Electronic and Atomic Collisions

ICPEAC XVII gratefully acknowledges the sponsorship and financial support of the various corporate and non-corporate bodies listed below. In particular the Conference received:

- Sponsorship from the Australian Academy of Sciences

- Sponsorship and financial assistance from the International Union of Pure and Applied Physics

- Financial support from the following non-corporate bodies:

  Australian Government
  Queensland Government
  Brisbane City Council
  Griffith University
  Ian Potter Foundation
  FOM Institute for Atomic and Molecular Physics
  Lawrence Livermore National Laboratory

- Financial support from the following companies:

  Qantas
  Australian Airlines
  Ortec
  Spectra Physics
  Balzers
  Coherent
  Oriel (Lastek)

Electronic and Atomic Collisions

Proceedings of the Seventeenth International Conference on
the Physics of Electronic and Atomic Collisions held in
Brisbane, Australia, 10–16 July 1991

Edited by W R MacGillivray, I E McCarthy and M C Standage

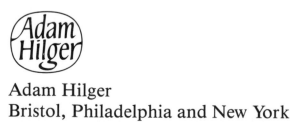

Adam Hilger
Bristol, Philadelphia and New York

British Library Cataloguing in Publication Data

CIP catalogue record for this book is available from the British Library

Library of Congress Cataloging-in-Publication Data are available

ISBN 0-7503-0167-8

Published under the Adam Hilger imprint by IOP Publishing Ltd
Techno House, Redcliffe Way, Bristol BS1 6NX, England
335 East 45th Street, New York, NY 10017-3483, USA
US Editorial Office: 1411 Walnut Street, Philadelphia, PA 19102, USA

Printed in Great Britain by J W Arrowsmith Ltd, Bristol

Preface

The Seventeenth International Conference on the Physics of Electronic and Atomic Collisions was held at the Sheraton Brisbane Hotel in Brisbane, Australia from July 10 to 16, 1991. This was the first ICPEAC conference to be held in the Southern Hemisphere. The Local Organising Committee was drawn from universities throughout Australia.

The conference was attended by about 460 delegates from 29 countries. The majority of these delegates also attended one or more of the five satellite meetings with more specific themes. Approximately 600 contributed papers were presented at the conference in lively poster sessions. Seven of these contributed papers were selected by the Programme Committee for oral presentation in Hot Topics sessions which concluded the conference.

This book contains written versions of the invited talks delivered at ICPEAC XVII, including the Hot Topics papers. Of the 71 invited speakers, 3 were unable to attend and 5 failed to submit manuscripts.

The opening keynote lecture was given by R W Crompton from the Australian National University on the foundations of atomic collision physics in Australia. This talk traced the development of this topic from its early beginnings to the now prominent role it has in Australian science. Plenary lectures were given by C L Cocke on ion–atom collisions, B Crasemann on utilising synchrotron radiation in atomic and molecular studies, F H Read on two electron correlations and A T Stelbovics on the Coulomb three body problem. The rest of the programme consisted of review talks and progress reports given in two parallel sessions.

The scope of the conference remains broadly based on its central theme of electronic and atomic collisions with such topics as photoionisation and detachment and ion–surface interactions included in the programme. The interconnections between areas of the conference may be best illustrated by the reported manifestations of two electron correlations in post collision interactions, two electron capture by ions and single photon double ionisation studies.

The editors wish to acknowledge the contributions in preparing this book of Dr Birgit Lohmann who coordinated the receipt of the manuscripts and Maureen Clarke and Graham Douglas of IOP Publishing for their assistance.

W R MacGillivray
I E McCarthy
M C Standage

International Conference on the Physics of Electronic and Atomic Collisions Organisation 1989–1991

Executive Committee

Chairman:	F A Gianturco (Italy)
Vice Chairman:	S Datz (USA)
Secretary:	J N Bardsley (USA)
Treasurer:	R Morgenstern (The Netherlands)
Members:	T Andersen (Denmark)
	A Dalgarno (USA)
	L J Dubé (Canada)
	B Fastrup (Denmark)
	R S Freund (USA)
	H B Gilbody (United Kingdom)
	Y Itikawa (Japan)
	C J Joachain (Belgium)
	M S Lubell (USA)
	T Lucatorto (USA)
	I E McCarthy (Australia)
	E Merzbacher (USA)
	J B A Mitchell (Canada)
	M C Standage (Australia)
Administrative Secretary:	L Roos (The Netherlands)

General Committee

Argentina
J E Miraglia

Australia
M C Standage
I E McCarthy
J F Williams

Austria
O Benka

Canada
L J Dubé
J B A Mitchell

Denmark
N O Andersen
T Andersen
B Fastrup

Federal Republic of Germany
K Bergmann
U Buck
W Domcke
B Fricke
U Heinzmann

France
J P Gauyacq
A Pesnelle
J Remillieux

India
D Mathur
A N Tripathi

Italy
F A Gianturco
G Stefani

Japan
Y Itikawa
N Kobayashi
H Nakamura

Mexico
C Cisneros

Romania
V Zoran

Spain
L F Errera

Sweden
A Barany

The Netherlands
R Morgenstern
B van Linden

United Kingdom
H B Gilbody
G King
A E Kingston
A C H Smith

USA
J N Bardsley
C Bottcher
A Chutjian
A Dalgamo
S Datz
R S Freund
H Helm
P S Julienne
P M Koch
C D Lin
M S Lubell
T Lucatorto
E Merzbacher
F W Meyer
R S Risley
A F Starace

USSR
J P Demkov
V P Shevelko
L A Shmaenok

ICPEAC XVII Local Organising Committee

Co-Chairmen:	I E McCarthy (The Flinders University of South Australia)
	M C Standage (Griffith University)
Secretary/Treasurer:	W R MacGillivray (Griffith University)
Members:	S J Buckman (Australian National University)
	R W Crompton (Australian National Univeristy)
	M T Elford (Australian National University)
	S C Haydon (University of New England)
	B Lohmann (Griffith University)
	R J MacDonald (University of Newcastle)
	L A Parcell (Macquarie University)
	J B Peel (La Trobe University)
	A T Stelbovics (Murdoch University)
	P J O Teubner (The Flinders University of South Australia)
	C J Webb (Griffith University)
	E Weigold (The Flinders University of South Australia)
	J F Williams (University of Western Australia)

Contents

x   *Contents*

Section 15: Reactive Collisions

Section 16: Clusters

Section 17: Exotic Systems

Section 18: Hot Topics

The foundations of the physics of atomic collisions in Australia

R.W. Crompton

Atomic and Molecular Physics Laboratories, Research School of Physical Sciences and Engineering, Australian National University, Canberra, ACT, 2601

ABSTRACT: Atomic collision physics began in Australia before 1910 with the work of W.H. Bragg in Adelaide on recombination, and J.A. Pollock in Sydney on the mobilities of atmospheric ions. Between the two World Wars ionospheric and atmospheric research motivated much of the laboratory work, and this influence continued to some degree after the Second World War. The scope of present-day research in atomic collisions in Australia is far wider, but the origins of some of the research groups can be traced to the personalities and themes of the first half of the Century.

1. INTRODUCTION

In 1895 Röntgen discovered X-rays, a discovery that was followed rapidly by numerous investigations of the ionization of gases by X-rays themselves and by the emissions from radioactive substances. By 1897, E. Wiechart, W. Kaufmann, and J.J. Thomson, in separate investigations, had determined the ratio of the charge to the mass of the electron, and therefore shown it to be a separate entity and not simply a negatively charged atom or molecule. Thus in a few short years in the last decade of the nineteenth century, the ground was prepared for the start of research into electronic and atomic collisions.

Not surprisingly, there were few physicists of any description in Australia in those years when there was so much excitement and activity in European laboratories. Some physicists were to be found in the Universities of Sydney, Melbourne, Adelaide, and Tasmania, and in the State observatories, while a few mathematicians in one or two of the universities were working on theoretical problems in physics. In all there were less than thirty scientists engaged in physics or related activities, and fewer than a dozen physicists working in universities where there was some opportunity of doing significant research in experimental physics. In addition, as we shall see later, there was one man who would have classified himself as a theoretical physicist, but he chose to work alone.

Despite their isolation, these early Australian physicists were well aware of the work of their colleagues overseas, and within a few years some of them had begun their own research in atomic collision physics.

2. THE BEGINNINGS: RECOMBINATION STUDIES IN ADELAIDE

It may not be generally known that W.H. Bragg could be said to have been the first to work in this field in Australia. Bragg was appointed to the Chair of Mathematics and Experimental Physics at the University of Adelaide in 1885 as the successor to the first Professor of Mathematics, Horace Lamb. However, it was not until many years later that Bragg began research work of any significance. Though this work, once begun, brought him rapidly into prominence, it did not involve the study of X-rays, the field of physics with which his

name is invariably associated these days.
The work I shall describe briefly here was
a precursor to that work. Nevertheless, it
played a very important part in an extraordinary scientific career - extraordinary
not only for its brilliance but also for its
origins and the course it ran in its early
years.

Bragg's Cambridge degree was in
mathematics, and legend has it (apparently
well substantiated) that much of the
physics he knew was learnt on the voyage
from England to Australia - and that he
learnt it by reading a text on electricity and
magnetism! His first years in Adelaide
were almost solely occupied with his
university teaching, but that soon spilled
over to reach, or affect, a far wider
audience - through lectures to public
audiences and scientific societies, and
through his influence on secondary
education, at that time dominated by
classics and mathematics.

During his first eighteen years in Adelaide,
Bragg's life proceeded profitably and en-
joyably, both within the University and
beyond, with as yet little indication of the
brilliant career in research that was to
follow. Once it began, however, in 1904,

William Henry Bragg 1862-1942

Bragg rose so rapidly to prominence that within three years he was elected a Fellow of the
Royal Society. Another year later he received an invitation to a chair in England which he
could not refuse, since it would enable him to be nearer the centre of action of his newly
found research interests. Reluctantly, he and Adelaide said farewell (Jenkin 1986).

At the age of 42, what precipitated the transformation in Bragg's career? It was, apparently,
the preparation of a major paper that he was to present at the Dunedin meeting of the
Australasian Association for the Advancement of Science (AAAS) as President of Section A
(Astronomy, Mathematics and Physics). In a paper entitled "Some recent advances in the
theory of ionisation of gases" (Bragg 1904), he gave a comprehensively researched survey
on what was then known about the ionization of gases by α-, β- and γ-rays.

On returning to Adelaide Bragg immediately began a series of experiments on the range of α-
particles, with the able assistance of a young man, R.D. Kleeman, who had previously been
employed as a blacksmith and had consulted him earlier on a problem in mathematics. In the
five years that followed before he left Adelaide Bragg published twenty-four papers on the
results of his experiments and on his theories of the nature of radioactive emanations. It was
primarily this work that earned him his FRS, but it was his subsidiary work on
recombination that I believe justifies him being named as the first to work on atomic collision
physics in Australia.

The work on recombination was begun with Kleeman, but it was pursued considerably
further by J.P.V. Madsen whose name will appear in several contexts in this paper. At the
time Bragg and Kleeman began their work little was known about recombination. It was
generally assumed that it was a simple volume neutralization process between charge carriers

of opposite sign, and that the loss rate/unit volume would be simply proportional to the number density of the ions, i.e.

$$\frac{dn}{dt} = -\alpha \, n \, p$$

where n and p are the number densities and α the recombination coefficient.

One way of measuring α was to look at the competition between the rate of production of ions in an ionization chamber by, say, α-particles, and the loss rate due to recombination and to ions being swept from the chamber by an electric field between its plates. However, the results of these experiments posed some questions, and Bragg and Kleeman (1905) set out to find answers to them. What they found was that recombination is more complicated than the simple process just described.

As the field strength in an ionization chamber is increased the current eventually saturates. This is shown in Figure 1, which is taken from Bragg and Kleeman's 1905 paper. Every ion produced is then collected, and from the current the production rate can be calculated.

Having found the saturation field strength, Bragg and Kleeman reduced the field to a smaller value such that only about 80% of the current was recorded, that is, to a value such that 20% of the ions produced originally were lost by a recombination process of some kind (ignoring diffusion losses). From the value of the current recorded at the lower field strength, they then calculated the ion density using values for the ion-mobility taken from the work of others. Then, using published values of α and the known volume V between the plates of the ionization chamber, they calculated the loss rate ($\alpha n^2 V$) due to volume recombination. When they compared this with the total production rate, calculated from the saturation current, they found that it accounted for only about 1/3000 of the 20% loss recorded at the lower field strength.

Following an earlier suggestion of Rutherford's, Bragg and Kleeman then went on to explore the implications of what they called 'initial recombination' whereby an electron released by ionization is recaptured before becoming attached to a neutral molecule to become a negative ion, thus adding to the pool from which the normal process, which they called 'general recombination', takes place.

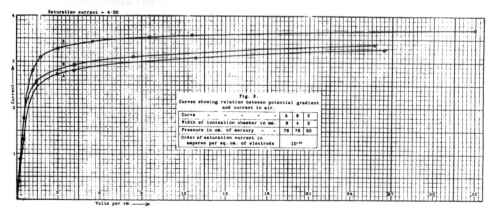

Fig. 1. Plots of ionization current versus voltage, showing saturation at high field strengths (Bragg & Kleeman 1905).

It is important to remember that this work was carried out less than ten years after the electron had been identified. The much higher mobility of free electrons was not to be known until the work of Franck (1910), Lattey (1910), and Townsend and Tizard (1912), more than five years later.

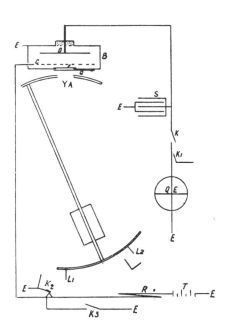

John Percival Vissing Madsen 1879-1969   Fig. 2. Madsen's apparatus for investigating recombination (Madsen 1908).

Before these experiments were completed Kleeman left to take up a scholarship at Cambridge, and shortly after Madsen took his place as Bragg's collaborator. Madsen had been trained as a physicist, mathematician, and engineer and was appointed to Bragg's staff in 1901. Though it can be claimed that Bragg initiated work on 'atomic collisions' as a complement to his all-absorbing early interest in naturally-occurring ionizing radiations, it was Madsen who was to have a greater influence on its subsequent development. But this was to be indirectly through the central role he played in science administration in Australia rather than through his direct involvement in research. In these early years, however, Madsen joined Bragg in the frenetic activity that first brought Bragg to prominence. Between 1906 and 1909 he published eight research papers, mostly with Bragg. One of them, on recombination (Madsen 1908), was a distillation of the thesis he submitted to the University of Adelaide for which he was awarded a DSc. In it he described an ingenious set of experiments designed to throw further light on the subject of 'initial' and 'general' recombination. Essentially what he sought to determine was the time interval during which the abnormal recombination process took place. Since there were no fast electronics, and he wished to examine phenomena on a time scale of the order of 50 milliseconds, this was a formidable challenge.

A much simplified diagram of Madsen's apparatus is shown in Figure 2, taken from his 1908 paper. There were in fact two pendulums, but only one is shown here. The second pendulum was released to commence its swing when the pin L_2 on the first pendulum tripped a relay to de-energize a holding magnet.

The sequence of events in the experiments was as follows. The gas in the ionization chamber was ionized by a short, but reproducible, pulse of α-particles from the radium bromide source, A. The radiation could enter the chamber only when the hole in the curved lead shutter, attached to the top of the pendulum, coincided with the aperture in the chamber floor. After a delay of 50 to 100 milliseconds (the duration of the delay being determined by the position of the pin L_1) the earthing switch K_2 was tripped. The potential of the previously-earthed electrode C then rose to that of the accumulator, T, establishing an electric field in the chamber and sweeping the ions from the ionized region between its electrodes. The field was removed by the action of a contact activated by the second pendulum.

The experiment entailed measuring the charge recorded by the electrometer, due to the collection of the ions, as a function of the applied field. Figure 3 shows the measured values, plotted as points, together with curves computed on the assumption that the only loss process was 'general' recombination. Since these results, taken with delay times as short as 50 milliseconds, were entirely consistent with the loss being solely due to 'general' recombination, Madsen concluded that the other ('initial') process took place on time scales of less than 50 milliseconds.

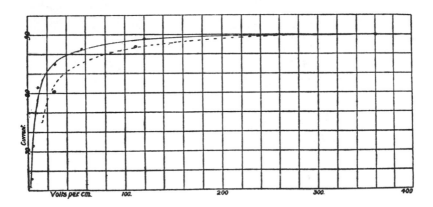

Fig. 3. Madsen's calculated and experimental data from which he inferred an upper limit to the time-scale of 'initial' recombination (Madsen 1908).

Unlike the majority of the papers of that period from Adelaide, which first appeared in the *Transactions of the Royal Society of South Australia* and were later reproduced in the *Philosophical Magazine*, Madsen's paper appeared only in the *Transactions*. Whether it was rejected by the prestigious English journal, or just not submitted, is not recorded, but its publication only in an obscure colonial journal probably explains why it is not referred to in later monographs describing the early development of the subject of recombination.

We shall return to Madsen later, but for the moment we leave Adelaide and look briefly at what was happening in Sydney and Melbourne at about the same time.

3. EARLY ION MOBILITY STUDIES: EXPERIMENT AND THEORY IN SYDNEY AND MELBOURNE

The names of three other Australians are to be found in the international scientific literature in the period 1905 to 1914, those of J.A. Pollock, E.M. Wellis(c)h, and W. Sutherland. Of the three, Sutherland's name is probably the best remembered these days for the model and potential in molecular dynamics named after him, and for the 'Sutherland correction' and 'Sutherland constant' in a formula for the temperature dependence of gaseous viscosity.

Pollock was the second professor of physics at the University of Sydney. The University was founded in 1862, and Pollock was appointed in 1899. Unlike other appointments to chairs in the Australian universities in the late nineteenth century, Pollock was not one of J.J. Thomson's students, although 'J.J.' was one of the committee who appointed him. It appears that Pollock shunned publicity, and the photograph shown here is the only one a biographer could find of him. He describes it as 'an enlargement from a photograph of a group of officers in the Tunnelling Battalion (1st World War), his presence no doubt being the result of orders from his commanding officer' (Branagan & Holland 1985a).

Pollock also worked on one of the fashionable topics of the day - ion mobilities. It is tempting to think that the initial motivation for his research was similar to Bragg's. Like Bragg (but about five years later), Pollock was called upon to deliver a presidential address to Section A of AAAS, and his first published research paper on ion mobilities appeared in the same year (Pollock 1909). However, his AAAS paper, entitled 'The ions of the atmosphere', seems to have been the consequence of his research rather

James Arthur Pollock 1865-1922

than its initiator, because in that paper he refers to work at Sydney that had been under way 'for some time'. Unlike Bragg's work on recombination, which was a byproduct of his main interests, Pollock's work on the nature of heavy ions in the atmosphere was his main interest for many years, and it has been judged to be his most important (Branagan & Holland 1985b).

At that time there was considerable interest in a class of heavy ions with very small mobilities - the so-called Langevin ions - which are now known to comprise charged aggregates of some 10^6 water molecules. Between 1909 and 1915 Pollock conducted a careful series of experiments to investigate the dependence of the formation and relative abundance of these ions on both the cleanliness of the air and its water vapour content. In the course of his work he discovered a new class of ion with intermediate mobilities and, using thermodynamic arguments, he concluded that the intermediate ions comprised a solid nucleus surrounded by an envelope of water vapour, while the envelope of the heavier Langevin ions was liquid water. Although the conclusions he drew from his experiments with filtered and unfiltered air concerning the nature of the nuclei on which the aggregation took place were later disproved, his work was nevertheless an important milestone in the development of an understanding of the nature of atmospheric ions.

The second of the three, E.M. Wellisch (who called himself Wellish in later years) was a student of Pollock's. Before gaining a scholarship in 1907 to work with J.J. Thomson he collaborated with Pollock in a paper on the relighting of the carbon arc. His work with Thomson cannot be said to have contributed to the *Australian* foundations of our subject, and unfortunately by the time he returned home in 1915 his appointment was in the Mathematics Department of the University of Sydney rather than the Physics Department. In the meantime he had done both experimental and theoretical research at the Cavendish Laboratory, predominantly on ion mobilities, before taking an appointment as Assistant Professor of Physics at Yale between 1911 and 1915. Following his return to Australia 'his research interests were surely sacrificed to the general needs of students and the Department (of Mathematics)' (Branagan & Holland 1985c), although he published two papers in 1931 on photoelectrons and negative ions. It was Wellish's work, however, that sparked some politely conducted argument in the literature between him and the third member of this group, William Sutherland.

As has already been noted, Sutherland is the best remembered of the three and he undoubtedly would have been more prominent still had he trodden the usual academic path. He was born in 1859 in Scotland but his family migrated to Sydney in 1864 then finally settled in Melbourne in 1870 where William eventually took first-class honours in 'natural science' in 1879 while at the same time completing coursework for the University's certificate of engineering. He won a scholarship to University College, London where he was awarded a BSc with first-class honours and a scholarship in experimental physics. He returned to Australia in 1882 but, to quote a biographer:

> seemingly without ambition for wealth or success, he devoted his life to reading and research. To support his modest needs, he did some private coaching and served as examiner at the university (Home 1990a).

Apart from temporary appointments at the University of Melbourne, where he responded to invitations on two occasions, Sutherland never held an academic appointment. This unconventional career seems to have been entirely to his liking, though he admitted that such positions did attract him because:

> The chief inducement with me (would be) the chance of a laboratory for research. I have tried to work at one or two things with the published results of other men's experiments, but have invariably stopped short at some point through want of experiments of my own (Osborne 1920a).

And he eschewed academia not because he was a man of means who could afford to do so (which he was not) but because 'the duties of it would interfere for sometime with my original work which I am eager to resume with vigour' (Osborne 1920b). (This was during one of the interludes at the University of Melbourne already referred to.) He did, in fact, apply for the Melbourne chair at that time, but through the Agent General in London so as not to obtain, as he felt, an unfair advantage through being personally known. But after the appointment was made 'came a letter from London stating that his application for the chair of Music (sic) had been received! Such are, or were, the ways of Agents General' (Osborne 1920b).

Sutherland submitted his first paper, to the *Philosophical Magazine*, in 1885 and there followed a constant flow totalling seventy-eight in all which ceased only in the year of his death in 1911. The fourth paper entitled 'The law of attraction amongst the molecules of gas' (Sutherland 1886) was the first of a series which, by 1893, had enabled him to account successfully for the unexpected temperature dependence of the

William Sutherland 1859-1911

viscosity of gases. He did so by modifying the then existing description of a gas as an assembly of colliding hard spheres with the molecular trajectories being straight lines terminated at collisions with a sudden change of direction. Such a model predicts a $T^{1/2}$ dependence of the viscosity on the absolute temperature T. Sutherland proposed that there was a weak attractive force between the molecules and showed that the viscosity should vary as $T^{1/2}/[1 + (S/T)]$. Both the model and the constant S bear his name. While the model is oversimplified it had remarkable success in predicting the observed temperature dependence for many gases over a wide range of conditions.

Though this is the work for which he is primarily remembered, Sutherland worked on many problems that could be tackled by examining gases, liquids and solids at the microscopic level. Thus his papers deal with gaseous diffusion, surface tension, the rigidity of solids, the properties of solutions, and an important series dealing with the properties of water — and this list is still by no means exhaustive. It does not include, for example, the topic which led to the exchange of views with Wellish in the *Philosophical Magazine* of 1910.

In 1909 Sutherland had written a long paper, containing thirty pages of detailed mathematical and physical argument, entitled simply 'The ions of gases' (Sutherland 1909). It was his major work for that year, and in it he applied to the subject of ion mobilities the ideas he had so successfully developed to explain the characteristics of the viscosity of gases.

In order to explain observed mobilities on a hard sphere model, it had been necessary to postulate that the ions were larger than a normal molecule and were thus comprised of an aggregate of molecules. By taking account of polarization forces both Sutherland and Wellish had been able to show that the enhancement of the collision frequency of the ions as a consequence of these forces rendered the assumption of clustering unnecessary. In his AAAS address to which I have already referred, Pollock made direct reference to the work of both Wellish and Sutherland in establishing that the experimental observations of the mobilities of 'small' ions could be accounted for on the assumption that they were charged *single* atoms or molecules. He summarized the new position thus:

> The idea of the small ion as a cluster of a few molecules, founded on insecure assumptions, was perhaps chiefly characterized by its vagueness; its replacement by a definite theory cannot but be regarded as making a great advance in our knowledge of ionic structure (Pollock 1909).

But while Wellish and Sutherland could agree that there was no need to invoke a cluster hypothesis to explain many of the experimental observations of the mobilities of small ions (some of the measurements having been by Wellish himself) they could not agree on some of the details. There followed an extremely polite exchange of letters in the *Philosophical Magazine* for 1910 (Wellisch 1910, Sutherland 1910) which did not appear to reach a mutually satisfactory conclusion. It was a pity they could not have met at that AAAS meeting in this city (Brisbane) ninety-two years ago. Wellish was still working away in Cambridge, Sutherland was in Melbourne. The tyranny of distance was even greater then than now. Pollock remarked in his AAAS address that 'Mr. Wellish communicates a paper on (the) subject to this section', and 'Mr. Sutherland, to our regret, is unable to be present at this meeting ... (but) permits me to mention the results of his investigation at this stage of our proceedings'. One suspects that this was one occasion on which Sutherland's unconventional and extremely frugal lifestyle counted against the advancement of science to which he was so totally committed.

Following these letters, Wellish's and Sutherland's interests seem to have turned elsewhere. There are one of two more papers on ions and free electrons by Wellish, published after he returned to Australia to take up his appointment as a lecturer in mathematics. Then the First World War intervened, and afterwards what research he was able to accomplish was in mathematics rather than physics. Sutherland was to live only one more year. He died suddenly in 1911 while still in full flight; in the last two years of his life he published no less than ten papers on a variety of topics (Home 1990b). Nevertheless, despite his relatively short life and unorthodox scientific career, Sutherland had made his name. 'In the Festschrift for Boltzmann's 60th birthday, of the 116 authors five only were from outside Europe and Sutherland was one of these five ... ' (Bolton 1988).

Pollock continued his research on atmospheric ions until, in 1915, at the age of 48, he enlisted for active service. After his return in 1919 there were only three short years before he, too, died prematurely. So concluded the first phase of Australia's involvement in atomic collisions research. In fifteen years the very small group of physicists in Australia at that time had, together with expatriates, made some significant contributions to a field that was just emerging.

4. BETWEEN THE WARS: ELECTRONS IN THE LABORATORY AND IN THE IONOSPHERE

The second phase commenced with the arrival of V.A. Bailey from England in 1924 to fill a vacancy caused by Pollock's death and, a few years later, D.F. Martyn's arrival, also from England, to join the newly-established Radio Research Board. Though both were excellent mathematicians as well as physicists their backgrounds were rather different. Bailey had done his DPhil research on low energy electrons in gases with J.S. Townsend, while Martyn, studying at the Royal College of Science in London, had specialized in electronics.

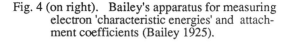
Victor Albert Bailey 1895-1964

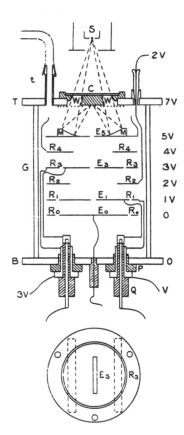

Fig. 4 (on right). Bailey's apparatus for measuring electron 'characteristic energies' and attachment coefficients (Bailey 1925).

It is not clear that Bailey's appointment was in any way connected with Pollock's erstwhile interests, but after his arrival in Sydney Bailey soon established an active group working on electron and ion physics. With Townsend he had published many papers exploiting Townsend's techniques for measuring electron diffusion and drift and extracting from the data the energy-averaged electron mean free paths and energy losses as functions of the 'characteristic energy' ε_k of the electrons, a quantity closely related to the mean energy. In Sydney he developed new techniques that enabled him to determine not only these quantities but also the rate at which electrons attach to form negative ions in electronegative gases.

Within two years he had published his first results using the new technique he had devised (Bailey 1925). A schematic diagram of his apparatus is shown in Figure 4. The principle of the method relies on the difference between the rate of lateral spread of *non-thermal* electrons and *thermal* negative ions due to diffusion as they traverse the apparatus filled with gas at a low pressure, p, under the action of an applied electric field E. The electrodes E_1 and E_3 contain narrow slits. As the mixed stream of electrons and ions drifts a distance c from E_5 to E_3, then from E_3 to E_1, the two components spread at different rates. By measuring the currents to the electrodes E_0, E_1, and E_3 Bailey was able to determine the product $R_e e^{-\eta c}$, where $R_e \equiv R_e(E,\varepsilon_k)$ would be the ratio of the transmitted to total current arriving at either E_1 or E_3 *if the stream were comprised solely of electrons,* and η is the attachment coefficient. From measurements at two pressures, keeping the ratio E/p and hence ε_k and η/p constant, both ε_k and η/p could be determined.

The method was typical of Bailey's ingenuity of which we shall see another example later. Using it, he and a succession of students provided a great deal of information on electron behaviour in attaching gases where no previous data existed.

At this point another man must be mentioned although it is unlikely that his name will be known to more than a handful of people in this audience. The name is that of H.H.L.A. Brose. Curiously, Brose links Bragg and Bailey. He was one of Bragg's last undergraduate students; there is, in fact, a record of him having given a piano performance at Bragg's farewell dinner in Adelaide!* And he visited Bailey's laboratory for a very brief period in 1926, publishing one paper from his work there on low energy electron collisions in ethylene. Between these two dates he had had rather a chequered career.

In 1913, some time after completing his first degree, he was awarded a Rhodes Scholarship to Oxford, and it appears that he was visiting family connections in Germany, shortly after arriving in England to take up his scholarship, when war broke out. He was interned as a civilian prisoner of war for its duration. Eventually he proceeded to a DPhil under Townsend and published a single-author paper on electrons in oxygen in which, surprisingly, he claimed he saw no evidence of attachment. His year in Australia in 1926 followed almost immediately, and in 1927 he took up an appointment at the University College of Nottingham.

Brose's unfortunate wartime experience had at least one consequence. As a result of chance events during his internment, he subsequently established contact with Einstein, and on several occasions when Einstein gave lectures in England, where he spoke in German, Brose was invited to attend to provide a translation. It is also possible that events then were responsible for him being the first member of the Townsend school to try to interpret the results of his experiments on low energy electron collisions by using the new quantum theory, for he began translating German articles on relativity during his internment and he continued to translate German texts after the war. Some of these were on quantum theory (Jenkin 1993). Thus we find a paper with E.H. Saayman in *Nature*, then two more in *Annalen der Physik,* in which the authors used the theory to try to explain the differences between electron collisions with hydrogen and with the inert gases (Brose & Saayman 1930).

---

* Brose family papers, currently in the possession of J. Jenkin, La Trobe University, Melbourne.

In the early history of Australian physics two names are invariably coupled together, those of V.A. Bailey and D.F. Martyn, even though the work they did together was only a very small part of the scientific output of each.

Martyn was a Scot who received his education in the UK, but he did essentially all his scientific work in Australia. It was Madsen, whose early work with Bragg has already been referred to, who was responsible for Martyn's move to this country. At the time to which I am referring (the late 1920s) Madsen was the professor of Electrical Engineering at Sydney University and had long ceased to be active in research in physics. But he retained strong interests in fundamental science which he continued to promote vigorously. During his long tenure at Sydney he demonstrated great foresight and commitment to both fundamental science and technology. His persistent advocacy was largely responsible for the formation of the Radio Research Board (RRB) and the establishment of a national standards laboratory, the cases for which he argued persuasively at only the second meeting of the Council for Scientific and Industrial Research (CSIR - now the Commonwealth Scientific and Industrial Research Organization - CSIRO). This was in 1926, the year of inauguration of the CSIR.

David Forbes Martyn 1906-1970

It took many years for the standards laboratory to be built, but Madsen managed to get agreement for his Radio Research Board much more quickly. By 1929 the CSIR and the Post Office had agreed to provide funds for research in the rapidly expanding field of radio science - potentially so important for a sparsely populated and isolated country - and Madsen went overseas to recruit staff. D.F. Martyn was one he recruited, L.G.H. Huxley was another.

Bailey's background was in gaseous electronics and the macroscopic and microscopic behaviour of electrons in gases, Martyn's was in electronics which, under the terms of reference of his RRB appointment, he now applied to problems of radiowave propagation. His initial work was on the characteristics of the ionosphere and on wave propagation that depended on these characteristics, and it was during this period that Bailey and Martyn joined forces to provide a satisfactory explanation of a baffling phenomenon that had become known as the 'Luxembourg effect'.

Early in 1933 a new high-power long-wave transmitter began broadcasting from Luxembourg. Almost immediately there were first reports, by Tellegen (1933) in *Nature,* of the superposition of one radio transmission on that of another. Tellegen, for example, who was based in Eindhoven, reported hearing a superposition of the Radio-Luxembourg program on the transmissions of eleven stations whose carrier frequencies were very different from that of Radio-Luxembourg.

vessels each of which was connected to a gas burette by means of an outlet tube. The upper vessel contained the corroding medium which was in contact with the metal and the lower one contained air. The air in the headspace of the upper chamber and in the solution was replaced by nitrogen and the whole apparatus was set up in a room held at 25° C.

Both the upper and lower burettes soon began to register an increase in volume. That in the upper one was the more rapid at first, but slackened later, and when the apparatus was taken down after two days, there was about 30 c.c. of hydrogen on either side, leaving about 9 c.c. (calculated from the loss in weight of the sheet) to be accounted for by absorption into the metal and by solution in the citric acid. I should be glad to know whether the passage of hydrogen through steel under similar conditions to the above has been recorded.

T. N. MORRIS.

Low Temperature Research Station,
    Downing Street,
    Cambridge.
    Jan. 12.

Interaction of Radio Waves

THE phenomenon recently reported by Tellegen[1] whereby the new broadcasting station at Luxembourg appears to interact with that portion of the carrier wave of the Beromunster station which is received in Holland, can be explained by taking into account the effect of such a powerful station (200 kw. and $\lambda = 1190$ m.) on the mean velocity of agitation (u) of the electrons in the ionosphere. Any change in u will produce a change in ν, the frequency of collision of an electron with molecules, and hence a change in the absorbing power of that part of the ionosphere in the vicinity of the station. Since this change depends on the magnitude of the electric vector in the disturbing wave, it follows that the absorbing power of this part of the ionosphere will vary in accordance with the modulation frequency of the modulation, and so the modulation will be impressed in part on any other carrier wave which may traverse this region.

We have examined these points quantitatively with the help of data obtained by Townsend and Tizard[2] on the motions of electrons in air, and have arrived at the following conclusions.

The amount of modulation of a carrier wave produced by a disturbing station of power P and modulation frequency f is approximately proportional to P and inversely proportional to f. There is thus introduced a distortion of the original modulation, at the expense of the higher frequencies of modulation.

The variation of the impressed modulation with the wave-length of the disturbing station is more complicated, being roughly proportional to $1/\{\nu^2 + (p-\omega)^2\}$ where $p = 2\pi c/\lambda$, $\omega = H_p e/cm$ and H_p is the component of the earth's magnetic field perpendicular to the electric vector of the disturbing wave. It is clear that the quasi-resonant state ($p = \omega$) can exist only in very localised regions of the ionosphere, and will contribute little to the total impressed modulation, which may be received over the whole path of the wave in the absorbing regions of the ionosphere. The disturbance will therefore be greatest when ω is small, that is, when the entire electric vector of the wave lies in the direction of the earth's magnetic field. The magneto-ionic

theory shows that under European conditions this can occur only for that part of the wave's path which is roughly horizontal. In such circumstances ω will always be small for waves much longer than 214 m.

We have examined the magnitude of the disturbance which would be experienced at Eindhoven when listening to the Beromunster station, and find that it would become appreciable for values of air pressure in the absorbing regions near those generally accepted. The disturbance experienced is proportional to ν^2, so that we should expect increased disturbance at times when the sky wave is weakened by increased absorption, for example, around sunrise and sunset, and in the daytime if signals be audible.

It is to be anticipated that the Warsaw station will also exhibit the effect in just appreciable intensity if careful investigation be made. It is not to be expected, however, that the very long wave high-power telegraph stations, such as Rugby or Nauen, could produce the effect, for such long wave-lengths are probably reflected at a level in the ionosphere below that which absorbs waves of broadcasting frequencies. Neither would such a station appreciably influence the reception of other very long wave stations, since most of the received signal on these wave-lengths is due to the ground wave.

The details of our investigation will be published elsewhere in the near future, together with a discussion of the possibility of utilising the phenomenon to derive further information about the ionosphere.

V. A. BAILEY.

Department of Physics,
    University of Sydney.

D. F. MARTYN.

Commonwealth Radio Research Board,
    Sydney.
    Nov. 29.

[1] NATURE, **131**, 840, June 10, 1933.
[2] Proc. Roy. Soc., A. **88**, 336 ; 1913.

Audibility of Auroras and Low Auroras

I WAS much interested in the article "Audibility of Aurora and Low Aurora" by F. T. Davies and B. W. Currie which appeared in NATURE of December 2, because I once witnessed an aurora and heard the swishing sounds referred to.

During the winter of 1908–1909, while attending Trinity College at Hartford, Conn., I observed a magnificent aurora. The light effects gave me the impression that the atmosphere was filled with fog, and that someone was illuminating it by playing a searchlight back and forth. The effect was very striking because the display was so close to the ground that I seemed to walk right through the illuminated fog.

The sound which I heard is exactly described by the word swishing. I do not believe I could say the swishing sound was in unison with the flickering of the lights because the sight was so new and strange that I did not observe it from the point of view of a scientist. All that I can say is the swishing sounds were heard while the lights were changing.

FLOYD C. KELLEY.

Research Laboratory,
    General Electric Company,
    1 River Road,
    Schenectady, N.Y.
    Jan. 4.

Bailey and Martyn's letter to *Nature* giving a preliminary account of their explanation of the Luxembourg effect

In a popular article Bailey published in *Wireless World* several years later (Bailey 1937a), he states that Tellegen's report was preceded several months earlier by a similar report by A.G. Butt in *World Radio*. However, in this case, apparently, it was the interference of the Radio-Luxembourg transmission by the transmission from Radio-Paris.

Some months later (Bailey wrote subsequently):

... a group of physicists was gathered round a tea-table in Professor Madsen's Department of Electrical Engineering at Sydney University discussing over their cups the remarkable phenomenon described by Tellegen.

The discussion gave no definite illumination, as it was generally agreed that the information in Tellegen's letter was inadequate for this purpose (Bailey 1937a).

Fortuitously, a few months later again, Bailey was examining a thesis concerned with radiowave propagation through the ionosphere which caused him to question the then standard assumption that the ionosphere was undisturbed by the passage of the radio wave. In the transmission of radio waves through the ionosphere the wave amplitude is attenuated by an amount that depends on the collision frequency of the free electrons in the ionosphere. Bailey recognized that the (oscillatory) field due to the 'disturbing' transmitter would change the collision frequency, ν, of the electrons, and hence the local absorption coefficient, along the path of the 'wanted' signal through the ionosphere. Using the experimental data of Townsend and Tizard for the dependence of ν in air on the average electron energy, he was able to show that, integrated along the path of the wanted wave, heating of the electrons by the 'disturbing' signal would lead to a significant change in the attenuation of the 'wanted' signal. But the transmission from Radio-Luxembourg was amplitude modulated, so the attenuation of the transmission from another station would have a time dependence approximating that of the amplitude modulation of the disturbing transmitter. Hence the amplitude of the 'wanted' signal would be further modulated due to the time-dependent absorption, and the program of Radio-Luxembourg would thus be superimposed on that of the other station.

The reproduction of the modulation of Radio-Luxembourg on the transmission of another station was not entirely faithful, it being attenuated more at the higher frequencies. This, too, could be explained. Provided the modulation frequency is sufficiently low, the variation of the 'electron temperature' can follow faithfully the slow variation of the effective field due to the RF carrier wave. If the frequency is high, however, the electrons have insufficient time to heat or cool as the amplitude changes, and the variation in the absorption coefficient is correspondingly reduced.

In their now classic paper, Bailey and Martyn (1934) showed that the depth of modulation of the wanted wave is proportional to $[(2\pi f)^2 + G^2\nu^2]^{-1/2}$ where f is the frequency of modulation of the disturbing transmitter and $G\nu$ is the product of the electron collision frequency ν at the site of the interaction and a constant G. This constant is defined such that the mean fraction of its energy lost by an electron per collision is the product of G and the mean excess of the electron energy above the mean thermal energy of the gas molecules. G is thus a characteristic of the gas.

The reason for the dependence of the depth of modulation on frequency has been discussed in the previous paragraph. However, the relation describing the depth of modulation had more significance than simply quantifying the dependence on f. Bailey and Martyn compared experimental data for the modulation depth with the dependence predicted by their formula and adjusted the parameters to obtain the best fit. Figure 5 shows their first attempt to do this (Bailey & Martyn 1935). Such fits enabled them to determine the product $G\nu$, and from independent estimates of ν, which had already been made by Appleton and Chapman and were made later by Martyn himself, they obtained G. Summarizing their conclusion in the popular article already referred to, Bailey wrote:

We thus conclude that G is equal to ... 0.0026.

The observations of Townsend and Tizard on electrons in air at ordinary temperatures lead to the value 0.0026 for G and those of Townsend and Bailey on electrons in hydrogen lead to the value 0.0031. Other gases like argon, helium and carbon-dioxide lead to values very different from these.

We may therefore conclude that the observations are consistent with the view that the gas in which the interaction occurs is either air or hydrogen at ordinary temperatures. Dr. Martyn, who has made a special study of such matters, informs me that hydrogen is completely excluded by the observations on polar and non-polar aurorae of Vegard and Cabannes. Hence we are left with air as the only gas which may be supposed present at ordinary temperatures in the region of interaction (Bailey 1937b).

Clearly, the reliability of conclusions about the composition of the ionosphere depended on the accuracy of the laboratory determination of G, which had been deduced from measurements of the characteristic energy ε_k and drift velocity. In air such measurements are complicated by the presence of attachment. Bailey had himself repeated measurements in air to confirm Townsend and Tizard's data using the technique already described to measure simultaneously the attachment rate and ε_k. His results for ε_k differed from the earlier ones by as much as 30%. It was obvious that more reliable data were required for any detailed interpretation of the results of cross modulation experiments, and it was this need that later led Huxley to initiate laboratory experiments on low-energy electron diffusion and drift in air. I shall return to Huxley and his work later.

In the ensuing years Bailey wrote prolifically on the interaction of radio waves in the ionosphere with particular emphasis on the effects to be expected at the electron gyro frequency.

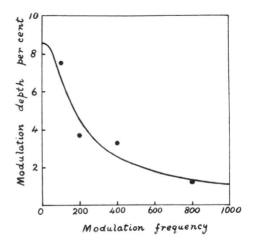

Fig. 5. Bailey and Martyn's first attempt to determine Gν by fitting their theoretical formula to measured data for the depth of modulation of a 'wanted' signal (Bailey & Martyn 1935).

During the Second World War several of his students, headed by J.M. Somerville, established a department of physics at Armidale as part of a new college of the University of Sydney, later to become the University of New England. The work on ionospheric cross modulation continued, but Somerville's interests turned to the study of arc discharges. Somerville recruited S.C. Haydon for the department in 1954. Haydon established the group that has made extensive studies of electrical breakdown in gases and its interpretation in terms of collision processes.

Martyn's collaboration with Bailey was short-lived. About that time he widened the scope of his researches to include a study of the properties of the upper atmosphere from the stratosphere upwards, and in a paper published in 1936 with O. Pulley (Martyn & Pulley 1936) he put forward a number of revolutionary conclusions and hypotheses concerning the properties of the upper atmosphere.

In the mid-1930s, it was generally believed that the outermost region of the atmosphere consisted of hydrogen that was very cold and still. The structure of the upper atmosphere proposed by Martyn and Pulley was radically different. From the data obtained from a number of sources they concluded:

(1) that there must be a second still colder stratosphere at about 80 km - the coldest region in the whole atmosphere;

(2) that above 80 km the temperature rose steadily to values of the order of 1000°C;

(3) that the atmosphere at heights of up to at least 300 km consisted mainly of nitrogen, with a substantial proportion of oxygen;

(4) that these upper regions were not still, but were subject to high velocity winds and turbulence.

These conclusions were vigorously contested at the time, but rocket and satellite observations subsequently have proved them to be largely correct (Piddington & Oliphant 1972).

Another of the conclusions of Martyn and Pulley is of particular relevance to the theme of this paper. At that time there was considerable argument concerning the process of electron loss in the ionosphere. To quote Massey:

Whereas some considered it (the electron loss) to be one of recombination between electrons and positive ions others, with the strong convictions which are apt to be held when there is little real knowledge of the subject, firmly believed that the electrons lost their freedom to participate in the reflection of radio waves through attachment to neutral molecules thereby forming negative ions too massive to produce any appreciable effect (Massey 1971).

Martyn and Pulley suggested, for the first time, that it was not sufficient to regard electrons as lost as soon as attachment occurs, and invoked the process

$$O^- + O \rightarrow O_2 + e,$$

(that is, what is now known as associative detachment) to explain why the electron concentration in the E-region remains high at night. Many years later, Fehsenfeld *et al.* (1966), using their flowing afterglow technique, were to show that this reaction is so fast that, at night time, negative ion formation contributes negligibly to electron loss even in the E-region.

Martyn made many important contributions towards advancing our knowledge of the ionosphere and upper atmosphere and in doing so, to quote Massey again, he made 'stimulating and important suggestions about atomic and molecular reactions in the ionosphere' (Massey 1971).

We come now to another influential figure of these years, L.G.H. Huxley. Martyn and Huxley were contemporaries in that they were both appointed by Madsen in 1930 as the first research officers of the Radio Research Board. Huxley's background and interests, however, were much closer to Bailey's than to Martyn's. Huxley went to Oxford as a Rhodes Scholar from Tasmania and took his DPhil under Townsend as Bailey had done five years earlier. His appointment to the RRB followed almost immediately, and although the financial circumstances of the Depression in the 1930s led to his return to England within two years, the work he did during that time kindled his interest in ionospheric and atmospheric physics.

Huxley went back to England to take up academic positions, first at the University College of Nottingham and later at the University College of Leicester, and in the period prior to the Second World War most of his research was concerned with theories of electron transport in electric and magnetic fields and radio-wave propagation. However, this work and his involvement with radar during the war were to set the stage for his work in England in the immediate post-war years, and later in Australia. Following his appointment after the war to the Department of Electrical Engineering at the University of Birmingham, he and two colleagues began an extensive investigation of the Luxembourg effect. Figure 6 shows an example of their observations of modulation depth and the fit that could be obtained between their data and the theoretical expression for the modulation depth as a function of modulation frequency (Huxley *et al.* 1948). With data of this quality it was possible to determine the value of Gv to 5% or better, and thus obtain an accurate value of the collision frequency at the site of the interaction, provided the value of G could be relied upon. (By this time the composition of the ionosphere was better known, and the aim was to determine v rather than infer the composition from a comparison of the laboratory value of G with the value derived from Gv and a 'known' value of v.) However, as we have seen, there was still some question as to the value of G. Thus, in parallel with the work on radio-wave interaction, Huxley initiated laboratory experiments on the diffusion and drift of low energy electrons in air with the aim of determining its value more accurately.

Leonard George Holden Huxley 1902-1988

While the fundamental concepts of these laboratory experiments were the same as those devised by Townsend, Huxley now made an important modification to the method of measuring the characteristic energy. This modification was based on theoretical work he had done before the war (Huxley 1940). The results of these experiments, when combined with Nielsen and Bradbury's (1937) time-of-flight measurements of electron drift velocity, gave a value of G that was about one half of the earlier value (Huxley & Zaazou 1949).

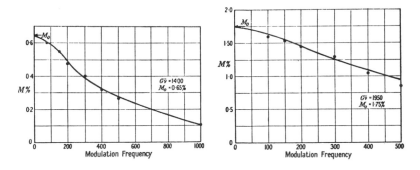

Fig. 6. Results of the 1948 observations of ionospheric cross modulation by Huxley, Foster and Newton.

5. POST-WORLD WAR II: A PERIOD OF RAPID GROWTH

In 1949 Huxley accepted the Chair of Physics at the University of Adelaide and opened a new chapter in atomic collisions research in Australia. At the time of his arrival, essentially no experimental laboratory physics of this kind was being carried out in the country. Bailey's laboratory studies of electrons and ions had ceased with the war, and his interests now lay elsewhere.

Apparently still not satisfied that the value of G was established reliably, Huxley now resumed, in Adelaide, the work he had started in Birmingham. It turned out that his reservations were well founded. His graduate students began a long series of investigations, first to establish the reliability of the experimental techniques in nitrogen and hydrogen, where the experiments were not complicated by attachment, and later in air and oxygen. A little later, at Huxley's instigation, one of these students, M.T. Elford, started a long series of ion mobility measurements, initially with the aim of providing data for calculating the spreading of meteor trails and hence their persistence (Huxley 1952). Ultimately the mobility data were to provide stringent tests of ion-atom interaction potentials.

The quest for a more accurate value of G had important consequences. In parallel with the development of reliable techniques for determining characteristic energies by the technique now known as the Townsend-Huxley method, and time-of-flight techniques for measuring drift velocities (the Bradbury-Nielsen method), came the development of computer-based techniques for analyzing the data. By this time, with the support of M.L.E. Oliphant, the Director of its Research School of Physical Sciences, Huxley had moved his laboratory to the Australian National University (ANU) in Canberra, where he had been appointed Vice-Chancellor, and new data and new techniques of analysis were coming out from both the ANU laboratory and from A.V. Phelps' group at the Westinghouse Research Laboratories (Frost & Phelps 1962). Phelps and his colleagues showed how it was possible to analyze the data using numerical techniques based on the previous work of Allis and Holstein, thus obtaining for the first time sets of energy-dependent elastic and inelastic cross sections that predicted the measured drift and diffusion data. The ANU laboratory had meanwhile also been developing numerical techniques for analyzing the data, but for the simpler case of elastic scattering only - a special case of the more general problem attacked by Phelps and his colleagues. The Westinghouse work was based on a solution of the Boltzmann equation; the ANU analysis was developed from the method of free paths of which Huxley was a master. Though the latter could be applied in this way only to the determination of energy-dependent *elastic* cross sections, Huxley showed that the two approaches led to the same analytical expressions. The first results from the ANU laboratory based on this new approach were reported at the 1965 ICPEAC in Quebec (Crompton & Jory 1965).

The application of computer-based techniques to the analysis of precision measurements of electron transport coefficients opened another era in the application of the Townsend 'swarm' techniques. In special circumstances these techniques now offered the possibility of providing benchmark cross-section data for low-energy electron-atom and electron-molecule collisions. However, in some situations more sophisticated transport theory was required. K. Kumar at the ANU and R.E. Robson, first at the ANU and later at James Cook University, took leading parts in developing it.

When Huxley left Adelaide in 1960 to take up his new appointment in Canberra, his successor, J.H. Carver, began a new line of atomic physics. Carver's initial career, at the ANU, was in nuclear physics, but he changed his interests when he went to Adelaide, prompted by the opportunities he saw for doing atmospheric physics in conjunction with the rocket facilities at the nearby Weapons Research Establishment. Under his direction a program of VUV and photoelectron spectroscopy was developed which integrated laboratory work with a program of rocket-borne experiments. E. Weigold, one of Carver's former students at the ANU, was appointed to his staff in 1962 to start a program on electron scattering. He left after a short time to take up an appointment with the US Air Force in Washington but returned in 1970 to join I.E. McCarthy and P.J.O. Teubner at the newly-established Flinders University of South Australia. Both were Adelaide graduates, and they too had returned from the U.S. to take up their appointments at Flinders, McCarthy from the

University of Oregon, and Teubner from the University of Pittsburgh. Together, McCarthy, Teubner and Weigold laid the foundations of the electron scattering group at that university.

My account thus far has been thematic rather than systematic in an attempt to describe the origins of Australian groups whose work is regularly represented at these conferences, and it is necessary to bring it to a close at a point many years ago to avoid the stage becoming too crowded. But even now there are important omissions because some of the groups are of relatively recent origin. One such is that of J.F. Williams at the University of Western Australia. Williams returned to Australia after many years overseas where he became well known for his work on electron scattering first at Gulf General Atomic in San Diego then at the Queen's University, Belfast. Williams was a student of Haydon's at the University of New England, then of D.N.F. Dunbar at the ANU where he worked on ion-beam scattering in gases. Dunbar's early career was also in nuclear physics, but he entered a new field on his appointment to the Department of Physics at the Canberra University College (later to become part of the ANU). In later years the work of Dunbar's group swung towards ion beam-solid interactions, so that the work of others who began their careers with him are now reported in other forums.

Another group that is relatively young is that of our co-chairman, M.C. Standage. Established shortly after the foundation of Griffith University in 1975 it has quickly established its reputation, but its roots are not to be found in the period on which most of this talk has focused.

Finally I return to the best known Australian in our field, H.S.W. Massey, and two of his distinguished colleagues who were fellow students in Melbourne and Cambridge.

Massey played an influential role in Australian science, though perhaps more through his interests in astronomy and upper atmosphere research than in atomic collisions as such. Nevertheless, to my knowledge he never published a paper that names an Australian institution in its by-line. He was born in Melbourne and received his first degree from the University of Melbourne. It seems that as a student in Melbourne he 'discovered' the new field of quantum mechanics for himself, and by 1928 was sufficiently familiar with it that he gave a paper to the first conference of Australian physicists which was held in Canberra in that year (Massey 1928). One year later he was on his way to Cambridge.

While Massey was laying the foundations of his distinguished career in England, theoretical atomic collision physics was languishing somewhat in Australia. There were, however, two prominent Australian theorists who must be included in this account: E.H. Burhop and C.B.O. Mohr. Both Burhop and Mohr were contemporaries of Massey's in Melbourne, and both collaborated with Massey when all three of them were PhD students in Cambridge in the early '30s, These early

Courtney Balthazar Oppenheim Mohr
1906-1986

collaborations continued after Massey left Cambridge in 1933, and even after Burhop and Mohr also left in 1936, Burhop for a lectureship in the Department of Natural Philosophy at the University of Melbourne, and Mohr for a similar appointment at the University of Cape Town.

Burhop remained at the University of Melbourne until 1942, when he joined the CSIR for a short period, and in 1945 he rejoined Massey at University College, London. Mohr returned to his old department in Melbourne in 1947. Thus Burhop and Mohr did not overlap in Melbourne and they never published together. In 1961 Mohr was appointed Professor of Theoretical Physics, one of the first such appointments in any Australian university. He was a man of great modesty whose work and influence were to continue well beyond his retirement in 1971. Nevertheless, it was perhaps not until McCarthy's appointment at Flinders that it could be claimed that Australia had a group, albeit small, of theoretical physicists specialising in atomic collisions.

6. CONCLUSION

In this talk I have tried to trace the origins of the Australian groups that have, over the years, contributed to the ICPEAC series. Their work does not, of course, by any means constitute all of the work in atomic physics that has been carried out in departments of physics and chemistry at the Australian universities, and in divisions of CSIRO, notably the Division of Chemical Physics. The emphasis has necessarily been on the origins of these groups - often related through relatively few influential people - rather than on the details of the work that has sprung from these origins. If I have given you some feel for the background against which the early scientists did their work, and how their interests and opportunities moulded the future development of research in electronic and atomic collisions in Australia, I will have succeeded.

ACKNOWLEDGEMENTS

I wish to thank many people who have supplied material that would otherwise have been difficult to trace, especially C.A. Bates, H.C. Bolton, R.H. Clayton, R.W. Home, J. Jenkin, M.L. Oliphant and R.P. Robertson. I also thank D.E. Boyd, A. Duncanson, J. Gascoigne and L. Robin for their invaluable assistance in the preparation of this manuscript.

REFERENCES

Bailey V A 1925 *Phil. Mag.* S 7 **1** 825
Bailey V A 1937a *Wireless World* February **26** 204
Bailey V A 1937b *Wireless World* March **5** 224
Bailey V A and Martyn D F 1934 *Nature* **133** 218
Bailey V A and Martyn D F 1935 *Nature* **135** 585
Bolton H C 1988 *Aust. Phys.* **25** 43 p 47
Bragg W H 1904 *Rept. Mtg. AAAS* **10** 47
Bragg W H and Kleeman R D 1905 *Trans. & Proc. Roy. Soc. S.A.* **29** 187; *Phil. Mag.* S 6 **11** 466
Branagan D and Holland G (eds) 1985a *Ever Reaping Something New* (Sydney: University of Sydney) p 93
Branagan D and Holland G (eds) 1985b ibid p 91
Branagan D and Holland G (eds) 1985c ibid p 78
Brose H L and Saayman E H 1930 *Nature* **126** 400
Crompton R W and Jory R L 1965 *IVth Int. Conf. on the Physics of Electronic and Atomic Collisions, Abstacts of Papers* ed B Bederson (New York: Science Bookcrafters) p 118
Fehsenfeld F C, Ferguson E E and Schmeltekopf A L 1966 *J. Chem. Phys.* **45** 1844
Franck J 1910 *Deutsch. Phys. Gesellsch.* **12** 291 613

Frost L S and Phelps A V 1962 *Phys. Rev.* **127** 1621
Home R W 1990a *Aust. Dict. Biog.* ed J Ritchie (Melbourne: Melbourne University Press) **12** p 141
Home R W 1990b *Physics in Australia to 1945* (Melbourne: University of Melbourne) p 190
Huxley L G H 1940 *Phil. Mag.* S 7 **30** 396
Huxley L G H 1952 *Aust. J. Sci. Res. A* **5** 10
Huxley L G H Foster H G and Newton C C 1948 *Proc. Phys. Soc.* **61** 134
Huxley L G H and Zaazou A A 1949 *Proc. Roy. Soc.* A **196** 402
Jenkin J 1986 *The Bragg Family in Adelaide* (Adelaide: The University of Adelaide Foundation)
Jenkin J 1993 *Aust. Dict. Biog.* ed J Ritchie (Melbourne: Melbourne University Press) **13** (to be published)
Lattey R T 1910 *Proc. Roy. Soc.* A **84** 173
Madsen J P V 1908 *Trans. & Proc. Roy. Soc. S.A.* **32** 12
Martyn D F and Pulley O 1936 *Proc. Roy. Soc.* A **154** 455
Massey H S W 1928 Conf. of Aust. Physicists, Canberra August 1928, Proceedings and Abstracts of Papers p 4
Massey H S W 1971 *Biographical Memoirs of Fellows of the Royal Society* **17** 497
Nielsen R A and Bradbury N (1937) *Phys. Rev.* **51** 69
Osborne W A 1920a *William Sutherland* (Melbourne: Lothian) p 40
Osborne W A 1920b ibid p 45
Piddington J H and Oliphant M L 1972 *Records of the Australian Academy of Science* **2** 47
Pollock J A 1909 *J. & Proc. Roy. Soc. N.S.W.* **43** 61
Sutherland W 1886 *Phil. Mag.* S 5 **22** 81
Sutherland W 1909 *Phil. Mag.* S 6 **18** 341
Sutherland W 1910 *Phil. Mag.* S 6 **19** 817
Tellegen B D H 1933 *Nature* **131** 840
Townsend J S and Tizard H T 1912 *Proc. Roy. Soc.* A **87** 357
Wellisch E M 1910 *Phil. Mag.* S 6 **19** 201

The Coulomb three-body problem

A T Stelbovics

School of Mathematical and Physical Sciences, Murdoch University, Perth, Australia 6150

Abstract.   Theoretical and experimental progress over the past decade is reviewed with emphasis on the electron-hydrogen and positron -hydrogen systems.

1. Introduction

The Coulomb three-body scattering problem is, perhaps surprisingly to the general scientific community, still a subject of intense interest for theorists and experimentalists alike. There are several reasons for this. First and foremost there is no complete theory of how to solve a three-body Schrödinger equation for arbitrary charges which incorporates all the boundary conditions relating to three charged particles. A second difficulty for the theorist is that in collision experiments our target atom, which for definiteness I take to be hydrogen, comprises a spectrum of an infinite denumerable number of excited bound states together with a continuous spectrum of non-localised states. Over the past twenty years it has become increasingly obvious that in order to model scattering processes to fine detail it is necessary to consider including representation of target states over a wide range of the energy spectrum. A further aspect which makes the theorist's job hard is that he also has to tailor his approach to modeling so as to suit the kinematic region of the scattering process when three free particle final states are produced.

From an experimentalist's point of view there are also a large number of difficulties that arise in performing the various scattering experiments. For example in carrying out an absolute electron-hydrogen measurement one is faced with the problem of determining to a high degree of accuracy, the ratio of atomic and molecular hydrogen in the scattering region. The measurement of differential cross sections requires long counting times especially at large scattering angles where the cross sections fall some several orders of magnitude. Greater challenges are posed in performing polarised-electron polarised-target scattering to study electron exchange effects and in replacing electrons with positrons in the incident beam.

The talk will emphasise the electron-hydrogen system for which many recent calculations and experiments have been performed; positron scattering will be mentioned briefly. Because of the vast nature of work done in recent years I will not attempt to cover every aspect of research but will concentrate on a few particular areas.

2. Electron-hydrogen scattering

I will begin with an analysis of electron-hydrogen scattering at an intermediate energy of 54.42 eV. My reason for choosing this energy as a starting point is two-fold.

Firstly there are no simplifying features and it provides the most stringent test of all our model assumptions. It is high enough to require solution of close-coupling equations for total orbital angular momentum up to J=20; electron-exchange effects are important, as is the coupling to the ionisation channels. Only then can asymptotic approximations such as second-Born, unitarised Born and finally Born approximations suffice. Secondly, experimentally there are several sets of data to compare with and tantalisingly there are many areas of disagreement with the theories.

2.1. The experimental situation

Elastic differential cross-sections (at 50eV) have been measured by Williams (1975) and absolute inelastic differential cross sections to the 2s and 2p levels by Williams (1981). Frost and Weigold (1980) carried out a relative measurement of the 2s and 2p excitation, normalising their results to a theory model at 10 degrees. Angular correlations for the 2p level have also been measured. They give information about the modulus and relative phases of the scattering amplitudes for scattering into the $m = -1, 0, 1$ magnetic sublevels. Three parameters λ, R and I defined as

$$\sigma = \sigma_0 + 2\sigma_1, \quad \lambda = \frac{\sigma_0}{\sigma}, \quad R = \frac{\mathrm{Re}\langle f_1 f_0^* \rangle}{\sigma}, \quad I = \frac{\mathrm{Im}\langle f_1 f_0^* \rangle}{\sigma}, \tag{1}$$

where σ_m is the cross section averaged over singlet and triplet scattering :

$$\sigma_m = \langle f_m f_m^* \rangle = \frac{1}{4}\langle f_m^S f_m^{S*} \rangle + \frac{3}{4}\langle f_m^T f_m^{T*} \rangle. \tag{2}$$

The first measurements were carried out by Hood *et al* (1979) for forward scattering angles. Measurements extending to angles of 140 degrees are given by Weigold *et al* (1980), Williams (1981) and again by Williams (1986). The 1986 measurements repeated the earlier ones, but also reported the first measurement of the I parameter. Further the experimental apparatus used in 1981 was recalibrated and absolute measurements of the 2s, 2p, 1s and summed n=2 cross sections were repeated leading Williams to report "the agreement is within 10% for all quantities so present measured values are not given here". These experiments have served as the basis for testing of all theory models up to the present day. Since absolute measurements of cross sections are difficult it should be mentioned that a further check of the accuracy of the measurements performed by the above workers was carried out by Lower *et al* (1987). They reported measuring accurate ratios of differential cross-sections for the inelastic (n=2) and elastic scattering of electrons by atomic hydrogen at several scattering angles at 100 and 200 eV. Their conclusion was thas "there is agreement essentially between all experiments...".

2.2. Theory models

Having established our experimental base let me turn to some observations of the status of theoretical calculations. I will confine my discussion to three types of non-perturbative models which are in use in the present day. They are the close-coupling optical model (CCO), pseudo-state close-coupling equations and intermediate energy

R-matrix (IERM). Each method seeks to solve the three-body Schrödinger equation by making an expansion of the three-body wave function $\Psi^{\pm}(1,2)$ in terms of a complete set of states for either one or both coordinates. If one uses the hydrogen target states and writes

$$\Psi^{\pm}(1,2) = \frac{1}{2}\sum_i (\phi_i(1)F_i^{\pm}(2) \pm \phi_i(2)F_i^{\pm}(1)), \tag{3}$$

one obtains the close coupling equations:

$$(-h_0 + E - \epsilon_j)F_j^{\pm} = \sum_k V_{jk}^{\pm}F_k^{\pm}. \tag{4}$$

The channel potentials are

$$V_{jk}^{\pm} = U_{jk}^{\pm} + W_{jk}^{\pm},$$

where

$$U_{jk}^{\pm} = \int d\mathbf{r}_1 \int d\mathbf{r}_2 |\mathbf{r}_1\rangle\phi_i(\mathbf{r}_2)\left[-\frac{1}{r_1} + \frac{1}{|\mathbf{r}_1 - \mathbf{r}_2|}(1 \pm P_{12})\right]\phi_k(\mathbf{r}_2)\langle\mathbf{r}_1| \tag{5}$$

and

$$W_{jk}^{\pm} = \pm|\phi_k\rangle(\epsilon_j + \epsilon_k - E)\langle\phi_j|. \tag{6}$$

The sum is used in the sense that it represents a discrete summation over bound states and an integration over the continuum. A direct solution of these equations using the continuum functions has not been attempted to date. Instead we can proceed by taking a basis of L^2 functions in which to diagonalise the target Hamiltonian. In practice one is restricted to a finite number of functions and includes in the basis set, those functions which describe the target channels of interest (P-space) exactly. One obtains an approximate representation of the remaining target states, discrete and continuum in terms of normalizable pseudo-states. These states are not true eigenstates and their pseudo-energies and form vary with number and type of basis functions.

The close-coupling optical models may be derived by making a Feshbach projection of the close-coupling equations onto the P-space. This is only one particular way of choosing the P and Q operators. Other choices are possible but I merely want to illustrate the form of the resultant optical potential so I will stick to this one. Then the close-coupling equations can be written as

$$(-h_0 + E - \epsilon_j)F_j^{\pm} = \sum_{k\in P}(V_{jk}^{\pm} + V_{jk}^{(Q)\pm})F_k^{\pm}, \quad j \in P, \tag{7}$$

where $V^{(Q)}$ is a complex energy dependent optical potential and is also known as the polarisation potential. The price one pays for restricting the sum in the channel indices to the P-space channels is that the full distorted wave function over all Q-space channels is embedded in the polarisation potential itself since

$$V_{jk}^{(Q)\pm} = \sum_{l,m\in Q} V_{jl}^{\pm}G_{lm}^{(Q)\pm}V_{mk}^{\pm}, \quad j,k \in P. \tag{8}$$

Here $G_{lm}^{(Q)\pm}$ is defined by the equation

$$\sum_{l \in Q} \left[(-h_0 + E - \epsilon_j)\delta_{jl} - V_{jl}^{\pm} \right] G_{lm}^{(Q)\pm} = \delta_{jm}, \quad j, m \in Q. \tag{9}$$

The optical potential contains contributions from discrete and continuum parts of Q-space. All applications of this method therefore rely on the nature of the approximations involved in calculating the Q-space sum. In the past decade optical potentials have been widely used by the Flinders group. The original model of the polarisation potential (McCarthy and Stelbovics 1980,1983b) used several approximations. Contributions to the polarisation were neglected for all but the ionisation continuum states. The Q-space three-body wave was modelled as a product of a Coulomb wave and a plane wave, and a screening approximation was adopted so that the plane-wave was always chosen for the faster electron. The integration over the continuum was carried out directly by performing the integration over both continuum electrons. Further approximations were made which included a localisation ansatz and spherical averaging. One overall check that the approximations are reasonable can be made by computing the total ionisation cross-section which is essentially the imaginary part of the optical potential (McCarthy and Stelbovics 1983a). The first optical potential calculation using this model with a P-space comprising 1s, 2s and 2p channels included polarisation potentials coupling only the diagonal channels. The results compared very favourably with the experimental differential cross-sections for 1s, 2s and 2p scattering of Williams(1981). van Wyngaarden and Walters (1986a) emphasised that neglecting off-diagonal coupling could lead to larger errors than McCarthy and Stelbovics had suggested. They studied electron scattering from hydrogen using a 21 pseudo-state basis which they solved in a hybrid pseudo-state close-coupling for a subset of the states and applied corrections due to the other Q-space contributions through a distorted-wave second-Born approximation. Their results for the differential cross-sections tended to fall below the experimental values, especially for scattering to the 2s channel. In the optical model calculation of Lower *et al* (1987) P-space comprising n=1,2 and 3 states was used and non-diagonal optical potentials for 1s-2s and 1s-2p coupling were included. The results for inelastic scattering were now in closer agreement with those of van Wyngaarden and Walters. In the most recent application of the model (Ratnavelu and McCarthy 1990) the locality approximation was removed for the most important contributions to the continuum polarisation, namely the 1s-1s, 1s-2s and 2s-2s potentials and retained for the others. In addition P-space was extended to include all n=1 to 4 channels except the 4f. Their calculations were now much closer to the van Wyngaarden and Walters results. Differences between the models were still noted for scattering to the 2p level.

The Ratnavelu and McCarthy calculation marked the end of the optical model in the original form. Other optical potential models using pseudo-states to approximate Q-space were also considered during this period (see Bransden *et al* 1985 and Callaway and Oza 1985 for details). Since that time important developments have been made to the polarisation potential representation. They center around the use of distorted waves to represent the Q-space in the polarisation potential (Bray *et al* 1990). This has been followed by a further calculation of the internal consistency of the distorted wave method (Bray *et al* 1991a) and a modified prescription for choice of P and Q space which I Bray will be reporting on (Bray *et al* 1991b) at this conference. The distorted waves are calculated until convergence is achieved. For Q-space over the bound states this means including target states with l=0 to 4, and for the Q-space continuum distorted waves up to l=15 are included. The major approximation that is made in the new generation of calculations is that the distorted waves are calculated in a weak coupling approximation. It is assumed the distortion due to coupling between different Q-space channels is not significant. This means that the $j \neq l$ potentials are set to zero in equation (9). Bray *et al* (1991a) have verified that it is a reasonable approximation for that part of Q-space comprising the discrete channels.

In parallel with the latest CCO calculations the R-matrix method has been improved in the past few years. The latest version of the theory developed is the intermediate energy R-matrix (IERM). The method was introduced by Burke *et al* (1987) and has been analysed by Scholz (1991). As in the standard R-matrix method (Burke and Robb 1975) the configuration space of the full wave function is partitioned into an internal and external region. The internal region is a sphere centered on the atom with boundary radius large enough to contain all the P-space channels. In this region the three-body wave function is expanded in a set of basis states as

$$\Psi^\pm(1,2) = \sum_{i,j} \phi_i(1)\phi_j(2)a_{ij}^\pm.$$ (10)

The orbitals ϕ_i include the P-space channels. The remainder are numerical orbitals defined only in the internal region and calculated as for the standard R-matrix method in a model potential. The IERM expansion includes continuum-continuum contributions in the sum (10) in contrast to the earlier R-matrix method. The a_{ij} coefficients are obtained by diagonalising the full three-body Hamiltonian within the internal region. In the external region the wave function is expanded in the form of an unsymmetrised close-coupling expansion, because the assumption is made that exchange effects are important only in the internal region (Scholz 1991). The total wave function over all space comprises a linear combination of the internal and external forms matching smoothly at the boundary. The wave function partitioning is formally expressed in terms of Q-space (the internal region) and P-space (the external region). Then employing a Feshbach projection one obtains an equation similar in form to that noted earlier for the CCO formulation with the main difference being that the optical potential term here contains poles in the three-body energy at the energies given by the diagonalisation of H in the internal region. This leads to amplitudes which exhibit pseudo-resonance structure. The resonance structure is removed by numerical averaging. An analytic justification for the procedure is discussed by Slim and Stelbovics (1988).

The present generation of IERM calculations use a hybrid formulation. The R-matrix is solved only for the first four partial waves in total angular momentum of the three-body wave function. Typically one is diagonalising a matrix of dimension of order 3,500 by 3,500 to construct the optical potential. This is time consuming and so a change is made at J=5. There the IERM results match smoothly over to a smaller R-matrix caculation using a set of pseudo-states due to Fon *et al* (1981).

The results of the most recent calculations are given by Scott *et al* (1989) and Scholz *et al* (1991).

2.3. Comparison of theory with experiment

Let me now summarise the most recent results of the Flinders and Belfast groups for scattering at 54.4 eV. I will mention only some of the results. In figures 1a and 1b the differential cross sections for scattering to the 2s and 2p levels are shown. Both theories produce a similar 2s cross section. The trend for model predictions to lie below the experimental results is continued. For the 2p channel the theories are in accord as well, especially at forward angles. In the angular range from 40 to 100 degrees, the Scholz *et al* (1991) values lie above and below the Bray *et al* (1991b). The decrease in the cross-section of Scholz *et al* at large backward angles may be an artifact of the caculational method. The IERM numbers for the 2p cross section include a distorted-wave Born correction to take account of the higher partial wave orbitals omitted in the expansion (10).

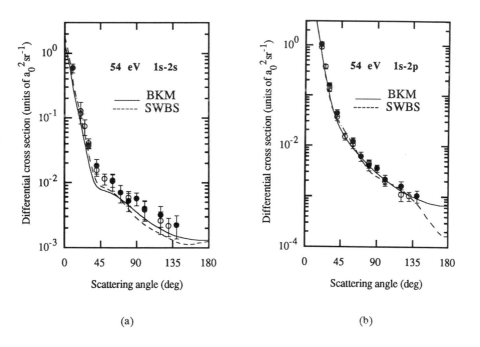

(a)   (b)

Figure 1. The differential 2s and 2p cross sections for electron scattering on atomic hydrogen. BKM denotes the CCO model of Bray *et al* (1991b). SWBS denotes the IERM model of Scholz *et al* (1991). The open and closed circles denote the measurements of Frost and Weigold (1980), and Williams (1981), respectively.

Looking at the angular correlations in figures 2a, 2b and 3a, one again observes a good deal of similarity between the results. The IERM results have more angular structure which appears to average around the smoother CCO curves. For all parameters there is still a marked discrepancy with the experiments. Since the three different sets of results (Williams 1981, 1986 and Frost and Weigold 1980) concur about the details of the structure in λ and R at backward angles one has to say that our most sophisticated models at intermediate energy still cannot explain some fundamental aspects of inelastic scattering to the n=2 levels of hydrogen by electron impact.

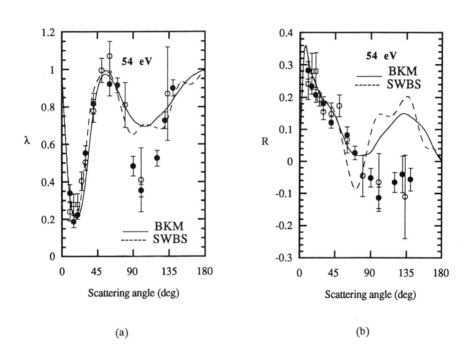

(a)	(b)

Figure 2. Angular correlation parameters λ and R for the 2p excitation level of atomic hydrogen. The labels are the same as for figure 1. The open squares denote the measurements of Hood *et al* (1979).

2.4. Spin asymmetry

Exchange effects are important at 54.4 eV. To date, there have been no measurements of inelastic scattering to the n=2 levels using polarised electrons on polarised hydrogen atoms. Progress in polarised-electron scattering has been reviewed recently by Kessler (1991). When we do have this data it will provide theorists with a more stringent test of their models. In order to indicate the magnitude of exchange effects let me introduce the spin asymmetry parameter A. For each cross-section measured it can be defined as

$$A(\theta) = \frac{\sigma^S - \sigma^T}{\sigma^S + 3\sigma^T}. \tag{11}$$

If singlet scattering dominates A is near 1 while for pure triplet scattering it is equal to -1/3. In the high energy limit where exchange effects are negligible it is 0.

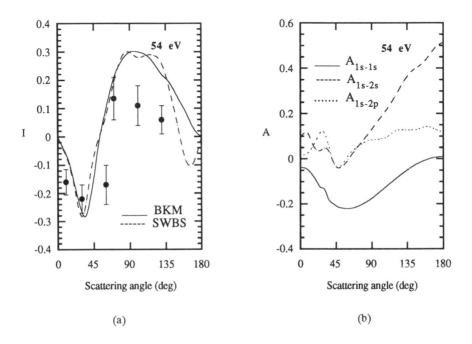

(a)   (b)

Figure 3. Angular correlation parameter I and spin asymmetries for electron scattering by atomic hydrogen. For I the theory curves are as in figure 1, the measurements are due to Williams (1986). The asymmetries are due to the theory of Bray *et al* (1991b).

At 54.4 eV the exchange asymmetries for elastic and n=2 inelastic cross sections are shown in figure 3b for the data of Bray *et al* (1991b). The elastic scattering in the 1s channel is predominantly triplet at intermediate angles and exchange is small at forward and backward angles. Singlet scattering on the other hand is more important for inelastic scattering to the n=2 levels. van Wyngaarden and Walters (1986b) carried out a study of the spin asymmetry over a range of energies using their pseudo state close-coupling caculation (van Wyngaarden and Walters (1986)). For the 2s scattering there is some nodal structure at forward angles dipping in the neighbourhood of 45 degrees and rising steadily large angles. The 2p cross section also has a node at forward angles, followed by a similar minimum as the 2s and rises slowly in the backward direction. The general trend of the features shown in figure 3b is similar to that exhibited by the asymmetries of van Wyngaarden and Walters although the magnitude and position of nodes of the inelastic asymmetries is somewhat different. It will be interesting to see in the coming years if the forthcoming experimental measurements will provide accurate enough results to permit yet another assessment of various models. At what energies do spin asymmetries become negligible? From van Wyngaarden and Walters results it appears that the answer is at least several hundred electron volts since at their maximum energy of 300 eV the inelastic asymmetries still reach values in the order of .1 in a node below 30 degrees.

2.5. Close-coupling at higher energies

The question of what constitutes high energy is interesting from the point of view of the close-coupling formalism. At a "high enough" energy, we know that our perturbative methods, such as the distorted-wave Born approximations (see for example Madison 1989) and eikonal- Born series, summing higher-order Born terms give reliable answers. (These methods are summarised by Joachain 1989.) However it is not obvious from inspection of the energy-dependent term W_{jk} (equ. 6) in the potentials of the close-coupling equations how this can be achieved since it is increasing with energy and it is unclear that its contribution to exchange will diminish. This resolution of the role of W_{jk} turns out to be subtle. The mathematical details are discussed by Stelbovics(1990) but the results can be stated as follows. If one does calculate the iterated Born series using the full close-coupling potentials, the sum will eventually diverge no matter how high the energy is. But if the close-coupling equations are solved first for that part of the potential excluding W_{jk}, then it can be shown that the T-matrix amplitudes of the full close-coupling equations are the sums of the amplitudes for the system with the restricted potential U_{jk}, plus a remainder term involving a combination of terms built from W and the restricted T-matrix. The remainder term can be shown to die off with increasing energy and the general off-shell T-matrix can be proven to diverge in the operator-norm sense. Any perturbative treatment of the close-coupling equations must exclude the W_{jk} part of the potential, which since it is of separable form can be summed to all orders in closed form. Thus the close-coupling equation amplitudes do approach those derived from the Born series summation methods in the limit of high energies.

At other energies discrepancies between theory and experiment are not as marked. For example in the near threshold region which is dominated by resonances Williams (1989) has carried out differential cross-section measurements to the 2p level as well as measuring total cross sections to the n=3 levels. Within experimental error there is substantial agreement with the large-scale pseudo-state calculations of Callaway (1988,1990).

3. Ionisation

Aside from inelastic scattering to discrete states, ionisation processes are of considerable interest. Recent reviews of this area have been given by McCarthy and

Weigold (1991) and Lahmam-Bennani (1991). Here I wish to touch on three topics to indicate recent progress. They are total ionisation cross-section, threshold behaviour of ionisation cross-section and three-body Coulomb scattering boundary conditions.

3.1. Total cross-sections

R-matrix and CCO theories can be used to calculate total ionisation cross-sections by using the optical theorem to obtain the total inelastic cross-section and then subtracting off contributions from all the discrete states. Scholz *et al* (1990) have calculated an ionisation-cross-section from threshold up to 3.7 Ryd. using total cross-sections obtained from the IERM calculation of Scott *et al* (1989). Estimates of the n=3 total cross section were obtained from experiment while the remaining inelastic contributions were parameterised by Born approximations. The results are uniformly about 15% above the experimental values of Shah *et al* (1987) over the whole range of energies. It is reasonable to attribute this discrepancy to the approximate treatment of the inelastic cross-sections. Incorporating more of the excited states into the IERM P-space can be expected to improve the estimate for the ionisation cross-section. An earlier pseudo-state close coupling caculation of Callaway and Oza (1979) tended to underestimate the cross-section by varying amounts.

Asymmetry measurements for the total ionisation cross-section have been presented by Fletcher *et al* (1985) and more recently by Crowe *et al* (1990) from threshold up to several hundred eV. The measurements show singlet scattering is dominant at threshold with a limiting value of $A \approx 0.45$ near threshold. The asymmetry decreases smoothly to zero around 500 eV. At 200 eV it still has a value of 0.1. This is very similar to the higher energy behaviour of asymmetries for the elastic and inelastic scattering to n=2 levels mentioned earlier. With regard to theoretical models, probably the most sophisticated caculation is that of Bray *et al* (1990). Using a one-channel CCO model they described the ionisation continuum in their polarisation potential by means of a Coulomb wave for each electron. This symmetric treatment leads to an impressive agreement with the experimental results. However a word of caution should be added; this choice of distorted wave leads to an overestimate of the total ionisation cross section. It would be prudent to assess the results of their model with an extended P-space and with the latest varient of symmetric projection operators (Bray *et al* 1991b).

3.2. Threshold ionisation

Recently Guo *et al* (1990) have presented results of asymmetry measurements in the vicinity of the ionisation threshold. Their measurements covered the range of energies from 13.57 to 15.27 eV. They conclude that there is clear evidence of structure with possible sharp features being masked by the 75 meV energy spread of the electron beams. For a restricted region extending 0.5 eV above threshold they find that a chi-squared fit to a constant asymmetry has only a 3% confidence level while a fit to a linear variation in energy has a 56% confidence level. They state that the Wannier theory is consistent with a constant asymmetry near threshold.

However Rau (1991) notes that by applying perturbation techniques to the Wannier formalism a linear energy dependence is permitted. The observation of threshold structure implies a departure from the Wannier model of ionisation which has been tested successfully for other atoms. An alternative model has been proposed by Temkin and is referred to as the Coulomb-dipole (CD) approximation. In the Wannier model the most probable mode for near threshold ionisation is assumed to be where the electrons escape on opposite sides of the ion core and where the total energy is uniformly shared between the electrons. Temkin (1982) argued that the most important configuration is that for which the two electrons have different energies. Bottcher (1988) has carried out numerical simulations which suggest Wannier theory is highly accurate unless the ratio of electron energies exceeds a factor of ten in which case Temkin's model is applicable. The application of the CD ansatz leads to cross-sections and asymmetries which have a rapid variation with energy near threshold. Temkin (1991) has noted that the precise form of this variation depends not only on a knowledge of the CD wavefunction which is accurate remote from the interaction region but also on the form of the wave function in the interaction region. Unfortunately in seeking to match the CD wavefunction onto an approximate form in the interaction region Temkin succeeded only in obtaining a linear dependence in energy for the ionisation cross-section.

The suggestion has been made also that the details of the final state wave function in the interaction region are important in interpreting the coplanar (e,2e) experiments of Ehrhardt *et al* (1989) and Schlemmer *et al* (1989). In the experiments, the ionised electrons with equal energy of 2eV and a fixed angle between them of 150 or 180 degrees were measured in coincidence for varying scattering angles using hydrogen and helium targets. The angular correlations for each target are substantially different. As far as Wannier theory is concerned, only the form of the wave function in the asymptotic region is important. Since the hydrogen and helium correlations are so different the suggestion has been made that this difference must be due to the details of the wave function in the interaction region. However Pan and Starace (1991) have shown that a calculation of the ionisation amplitude using distorted wave for the initial state and Coulomb waves for the exiting electrons with effective charges determined from a screening argument is sufficient to explain the observed differences.

3.3. Three-body Coulomb boundary conditions

e-2e experiments probe the three-body wave function for three asymptotically free Coulomb particles. Most of the total ionisation cross-section comes from the kinematic region where the momentum transfer of the incident electron is small. There are a number of theory models to describe the triple differential ionisation cross-section. They include a pseudo-state close-coupling model of Curran and Walters (1987), the unitarized eikonal-Born series of Byron *et al* (1985) and a Distorted-wave Born approach (McCarthy and Zhang 1989). All the above mentioned methods have treated the ionised electrons in terms of a product of screened Coulomb waves. All are reasonably successful in describing the experimental data of Ehrhardt *et al* (1985). Recently a different approach to the cross-section modelling has been presented by Brauner *et al* (1989) in which they stress the importance of incorporating the correct three-body boundary conditions. They use an approximate form of the three-body wave function which satisfies the boundary conditions for the final state and are able to compute ionisation amplitudes without resorting to partial-wave decompositions. They also obtain very good agreement with the experimental data at higher energies. Their approach is satisfying from the point that electron-electron correlations are included in a physically sensible way. On the other hand Curran *et al* (1991) have pointed out that the ionisation amplitude depends on the overlap with the initial ground state of hydrogen which is localised, so the approximate nature of the three-body wave function in the interaction region may lead to an error which is difficult to assess. This of course is the same argument that is leveled at the other approaches which include distortion in the entrance channel at the expense of an approximate treatment of the three-body asymptotic state.

For future work it would be interesting to see if the correct boundary conditions could be implemented in a close-coupling formalism when continuum target states are included in the target expansion. Gailitis (1990) points out that this may be impractical for methods such as close-coupling which expand over target states of all angular momenta since the effect of the electron-electron correlation in the three-particle asymptotic states translates into partial-wave expansions whose terms do not die off with increasing angular momentum.

4.   Positron scattering

Positron-atom scattering experiments are a valuable additional tool in our understanding of the collision reaction mechanism (see Kauppila and Stein 1990). To date there has been only one type of experiment reported for positron-hydrogen scattering, that which measures the total ionisation cross-section for positron impact. Spicher *et al* (1990) measured positron and electron impact ionisation cross sections from 17.5 to 600 eV. The positron cross-section is larger over the whole range of energies, roughly by factor of two at lower energies and still significantly larger at 400 eV. The positron and electron cross-sections appear to be converging above 500eV where one expects the Born approximation to be good. The existing models underestimate the positron cross-section significantly, lying much closer to the electron ionisation cross-section results. It is interesting to note that triple differential cross-section models (for example Joachain and Piraux 1986 and Brauner *et al* 1989) are larger by about the right order of magnitude. It would be desirable to see a calculation of the integrated cross-section using these more sophisticated models; one suspects that it could come close to reproducing the measured cross-section above 100eV.

There have been several recent theoretical investigations in anticipation of future experiments. Elastic scattering T-matrix elements as well as elastic and total cross sections up to 3.7 Rydbergs have been reported by Higgins *et al* (1990) using the IERM model. Positronium formation to the n=1 and n=2 levels has been investigated by Hewitt *et al* (1990) up to 68eV using a close-coupling formulation. They find an enhancement up to a factor of four for the positronium formation cross-section in the 2s channel compared to earlier distorted-wave calculations below 30 eV incident energy. Similar enhancements are observed for the 2p cross-sections up to 50 eV. This is due to the polarisation in the incident and rearrangement channels introduced by their basis sets.

5. Concluding remarks and future directions

I have taken a quick journey through the jungle that is the three-body Coulomb problem. I have deliberately selected areas where there are good experiments to model. I have also tried to emphasise models which attempt to include as much of the physics as possible. What type of experiments would a theorist wish for in the near future? Without doubt, measurements of asymmetry parameters for elastic and inelastic scattering would be of interest especially since the asymmetries seem to behave differently for different types of excitations. With the increasing number of polarized electron experiments coming on stream perhaps this will be a reality in the not too distant future. Next it would also be desirable to have a fresh set of absolute differential and total cross-sections at intermediate energies for elastic and inelastic scattering. Both IERM and CCO models are reaching similar conclusions and it is time for smaller error bars in the experimental measurements. Further experiments measuring excitation to $n=3$ levels would be of interest over a whole range of energies; they would make additional demands on the sophistication of the coupled channels and R-matrix methods. Finally any information concerning positron scattering of hydrogen targets to elastic or inelastic channels will be very welcome.

Acknowledgments

I wish to thank the Australian Research Council for supporting my research. I would also like to thank T. Scholz and I. Bray for access to unpublished results.

References

Bottcher C 1988 Adv. Mol. Phys. **25** 303
Bransden B H, McCarthy I E, Mitroy J D and Stelbovics A T 1985 Phys. Rev. A **32** 166
Bray I, Madison D H and McCarthy I E 1990 Phys. Rev. A **41** 5916
Bray I, Konovalov D A and McCarthy I E 1991a J. Phys B: At. Mol. Opt. Phys. **24** 2083
Bray I, Konovalov D A and McCarthy I E 1991b Phys. Rev. A (in press)
Burke P G, Noble C J and Scott P 1987 Proc. R. Soc. A **520** 289
Burke P G and Robb W D 1975 Adv. At. Mol. Phys. **11** 143
Brauner M, Briggs J S and Klar H 1989 J. Phys B: At. Mol. Opt. Phys. **22** 2265
Byron F W, Joachain C J and Piraux B 1985 J. Phys B: At. Mol. Phys. **18** 3203
Callaway J 1988 Phys. Rev. A **37** 3692
Callaway J 1990 J. Phys B: At. Mol. Opt. Phys. **23** 751
Callaway J and Oza D H 1979 Phys. Lett. **72A** 207
Callaway J and Oza D H 1985 Phys. Rev. A **32** 2628
Crowe D M, Guo X Q, Lubell M S, Slevin J and Eminyan M 1990 J. Phys B: At. Mol. Opt. Phys. **23** L325
Curran E P and Walters H R J 1981 J. Phys. B: At. Mol. Opt. Phys. **20** 337
Curran E P, Whelan C T and Walters H R J 1991 J. Phys B: At. Mol. Opt. Phys. **24** L19

Ehrhardt H, Knoth G, Schlemmer P and Jung K 1985 Phys. Lett. **110A** 92

Ehrhardt H, Rosel J, Sclemmer P, Agricola R and Jung K 1989 In 16th ICPEAC New York Abstr. Cont. Papers 232

Fletcher G D, Alguard M J, Gay T J, Hughes V W, Wainwright P F, Lubell M S and Raith W 1985 Phys. Rev. A **31** 2854

Fon W C, Berrington K A, Burke P G and Kingston A E 1981 J. Phys. B: At. Mol. Phys. **14** 1041

Frost L and Weigold E 1980 Phys. Rev. Lett. **45** 247

Hood S T, Weigold E and Dixon A J 1979 J. Phys. B: At. Mol. Phys. **12** 631

Gailitis M 1990 J. Phys B: At. Mol. Opt. Phys. **23** 85

Guo X Q, Crowe D M, Lubell M S, Tang F C, Vasilakis A, Slevin J and Eminyan M 1990 Phys. Rev. Lett. **65** 1857

Hewitt R N, Noble C J and Bransden B H 1990 J. Phys B: At. Mol. Opt. Phys. **23** 4185

Higgins K, Burke P G and Walters H R J 1990 J. Phys B: At. Mol. Opt. Phys. **23** 1345

Joachain C J 1989 Invited Papers of the 16th ICPEAC (Eds. Dalgarno A *et al*) AIP Conf. Proc. 205 68

Joachain C J and Piraux B 1986 Comm. At. Mol. Phys. **17** 261

Kauppila W E and Stein T S 1990 Adv. At. Mol. Opt. Phys. **26** 1

Kessler J 1991 Adv. At. Mol. Opt. Phys. **27** 81

Lahmam-Bennani A 1991 J. Phys B: At. Mol. Opt. Phys. **24** 2401

Lower J, McCarthy I E and Weigold E 1987 J. Phys B: At. Mol. Phys. **20** 4571

Madison D H 1989 Invited Papers of the 16th ICPEAC (Eds. Dalgarno A *et al*) AIP Conf. Proc. 205 149

McCarthy I E and Stelbovics A T 1980 Phy. Rev. A **22** 502

McCarthy I E and Stelbovics A T 1983a Phys. Rev. A **28** 1322

McCarthy I E and Stelbovics A T 1983b Phys. Rev. A **28** 2693

McCarthy I E and Weigold E 1991 Adv. At. Mol. Opt. Phys. **27** 201

McCarthy I E and Zhang X 1989 J. Phys. B: At. Mol. Opt. Phys. **22** 2189

Pan C and Starace A F 1991 Phys. Rev. Lett. **67** 185

Ratnavelu K and McCarthy I E 1990 J. Phys B: At. Mol. Opt. Phys. **23** 1655

Rau A R P 1991 Private Communication

Scholz T T 1991 J. Phys B: At. Mol. Opt. Phys. **24** 2127

Scholz T T, Walters H R J and Burke P G 1990 J. Phys B: At. Mol. Opt. Phys. **23** L467

Scholz T T, Walters H R J, Burke P G and Scott M P 1991 J. Phys B: At. Mol. Opt. Phys. **24** 2097

Scott M P, Scholz T T, Walters H R J and Burke P G 1989 J. Phys B: At. Mol. Opt. Phys. **22** 3055

Schlemmer P, Rosel T, Jung K and Ehrhardt H 1989 Phys. Rev. Lett. **63** 252

Shah M B, Elliot D S and Gilbody H B 1987 J. Phys B: At. Mol. Phys. **20** 3501

Slim H A and Stelbovics A T 1988 J. Phys B: At. Mol. Opt. Phys. **21** 1519

Spicher G, Olsson B, Raith W, Sinapius G and Sperber W 1990 Phys. Rev. Lett. **64** 1019

Stelbovics A T 1990 Phys. Rev. A **41** 2536

Temkin A 1982 Phys. Rev. Lett. **49** 365

Temkin A 1991 J. Phys. B: At. Mol. Opt. Phys. **24** 2147

Weigold E, Frost L and Nygaard K J 1980 Phys. Rev. A **21** 1950

Williams J F 1975 J. Phys B: At. Mol. Phys. **8** 2191

Williams J F 1981 J. Phys B: At. Mol. Phys. **14** 1197

Williams J F 1986 Aust. J. Phys. **39** 621

Williams J F 1989 Invited Papers of the 16th ICPEAC (Eds. Dalgarno A *et al*) AIP Conf. Proc. 205 115

van Wyngaarden W L and Walters H R J 1986a J. Phys B: At. Mol. Phys. **19** 929

van Wyngaarden W L and Walters H R J 1986b J. Phys B: At. Mol. Phys. **19** 1817; ibid 1827

Recent experimental work on two-electron correlations

Frank H Read

Department of Physics, Schuster Laboratory, University of Manchester, Manchester, M13 9PL, U.K.

ABSTRACT: The experimental evidence for the existence of strong electron correlations in three areas of atomic physics will be discussed. These areas are (1) double–Rydberg atomic states, (2) threshold ionisation processes, and (3) capture processes in which doubly-excited states are produced.

1. INTRODUCTION

Electron–electron correlations play an important and occasionally dominant role in several areas of atomic and molecular physics. Three such areas will be discussed here. The first concerns those atomic states in which two electrons are excited to Rydberg orbitals. This area offers a rich and diverse set of examples of strongly correlated systems, the full extent of which has only just begun to be explored. The second area is that of threshold ionisation processes in which two slowly moving electrons are produced. Several processes of this type have been well studied, but some important issues remain unresolved. Finally processes in which slowly moving, highly charged ions capture two electrons into doubly-excited states are discussed, both because these processes offer one of the few means of creating such states and because the processes may themselves involve correlation effects.

2. DOUBLE-RYDBERG ATOMIC STATES

The double–Rydberg states are a sub-set of the doubly-excited states, which include states such as He $2\ell n\ell'$, in which the principal quantum number of one or both electrons is only one higher that that of the ground state, so that these electons are then not usually considered to be Rydberg electrons. Electron–electron correlations appear in the strongest and purest form only when <u>both</u> electrons are Rydberg electrons, because the electron–core interactions then lose their dominance.

The set of double–Rydberg states $n\ell n'\ell'$ can itself be divided into further, sometimes overlapping, sub-sets, three of which will be discussed here. These are (i) the intrashell states ($n = n'$), (ii) the "planetary" states ($n, n' \gg 1$), and (iii) the "circular" states ($n = \ell + 1$, or $n' = \ell' + 1$).

At the present time the structure and description of the intrashell double–Rydberg states is far from being understood, but rapid progress is being made (e.g. Müller et al 1991). It seems safe to assume that the state with the lowest potential and total energy in an intrashell manifold of states is that in which the two electrons are localised to the greatest extent in the region of the Wannier point ($\underline{r}_1 = -\underline{r}_2$), with r_1 approximately constant and with no overall rotation (L =0). These will be given the provisional description here of "Wannier-point" states, and this description will also include states of the same type but with an overall rotation, as depicted in figure 1.

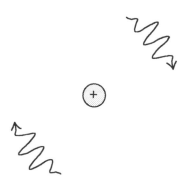

Figure 1.   Schematic of a rotating "Wannier-point" state, a particular example of an intra-shell (i.e. n = n´) double-Rydberg state.

Higher internal energy can arise from (i) symmetric stretching along the inter-electron axis ($r_1 \approx r_2$, varying; $\theta_{12} \approx 180°$), (ii) asymmetric stretching along the same axis ($r_1 \neq r_2$; $\theta_{12} \approx 180°$) or bending about the axis ($r_1 \approx r_2 \approx$ const; $\theta_{12} \neq 180°$), or combinations of these. Overall rotation also provides additional energy. The bending mode is doubly degenerate and so can give rise to internal angular momentum. These possibilities are illustrated in figure 2.

The region for which $r_1 \approx r_2$ is known as the Wannier ridge, because a local maximum exists in the potential energy when this is expressed in terms of hyperspherical coordinates ($\alpha = \tan^{-1}(r_2/r_1)$, $R = (r_1^2 + r_2^2)^{\frac{1}{2}}$, θ_{12}) (Wannier 1953), and the lower-lying intrashell states have often been described in recent years as "Wannier-ridge" or "Wannier ridge-riding" states. The symmetric stretching and bending modes are covered by this description, but not the asymmetric stretching mode. This description may however be misleading or wrong. The work of Watanabe (1987), Rost et al (1991) and Kim and Ezra (1991) suggests that it is the asymmetric stretching mode that is excited in the lower-lying states. Furthermore Watanabe (1987), Richter and Wintgen (1991) and Kim and Ezra (1991) find from studies of the classical electronic motion that the symmetric stretching mode is significantly less stable than the antisymmetric mode.

Another type of motion is that along the Langmuir orbit (Langmuir 1921), in which $r_1 = r_2$ but r_1 and θ_{12} vary. The motion here is on the Wannier ridge, but is not one of the simple bending modes referred to above because r_1 is not constant and θ_{12} has a large and fixed excursion (to $\theta_{12} = 24°$). Richter and Wintgen (1991) find that the Langmuir orbit of helium is stable, as do Müller et al (1991), who considered semiclassical quantization of the L = 0 intrashell states of the helium atom. The "Langmuir states" lie at the top of the manifold of intrashell states. Müller et al (1991) have derived a formula for the energies of the Langmuir states in terms of the principal quantum number n and the double-electron quantum number K (see below), and by generalising this to include all the L = 0 states of a manifold have been able to reproduce, to within a few percent, all the measured or calculated energies of such states in helium, including the ground state! The reason for the accuracy of this extrapolation is not understood.

Yet another type of motion that is classically stable (Klar 1987) is the "double circle mode", also first proposed by Langmuir (1921). The two Langmuir motions are illustrated in figure 2. Although at first sight figure 2(f) appears to be related to 2(e) as 2(d) is to 2(c), this is not so because the Langmuir motion is not degenerate.

To summarise we may provisionally distinguish between (i) the Wannier point states with minimum internal motion and energy, (ii) the symmetric stretching states, which may be the most unstable, (iii) the asymmetric stretching states, which seem to include the lower-lying states of an intrashell manifold, (iv) the bending states, (v) the Langmuir states, which seem to be stable (at least for Z = 2) and which have a high energy, and (vi) the double circle states. Combinations can occur for the non-Langmuir modes, and external angular momentum can be added.

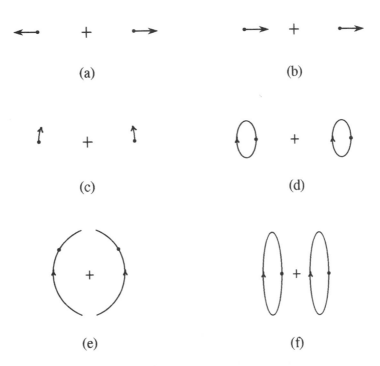

Figure 2. Classical modes of correlated motion of a two-electron atom. (a) symmetric stretching, (b) asymmetric stretching, (c) bending, (d) bending with internal rotation, (e) Langmuir motion, (f) double circle motion.

The intrashell double–Rydberg states have been extensively studied theoretically (see for example Fano 1983, Lin 1984, 1986, 1989, Ho and Callaway 1986, Feagin and Briggs 1988, Molmer and Taulbjerg 1988, Aymar 1989, Makarewicz 1989, Rost and Briggs 1989, Rau 1990, Rost et al 1991 and Macías and Riera 1991). Most of this work is concerned with the simplest atoms (such as H$^-$ and He) in which the ion core is a bare nucleus. There have also been many experimental observations of $n\ell n\ell'$ states for which n has the lowest Rydberg value, but few for higher values of n (see for example Brunt et al 1976, Zubek et al 1989, Domke et al 1991). The only observation of an extended Rydberg series of Wannier–ridge states remains that of Buckman et al (1983), shown in figure 3 (see also Buckman and Newman 1987). Here the $1sn\ell n\ell'$ states of He$^-$ are seen as resonances in e–He inelastic scattering. The lowest resonances in each group can be identified as $1sns^2$ states and are visible in this spectrum for values of n up to 7 (and up to 8 in the spectrum of Buckman and Newman 1987).

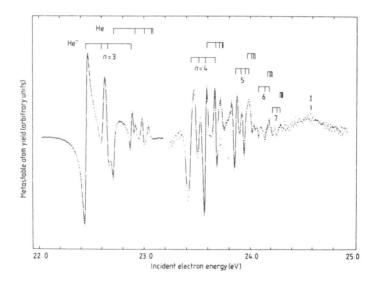

Figure 3. Double–Rydberg states of He$^-$, appearing as resonances in e–He inelastic scattering (Buckman et al, 1983). The energies of the lowest intrashell states of He$^-$ in each manifold are indicated, as well as the minimum and maximum energies of the manifolds of single-electron Rydberg states of He.

The independent particle classifications used above are of course incorrect, although they retain some usefulness due to their familiarity. For systems in which the electron–electron correlations are strong enough to cause localisations such as that illustrated in figure 1, it is clearly inappropriate to describe the two individual electrons as ns electrons or more generally to attach any single-particle quantum numbers to them. "Double-electron" quantum numbers must therefore be used. Building on the early group-theoretical work of Herrick and collaborators (see references in Herrick 1983), Lin (1984, 1986, 1989) has developed a suitable model and classification scheme. Other models have also been proposed (for example Nikitin and Ostrovsky 1985, Feagin and Briggs 1988, Molmer and Taulbjerg 1988,

Markarewicz 1989, Rost et al 1991), each with its own set of approximately conserved quantities and associated quantum numbers. These models are closely related to each other, being all based essentially on the assumption that the inter-electron axis (see figures 1 and 2) has some integrity and stability.

In Lin's classification scheme a correlated state is specified as $_n(K, T)_n^A$ $_{2S+1}L^\pi$ where K, T and A are the new "double electron" quantum numbers. These numbers can be loosely associated with the vibrational and rotational motion with respect to the inter-electron axis, as described above. The bending motion gives rise to a vibrational quantum number v, an energy $\hbar\omega(v+1)$ and a possible internal angular momentum $\hbar\ell$. K and v are related by

$$K = n - 1 - v \tag{1}$$

and T can be identified with ℓ. The internal angular momentum combines with the angular momentum of the inter-electron axis in the laboratory frame to produce the total orbital angular momentum L of the electron pair. The rotational energy has the form $B[L(L+1)-\ell^2]$. Finally A can be associated with the symmetric and anti-symmetric stretching motion along the inter-electron axis, which in turn leads to A having the value +1 when there is a node in the wavefunction at the Wannier point (i.e. at $r_1 = -r_2$), −1 when there is an antinode and 0 when the electron motion is not correlated. Thus the most localised states, the "Wannier-point" states, are those that have $(K, T)^A = (n-1, 0)^1$.

A justification of this geometrical interpretation of the double-electron quantum numbers is that the accurately calculated energies of the lower-lying states of a manifold of intrashell states (see e.g. Lin 1986) do indeed display vibrational and rotational energy spacings. The full group-theoretical treatment provides the restrictions on the quantum numbers, which are:

T = 0, 1, 2, ..., min (L, n−1), if $\pi = (-1)^L$,

T =   1, 2, ..., min (L, n−1), if $\pi = (-1)^{L+1}$,

K = n−T−1, n−T−3, ..., −(n−1−T),

A = $\pi(-1)^{S+T}$, if K > L−n,

A = 0,   if K ≤ L−n   (2)

When the two electrons have different values of n, the lower one (for the inner electron) should be used in all the above formulae.

The experimental evidence quoted so far on Wannier-ridge states has involved electron impact excitation, with its inherent limitations in energy resolution. Laser excitation would provide a considerable improvement in resolution, but despite considerable effort to this end (e.g. Freeman et al 1985, Gallagher 1988, Morita et al 1988, Morita and Suzuki 1988, Boulmer et al 1988, Eichmann et al 1990) it has not yet been shown that Wannier-ridge states can be excited in this way. As an example of such a study figure 4 shows the excitation scheme used by Morita et al (1988). As in all such schemes the two electrons are excited sequentially. Here one of the 4s electrons of calcium is excited (using 2 separate lasers) to a n's state, where n′ = 8 to 18, and then this electron acts as a "spectator" while the other electron is excited (using frequency-doubled two-photon excitation) to a ml state, such as 9s. The n-value of the spectator electron increases by 1 or 2 during the second excitation, so the final states reached should include the 9sns states where n = 9 to 20. The lowest of these, the Wannier-point $9s^2$ state, cannot however be positively identified in the spectra.

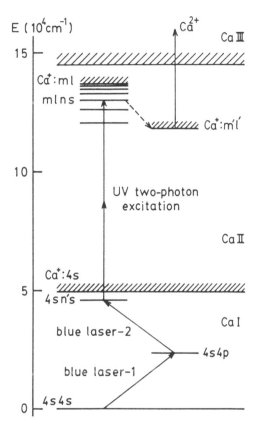

Figure 4. Schematic of the excitation scheme used by Morita et al (1988) to populate double-Rydberg states of calcium.

One of the many problems that beset such laser studies is that it is not known at what wavelengths the transitions to intrashell states should occur because of the difficulty of predicting the energies of these states for atoms that have multi-electron cores. It is not possible to estimate the energy of a ns^2 state by extrapolation of measured energies of nsms (m>n) states using a single-Rydberg formula, because the correlations cause the ns^2 state to have a radically different structure. The few calculations of the energies of intrashell states of such atoms include those for the $1s^2ns^2$ state of Li^- and of beryllium-like ions (e.g. Martin et al 1988, Dulieu 1989, Vaeck and Hansen 1989, Chen and Lin 1990) and the ns^2 states of the alkaline-earth atoms Be through to Ba (Aymar 1989). The even fewer experimental investigations include those cited for figure 3 above and those to be mentioned in section 4 below. Simpler atoms, for which the core is a bare nucleus (such as H^- and He), have by contrast been well studied, and the energies of their double-Rydberg states have been comprehensively and accurately calculated (see Lin 1986 for earlier references) and less-comprehensively measured (e.g. Williams 1988, Zubek et al 1989 Domke et al 1991).

To help in predicting the energies of Wannier-ridge states of two electrons outside a complex ion core, double-Rydberg formulae have been developed (Read 1977, Read 1982, Rau 1983, Wang 1986), the typical form of which is

$$E(ns^2) = I - \frac{2R(Z-\sigma)^2}{(n-\mu)^2} \qquad\qquad (3)$$

where R is the Rydberg energy, Z is the charge of the core, σ is a screening parameter and μ is a "double-electron" quantum defect. If σ is given the value of 0.25 appropriate to the Wannier point and μ is taken to be that of a single ns electron outside the same core, the formula reproduces the known energies of the lowest ns^2 states of atoms surprisingly well, even although there are then no adjustable parameters. For the higher ns^2 states the available evidence suggests (Read 1990) that μ decreases as n increases. This in turn suggests that at the centre of the atom the amplitude of a Wannier-point state decreases with n more strongly than does the amplitude of a single-Rydberg state (which is proportional to $n^{-3/2}$), as had been argued earlier by Macek and Feagin (1985). This partial "hole" in the wavefunction at the centre of the atom also helps to explain why the cross-sections for electron-impact excitation of the $3\ell3\ell'$ states of helium are nearly 2 orders of magnitude smaller than those of the $2\ell2\ell'$ states (Brotton et al 1991).

Clearly much remains to be done, experimentally and theoretically, on the study of intrashell double-Rydberg states.

A second sub-set of the double-Rydberg states consists of the "planetary" states for which n and n' are both large. States of this type have been excited (with n ≠ n') using multi-laser techniques (e.g. Camus et al 1989, Eichmann et al 1990, and Gallagher 1988 for earlier references). As an example, Eichmann et al (1990) have used 6 different lasers to excite planetary states of barium, such as the 40g 78d state. The mean distances of the electrons from the ion core in such a state ($\bar{r}_1 \sim 360\text{Å}$, $\bar{r}_2 \sim 3200\text{Å}$) are much larger than their wavelengths at these distances ($\lambda_1 \sim 10\text{Å}$, $\lambda_2 \sim 40\text{Å}$), hence justifying the description "planetary". The ratio of the classical orbital periods is approximately 30. The experimental results are explained well by a model in which the inner electron feels a nearly static electric field produced by the slower outer electron, giving a Stark manifold of states for the inner electron. The wavefunctions of the extreme states of this manifold are highly localised, with the inner electron being either on the same or the opposite side of the core as the outer electron. It is therefore clearly incorrect to label such states using the angular momentum quantum numbers of the individual electrons. This underlines the high degree of correlation that can, and usually must, exist in planetary atoms. Another manifestation of strong corrections is the colinear motion found by Richter and Wintgen (1990, 1991) in which the two electrons oscillate with the same frequency on the same side of the nucleus.

Another interesting type of double-Rydberg state is that in which one or both electrons are in "circular" orbits, for which $\ell = n-1$ (e.g. Roussel et al 1990, Jones et al 1991). These orbits give decreased autoionisation rates and hence greater stability. An example is the double circular 4f5g (J = 3) configuration of Ba, excited with the aid of 3 dye lasers (Jones et al 1991). One state of the configuration has a lifetime that is particularly long, corresponding to more than 300 classical orbital periods of the 5g electron.

A rich diversity of correlations and coupling schemes in double-Rydberg atoms can be confidently expected, and will offer a rewarding field of study for many years to come.

3. THRESHOLD IONISATION PROCESSES

Electron correlations play a dominant role in near-threshold ionisation processes such as

$$e + He \rightarrow He^+ + e + e,$$

$$h\nu + He \rightarrow He^{++} + e + e,$$

$$h\nu + H^- \rightarrow H^+ + e + e \tag{4}$$

According to the Wannier model (Wannier 1953, for a review see Read 1985) two free electrons are produced in such processes only when the electron pair has an initial position on or near the Wannier ridge.   If there is too large a component of the asymmetric motion depicted in figure 2(b), then one of the electrons will remain bound.   This happens because there is a local maximum in the potential on the ridge, as mentioned above, and so electron pairs that start sufficiently far off the ridge move progressively further from the ridge, finishing in a region that corresponds to one of the electrons being bound and the other free.   This limitation causes a reduction in the yield of processes in which the two electrons are free.   The reduction is most pronounced at threshold, and usually ceases to have any observable effect at energies more than a few eV above threshold.

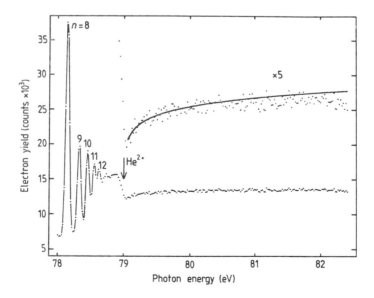

Figure 5.   Yield of low energy (<5meV) photoelectrons from helium (from Hall et al 1991).   The peaks below the double ionisation threshold correspond to photoionisation into Rydberg state of He^+, while the yield above the threshold corresponds to photo-double-ionisation. The curve through the latter yield has the Wannier dependence $E^{0.056}$ on the excess energy E.

A recent example of the reduction in yield is shown in figure 5. A dip or cusp, first seen in electron impact ionisation (Cvejanovic and Read 1974), is clearly present at the double ionisation threshold, although it is so sharp that it is severely attenuated here by the instrumental energy resolution. Above the double ionisation threshold the reduction in yield manifests itself as a power-law dependence ($E^{0.056}$) on the excess energy E, as shown in figure 5. Below that threshold resonant photoionisation contributes

$$h\nu + He \rightarrow He^{**} \rightarrow He^{+} + e \qquad (5)$$

as well as direct photoionisation. The Wannier model predicts (Read and Cvejanovic 1988) that there is an approximately symmetric dip or cusp in the direct contribution below and above the threshold. The additional resonant contribution can be clearly seen in figure 5. The resonances that are excited here must be double-Rydberg states of He, and work is progressing (e.g. Rost and Briggs 1991) on elucidating the types of such states that are involved.

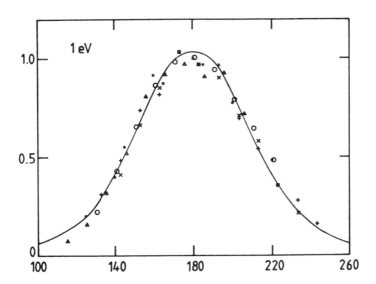

Figure 6. Unnormalised differential cross-section for helium (e,2e) scattering in the perpendicular plane (ie for $\theta_1 = \theta_2 = 90°$) at an excess energy E of 1eV (Jones et al 1991). The individual electron energies are 0.5eV and the yield is plotted as a function of the mutual angle ϕ_{12}.

More detailed information on near-threshold electron correlations is supplied by (e, 2e) coincidence experiments. As mentioned above, the electrons start with $r_1 \approx r_2$, and θ_{12} is nearly unrestricted. In the asymptotic region the ratio r_1/r_2 becomes unrestricted (i.e. the partitioning of the excess energy E is unrestricted) but θ_{12} is concentrated around 180°, with the angular width being given by $\theta_0 E^{\frac{1}{4}}$. The purpose of a recent set of experiments (Jones et al 1991) was to measure the key Wannier parameter θ_0 for helium, and also to help to establish the range of excess energy E over which the Wannier model is valid. Some of the results are shown in figure 6. θ_0 was found to be 74 ± 7° (for E in eV), which is consistent with values established theoretically (Feagin 1986, Altick 1985, Crothers 1986). The energy range of validity was found to be approximately 1.5 eV.

The dominance of electron–electron correlations in the near–threshold processes (4) and the validity of the Wannier model are now reasonably well established.   For example, no other model had been adduced to explain the origin of cusps such as that seen on figure 5.   But several problems are yet to be resolved, such as those concerning the influence of the particular atom and of its non–Wannier scattering parameters on the form of (e, 2e) angular distributions (Rosel et al 1990) and on the value and constancy of θ_0 (Gailitis and Peterkop 1989).

4. TWO–ELECTRON CAPTURE BY HIGHLY CHARGED IONS

There have been many studies of the capture of two electrons into intrashell doubly–excited states by slowly–moving highly–charged ions.   (see for example Crandall et al 1976, Barany et al 1985, Niehaus 1986, Bordenave–Montesquieu et al 1987, Barat et al 1987, Mann and Schulte 1987, Mack et al 1989, Kilgus et al 1990, Stohltefoht et al 1990, Chetioui et al 1990, Harel and Jouin 1990, Giese et al 1990, Boudjema et al 1991).   As an example, a spectrum of Mack et al (1989) is shown in figure 7.   Here O^{6+} ($3\ell 3\ell'$) states are formed in the process

$$O^{8+} + He \rightarrow O^{6+} + He^{++} \tag{6}$$

using ions of energy 96keV, and are detected through the electrons that they eject.   An analysis by Bordenave–Montesquieu et al (1987) of a similar spectrum

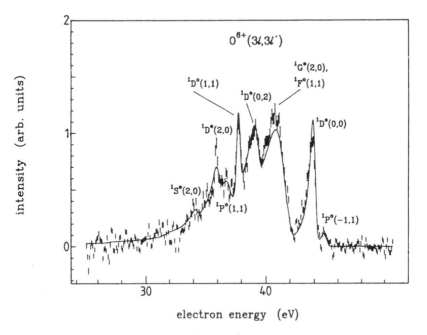

Figure 7.   Yield of O^{6+} ($3\ell 3\ell'$) states resulting from two–electron capture from He by 96keV O^{8+} ions (Mack et al 1989).   The states are labelled by the double–electron quantum numbers K and T, and are as identified by the calculations of Ho (1987).

shows that the state that is populated with the highest probability is the $(2,0)^+$ $^1G^e$ state, in the (K,T) notation (see above). This is a rotating "Wannier-point" state, as illustrated in figure 1. The quantum numbers v and ℓ (see above) are zero, giving the maximum possible localisation. The "external" angular momentum quantum number L has the maximum value (4) for this manifold.

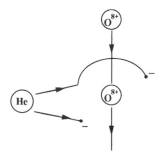

Figure 8. Schematic of the classical model of double capture.

An insight into the mechanism by which this type of rotating Wannier point state is created can be gained from the classical model developed by Mann et al (1981), Barany et al (1985), Niehaus (1986) and Mack et al (1989). The two electrons of the target He atom experience the additional Coulomb potential of the approaching O^{8+} ion, and in this model they are successively captured when the ion is near enough to lower the potential energy to the value required to ionize them. At the moment of capture each electron is at rest in the laboratory frame but has angular momentum in the frame of reference of the ion.

In this model the first electron is captured when the O^{8+} ion is approximately 3.9Å away, into a state that has $n \approx 4$ and $\ell \approx 2$ (depending on the impact parameter that is assumed), while the second electron is captured at a distance of 2.6Å into a state that has $n \approx 3$ and $\ell \approx 2$. The two angular momenta are in the same direction, so it is not surprising that the states of high total L are preferentially populated.

Two further features of the classical capture model help to explain why the two electrons are preferentially captured in Wannier-point states. The first (Mack et al 1989) is that the second capture occurs more easily when the first electron is on the opposite side of the ion core, as illustrated in figure 8. The second feature is that the time taken for the first electron to travel to the opposite side is approximately equal to the time interval between the two capture events, for the experimental conditions used by Mack et al (1989) and Bordenave-Montesquieu et al (1987). There is thus a (perhaps fortuitous) "resonance" between the two time spans.

These two features imply that an element of correlation exists in the capture events. It seems that this has not yet been looked for experimentally or been considered in the conventional quantum-mechanical curve-crossing treatments of double-electron capture processes.

ACKNOWLEDGEMENTS

The author wishes to thank those who have given permission to reproduce figures and the many authors who have supplied preprints and other information.

REFERENCES

Altick P L, 1985, J.Phys.B: At.Mol.Phys. **18**, 1841–6

Aymar M, 1989 J.Phys.B: At.Mol.Opt.Phys. **22**, 2359–67

Barany A, Astner G, Cederquist H, Danared H, Huldt S, Hvelplund P, Johnson A, Knudsen H, Liljeby L and Rensfelt K–G, 1985 Nuclear Instruments and Methods in Physics Research B9, 397–9

Barat M, Gaboriaud M N, Guillemot L, Roncin P, Laurent H and Andriamonjes, 1987, J.Phys.B: At.Mol.Phys. **20**, 5771–83

Bourdenave-Montesquieu A, Benoit–Cattin P, Boudjema M, Gleizes A and Bachau H, 1987, J.Phys.B: At.Mol.Phys. **21**, L695–703

Boudjema M, Cornille M, Dubau J, Moretto–Capelle P, Bordenave-Montesquieu A, Benoit–Cattin P and Gleizes A, 1991 J.Phys.B: At.Mol.Opt.Phys. **24** 1713–37

Brotton S J, Srigengan V, Currell F J and Read F H, 1991, to be published.

Brunt J N H, King G C and Read F H, 1976, J.Phys.B: At.Mol.Phys. 9, 2195–2207

Buckman S J, Hammond P, Read F H and King G C, 1983, J.Phys.B: At.Mol.Phys. **16**, 4039–47.

Buckman S J and Newman D S, 1987, J.Phys.B: At.Mol.Phys. **20**, L711–5

Camus P, Gallagher T F, Lecomte J–M, Pillet P, Pruvost L and Boulmer J, 1989 Phy.Rev.Lett. **62**, 2365–8

Chen Z and Lin C D, 1990 J.Phys.B: At.Mol.Opt.Phys. **23**, L87–8

Chetioui A, Martín F, Politis M F, Rozet J P, Touati A, Blumenfeld L, Vernhet D, Wohrer K, Stephan C, Barat M, Gaboriaud M N, Laurent H and Roncin P, 1990 J.Phys.B: At.Mol.Opt.Phys. **23**, 3659–75

Crandall D H, Olson R E, Shipsey E J and Browne J C, 1976 Phys.Rev.Lett. **36**, 858–60

Crothers D S F, 1986, J.Phys.B: At.Mol. Phys. **19**, 463–83

Cvejanović S and Read F H, 1974, J.Phys.B: At.Mol.Phys. **10**, 1841–52

Domke M, Xue C, Puschmann A, Mandel T, Hudson E, Shirley D A, Kaindl G, Greene C H, Sadeghpour H R, and Petersen H, 1991 Phys.Rev.Lett. **66**, 1306–9

Dulieu O, 1989, Z.Phys.D. **13**, 17–24

Eichmann U, Lange V and Sandner W, 1990, Phys.Rev.Lett. **64**, 274–7

Fano U, 1983, Rep.Prog.Phys. **46**, 97–169

Feagin J M, 1984, J.Phys.B: At.Mol.Phys. **17**, 2433–51

Feagin J M and Briggs J S, 1988, Phys.Rev.A **37**, 4599–4613

Freeman R R, Bloomfield L A, Boker J and Cooke W E, 1985, Laser Spectroscopy <u>VII</u> ed T W Hänsch and Y R Shen (Berlin:Springer) p77

Gailitis M and Peterkop R, 1989, J.Phys.B: At.Mol.Opt.Phys. **22**, 1231-9

Gallagher T F, 1983, Rep.Prog,Phys. **51**, 143-88

Giese J P, Schulz M, Swenson J K, Schöne H, Benhenni M, Varghese S L, Vane C R, Dittner P F, Shafroth S M and Datz S, 1990, Phys.Rev.A **42**, 1231-44

Hall R I, Avaldi L, Dawber G, Zubek M, Ellis K and King G C, 1991, J.Phys.B: At.Mol.Opt.Phys. **24**, 115-125

Harel C and Jouin H, 1990 Europhys.Lett. **11** (2), 121-6

Herrick D R, 1983, Adv.Chem.Phys. **52**, 1-115

Ho Y K, 1986, Phys.Rev.A **35**, 2035-43

Ho Y K and Callaway J, 1986, Phys.Rev.A **35**, 130-7

Jones R R, Panming F and Gallagher T F, 1991, preprint

Jones T J, Read F H, Cvejanović S, Hammond P and King G C, 1991, J.Phys.B: At.Mol.Opt.Phys. submitted

Kim J H and Ezra G S, 1991, preprint

Kilgus G, Berger J, Blatt P, Grieser M, Habs D, Hochadel B, Jaeschke E, Krämer D, Neumann R, Neureither G, Ott W, Schwalm D, Steck M, Stokstad R, Szmola E, Wolf A, Schuch R, Müller A, Wagner M, 1990 Phys.Rev.Lett. **64**, 737-40

Klar H, 1987, Z.Phys.D. **6**, 107-111

Langmuir I, 1921, Phys.Rev. **17**, 339-353

Lin C D, 1984, Phys.Rev.A **29**, 1019-33

Lin C D, 1986, Adv.At.Mol.Phys. **22**, 77-142

Lin C D, 1989 Phy.Rev.A **39**, 4355-61

Macias A and Riera A, 1991, J.Phys.B: At.Mol.Opt.Phys. **24**, 77-90

Macek J and Feagin J M, 1985, J.Phys.B: At.Mol.Phys. **18**, 2161-79

Mack M, Nijland J H, Straten P v d and Niehaus A, 1989, Phys.Rev.A **39**, 3846-54

Makarewicz J, 1989 J.Phys.B: At.Mol.Opt.Phys. **22**, L235-40

Mann R, Folkmann F and Beyer H F, 1981, J.Phys.B: At.Mol.Phys. **14**, 1161-81

Mann R and Schulte H, 1987, Z.Phys.D **4**, 343-9

Martín F, Mó O, Riera A and Yáñez M, 1988, Phys.Rev.A **38**, 1094-7

Molmer K and Taulbjerg K, 1988, J.Phys.B: At.Mol.Opt.Phys. **21**, 1739–49

Morita N and Suzuki T, 1988, J.Phys.B: At.Mol.Opt.Phys. **21**, 439–44

Morita N, Suzuki T and Sato K, 1988, Phys.Rev.A **38**, 551–4

Müller J, Burgdöfer J and Noid D, 1991 preprint

Niehaus A, 1986, J.Phys.B: At.Mol.Phys. **19**, 2925–37

Nikitin S I and Ostrovsky V N, 1985, J.Phys.B: At.Mol.Phys **18**, 4349–69

Rau A R P, 1983, J.Phys.B: At.Mol.Phys. **16**, L699–705

Rau A R P, 1990, Rep.Prog.Phys. **53**, 181–200

Read F H, 1977, J.Phys.B: At.Mol.Phys. **12**, 449–58

Read F H, 1982, Aust.J.Phys. **35**, 475–99

Read F H, 1985, Electron Impact Ionisation, ed T D Märk and G H Dunn (New York: Springer) pp 42–88

Read F H, 1990, J.Phys.B: At.Mol.Opt.Phys. **23**, 951–8

Read F H and Cvejanović S, 1988, J.Phys.B: At.Mol.Opt.Phys. **21**, L371–5

Richter K and Wintgen D, 1990, J.Phys.B: At.Mol.Opt.Phys. **23**, L197–210

Richter K and Wintgen D, 1991, preprint

Rösel T, Bär, Jung K and Ehrhardt H, 1990, private communication

Rost J M and Briggs J S, 1989 J.Phys.B: At.Mol.Phys. **22**, 3587–3602

Rost J M, Gersbacher R, Richter K, Briggs J S, and Wintgen D, 1991 J.Phys.B: At.Mol.Opt.Phys. **24**, 2455–66

Rost J M, and Briggs J S, 1991, J.Phys.B: At.Mol.Opt.Phys. **24**, L393–6

Roussel F, Cheret M, Chen L, Bolsinger T, Spiess G, Hare J and Gross M, 1990 Phys.Rev.Lett. **65**, 3112–5

Stolterfoht N, Sommer K, Swenson J K, Havener C C and Meyer F W, 1990 Phys.Rev.A **42**, 5396–5405

Vaeck N and Hansen J E, 1989 J.Phys.B: At.Mol.Opt.Phys. **22**, 3137–53

Wang H, 1986, J.Phys.B: At.Mol.Phys. **19**, 3401–10

Wannier G H, 1953, Phys.Rev. **90**, 817–25

Watanabe S, 1987, Phys.Rev.A **36**, 1566–74

Williams J F, 1988, J.Phys.B: At.Mol.Opt.Phys. **21**, 2107–16

Zubek M, King G C, Rutter P M and Read F H, 1989, J.Phys.B: At.Mol.Opt.Phys. **22**, 3411–21

Recent trends in ion atom collisions

C L Cocke

J.R. Macdonald Laboratory, Physics Dept., Kansas State University, Manhattan, KS 66506

ABSTRACT: Recent experiments in ion-atom collisions which involve binary encounters between electrons and electrons or heavy particles and electrons are discussed.

I. INTRODUCTION

The terms "recent" and "trends" are in some sense incompatible with each other. To an ICPEAC audience, the word recent could reasonably be taken to mean "since the last ICPEAC". It is doubtful that developments on such a time scale could be classified as "trends," however.   This talk will therefore necessarily incorporate work which has developed on a longer time scale but will include the most recent results known to this author.

The impossibility of any comprehensible coverage of the activities within ion-atom collisions is partially indicated in Figure 1, where the geographic distribution of activities in this field (based mainly on papers submitted to ICPEAC XVI) is shown.   There are 65 laboratories and institutes on the map, representing a slightly smaller number of topics. Common to the work of these laboratories is the bombardment, either in the laboratory or on paper, of neutral atom targets with ions.

The last four ICPEACs have seen reviews of the history of ion-atom collisions by Merzbacher (1983) and Briggs (1988), and a perspective over the wider field of atomic physics by Datz (1989). My own prejudice is to characterize the 70's and early 80's as a time during which one electron processes, namely the capture, ionization and excitation of a single target electron by the mean field of the projectile, were emphasized and, to a large extent, understood. For fast collisions with low charged projectiles, perturbation expansions were shown to be successful, while for slow collisions the Fano-Lichten model for inner shells and interacting orbital models for outer shells provided the basic physical framework.   During the last half of the 80's, the technical complexity of experiments grew enormously. Multiparameter computer-based data taking has become common, highly charged ions have become routinely available from both ECR and EBIT/EBIS sources (Bliman 1989, Salzborn et al. 1991) [only a few recent references are given] laser-polarized (Liu 1989) and/or excited targets (Dowek 1991) are available. Merged and crossed beam experiments have become common (Gregory 1990, Wåhlin 1991, Schön et al. 1987, Havener et al. 1989) storage rings (Kilgus et al. 1990) and electron coolers (Andersen et al. 1990, Müller et al. 1990) for high resolution electron-ion collisions have become

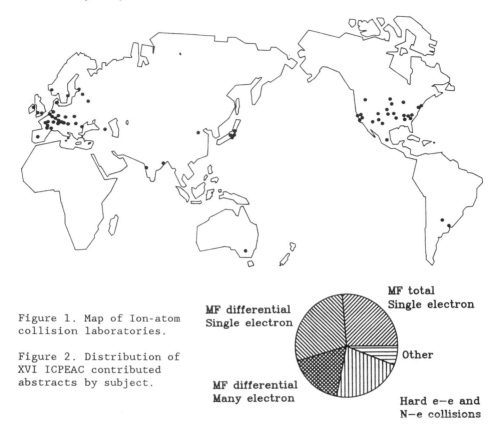

Figure 1. Map of Ion-atom collision laboratories.

Figure 2. Distribution of XVI ICPEAC contributed abstracts by subject.

available, the use of antimatter beams (Andersen et al. 1987, Humberson et al. 1989) and the very highest energy heavy ions available anywhere (Datz et al. 1991) have been turned to atomic physics. It has become relatively rare that a total cross section is sufficient. Alignment and orientation of final states, high resolution electron and photon spectroscopy, and experiments which differentiate the final products in angle, energy and charge state demonstrate our growing ability to examine the final product space in ever more detail to reveal the more inner workings of the reaction processes. Channels in single crystals have come to be used almost routinely as quasi free electron targets for resonant coherent excitation (Y. Iwata et al. 1990) electron ionization (Claytor et al. 1988) and convoy electron production (Kimura et al. 1991). On the theoretical side, codes with very large basis sets (Fritsch and Lin 1991) have been able to provide good descriptions of a wide range of single electron processes, and CTMC codes (Reinhold et al. 1991a) have addressed both single and multielectron processes. Very relativistic collisions have been addressed (Deco et al. 1990, Bottcher and Strayer 1989). It can be argued that the theoretical framework is in place to analyze most collision processes accessible to experimentalists today. Of course, there is always the little detail of actually doing the calculation for any particular collision system.

Figure 2 shows a rather subjective classification of the contributed papers given at ICPEAC XVI into categories which I label as "mean field" (MF) single electron processes (subdivided into total and differential cross section work, and further subdivided to distinguish experiments focussing on single or multielectron processes) "binary collision" (BC) processes (including "hard" electron-heavy particle and electron-electron collisions) and other. I discuss further below what I mean by the MF and BC designations. This figure shows that my prejudice is little more than that. The center of activity of the field still concentrates on the investigation of one electron MF transitions. Nevertheless, multiple electron processes and ones involving quasi binary encounters between components of the colliding systems represent a growing segment of activity in ion-atom collisions.

2. THE ROLES OF COMPONENT ELECTRONS IN ION-ATOM ENCOUNTERS

The interaction of a projectile ion with a target atom must ultimately be explained on the basis of the individual Coulomb interactions among the charged components of the two systems, comprised of both electrons and nuclei. Since this is a complicated multi-body problem, approximations must be made. A common one is to focus on the motion of one active electron and to describe its ionization, excitation or capture as due to the action of the average potential of the remaining components of the system. The exchange of energy and momentum in such a process is between the entire collision partners. Such processes I label as MF.

Many recent experiments in ion-atom collisions have been just those for which this picture does not work. They have been ones for which the intrinsic lumpiness of the exciting potential, i.e., that it incorporates not electron clouds but point electrons which act as individual particles, is essential. In this case, the exchange of momentum and energy is between the electron and the striking particle only and the process can be identified on the basis of this two body kinematics if the collision is a "hard" one. Such processes are ones which I classify as BC in Figure 2 and it is on these processes I have chosen to concentrate in this talk. The reader is referred to recent complimentary reviews of this subject by Sellin (1991), Richard (1991) and McGuire (1991) for further coverage. I will restrict my attention to two types of hard collision BC process. The first involves a hard collision between a heavy projectile and a target electron and is identified by a large momentum transfer to either the heavy particle or the struck electron. The second involves hard collisions between an electron carried by the projectile and one on the target. In this case, the target electron serves as an exciting agent of the projectile and *vice versa*. Such a process can be identified by the binary encounter kinematics for free electron-electron scattering or by simultaneous electronic transitions on target and projectile, and are closely related to corresponding electron-ion collision processes.

When speaking of a hard two-particle Coulomb encounter between components of an ion-atom system, it is convenient to draw a classical diagram of the process. In deciding how seriously to take such a diagram, one should bear in mind the criterion set forth by Bohr (1948) which must be met for such an encounter to follow a full classical trajectory, namely that

$$K = q_1 q_2 / \hbar v$$

where q_1 and q_2 are the charges of the collision partners and v the relative velocity, be large compared to unity. For the cases I will discuss here, always involving an electron as one of the collision partners, this condition is frequently not met. Then why insist on a classical picture in the discussion? There are two answers to this. First, classical terms are very useful to identify the basic physical process, even if quantitative calculations must be done with a quantal framework. There is generally a one-to-one correspondence between classical hard scatterings and terms in a quantal perturbation expansion. Second, since the differential scattering cross sections calculated quantum mechanically are identical with the classical result for distinguishable particles for any value of K. Thus quantitative classical calculations may give correct results whether K is large or not. Whether this should be viewed as lucky or reveals some more profound aspect of atomic collision physics is somewhat a matter of taste.

2.1 Hard Collisions Between Heavy Projectiles and Quasi-Free Electrons

2.1.1 Elastic Scattering

2.1.1.1 Binary Encounter Electrons

Fast electrons generated from hard collision collisions of the projectile with target electrons, referred to as "binary encounter electrons," are a well recognized and researched feature the spectra of electrons from ion atom encounters and have been studied for many years (Rudd and Macek 1972, Stolterfoht 1978). They appear as a peak in the spectrum at an angle determined by the kinematics of a collision between the projectile and a free electron, spread out by the Compton momentum distribution of the target electrons which are not truly at rest before the collision. A typical spectrum is shown in Figure 3a, with theoretical curves calculated from both an impulse approximation and a Born approximation. Figure 3b shows that the expected Z^2 scaling expected for such a collision is well satisfied for bare projectiles. The agreement is excellent. Can there be anything new happening with such a well understood process? Two recent experiments which presented features which were not immediately understood suggest that there are.

The first of these was a study of the binary encounter peak seen at zero degree scattering in the laboratory (180 degree scattering in the electron's center of mass system) with non-bare projectiles. Such a scattering is indeed a very hard electron-ion encounter, and one might expect that the scattering would be very well described by the scattering of quasi-free electrons by the nuclear charge of the projectile. For clothed projectiles, one would expect that the process would be described by a screened projectile potential for large impact parameters, but that close collisions would be sampling again mainly the projectile nuclear potential. It was therefore originally surprising when Richard et al. (1990) found that the differential cross section for the binary encounter peak for such a collision depended very much on the number of projectile electrons carried into the collision and in a direction opposite that which screening would produce for large impact parameters. Their results are shown in Figure 4. Although one's first intuition might be that screening would reduce rather than increase the differential cross section, that intuition is simply wrong, as has now been well demonstrated by Reinhold et al. (1990), Taulbjerg (1990) and Shingal et al. (1990). The deflection function for free electrons scattering from a screened potential is quite different from that for a Coulomb potential, and in such a way that the differential cross section at backward electron angles is enhanced over the bare nucleus case. A physical picture given by Reading (1990) is that the major effect of the

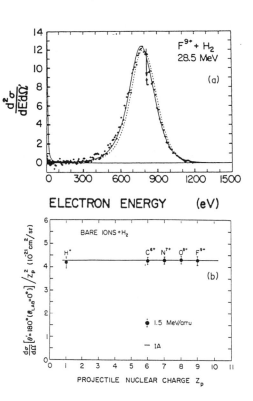

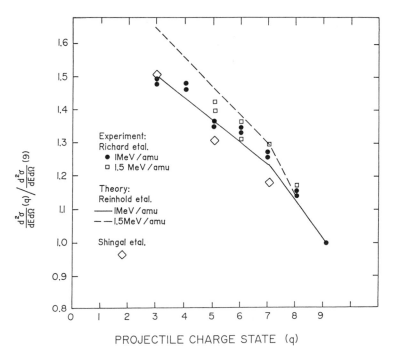

Figure 3. a) Binary encounter electron spectrum in projectile frame. b) Differential cross section at zero degrees lab. for BE peak (Lee, 1990).

Figure 4. Ratio of differential cross section at zero degrees lab. (180° electron) to that for a bare projectile (Richard 1990). The theory is from Reinhold (1990) and Shingal (1990).

screening for small impact parameters is to reduce the effective collision energy with which the electron arrives at the inner region where it scatters from the projectile nucleus. Thus for such collisions the differential cross section is characteristic of a lower collision energy than the asymptotic one, and is thus enhanced. For larger b the screening acts the other way to reduce the differential cross section and there is a cross-over at a backward scattering angle for which the differential cross section accidentally matches that for scattering from the nucleus alone. In the calculations of Reinhold et al. and Shingal et al., both shown in Figure 4, a static screened potential was assumed. Taulbjerg has found that better agreement with the data is achieved if exchange effects are included, and more sophisticated treatments are underway.

The second recent puzzle in binary electron spectra was the observation by Kelbch et al. (1989) that for very highly charged but also heavily clothed projectiles on multielectron targets, the binary encounter peak showed very strange behavior. Over a very small angular range, the binary encounter peak was found to break into two peaks, neither of which was located at the normal kinematically required position, an then emerge again to a single peak at larger angles (see Figure 5a). After a number of more exotic proposals, a convincing explanation has been provided by Reinhold et al. (1991), and again finds it origins in the non Coulomb character of the potential presented by a clothed projectile. The deflection function for a free electron scattering from a heavily screened projectile ion not only differs dramatically from that for a Coulomb potential, but, unlike the lighter ion case discussed above, possesses rainbow structures which give rise to very non-monotonic behavior of the differential cross section at backward electron scattering angles. When applied to ion atom collisions by folding in the Compton profile of the target electrons, the free electron scattering features, shown in Figure 5b, survive and give rise to the bifurcation of the binary encounter peak as observed in the experiment. Comparison between experiment and theory is shown in Figure 5a, where near quantitative agreement is seen. The cross section at backward electron angles for the clothed projectile can be several orders of magnitude larger than that for bare projectiles.

Miraglia and Macek (1991) have recently revisited the whole impulse approximation for the production of binary encounter electrons and have developed a formalism for dealing with scattering when the projectile charge exceeds the collision velocity, for which a first Born approximation is not valid. They find that for slightly higher Z than shown in Figure 3, the cross section does not even scale as Z^2! Clearly, surprises still await in hard electron-nuclear scattering in ion-atom collisions.

2.1.1.2 Heavy Particle Scattering

It was shown by Rutherford (1911), Geiger and Marsden (1909) that the atom possessed a hard nuclear core. Evidence for this was the observation of large-momentum-transfer collisions between alpha particles and the nuclei of atoms. A similar phenomenon is seen in scattering protons from a helium target. While the differential cross section for total scattering is rather featureless, reflecting the deflection of the proton by the screened Coulomb potential of the He atom, the differential scattering cross section for those protons which ionize the He shows evidence for the existence of electrons within the atom. (There is other evidence for this.) Figure 6a shows the large angle differential cross section for the scattering of 6 MeV protons from He for those collisions which ionize the He. For proton

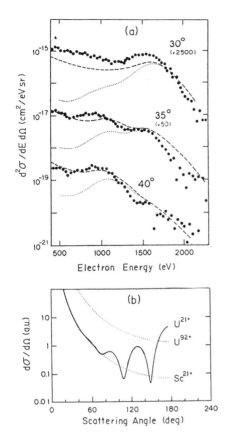

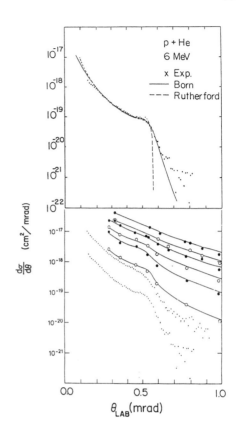

Figure 5. a) Differential cross section for electron production near the binary encounter peak. The lines are calculated in the impulse approximation using a static potential for electron scattering from a clothed U^{21+}. b)Differential cross section for free electron scattering from the U^{21+} potential (from Reinhold et al. 1991).

Figure 6. a) Differential cross section for proton scattering from He in which the He is ionized (Kamber 1988a). The solid curve is a Born calcula- tion by McGuire; the dashed curve represents Coulomb scattering. b) Similar to (a) but for proton energies of (top to bottom) .3,.4,.5, .7,1,2,3 and 6 MeV (Kristensen and Horsdal Pedersen 1990).

scattering from free electrons, a rainbow appears in the laboratory at an angle of 0.55 mrad, a "large" angle collision in the laboratory for a bowling ball (proton) hitting a ping-pong ball (electron), because the proton cannot be scattered to a larger angle by a free electron. This rainbow, which is reduced to a shoulder in the data by the Compton momentum distribution of the He target and by finite resolution of the experiment, is an unambiguous signature of a hard collision between the proton and a quasi-free target electron. The requirement that the He be ionized pre- selects hard electron-proton collisions in such a way as to isolate the binary encounters with the target electrons. The angular distribution is very well accounted for by the Born calculation of McGuire (in Kamber 1988a) and by several other calculations (Salin 1989, Olson et al. 1989). In this case, it is also rather well described by a simple Rutherford scattering calculation folded into the experimental resolution. Figure 6b displays further data from Kristensen and Pedersen (1990) which shows the evolution of this feature over a wider range of bombarding energy. The feature is gradually dissolved by the Compton momentum profile as the projectile velocity ceases to exceed that of the target electron sufficiently.

2.1.2 The Double to Single Ionization Ratio for He: Sudden Single Ionization and Photoionization

The above described collisions between projectiles and the electrons in He suggest a means of probing the double ionization of He which occurs when one electron is suddenly removed from the atom. This subject is of interest partially because it probes the correlation structure of He in both initial and final states, and because it has stimulated an examination of the relationship between photon and charged particle impact ionization processes. Attention is focused on the ratio of double to single ionization, hereafter designated as R. For photoionization, the incident photon interacts with only one electron, and double ionization comes about either through shakeoff or by interaction of the departing photoelectron with the remaining electron. Shakeoff is not consistently defined in the literature (Carlson 1967, Åberg 1973, Vegh and Burgdörfer 1990). Physically it is usually described as the result of a sudden change in the potential attending sudden removal of the primary electron, which results in an overlap between the wave function of the second electron and continuum states of the new hamiltonian. Summed over final continuum states, the square of this overlap is the probability for double ionization, R, and is independent of the method of removal of the first electron. This description is not adequate for He, since the monopole potential change calculated in a Hartree Fock approximation gives a very small R (Åberg 1973) and correlations in the initial state wave function not describable by a mean potential change give the main contributions to the double ionization in the high energy limit.

In photoionization, the energy of the outgoing electron is essentially that of the incoming photon minus the ionization energy, and it is easy to ensure that the first electron leaves fast by using large photon energy. The experimental and theoretical situation is shown in Figure 7 where inelastic electron scattering data is included and taken to be equivalent to photoionization data. The experimental ratio rises from zero at threshold to nearly five percent. Except for one data point taken before synchrotron light was available, the pre-1990 data stopped at 300 eV. Several theoretical curves are shown.

How can one investigate this using proton impact? First, one must be sure that the impact remove a fast primary electron, so as to reduce the probability that this departing electron interact with the second electron. Second, the charged particle itself must be fast enough that in any single pass through the He the probability P that it ionize either electron be much smaller than R. If two independent encounters with the projectile dominates, R is just given by P/2, with P suitably averaged over impact parameter.

The first condition requires that hard ion-electron collisions be selected. Total cross sections, which are dominated by soft collisions producing slow primary electrons, never do this, and instead give values of R of only a few parts in a thousand. This subject has itself been the center of much activity(Andersen et al. 1987) since R is found to depend on the charge of the projectile, different for protons and antiprotons for example, and several theoretical explanations have been offered, but it is not the topic of this discussion. One can isolate hard collisions by either demanding kinematically that a hard collision occur, or by looking at electron capture at high energy which requires that the primary (captured) electron depart fast.

The second condition is not easy to evaluate rigorously, but at least requires that P be less than a percent or so, which means that the ratio of the projectile velocity to its charge be at least above ten. Interestingly, this condition is incompatible with the assumption that the scattering of the electron be classical, although this does not prevent either experimentalists or theorists from discussing the collision classically. No experiment has been done to date with a projectile of charge above unity at sufficiently high velocity to satisfy this condition even in the total cross section.

The recent experimental results for proton impact are shown in Figure 7, where they are put on the photoionization figure by matching the outgoing electron energy with the photoelectron energy. The values of R for large angle proton scattering lie well below the suggested trend of the pre-1990 photon data, while the capture data fall in between. This data would suggest that even well above the He double ionization energy, there is no common value of R. As discussed by several authors (Åberg 1973, Cocke 1989, Vegh and Burgdörfer 1990) the limiting value of R should not necessarily the same for the three cases, even if a certain simplistic physical reasoning suggests that all three should approach a common shakeoff limit in the limit of infinitely fast primary electron removal. The basic physical reason for the difference is that the limiting value of R depends not just on the energy of the fast electron but on the momentum transfer imparted to the recoiling He ion during the primary ionization event. For a fixed primary electron energy, this momentum is much smaller in the case of heavy particle impact than it is in the case of photoionization, and therefore one might expect different limiting values of R.

The most recent developments thicken the plot. A recent photoionization measurement by Levin et al. (1991) gives a much lower value of R at a large photon energy than the trend of the earlier data would suggest. One might conclude that R for photoionization is simply dropping fast with E, but the most recent theoretical many body perturbation theory calculation by Isihara et al. (1991) which is in good agreement with the new data point, is not in agreement with the large value of R found in earlier photoionization work. It is in agreement with the charged particle data, however. Is this luck? On whose part? No full *ab initio* calculation for the charged particle case has been carried out. Neither the energy dependence of the photoionization ratio nor the connection between this and the charged particle results can be considered well understood.

2.1.3 Angular Dependence of R: Mechanism Differentiation

Outside a scattering angle of 0.55 mrad, protons must be scattering from the He nucleus, and thus one might expect some kind of qualitative change in ionization/scattering mechanisms as reflected in the behavior of R at around this scattering angle. Such a phenomenon has been identified by Giese and Horsdal (1988), as shown in Figure 8. The value of R shows a

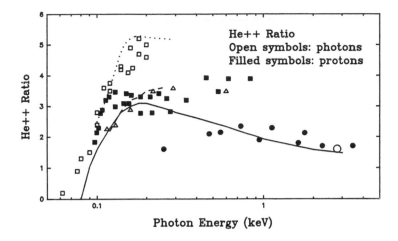

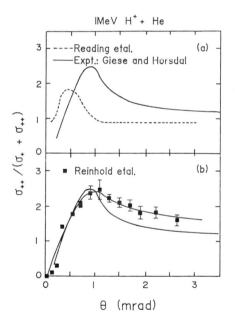

Figure 7.   The ratio R of double to
single ionization for He.   The data
are: photon and electron: open
triangles: Carlson 1967; open
squares: Schmidt et al. 1976, Wight
et al. 1976, Holland et al. 1979;
large open circle, Levin et al.
1991.   Capture data: filled squares,
Kristensen and Horsdal-Pedersen
1990.   Filled circles, Kamber 1988.
Theory: dotted line: Carter and
Kelly 1981; dashed line, Byron and
Joachaim 1967; solid line, Ishihara
et al. 1991.

Figure 8.   The fraction of double
ionization versus proton scattering
angle.   Data (solid line) are from
Giese and Horsdal 1988.   Theory:
(a) Reading et al. (1989), dashed
curve; (b) Olson et al. (1990),
CTMC, divided by three (points).

peak, not at 0.55 mrad, but at nearly twice this angle.   Classically, such
a large scattering angle can only occur if the projectile either scatters
from the He nucleus or suffers two scatterings from the target electrons.
Several explanations have been offered for this behavior, but a complete
calculation to evaluate the models is very difficult to carry out even for
so small number of particles.   Reading et al. (1989) suggest that the double
ionization near the peak is dominated by a hard electron-proton collision
followed by shakeoff, while to either side of the peak the proton is
deflected by the nucleus and the second electron is removed in a soft
collision.   The peak position at 1 mrad is not reproduced, however.   Vegh
(1989) suggested a multiple step hard collision process, but without

quantitative evaluation. Olson et al. (1989) did a CTMC calculation which, while too high in magnitude by a factor of three, gives the angular dependence very well. Analysis of their calculation reveals that the double ionization near the peak is dominated by independent hard scatterings of the proton off the two electrons, which result in a net proton deflection beyond the 0.55 mrad limit. Single ionization by such a hard collision drops off at 0.55 mrad, and thus R rises outside this angle. Beyond about 1.1 mrad the proton must be scattered by the nucleus, with both single and double ionization resulting from accompanying soft collisions with the electrons, and R again decreases.

This description and recent measurements of P by Kristensen and Horsdal-Pedersen (1990) call into question whether the energy of the outgoing primary electron can be identified on the basis of proton scattering angle as was done by Kamber et al. In addition, Kristensen and Horsdal-Pedersen found rather large values of P, sufficiently large that ignoring double ionization for hard collisions with the He nucleus seems to be hardly justified. For example, at 3 MeV, P was found to be nearly 6%, which would account for all of R (and more) by double interactions with the projectile for both the capture and large angle scattering case. This result suggests that none of the charged particle experiments are yet at high enough energy to see any limiting value for R. On the other hand , the data for both of the charged particle experiments seem to behave as if they have reached some kind of limit for R. Substantial questions remain unanswered.

2.1.4 Hard Collisions in Electron Capture Mechanisms

It is well known that electron capture by fast point projectiles requires the action of both the projectile potential and a third body potential, usually that of the target nucleus. A mechanism involving two hard collisions for electron capture was made by Thomas (1927) and seen experimentally in 1983 (Horsdal-Pedersen et al. 1983, Vogt et al. 1986). The normal Thomas scattering for capture by protons involves a hard collision by the projectile on a target electron and a second one by this electron off the target nucleus. A second similar kind of Thomas scattering (T2) in which the second scattering is from a second target electron was discussed by Shakeshaft and Spruch (1979) and by Briggs and Taulbjerg (1979). In a quantal treatment, such scatterings are represented by corresponding terms in a perturbation expansion. Evidence for detection of this process was reported by Horsdal et al. (1986) who used the measurement of R as a function of scattering angle for single capture to identify the process. If an electron is captured a T2 process, the proton is expected to be scattered at 0.55 mrad, and both the captured electron and the scattering electron are removed from the He. As shown in Figure 9, a large enhancement of R for the expected scattering angle is seen, with R becoming as large as 0.15! Since 0.55 mrad is a kind of magic angle for many imaginable processes involving hard proton-electron collisions, one should be careful that there could not be other mechanisms which could produce such behavior. Gayet and Salin (1990) have suggested one such process which has nothing to do with billiard ball physics, but which attributes the effect entirely to the behavior of the phases of the amplitudes for the two electron process. Their calculation, shown in Figure 9, is in rather good agreement with experiment.

Evidence for the T2 process was more recently found by Pálinkás et al. (1989) as shown in Figure 10. In a T2 scattering, the second electron leaves the collision at 90 degrees in the laboratory, with an energy which

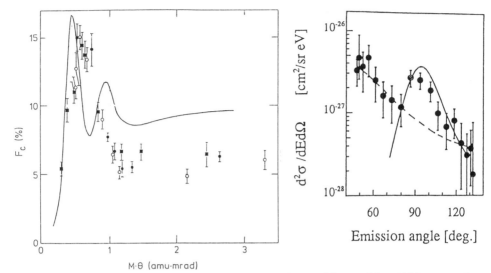

Figure 9. The fraction of double
ionization versus hydrogen scattering
angle for capture. The data are from
Kristensen and Horsdal-Pedersen 1990.
The theory is from Gayet and Salin
1990.

Figure 10. Double differential
cross sections for electron
emission in transfer ionization
of He by 1 MeV protons. The
solid curve is the calculation
by Briggs and Taulbjerg 1979
for the T2 process (from
Pálinkás et al. 1989).

can be deduced from the kinematics of the collision with two quasi-free
electrons. However, it is interesting to note that conservation of energy
and momentum alone, for any process in which the target nucleus does not
participate, will require that a 90 degree electron have the same energy,
lying on a "kinematic ridge," as discussed by McGuire et al. (1989), and
thus will produce a peak in angle at this energy. The double scattering
mechanism is very likely to be the leading contributor to the population
of this ridge, but is not required to be the only one. The data agree
rather well with the theoretical expectations. Exploration of the whole
ridge structure would be even more informative, however.

2.1.5 Recoil Momentum Spectroscopy of The p-He System

One way to isolate what is colliding with what in the ionization of He by
protons is to measure both the proton deflection angle and the He recoil
momentum. This has been done by Dörner et al. (1989) for .5 MeV protons on
He and more recently for 3 MeV protons on He. They use a He gas target
cooled to 30 K, which reduces the thermal target momentum distribution to
1.2 au, and detect the He ion in coincidence with the proton. As shown in
Figure 11 the major contribution to the transverse momentum exchange for
singly ionizing collisions for an angle of 1 mrad (outside 0.55 mrad), is
a hard encounter between the nuclei. However, the distribution in momentum
of the projectile around the value which would be expected from just a
proton-He nuclear encounter is due to momentum carried off by the ionized
electron. Even if the ionization were due to a hard proton-electron
encounter, the He nucleus carries away at least its Compton momentum

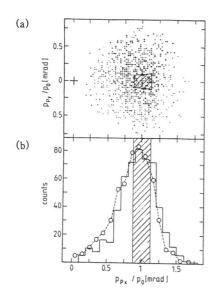

Figure 11. The distribution in the transverse plane of projectile deflections for a fixed He$^+$ recoil momentum of 10^{-3} times the projectile momentum. The square denotes the location of the heavy particle transverse momentum conservation. The lower curve is a CTMC calculation of the projection of p_x compared to experiment (from Dörner et al. 1989).

distribution, which amounts to a quite observable 2 au or so. Targets under construction for such experiments may go as low as 1 K, and allow recoil momentum spectroscopy with a resolution near .2 au. This subject was discussed by Schmidt-Böcking in his talk at this conference.

2.2 Electrons as Exciting Agents in Ion-Atom Collisions

From the projectile's reference frame, the electrons carried in by the target can act coherently to screen the atomic nuclear charge, thereby reducing the exciting power of the target potential, a MF process. They can also act incoherently, as independent agents with the ability to electronically excite the projectile in BC processes. In the latter capacity, they cause to occur all the same processes as if the projectile ion were subjected to a flux of free electrons. The main difference is that the ion-atom case is intrinsically "low resolution," because of the Compton distribution of electron momenta in the atom, and that processes which involve double interactions with the atomic nucleus can reach the same final states as those involving electron-electron collisions. We review here examples of electron-ion collisions studied using atomic targets as the source of electrons.

2.2.1 Dielectronic Recombination/ RTE

Classically, a dielectronic recombination event is one in which an electron collides with an electron of an ion, losing enough energy that the ion's electron is excited while the projectile electron drops into a bound state of the ion. Such a process is resonant in electron energy when it goes through a doubly excited state of the ion embedded in the continuum. When the free electron is replaced by one carried in by an atom, the process, labelled resonant transfer and excitation (RTE) is still resonant in ion velocity (corresponding to the velocity of the resonant free electron) but broadened by the Compton profile of the electrons. When one is dealing with an inner (e.g., K shell) excitation, the electron-electron collision must be a hard one in order that the excitation occur.

It was not until 1983 that DR was finally observed under single collision conditions in the laboratory (Dittner et al. 1983, Mitchell et al. 1983, Belic et al. 1983). RTE was observed earlier by Tanis et al. (see Tanis

1989 for a review), but there was some reluctance to grant that this was equivalent to DR. Indeed, although the case for RTE being nearly the same as DR is now overwhelming, there is the fundamental complication in the former case that the electron collision occurs in the presence of the strong Coulomb field of the target atom. In exactly what way this changes the process is still not fully explored.

Both DR and RTE are now becoming quite mature subjects, and we give only a brief review. A comparison between resonant dielectronic capture observed in several ways is shown in Figure 12. The energy scales have been adjusted to align common features in the spectra. For the Ge case, an RTE result (Mokler et al. 1990), the resonant capture was observed by detecting in coincidence the two photon decay of the $(1s2s)^1S$ state fed in cascade following the stabilizing K x ray from the $(2s2p)^1P$ doubly excited state. The resulting cross section is in good agreement with theory, but limited in resolution by the Compton width of the target. The Ar result, from an EBIT source with 50 eV resolution (Ali et al. 1990), shows the same features for true DR, again in excellent agreement with theory. The Ti result is for an ion channeled through a single crystal of Si (Datz et al. 1990). A channelled ion is kept away from the nuclei of the target, and thus mainly samples a free electron gas at the center of the crystal channel whose Compton profile is narrower than that for a single target atom. Because of the large projectile velocity, such an ion sees an enormous electron current density (approximately 10^{14} A/cm^2 for the case of Figure 12), which is much lower than the typical current density of 1000 A/cm^2 which an ion in an EBIS/T source sees. However, the product of current density times containment time for the EBIS/T is typically 300 C/cm^2, larger than the 3 C/cm^2 seen by the ion traversing the channel. The latter case presents the opportunity to study electron ion collisions under conditions such that radiative relaxation is comparable or slower than electronic excitation times. The oxygen case represents the highest resolution seen so far in K-shell DR, done with the storage ring at Heidelberg (Kilgus et al. 1990). In all of these cases the agreement with theory is excellent.

For a relativistic U ion, normal electron capture from a low-Z target mediated by the momentum distribution of the target becomes so weak that radiative electron capture becomes the major way in which an ion can capture an electron from a target. RTE is then observable as a resonant version of this process in an experiment which detects only the total electron capture cross section. The results of Graham et al. (1990) are shown in Figure 13, where again good agreement between experiment and theory is found.

The above data show that theoretical calculations for DR, at least for K excitation, are on excellent footing. The RTE data, except for low Z where the impulse approximation is questionable, are in rather good agreement with theory. Surprises still occur, however. Figure 14 shows the results for RTE from a single crystal of Au (Belkacem 1990), which should be compared with those for Si in Figure 12. The DR resonance is so narrow as to be almost easy to miss, and corresponds to a Fermi energy below 1 eV at the channel center. This value is much narrower than the 10 eV found for the Si case, and no explanation for this remarkably narrow width is yet at hand. In addition, the resonances are shifted from their normal positions. This problem is still under investigation.

If the DR resonances are observed in the elastic scattering channel, they appear as coherent scattering amplitudes interfering with normal elastic

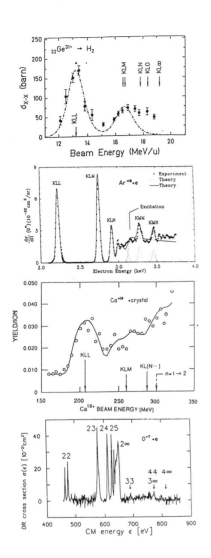

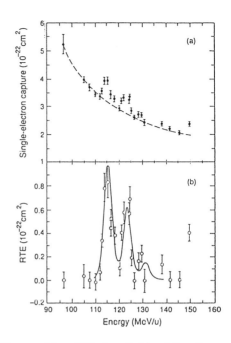

Figure 13. RTE for Heliumlike U in H_2 (Graham et al. 1991).

Figure 12. a) RTE for Ge^{31+} on H_2 (Mokler et al. 1991); (b) DR for electrons on Ar^{+16} in an EBIS (Ali et al. 1990); (c) RTE for Ca^{19+} channeled through a thin Si crystal (Datz 1990); (d) DR for electrons on O^{+7} in a heavy ion storage ring (Kilgus et al. 1990).

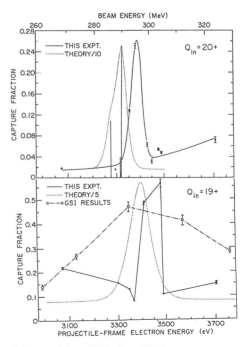

Figure 14. RTE for Ti ions channeled in a thin Au crystal (Belkacem et al. 1990).

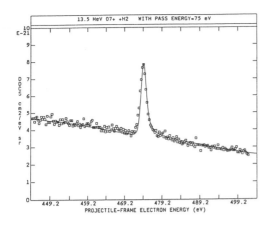

Figure 15. High resolution electron spectrum of the Auger line from the O^{+6} $(2p^2 \ ^1D)$ state formed in RTE of $O^{+7}(1s)$ on H_2. The solid line is a best fit to elastic (binary encounter) plus elastic scattering. The experimetal resolution is 1.5 eV (Richard 1990; Lee 1990).

scattering. This also occurs for the ion-atom case, RTEA, where the Auger lines observed in high resolution are seen to display Fano profiles due to their interference with the binary encounter amplitude. An example is shown in Figure 15 for the $O^{6+}(2p^2 \ ^1D)$ resonance excited in the collision of $O^{+7}(1s)$ with H_2 (Richard 1990). Since the Auger lines are observed in high resolution, the effective resolution of the experiment is given by the electron spectrometer, not by the Compton spread of the target. The effect of the latter spread is roughly to allow the experiment to study a large range of incident electron energies without ever changing the ion beam velocity. The Compton spread does the "scanning." A detailed analysis of the shape and strength of the Auger profile can yield the resonance parameters, including the resonance strength without ever measuring an absolute cross section or scanning the ion energy.

In spite of the maturity of DR and RTE, features such as angular distributions of the electrons and photons and the above interference structure are still active targets of investigation (Bhalla 1991).

2.2.2 Excitation

Electron collisions can contribute to projectile electron excitation without capture as well. The high resolution data of Zouros et al. (1989), shown in Figure 16, show the cross section for the creation of the $1s2s2p^4P$ state in Li-like O as a function of projectile velocity. This state cannot be excited by a single interaction with the projectile nuclear charge, since it requires a spin flip for its creation. A threshold for its creation is seen near the free electron threshold velocity, and its population is attributed to electron exchange whereby an electron from the atomic target collisionally replaces the projectile electron but ends up in an excited state. The data is reproduced in shape by folding the free electron cross section into the target Compton profile. The corresponding process without exchange also exists for the $1s2s2p^2P$, but is harder to see since it competes with normal excitation of the projectile by the target nucleus.

2.2.3 Ionization

A corresponding process exists for ionization. Claytor et al. (1988) used thin solid targets to present an electron flux to very relativistic U ions,

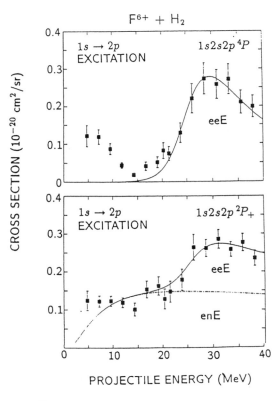

Figure 16. (a) Cross section for the formation of the $1s2s2p\,^4P$ state for O^{+5} ($1s^2\,^2S$) on H_2. The solid curve is calculated using the impulse approximation from the free electron excitation cross section for $1s$ to $2p$ excitation, multiplied by 1.5. (Zouros 1989; Richard 1990; Lee 1990). (b) Similar, but for the $1s2s2p\,^2P$ state.

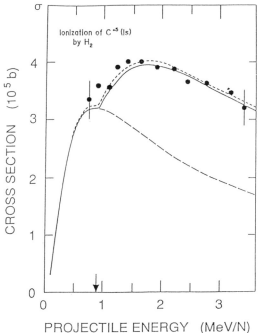

Figure 17. The cross section for ionization of C^{+5} by H_2. The lower long dashed curve is calculated taking into account only nuclear ionization, while the solid and short dashed curves include the excitation due to quasi-free electrons of the H_2. The arrow indicates the threshold energy for the latter process (Meyerhof et al. 1991).

and deduced ionization cross sections about a factor of two higher than theory, a discrepancy which has not yet been explained. Hülskötter et al. (1989) showed evidence for electron ionization in collisions of one electron C and O ions with He.   Figure 17 shows the cross section for collisional removal of the lone K electron of C in collision with H_2. Below threshold, the target nucleus does all of the work. Above threshold, there is a rise in the cross section attributed to the action of the H_2 electrons as ionizing agents of the C 1s electron. Of course, the theoretical curves are important, since without them one would have some difficulty seeing the effect. In this case the theoretical calculation must incorporate the possibility that the two scattering centers of the H_2 act coherently, and amplitudes must be added. A more unambiguous identification of the electron ionization might be possible through transverse momentum spectroscopy of either the projectile or the target.   For example, in hard collisions between C^{+5} and a He nucleus, the carbon should be scattered at much larger angles, and the He receive much larger transverse momentum, than would be the case in a collision which involved only the C^{+5} and one of the He electrons. Preliminary results of Meyerhof and Montenegro (1991) for the projectile angular distributions appear to show an enhancement of large impact parameter contributions above threshold for the electron excitation process. Further work on the momentum transfer question would be valuable.

3. CONCLUSIONS

Our ability to better determine what is happening during a fast ion-atom collision is clearly improving, and one can look forward to a continuing investigations which reveal the roles of both electron-electron collisions and electron-heavy particle collisions. The experiments reviewed here represent   only   a   smattering   within   the   chosen   subject   of   hard electron-electron and electron-nucleus collisions, and this subfield itself represents well less than half of the work presently going on in the vigorous subject of ion-atom. Many trends are left for the next reviewer to identify.

Acknowledgment

This work was supported by the Chemical Sciences Division of Basic Energy Sciences, US DOE.

REFERENCES

Åberg T 1973 USAEC Rept. CONG-720404 1509 (Atlanta Mtg.)
Ali R et al. 1990 Phys. Rev. Lett. **64** 633
Amusia M Ya and Kheifets A S 1985 J. Phys. B18 364
Andersen L H et al. 1987 Phys. Rev. A36 3612
Andersen L H et al. 1989 Phys. Rev. Lett. **62** 1731
Andersen L H, Bolko J and Kvistgaard P 1990 Phys. Rev. Lett. **64** 729
Belic D S et al. 1983 Phys. Rev. Lett. **50** 339
Belkacem A E et al. 1990 Phys. Rev. Lett. **64** 380
Benhenni M et al. 1990 Phys. Rev. Lett. **65** 1849
Bhalla C P 1991 Nucl. Inst. Meth. B56/57 140
Bhalla C P 1991a Phys. Rev. Lett. **64** 1103
Bliman S 1989 Journal de Physique C1, V 50
Bohr N 1948 Mat.-Fys. Medd. Dan. Vidensk. Selsk. 18 No. 8
Bottscher C and Strayer M R 1989 Phys. Rev. D39 1330
Briggs J 1988 Proc. XV ICPEAC ed H G Gilbody et al. (N. Holland) 13

Briggs J and Taulbjerg K 1979 J. Phys. B12 2569
Brown R L 1970 Phys. Rev. A1 586
Byron F W and Joachain C J 1967 Phys. Rev. A8 2887
Cocke et al. 1989 Nucl. Inst. Meth. B42 545
Carlson T A 1967 Phys. Rev. 156 142
Carter S L and Kelly H P 1981 Phys. Rev. A24 170
Claytor N et al. 1988 Phys. Rev. Lett. 61 2081
Datz S 1989 Proc. XVI ICPEAC ed A Dalgarno et al. (AIP: New York) 2
Datz S et al. 1990 Phys. Rev. Lett. 63 742
Datz S et al. 1991 ORNL Progress Rept. (unpublished) 124
Deco G, Momberger K and Grün N 1990 J. Phys. B23 2091
Dittner P F et al. Phys. Rev. Lett. 51 31
Dörner R, Ullrich J, Schmidt-Böcking H and Olson R E 1989
    Phys. Rev. Lett. 63 147
Dowek D et al. 1990 Phys. Rev. Lett. 64, 1713
Fritsch W and Lin C D 1991 Physics Reports 202 1
Gayet R and Salin A 1990 in "Lecture Notes in Physics" 376
Geiger H and Marsden E 1909 Proc. Roy. Soc. (London) A82 495
Giese J P and Horsdal E 1988 Phys. Rev. Lett. 60 2018
Graham W G et al. 1990 Phys. Rev. Lett. 65 2773
Gregory D C et al. 1990 Phys. Rev. A41 106
Hahn Y 1989 Proc. XVI ICPEAC (AIp: New York) 550
Havener C C et al. 1989 Phys. Rev. A39 1725
Holland D M P, Codling K, West J B and Marr G V 1979 J. Phys. B12 2465
Horsdal-Pedersen, Cocke C L and Stöckli M 1983 Phys. Rev. Lett. 50 1910
Horsdal-Pedersen, Jensen B and Nielsen K O 1986 Phys. Rev. Lett. 57 1414
Hulskotter H-P, Meyerhof W E, Dillard E and Guardala N 1989 Phys. Rev.
    Lett. 63 1938
Humberston J W 1989 Proc. XVI ICPEAC (AIP: New York) 614
Ishihara T, Hino K and McGuire J H 1991 private comm.
Iwata Y et al. 1990 Nucl. Inst. Meth. 48 97
Kamber E Y, Cocke C L, Cheng S and Varghese S L 1988 Phys. Rev. Lett. 60
    2026
Kamber E Y, Cocke C L, Cheng S and Varghese S L 1988a J. Phys. B21 L455
Kelbch C et al. 1989 Phys. Lett. A139 304
Kilgus G et al. 1990 Phys. Rev. Lett. 64 737
Kimura K et al. 1991 Phys. Rev. Lett. 66 25
Knapp D A et al. 1989 Phys. Rev. Lett. 62 2104
Kristensen F G and Horsdal-Pedersen E 1991 J. Phys. B23 4129
Lee D H et al. 1990 Phys. Rev. A41 4816
Lee D H 1990a Ph.D. dissertation Kansas State Univ. (unpublished)
Levin J et al. 1991 Bull. Am. Phys. Soc. 36 1328 and private comm.
Liu C J et al. 1989 Contr. Abs. XVI ICPEAC ed A Dalgarno et al. 524
Schulz M et al. 1989 Phys. Rev. Lett. 62 1738
McGuire J H 1991 Advances in Atom. Mol. and Optical Physics (to appear)
McGuire J H et al. 1989 Phys. Rev. Lett. 62 2933
McGuire et al. 1989 Phys. Rev. Lett. 62 2933
Meyerhof W E et al. 1991 Phys. Rev. A43 5907
Meyerhof W E and Montenegro E C 1991 private communication
Merzbacher E 1983 Proc. XVI ICPEAC ed. J Eichler et al. (N Holland) 1
Miraglia J E and Macek J 1991 Phys. Rev. A43 5919
Mitchell J B A et al. 1983 Phys. Rev. Lett. 50 335
Mokler P H et al. 1990 Phys. Rev. Lett. 65 3108
Mueller A et al. 1991 Zeit. fur Phys. D (to appear)
Olson R E, Reinhold C O and Schultz D R 1990 J. Phys. B23 L455

Olson R E, Ullrich J, Dorner R and Schmidt-Bocking H 1989 Phys. Rev. A40 2843

Pálinkás J, Schuch R, Cederquist H and Gustafsson O 1989 Phys. Rev. Lett. 63 2464

Reading J F, Ford A L and Fang X 1989 Phys. Rev. Lett. 62 245

Reading J 1990 Japanese American Seminar, Nucl. Inst. Meth. (to appear)

Reinhold C O et al. 1991 Phys. Rev. Lett. 66 1842

Reinhold C O, Olson R E and Schultz D R 1991a Physics Reports

Reinhold C O, Schultz D R and Olson R E 1990 J. Phys. B23 L591

Richard P 1980 "Methods of Expt. Phys.: Atomic Phys." (Academic: New York)

Richard P 1990 X-90, 15th Intl. Conf. X-Ray and Inner Shell

Richard P, Lee D H, Zouros T J M, Sanders J and Shinpaugh J 1990 J. Phys. B23

Rudd M E and Macek J H 1972 Case Stud. At. Phys. 3 47

Rutherford E 1911 Phil. Mag. 21 669

Salzborn E et al. 1991 "Proc. Intl. Conf. Highly Charged Ions" Zeit. fur Phys. D (to appear)

Salin A 1989 J. Phys. B22 3901

Samson J A R 1990 Phys. Rev. Lett. 65 2861

Schmidt V, Sandner N and Kuntzemuller 1976 Phys. Rev. A13 1748

Schön W et al. 1987 J. Phys. B20 L759

Sellin I 1978 "Topics in Current Phys: Structure and Collisions of Ions and Atoms" (Springer: Berlin)

Sellin I 1990 in "Lecture notes in Phys." 376 ed. D Berenyi and G Hock (Springer: Berlin) 4

Shakeshaft R and Spruch L 1979 Rev. Mod. Phys. 51 369

Shingal R, Chen Z, Karim K R, Lin C D and Bhalla C P 1990 J. Phys. B23 L637

Stolterfoht N 1978 in "Structure and Collisions of Ions and Atoms" ed I A Sellin (Springer: Berlin) 155

Tanis J A 1989 Proc. XVI ICPEAC (AIP: New York) 538

Taulbjerg K 1990 J. Phys. B23 L761

Thomas L H 1927 Proc. Roy. Soc. London 114 561

Tinschert K, Müller A, Phaneuf R A, Hofmann G and Salzborn E 1989 J. Phys. B23 1241

Ullrich J 1991 private comm.

Vegh L 1989 J. Phys. B L35

Vegh L and Burgdörfer J 1990 Phys. Rev. A42 655

Vogt H et al. 1986 Phys. Rev. Lett. 57 2256

Wåhlin E K et al. 1991 Phys. Rev. Lett. 66 157

Wight G R and Van der Wiel M J 1976 J. Phys. B9 1319

Zouros T J M, Lee D H, and Richard P 1989 Phys. Rev. Lett. 62 2261; and Proc. XVI ICPEAC (AIP: New York) 568

Atomic and molecular physics with synchrotron radiation

Bernd Crasemann

Department of Physics, University of Oregon, Eugene, Oregon 97403 USA

Abstract. Tunability, intensity, polarization and time structure make synchrotron radiation a powerful tool for the exploration of atomic and molecular structure and dynamics. Some applications in this field are reviewed here, with special attention to inner-shell processes and to the potential of the new "third-generation" sources currently coming on-line.

1. Introduction

The electromagnetic radiation emitted by electrons or positrons that move at relativistic velocities under transverse acceleration in storage rings has unique properties that make it a valuable tool in scientific research and technology. Applications in materials, surface and interface science, biophysics, lithography and microscopy, for example, have been very fruitful. In atomic and molecular science, many interesting results have been obtained but the tenuity of gas-phase targets has bounded the range of possible experiments with the older rings. This limitation is now being lifted with the advent of far more brilliant "third-generation" sources that optimize the use of insertion devices to generate radiation. As these new facilities are becoming operational, this is a propitious time to survey recent advances in the study of atoms and molecules with synchrotron radiation and to venture a look ahead at some of the new possibilities likely to arise.

1.1. History

It is well-known how critical a role x rays have played in the development of contemporary science and technology, during the relatively brief period since their discovery by Wilhelm Conrad Röntgen in the Christmas holidays of 1895. The applications of x rays have been hampered, however, by the fact that most of the power radiated by a Röntgen tube is concentrated in a few characteristic lines of the target material, while the accompanying continuous bremsstrahlung is relatively weak. This continuum is emitted as the electron beam in the tube is decelerated upon penetrating the target. The electrons radiate in accordance with Larmor's formula, which does not involve the *direction* of the acceleration vector—whence it seems curious that for half a century following Röntgen's discovery man-made x-ray continuum sources were limited to devices in which the electron acceleration is essentially colinear with the velocity. All along, Nature held up beautiful examples of the radiation from transversely accelerated electrons, as visible (since 1054 A.D.) in the Crab nebula.

In fact, Liénard (1898) already derived the relativistic generalization of Larmor's formula and showed that a particle of charge e and mass m undergoing centripetal acceleration in a circular orbit of radius ρ radiates at the rate

$$P = \frac{2e^2c}{3\rho^2}\beta^4\gamma^4 \tag{1}$$

where P is the instantaneous power, γmc^2 is the total energy of the electron, and β is the ratio of its velocity to that of light, c. Interest in this radiation was stimulated in the 1940's when it was realized that radiative power dissipation is of significance for the design of particle accelerators of higher energies (Ivanenko and Pomeranchuk 1944, Blewett 1946). The classical radiation theory for an arbitrary electron trajectory was derived by Schwinger (1946, 1949) who subsequently examined the importance of quantum mechanical corrections (Schwinger 1954). Parallel work was conducted independently in the U.S.S.R. by Ivanenko and Sokolov (1948) and Sokolov *et al* (1953), corroborating Schwinger's results, *viz.*, primarily, that the radiation is sharply peaked in the direction of the electron's motion and that the radiated power extends to ever higher harmonics of the orbital frequency as the electron energy increases. The actual observation of synchrotron radiation did not take place until 1947, through accident, on a 70-MeV synchrotron in the General Electric Research Laboratories in Schenectady. The early history of the subject is described by Lea (1978) and by Winick and Doniach (1980); Hartman (1988) has given an eyewitness account of early experimental work.

1.2. Properties of synchrotron radiation

The physical properties of synchrotron radiation are lucidly derived in the textbook by Jackson (1975); several informative reviews exist that also cover practical aspects, e.g., by Krinsky *et al* (1983). In addition to a high degree of polarization and sharp time structure, the two properties of synchrotron radiation that help make it an ideal probe for atomic and molecular investigations are high intensity and wide energy tunability. Considerable power is radiated in the bending magnets of storage rings, according to Eq. (1); the energy radiated per turn is proportional to the fourth power of the electron energy and inversely proportional to the radius of the orbit. In a circular path of 10-m radius, 100 mA of 3-GeV electrons emit more than 70 kW of electromagnetic radiation. The electrons radiate in a tight cone in the forward direction, of root-mean-square aperture $1/\gamma = mc^2/E$, or 0.2 mrad for $E=3$ GeV. Substantial flux can thus be brought to experimental apparatus well removed from the ring. From a 5-mrad segment of the orbit in the foregoing example, an amount that can readily be used in a single experiment, more than 3×10^{12} photons per second are contained in a 2-eV bandwidth at 1.8 keV.

The spectrum of synchrotron radiation from the bending magnets of a storage ring is produced by the superposition of many harmonics of the fundamental frequency of revolution of the electrons; the individual harmonic lines are smeared out into a continuum that extends from the visible all the way into the hard x-ray regime (Figure 1). The spectrum is characterized by a *critical photon energy* ε_c; half the radiated power is contained in photons of less than ε_c, and half above. The critical photon energy

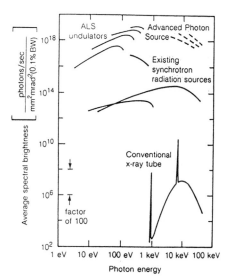

Figure 1. Spectral brightness of various synchrotron-radiation sources, compared with that of a typical conventional x-ray tube. (Courtesy A. L. Robinson, Lawrence Berkeley Laboratory.)

is proportional to the magnetic field and to the square of the stored-electron energy. The photon flux peaks at $0.4\varepsilon_c$, and falls off nearly exponentially above ε_c. For most experimental applications, a slice of the synchrotron-radiation spectrum is selected by monochromatization—with diffraction gratings at the longer wavelengths and by Bragg-crystal diffraction for x rays. The photon energy at the target is readily tuned by adjusting the monochromator, usually under computer control.

The intensity of synchrotron radiation can be further enhanced and its spectrum compressed through the use of insertion devices, i.e., wigglers and undulators, installed in straight sections between the bending magnets. These devices are linear arrays of magnetic dipoles with alternating polarities that produce magnetic fields perpendicular to the plane of the ring (Kim 1987). As the electrons or positrons in the beam traverse the magnetic regions, their trajectory wiggles in the orbital plane and they constantly emit photons. The electron motion in an insertion device is characterized by a deflection parameter K, which measures the maximum angular deflection of the orbit in units of the opening angle $1/\gamma$ and depends upon the magnetic field strength and the distance between alternating magnet poles. In wigglers, K is larger than 1, the electron path swings outside the radiation cone, and the radiation adds incoherently. The spectrum is similar to that from a bending magnet, but brighter by a factor of approximately the number of magnet poles. The photon energy at which the spectrum peaks can be tuned by changing the magnet field strength or gap. In contrast to wigglers, undulators are characterized by a deflection parameter $K \leq 1$; the electron beam remains within the radiation cone, and light emitted from successive beam excursions interferes coherently. Undulator radiation is concentrated in sharp peaks at a fundamental energy, with less intense peaks at higher harmonics. Undulators thus produce exceedingly intense, partially coherent photon beams at specific energies, which can be tuned by varying the gap between the magnets.

1.3. Synchrotron-radiation facilities for research

For early experiments, synchrotron radiation was derived from accelerators or storage rings that had been designed for other purposes; the experiments were "parasitic" on these machines which are referred to as "first-generation." As the value of the radiation became more widely recognized, some of these rings were turned over to dedicated use as synchrotron-radiation sources and others were specifically designed and built as light sources. During the last decade, impressive growth of these "second-generation" machines has occurred throughout the world (Winick 1989).

Currently we are seeing the initiation of "third-generation" light sources which are designed specifically to optimize the output from undulators. They are based on electron or positron storage rings that have very small beam emittance (cross section × divergences) and large current (up to 1 ampere) creating high brightness, as well as long ($\simeq$ 6 m) dispersion-free straight sections to accomodate undulators (Jackson 1990). Two classes of machines are being planned, distinct because the photon energy depends on the undulator period and electron (or positron) beam energy. To cover the ultraviolet and soft x-ray spectrum, Super-ACO has been built in France, and other 0.8-2 GeV rings are at various stages of design or construction in Berkeley and in Brazil, the People's Republic of China, Germany, India, Italy, Korea, the United Kingdom, Sweden, Taiwan, and the U.S.S.R. Larger rings (6-8 GeV) to provide 1-100 keV hard x rays are under construction in the Argonne National Laboratory and by the European countries in France; one is in the design stage in Japan (Levi 1991). The new facilities will produce synchrotron radiation that is far brighter than available sources can provide (Figure 1). Complementing existing machines, the new sources will offer drastically new opportunities in fields of science and technology where experiments have been intensity-starved. Atomic and molecular science falls into this category.

1.4. Opportunities in atomic and molecular science

The potential of synchrotron radiation in atomic physics research was first revealed quite strikingly through classic experiments conducted on the National Bureau of Standards storage ring SURF in the early 1960's. This work made great impact as it led to the discovery of a wealth of discrete resonances in the photoionization continua of rare gases. The first atomic absorption spectrum obtained with synchrotron radiation, that of He in the 200-Å region, revealed series of autoionizing states (Madden and Codling 1963). This discovery and its explanation (Cooper *et al* 1963, Fano 1983) were of crucial importance for the understanding of electron correlation effects; they have led to the formulation of new correlation quantum numbers (Herrick and Sinanoğlu 1975, Lin 1984). Only very recently, an analogous high-resolution photoionization experiment has been performed on the BESSY source in Berlin. With a resolving power of $E/\Delta E \simeq 10,000$, more than 50 states as narrow as 0.1 meV were observed, with evidence for interchannel interference that could be interpreted in terms of multichannel quantum-defect theory (Domke *et al* 1991).

This example of fundamental insights generated by the study of He double-excited states coupling to the continuum illustrates the significant evolution that has occurred in the use of synchrotron radiation in atomic and molecular physics over the last quarter century (see, e.g., McKoy *et al* 1984, Crasemann and Wuilleumier 1985, Dehmer *et al* 1987, Nenner and Beswick 1987).

Major frontiers of atomic and molecular science, where knowledge is bounded by lack of experimental information and theoretical understanding, can be crudely characterized as falling into some broad categories which include

- The *structure* of atoms and molecules and the *dynamics* of their interactions with radiation and matter—especially near energy thresholds,

- *Many-body effects*: electron-electron correlations and other phenomena that transcend independent-particle models,

- *Relativistic and quantum-electrodynamic effects* including those due to the Breit interaction, self energy and vacuum polarization.

In each of these categories, important new classes of investigations will become possible with the new synchrotron-radiation sources. To guide planning of upcoming facilities, workshops and panel sessions have been devoted to the study of these research opportunities (Berry *et al* 1990, Baer *et al* 1991). Some of the most clearly evident topics can be grouped according to experimental research techniques as follows:

Absorption spectrometry: Absolute cross sections, edge structure, x-ray circular dichroism

X-ray scattering: Absolute scattering probabilities, depolarization of resonance fluorescence, angular distributions

X-ray fluorescence: Polarization and angular distributions, chemical shifts, multi-electron effects

Visible-uv fluorescence: molecular fragmentation, molecular vibration/rotational resolution

Photoelectron spectrometry: Cross sections, angular and spin distributions, multi-electron effects, shape and autoionizing resonances, post-collision interaction, coincidence studies, two-color experiments

Auger-electron spectrometry: Auger yields, energy levels of multiply charged ions, satellites and many-electron effects, time-resolved studies, threshold resonances, cascade effects, post-collision interaction, angular distributions and alignment, spin-resolved spectrometry, coincidence studies

Ion spectroscopy: Molecular fragmentation, multiple ionization, ion coincidence studies, studies of trapped ions, two-color experiments

In the following we elaborate on some of these topics, to illustrate what can be done and what may lie ahead.

2. Molecular structure

Fluorescence studies of molecules benefit from the brightness as well as the polarization of insertion-device generated radiation (Poliakoff *et al* 1986). The geometrical and electronic structure of molecules has long been studied through x-ray absorption and emission spectrometries, but identification of the associated molecular symmetries was achieved only indirectly by comparing observed energies and relative intensities with molecular-orbital calculations. Now direct experimental determination of the symmetries associated with x-ray features in small molecules has become possible owing to the discovery of strongly polarized x rays emitted from randomly oriented molecules following selective near-threshold excitation (Lindle *et al* 1988). By tuning a monoenergetic polarized x-ray beam so that a K-shell electron is promoted to an unoccupied valence orbital, molecules aligned with respect to the incident-beam polarization can be selected from a random ensemble of free molecules. Fluorescent x-ray emission occurs quickly as compared with molecular rotation, before any significant dealignment of the molecules. If the core hole is filled by an electron from a valence molecular orbital, it can result in polarization of a fluorescence component. This phenomenon has been used to study CH_3Cl (Lindle *et al* 1988) and, most recently, the molecule H_2S, which has a subthreshold absorption resonance predicted to be composed of excitations to two ($2b_2$ and $6a_1$) orbitals; calculations were verified by the observed polarization behavior (Mayer *et al* 1991). This promising technique may eventually serve to determine the symmetries of atomic sites in larger, less well characterized molecules, and to study oriented molecules on crystal surfaces.

Other synchrotron-radiation-based approaches to the study of molecules include high-resolution spectrometry of the decay dynamics of selectively excited molecular states (Nenner *et al* 1980), studies of shape resonances (Dehmer *et al* 1987, Nenner and Beswick 1987), fragmentation of selectively core-excited molecules (Nenner and Beswick 1987, Larkins *et al* 1988), and the spectroscopy of excited molecular species through two-color double-resonance experiments in which synchrotron radiation and laser light are used in a pump-probe arrangement (Wuilleumier *et al* 1987). Combined with advances in the theory of molecular transitions (Larkins 1990, Chen *et al* 1990), these types of experiments can be expected to lead to significant new insights in molecular science.

3. Many-body aspects of atomic processes

Many-electron processes induced by photon impact epitomize the limitations of the conventional, most tractable models of atomic structure. In first-order perturbation theory, the photon-electron interaction is described by a one-electron operator. The traditional frozen-core, central-field model consequently does not predict such phenomena as multiple excitations, Raman processes and post-collision interaction. These effects, some of which occur mostly near thresholds, can be studied in detail with tunable highly monochromatized synchrotron radiation; they offer some of the most interesting and fruitful approaches to a deeper understanding of atomic and molecular processes.

3.1. Dynamics of atomic core hole states

Synchrotron radiation is making it possible to explore the properties of atomic inner shells, which differ greatly from those of outer electrons involved in optical transitions.

Deep inner-shell hole states are deexcited primarily by radiationless processes; they can have lifetimes as short as femtoseconds or even attoseconds and correspondingly large widths, up to tens of electron volts.

We consider the creation of a core vacancy by photoexcitation or ionization and the subsequent rearrangement of the atomic electron cortège leading to deexcitation until the daughter ion finds itself in its ground state. The escape of the photoelectron from an inner shell of a many-electron atom involves complex dynamics of electron excitations with mutiple correlational aspects. Perhaps most interestingly, the entire process of inner-shell photoionization and subsequent deexcitation has drastically different characteristics near threshold (Krause 1987, Crasemann 1987, Becker and Shirley 1990) from those in the high-energy limit. In the latter case, if an x-ray photon promotes an inner-shell electron to an energetic continuum state, the atom first relaxes in its excited (hole) state. In a practically distinct second step, the hole is then filled—most often by a radiationless transition, under emission of an Auger electron with *diagram* energy that can readily be calculated from the wave function of the stationary, real intermediate state (Mehlhorn 1985).

In the vicinity of core-level energy thresholds, on the other hand, atomic photoexcitation and the ensuing x-ray or Auger-electron emission can occur in a single second-order quantum process, the *resonant Raman effect*. Here, the intermediate states are *virtual* and there is no relaxation phase. The width of the emitted x-ray or Auger line reflects that of the incident radiation and hence can be much *narrower* than the natural lifetime width of the initial hole state. As one tunes the energy of the incident x rays through a broad core level and observes a characteristic resonant x-ray or Auger line (Figure 2), the energy of the line displays linear *dispersion* with incident x-ray energy; the intensity of the line traces out the Lorentzian shape of the core-hole state (Brown *et al* 1980).

X-ray and Auger resonant Raman transitions in atomic inner shells have been observed with existing synchrotron-radiation sources (Eisenberger *et al* 1976, Brown *et al* 1980, Krause and Caldwell 1987), but signal-to-noise ratios have been so poor as to prevent a thorough investigation of these phenomena beyond the original pilot studies. One of the most interesting uses of next-generation sources will probably be in studies of threshold Raman processes, both for the crucial role that they play in the evolution of atomic inner shell-dynamics and for the sake of applications.

Among applications, one obvious possibility is to use the resonant Raman effect in a source of sub-natural-linewidth, narrowly tunable radiation. A promising use of the Auger resonant Raman effect is the decomposition of x-ray absorption edges into their individual components due to discrete bound-bound electron transitions to valence or Rydberg states and to transitions into the continuum (Breinig *et al* 1980) by tracing the intensities of spectator-electron Auger satellites, as illustrated in Figure 2(a) (Brown *et al* 1980). In this way, single-electron contributions to the anatomy of x-ray absorption edges could be distinguished from poorly understood many-body effects which, especially in metals, often still elude quantitative understanding (Wendin and Del Grande 1985, Del Grande 1990). Finally, since a one-step Raman process is mediated, in principle, by a complete set of intermediate discrete and continuous hole states (Åberg 1980, Tulkki *et al* 1987), it should be sensitive to the effect of extraatomic influences. For example, it is well-known that atomic energy-level shifts caused by metallic environments can open intense radiationless channels that grossly alter hole-state widths (Yin *et al* 1973, Sorensen *et al* 1991); such effects could be studied in detail, especially near thresholds, and traced to

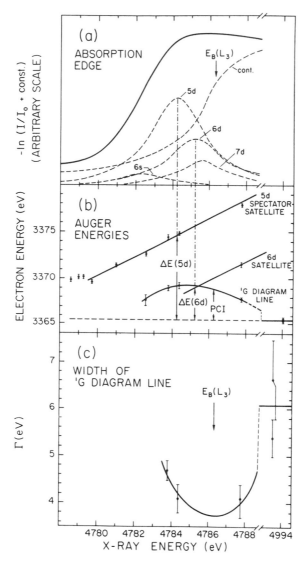

Figure 2. (a) Measured Xe L_3 absorption edge, decomposed according to Breinig *et al* (1980). (b) Energies of the $L_3 - M_4M_5(^1G_4)$ Auger line and its satellites. Near threshold, the ^1G diagram line is shifted by PCI. The satellites are shifted due to screening by $5d$ and $6d$ spectator electrons, respectively; their energies exhibit linear (Raman) dispersion. (c) Width of the measured ^1G diagram-line spectrum, as a function of excitation energy. From Brown *et al* (1980).

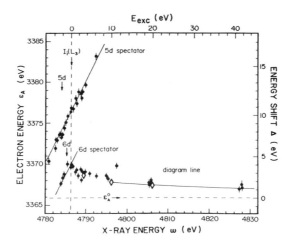

Figure 3. Energy of the Xe $L_3 - M_4 M_5(^1G_4)$ Auger-electron line as a function of exciting x-ray energy. Near-threshold Auger satellites, caused by the spectator photoelectron in bound states, exhibit linear Raman dispersion. The post-collision interaction shift (right-hand scale) vanishes in the asymptotic limit. From Armen *et al* (1985).

modifications of the wave functions in the pertinent Auger matrix elements. One can, in fact, speculate that resonant Raman spectrometry in the x-ray regime may become as useful a tool in atomic physics and materials science as optical Raman techniques now are in biophysics and chemistry.

3.2. Post-collision interaction

Two fundamental questions arise from the foregoing discussion. How does Nature link the single-step process of vacancy creation and decay that takes place in the threshold Raman regime with the drastically different transition dynamics in the high-energy limit, where ionization and subsequent deexcitation are distinctly separated by the lifetime of the hole state during which the atom relaxes, making the characteristics of the decay nearly independent of the mode of excitation? What theoretical framework can lead to a unified, consistent description of these two different mechanism?

The answer to the first question is that the bridge which links the extreme Raman and asymptotic two-step regimes is *post-collision interaction* (PCI) (Schmidt 1987). In near-threshold inner-shell photoionization followed by Auger decay, the Auger electron initially screens the ionic Coulomb field in which the photoelectron recedes. The screening subsides when the faster Auger electron passes the slow photoelectron. The Auger line shape is distorted by this effect and energy is transferred from the slow photoelectron to the fast Auger electron (Figure 3). Intuitive insight and surprisingly accurate predictions can be derived from a semiclassical potential-curve model of PCI (Niehaus 1977, Niehaus and Zwakhals 1983). The attractive potential well in which the slow photoelectron moves is considered to deepen suddenly when the residual singly ionized atom undergoes Auger decay and its net charge changes from $+e$ to $+2e$. There is no time for the photoelectron to speed up; it falls into the deeper well and the energy it loses is transferred to the Auger electron. Not only does the photoelectron slow down—it may be *recaptured* by the atom from which it was emitted (Van der Wiel *et al* 1976, Eberhardt *et al* 1988, Tulkki *et al* 1990). Semiclassical PCI theory has been refined to take account of the time required for the Auger electron to overtake the photoelectron (Ogurtsov 1983, Russek and Mehlhorn 1986).

A *resonant-scattering* theoretical formulation of the decay of metastable states (Åberg 1980, 1981), within the framework of relativistic quantum mechanics, permits development of a unified theory of photoionization and Auger decay that includes both the resonant Raman and PCI effects near threshold and smoothly blends into the high-energy limit. In this scheme, a transition matrix element has been derived that treats photoionization followed by Auger decay as a one-step process which results from the photon-electron interaction: a resonance in the double photoionization cross section. The PCI effect then appears as a property of the resonance behavior of double photoionization (Tulkki *et al* 1987, Armen *et al* 1987b, Tulkki *et al* 1990). Results agree well with experiments (Borst and Schmidt 1986, Armen *et al* 1987a).

3.3. Shake-modified resonant autoionization

The powers of resonant scattering theory and of photoelectron spectrometry with synchrotron radiation are epitomized by a recent study of Rydberg electrons subject to the radiationless decay of a resonantly excited inner-shell hole state (Whitfield *et al* 1991a). These authors used the relativistic formulation of PCI (Tulkki *et al* 1987, Armen *et al* 1987b, Tulkki *et al* 1990) in multichannel multiconfiguration Dirac-Fock calculations (Tulkki 1989) to treat autoionization following photoexcitation as a one-step process mediated by a complete set of intermediate discrete and continuous one-hole states. They noted, however, from the general transition-matrix element that the major contribution to the cross section must come from a final-state phase-space part in which one of the electrons moves with nearly the characteristic autoionizing energy. Under realistic assumptions they were hence able to factorize the many-electron interaction amplitude into an overlap element and the autoionization probability amplitude. They could thus show that previously unexplained features of the Mg $2p$ autoionization spectrum (Whitfield *et al* 1991b) are due to shake transitions of a spectator electron subject to the decay of resonantly excited autoionizing states—revealing for the first time the relationship between shake transitions and the post-collision interaction. These findings lead to important unifying insights; they point the way toward future synchrotron-radiation experiments on the dynamics of resonantly excited sub-threshold inner-shell hole states.

3.4. Coster-Kronig transitions

A unique role in atomic inner-shell dynamics is played by Coster-Kronig processes—radiationless transitions through which a deep core vacancy "bubbles up" to a higher level with the *same* principal quantum number (McGuire 1975). The fastest transitions in atomic deexcitation, Coster-Kronig processes can take place in mean times of the order of femtoseconds to attoseconds, shorter than the Bohr period of the initial vacancy; their strength strains perturbative approaches to the limit. When energetically possible, Coster-Kronig transitions are the principal means by which ionized atoms lose energy; they dominate the characteristics of the vacancy cascade that ensues inner-shell excitation.

Despite their importance, Coster-Kronig transitions prove elusive in both theory and experiment. The matrix elements are exceedingly sensitive to the atomic model; they involve the overlap of three bound-state wave functions with a long-wavelength continuum function. Pronounced many-body features come to play in these transitions. Predictions from single-configuration independent-particle calculations disagree strikingly with observations: for the much-studied case of Ar $L_1 - L_{2,3}M_1$, for example, single-configuration calculations overestimate the rate by a factor of 4 and the $^1P/^3P$ intensity ratio by a factor of 120 (Karim *et al* 1984). Multiconfiguration calculations including the effects of relativity (Bruneau 1983) and of exchange and relaxation (Karim *et al* 1984) have reduced these discrepancies, as have diagrammatic many-body calculations (Wendin 1982). An important correlation effect that influences Coster-Kronig transitions is produced by dynamic relaxation processes or interaction with radiationless continua, in which the core hole fluctuates to intermediate Coster-Kronig levels in addition to creating electron-hole pair excitations (Beck and Nicolaides 1978, Wendin 1982); this effect causes a shift in atomic energy levels (Chen *et al* 1981a, 1985, Indelicato 1991).

Experimental data on Coster-Kronig rates are sorely needed to guide theoretical efforts and for a multitude of practical applications, but they are exceedingly scarce: spectra of the continuum electrons are difficult or impossible to measure because of their low energies, and rate determinations by coincidence techniques are limited to special cases (Bambynek et al. 1972), so that vast *lacunae* remain in the available information (Krause 1979).

A new method of measuring Coster-Kronig transition probabilities is based on selective subshell ionization with narrowly monochromatized synchrotron radiation (Jitschin *et al* 1985). The approach relies on the fact that ionization of a particular subshell can be turned on or off by tuning the incident photon energy across the respective ionization threshold. Detection of the induced fluorescence (Werner and Jitschin 1988) or Auger-electron emission (Jitschin *et al* 1989, Sorensen *et al* 1989) permits determination of Coster-Kronig transition probabilities and subshell fluorescence yields. Present-day facilities have made it possible to demonstrate the feasibility of the method, but its more widespread use with good statistics must await the availability of dedicated atomic-physics stations on third-generation light sources. Yet, at least one interesting result illustrates the power of the technique.

According to *free-atom* energy calculations, $L_2 - L_3N_1$ is the only available Coster-Kronig channel for an L_2 hole in copper; it contributes a partial width of only ~ 0.02 eV (Chen *et al* 1977). On the other hand, there has been indirect evidence that ~ 30 times more intense $L_2 - L_3M_{4,5}$ transitions, cut off in free Cu atoms (Figure 4), are possible in *metallic* Cu (Mårtensson and Johansson 1983). Synchrotron-radiation experiments (Wassdahl *et al* 1990, Sorensen *et al* 1991) have now shown incontrovertibly that this is in fact the case, confirming a unique situation in which *extraatomic* relaxation can switch on an intense inner-shell deexcitation channel.

4. Relativity and quantum electrodynamics in inner-shell processes

It is well known that relativity substantially affects the properties of atoms and molecules, particularly but not limited to the heavier species. Three major factors can be distin-

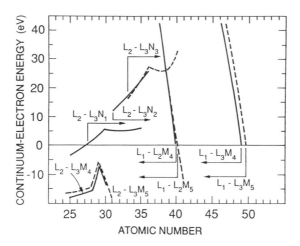

Figure 4. Near-threshold Coster-Kronig energies for some dominant transitions, calculated for *free* atoms (Chen *et al* 1977), as functions of atomic number. Thresholds of relevant transitions are indicated by vertical lines, with horizontal arrows pointing into regions where these transitions can take place. From Sorensen *et al* 1991.

guished. Notably, relativity *redistributes the electronic charge* in an atom. The relativistic mass increase of fast-moving s and p electrons close to the nucleus reduces the mean distance of these electrons from the origin. A simple hydrogenic calculation shows that the electron-mass increase, and hence the change in orbital radius, reaches 0.5% for $Z/n = 14$ and 5% for $Z/n = 43$, where Z is the atomic number and n, the principal quantum number of the orbital. The 1s binding energy is affected by 0.5% for $Z = 19$ and by 5% for $Z = 61$ (Dyall 1986). The increased electronic charge density near the nucleus produced by the relativistic contraction of s and p orbitals causes increased screening of the nuclear charge, whence outer electrons move farther away from the origin. Thus even the wave functions of *per se* nonrelativistic outer electrons, e.g. in the d and f orbitals, can be noticeably affected by relativity (see e.g. Crasemann *et al* 1984, Crasemann 1989). Other major relativistic factors in atomic structure and transitions are *spin-orbit splitting* and the *Breit interaction*, the first dynamic correction to the electrostatic Coulomb interaction in the relativistic Hamiltonian, to which we return below.

The radiative corrections of quantum electrodynamics can be of comparable magnitude as the Breit energy; they comprise the *self energy* and *vacuum polarization*, of opposite signs, which together cause the Lamb shift. Calculation of the self energy in many-electron atoms is a formidable problem; as a first approximation, an effective-charge screening procedure (Chen *et al* 1985) can be applied to point-Coulomb results (Mohr 1982); more sophisticated work is emerging (Mohr 1991, Cheng *et al* 1991). Figure 5 illustrates the relative contributions of Breit energy and radiative corrections to the atomic K-shell binding energy.

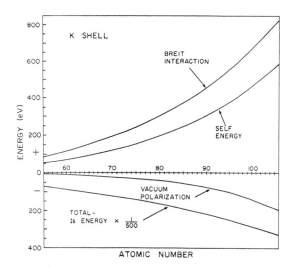

Figure 5. Shifts produced in the 1s energy level by the Breit interaction, self energy, and vacuum polarization, as functions of atomic number. From Chen *et al* 1981.

4.1. Breit interaction

The Breit operator in the relativistic Hamiltonian contains the correction to Coulomb's law to lowest order in the fine-structure constant arising from the exchange of a single transverse photon (see e.g. Armstrong 1978). In the Pauli limit, this corresponds to the orbit-orbit, spin-spin, and spin-other-orbit interactions between two electrons (or two holes). The operator can be written as the sum of a Gaunt term that expresses the magnetic interaction between two electrons and a retardation term that takes account of the effect of the finite velocity of light on both the Coulomb and magnetic interactions. Breit's original formulation of the operator holds in the long-wavelength limit, for electrons that move with much less than the velocity of light (Breit 1932). Mann and Johnson (1971) derived a frequency-dependent form of the Breit operator, suitable for highly relativistic electrons, which is now universally used:

$$H_{Breit}(\omega) = -\frac{1}{r_{ij}}[\alpha_i \cdot \alpha_j \cos \omega r_{ij} + (1 - \cos \omega r_{ij})] \qquad (2)$$

Here, the α_i are Dirac matrices and r_{ij} is the distance between the two interacting electrons; ω is the energy of the virtual photon. The first term in square brackets in Eq. (2) pertains to the retarded Gaunt part, and the second term, the retardation correction to the charge-charge interaction.

While the Breit interaction contributes only a small fraction of atomic binding energies (Figure 5), it can have an impressively large effect in circumstances where the Coulomb interaction cancels out, such as in multiplet splitting. For example, the fine-structure splitting between the $J = 0$ and $J = 1$ multiplet states of the $[1s2p_{1/2}]$ two-hole configuration is greatly increased by the Breit interaction at high Z, and the order of the two levels is inverted. Splitting between the $J = 1$ and $J = 2$ states of the $[2p_{3/2}2p_{1/2}]$

two-hole configuration, on the other hand, is reduced by the Breit interaction and the two levels actually cross at $Z = 93$, according to theory (Chen *et al* 1982). Another extreme example is found in the high-spin metastable states of Li-like ions, such as $1s2s2p^4 P_{5/2}$, which do not decay at all in the nonrelativistic approximation but undergo spin-forbidden Auger decay due to spin-orbit mixing with other states, mediated by the current-current part of the Breit interaction (Cheng *et al* 1974, Chen *et al* 1983).

4.2. Experimental opportunities

Possibilities exist for synchrotron-radiation experiments to explore relativistic and QED effects in few-electron and few-hole systems, although work in this field is in its infancy. Efforts to explore *multiple core-electron photoexcitation* or ionization by absorption spectrometry with synchrotron radiation have recently been reviewed by Kodre *et al* (1990). The signals are weak and readily confused with extraneous effects such as glitches and *umweganregung* reflections in the monochromator crystals; thus, interpretation must be guided by reliable theoretical predictions (Deutsch *et al* 1991, Schaphorst *et al* 1991). Of particular interest would be identification of the splitting between $^3S_1$ and $^1S_0$ double-hole states, which arises from the spin-dependent relativistic exchange integral of the Coulomb interaction and from the Breit interaction (Chen *et al* 1982). Absorption "edges" at double-ionization thresholds are not to be expected, because the onset of shakeoff cross sections is gradual, but thresholds of relevant bound-bound shakeup processes may lead to more readily identifiable features, according to theory (Tulkki and Åberg 1985, Schaphorst *et al* 1991) and results from synchrotron-radiation-excited fluorescence (Deslattes *et al* 1983) and Auger-satellite spectrometry (Fig. 6; Armen *et al* 1985).

4.3. Hypersatellites

X-ray hypersatellite shifts are another quantity, measurable in principle, in which the effect of the Coulomb interaction largely cancels and that of the Breit interaction becomes very pronounced. *K hypersatellites* are x rays emitted in transitions to a completely vacant K shell; they have higher energies than the normal, diagram K x-ray lines. At intermediate atomic numbers, $\simeq 10\%$ of the hypersatellite shift is due to the Breit interaction; at $Z=90$, the fraction is 20% (Chen *et al* 1982). An interesting point to note is that for low-Z elements the $[1s^2] \rightarrow [1s2p]^3P_1$ transition (where square brackets indicate hole states) is *dipole-forbidden* in *LS*-coupling; photon emission then arises from mixing between $^3P_1$ and $^1P_1$ states of the final $[1s2p]$ configuration, which makes the $K\alpha_1^h / K\alpha_2^h$ x-ray hypersatellite intensity ratio very sensitive to the Breit interaction (Åberg and Suvanen 1980, Chen *et al* 1982). Brown *et al* (1991) are yielding to the temptation to search for this effect at the National Synchrotron Light Source.

5. Experiments on ions

Even though stripped atoms vastly preponderate over neutral species in the universe, experimental data on the structure and dynamics of ions are exceedingly scarce. Information on the properties of ions required for the understanding of plasmas and astrophysical phenomena, for example, must largely be derived from rarely verified theory.

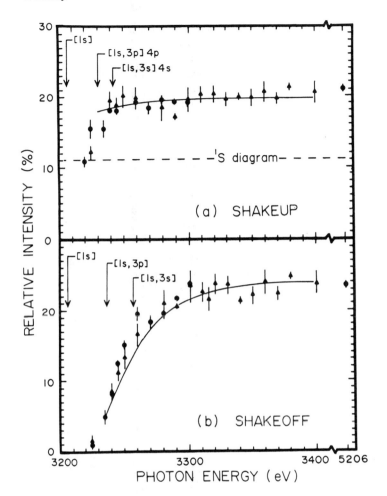

Figure 6. Relative intensities of satellite groups in the Ar $K - L_{2,3}L_{2,3}$ Auger spectrum, showing the sharp threshold onset of shakeup transitions accompanying $1s$ ionization, as against the gradual onset of shakeoff transitions. From Armen *et al* 1985.

Few-electron ions, moreover, provide an ideal testing ground for the theory of relativistic and quantum electrodynamic effects, as discussed above in Section 3. Experiments in this area are therefore of great interest. Highly precise x-ray spectra of selected few-electron species are just becoming available from electron-beam ion traps (EBIT's) (e.g., Cowan *et al* 1991). Some photoabsorption experiments (Sonntag *et al* 1986, Wuilleumier 1989) and photoion spectrometry measurements (Lyon *et al* 1986, 1987a,b) on positive ions have been conducted with synchrotron radiation. Yet it has long been recognized that the most detailed information on the photoionization process is derived from photoelectron spectrometry (e.g., Wuilleumier and Krause 1974, Wu *et al* 1990). Only third-generation synchrotron-radiation sources offer a possibility to measure photoelectron spectra from ion beams, and the long-expected first experiment of this nature has just been completed (Bizau *et al* 1991). With synchrotron radiation from a Super ACO storage-ring undulator, a Ca^+ ion beam was resonantly photoionized and the photoelectron spectrum recorded. This pioneering experiment constitutes the threshold for future work on photoionization and Auger studies of ion beams.

At the same time, the first successful effort to photoionize trapped ions with synchrotron radiation was reported (Kravis *et al* 1991). The distribution of ion charge states following K-shell ionization of Ar^{2+} ions was measured with unmonochromatized radiation from a bending magnet on the x-ray ring of the National Synchrotron Light Source. This work, as well, demonstrates that one can anticipate increased activity in this area, however difficult the experiments may be, because of the importance of potential results.

Acknowledgments

In parts of this paper, the author has drawn upon a lecture he gave in the Argonne National Laboratory (Berry *et al* 1990) and the Berkeley Workshop Report by Baer *et al* (1991). Thanks are due to T Åberg, M O Krause, D W Lindle, J Tulkki, S B Whitfield and F Wuilleumier for sharing results in advance of publication. Honghong Wang kindly assisted with the preparation of this work. This research was supported in part by the National Science Foundation under Grant PHY-9014517.

References

Åberg T 1980 *Phys. Scr.* **21** 495

——1981 *Inner-Shell and X-Ray Physics of Atoms and Solids* ed D J Fabian, H Kleinpoppen and L M Watson (New York: Plenum) pp 251–260

Åberg T and Suvanen M 1980 *Advances in X-Ray Spectroscopy* ed Y Cauchois, C Bonnelle and C Mondé (New York: Perganmon)

Armen G B, Åberg T, Karim K R, Levin J C, Crasemann B, Brown G S, Chen M H and Ice G E 1985a *Phys. Rev. Lett.* **54** 182

Armen G B, Åberg T, Levin J C, Crasemann B, Chen M H, Ice G E and Brown G S 1985b *Phys. Rev. Lett.* **54** 1142

Armen G B, Sorensen S L, Whitfield S B, Ice G E, Levin J C, Brown G S and Crasemann B 1987a *Phys. Rev. A* **35** 3966

Armen G B, Tulkki J, Åberg T and Crasemann B 1987b *Phys. Rev.* A **36** 5606

Armstrong L Jr 1978 *Structure and Collisions of Ions and Atoms* ed I A Sellin (Berlin: Springer-Verlag) pp 69-103

Baer T, Crasemann B, Dehmer J L, Kelly H P, Krause M O and Lindle D W 1991 *Future Research Opportunities in Atomic, Molecular, and Optical Physics* Lawrence Berkeley Laboratory Report PUB-5305 pp 3–21

Bambynek W, Crasemann B, Fink R W, Freund H-U, Mark H, Swift C D, Price R E and Venugopala Rao P 1972 *Rev. Mod. Phys.* **44** 716

Beck D R and Nicolaides C A 1978 *Excited States in Quantum Chemistry* ed C A Nicolaides and D R Beck (Dordrecht: Reidel)

Becker U and Shirley D A 1990 *Phys. Scr.* **T31** 56

Berry H G, Azuma Y and Mansour N B 1990 eds *Atomic Physics at the Advanced Photon Source: Workshop Report* Argonne National Laboratory Report ANL/APS/TM-8

Bizau J M, Cubaynes D, Richter M, Wuilleumier F J, Obert J, Putaux J C, Morgan T J, Källne E, Sorensen S L and Damany M *this Conference* and to be published

Blewett J P 1946 *Phys. Rev.* **69** 87

Borst M and Schmidt V 1986 *Phys. Rev.* A **33** 4456

Breinig M, Chen M H, Ice G E, Parente F, Crasemann B and Brown G S 1980 *Phys. Rev.* A **22** 520

Breit G 1932 *Phys. Rev.* **39** 616

Brown G S, Chen M H, Crasemann B and Ice G E 1980 *Phys. Rev. Lett.* **45** 1937

Brown G S, Cowan P L, Crasemann B, Kodre A F, Ruscheinski J and Schaphorst S J 1991 unpublished

Bruncau J 1983 *J. Phys. B: At. Mol. Phys.* **16** 4135

Chen M H, Crasemann B, Huang K-N, Aoyagi M and Mark H 1977 *At. Data Nucl. Data Tables* **19** 97

Chen M H, Crasemann B and Mark H 1981a *Phys. Rev.* A **24** 1158

Chen M H, Crasemann B, Aoyagi M, Huang K-N and Mark H 1981b *At. Data Nucl. Data Tables* **26** 561

Chen M H, Crasemann B and Mark H 1982 *Phys. Rev.* A **25** 391

1983 *Phys. Rev.* A **27** 544

Chen M H, Crasemann B, Mårtensson N and Johansson B 1985 *Phys. Rev.* A **31** 556

Chen M H, Larkins F P and Crasemann B 1990 *At. Data Nucl. Data Tables* **45** 1–205

Cheng K T, Lin C-P and Johnson W R 1974 *Phys. Rev.* A **48** 437

Cheng K T, Johnson W R and Sapirstein J 1991 *Phys. Rev. Lett.* **66** 2960

Cooper J W, Fano U and Prats F 1963 *Phys. Rev. Lett.* **10** 518

Cowan T E, Bennett C L, Dietrich D D, Bixler J V, Hailey C J, Henderson J R, Knapp D A, Levine M A, Marrs R E and Schnmeider M B 1991 *Phys. Rev. Lett.* **66** 1150

Crasemann B 1987 *J. Physique* **48** C9-389

——1989 *Acta Phys. Hungarica* **65** 171

Crasemann B, Chen M H and Mark H 1984 *J. Opt. Soc. Am.* B **1** 224

Crasemann B and Wuilleumier F 1985 *Atomic Inner-Shell Physics* ed B Crasemann (New York: Plenum) pp 281–315

Dehmer J L, Parr A C and Southworth S H 1987 *Handbook on Synchrotron Radiation* vol 2 ed G V Marr (Amsterdam: North Holland) chap 5

Del Grande N K 1990 *Phys. Scr.* **41** 110

Deslattes R D, LaVilla R E, Cowan P L and Henins A 1983 *Phys. Rev. A* **27** 923

Deutsch M, Brill G and Kizler P 1991 *Phys. Rev. A* **43** 2591

Domke M, Xue C, Puschmann A, Mandel T, Hudson E, Shirley D A, Kaindl G, Greene C H, Sadeghpour H R and Petersen H 1991 *Phys. Rev. Lett.* **66** 1306

Dyall K G 1986 *Aust. J. Phys.* **39** 667

Eberhardt W, Bernstorff S, Jochims H W, Whitfield S B and Crasemann B 1988 *Phys. Rev. A* **38** 3808

Eisenberger P, Platzman P M and Winick H 1976 *Phys. Rev. Lett.* **36** 623

Fano U 1983 *Rep. Prog. Phys.* **46** 97

Hartman P L 1988 *Synch. Rad. News* **1** No 4 28

Herrick D R and O Sinanoğlu 1975 *Phys. Rev. A* **11** 97

Indelicato P 1991 Private communication

Ivanenko D and Pomeranchuk I 1944 *Phys. Rev.* **65** 343

Ivanenko D and Sokolov A A 1948 *Dokl. Akad. Nauk (USSR)* **59** 1551

Jackson A 1990 *Synch. Rad. News* **3** No 3 13

Jackson J D 1975 *Classical Electrodynamics* (New York: Wiley) Second Edition, Sec. 14.6

Jitschin W, Materlik G, Werner U and Funke P 1985 *J. Phys. B: At. Mol. Phys.* **18** 1139

Jitschin W, Grosse G and Rohl P 1989 *Phys. Rev. A* **39** 103

Karim K R, Chen M H and Crasemann B 1984 *Phys. Rev. A* **29** 2605

Kim K-J 1987 *Physics of Particle Accelerators* AIP Conference Proceedings No 184 ed M Month and M Dienes (New York: American Institute of Physics) pp 565–632

Kodre A F, Schaphorst S J and Crasemann B 1990 *X-Ray and Inner-Shell Processes* AIP Conference Proceedings No 215 ed T A Carlson, M O Krause and S T Manson (New York: American Institute of Physics) pp 582-590

Krause M O 1979 *J. Phys. Chem. Ref. Data* **8** 307

Krause M O 1987 *Phys. Scr.* **T17** 146

Krause M O and Caldwell C D 1987 *Phys. Rev. Lett.* **59** 2736

Kravis S D, Church D A, Johnson B M, Meron M, Jones K W, Levin J, Sellin I A, Azuma Y, Berrah Mansour N, Berry H G and Druetta M *Phys. Rev. Lett.* **66** 2956

Krinsky S, Perlman M L and Watson R E 1983 *Handbook on Synchrotron Radiation* ed E-E Koch vol 1A (Amsterdam: North-Holland) pp 65–171

Larkins F P, Eberhardt W, Lyo I W, Murphy R and Plummer E W 1988 *J. Chem. Phys.* **88** 2948

Larkins F P 1990 *J. Electron Spec. Relat. Phenom.* **51** 115

Lea K R 1978 *Phys. Rep.* (Section C of Physics Letters) **43** 338

Levi B G 1991 *Physics Today* **44** No 4 17

Liénard A 1898 *L'Éclairage Électr.* **16** 5

Lin C D 1984 *Phys. Rev.* A **29** 1019

Lindle D W, Cowan P L, LaVilla R E, Jach T, Deslattes R D, Karlin B, Sheehy J A, Gil T J and Langhoff P W 1988 *Phys. Rev. Lett.* **60** 1010

Lyon I C, Peart B, West J B and Dolder K 1986 *J. Phys. B: At. Mol. Phys.* **19** 4137

——1987a *J. Phys. B: At. Mol. Phys.* **20** 5403

    bibitem Lyon I C, Peart B, Dolder K and West J B 1987b *J. Phys. B: At. Mol. Phys.* **20** 1471

Madden R P and Codling K 1963 *Phys. Rev. Lett.* **10** 516

Mann J B and Johnson W R 1971 *Phys. Rev.* A **4** 41

Mårtensson N and Johansson B 1983 *Phys. Rev.* B **28** 3733

Mayer R, Lindle D W, Southworth S H and Cowan P H 1991 *Phys. Rev.* A **43** 235

McGuire E J 1975 *Atomic Inner-Shell Processes* ed B Crasemann vol 1 (New York: Academic) pp 293–330

McKoy V, Carlson T A and Lucchese R R 1984 *J. Phys. C: Solid State Phys.* **88** 3188

Mehlhorn W 1985 *Atomic Inner-Shell Physics* ed B Crasemann (New York: Plenum) pp 119–180

Mohr P 1982 *Phys. Rev.* A **26** 2338

——1991 Private communication

Nenner I and Beswick J A 1987 *Handbook on Synchrotron Radiation* vol 1 ed G V Marr (Amsterdam: North-Holland) chap 6

Nenner I, Guyon P M and Baer T 1980 *J. Chem. Phys.* **72** 6587

Niehaus A 1977 *J. Phys. B: At. Mol. Phys.* **10** 1845

Niehaus A and Zwakhals C J 1983 *J. Phys. B: At. Mol. Phys.* **16** L135

Ogurtsov G N 1983 *J. Phys. B: At. Mol. Phys.* **16** L745

Ohno M and Wendin G 1979a *J. Phys. B: At. Mol. Phys.* **11** 1557

——1979b *J. Phys. B: At. Mol. Phys.* **12** 1305

Poliakoff E D, Ho M H, Leroy G E and White M G 1986 *J. Chem. Phys.* **85** 5529

Russek A and Mehlhorn W 1986 *J. Phys. B: At. Mol. Phys.* **19** 922

Schaphorst S J, Kodre A F, Ruscheinski J, Crasemann B, Åberg T, Tulkki J, Chen M H, Azuma Y and Brown G S 1991 *Bull. Am. Phys. Soc.* **36** 1329 and to be published

Schmidt V 1987 *J. Physique* **48** C9-401

Schwinger J 1946 *Phys. Rev.* **70** 798

——1949 *Phys. Rev.* **75** 1912

——1954 *Proc. Nat. Acad. Sci. (Washington)* **40** 132

Sokolov A A, Klepikov N P and Ternov I M 1953 *Dokl. Akad. Nauk (USSR)* **89** 665

Sonntag B F, Cromer C L, Bridges Y M, McIlrath T J and Lucatorto T B 1986 *Optical Science and Engineering Series* 7 AIP Conference Proceedings No 147 ed D T Attwood and Y Bokor (New York: American Institute of Physics) pp 412-422

Sorensen S L, Carr R, Schaphorst S J, Whitfield S B and Crasemann B 1989 *Phys. Rev.* A **39** 6241

Sorensen S L, Schaphorst S J, Whitfield S B, Crasemann B and Carr R 1991 *Phys. Rev.* A **44** 350

Tulkki J 1989 *Phys. Rev. Lett.* **62** 2817

Tulkki J and Åberg T 1985 *J. Phys. B: At. Mol. Phys.* **18** L489

Tulkki J, Armen G B, Åberg T, Crasemann B and Chen M H 1987 *Z. Phys.* D **5** 241

Tulkki J, Åberg T, Whitfield S B and Crasemann B 1990 *Phys. Rev.* A **41** 181

Van der Wiel M J, Wight G R and Tol R R 1976 *J. Phys. B: At. Mol. Phys.* **9** L5

Wassdahl N, Rubensson J-E, Bray G, Glans P, Bleckert P, Nyholm R, Cramm S, Mårtensson N and Nordgren J 1990 *Phys. Rev. Lett.* **64** 2807

Wendin G 1982 *X-Ray and Atomic Inner-Shell Physics–1982* AIP Conference Proceedings No 84 ed B Crasemann (New York: American Institute of Physics) pp 495–516

Wendin G and Del Grande N K 1985 *Phys. Scr.* **32** 286

Werner U and Jitschin W 1988 *Phys. Rev.* A **38** 4009

Whitfield S B, Tulkki J and Åberg T 1991a submitted to *Phys. Rev.* A

Whitfield S B, Caldwell C D and Krause M O 1991b *Phys. Rev.* A **43** 2338

Winick H 1989 *Proceedings of 1989 IEEE Particle Accelerator Conference* (IEEE Cat No 89CH2669-0) pp 7–11

Winick H and Doniach S 1980 *Synchrotron Radiation Research* (New York: Plenum) Chap 1

Wu J Z, Whitfield S B, Caldwell C D, Krause M O, van der Meulen P and Fahlman A 1990 *Phys. Rev.* A **42** 1350

Wuilleumier F J 1989 *Proceedings of the Workshop on Photon-Ion Interactions* ed F J Wuilleumier and E Källne (Grenoble: ESRF) pp 5 and 231

Wuilleumier F J and Krause M O 1974 *Phys. Rev.* A **10** 242

Wuilleumier F, Ederer D L and Picqué 1987 *Adv. At. Mol. Phys.* **23** 197

Yin L I, Adler I, Chen M H and Crasemann B 1973 *Phys. Rev.* A **7** 897

Interaction of multi charged ions at low velocities with surfaces and solids

H J Andrä[1], A Simionovici[1], T Lamy[1], A Brenac[1], G Lamboley[1], A Pesnelle[2], S Andriamonje[3], A Fleury[4], M Bonnefoy[4], M Chassevent[4] and J J Bonnet[4]

[1] LAGRIPPA, CEN-G, BP 85X, 38041 Grenoble Cedex, France
[2] SPAM, C.E. Saclay, 91190 Gif-sur-Yvette Cedex, France
[3] CEN Bordeaux-Gradignan, Le Haut-Vigneau, 33175 Gradignan Cedex, France
[4] Lab. de Phys. des Coll. Atomiques et Moléculaires, CNAM,
    292, rue St. Martin, 75141 Paris Cedex 03, France

ABSTRACT: The interaction of multicharged ions with conducting surfaces and solids is reviewed for two energy regimes. At high energy the ions penetrate into the solid and are neutralized inside the bulk via direct electron capture into inner shells and subsequent cascades, the dynamics of which can be observed with X-ray spectroscopy. At low and very low energies, completely inverted, neutral atoms may be formed in front of the surface by resonant electron capture from the conduction band into high-lying Rydberg levels while the inner shells stay empty. Secondary electrons yields, X-ray and Auger electron spectra give experimental hints for the transient existence of these exotic atoms.

1. INTRODUCTION

An ion approaching a surface at high energy (>1 keV/u) is known to penetrate into the solid with a probability close to one and eventually to be stopped in the bulk. However, when it dives at low energy into the electron density of the surface, which reaches out decreasing exponentially into the vacuum, its nucleus starts experiencing the collective repulsion of all nearby surface nuclei, as summarized by Andrä (1989). This repulsion is transformed into the so called planar surface potential which depends on the nuclear charges Z_1 and Z_2 of the ion and the target atom, respectively, and on effective charge of the projectile near the surface. A Multi Charged Ion (MCI) may thus be totally reflected on this planar potential at energies between 20 and 200 eV.

At such low energies a MCI is considered to resonantly capture electrons from the conduction band of a conducting target (defined by the work function ϕ_w, the depth of the potential well U_0, and the Fermi energy ε_F) into high lying Rydberg manifolds $n \approx q/\sqrt{2}$ ϕ_w at a great distance $z_c \approx 1.6 <r_{ns}> = 1.2 \, q/\phi_w$ in front of the surface, without taking the charge image into account. z_c is related to the spatial extension of the wavefunction of the Rydberg manifold n, since the overlap of its exponential tail with the exponential tail of the electronic wavefunction of the solid at ε_F yields a sufficient transition rate from the conduction band towards the MCI to populate this manifold n at the distance z_c. z_c may be modified somewhat due to the image charge which has to be derived under the condition that the potential is zero at the surface. This condition implies a potential barrier between the potential wells of the solid and the ion. It leads to a slight expansion of the exponential tails of the wavefunctions of the electrons of the solid and to a considerable spatial reduction of the tail of the ionic wavefunction so that their net overlap is reduced. With the lack of a complete quantum treatment of this situation, the pure geometric shape of the net potential suggests a reduction of z_c to an estimated $z_c \approx \sqrt{(10 \cdot q \cdot 12 + 144)}$. This problem has been treated by Snowdon (1988) and very recently by Burgdörfer et al.(1991)

and is thus still a matter of discussion and introduces a great uncertainty into the value of the real z_c to be used. It will, however, become evident that it is only its order of magnitude which is of significance at the present sensitivity of the experiments.

Completely inverted neutral atoms may thus be formed in this way in front of the surface with q electrons in high lying Rydberg manifolds while the inner shells stay empty. From these Rydberg manifolds the electrons may eventually drop down via Auto-Ionization (AI) cascades to lower Rydberg manifolds accompanied by the emission of low energy electrons which are immediately replaced from the surface by further resonant capture. These atoms of an absolutely new and unexpected species in atomic physics evidently create a great interest in the study of their transient behavior before they hit the surface. There is clearly also an obvious practical interest in the comprehension of these highly inverted atoms since they could be the ideal medium for X-ray lasers in case one could produce them in the appropriate densities.

The adiabatic approximation used in the considerations for the resonant electron capture sets the limit of vertical velocity v_v for the application of this model. One has to require that the period $T_n = \pi \cdot q / \sqrt{(2 \cdot E_n^3)}$ of an orbital revolution in the manifold n is about one tenth of the time ΔT of approach from z_c to the surface. Setting $E_n = \phi_w$, one can readily deduce from this condition that very low energies, 288 eV for Ne^{9+} and 576 eV for Ar^{17+}, are required for the observation of this adiabatic situation with well defined energy levels and transition rates.

At high energies, as defined by $\Delta T \leq T_n$ and corresponding to ≥ 28 keV for Ne^{9+} and ≥ 57 keV for Ar^{17+}, individual energy levels are washed out by the uncertainty principle, except for lower lying shells. Resonant electron capture will nevertheless take place from the conduction band into global regions of low binding energy and global cascading from this region can occur. All Rydberg manifolds with $T_n \leq 0.1 \cdot \Delta T$ can, however, still be treated individually. As will be used later, this particularly applies to the manifolds with $n \leq 5$, the cascading of which seems to dominate the spectroscopic information obtained as yet.

2. EXPERIMENTS WITH INCIDENT MCI AT HIGH ENERGY

When a hydrogenic MCI approaches a surface with high energy, its approach time ΔT is so short that the population of the n = 2 manifold via cascading from $n \leq 5$ becomes negligible. The MCI thus penetrates into the solid with high probability and with a nearly empty L-shell. The electrons in manifolds $n \geq 3$ are considered to be shaken off when they experience a sudden change of their effective central potential when the projectile is diving into the screening electron density of the solid.

In these conditions, the X-ray emission can yield information on what happens to the K and L shells of the MCI deep inside the bulk, while the information carried by Auger electrons is limited to the history of the MCI in the first few atomic layers due to the short range escape probability of such Auger electrons from the bulk as summarized by Ley *et al.* (1979). Donets (1983) was the first to observe with a Ge detector the X-ray emission of Ar^{17+} impinging at 18 keV on Be. His low resolution spectrum is shown in Fig.1a compared to the K_α, K_β lines of singly ionized Ar and the "Ly_α", "Ly_β", "Ly_γ" lines of He-like Ar^{16+}. The dominant peak is attributed to K_α satellites while the high energy tail is assumed to stem from K_β, and K_γ satellites. As shown in Fig.1b, the region between K_α and "Ly_α" can so be decomposed in its 8 K_α satellites which differ in energy due to the screening by different numbers of L electrons. The K_α line corresponds to a completely filled L shell while the "Ly_α" line is due to a single 2p electron in the L shell. Neglecting the M shell populations, one can readily deduce from the peak position in the spectrum that an average of 5 electrons are in the L shell when these X-rays are emitted. An analysis with correct fluorescent yields by Zschornak *et al.* (1983) yields 4-5 electrons. This interpretation has been perfectly well confirmed by a recent high resolution measurement by Briand *et al.*(1990) in Fig.1c where the X-rays from Ar^{17+} at 340 keV on Ag were analyzed by a plane, mosaic graphite crystal in conjunction with a position sensitive detector. Besides the non-spectroscopic notation and inconsistencies in the energy scale, their spectrum shows as essential feature the 8 resolved K_α satellites with an intensity distribution which corresponds very well to the interpretation of Donets' result.

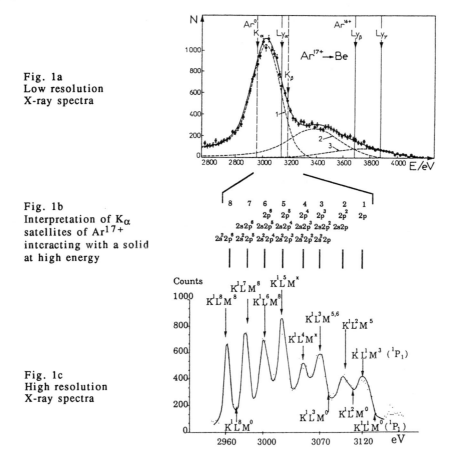

Fig. 1a
Low resolution
X-ray spectra

Fig. 1b
Interpretation of K_α
satellites of Ar^{17+}
interacting with a solid
at high energy

Fig. 1c
High resolution
X-ray spectra

In order to understand the filling dynamics of the L shell which is responsible for the emission of the spectrum in Fig.1c we have modelled it with a set of coupled rate equations for the K, M shells and for the 2s and 2p L subshells under the following constraints:

(i) Inside the bulk these shells are directly populated from inner shells of the target atoms with constant cross sections σ_K, σ_L (for 2s and 2p), and σ_M, respectively, which transform into filling rates via $R_j = \rho \cdot \sigma_j \cdot v$, where ρ is the density of the target atoms and v the constant velocity of the projectile. The σ_j cross sections, comprise quasi-resonant capture, molecular orbital promotion-type capture, and interatomic Auger capture. Due to this complexity, no *ab initio* estimates exist as yet so that the σ_j have to be considered as free parameters, which may be further related to the geometric size $<r_{nl}>^2$ of the capturing orbits.

(ii) The MCI can accumulate a maximum of q electrons only to become at best neutral but not negative.

(iii) The X-ray and Auger transition probabilities are extrapolated as linear functions of the populations in the upper shells involved in the transitions starting from the values in the literature.

(iv) The rate equations are solved numerically for an initial $Ar^{17+}(1s)$ ion which enters at distance $z = 0$ and time $t = 0$ into the solid.
(v) The intensity of every K_α satellite at its average energy position is accumulated for all time increments until the initially empty K hole is filled.

The result so obtained for the populations of the shells as a function of the time spent inside the bulk and for the K_α satellite spectrum is shown in Fig.2a,b, respectively, for the cross sections indicated. While the populations seem to show a predominance of the direct population of the M shell, it can be shown by systematic variations of the σ_j and of the transition probabilities used in the model that the direct population of the L shell is decisive for the close reproduction of the spectrum observed. The value of σ_L is in addition very close to the geometrical cross section $\pi \cdot <r_{2s}>^2$ of Ar^{17+} so that one may accept our model with the indicated cross sections as a reasonable comprehension of the neutralization process of a MCI inside the bulk.

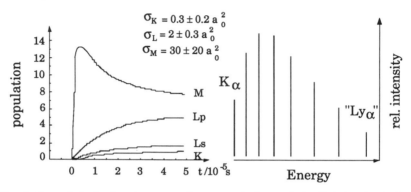

$$\sigma_K = 0.3 \pm 0.2 \, a_0^2$$
$$\sigma_L = 2 \pm 0.3 \, a_0^2$$
$$\sigma_M = 30 \pm 20 \, a_0^2$$

Fig.2 Simulation of a) the development of the populations of the K,L,M shells inside the solid and b) of the resulting X-ray spectrum of Fig.1c.

In order to make sure that Donets' experiment and its interpretation represent indeed a pure Ion-Bulk Interaction (IBI) it was necessary to give experimental evidence for the existence of an interaction of the MCI in front of the surface and to demonstrate the transition from a pure IBI towards an Ion-Surface Interaction (ISI). To this end we have taken advantage of the high optical quality of the MCI beams of the NEW LAGRIPPA FACILITY, see Lamy *et al.* (1990). We have systematically varied ΔT by varying the vertical velocity $v_v = v \cdot \sin \phi_{in}$ of Ar^{17+} at 340 keV via the angle of incidence ϕ_{in} of the trajectory with respect to the surface planes of highly polished, polycrystalline Ta or monocrystalline doped Si surfaces. While varying ϕ_{in} from 60° to 2°, the emitted X-rays were analyzed with a Si(Li) detector to yield the low resolution spectra for the two extreme angles in Fig.3 which have been observed with the same result also by Schulz *et al.* (1991).

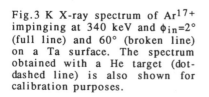

Fig.3 K X-ray spectrum of Ar^{17+} impinging at 340 keV and $\phi_{in}=2°$ (full line) and 60° (broken line) on a Ta surface. The spectrum obtained with a He target (dot-dashed line) is also shown for calibration purposes.

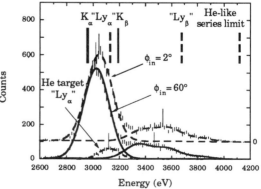

The spectrum at 60° corresponds very well to the one shown in Fig.1a while the average energy $<E_{K\alpha}>$ of the K_α satellites in the spectrum at 2° is clearly shifted towards higher energy, a result also recorded by Schulz *et al.* (1991). When plotting $<E_{K\alpha}>$ as a function of ϕ_{in} for two different energies of Ar^{17+}, one obtains the systematic increase of $<E_{K\alpha}>$ with decreasing ϕ_{in} in Fig.4a. In order to understand this behavior, one has to assume some sort of ISI because pure IBI would result in straight horizontal lines at the level of $<E_{K\alpha}(60°)>$.

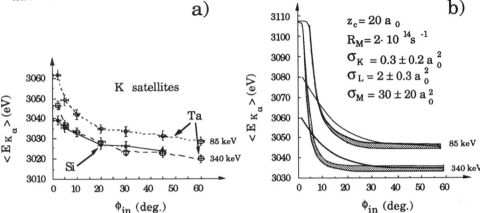

Fig.4 Measured a) and simulated b) average energy of the K_α satellites as a function of the angle of incidence of Ar^{17+} at two energies.

Anticipating a later discussion, one can assume that at a distance of about $z = 20\ a_0$ all electrons are in low enough Rydberg manifolds ($n \leq 5$) so that Auger cascades dominate the further transient development of the ion. With the preference of $\Delta n = 1$ steps of Auger cascading, one can thus approximate the ISI by a single and constant feeding rate $R_M(ISI)$ into the M shell for $z \leq 20\ a_0$. This approximation has been used to reproduce the data of Fig.4a in Fig.4b with the former set of σ_j, with the indicated R_M, and with the constraint that all M shell population is shaken off when the projectile dives into the bulk. Excellent agreement is obtained for the relative change of $<E_{K\alpha}(60°)>$ with the energy of the MCI. This strongly suggests that the linear dependence of the direct feeding rates on v with constant σ_j is a good description of the physics involved inside the bulk. For the ϕ_{in} dependence, the general trend of the data is reproduced but at very small angles the increase of $<E_{K\alpha}>$ is evidently overestimated. This is, however, with great probability not a defect of the model which is based on the use of atomically flat surfaces while the experiment deals with realistic surfaces. An approximate mathematical description of the surface roughness has therefore to be used for a convolution with the calculated $<E_{K\alpha}>$ dependence. To this end one may subdivide the surface into micropatches with a distribution of inclinations with respect to the average surface and then describe the surface by a Gaussian distribution of these inclinations with a maximum of 1 and a halfwidth of 10° resting on a constant background of 0.05 (in order to account for surface steps) which extends from -90° to +90°. A convolution of the calculated dependence with this distribution yields the full lines in Fig.4b which are satisfactory reproductions of the data.

One can therefore conclude that :

(i) an ISI exists,

(ii) the description of this ISI by a single feeding rate $R_M(ISI)$ in the regions $z \leq 20\ a_0$ is an acceptable approximation when the surface roughness is taken into account,

(iii) R_M may be varied to a large extent without significantly modifying the results of the convoluted simulations,

(iv) the surface roughness has such a significant influence on the data that no finer details of the ISI can be deduced.

In order to improve this situation, we have carried out high resolution measurements of X-rays emitted from Ne^{9+} impinging at 35 keV and ϕ_{in} = 5° and 50° on Ta using a standard plane RbAP(Beryl)-crystal spectrometer with Soller slits described by Fleury *et al.* (1986) The results are shown in Fig.5a,b and compared to our model calculation in Fig.5d with the indicated parameters. The spectrum is dominated by the K_α line at ϕ_{in} = 50° which confirms the picture of a rapid filling of the L and M shells inside the bulk. It is very well reproduced by the model which at 35 keV and 50° does not depend at all on an ISI. At 5°, however, the ISI becomes very important and is here approximated by a R_M(ISI) for z ≤ 12 a_0. With this assumption the spectrum in Fig.5e is expected, in evident contradiction with the experimental result in Fig.5b. The measurement clearly shows a dramatic change of the spectrum from Fig.5a towards a predominance of "Ly_α" at 5° in Fig.5b. There is, however, a considerable intensity left in the K_α line at 5° while the model predicts near-zero intensity. One has to suspect therefore that the surface roughness again plays a very perturbing role in the experiment such that the ions approaching the surface at 5° penetrate into the bulk with high probability due to a great fraction of surface patches with inclinations > 5°. As a consequence, one has to consider the spectrum in Fig.5b as a superposition of Fig.5a and of Fig.5c.

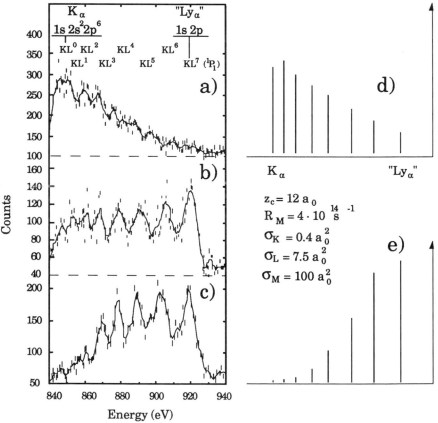

Fig.5   K X-ray spectrum in the wavelength range of 1.3 to 1.5 nm emitted from Ne^{9+} impinging a) at 35 keV and ϕ_{in}= 50°, b) at 35 keV and ϕ_{in}= 5° on a Ta surface, and c) at 90 eV and ϕ_{in}= 89° on a doped Si surface. The full lines are 5 point smooths to guide the eye. d) and e) are simulations of the spectra in a) and c), respectively.

In order to prove this interpretation, a high resolution X-ray spectrum of Ne^{9+} interacting at only 90 eV (!) and at ϕ_{in} ≈ 89° with a monocrystalline, doped Si surface has been recorded by Brenac *et al.* (1991) and is shown in Fig.5c, conditions at which the penetration into the bulk becomes negligible. This spectrum convincingly demonstrates that the K_α intensity becomes indeed negligible when the ion does not penetrate. Instead, the spectrum is dominated by the emission due to excited configurations with only a few electrons in the L shell. It is, however, also obvious from the comparison with the model calculation in Fig.5e that the fraction of intensity with 3-5 electrons in the L shell is clearly underestimated by the model so that improvements which go beyond the simple cascading with a constant R_M (ISI) have to be considered later on. It should be noted that even a completely unrealistic R_M (ISI) $>> 4 \cdot 10^{14} s^{-1}$ cannot reproduce the spectrum in Fig.5c and that the high value of $4 \cdot 10^{14} s^{-1}$ has been chosen in Fig.5e in order to enhance the intensity of KL^6 (2 electrons in the L shell) with respect to "Ly$_\alpha$" which becomes dominant with the decrease of R_M (ISI).

From the ensemble of X-ray measurements available as yet one may therefore conclude that :
(i) the MCI interaction at high energy and at $\phi_{in} > 30°$ is a pure IBI and has nothing to do with a surface interaction
(ii) the MCI interaction at high energy and $2° \leq \phi_{in} \leq 20°$ clearly indicates the existence of an interaction in front of the surface which is, however, heavily masked by the perturbation induced by the surface roughness
(iii) the MCI interaction at very low energy and $\phi_{in} \leq 90°$ gives clear evidence not only for the existence of an ISI but also for further details of it, which have to be rediscussed later on
(iv) the deceleration of MCIs seems to be a much easier experimental approach for the study of the ISI compared to the enormous effort necessary to prepare near-atomically flat monocrystalline surfaces in UHV for measurements at high energies and grazing angles.

3. EXPERIMENTS WITH INCIDENT MCI AT LOW ENERGY

Low energies, as defined above, allow to discuss discrete, high lying Rydberg manifolds which are populated with less than q electrons and may eventually AI-cascade down to lower Rydberg manifolds. This cascading process was initially suggested by Hagstrum (1954) for the interpretation of his first measurement of the low energy electron emission spectra from an MCI-surface interaction which peaks at very low energies ≤ 5 eV. It was mathematically formulated by Arifov *et al.* (1973) to explain their surprising experimental result that the total secondary electron yield per ion Γ is – for ions up to q = 7 – proportional to the total potential energy $W_q = \Sigma_i (E_{bi} - 2 \phi_w)$ stored in the ion, where E_{bi} are the ionization energies of each charge state. They concluded that the lowest energies of the emitted free electrons in AI-cascades are favored so much that the spectrum indeed peaks at ≤ 5 eV and that the average energy emitted is $\varepsilon \approx 20$ eV, a value which has to be revised to $\varepsilon \leq 15$ eV due to improved spectroscopic experiments by Delaunay *et al.* (1987a). With this ε, Γ becomes indeed proportional to W_q when setting $\Gamma = k \cdot W_q / \varepsilon$. This relation has in the meantime been experimentally verified by Delaunay *et al.* (1987b) up to q = 12 at a velocity of the MCIs of only $v_v = 4 \cdot 10^4$ m/s as shown in Fig.6a. At $v_v = 2 \cdot 10^5$ m/s, however, a drastic deviation from the proportionality on W_q is observed in Fig.6b which is attributed to the lack of time available to complete the cascading of q electrons to the ground state. Surprisingly enough, the result at this and even higher velocities can be represented by $\Gamma \propto q^2$ whereas $\Gamma \propto W_q$ at the low velocity of Fig.6a implies $\Gamma \propto q^a$ with a > 2.

These results on the shape of the low energy spectra and on the yield Γ of secondary electrons seem to support the AI-cascading model quite well. It is, however, accepted by now that the AI transition rates of the high lying Rydberg manifolds with n > 10 are too small ($\gamma < 10^{13}$ s^{-1}) for ever completing an AI-cascade in the time ΔT available even at the lowest velocities. This theoretical result of incomplete AI-cascading is corroborated by the small factor of proportionality k = 0.17 observed in Fig.6a which should be 0.7 < k <

1 for a complete cascade. The uncertainty is due to the unknown losses among the electrons emitted towards the surface and also to the AI-cascading electrons emitted into the empty continuum of the solid above the conduction band. A promising new experimental technique has been developed by Lakits *et al.* (1989) which may allow to measure the secondary electron emission statistics and in particular Γ, with high precision as outlined by Winter (1991).

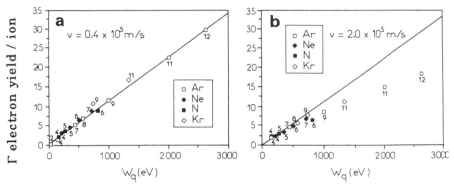

Fig.6 Secondary electron yield per ion vs W_q at a) $4 \cdot 10^4$ m/s and b) $2 \cdot 10^5$ m/s MCI velocities.

If the AI-cascading is not fast enough, one has to find the missing link between the resonant population of high lying Rydberg manifolds and the observation of an ISI as detected via K_α satellites which implies a sufficient population of the L shell in front of the surface. How do some of the q electrons initially in the high Rydberg manifolds descend to the L shell ? In order to develop an insight on this process Andrä *et al.* (1991) have included the mutual screening of the q electrons in the Rydberg manifolds.

As soon as the first electron is resonantly captured into a manifold n of a MCI with charge q at a binding energy $E_{n1} > \phi_w$, the charge q is partially screened for the capture of the second electron into the same n. The n manifold is thus slightly shifted upwards in binding energy to $E_{n2} = E_{n1} - \Delta e_n$ for the capture of the second electron as sketched in Fig. 7. For the capture of the third electron into n the charge q is now screened by two n electrons and the n manifold is thus again shifted upwards in a linear approximation to $E_{n3} = E_{n1} - 2 \cdot \Delta e_n$. This shifting of the n manifold continues with the further capture of electrons until the kth electron brings it to E_{nk} just below the Fermi edge such that the energy of the manifold for the capture of the $(k+1)^{th}$ electron would be just above the Fermi edge. Since an electron in a level above the Fermi edge is rapidly reionized into empty states of the continuum of the solid, an equilibrium population of k electrons in the n manifold just <u>below</u> the Fermi edge is maintained.

Fig.7 Screening dynamics of Rydberg manifolds, see text.

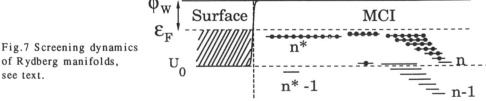

Simultaneously with the energetic shift of the n manifold the n-1 manifold is also affected by the screening due to the electrons in the n manifold. MCDF calculations allow to calculate the average shift Δe_n in the n manifold and Δe_{n-1} in the n-1 manifold per electron for k electrons in the n manifold, so that $E_{n-1}(k) = E_{n-1}(0)-k\cdot\Delta e_{n-1}$. The n-1 manifold may so be shifted into resonance with the conduction band by the presence of electrons in the n manifold such that resonance capture into the n-1 manifold can take over the neutralization of the projectile at a somewhat shorter distance from the surface. This distance is approximately given by $z_{n-1} \approx 1.6 <r_{n-1,s}(E_{n-1} \approx U_0)> \approx 2.7\cdot(n-1)\ a_0$ for a resonance energy of about 0.4 a.u. near the bottom of the conduction band, where $r_{n-1,s}$ is the radius of the (n-1)s subshell. While filling up the n-1 manifold, the n manifold shifts successively above the Fermi edge and so looses its electrons by re-ionization into the continuum of the solid. The n-1 manifold thus completely takes over the initial role of the n manifold at shorter distance from the surface but in resonance with the conduction band.

We may call this the first cycle of Screening Dynamics (SD) which transfers the initial resonant population of the n manifold into the n-1 manifold. These SD cycles continue towards successively lower Rydberg manifolds until the n* manifold is reached. n* is defined by the fact that the screening of the n*-1 manifold by q electrons in the n* manifold is not anymore sufficient to shift the n*-1 manifold into resonance above the bottom of the conduction band. A MCDF calculation yields for Ne^{9+} n* = 6 and for Ar^{17+} n* = 9. These results of the SD corroborate very well the semiclassical treatment of the capture of electrons by a MCI by Bardsley *et al.* (1991).

Under the action of the image charges, which is negligible compared to the screening at distances $z \geq 2.7\cdot n*$, manifolds with binding energy $\leq 2\cdot U_0$ may be pushed into resonance at distances $z < 2.7\cdot n*$. The combined action of SD and image charges may thus create an atom with q electrons in the n'<n* manifold at a distance of about $z \approx 2.7\cdot n'$, i.e. for Ne^{9+} n'=5 and for Ar^{17+} n'=7. For these manifolds, the AI-cascading rates have been shown by Vaeck *et al.*(1991) to be sufficiently high to allow a rapid filling of the M,L,K shells during the further approach of the projectile from the surface.

AI-cascading takes place during this whole process of SD from z_c towards $z \approx 2.7\cdot(n*-1)$ but it is evidently cut short by the SD and can so produce only a reduced number of low energy electrons in good agreement with the observations. The model of SD thus provides a satisfactory explanation for the yield of low energy electron emission as well as for the transfer of population from high towards low lying Rydberg manifolds.

The populations in these low lying Rydberg manifolds are essential for the understanding of the emission of relatively high energy Auger electrons which was discovered by Zehner *et al.*(1986) and further studied by de Zwart (1987), and Delaunay *et al.*(1987a). These Auger transitions are now known to be observable whenever innershell holes exist in the incoming MCI which allow for Auger electron energies above $\approx$ 30 eV. This is the case in Ar^{9+} for example with its $1s^2 2s^2 2p^5$ configuration where LMX Auger transitions with X = M,N,O... fill the empty 2p hole and emit Auger electrons in the region around 200 eV. These transitions were extensively studied by de Zwart (1987) using a high resolution spherical electron spectrometer in an UHV apparatus which allowed to measure the spectrally resolved yield per ion with incoming Ar^{9+} ions decelerated to systematically varied energies as low as 45 eV (!) as shown in Fig.7 of de Zwart (1989). Taking into account the small escape depth of 200 eV electrons from the solid, one can interpret these spectra in good agreement with the conclusions on the X-ray data above. From 5 to 250 keV the few electrons observed stem from the trajectory in the first atomic layers inside the bulk of deeply penetrating ions. From 2 keV to 500 eV not only the time spent in these first atomic layers increases so that more electrons per ion are observed from the bulk but also an increasing fraction of the electrons stem from an ISI in front of the surface. From 200 eV to 45 eV, the emission is supposedly dominated by an ISI so that de

Zwart (1987) could analyze the spectral structure in terms of LMX Auger transitions with X = M,N,O..., on the basis of calculations by Folkmann *et al.* (1983) on doubly excited 3snl configurations of Ar^{7+} neglecting the presence of the other 7 electrons in the n = 3-12 shells.

An alternative to this interpretation was proposed by Folkerts *et al.* (1989) by discussing the emission from neutral Ar with one 2p electron excited to 4s and taking into account an energy shift due to the image charge. LMM transitions from such an atom are indeed expected in the region of 200 eV so that a contradiction seemed to be established for the existence of an ISI since an ISI cannot produce such an atom. We have therefore repeated these measurements with the apparatus described by Andrä *et al.*(1991). The result for two energies of Ar^{9+} interacting at 45° with a doped monocrystalline Si surface is shown in Fig.8. The general spectral structure of de Zwart is reproduced but one notes that
(i) for Ar^{9+} at 347 eV the energetic position of the peak of the spectral feature is found at 203 eV in comparison to 212 eV for de Zwart
(ii) this peak shifts to the still lower energy of 196 eV when Ar^{9+} is decelerated to 153 eV
(iii) the shape of the spectrum also changes such that its average energy shifts by even more than the 7 eV of the peak.
For verification we have measured these same features with Cl^{8+}.

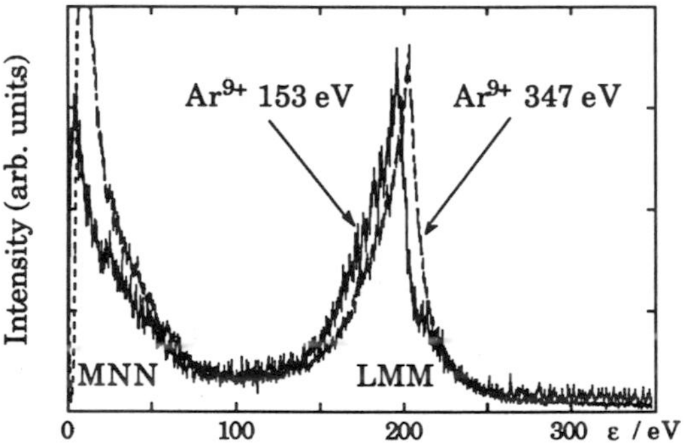

Fig.8 L Auger spectra of Ar^{9+} interacting with a doped, mono-crystalline Si surface at two very low energies.

MCDF calculations show that the average energy of the dominant LMM Auger transitions shifts in the same manner when assuming for 347 eV Ar^{9+} a $2p^5 3s^2 3p^6 3d$ upper configuration with 9 electrons in the M shell and for 153 eV Ar^{9+} a $2p^5 3l3l'4s^2 4p^5$ upper configuration with 2 electrons in the M shell and 7 electrons in the N shell. From this analysis one may conclude that Ar^{9+} at >300 eV has not enough time for sufficient cascading into the M shell via ISI and thus comes close to the surface with only 0 or 1 electron in the M shell. The M shell is then extremely rapidly filled from inner shells of the target atoms either at the closest approach during a total reflection of the projectile or in the first atomic layers when the projectile penetrates. A spectrum is thus emitted which corresponds indeed to close to 9 electrons in the M shell. Ar^{9+} at <160 eV has more time for cascading into the M shell via ISI and thus emits a partial spectrum with about 2 electrons in the M shell and 7 other electrons in higher shells superimposed on a background stemming from 3 to 9 electrons in the M shell. More detailed MCDF calculations including the transition probabilities are, however, necessary for a close reproduction of the spectra under these assumptions.

In order to arrive at more structure in the spectrum we have repeated the same measurements with Fe^{17+} on polycrystalline Ta with the same initial configuration $1s^2 2s^2 2p^5$. For this case, the LMM electrons are expected around 600 eV so that any spectral features are also expected to be separated by a factor of 3 more than for Ar^{9+}. In Fig.9 one can indeed see for Fe^{17+} at 2260 eV a separation into two peaks which can be assigned to LMM transitions with close to 17 electrons in the M shell. With Fe^{17+} at only 85 eV (!) one observes again a significant shift of the average energy towards lower energies and a pronounced change of the spectral shape. Without going into the details of our MCDF calculations both effects seem to be accounted for by an emission from neutral Fe with 2 electrons in the M shell and 15 electrons in the N or higher shells. These measurements thus corroborate very well the interpretations of the Ar^{9+} results. They indicate for fluorine-like MCI a transition from the IBI to the ISI at only very low MCI energies.

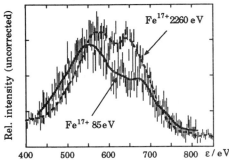

Fig.9 L Auger spectra of Fe^{17+} interacting with a polycrystalline Ta surface at two very low energies.

When using hydrogenic MCIs as projectiles, KLL Auger electrons should show up in addition to LMM electrons. The first inner shell Auger electrons observed by Zehner *et al.*(1986) were essentially stemming from such KLL transitions which have been extensively studied under grazing incidence conditions by Meyer *et al.*(1987), Zeijlmans van Emmichoven *et al.* (1991), Meyer *et al.*(1991) and Köhrbrück *et al.* (1991). It was only recently that the LMM transitions of hydrogenic MCIs were clearly identified and analyzed by Folkerts *et al.*(1990) and Andrä *et al.*(1991) when very low energy MCIs were used. The spectra obtained by Folkerts *et al.*(1990) for C^{5+}, N^{6+}, O^{7+}, and Ne^{9+} impinging at 45° and energies of 150, 90, 150 and 600 eV respectively on clean, polycrystalline tungsten in an UHV of $2 \cdot 10^{-8}$ Pa are shown in their Fig.1. They clearly show a spectral feature at low energies which can be assigned to L Auger emission and another spectral feature at higher energies which can be assigned to K Auger emission. One notes that spectral structures are better resolved with increasing charge and with decreasing energy of the MCI, as shown again by Folkerts *et al.* (1991).

In order to demonstrate this latter aspect even further, we have investigated spectra of the same species at still lower energies. Although they were recorded with a monocrystalline, doped Si target at 45° under less favorable vacuum conditions of 10^{-5} Pa only, they yield the same spectra as Folkerts *et al.* (1990) at their energies. We only present here the most striking modifications of the spectral structures of the L and K Auger emission in Fig.10 when reducing the energy of O^{7+} to 54 eV and equivalently in Fig.11 when reducing the energy of Ne^{9+} to 145 eV. These measurements clearly demonstrate the importance of the reduction of the velocity by another factor of 2 (*e.g.* Ne^{9+} from 600 to 145 eV) and strongly suggests to increase the experimental effort to gain another factor of 2 in the future.

They evidently show for the K Auger emission that the "left" peak becomes more and more dominant with smaller energies and that a well separated peak develops on the high energy side between 800 and 900 eV with Ne^{9+} which can only be seen as a weak shoulder with O^{7+}. Our intermediate spectra with F^{8+}, not shown here, clearly demonstrate this transition from O^{7+} to Ne^{9+}. The L Auger emission of O^{7+} at 54 eV and of Ne^{9+} at 145 eV split into two contributions when compared to unresolved peaks at the energies of Folkerts *et al.* (1990)

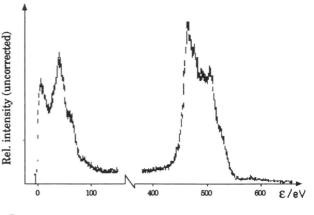

Fig.10 Auger electron spectrum emitted from O^{7+} interacting at 54 eV and $\phi_{in}= 45°$ with a doped, monocrystalline Si target.

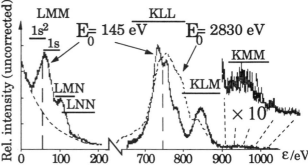

Fig.11 Auger electron spectrum emitted from Ne^{9+} interacting at 2830 eV (broken line) and 145 eV (full line) and $\phi_{in}= 45°$ with a doped, monocrystalline Si target.

For the further interpretation we shall discuss the data with Ne^{9+} at 145 eV in Figs.11 since they show the most spectral features. To the K Auger spectrum obtained with Ne^{9+} at 145 eV we have superimposed as a broken line the spectrum obtained with Ne^{9+} at 2830 eV. We also indicate the region of known KLL Auger emissions from Li-like Ne ions to singly K-ionized Ne where the latter corresponds to the upper configuration of the K_α X-ray emission.

At 2830 eV Ne strongly penetrates into the solid as seen in the low energy tail of the KLL spectrum, a situation for which the σ_j used in Fig. 5c yields an average population of the L shell of 4 electrons only. Taking into account the finite escape depth of about 30 a_0 for 800 eV electrons from the bulk of Si, one expects to observe only the incomplete filling of the L shell and consequently a K Auger emission from upper configurations with 3-4 electrons in the L shell. Globally, this corresponds well to the broken line spectrum which evidently does not peak at the upper KLL limit corresponding to a full L shell.

When changing from 2830 eV to 145 eV, the higher energy part of the KLL region reduces considerably in favor of a peak at 732 eV while still exhibiting two peaks which continue to decrease as the energy of the MCI is further decreased. Above the KLL region a KLM peak shows up already at 2830 eV which becomes more and more pronounced and shifts to higher energies when the MCI energy is decreased. A well separated KLM peak has also been reported by Schulz *et al.* (1990) who used 51 keV Ar^{17+} on Ge at 60° and 3°. Between 900 and 1000 eV, a further, very weak peak appears with a maximum at about 360 eV MCI energy and then reduces towards 190 eV MCI energy. The ensemble of these observations are significant indications for a transition from IBI to ISI when decelerating the MCI.

The analysis of the spectral features in Fig.11 is based on MCDF calculations which show that below the broken vertical line no other K Auger emission from neutral Ne is possible than from configurations with 2 electrons in the L shell and various distributions of 7 electrons in the M and N shells. These configurations are typical results of cascading in agreement with an ISI. The two peaks in the KLL region above the broken line stem from K Auger transitions with upper configurations in neutral Ne with 3 and more electrons in the L shell. Their K Auger electron energies seem to accidentally concentrate into two groups, the higher one with more electrons in the L shell. The appearance of more than 2 electrons in the L shell via cascading is very unlikely so we assume a direct capture of inner shell target electrons into the L and M shells during the closest approach (1-2 a_0) of the projectile when it is reflected from the first or second atomic layer of the surface. This ISI during Reflection (ISIR) may explain the behavior of the weak peak above 900 eV as a function of the MCI energy. The MCDF results show that it may be due to KMM transitions from the same upper configurations which become more and more abundant during the transition from IBI to ISIR and then less abundant when the ISIR decreases due to increasing distance of closest approach at the very low MCI energy. The KLM peak carries besides its very existence only little information on the interaction processes involved since it stems, according to the MCDF results, from a superposition of KLM Auger transition energies from all upper configurations of neutral Ne with 1 to 7 electrons in the L shell.

The double peak structure of the L Auger transitions in Figs.10 and 11 may at first sight be interpreted as a LMM and LMN peak. After extensive MCDF calculations it emerges, however, that in the presence of one K hole all L Auger transition energies fall above the vertical broken line. The "left" half of the "left" peak has therefore to be assigned to L Auger transitions with a completely filled K shell. These transitions are of course expected to occur since the observation of K Auger transitions automatically implies that the initial K hole will be filled at some time of the interaction leaving behind configurations with one or more holes in the L shell. The LMM transitions with full K shell have actually been observed separately by Folkerts *et al.* (1991) when using O^{6+} with a full K shell as incident MCI at low energy. The double peak LMM and LMN structure of these L Auger electrons could even be resolved by Köhrbrück *et al.* (1991) when using 320 keV Ar^{16+} with full K shell as incident MCI at 10° on Cu which is, however, most likely due to a sub-surface interaction.

With this reasoning, the double peak structure in Figs.10 and 11 has to be interpreted as a superposition of a first spectrum emitted by configurations with one K hole and a second spectrum emitted by configurations with a full K shell. Both of them are composed of two dominant groups of LMM and LMN transitions and a third, very weak group of LNN transitions as indicated in Fig.11 for the configurations with one K hole. Also indicated is the group of expected LMM transition energies for the configurations with full K shell which thus readily explains the low energy part of the spectrum below the broken line. The complexity of the composition of the spectrum in Fig.11 does unfortunately not allow to draw further conclusions on the interaction mechanisms responsible for their appearance.

4. SCENARIO FOR THE INTERACTION OF NE^{9+} AT VERY LOW ENERGY

When summarizing the informations on the interaction of MCIs with surfaces or solids presented as yet, one may sketch a scenario on what a very slow Ne^{9+} ion at 145 eV experiences during its interaction with a conducting surface in Fig.12. At rather long distances $38 < z_c < 54$ a_0 in front of the surface, a hyperexcited neutral Ne with 9 electrons in high lying Rydberg manifolds is created by resonant capture of electrons from the conduction band of the surface. During the further approach of this Ne atom towards

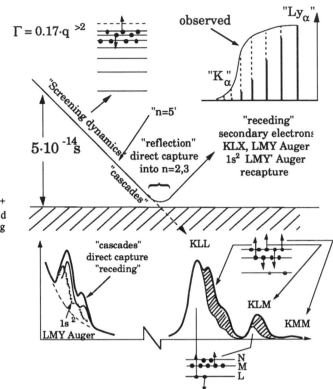

Fig.12 Scenario of Ne^{9+} interacting at 145 eV and 45° with a conducting surface.

the surface, the Screening Dynamics (SD), as described above, transfers these electrons to the n = 5 Rydberg manifold at a distance $z_{SD} \approx 10.8 \ a_0$. This SD is accompanied by AI-cascading which produces an important fraction of the secondary electron yield Γ at very low energies.

At distances z < z_{SD}, Auger cascading becomes the dominant channel of population of still lower shells which are accessible to observation via Auger electron or X-ray spectroscopy. It is during this phase that the typical ISI spectra are emitted with small population in the L shell and the rest of the q electrons in M and N shells. Representative for this phase are the blank part of the K Auger spectrum on the lower right of the figure and the full line part of the X-ray spectrum on the upper right of the figure.

At still smaller distances the combined action of an effective image charge and of the repulsive planar potential, modify the trajectory of the projectile such that at 145 eV a reflection on the first or second atomic layer of the surface takes place with a probability close to one. During the closest approach on the order of 0.5-2 a_0, direct capture of electrons due to inter-atomic Auger transitions or via quasi-resonant capture from inner shells of the surface atoms into the L and M shells of Ne takes place abundantly due to the relatively long time spent close to these atoms (head on collision !). This ISIR leads during the reflection and while the projectile is receding from the surface to the emission of spectra with a medium population of the L shell and the rest of the q electrons in the M or N shells. Representative for this phase are the hatched part of the K Auger spectrum on the lower right of Fig.12 and the dashed line part of the X ray spectrum on the upper right of Fig.12.

All along the sections of the trajectory called "cascades", "reflection", and "receding" in Fig.12, emission of L Auger electrons takes place. In the section "cascades" the emission from configurations with one K hole dominates, but its contribution is exponentially reduced in favor of the emission from configurations with full K shell which probably dominates in the "receding" section . The very observation of L Auger emission with full K shell implies that resonant electron capture or near resonant Auger capture from the conduction band to continuously lowering n manifolds takes place in the "receding" section. The distance from the surface up to which this neutralization will take place is, however, small compared to z_c. The capture of a final $n = 2$ electron, the only channel for the completion of the neutralization, can take place up to $z \leq 1.6 \cdot <r_{2s}> \approx 10\ a_0$ only. It is thus very likely that the neutralization will not be completed on a receding trajectory.

5. CONCLUSIONS

The interpretations of the different groups seem to converge towards a gross comprehension of the X-ray and Auger electron spectra emitted during the neutralization of a MCI inside the solid at the "high" energies defined above. However, this comprehension remains open for detailed study of all the physics involved in cross sections or filling rates of inner shells of the MCI during the binary or multi-center collisions inside the sea of conduction electrons and inside the average density of more tightly bound target electrons. The near future has to show whether still better resolved X-ray spectra can corroborate theoretical efforts in this domain.

At very low MCI energies, a scenario of the physics taking place in front of the surface begins to emerge. It needs, however, multiple experimental confirmations with further decelerated beams of MCIs impinging on clean surfaces in UHV in order to disentangle many still puzzling details. These experimental efforts require theoretical corroboration not only concerning X-ray and Auger transition energies and transition probabilities but particularly on the very concept of the physics of the interaction of a MCI with a conducting surface.

Such combined experimental and theoretical efforts could force us to modify the presented scenario in a number of details. This should not to be considered as a weakness neither of the present situation nor of this review. It should instead be a measure of the great excitement felt by all groups on the possibility of living yet to come surprises in a still widely open field.

This work has been performed at LAGRIPPA, a joint laboratory of the CNRS and of the CEA, the support of which is gratefully acknowledged. H J A is very grateful for the support by the SFB 216 der Deutschen Forschungsgemeinschaft and indebted to his generous colleagues of the Institut für Kernphysik der Westfälischen Wilhelms Universität, Münster and A S is indebted to Dr. J P Desclaux for numerous helpful discussions concerning the MCDF calculations.

REFERENCES

Abstracts of Contributed Papers of the XVII ICPEAC 1991, eds. McCarthy I E, MacGillivray W R, and Standage M C (1991) are referred to as: XVII ICPEAC Abstr.
Invited Papers presented at the XVII ICPEAC 1991, eds. McCarthy I E, MacGillivray W R, and Standage M C (1991) are referred to as: XVII ICPEAC Inv.

Andrä H J 1989 "Electronic Interaction of Multi-Charged Ions with Metal Surfaces at Low Velocities", in Proceedings of the NATO Advanced Study Institute on "Atomic Physics of Highly Charged Ions", ed. R. Marrus, Plenum Press p. 377, see references therein; see also Nucl Instr Meth **B43**, 306 (1989)

Andrä H J, Simionovici A, Lamy T, Brenac A, Lamboley G, Bonnet J J, Fleury A, Bonnefoy M, Chassevent M, Andriamonje S, and Pesnelle A 1991 Proceedings of the Vth Int Conf Phys Highly Charged Ions, Sept 1990, Giessen-FRG, ed Salzborn E, to be published in Z.Phys.**D**.

Andrä H J, Pesnelle A, Simionovici A, Lamy T, Brenac A, Lamboley G 1991 XVII ICPEAC Abstr p 620

Arifov U A, Kishinevskii L M, Mukkamadiev E S, and Parilis E S 1973 Zh.Tekh.Fiz.**43**, 181 ; Sov.Phys.Tech.Phys.18, 118 (1973)

Bardsley J N and Penetrante B M 1991 in XVII ICPEAC Abstr p 615 and private communication.

Bonnet J J, Fleury A, Bonnefoy M, Chassevent M, Lamy T, Brenac A, Simionovici A, Andrä H J, and Andriamonje S 1991 Proceedings of the Vth Int. Conf. Phys. Highly Charged Ions, Sept. 1990, Giessen-FRG, ed. E. Salzborn, to be published in Z.Phys.D

Brenac A, Simionovici A, Andrä H J, Lamy T, Lamboley G, Bonnet J J, Fleury A, Bonnefoy M, Chassevent M, and Andriamonje S 1991 XVII ICPEAC Abstr. p 621. Please note the error in the caption of Fig.1c which should read Si instead of Ta.

Briand J P, de Billy L, Charles P, Essabaa S, Briand P, Geller R, Desclaux J P, Bliman S, and Ristori C 1990 Phys.Rev.Lett. **65**, 159

Burgdörfer J, Lerner P, and Meyer F 1991 XVII ICPEAC Abstr. p 611

Delaunay M, Fehringer M, Geller R, Varga P, and Winter H 1987a Europhys.Lett. **4**, 377

Delaunay M, Fehringer M, Geller R, Hitz D, Varga P, and Winter H 1987b Phys.Rev. **B35**, 4232

de Zwart S T 1987 Thesis, Groningen; ibid. Nucl.Instr Meth. **B23**, 239 (1987)

de Zwart S T, Drentje A G, Boers A L, and Morgenstern R 1989 Surface.Sci. **217**, 298

Donets E D 1983 Phys.Scr. **T3**, 11; ibid. Nucl.Instr.Meth. **B9**, 522 (1985)

Emmichoven P A, Havener C C, and Meyer F W 1991 Phys.Rev. **A43**, 1405

Fehringer M 1987 Thesis T U Wien

Fehringer M, Delaunay M, Geller R, Varga P, and Winter H 1987 Nucl.Instr.Meth. **B23**, 245

Fleury A, Debernardi J, Bonnefoy M, Bliman S, Bonnet J J, and Chassevent M 1986 Nucl.Instr.Meth. **B14**, 353

Folkerts L and Morgenstern R 1989 J.Phys.(Paris) **50**, C1-541

Folkerts L and Morgenstern R 1990 Europhys.Lett. **13**, 377

Folkerts L, Das J, and Morgenstern R 1991 XVII ICPEAC Abstr. p 619

Folkman F and Cramon K M 1983 Phys.Scr. **T3**, 166

Hagstrum H D 1954 Phys.Rev. **96**, 325 ; ibid. **96**, 336 (1954); in "Electron and Ion Spectroscopy of Solids", edited by L.Fiermans, J.Vennik, and W.Dekeyser (Plenum, New York, 1978)

Köhrbrück R, Lecler D, Fremont F, Roncin P, Sommer K, Bleck Neuhaus J, and Stolterfoth N 1991 XVII ICPEAC Abstr. p 616

Lakits G, Aumayr F, and Winter H 1989 Rev.Sci.Instr. **60**, 3151

Lamy T, Lamboley G, Hitz D, and Andrä H J 1954 Rev.Sci.Instr. **61**, 336

Ley L et al. 1979 in "Photoemission in Solids II", eds. Ley L and Cardona M, Springer Verlag

Matthews D L, Johnson B M, and Moore C F 1975 At.Data Nucl.Data Tables **15**, 41

Meyer F W, Havener C C, Snowdon K J, Overbury S H, and Zehner D M 1987 Phys.Rev. **A35**, 3176

Meyer F W, Overbury S H, Havener C C, Zeijlmans van Emmichoven P A, and Zehner D M 1991 XVII ICPEAC Abstr. p 617, and to be published in Phys.Rev.Lett. (1991)

Schulz M, Cocke C L, Stöckli M, Hagmann S, and Schmidt-Böcking H 1990 Proceedings of the Vth Int. Conf. Phys. Highly Charged Ions, Sept., Giessen-FRG, ed. E. Salzborn, to be published in Z.Phys.D (1991)

Schulz M, Cocke C L, Hagmann S, Stöckli M, and Schmidt-Böcking H 1991 to be published in Phys.Rev.A.

Snowdon K J 1988 Nucl.Instr.Meth. **B34**, 309

Vaeck N and Hansen J E 1991 XVII ICPEAC Abstr. p 613; ibid to be published in J.Phys.B and private communication.

Winter H 1991 XVII ICPEAC Inv.

Zehner D M, Overbury S H, Havener C C, Meyer F W, and Heiland W 1986 Surf.Sci. **178**, 359

Zschornak G, Musiol G, and Wagner W 1983 Phys.Scr. **T3**, 194

Collision processes in charge exchange recombination spectroscopy of plasmas

A. N. Zinoviev

A.F.Ioffe Physical–Technical Institute of the Academy of Sciences of the USSR, 194021 Leningrad, Polytechnicheskaya 26, USSR

Abstract.   Charge exchange recombination spectroscopy is a method of plasma diagnostics which is based on a study of the radiation produced by atomic beams injected into plasma. This method allows us to measure the local concentration of impurity and hydrogen ions, their temperatures, velocities of plasma rotation, values of magnetic fields etc. The report deals with the discussion of collision processes which are important for plasma spectroscopy using atomic beams. Special attention is given to the charge exchange reaction in collisions between nuclei of elements and hydrogen atoms, to processes of excitation by ion impact and to collisional state mixing in plasmas. In conclusion atomic data needed for future development of these diagnostics are discussed.

1. Introduction

Charge Exchange Recombination Spectroscopy (CHERS) is widely used as a special diagnostic technique to investigate plasma ion parameters. This method is based on the observation of radiation produced during injection of fast hydrogen atom beams into plasmas and was proposed by Afrosimov *et al* (1977) to measure impurity ion concentrations. In the collisions between beam atoms and plasma ions highly excited states of multicharged ions are populated with large cross sections. The following decay of the excited states produces characteristic radiation:

$$I(q) + H \rightarrow I(q-1,*) + H(+) \rightarrow I(q-1) + H(+) + \hbar\omega.$$

If the beam intensity, geometry of the experiment, detection efficiency of spectroscopic channel and emission cross section are known the impurity concentration could be measured.

This method has several advantages as compared with traditional plasma spectroscopy. First local measurements of plasma parameters by using CHERS are possible because the signal is detected from the small volume formed by the intersection of the beam and spectrometer viewing line. By observing different parts of the beam trajectory the spatial profile of plasma parameters could be measured.

Second, because the excitation reaction is known, the method produces independent measurements without using other plasma parameters. Beam intensity and the injection time can be chosen in such a way that investigated plasma parameters will not be disturbed.

Finally, this method allows us to measure the concentration of nuclei of elements. During a charge exchange reaction nuclei are transformed into hydrogen-like ions whose characteristic radiation is observed. Under present tokamak conditions nuclei of light elements (carbon, oxygen, sometimes beryllium or boron) are the dominant impurity ions in the plasma. Impurities determine the major energy loss channel from the plasma. If their content exceeds ten percent the radiation losses become so large that plasma parameters needed for a self-supporting thermonuclear reaction never will be achieved. The presence of impurities influences also the plasma stability and the efficiency of the additional heating methods. Therefore control of the impurity content and especially the nuclei concentration measurements are extremely important.

The increase of impurity radiation of the O(7+) lines was first observed by Isler(1977)in neutral beam injection experiments on the ORMAK tokamak but this observation was not used for impurity diagnostics. The first diagnostic experiment where the concentration of carbon nuclei was really measured has been performed in our work on the tokamak T-4 (Afrosimov 1978, 1979). These experiments were extended to measure the radial profile of oxygen impurities in our measurements on the T-10 (Zinoviev *et al* 1980) and later by Fonck *et al* (1982) on the PDX.

An analysis of the form of the observed line allows us to obtain the ion temperature and the velocity of plasma rotation. The process of electron capture is occurring at rather large impact parameters. As a result an incoming particle practically does not disturb the initial energy distribution of plasma ions.The resulting excited particles have practically the same energy distribution. Because of the motion of plasma ions the observed line could be Doppler-broadened or shifted. The line width and the line shift give information about plasma ion temperature and plasma rotation. This method of ion temperature measurements was first considered by Hess (1976) and the first measurements were made on the tokamak T-10 by Berezovsky *et al* (1982, 1985). For line shape analysis both impurity and hydrogen lines are used. In the p+H case the main channel is resonant charge exchange where hydrogen atoms in the ground state are produced. These atoms could escape from the plasma and be detected. This method allows us to measure the tail of the energy distribution function. In the charge exchange reaction excited hydrogen atoms are also produced. The analysis of the line shape of the radiation which is emitted by these atoms allows us to study the energy distribution function of hydrogen in the region $0-2kT_i$. Using both CHERS and analysis of hydrogen atoms escaping from the plasma the first measurements of the local energy distribution function were made in our experiments (Grigoryev *et al* 1984) on the tokamak 'TUMAN'-3. This function was shown to be very close to a Maxwellian one.

Additional possibilities are provided by observation of radiation connected with the beam atom excitation. The registration of these signals allows us to measure plasma density (Kadota *et al* 1978, Hellermann *et al* 1988), effective plasma charge and magnetic field in the plasma (Boileau *et al* 1989a). Now this method of plasma diagnostics is widely used on all large tokamaks (Gordeev, Zinoviev 1980, Cotrell 1983, Groebner 1983, Post *et al* 1983, Duval *et al* 1985, Hawkes, Peacock 1985, Fonck 1985, Isler 1985, Seraydarian *et al* 1986, Carolan *et al* 1987, Fonck et al 1987, Boileau *et al* 1989b, Hellerman *et al* 1990, Zinoviev, Afrosimov 1990 etc)

2. Charge exchange capture in collisions between light impurities and hydrogen atoms

For measurements of impurity concentration well-known values of the cross sections for electron capture into states of multicharged ions with different n and l quantum numbers are needed. The required accuracy of the cross sections must be 20–30% or better. At present the total capture cross section for such elements as helium, carbon and oxygen are well-known. There are different theoretical calculations using strong coupling methods (Green *et al* 1982, Shipsey *et al* 1983, Fritsch and Lin 1984) and many measurements (Panov *et al* 1983, Bendahmen 1985, Meyer *et al* 1979, 1985, Phaneuf *et al* 1982, Afrosimov 1983, Dijkkamp *et al* 1985, Shah and Gilbody 1978, Ciric *et al* 1975 etc) which agree rather well.

State populations due to an electron capture are usually studied spectroscopically (Afrosimov *et al* 1983, Dijkkamp *et al* 1985, Bliman *et al* 1983, Hvelplund *et al* 1983, Gordeev *et al* 1983, Zinoviev 1986 etc) or using energy gain spectroscopy(Afrosimov *et al* 1980, Iwai *et al* 1982), where energy spectra of particles which captured an electron are studied (separate peaks in such spectra correspond to the capture into different n-shells). A special program of cross-section measurements needed for CHERS diagnostics has been carried out in cooperation between KVI (Gronigen) AMOLF(Amsterdam) and Ioffe Institute in Leningrad (Hoekstra 1987 a,b,c, 1988, Ciric *et al* 1988). Some results for collisions He(2+), C(6+), O(8+)–H are presented in figures 1–3.

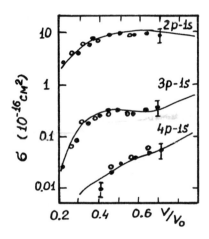

Figure 1. Emission cross sections for He(2+)–H collisions: the open and full circles are from the experiment by Ciric *et al* (1985 and 1988), and the curves are from the calculations by Fritsch (1988).

States with $n = Z^{\frac{3}{4}}$ are dominantly populated in accordance with the theoretical predictions. For these states agreement between experimental data and the results of strong coupling calculations (Green *et al* 1982, Shipsey 1983, Fritsch, Lin 1984) is quite good. With increasing n the agreement between different theoretical calculations becomes worse. This fact could be connected with an insufficient basis of states being taken into account. Experimental data agree better with the results of calculations using the molecular basis (Green *et al* 1982, Shipsey *et al* 1983).

In the case of optical lines, where rather high n are considered, the agreement of our measurements with calculations by Fritsch (1984,1988) and Ryufuku(1982) is very

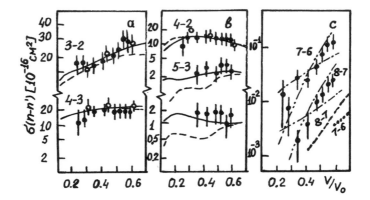

Figure 2. Emission cross sections for C(6+)–H collisions. Open circles—experiments by Dijkkamp *et al* (1985); full circles—experiments by Hoekstra *et al* (1988). Theoretical calculations:full curve—Green *et al* (1982); broken curve—Fritsch and Lin (1984);dash-dot curve—Fritsch (1988); dotted curve—Ryufuku (1982);dash-dot-dot curve—Ovchinnikov (1989).

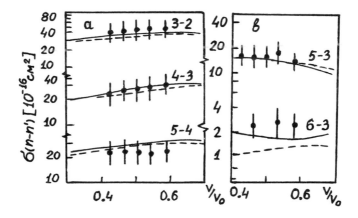

Figure 3. Emission cross sections for the Ö(8+)–H case. Full circles—experimental data by Hoekstra *et al* (1987 c); full curves—calculations by Shipsey *et al* (1983); broken curves—calculations by Fritch and Lin (1984).

bad. Recently a new approach based on consideration of hidden crossings in a complex plane of internuclear distance was developed by Ovchinnikov and Solov'ev (1988). In this technique the accuracy of results obtained for high n is nearly the same as for dominantly-populated states. Results obtained by Ovchinnikov (1989) agree quite well with our measurements.

These experiments provide a good test for the theoretical calculations in the most complicated quasimolecular region of collision velocities and prove that results of strong coupling calculations can be considered as quite a reliable basis for CHERS diagnostics. Of course it will be very desirable to extend experiments into the collision energy region $E = 10 - 200$ keV/amu and perform also measurements for B and Be nuclei.

Lines excited by charge exchange recombination have wave lengths in a wide spectroscopic region: from soft X-ray to the optical region. Signals produced by beam injection for different C(5+) lines are proportional to the emission cross sections which are listed in table 1. The same lines are excited by collisions of plasma electrons and hydrogen-like ions and this background is characterized by corresponding rate coefficient $\langle\sigma_{ex}v_e\rangle$. The ratio $\sigma_c/\langle\sigma_{ex}v_e\rangle$ determines the contrast of the signal in the CHERS experiment and is extremely important. It is possible to see that the use of ultraviolet lines gives by a factor 30 a higher contrast ratio as compared with the L_α lines detected in the first experiments. The optical lines have approximately the same contrast ratio. The lack of intensity of light connected with decreasing cross section could be easily compensated by the high efficiency of optical spectrometers and the possibility of using focusing optics. In this region it is also easier to organize the spectral analysis of line shapes. In future fusion devices intensive X-ray and neutron irradiation requires us to remove diagnostic spectrometers far away from the tokamak. Again the use of the optical region gives advantages because the existence of optical fibers and focusing optics helps to solve this problem very easily.

Table 1. Emission cross sections $\sigma_c(n - n')$ for collisions C(6+)–H, 50 keV/amu (Green *et al* (1982), Shipsey *et al* (1983)), rate coefficients for excitation of the same lines of C(5+) ions by electron impact and their ratios which determine the signal contrast in CHERS.

Transition	2–1	3–2	4–3	5–4	6–5	7–6	8–7
$\sigma_c(n - n')$ (10^{-15}cm^2)	1.95	1.5	1.4	0.56	0.078	0.032	0.015
$\langle\sigma_{ex}v_e\rangle$ $10^{-10}\text{ cm}^3\text{ s}^{-1}$	6.3	0.3	0.11	0.053	0.030	0.019	0.013
$\sigma_c(n - n')/\langle\sigma_{ex}v_e\rangle$ $(10^{-5}\text{ s cm}^{-1})$	0.31	5.0	13	10.6	2.6	1.7	1.2

Helium nuclei—alpha particles—are another subject of CHERS diagnostics. They would be produced via thermonuclear reactions or sometimes are added to the plasma to increase the efficiency of additional heating methods. In thermonuclear reactions fast alpha particles with energies 3.5 MeV are produced. More then 90% of these particles will be captured by the plasma, and because of Coulomb collisions with plasma electrons and ions they will be thermalized with characteristic time 0.1–1 s. Therefore slowing down and thermalized alpha particles could be present in the plasma.

The most suitable line for alpha-particle diagnostics corresponds to the optical transition 4–3. 4l-states could be populated via charge exchange in He(2+)–H collisions. These states have rather long lifetimes and they are mixed by collisions in the plasma.

In the gas phase experiment 4s and 4p states are expected to be prominently populated. Therefore the cross section for emission of the 4–3 line measured in gas phase experiment cannot be directly used for CHERS. To obtain the required cross section for collisions in plasmas we need to measure electron capture cross sections for all 4l-states and then possible mixing effects must be taken into account.

In the gas-phase He(2+)–H collisions excited ions of He(+) are produced from the beam particles. They have large velocities and could leave the registration region without emitting light. We used the fact that the lifetimes for 4s,4p,4d and 4f states τ_i differ very much. Correspondingly characteristic distances for 4l-state decay $l_i = \tau_i v$ will be also quite different. Using a scanning spectrometer the spatial profile of the emission could be measured. By fitting this profile capture cross sections to all 4l-states can be obtained (figure 4).

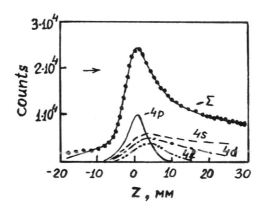

Figure 4. Spatial profile of emission of the 4–3 line in collisions He(2+)–H, $E = 100$ keV/amu (Frieling *et al* 1991). The beam direction is indicated by an arrow. Fitted profile and contributions of 4l-states are shown.

Such measurements have been performed by Hoekstra (1989) for collision energies 1–13 keV/amu and were recently extended to energies 27–130 keV/amu in our experiments (Frieling *et al* 1991). The cross sections for capture into $n = 4$ shell and relative populations of 4l-states are shown in figures 5 and 6. At small energies these results agree quite well with strong-coupling calculations by Fritsch (1989). At large energies the best agreement is achieved with the calculations by Belkic *et al* (1990), where the Born method with continuum disturbed wave functions was used.

Detailed analysis of the possibilities of investigating alpha particles using the CHERS method was given by Petrov (1991) and by Frieling (1990). The main conclusion is that for diagnostics of alpha particles with energies less then 200 keV CHERS has good prospects. For larger energies the injection of a special diagnostic beam of helium atoms is required. Alpha particles could capture two electrons from beam atom and escape

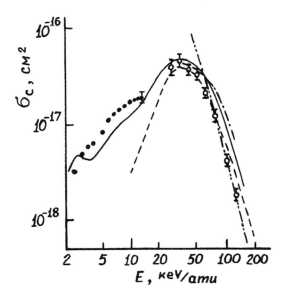

Figure 5. Cross sections for electron capture into the $n = 4$ shell in He(2+)–H collisions: full circles—experiment by Hoekstra *et al* (1989);open circles— our measurements (Frieling *et al* 1991). Calculations:full curve— Fritsch (1988);broken curve—Ryufuku (1982);dash-dot curve—Olson and Schultz (1988);dash-dot-dot curve—Belkic *et al* (1990).

from the plasma. Then using neutral-particle energy analyzers these atoms could be detected.The CHERS method has in this case a factor 100 lower sensitivity as compared to thermal alpha particles.

3. Scaling law for filling nl states during charge exchange

In many cases which are of interest for CHERS diagnostics quite reliable experimental data or results of theoretical calculations are available. But still it is necessary to have some scaling which would give a possibility of determining the scheme for filling the excited states with different quantum numbers n and l in the charge-exchange reaction with participation of various ions and over a broad range of collision velocities.

In the collisions A^{Z+}–$H(n_0)$ the most populated level is known to be one with $n_m = n_0 Z^{3/4}$ (here n is an initially populated state of the hydrogen atom). We proposed (Zinoviev,Korotkov 1989)to describe the filling of different n-states using the function $F(t, s)$ with universal coordinates $t = n/n_m$ and $s = vn_0/Z^{1/4}$. The parameter s is the ratio of the collision velocity v to the orbital velocity of an electron for the state with $n = n_m$. At fixed collision velocity the function $F(t, s)$ will describe the ratio of the state

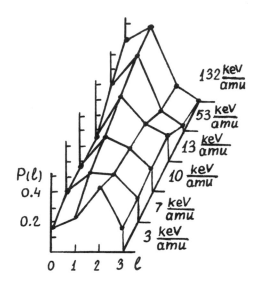

Figure 6. Relative populations of 4l-states of He(+) due to electron capture in He(2+)–H collisions. Data at 3–13 keV/amu are taken from Hoekstra *et al* (1989), at larger energies - results of measurements by Frieling *et al* (1991).

populations with quantum numbers n and n_m. The function $F(t, s)$ could be normalized to 1.0 at maximum. We can choose certain values of $F = 1, 1/2, 1/5, 1/10...$ and plot dependencies $t(s)$ for these values of F in figure 7 using the results of existing theoretical calculations for the different Z and n_0 listed in the figure caption. All points are situated very close to the universal dependence $t(s)$. This allows us to find an empirical form of the function $F(t, s)$. The data on figure 11 correspond to the collision velocity region near the maximum of the charge capture cross section when many states interact. In our opinion this is a reason why we have the universal form of $F(t, s)$. At lower velocities when only few states are strongly coupled such universal scaling fails. At higher velocities ($v > 2$ a.u.) the dependencies of parameters t and s on Z should be changed.

Using the universal function $F(t, s)$ shown in figure 7 we can determine n-state filling for the case needed. An example of using the scaling for the collision He(2+)–H, 25 keV/amu is shown in figure 8. This case corresponds to the values of $F(t, s = 0.84)$. The n-state populations agree well with the results of strong-coupling calculations by Fritsch (1988), which were not known when this scaling was proposed.

Usually the region $v \geq 1$ a.u. is of interest for CHERS diagnostics. At such collision velocities a function $P(n, l)$ can be used to describe the filling of different l-states which approximates well the results of classical calculations (Olson,1981):

$$P(n, l) = 3(l^2 + 1.7 \times l + 1.1)/\{k(k^2 + 1.05k + 1.25)\}$$

where $k = n$ if $n \leq n_m$ and $k = n_m + 1$ if $n > n_m$ and $l \leq n_m$. For $n > n_m$, $l > n_m$

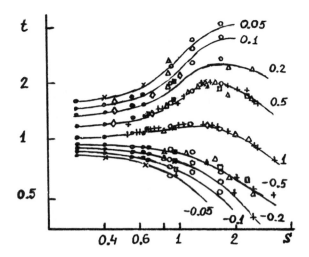

Figure 7. Universal dependence of the parameter $t = n_k/n_m$ on the reduced velocity s, where n is the principal quantum number of the ion state for which $F(t, s) = k$ is constructed on the basis of calculations of cross sections for charge exchange nuclei in collisions with ground state hydrogen atoms: full circles—$Z = 6$ (Green *et al* 1982); crosses—$Z =2,4$–8 (Fritsch and Lin 1984);+—$Z =2$–6,8,9 (Ryufuku 1982);diamonds—$Z =3$–5 (Bransden *et al* 1980); stars—$Z =2,6,8$ (Hardie and Olson 1983) and with excited hydrogen atoms (Isler and Olson 1988): open circles— $Z =6$, $n_0 =2$; full triangles—$Z =6$, $n_0 =3$; open triangles—$Z =8$, $n_0 =2$; full squares—$Z =8$, $n_0 =3$. The labels on the curves give the values of the function $F = k$, where $k < 0$ for $n < n_m$, $k > 0$ for $n > n_m$.The figure is taken from Zinoviev and Korotkov (1989).

$P(n, l) = 0$ and $\Sigma_l P(n, l) = 1.0$.

To determine the cross section the sum of the values of the function F over all n must be normalized to the total capture cross section σ_c. This value of σ_c could be taken from experiment or determined from the universal dependence of the reduced cross section $\bar{\sigma}$ on the reduced collision velocity $\bar{v}$ from work by Ryufuku *et al* (1979) if we assume $\bar{\sigma} = \sigma_c/(Z^{1.07}n_0^4)$ and $v = \bar{s}$.

Values of emission cross sections calculated using this scaling are compared in table 2 with an experiment by Hoekstra *et al* (1988) and seem to be accurate within the uncertainties of 30%.

4. Collisional state mixing

The excited states formed during the charge-exchange reaction in beam-atoms–plasma-ion collisions could be mixed if their life times exceed the characteristic time for collision state mixing. This problem was first mentioned by Fonck *et al* (1984). Now

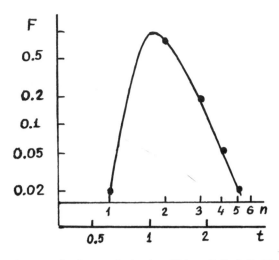

Figure 8. The normalized n-population in collisions He(2+)–H, E =25 keV/amu. The curve is a function $F(T, s = 0.84)$ constructed from the data shown in figure 7, the points are results of calculations by Fritsch (1988). The figure is taken from Zinoviev, Korotkov (1989).

Table 2. Emission cross sections (in 10^{-15}cm^2) obtained using our scaling (Zinoviev, Korotkov, 1989) and experimental results by Hoekstra *et al* (1988).

Line	Scaling data	Experiment
3–2	23.6	26.7+6.5
4–3	23.7	18.0+4.5
4–2	8.8	12.6+3.4
5–3	2.33	2.6+0.7
5–2	1.14	1.5+0.5
7–6	0.11	0.13+0.045
8–7	0.017	0.025+0.006

dedicated collision-mixing codes have been developed by Boley *et al* (1984), Spence and Summers(1986), Korotkov and Zinoviev (1988). In these codes a collisional radiative model of the plasma is used where radiative transitions are considered to compete with collisional excitation, deexcitaion, ionization, recombination and state mixing processes. This model of state populations in plasma has been described in many papers (see Johnson, Hinnov (1973), Sobelman *et al* (1981) and references therein). Special features of present codes include an accurate consideration of level population via the charge capture reaction which is dominant in this case.

Using existing experimental and theoretical data the data files were constructed which describe the populations of different states with $n < 13$ for collisions D(+),He(2+),C(6+),O(8+)-D. For an electron capture in the collisions Be(4+),B(5+)-D and for electron capture from excited beam atoms in all cases the proposed scaling was used.

Special attention was given to the collisional mixing of states with the same principal quantum number where collisional mixing rates are extremely large. In hydrogen-like ions these states are almost degenerate. Therefore an accurate consideration of possible state energy-splitting effects must be made. For such ions as O(7+) and C(5+) the fine structure and Lamb shift are very important. Because of the presence of electric and magnetic fields in a plasma both Stark and Zeeman splitting must also be taken into account. The electric field could be connected with the plasma potential $\varepsilon = T_e/a$ (here T is an electron temperature and a is a minor plasma radius) or with the Lorentz electric field for a particle moving across the magnetic field $\vec{\varepsilon} = \vec{v} \times \vec{B}$ ($\vec{v}$ is the particle velocity, $\vec{B}$ is the magnetic field).

For excitation processes the dipole transitions usually dominate. Sobelman *et al* (1981) proposed a simple analytical formula for dipole transitions based upon the solution of the two-state strong- coupling equations:

$$\sigma_{ij} = 2\pi(\lambda_{ij}/v)^2 \int_0^\infty \sin^2(C/y) \exp\{-2[b(1 + by^2)]^{1/2}\}y\,dy \tag{1}$$

where

$$C = \int_0^\infty V[R(t)]dt.$$

Cross section (1) depends only on two parameters: λ_{ij} is a dipole matrix element between considered states, and $b = \lambda_{ij}\omega/v^2$ where ω is the transition energy, v is the collision velocity. This formula has the correct dependence on collision energy E in the Born region but with the wrong normalization factor ($\sigma = \pi^3 \ln E/E$ instead of $\sigma = 4\pi \ln E/E$). Probably this behaviour is connected with the lack of validity of the dipole approximation for the interaction potential in the Born region. To normalize empirically an expression (1) to the correct Born asymptote the value of the atom radius $C = 2$ was assumed. This formula describes reasonably well the cross section behaviour at small velocities if we take into account the existence of strong coupling of all states in the n-shell multiplet. An analysis of the structure of strong-coupling equations shows that a necessary correction can be introduced replacing b by a new parameter $b = \lambda\omega/v^2$ where $\lambda^2 = \Sigma_j\lambda_{ij}^2$. In this formula the existence of cutoff radii of interaction for nearly degenerate states can be taken into account. The first radius $R_1 = v/\omega$ is connected with the cutoff of the transition probability with increasing impact parameter. Radius $R_2 = 2v/(A_i + A_j)$ reflects the finite lifetime of the excited states. A_i, A_j are total decay probabilities for the states considered. Radius $R_3 = 7.12 \times 10^{-12} (T_e/n_e)^{1/2}$ is a Debye radius for Coulomb interaction in the plasma (T_e and n_e are electron temperature and density of the plasma. The total cutoff radius could be taken in the form $1/R_\Sigma = 1/R_1 + 1/R_2 + 1/R_3$. Replacing this parameter by $b_1 = \lambda/(vR_1)$ all kinds of cutoff radii can be taken into account since $b_1 = \lambda/(vR_1)$.

Finally we have the following form for excitation cross section

$$\sigma_{ij} = 2\pi(\lambda_{ij}/v)^2 \int_0^\infty \sin^2(2/y) \exp\{-2[b_2(1 + b_2y^2)]^{1/2}\}. \tag{2}$$

This formula for the excitation cross section satisfies the balance equation for direct and reverse transitions and within a certain limit this formula is equivalent to the well-known results obtained by Pengelly and Seaton (1964) for transitions between degenerate states.

In figure 9 results obtained using formula (2) are shown to be in good agreement with results of other authors. Formula (2) can be also extended for impurity-ion excitation if the parameter λ is replaced by λZ where Z is the initial ion charge.

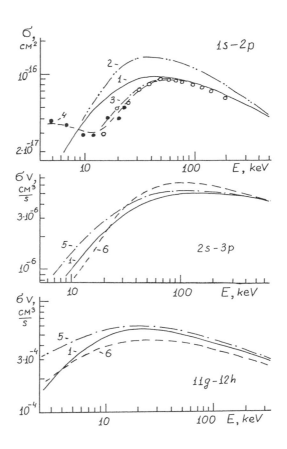

Figure 9. Excitation cross sections for p+H collisions. Curves marked by 1 are obtained using formula (2), open and closed points are measurements by Park *et al* (1976) and by Morgan *et al* (1973). Curves 2–6 are calculations by Janev, Presnyakov (1980), Shakeshaft (1979), Cheshire *et al* (1970), Vriens, Smeets (1980) and Lodge *et al* (1976) respectively.

For transitions from or to a ground state the dipole approximation is not valid. For such transitions we prefer to use an empirical formula suggested by Janev (1988)

$$\sigma_{12} = 3.23 \times 10^{-15} \exp(-42.36/E) \ln(2.69 + 0.68E)/E. \tag{3}$$

Here σ_{12} is in cm^2, E in keV/amu.

For the excitation by electron impact the Seaton formula with Gaunt factor proposed by Zhitkov *et al* (1979) can be used

$$\langle\sigma_{ex}v_e\rangle = 1.6 \times 10^{-5}G_{nm}f_{nm}\exp(-\omega/T_e)/(\omega T_e^{1/2})$$

$$G_{nm} = 0.19\{1 + 0.9[1 + 0.05m(m-n)(1 + (1-2/Z))]\}E - 1(\omega/T_e)\exp(\omega/T_e). \quad (4)$$

Here E is an integral exponent, m and n are principal quantum numbers for considered states.

For ionization by electron impact a formula by Lotz (1970) is considered as a good approximation:

$$\langle\sigma_i v_e\rangle = 2.68 \times 10^{-6}E_1(I_n/T_e)/(I_n T_e^{1/2}). \quad (5)$$

Here I_n is an ionization potential.

For ion impact this formula can be rewritten in the following form (Korotkov, Samsonov, 1989)

$$\sigma_i v_i = 3.21 \times 10^{-6}\ln\{0.545E/I_n[1 + 0.459(I_n/E)^2]\}/(I_n E^{1/2}). \quad (6)$$

For impurity impact formulae proposed by Janev (1988) could be used.

5. Effective cross sections for CHERS spectroscopy

The set of equations for state population coefficients in the steady state collisional radiative model of a plasma was solved twice. First the state populations for an atomic beam passing through the plasma were found. Then the contributions for charge capture into specific nl states can be calculated taking into account the collisions with the beam atoms in the ground and all excited states. Afterwards, the set of equations rewritten for plasma ions can be solved again and coefficients for state populations of these ions obtained. By normalizing these data to one considered particle, effective cross sections for line emission can be found.

These values are of interest for CHERS spectroscopy and reflect all collisional-mixing effects. Some results obtained using our mixing code are shown in figure 10. For carbon ions (the same for oxygen ions) and plasma densities 10^{13}–10^{14} cm^{-3} which are typical for tokamaks, cross sections for ultraviolet and soft X-ray lines are practically independent of plasma density, but optical lines are very sensitive to plasma conditions.

Very important contributions to optical line emission connected with the electron capture from exited atoms of the beam (Isler, Olson 1988, Zinoviev, Korotkov 1989). Corresponding cross sections are very large. They increase as n where n is the principal quantum number of the excited state of the hydrogen atom, but fall very rapidly with increasing beam energy. At energies larger than 40 keV/amu the contribution of excited atoms become negligible. At high plasma densities effective cross sections begin to fall as n. This fact reflects the transition from the corona plasma model to the local thermodynamic equilibrium of the plasma.

Analogous data were obtained for all lines with $n \leq 10$ emitted in collisions D(+), He(2+), Be(4+), B(5+), C(6+), O(8+)–D for beam energies 10–120 keV/amu. The

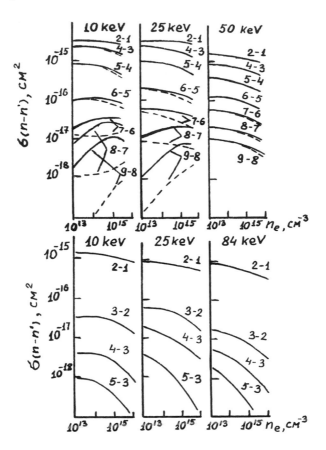

Figure 10. Effective cross sections for line emission in collisions C(6+)–H (a) and He(2+)–H (b) (Zinoviev, Korotkov, 1989).

accuracy of these data is believed to be better then 20–30% in all cases where we have calculations of cross sections for charge capture into specific nl-shell checked by experiment and 30–50% in other cases when results of the proposed scaling were used (Zinoviev, Korotkov,1991).

The problem of the accuracy of cross section data is extremely important for analysis of impurity transport. Let us consider as an example the ionization balance equation for nuclei:

$$dn(q)/dt + \mathrm{div}j(q) = \langle \sigma_i v_e \rangle n(q-1)n_e - \{\langle \sigma_R v_e \rangle n_e + \langle \sigma_c v_i \rangle n_H\}n(q). \qquad (7)$$

The first term in the right hand part of the equation (7) is connected with ionization of hydrogen-like ions. The second term includes all kinds of recombination processes (radiative recombination, electron capture etc) and the second term on the left side is associated with impurity transport. If we can measure the spatial profiles of concentrations of nuclei and hydrogen-like ions and plasma conditions are close to a steady state we can rewrite this equation and extract the contribution of transport processes as the difference between two atomic process terms. It can be clearly seen that even small errors in values of cross sections could greatly influence the results of such analysis.

For the case of nuclei cross sections are known quite well. This allows us to obtain experimentally the local values of the flux of oxygen nuclei (Korotkov 1988, Zinoviev, Afrosimov 1990) by analyzing results of our measurements on the tokamak T-10. For other elements the uncertainties of cross section data (especially in the rates of dielectronic recombination processes) could make such analysis of transport processes impossible.

6. Penetration of beam atoms into plasmas

Beams passing through the plasma are absorbed mainly because of resonance charge exchange process in collisions with plasma protons or because of ionization either by electron or proton impact. If the impurity concentration exceeds 5%, electron capture by impurity ions also must be taken into account. Beam attenuation cross sections can be calculated according to the formula

$$\sigma_\Sigma = \sigma_c + \sigma_{ip} + \langle \sigma_{ie} v_e \rangle / v_a \qquad (8)$$

where σ_C and σ_{ip} are cross sections for electron capture and ionization by proton impact, $\langle \sigma_{ie} v_e \rangle$ is the ionization rate coefficient for electron impact, v_a is the beam velocity.

As has been shown in many papers (Boley *et al* (1984), Korotkov, Samsonov 1989) the existence of multistep excitation processes leads to enhancement of beam attenuation. Let us introduce an enhancement factor as a ratio $\delta = (\sigma_{ef} - \sigma_\Sigma)/\sigma_\Sigma$. Values of enhancement factors are shown in figure 11. At collision energies 10–100 keV/amu enhancement factors are 10–30% but at higher energies they can exceed a factor of 2. Even small corrections in the beam attenuation cross sections are important because they appear in the exponent when the beam penetration into the plasma is calculated.

The beam density inside the plasma could be directly measured using the CHERS method by observing the excitation of beam particles. The corresponding cross sections are shown in figure 12. At typical tokamak plasma densities they are 3–5 times lower as compared the gas phase collisions because of the collisional deexcitation processes in the plasma.

At energies 40 keV and larger the excitation by impurity impact contributes. For these processes formulae proposed by Janev (1988) can be used which fit the results of strong-coupling calculations by Fritsch and Schartner (1987). These data agree well with results of other authors for p–H and He(2+)–H collisions (figure 13). In the cases of Be(4+)–H and C(6+)–H collisions these results could be less accurate because a very extensive basis of states is needed. Therefore an experimental check of these data is very desirable.

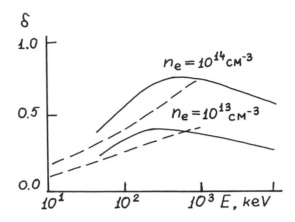

Figure 11. Enhancement factors for beam attenuation cross section calculated by Korotkov, Samsonov (1989) (full curves) and by Boley *et al* (1984) (broken curves).

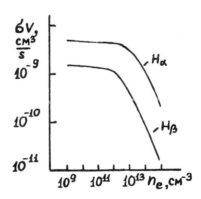

Figure 12. Effective cross sections for hydrogen beam atom excitation. $E = 40$ keV/amu.

7. Active Balmer alpha spectroscopy

A new step in the development of CHERS spectroscopy was made in our joint experiments on the largest tokamak JET (Boileau *et al*,1989). The part of the spectrum situated near the Balmer alpha line shows two intense well-separated features (figure 14). The first part is connected with charge exchange processes into excited states of deuterons in the bulk of plasma. The Doppler width and Doppler shift of the corresponding line gives the local deuterium temperature and velocity of plasma rotation.

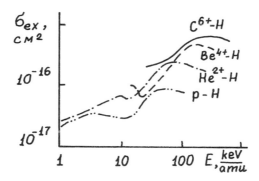

Figure 13. Excitation cross section into $n = 2$ in A(Z+)–H collisions: for the cases p+H and He(2+)–H, our compilations, based on the data shown in figure 9 for p+H case and calculations by Rapp (1974) and by Fritsch and Schartner (1987) are presented. For other cases the results by Fritsch and Schartner(1987) are given.

The absolute intensity of this line allows one to measure deuteron density.

The second part of the spectrum is associated with excited deuterium atoms in the beam. At an observation angle different from 90deg this feature is Doppler-shifted. Beam-atom excitation is produced mainly by collisions with plasma deuterons and fully-stripped impurity ions and to a lesser extent by electron impact. The ratio between these two features is sensitive to the impurity content in the plasma and can be used to obtain the effective plasma charge (Hellermann *et al* 1988). Because the injected beam has also atoms with one half and one third of the full energy, corresponding components are present in observed spectrum. If we look at the part of the spectrum belonging to the full energy component we can see structure which is typical of the Stark effect. This structure is associated with the Lorentz electric field which is induced in the moving frame of coordinates when particles cross magnetic field lines. In the case of the JET experiments when beam energy was 40 keV/amu and a typical toroidal field was 3.2 T the induced Lorentz field was $\vec{\epsilon} = \vec{v} \times \vec{B} = 10^7$ V/m, which produced a Stark pattern with the width 17 Å.

Using perturbation theory the problem of the observed spectrum description can be solved completely. In the beginning the eigenfunctions and eigenvalues of the Hamiltonian can be found. Then using new eigenfunctions we can calculate all needed dipole matrix elements and use them for the calculations of both radiative transition probabilities and cross sections for excitation and deexcitation processes. Afterwards the mixing code described above can be used to obtain absolute intensities of spectrum features. There is only one free parameter which is the local value of the magnetic field. Fitting calculated spectrum to the measured one we can obtain with the accuracy better then 0.2% the value of local magnetic field. Because the toroidal magnetic field produced by coils is known it is possible to obtain with accuracy 10–20% the value of the poloidal magnetic field connected with the plasma current (Boileau *et al* (1989). An

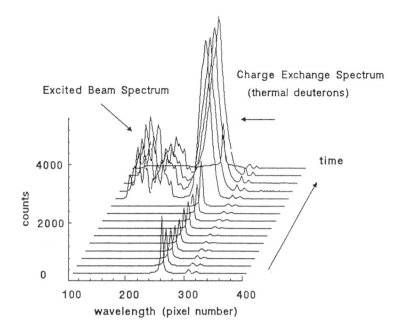

Figure 14. Emission spectrum near the D line produced during beam injection in JET plasma (Boileau *et al*, 1989).

analysis of the Stark multiplet polarization pattern (Wroblewski *et al* 1989, Levinton *et al*, 1991)can give the orientation of the magnetic field in space and improve the accuracy of measurements.

8. Conclusions

At first, let us list the possible future applications of the CHERS method:
1. investigations of thermonuclear alpha particles;
2. an analysis of impurity transport;
3. measurements of the complete velocity distribution function in the plasma when emitted light in three perpendicular directions is observed;
4. application of beam-atom excitation for the control of energy input into the plasma;
5. experimental determination of the enhancement factors.

Additional research on the following atomic processes seems to be needed:
1. charge-exchange reaction:
    a) measurements of H_α and H_β emission produced via charge exchange in collisions p+H with energies 10–150 keV/amu;

b) measurements of capture and line emission cross sections for Be(4+)–H and B(5+)–H collisions;

c) extension of measurements for cases C(6+)–H and O(8+)–H in the energy region 10–150 keV/amu;

2. Excitation processes. Here the measurements of H and H excitation in the collisions p,He(2+),Be(4+), B(5+),C(6+) and O(8+)–H, E =10–150 keV/amu are needed.

3. Collisions with excited atoms. In particular investigations of capture processes in collision between excited hydrogen atoms and multiply charged ions are important for the description of optical line emission. As a model, collisions with elements like Li and Na, where an electron is weakly bound, could be used.

4. Collisions in strong electric and magnetic fields. The presence of the fields removes a degeneration of hydrogen - like ion states and this fact extends possibilities for spectroscopic studies of highly-excited states. In addition such experiments could imitate collisions in plasmas where the fields are strong.

5. The theoretical approach based on analysis of analytical properties of quasimolecular terms in the complex plane of internuclear distance should be widely used for calculations of electron capture, excitation and ionization processes. The work by Khrebtukov published in this conference-book and based on the theoretical conception of Solov'ev,Vinitsky (1985) gives hope that by this approach an old problem of introducing translation factors and rotational coupling in the scheme of strong-coupling calculations would be solved.

Acknowledgments

The author is grateful to Drs F de Heer, M von Hellermann, A A Korotkov, D B Khrebtukov, R Morgenstern and H P Summers and for many fruitful discussions.

References

Afrosimov V.V., Gordeev Yu.S., Zinoviev A.N 1977, Pisma Zh Tekh Fiz 3, pp 97–101.

Afrosimov V.V., Gordeev Yu.S., Zinoviev A.N. 1978, Pisma Zh Eksp Teor Fiz 28, pp 540–543; 1979 Fiz Plasmy 5,pp 987–995.

Afrosimov V.V. *et al* 1980, Zh Tekh Fiz, 50, pp 519–529 .

Afrosimov V.V.*et al* 1983, Pisma Zh Eksp Teor Fiz 38, pp 70–73 .

Belkic D *et al* 1990 (to be published)

Bendahmen M., Bliman., Dousson S. *et al* 1985, J. de Physique, 46, pp.561–572.

Berezovsky E.L. *et al* 1982, Pisma Zh Tekh Fiz 8, pp 1382–1385; 1985 Nucl.Fusion 25, p.1495..

Bliman S., Bonneyfoy M., Bonnet J.J. *et al* 1983,Physica Scripta T3, pp.63–67.

Boileau A., Hellermann M.G., Mandl W. *et al* 1989a, J.of Phys. B22, pp. L145–152.

Boileau A., Hellermann M.G., Horton L. *et al.* 1989b, Plasma Phys. and Contr.Fusion 31, pp.779–804.

Boley C.D.,Janev R.K.,Post D.E. 1984, Phys.Rev.Lett.52, pp.534–537.

Bransden B.H., Newby C.W.,Noble C.J. 1980, J.Phys.B13, p.4245–4255.

Carolan P.Q., Duval B.P., Field A.R. *et al* 1987, Phys.Rev.A35, pp.3454–3471 .

Cheshire I.M., Gallaher D.F., Taylor A.J. 1970,J.Phys.B3, pp.813–32

Ciric D., Dijkkamp D., Vlieg E., de Heer F.J. 1985 J.Phys. B18, pp.L17–L22 , ibid., 18, pp.4745–4762.

Ciric D., Hoekstra R., de Heer F.J. and Morgenstern R. 1988 in Electronic and Atomic Collisions ed. Gilbody *et al*, Elsevier Science Publ. pp. 655–660.

Cotrell G.A. 1983, Nucl. Fus. 23, pp.1689–1696 .

Dijkkamp D., Ciric D., de Heer F.J. 1985 Phys.Rev.Lett. 54, pp.1004–1007 .

Duval B.P., Hawkes N.C., Fielding S.J. *et al* 1985 Nucl. Instr. Methods, B9. pp.689–697.

Frieling G.J. *et al* 1990 Proc.17th EPS Conf.on Contr.Fusion and Plasma Heating, Vol. 14–4, p. 1664, Amsterdam

Frieling G.J., Hoekstra R., Smulders E.,Dickson W., Zinoviev A.N., Kuupens S.J.,de Heer F.J. 1991 (to be published)

Fritch W., Lin C.D. 1984, Phys.Rev..A29, pp.3039–3051.

Fritch W. 1984, Phys.Rev. A30, pp.3324–3327.

Fritsch W., Schartner K.H. 1987, Physics Letters A126, pp.17–20.

Fritch W. 1988, Phys.Rev. A38, p.2664.

Fritsch W. 1989 J.Physique Coll. 50, p.87

Fonck R.J., Finkerthal M., Goldston R.J. *et al.* 1982 Phys.Rev.Lett. 49, pp.737–740.

Fonck R.J., Darrow D.S., Jaehnig K.P. 1984, Phys.Rev. A29, pp.3288–3309.

Fonck R.J. 1985 Rev.Sci.Instr. 56, pp.885–890.

Fonck R.J., Howell R.B., Jaehnig K.P, Knize R.J. 1987, Proc. 6th APS Conf.on At.Proc.in High Temp.Plasmas. Santa Fe.

Gordeev Yu.S. and Zinoviev A.N. 1980, The Physics of Ionized Gases Ed. M.Matic, Boris Kydric Inst. Nucl. Sci. Belgrad pp.215–250.

Gordeev Yu.S., Dijkkamp D., Drentje A.G., de Heer F.J. 1983, Phys. Rev. Lett. 50, pp.1842–1845.

Green T.A., Shipsey E.J., Browne J.C. 1982, Phys.Rev. A25, pp.1364–1373

Grigoryev A.V., Zinoviev A.N.,Kyslyakov A.I. *et al* 1984, Pisma Zh Tekh Fiz 10, pp 76–80 .

Groebner R.J., Brooks N.H., Burrel K.H., Rotter L. 1983, Appl. Phys. Lett. 43, pp.920–922.

Hardie D.J.B.,Olsen R.E. 1983, J.Phys.B16, p.1983.

Hawkes N.C., Peacock N.J. 1985 Nucl.Fus.25, pp.971–980.

Hellermann M.G., Boileau A., Mandl W. *et al* 1988 Abstr. 12 th Int.Conf. on Plasma Phys. and Contr. Nucl. Fusion Research. Nice, 1988.

Hellermann M.G., Mandl W., Summers H.P. *et al* 1990, JET P(90)17.

Hess W.R. 1976, Proc.Workshop Specific Diagnostic Heating Exsperim. Grenoble, Rep. A6HI.

Hoekstra R., Ciric D., Zinoviev A.N. *et al* 1987a, Proc.Int.Conf. 'Atomic Collisions in Fusion' Oxford, p.B1.

Hoekstra R., Ciric D., Zinoviev A.N., *et al* 1987b, Abstracts XII ICPEAC, North Holland Publ.Com., Brighton, p. 530.

Hoekstra R., Ciric D., de Heer F.J., Morgenstern R. 1987c, Phys.Lett. A124, pp.73–76.

Hoekstra R., Ciric D., Zinoviev A.N.,*et al* 1988 Z.Phys.D 1988, 8, pp.57–61.

Hoekstra R, Frieling J, Schlakmann B, *et al* 1989, Abstracts XVI ICPEAC, N.Y., ed. A.Dalgarno et el. p.562.

Hvelplund P., Samsoe E., Andersen L.H. *et al* 1983, Physica Scripta T3, pp.176–181.

Isler R.C. 1977, Phys.Rev.Lett. 38, pp.1359–1362.

Isler R.C. 1985 Nucl. Instr. Methods, B9, pp.673–678.

Isler R.C.,Olson R.E. 1988, Phys.Rev.A 1988, 37, pp.3399–3404.

Iwai T., Kaneko Y., Kimura M., *et al* 1982, Phys. Rev.A26,pp.105–114.

Janev R.K., Presnyakov L.P. 1980, J.Phys.B13, pp.4233–4244.

Janev R.K., 1988, Proc.Int.Conf.on Plasma Diag., Dubna.

Johnson L.C., Hinnov E. 1973,J.Quant. Spectr.Rad.Trans. 13, pp.333–358.

Kadota K, Tsuchida K. *et al* 1978, Plasma Physics 20, p.1011.

Korotkov A.A., Zinoviev A.N. 1988, Fiz Plasmy 15, pp. 223–236.

Korotkov A.A. 1988, J Tekh Fiz 58, pp. 1274–1282.

Korotkov A.A., Samsonov M.S. 1989, Prepr.Ioffe Institute-1351, Leningrad

Lodge J.G., Persival I.C., Richards D. 1976, J.Phys.B9, pp.239–254.

Lotz W. 1970, Z.Physik, Bd.232, pp.101–107.

McCormick K. *et al* 1977 Proc.8-th Europ.Conf.Contr.Fusion and Plasma Physics, p.140.

Meyer P.M., Phaneuf R.A., Kim H.J. *et al* 1979 Phys.Rev. A19, pp.515–521.

Meyer F.W., Howald A.M., Havener C.C., Phaneuf R.A. 1985, Phys.Rev. A32, pp.3310–3318.

Morgan T.J., Geddes J., Gilbody H.B. 1973, J.Phys.B.6, pp.2118–2138.

Olson R.E. 1981 Phys.Rev.A24, pp. 1726–1734.

Olson R.E., Schultz 1988 (private commmunication)

Ovchinnikov S.Yu., Solov'ev E.A. 1988 in Elecronic and Atomic Collisions ed. Golbody H.B. *et al*: Elsevier Science Publ. pp..439–450.

Ovchinnikov S.Yu. 1989, Abstractrs XVI ICPEAC,N.Y.,p.458

Panov M.N., Basalaev A.A., Lozhkin K.O. 1983,Physica Scripta, T3,pp.63–67.

Park J.T., Alday J.E., J.M.George, J.L.Peacher 1986, Phys.Rev.A14, pp.608–614.

Pengelly R.M., Seaton M.N. 1964, M.N.Royal Astr.Soc. 127, pp.165–175.

Petrov M.N. 1991, Proc.IAEA Meeting (to be published)

Phaneuf R.A., Alvarez I., Meyer F.W., Crandall D.H. 1982, Phys Rev.A26, pp.1892–1906 .

Post D.E., Grisham L.R., Fonck R.J. 1983 Physica Scripta.T3, pp.135–147.

Rapp D. 1974 J.Chem.Phys.61, pp.3777–3779.

Ryufuku H. 1982, JAERI-Report -M 82-031

Ryufuku H., Watanabe T. 1979, Phys.Rev. A20,p.1828–1837

Seraydarian R.P., Burrell K.H., Brooks N.H. *et al* 1986, Rev.Sci. Instrum. 57, pp.155–163 .

Shah M.B., Gilbody H.B. 1978, J.Phys. B11, pp.121–130.

Schakeshaft R. 1979, Phys.Rev. A18, pp.1930–1934.

Shipsey E.J.,Green T.A., Browne J.C.1983, Phys. Rev..A27, pp.821–832.

Sobelman I.I., Vainstein L.A., Yukov E.A. Excitation of atoms and broadening of spectral lines, Berlin, Springer, 1981.

Spence J., Summers H.P. 1986, J.Phys.B19, pp.3749–3776.

Vriens L., Smeets A.H.M. 1980, Phys.Rev.A22, pp.940–951.

Zhitkov A.G., Marchenko V.S., Yakovlenko S.Ya. 1979,Prep.IAE-3278/6,Moscow

Zinoviev A.N., Korotkov A.A.,Kzhizhanovsky E.R *et al* 1980, Pisma Zh Eksp Teor Fiz 32, pp 557–560.

Zinoviev A.N. 1986, I Soviet-Britain Symposium on the spectroscopy of multicharged ions. Troitsk, pp.273–274.

Zinoviev A.N., Korotkov A.A.1989 , Pisma Zh Eksp Teor Fiz 50, pp 276–279.

Zinoviev A.N., Afrosimov V.V. 1990, Plasma diagnostics (in Russian) , Moscow, Vol. 7, pp. 56–111, ed. M.I.Pergament

Zinoviev A.N., Korotkov A.A. 1991 (to be published).

Electron attachment to molecules at ultralow electron energies

Ara Chutjian

Jet Propulsion Laboratory, California Institute of Technology
Pasadena, CA 91109 USA

Experimental techniques, both single-collision and multiple-collision, for studying electron attachment to molecules at energies below 0.1 eV are reviewed. Recent attachment results in HI, DI, F_2, SF_6, CCl_4, and $CFCl_3$ are examined from the point of view of threshold behavior (s- or p-wave), threshold energetics, and the temperature dependence of the attachment rate constant. An interesting application of the s-wave phenomenon to trace-species detection is given.

1. INTRODUCTION

With the advent of several new techniques in the past decade, one has been able to explore the electron attachment properties of diatomic and polyatomic molecules at very low electron energies (below 100 millielectron volts) with resolutions of 5 meV and less. This work has highlighted the fact that for some diatomic molecules, and for several broad classes of polyatomic molecules, the attachment in this energy range is dominated by an s-wave threshold law. In this case, the attachment cross section $\sigma_A(\epsilon)$ is predicted to vary as $\epsilon^{-1/2}$, where ϵ is the electron energy. And hence the cross section diverges in the limit of zero electron energy. Such was noted by Wigner in a completely different application (thermal neutron capture by light nuclei) (see, for example, Bethe 1935, Wigner 1948). As is the strength of threshold-law analysis, the same behavior is applicable in dissociative, and nondissociative, electron capture by some molecules. s-wave electron attachment is one of the rare cases in molecular physics where one encounters an infinite cross section.

The individual diatomic molecules studied to date, in the energy range 0-100 meV have been HI, DI, HCl, and F_2. The polyatomic molecules have been SF_6, butanedione, pentandione, so-called "superacids", members of a class of perfluorinated carbon compounds, chlorohalocarbon compounds, and even explosives molecules like RDX, EGDN, PETN, and TNT. In fact, the s-wave phenomenon is at the heart of two new types of explosives detectors under study: the thermal neutron analyzer (TNA) now deployed at some airports, and the reversal electron attachment detector (READ) under development at JPL. One wonders if Bethe or Wigner knew he was touching on problems of explosives detection when pondering threshold laws!

We present herein a summary of several experimental techniques which have been extensively used in the study of low-energy electron attachment. These are, as single-collision methods, the high-Rydberg collisional ionization, and krypton photoionization techniques; and as multiple-collision methods the Cavalleri electron density sampling (CEDS), electron swarm, flowing-afterglow/Langmuir-probe (FALP) and microwave conductivity/pulsed radiolysis (MCPR) techniques. We present a brief summary of these methods, and show that results from these methods can be combined and compared to give new insights into the mechanism of electron attachment. We then give a practical application of the s-wave phenomenon to trace species detection.

2.   EXPERIMENTAL METHODS

A.   High-Rydberg Collisional Ionization

Consider the reaction

$$X(nl) \quad + AB \rightarrow X^+ + AB^- (\text{or } A + B^-) \tag{1}$$

where $X(nl)$ is usually a xenon or rubidium atom in a high-Rydberg state, and the target AB can be any thermal-energy electron-attaching species such as SF_6, CCl_4, $c-C_7F_{14}$, etc (Kalamarides et al 1990). The so-called "free-electron model" (Matsuzawa 1983) predicts that the attachment rate constant of AB in equation (1) will be equal to that for attachment of free (continuum) electrons having the same velocity distribution as the Rydberg electrons. In essence, the high-Rydberg atom, for sufficiently large n, is a carrier of a free electron, of energy equal to its ionization potential minus binding energy.

One form of apparatus using $Xe(nl)$ or $Rb(nl)$ atoms is shown in figure 1. The Rb atoms effuse from an oven, and are intersected at right angles by a frequency-stabilized, single-mode pulsed dye laser. Two-photon excitation populates $n\ ^2S_{1/2}$ and $n\ ^2D_{5/2}$ levels, where $10 \le n \le 110$. Collisions between the target and Rydberg atom are allowed to occur in a field-free region for a variable time t (1-10 μs) after the laser pulse. The number of atoms remaining after the delay time are sampled by selective field ionization (SFI). The SFI signal S(t) is proportional to the number N(t) of remaining atoms, and the rate constant k(v) for atom destruction is related to N(t) by

$$N(t) = N(0)\ e^{-bt} \tag{2}$$

where $b = A_{eff} + k(v)[AB]$, $A_{eff} = \tau^{-1}_{eff}$ is the effective spontaneous radiation rate, and [AB] denotes the density of target AB. A measurement of b vs [AB], extrapolated to zero pressure, gives both A_{eff} (intercept) and k(v) (slope). The rate constants k(v) are then converted to cross sections via the relation

$$k(v) = \int \sigma_A(v)\ vf(v)\ dv \tag{3}$$

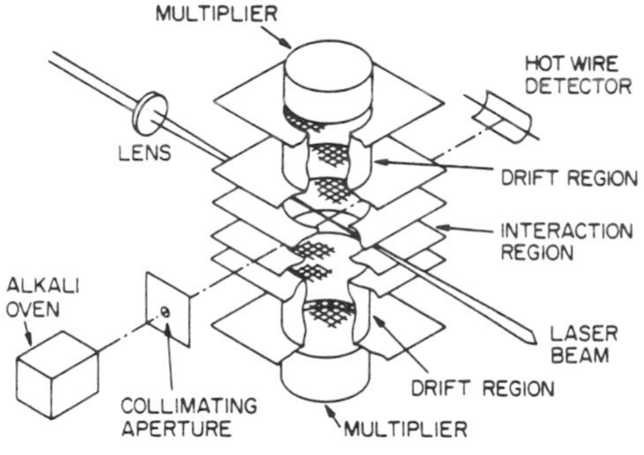

Fig. 1.   Schematic of the high-Rydberg attachment apparatus.

where f(v) is the Rydberg electron velocity distribution, and σ(v) the attachment cross section for the "quasi-free" electrons.

The electron energies accessed by this technique are about 0.5-20 meV. The higher energy is limited by post-attachment interactions (Kalamarides et al 1989), and the lower by a diminishing Rydberg population and the increasing effect of stray electric fields.

B.   The Krypton Photoionization Method

This technique utilizes the sharp $^2P_{1/2}$ photoionization threshold in atomic krypton to generate continuum ("truly free") electrons of energies in the range of 1-200 meV with resolutions of 4-6 meV (FWHM) (Ajello and Chutjian 1979, Chutjian and Alajajian 1985).   A schematic diagram of the apparatus is shown in figure 2.   Kr atoms in the presence of the target AB are photoionized within a field-free collision box by narrow-band photons from a He-Hopfield continuum lamp with a 1-meter VUV monochromator.   The ionization step

$$\hbar\omega + Kr(^1S_0) \rightarrow Kr^+ (^2P_{1/2}) + e(\epsilon) \qquad (4)$$

produces electrons **e** of energy ϵ where ϵ is given by the difference $\epsilon = \hbar\omega - E_t$ for photon energy $\hbar\omega$ and threshold ionization energy E_t (14.666 eV in Kr).   The electrons then attach to the admixed target by

$$e(\epsilon) + AB \rightarrow AB^- \text{ (or A + B}^-\text{)}. \qquad (5)$$

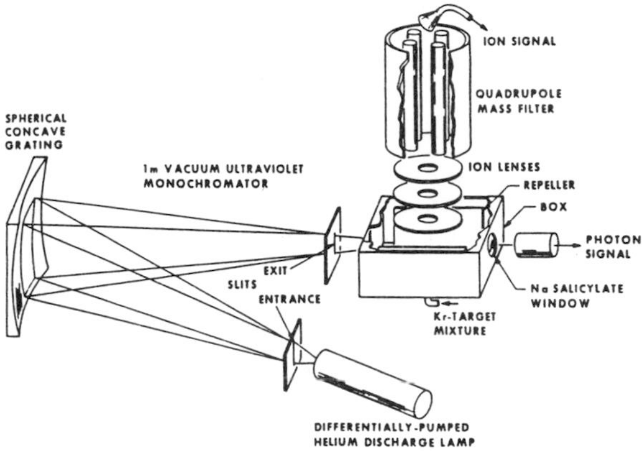

Fig. 2.   Schematic of the Kr photoionization apparatus.

The resulting ions are extracted from the collision region (Alajajian et al 1988), mass-analyzed, and the signal stored via multichannel scaling.   After deconvolution from the spectral slitwidth, the lineshape is parametrized in the form

$$\sigma_A(\epsilon) = N[a(\epsilon^{-1/2} e^{-\epsilon^2/\lambda^2}) + e^{-\epsilon/\gamma}]$$

where parameters a and λ are determined from the lineshape unfolding, γ from the high-energy ($\epsilon > 10$ meV) part of the lineshape, and N by normalization to the thermal attachment rate constant at 300 K.   For the latter, one uses the expression

$$k(\bar{\epsilon}) = (2/m)^{1/2} \int_0^\infty \sigma_A(\epsilon) \; \epsilon^{1/2} f(\epsilon) \; d\epsilon \qquad (6)$$

where $k(\bar{\epsilon})$ is the thermal attachment rate constant at mean electron energy $\bar{\epsilon} = 38.78$ meV (300 K), m the electron mass, and where $f(\epsilon)$ is the Maxwellian electron energy distribution function (EEDF). Values of $k(\bar{\epsilon})$ are obtained from other techniques, such as the multiple-collision Cavalleri, swarm, FALP, or MCPR methods outlined below. The normalization is only possible, of course, if the attachment channels described by equation (5), and measured in $k(\bar{\epsilon})$, are the same.

This approach has recently been extended by Klar et al (1991) to lower electron energies and sub-meV resolution.   Excitation is made from the metastable Ar(4s $^3P_2$) level by one laser to the Ar(4p $^3D_3$) level.   A second laser is used to excite from this level to either *ns* or *nd* Rydberg levels, or to the Ar$^+(^2P_{3/2})$ continuum (free electrons).   With attachment of free electrons to SF_6 one observes both the *s*-wave phenomenon, and vibrational excitation of $SF_6(\omega_1)$. In addition, Boumsellek et al (1991a) are currently extending their original VUV approach to lower electron energies and sub-meV resolution by direct laser ionization to the $Xe^+(^2P_{1/2})$ continuum in a single step with 90 nm radiation.

C.   The Cavalleri Electron Density Sampling Method

The CEDS technique has been extensively used by Crompton and co-workers to measure thermal attachment rate constants (Cavalleri 1969, Gibson et al 1973). A schematic diagram of the apparatus is shown in figure 3. A cylindrical cell contains a mixture of the target AB and N_2. The N_2 pressure is several kPa and, for strongly-attaching molecules, the relative concentration of AB is less than 10 ppm. An X-ray flash of a few μsec duration is transmitted through a thin beryllium window to produce electrons within the cell. After a delay time t (during which the electrons are depleted by attachment) an rf field is applied via external electrodes. The excited, residual electrons excite the N_2 to produce optical emissions which are then detected by the phototube. The intensity of emission is proportional to the number of unattached electrons. About 4×10^4 X-ray shots at a 1 Hz rate are required for good statistics. Electron and ion densities are kept small, and the total pressure high, so that effects of ambipolar diffusion are avoided. Measurements of the rate constant are made over a range of temperatures by heating and cooling both the target and electrons in the collision region.

D.   The Electron Swarm Method

The electron swarm technique measures $k(\bar\epsilon)$ as a function of mean electron energy and E/N, where E is the magnitude of the swarm electric field E and N the carrier gas density (Hunter and Christophorou 1984, Spyrou and Christophorou 1985). One solves the Boltzmann equation, in some degree of approximation, to obtain the EEDF $f(\epsilon, E/N)$ at the mean energy $\bar\epsilon$ of the swarm.

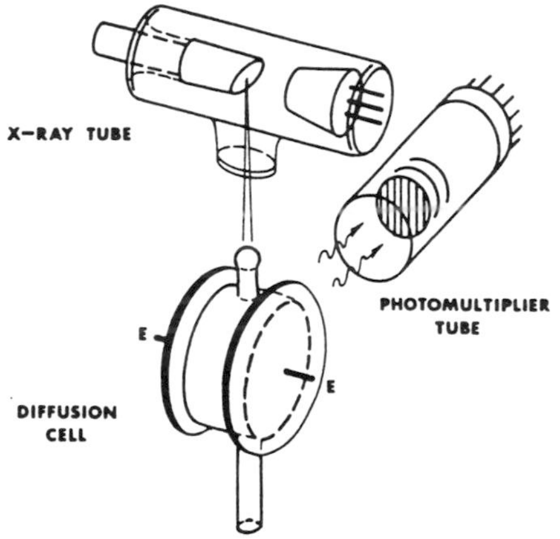

A schematic diagram of the apparatus used by the Oak Ridge group is shown in figure 4. Bursts of electrons from an alpha-particle emitter are allowed to drift to the

Fig. 3.   Schematic of the CEDS apparatus.

anode under the influence of a uniform E field (directed upward in figure 4). The alpha-particle trajectories are well collimated so that the starting plane of the swarm is well defined.

Each current burst induces a voltage pulse at the output amplifier at a specific E/N value. The pulse-height distribution is first recorded as a function of E/N with only buffer gas (N_2 or Ar) in the drift tube. A small quantity (usually 10^{-5} to 10^{-8} the buffer gas density) of the target AB is added, and the ratio of pulse heights recorded as a function of E/N. This ratio is then used to obtain the quantity $\eta(\bar\epsilon)/[AB]$, the attachment coefficient is units of cm^2. The rate constant $k(\bar\epsilon)$ is related to $\eta(\bar\epsilon)/[AB]$ by $k(\bar\epsilon) = \eta(\bar\epsilon)w/[AB]$. Here w is the electron drift velocity usually obtained in a separate measurement using pure buffer gas. The drift region is enclosed in a stainless steel pressure vessel capable of standing pressures to 10 MPa, and capable of being heated and cooled to measure temperature dependencies in $k(\bar{\epsilon})$.

E.   The Flowing-Afterglow/Langmuir-Probe Technique

The FALP technique measures thermal attachment and dissociative recombination rate constants

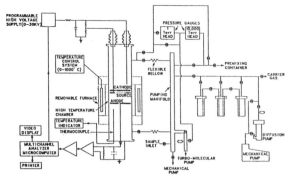

Fig. 4.   Drift-tube swarm apparatus.

(Smith et al 1984, Adams et al 1985).   In the FALP approach one generates an afterglow plasma in a fast-flowing helium carrier gas of ~0.1 kPa pressure by means of a microwave discharge located upstream.   Argon atoms are added to the plasma to quench any metastable He atoms, and a small quantity of the target gas (ppm concentrations) is added downstream.   A schematic of the apparatus is shown in figure 5.

At the point of addition of the target AB one generates a large axial gradient of the electron density due to continual depletion by attachment to AB. Neglecting ambipolar diffusion, the rate of electron loss (or negative-ion formation) can be written as

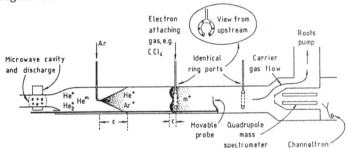

Fig. 5.   Schematic of the FALP apparatus.

$$\frac{\partial n_e}{\partial t} = -k(\bar{\epsilon}) \, n_e [AB] \tag{7}$$

where n_e is the electron density and $k(\bar{\epsilon})$ the attachment rate constant. However, because of the limited sensitivity of the Langmuir probe, FALP requires high ion and electron densities, so that ambipolar diffusion cannot be neglected.   In this case solution of the appropriate continuity equation leads to the expression (Smith et al 1984)

$$\frac{n_e(z)}{n_e(0)} = \left(1 - \frac{\nu_D}{\nu_a}\right)^{-1} \left[\exp\left(-\frac{\nu_a}{v_p} z\right) - \frac{\nu_D}{\nu_a} \exp\left(-\frac{\nu_D}{v_p} z\right) \right] \tag{8}$$

where $\nu_D = D_{ae}/\Lambda^2$, and $\nu_a = k(\bar{\epsilon})[AB]$. D_{ae} is the electron-positive ion (Ar^+) ambipolar diffusion coefficient, and Λ the characteristic diffusion length.

F.   The Microwave-Cavity/Pulsed-Radiolysis Technique

The MCPR technique has been developed by Hatano and coworkers to measure $k(\epsilon)$, thermal rate constants in a Bloch-Bradbury (BB) sequence (attachment followed by third-body stabilization), and for attachment to van der Waals molecules (Shimamori and Hatano 1976, Toriumi and Hatano 1985). In this technique a pulsed, high-energy electron beam of ~ 1 ns duration impinges on a thin tungsten foil. The collision area becomes a source of X-rays which in turn irradiate a high-pressure (10-100 kPa) gas mixture within a microwave resonator (figure 6). The resonator itself can be heated or cooled in the temperature range 77-373 K to measure temperature dependencies of the attachment rate constant (see below).

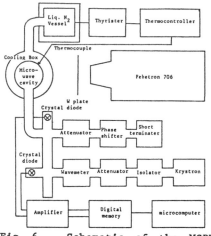

Fig. 6.   Schematic of the MCPR apparatus.

Electrons generated in the cavity are rapidly thermalized, and the cavity Q and resonant frequency are altered. The changes are proportional to the electron density, and are detected by differentially-amplified signals which monitor the klystron and cavity frequencies. The product of the observed electron lifetime and target density, τ_0 [AB], can be related to the direct (collisionless) rate constant k_1 and the three-body (quenching) rate constant k_M by the expression

$$\tau_0[AB] = 1/k_1 + 1/k_M[M] \tag{9}$$

where [M] is the pressure of the buffer gas. Measurements of τ_0[AB] vs [M]$^{-1}$ then yield k_M (slope) for each collision partner, and k_1 (intercept).

Extensive measurements of $k(\bar{\epsilon})$ in the energy range $0.03 \leq \bar{\epsilon} \leq 1.0$ eV have been recently reported by Shimamori et al (1991). The molecules studied include CCl_4, CH_3I, CF_3Br, $CHCl_3$, C_6H_5I, C_6H_5Cl, BCl_3 and c-C_7F_{14}. Unlike the Cavalleri, swarm or FALP techniques, the MCPR method causes electron heating, while leaving the target at the chosen temperature of the cavity. And hence, effects on $k(\bar{\epsilon})$ of electron temperature and target temperature can, potentially, be assessed independently.

3.   RECENT RESULTS IN THERMAL ELECTRON ATTACHMENT

Rather than review the vast amount of data in the area of low-energy electron attachment, we draw upon recent results which illustrate an interesting effect. In diatomic molecules, we discuss low-energy attachment to HI, DI and F_2; and in polyatomic molecules results in SF_6, CCl_4, and $CFCl_3$ which underscore the importance of measuring temperature dependencies of attachment rate constants. Finally, we combine our knowledge of s-wave attachment properties, with a bit of electrostatics, to point out an interesting application to trace species detection. And it's all done with mirrors (electrostatic ones, that is)!

A.   Electron Attachment in HI and DI

The spectroscopic constants of the diatomic molecules HI and DI are known to high accuracy. One easily calculates that the zero-energy dissociative attachment (DA) process in HI should be exothermic by 5.3 meV, and that in DI endothermic by 35.5 meV. And hence DA in DI is allowed via rotational levels J'' populated at elevated temperatures.   The DA cross sections for both HI and DI were measured using the Kr photoionization technique (Alajajian and Chutjian 1988). The HI cross sections were obtained by normalization to $k(\bar{\epsilon})$ of Smith and Adams (1987) using the FALP method; and DI cross sections obtained through an intercomparison technique (Alajajian and Chutjian 1987).

Measurements of the T-dependence of DA signal in DI relative to HI were then carried out (Chutjian et al 1990).   The effect of increasing I^- signal in DI, relative to I^- in HI, was accounted for by the increase in the Boltzmann population of DI(J'') levels with T. Shown in figure 7 is the experimental relative enhancement in I^- signal, along with the calculated expected increase in population of higher J'' levels. The effect in terms of potential energy curves is shown in figure 8.

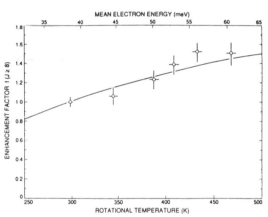

It would be interesting to exploit this effect to rapidly "turn on" a DA process by, for example, laser-exciting "cold" DI to a v,J level above its DA threshold.   This could be used as a gaseous opening switch: a plasma is made less conductive by converting electrons to negative ions.

B.   Electron Attachment in F_2

Fig. 7. Calculated (——) and measured (o) DA in DI relative to HI.

Zero-energy electron attachment to F_2 to produce F^- has been a source of controversy for nearly a decade. One expects the relevant electronic states to be F_2 ($X^1\Sigma_g^+$) and F_2^- ($X^2\Sigma_u^+$).   By symmetry, the negative ion cannot support an s-wave, and the first allowed is p-wave.   This would lead to a threshold cross section varying as $\epsilon^{1/2}$ (Wigner 1948), and hence vanishing at zero energy.

A rising cross section has been detected in every experiment on F_2. The results are summarized in figure 9.   Here one sees that beam (•, Chantry 1982), swarm (o, McCorkle et al 1986), and Kr photoionization data (——, Chutjian and Alajajian 1987) are in quite good agreement for such an experimentally intractable molecule. Possible reasons for this apparently s-wave behavior are that (a) some ungerade character from higher-lying negative-ion states is admixed with the F_2^- (X) state, (b) electron-rotation (coriolis) interaction mixes gerade states, and (c) nonadiabatic terms mix the $^1\Sigma_g^+$ and $^2\Sigma_u^+$ states when the Hamiltonian is expressed in electronic coordinates relative to the separated nuclei (Pack and Hirschfelder 1968). Or, the answer could be simply that (d) one has not gone low enough in energy, because threshold laws never specify the energy at which threshold behavior is attained! Measurements are currently in progress at JPL with a new approach (Boumsellek et al 1991a) to extend the energy limit to below 5 meV, at a resolution of 0.5 meV, to determine if the cross section is still rising, or falling.

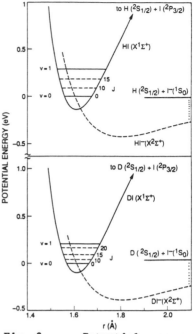

Fig. 8.   Potential energy curves for HI and DI.

C.   Attachment to SF_6, and Its Temperature Dependence

Cross sections and rate constants for electron attachment to SF_6 have been reported using all the techniques outlined in Sec 2. Attachment to form the parent ion SF_6^- is s-wave, as confirmed in four different cross-section measurements (Zollars et al 1985, McCorkle et al 1980, Chutjian and Alajajian 1985, Klar et al 1991). The s-wave behavior has also been treated theoretically in terms of electron capture through non-adiabatic coupling to the nuclear motion (Gauyacq and Herzenberg 1984). Theory also predicted the ω_1 vibrational excitation recently observed by Klar et al (1991).

The puzzle in SF_6 appeared to be rather in the rate constant, and its T dependence. The matter was resolved through combining results of a calculation (Chutjian 1982, Orient and Chutjian 1986), with the T-dependence of $k(\bar{\epsilon})$ as measured by four different techniques.

Using the theory of O'Malley (1966) one may write the attachment cross section $\sigma_A^{v\omega,J}(\epsilon)$ for the vth harmonic of the ωth normal mode, and rotational level J, as

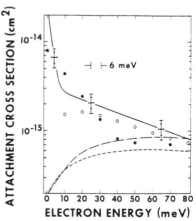

Fig. 9.   DA cross sections in F_2.

$$\sigma_A^{v\omega,J}(\epsilon) = (4\pi^2 g/k^2)(\Gamma_a/\Gamma_d)\left|\tilde{\chi}_{v\omega,J}(R_\epsilon - i\Gamma_a/\Gamma_d)\right|^2 \times \exp[-\rho(\epsilon)], \qquad (10)$$

where k is the electron momentum, Γ_a and Γ_d are the autoionization and dissociation widths, respectively, g is a spin degeneracy factor, $e^{-\rho(\epsilon)}$ the survival probability, and R_ϵ is the turning point in the final negative-ion state at energy ϵ. The quantity $\chi_{v\omega,J}$ is the complex overlap integral between the vth vibrational level in the ωth normal mode, and the quasicontinuum of SF_6^- states represented by a modified δ-function (see Orient and Chutjian 1986 for details). This cross section of each v,J level must in turn be summed, at each T, over all significantly populated states S of the system, or

$$\sigma_A(\epsilon) = \sum_S \sum_{v\omega} \sum_J b_{v\omega,J} \, \sigma_A^{v\omega,J}(\epsilon) \tag{11}$$

A particular state S consists of the set of excitations of one or more normal modes ω, with v quanta of excitation in each mode.

Comparisons of the calculated $\sigma_A(\epsilon)$ from equation (11) with swarm (McCorkle at al 1980), high-Rydberg ionization (Zollars et al 1985), and Kr photoionization results (Chutjian and Alajajian 1985) can be found in Orient and Chutjian (1986). We turn our attention herein rather to the rate constant as a function of T. This can be calculated by combining equations (11) and (6) at each T, and comparing that result to the large body of experimental data. This theoretical comparison is shown by the solid lines in figure 10. One sees that the rate constants are "well-behaved": a rate constant exhibiting s-wave behavior ($\ell=0$) becomes constant as T is lowered (put $\sigma_A(\epsilon) \sim \epsilon^{-1/2}$ in equation (6) to see this), while that varying as p-wave ($\ell=1$) vanishes as T → 0 K.

Experimentally, from examining beam-gas (▲, Spence and Schulz 1973), swarm (◉, Fehsenfeld 1970), and CEDS data (■, Petrović and Crompton 1985) one would conclude that a constant cross section had, in fact, been attained; but in a much higher temperature range, between 200 and 600 K, and not in the expected theoretical limit T → 0 K. An explanation for this constancy at the higher T came from the calculation, and from another piece of evidence. There are two channels open in the SF_6 attachment: one to produce SF_6^-, and the other to produce $SF_5^- + F$. When one adds to the calculated rate constant for the SF_6^- channel alone, the measured rate constant for the second channel (Smith et al 1984, Fehsenfeld 1970), one then obtains the curves (····) shown. To summarize, the constant rate constant in the range 200-600 K was not the true s-wave behavior, but only a compensation of a falling SF_6^- rate constant by a rising $SF_5^- + F$ one. The true s-wave behavior will be apparent if $k(\bar\epsilon)$ can be measured at T<200 K.

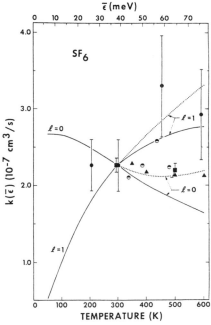

Fig. 10. Calculated and measured rate constants for attachment in SF_6 to produce SF_6^- and $SF_5^- + F$.

Finally, we point out that the FALP results (●) could not be reconciled with either theory or the other experiments.

D.   Attachment to $CFCl_3$ and CCl_4, and Their Temperature Dependencies

The discrepancy between FALP results and measured cross sections for SF_6 led to a similar examination of cross sections and rate constants in $CFCl_3$ and CCl_4. One's expectation is clearly that for these molecules the rate constant should not exhibit a peak at higher temperatures. The cross section in SF_6 is falling with energy, and a Maxwellian EEDF simply places more electron population towards higher energies (smaller $\sigma_A(\epsilon)$) as T is increased, and hence the rate constant should decrease uniformly.

To test this hypothesis, a series of calculations was carried out on $CFCl_3$ and CCl_4 (Orient et al 1989) to compare with earlier measurements (Chutjian et al 1984, Chutjian and Alajajian 1985). Simultaneously, a series of CEDS measurements of $k(\bar{\epsilon})$ *vs* T for $CFCl_3$ and CCl_4 were made by Crompton and co-workers.

Results for $CFCl_3$ are shown in figure 11. One sees again good agreement of the calculated (*s*-wave) cross section (—, the shading is the error in the underlying experimental cross section), CEDS experiments (■, Orient et al 1989), and swarm results (O, McCorkle et al 1980). FALP results (∇, Smith et al 1984) appear to give a peak in the rate constant near 450 K which is difficult to reconcile either with CEDS results or theory.

E.   The Reversal Electron Attachment Detector

The fact that an electron attachment cross section actually tends towards infinity at energies approaching zero can be ample fuel for the imagination. Conventional ion sources and glow discharges usually have mean electron energies well above this limit, and hence can provide only a small fraction of the available EEDF to attach to species such as SF_6, $CFCl_3$, CCl_4, etc. To utilize these large cross sections a concept was developed of reversing trajectories in an electron beam directed towards an electrostatic mirror. At the point of reversal, the electrons would have zero, or near-zero longitudinal and radial energies. Introduction of the target at this point would then lead to production of negative ions. This idea was first demonstrated in a simple geometry (Orient et al 1985). A second charged-particle optics system was then computer-designed from first principles, and tested. This design increased the space-charge limited currents in the beam, reduced the radial and longitudinal energies to zero, and increased the extraction efficiency of ions (Bernius and Chutjian 1989, 1990). A schematic diagram of the reversal electron attachment detector (READ) is shown in figure 12.

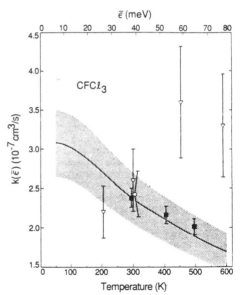

Fig. 11. Comparison of calculated and measured rate constants for attachment in $CFCl_3$ to produce Cl^-.

One practical use of this device is that explosives molecules also have the interesting property of attaching zero-energy electrons! And hence, the

READ can be used to detect some of the explosives-of-choice, such as RDX, PETN and TNT. Shown in figure 13 is the first zero-energy attachment spectrum of the explosive PETN (Boumsellek et al 1991b). Work is currently underway to measure the sensitivity of READ to detection of these species.

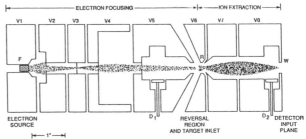

Fig. 12.   Schematic of the READ.

ACKNOWLEDGEMENTS

Thanks are due to all those who contributed to this work over the years: J Ajello, S Alajajian, M Bernius, S Boumsellek, K-F Man, and O Orient. This work was carried out at the Jet Propulsion Laboratory, California Institute of Technology, and was supported by the Department of Transportation through agreement with the National Aeronautics and Space Administration.

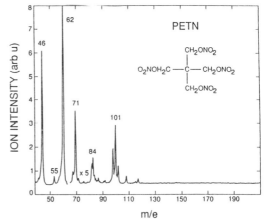

Fig. 13.   Mass spectrum of PETN using the READ.

REFERENCES

Adams N G, Smith D and Alge E 1985 *Chem Phys Lett 116* 460
Ajello J M and Chutjian A 1979 *J Chem Phys 71* 1079
Alajajian S H and Chutjian A 1987 *J Phys B: At Mol Phys 20* 2117
Alajajian S H and Chutjian A 1988 *Phys Rev A 37* 3680
Alajajian S H, Bernius M T and Chutjian A 1988 *J Phys B: At Mol Opt Phys 21* 4021
Bernius M T and Chutjian A 1989 *J Appl Phys 66* 2783
Bernius M T and Chutjian A 1990 *Anal Chem 62* 1345
Bethe H A 1935 *Phys Rev 47* 747

Boumsellek S, Alajajian S H and Chutjian A 1991a (work in progress)
Boumsellek S, Alajajian S H and Chutjian A 1991b *J Am Soc Mass Spectrom* (in press)
Cavalleri G 1969 *Phys Rev 179* 86
Chantry P J 1982 in *Applied Atomic Collision Physics* (eds H S W Massey, E W McDaniel and B Bederson New York: Academic) V 3, p 56
Chutjian A 1982 *J Phys Chem 86* 3518
Chutjian A, Alajajian S H, Ajello J M and Orient O J 1984 *J Phys B: At Mol Phys 17* L745; 1985 *J Phys B: At Mol Phys 18* 3025 (corrigendum)
Chutjian A and Alajajian S H 1985 *Phys Rev A 31* 2885
Chutjian A and Alajajian S H 1987 *Phys Rev A 35* 4512
Chutjian A, Alajajian S H and Man K-F 1990 *Phys Rev A 41* 1311
Fehsenfeld F C 1970 *J Chem Phys 53* 2000
Gauyacq J P and Herzenberg A 1984 *J Phys B: At Mol Phys 17* 1155
Gibson D K, Crompton R W and Cavalleri G 1973 *J Phys B: At Mol Phys 6* 1118
Hunter S R and Christophorou L G 1984 *J Chem Phys 80* 6150
Kalamarides A, Walter C W, Lindsay, B G., Smith, K A and Dunning, F B 1989 *J Chem Phys 91* 4411
Kalamarides A, Marawar R W, Ling X, Walter C W, Lindsay B G, Smith K A and Dunning, F B 1990 *J Chem Phys 92* 1672
Klar D, Ruf M-W and Hotop H 1991, *Joint Symposium on Electron and Ion Swarms and Low Energy Electron Scattering* (Bond Univ, Gold Coast, Australia), abstracts p 2
Matsuzawa M 1983 "Theoretical Studies of Collisions of Rydberg Atoms with Molecules" in *Rydberg States of Atoms and Molecules* (eds R F Stebbings and F B Dunning, New York: Cambridge)
McCorkle D L, Christodoulides A A, Christophorou L G and Szamrej I 1980 *J Chem Phys 72* 4049
McCorkle D L, Christophorou L G, Christodoulides A A and Pichiavella L 1986 *J Chem Phys 85* 1966
O'Malley T F 1966 *Phys Rev 150* 14
Orient O J, Chutjian A and Alajajian S H 1985 *Rev Sci Instr 56* 69
Orient O J and Chutjian A 1986 *Phys Rev A 34* 1841
Orient O J, Chutjian A, Crompton R W and Cheung B 1989 *Phys Rev A 39* 4494
Pack R T and Hirschfelder J O 1968 *J Chem Phys 49* 4009
Petrovic' Z Lj and Crompton RW 1985 *J Phys B: At Mol Phys 17* 2777
Shimamori H and Hatano Y 1976 *Chem Phys Lett 38* 242
Shimamori H, Nakatani Y and Ogawa Y 1991 *Joint Symposium on Electron and Ion Swarms and Low Energy Electron Scattering* (Bond Univ, Gold Coast, Australia) abstracts p 28
Smith D, Adams N G and Alge E 1984 *J Phys B: At Mol Phys 17* 461
Smith D and Adams N G 1987 *J Phys B: At Mol Phys 20* 4903
Spence D and Schulz G J 1973 *J Chem Phys 58* 1800
Spyrou S M and Christophorou L G 1985 *J Chem Phys 82* 2620
Toriumi M and Hatano Y 1985 *J Chem Phys 82* 254
Wigner E P 1948 *Phys Rev 73* 1002
Zollars B G, Higgs C, Lu F, Walter C W, Gray L G, Smith K A, Dunning F B and Stebbings R F 1985 *Phys Rev A 32* 3330

Atom cooling and trapping, and collisions of trapped atoms

Alan Gallagher[†]

Joint Institute for Laboratory Astrophysics, National Institute of Standards and Technology and Univ. of Colorado, Boulder, Colorado 80309-0440 USA

Abstract. Methods currently used to optically cool and trap atoms, and the resulting trap densities and temperatures are discussed. Next, the types of cold-atom collision processes that occur in these traps are discussed, with emphasis on their differences compared to more familiar atomic collisions at thermal and higher energies. A semiclassical model and calculations for trap-loss and associative ionization, due to excited state collisions of these very cold atoms, is then described.

Trapping and cooling of atoms with laser beams has now produced very cold and dense clouds of atoms. The collisions that occur between pairs of these atoms are important to trap properties as well as to progress toward Bose condensation. These very slow collisions also introduce fascinating new features to atomic collisions, as well as provide an opportunity for major light-induced modifications of collision dynamics. Here I will first provide a brief overview of current optical cooling and trapping methods, and the types of slow collision processes that occur in traps. I will then qualitatively discuss the major differences between these very slow collisions and normal thermal collisions. Finally, I will describe my semiclassical picture and calculations of two of these excited-state collision processes, done partly in collaboration with Dave Pritchard. Before starting, I also want to clarify the fact that I do not have any trapping experiments in my laboratory, although my colleagues Jan Hall and Carl Wieman do. To my knowledge, about a dozen laboratories are engaged in atom trapping. In the brief time available I will concentrate on phenomena rather than credits, and I refer you to Refs. 1 and 2 for additional information.

Light traps are energetically shallow, so thermal atomic beams or vapors must first be slowed down to load the trap. This is done by transferring resonance-line photon momentum to the atoms. Since fluorescence is isotropic on average, each absorption and emission event transfers one incident photon momentum $h\nu/c$ to the atom, and as this can be repeated each excited state lifetime τ this force can reach $h\nu/c\tau$, typically 10^5 times Earth's gravity. One photon recoil changes a Na velocity by ~ 10 cm/s, so $\sim 10^4$

[†] Staff member, Quantum Physics Division, National Institute of Standards and Technology

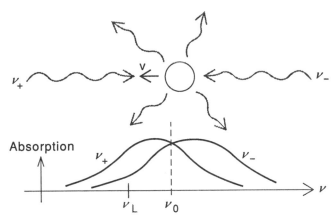

Fig. 1. Diagram of an atom moving to the left in counterpropagating beams of frequency ν_L, with the absorption spectrum for each beam shown below. The greater absorption of the ν_+ beam transfers net photon momentum that opposes the atom velocity. This slowing effect occurs for either velocity direction, cooling the atom motion.

recoils are needed to load a Na trap. The atomic resonance Doppler shifts as the atom slows down, so some trick must be used to maintain efficient excitation during slowing; the most common is laser-frequency chirping, resulting in intermittent trap loading.

When the laser frequency ν_L is red-detuned from an atomic resonance ν_o, an atom moving toward the laser is Doppler shifted closer to resonance, increasing the probability of absorption and the slowing force on the atom. The inverse holds for motion away from the beam, as shown in Figure 1. Thus, an atom in counterpropagating red-detuned beams feels a net retarding force in either direction and cools, as if moving in "molasses." (The average fluorescence is blue-shifted from ν_L, consistent with this cooling.) Six beams directed along three orthogonal axes are used to thus cool initially slowed atoms, using red detuning of $\sim$ one natural linewidth Γ. The random character of the fluorescence-recoil directions produces a diffusional heating mechanism that competes with the "molasses" cooling to produce a "Doppler cooling limit" of $kT = \hbar\Gamma/2$, which is $\sim 240\ \mu K$ for Na and corresponds to a mean velocity of ~ 50 cm/s. The same "Doppler cooling" method is being used to cool trapped ions.

An additional optical cooling mechanism, called "polarization-gradient" or "Sisyphus" cooling, allows much lower temperatures to be reached for atoms that can be optically pumped.[2] This results from the fact that the coherent fields of the six laser beams produce a polarized net field $\bar{E}_T$ at each spatial position, and the polarization ellipticity and direction oscillates every $\lambda/2$ in each spatial direction. Atoms with multiple ground states are optically pumped back and forth between ground state Zeeman levels $|M\rangle$ as they move slowly through this spatially varying $\bar{E}_T$. The near-resonant radiation causes different ac Stark shifts of the $|M\rangle$ levels, as shown in the left diagram of Figure 2 for a position with net σ^+ radiation. As $\bar{E}_T$ oscillates with spatial position, oscillating M-dependent potential wells are formed as shown in the right diagram of Figure 2. At each position optical pumping transfers atoms from the higher to the lower energy

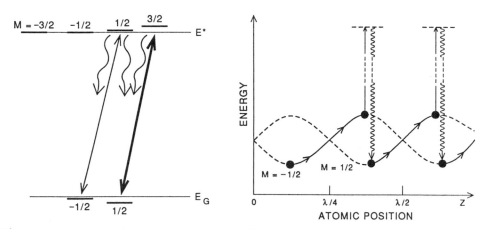

Fig. 2. The ac Stark-shifted potentials and radiative transitions responsible for polarization gradient cooling. Figure adapted from Ref. 2.

$|M\rangle$ level, as can be seen in the example in Figure 2. Thus, as shown the atoms climb potential hills as they move. This repetitive process of hill climbing cools the atoms to the velocity corresponding to a single photon recoil, or ~ 10 cm/s and 10 μK for Na. The velocity distribution is most commonly measured by switching off the trapping light and measuring atomic positions versus time. These highly cooled atoms are now being loaded into magnetic traps in some laboratories, and further cooling is then achieved in the absence of resonance radiation.[1,3]

I have so far described optical cooling, but not trapping into a small spatial region. Only with trapping plus cooling are densities and trap lifetimes large enough to produce noticeable consequences of collisions between the slowed atoms. Present optical traps utilize the Zeeman-level structure to achieve a relatively deep trap; ~ 1 K is typical. Although actual Zeeman structure is generally more complex, these "spontaneous force" traps utilize the principle shown in Figure 3 for a $J = 0$ to 1 transition.[4]

Here six red-detuned laser beams converge along three orthogonal axes, and oppositely directed beams have opposite circular polarizations. The trap is at the midpoint between reverse-current Helmholtz coils, where the magnetic field is zero and increasing radially outward. A field gradient then exists along all three orthogonal axes, producing level shifts as shown in the figure. When an atom moves away from the trap center, the Zeeman level shifts shown tune it closer to resonance for the position-restoring beam, and further from resonance for the oppositely-directed beam. The net photon momentum transfer then yields a restoring force and a quadratic potential well exists in each orthogonal direction, producing an anharmonic trap. The combination of atom temperature and trap depth typically confines the atoms to a ~ 0.1 mm volume for densities below $\sim 10^{10}$ cm^{-3}. Radiation forces between atoms cause bizarre, nonuniform spatial distributions at higher densities; these are partially understood but I will not describe them here.[5] Once atoms are trapped and cooled in these "deep" traps, they can be loaded into magnetic traps, which are much shallower. There they can be further cooled, without using resonance-time radiation which would heat the atoms back to the recoil or Doppler

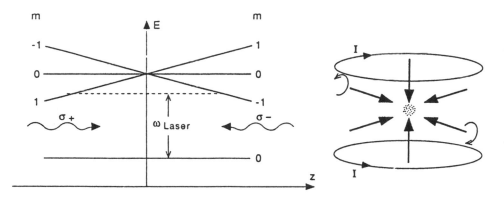

Fig. 3. Diagrammatic representation of Zeeman light trapping, adapted from Ref. 4. Energy levels versus position, due to the magnetic field, are shown in the upper part and the arrangement of field coils and six light beams are shown below.

limit.[1,3] However, resonance radiation is normally used to detect the quantities, velocities and positions of atoms in the trap after they have been manipulated in the dark. In this way atom pair collisions at temperatures well below 1 μK, are being investigated, and perhaps the final "collective" collision of condensation will be observed before long.

I will now discuss the different kinds of atomic collisional phenomena that play important roles in these traps. First, trapped atoms can be lost due to collisions with background-gas which desorbs from the walls. These typically limit trap lifetimes to $\sim$ 100 s at 10^{-10} Torr vacuum. Most experiments are done in room temperature enclosures, in which case these are ordinary thermal elastic collisions. These involve the somewhat unusual phenomenon of a very small momentum transfer, but the only interest in these so far has been to minimize their deleterious effect. Collisions between pairs of slow atoms are the dominant trap-loss mechanisms that limit densities in most traps. This loss results from electronic energy transfer, with the atom pair gaining a corresponding kinetic energy that ejects both from the trap. First, considering collisions between ground-state atoms, spin exchange in collisions of H or alkali pairs can transform the hyperfine energy splitting into atom-pair kinetic energy. This is sufficient energy to eject the heavier alkalis from weaker optical traps, and can eject H and all alkali atoms from magnetic traps. Next, elastic collisions between ground-state atoms produce relaxation toward a thermal phase-space distribution in magnetic traps, and this has a profound effect on trap dynamics and evaporative cooling.

Turning now to excited-state interactions, the long-range pair interactions due to the optically induced dipoles are profoundly complex and important to the pair dynamics. At $R > \bar{\lambda}$ spacing these forces depend on atomic positions relative to the driving fields and on optical frequency, power and polarizations, and it is not even clear if these are conservative forces in trap standing-wave fields. Excited-state energy-transfer collisions only occur in light traps, but they have much larger cross sections than ground state collisions and they generally dominate such trap losses. I will discuss them in detail below. Finally, all atoms studied so far have very strong close-range pair binding compared to trap energies. Thus, the rate of three-body association to a bound pair can be an important loss, particularly if this initiates agglomeration into a solid. The latter issue becomes particularly important in efforts to achieve Bose-Einstein condensation due to the long-range forces, without allowing the associated metastable state with an interatomic spacing of 100-1000Å to collapse to a tightly-bound cluster of 3Å spacing. In all cases these very slow collisions are highly quantum mechanical, and in most cases not studied theoretically. However, I will not discuss them further here.

With the exception of the background gas problem, all of the pair or three-body collisions I have just described occur for very slow, cold atoms. They generally do not behave like the thermal energy-transfer collisions we are accustomed to, and in many cases it is not even clear how to describe the phenomenology, much less calculate it. In addition to the intrinsic interest of these very unusual collision phenomena, it is obvious that the spectroscopy and behavior of trapped atoms and ions have enormous potential for new physical discoveries. Thus, I highly recommend that collision physicists take a good look at this fascinating field.

The remainder of my discussion will be about excited state collisions in atom traps that utilize resonance-line trapping fields. These are very important in atom traps, since in the 100-1000Å region of primary importance the excited state interaction, $V^*(R)$ is typically 10^5 times stronger than the ground-state pair interaction $V(R)$. Although the atomic deBroglie wavelengths are very large, I will utilize a semiclassical picture to describe atom-pair motion. This has an immense advantage in conceptualization and simplicity, and I will shortly show that it is also a reasonable representation of wave-packet dynamics. As in most physics, these semiclassical analogies are not the final word in accuracy, but they provide valuable insight and approximations to fully quantum calculations. I will start with a diagrammatic represenation, in Figure 3, of the most crucial differences between these very slow collisions and normal thermal collisions. In the thermal case collision times τ_c are typically 10^{-4} of radiative lifetimes τ_N, so the process is traditionally described as photoexcitation of a free atom followed by a pair collision without any influence of the radiation on the collisions. In addition, the deBroglie wavelength is normally smaller than the important pair-interaction distances and many angular momenta contribute, so that classical nuclear motion in the molecular potentials V(R) is a good approximation. Also, the interactions are only comparable to kT, thereby modifying the motion, at close range of typically ~ 5Å. Energy transfer occurs at somewhat larger R (20-50Å) in the present case, because of the very strong, long-range resonant-dipole interaction $V^*(R) = C_3/R^3$ between a ground-state/excited-state pair A-A*. However, this does not alter the $\tau_c \ll \tau_N$ condition or the validity of classical nuclear motion for normal-temperature collisions.

In the trap-collision case with $T \leq 1$ mK, the $V^*(R)$ interaction is typically comparable to kT at 300-1000Å and velocities are 0.1-1 m/s, so $\tau_c \geq \tau_N$. Here radiative transitions in mid-collision are the rule rather than the exception, and these transitions strongly modify collision orbits, induce nonconservative motion and have molecular rather than free atomic spectral dependences. One consequence of mid-collision radiation is shown in the middle diagram of Figure 4. There an atom is photoexcited to an attractive $V^*(R)$ at ~ 500Å, but it decays back to the ground state before the pair reach the small-R region where excited-states energy transfer could have occurred. An inverse

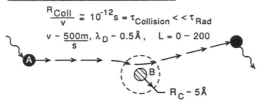

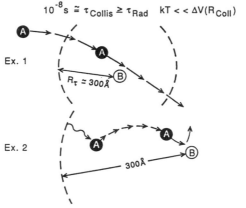

Fig. 4. Diagram of typical distance and velocity parameters, and their effects on excited-state collision orbits, in normal thermal vapors and in atom traps.

example of an optical transition in mid-collision is shown at the bottom of Figure 4. There a pair of unexcited atoms are passing by at ~ 200Å spacing until photoabsorption to an attractive excited state suddenly causes the pair to be drawn together in the very strong $V^*(R)$. In both examples, the critical distance is R_τ, the distance from which the pair are drawn into contact by $V^*(R)$ in one excited state lifetime. Excited-state energy transfer will be dominated by photoexcitation at $R \leq R_\tau$. As T decreases, the Langevin orbiting radius for an excited state collision in $V^*(R)$ will exceed R_τ. The Langevin rate will then exceed the actual collision rate because radiative decay will terminate the orbit in $V^*(R)$ before contact.

In Figure 5, I illustrate a semiclassical model for the internal dynamics of two types of $A^* - A$ energy-transfer collisions which can cause trap losses, using Na in this example. Figure 5a, taken from Ref. 6, is a traditional avoided-crossing electronic energy-transfer collision, where the Na-Na pair approach with kinetic energy E_i, are excited to V_2, an attractive $V^*(R)$, orbit together while exchanging electronic and nuclear energy, jump to the lower state while transversing R_c, and exit on V_1 with kinetic energy $KE_f = KE_i + \Delta E_J$. KE_f is shared equally by the pair, ejecting each with $\sim 12°$ K of energy. Figure 5 starts in the same way, but spontaneous emission occurs during the Na*-Na approach. The pair separate with kinetic energy $\Delta V_2 + KE_i$, where ΔV_2 is the $V_2(R)$ change shown, and if it exceeds twice the trap depth this also ejects the pair.

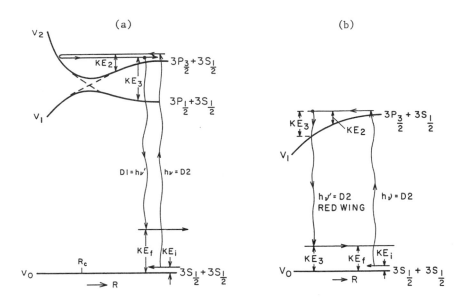

Fig. 5. (a) Diagrammatic representation of the nuclear kinetic energies and optical transitions involved in the Na, J-changing transition. (b) The same for kinetic energy gain by "radiative redistribution" or photo emission in mid-collision. Adopted from Ref. 6.

This also occurs primarily for excitation at $R < R_r$. This form of "energy-transfer" is not important in thermal collisions, but Figure 5b is also the semiclassical picture for "radiative redistribution," a familiar process in line broadening in which the absorbed and radiated photon energies are different (by ΔV_2, in the example).

Note that, since photoexcitation of a pair at $R < R_r$ dominates excited-state energy transfer, the absorption spectrum for inducing such losses is a molecular spectrum, not a free-atom spectrum. In fact, the process in Figure 5a can also be described as molecular absorption, internal conversion and predissociation, and that in 5b as a free-bound-free molecular Raman process.

I have already noted that describing the internuclear position and motion classically, as in Figures 4 and 5, must be justified as even an approximation, because the atomic deBroglie wavelengths are very large, $\sim 300\text{Å}$ for a 1 mk Na pair. Thus, elastic ground-state pair collisions are highly quantum mechanical at $T < 1$ mk, involving essentially pure s-wave scattering. In contrast, a picture of point nuclei undergoing classical motion has considerable validity for these excited state collisions, for the following reason. For a red-detuned laser wavelength, the free-bound Franck-Condon factors, combined with the natural Lorentzian shape of each molecular transition, produce excitation of a coherent superposition of bound states in $V^*(R)$. This produces a localized wavepacket of $\sim 0.1\ R_i$ width centered at R_i, as shown in Figure 6. For small T the size and subsequent motion of this excited-state wavepacket is independent of the initial, ground-state motion. This

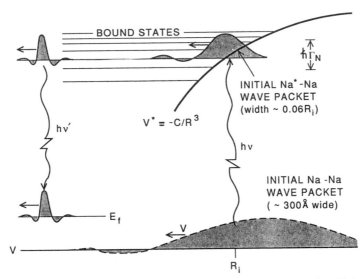

Fig. 6. Diagrammatic representation of photoexcitation to an attractive Na$_2^*$ state by red-detuned light, using a wave-packet picture of the "radiative redistribution" described classically in Figure 5.

initial excited-state wavepacket moves toward decreasing R at the classical rate, while radiatively decaying as for each state of the superposition. When the red detuning is increased until $R_i \ll R_r$ the vibrational spacing is much greater than the natural line width and primarily one bound state is excited. This excited state is then spread out from $R \cong 0$ to R_i as it decays radiatively. The classical picture remains consistent with this, since in this case many classical vibrations from $R = R_i$ to ~ 0 occur before radiative decay. Consistent with the classical approximation for nuclear motion, the classical Franck-Condon principle (CFCP) replaces this excitation of a wavepacket with excitation of point nuclei at R_i, defined by $V^*(R_i) - V(R_i) = h\nu$. These two approximations thus yield a quantitatively meaningful framework for analyzing these collisions. In the case of blue detuning, a much larger set of bound states in the attractive $V^*(R)$ are coherently excited in their far off-resonant Lorentzian wing. Without further discussion I will treat this as an incoherent superposition of classical-motion states, each starting at an R_i and decaying exponentially as the pair converge. The discrete splitting of the bound states produces a structured continuum in the red-wing absorption spectrum and complicated saturation behavior at high intensities, but the classical picture provides a valuable conceptual as well as quantitative basis for understanding the process.

For illustration I will briefly describe how Dave Pritchard and I used these two approximations to calculate the trap-loss rate $\Gamma_{coll}(I, \nu)$ due to Na*-Na collisions, where I and ν are the trap optical intensity and frequency.[6] In essence, we took

$$\Gamma_{coll} = \mathcal{V} \int_0^{\infty} \left[\frac{n^2}{2} 4\pi R_2^2 \partial R_i \right] B(I, R_i, w_L) \left[P_{\Delta J}(R_i) + P_{red}(R_i) \right] , \qquad (1)$$

where $\mathcal{V}$ is the trapped-atom volume and in the density, the first bracket is the probability of an Na pair at $R_i \rightarrow R_i + \partial R_i$ separation, $B(I, R_i, \nu)$ is the absorption rate for such

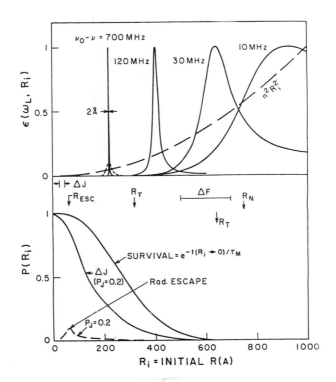

Fig. 7. Factors in the integrand of Eq. (1), versus red detuning, adapted from Ref. 6.

a pair, $P_{\Delta J}$ is the probability that an Na*-Na pair starting at rest at R_i undergo a ΔJ transition at small R, and P_{Rad} is the probability of trap loss due to radiative redistribution. For $B(I, R_i, \nu)$ we used a Lorentzian absorption line centered at $V^*(R_i) - V(R_i)$, corresponding to the CFCP, of width Γ and height $\lambda^2/2\pi$. These various factors in the integrand are shown diagrammatically in Figure 7, adapted from Ref 6.

In keeping with the deliberately simplified form of this calculation, the effects of hyperfine structure, multiple attractive Na*-Na potentials, optical saturation, bound-state spacing, and multiple excitations during a Na pair transit are ignored. That the calculation nonetheless describes the essence of this highly unusual excited-state collision process is demonstrated by the following measurement of Sesko et al.[7] Additional considerations that enter these trap-loss collisions can be found in Refs. 8-12.

Trap loss due to cold atom collisions was first reported by Prentiss et al. for Na, but using what is now recognized as highly saturated conditions,[14] such that all Na pairs that reach $\sim$ 500Å separation undergo ΔJ change. Sesko et al.[7] later measured Cs trap loss under conditions where the low-power limit is tested, and compared to the above classical calculation using Cs parameters. Their results and this comparison are shown in Figure 8, where β is a trap-loss rate coefficient per ground-state pair. In part (a) the laser is red detuned by $\Gamma/2\pi = 5$ mHz and the power is varied. At low power the effective trap depth is too shallow to confine Cs pairs that have gained 0.5 K through an $F = 4$ to

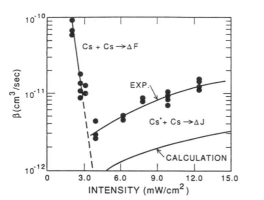 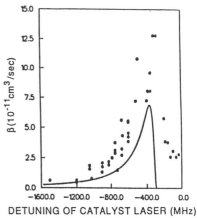

Fig. 8. Measured and calculated trap-loss rate coefficients β due to ΔJ change of Cs* in Cs pair collisions. $(dn/dt = -\beta n^2$ where n is the Cs density$)$. The catalyst laser beam intensity is 24 mW cm^{-2} in part b; it produces $n^*/n = 10^{-4}$ at $\sim$ 500 MHz detuning.

3 spin-exchange collision. This occurs because the optical trap operates on sublevels with the highest ground and excited-state F values, to avoid optical pumping to a "black" Zeeman state that does not absorb. As power increases in Figure 8a, this hyperfine loss is eliminated and the excited-state loss, proportional to I, dominates. The prediction of the theory is a factor of $\sim$4 below the data, but the excited-state hfs that was ignored in the calculation is expected to be very important for such small detuning. In order to test the theory more definitely, Sesko *et al.* next kept the trapping lasers fixed and induced additional losses with an additional "catalyst laser" beam. The results of this are compared to the calculation in Figure 8b. At the large detunings shown, hfs should be relatively unimportant; here the calculation is a factor of $\sim$2 too small. The calculation is sensitive to the probability of a ΔJ change when traversing the small R region, and uses an estimate based on room temperature alkali-pair data and adiabaticity parameters. In view of this and other approximations this level of agreement is considered excellent.

I now wish to discuss associative ionization (AI) of trapped Na, as another example of the very unique character of these very slow collisions. The effective rate coefficients and additional details of this process, formally written as Na* + Na* $\rightarrow$ Na$_2^+$ + e, has been measured by Gould *et al.*[15] I find that the actual process that dominates the formation of Na$_2^+$ is quite different than the collision of two excited Na* atoms. This is a consequence of the $\sim 10^4$ times stronger long-range attraction of an Na*-Na* pair compared to a Na* - Na* pair. This process is shown diagrammatically in Figure 9b, where it is contrasted with the normal thermal process shown in Figure 9a. Here the normal, thermal velocity process involves two free-atom excitations, followed by the third step of the Na* + Na* $\rightarrow$ Na$_2^+$ + e collision with negligible perturbation of the collision by radiation. In contrast, the trapped atom process is dominated by the four-step process in part (b). This involves (1) one photon absorption by the Na pair, (2) pair attraction in $V^*(R)$, (3) a second photon absorption to the $V^{**}(R)$ state connected to Na* + Na* at large R, (4) continued inward motion until the transition to Na$_2^+$ + e occurs at $R \sim 4$Å. The inward

Na*+Na* ASSOCIATIVE IONIZATION & ENERGY TRANSFER

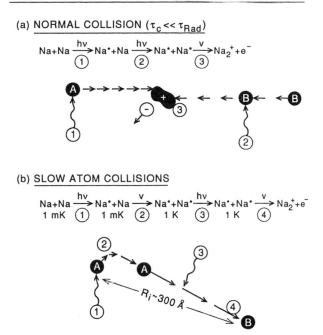

Fig. 9. Diagrammatic representation of associative ionization, as it occurs at normal temperatures in three isolated steps, and as it occurs in atom traps in four coupled steps.

velocity gained at very long range in step 2 makes this four-step process efficient for Na pairs initially within ~400Å. The Na* + Na* interaction alone will only draw Na* pairs together within an excited state lifetime if they are initially within ~50Å spacing. In addition, the four step process involves molecular absorptions, and for properly matched laser wavelengths this process is $> 10^4$ faster than the normal Langevin orbiting rate in $V^{**}(R)$.

The character of the detuning dependence of this four-step AI can be seen qualitatively in Figure 10, where the associated nuclear kinetic energies and interatomic potentials are shown diagrammatically. A 50-500 MHz red detuning (Δ_1) enhances step 1 absorption in the most effective 300-500Å region, as seen in Figure 10. A blue detuning (Δ_2) slightly greater than $|\Delta_1|$ then enhances step 3 absorption after the Na* − Na pair have gained some inward velocity.

Using the CFCP, classical nuclear motion, exponential radiative decay of Na_2^* and Na_2^{**} states, and natural-linewidth broadening of the CFCP relations $V^{**}(R_3) - V^*(R_3) = h\nu_2$, and $V^*(R_1) - V(R_1) = h\nu_1$, as in Figure 10, the four-step AI rate Γ_{AI} is given by

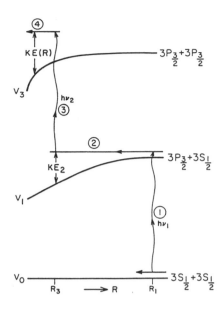

Fig. 10. The interatomic potentials and kinetic energies associated with the four-step associative-ionization process of Figure 9.

$$\Gamma_{AI}/\epsilon_{AI} = \mathcal{V} \int_0^\infty \frac{n^2}{2} 4\pi R_1^2 \, dR_1 B(I_1, \Delta_1, R_1) \int_0^{R_1} dR_3 \, \frac{e^{-\gamma t^*(R_1, R_3)}}{v(R_1, R_3)} \, B'(I_2, \Delta_2, R_3) \, e^{-\gamma t^{**}(R_1, R_3)}$$

(2)

Here B is the Na pair absorption coefficient, as previously defined, and B' is the equivalent for the $Na_2^* \to Na_2^{**}$ absorption, t^* is the time spent in step 2 moving from R_1 to R_3, and t^{**} is the time from R_3 to $R = 0$ at the velocity gained in step 2. The terms in the integrand represent, from left to right, the step 1 rate, the probability of surviving step 2 without radiative decay, the probability of step 3 within $R_3 \to R_3 + dR_3$, and the probability of surviving step 4 without radiative decay. The right side of Eq. (2) is the rate of Na_2^* pairs reaching $R \cong 0$, while the factor ϵ_{AI} on the left side is the probability that AI occurs as the Na* pair traverse the small R region; it includes the statistical probability of approaching on a molecular potential that can produce AI. This factor depends on the very complex character of the small-R molecule, and is a larger uncertainty in the calculation. The relevant potentials and kinetics have been studied by Henriet and Masnow-Seeuws,[16] and several higher-temperature experiments yield relevant results, but a discussion of these does not belong here. The predictions of Eq. (2), using a reasonable estimate of $\epsilon_{AI} \cong 0.003$, are shown in Figures 11 and 12 for the case of one laser frequency ($\Delta_1 = -\Delta_2$) and two laser frequencies as in a "catalyst laser" experiment.

In the single frequency case the laser is highly detuned from the molecular resonance in steps 1 or 3, but it induces the off-resonant transition through the far Lorentzian wing. This, of course, produces a transitory virtual state of the molecule, just as occurs

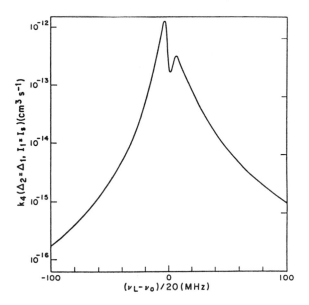

Fig. 11. Rate coefficient k_4 for Na* pairs converging to $R \sim 0$ in a trap, versus detuning of a single laser frequency for 80 mW/cm² intensity ($\Gamma_{AI}/\epsilon_{AI} = \mathcal{V}k_4 n^2/2$).

in off-resonant Raman or Rayleigh scattering, and only the long-lived final $Na_2^* + e$ state conserves energy. This aspect does not arise in the dominant regions of the two-wavelength case with red and blue detunings. The curves for this case, given in Figure 12, are normalized to unity for clarity; the peaks correspond to an AI rate coefficient per Na pair of $\sim 10^{-12}$ cm³ s⁻¹ when both laser beam intensities equal $I_s = 2\pi(h\nu)\gamma\lambda^{-2} = 80$ mW/cm², equal to the atomic two-level saturation intensity. However, if one refers the AI rate to the excited state density, as is normally done (i.e. $\Gamma_{AI} = k'_{AI}\mathcal{V}$ (Na*)²/2 rather than $k_{IA}\mathcal{V}$ (Na)²/2), then the peak of the $\Delta_1 = 100$ (2 GHz detuning) curve is an apparent rate coefficient $k'_{AI} \cong 10^{-2}$ cm³ s⁻¹. This is a rather spectacular rate coefficient at very low laser power.

It turns out to be rather difficult to measure the low-power limit of the AI rate coefficient, given above. The problem is that Na pairs are moving so slowly that they are very easily depleted by Na* − Na ΔJ change during transit of the $\sim$600Å diameter region where this occurs efficiently. The powers used so far in reported AI data are very many orders of magnitude greater than I_s,[14] causing trap loss of all Na pairs foolish enough to wander into this disaster zone; the observed AI rate then results from a small fraction of these pairs that manage AI before ΔJ change. This is evaluated quantitatively in Ref. 14.

I am grateful to Jinks Cooper and Carl Wieman for many discussions on these topics.

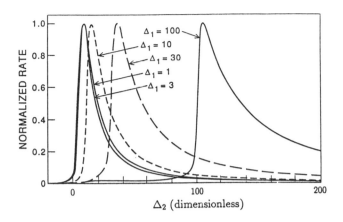

Fig. 12. Normalized Na associative ionization rate coefficients in a trap versus laser frequencies, for one red and one blue-detuned beam.

References

[1] *J. Opt. Soc. Am. B* **6** Nov. 1989 is a special issue on laser cooling

[2] Cohen-Tannoudji C N and Phillips W 1990 *Physics Today* (October) p. 33

[3] Monroe C, Swann W, Robinson H and Wieman C 1990 *Phys. Rev. Lett.* **65** 1571

[4] Raab E L, Prentiss M, Cable A, Chu S and Pritchard D E 1987 *Phys. Rev. Lett.* **59** 2631

[5] Sesko D W, Walker T G and Wieman C E 1991 *J. Opt. Soc. Am. B* **8** 946; also 1990 *Science* **248** 1076

[6] Sesko D W, 1990 "Interactions between atoms in an optical trap," (JILA Thesis #169) University of Colorado

[7] Gallagher A and Pritchard D E 1989 *Phys. Rev. Lett.* **63** 957

[8] Sesko D, Walker T, Monroe C, Gallagher A and Wieman C 1989 *Phys. Rev. Lett.* **63** 961

[9] Vigue J. 1986 *Phys. Rev. A* **34** 4476; Julienne P S, Heather R and J Vigue 1990 in *Atomic Physics* (American Institute of Physics: New York) **12**

[10] Pritchard D 1986 in *Electronic and Atomic Collisions* (North Holland: Amsterdam)

[11] Julienne P S, Pan S H, Thorsheim H R and Weiner J 1988 in *Advances in Laser Science* (American Institute of Physics: New York)

[12] Julienne P S 1988 *Phys. Rev. Lett.* **61** 698

[13] Julienne P S and Mies F H 1989 *J. Opt. Soc. Am. B* **6** 2257

[14] Prentiss M, Cable A, Bjorkholm J E, Chu S, Raab E L and Pritchard D E 1988 *Opt. Lett.* **13** 452

[15] Gallagher A *Phys. Rev.* (submitted)

[16] Gould P L, Lett P D, Julienne P S, Phillips W D, Thorsheim H R, and Weiner J 1988 *Phys. Rev. Lett.* **60** 788

[17] Dulieu O, Giusti-Suzor A and Masnou-Seeuws F 1991 *J. Phys. B: At. Mol. Opt. Phys.* (submitted); Henriet A and Masnou-Seeuws F 1987 *J. Phys. B: At. Mol. Phys.* **20**, 671

Circular and linear dichroism in atomic and molecular photoelectron emission

N A Cherepkov

Aviation Instrument Making Institute, 190000 Leningrad, USSR

ABSTRACT: Polarized atoms and oriented molecules reveal Circular Dichroism in the Angular Distribution (CDAD) and Linear Dichroism in the Angular Distribution (LDAD) of photoelectrons, which mean a difference between photoelectron currents, ejected at a definite angle by left and right circularly polarized light, or by linearly polarized light of two mutually perpendicular polarizations, respectively. Measurements of CDAD and LDAD enable one to obtain the most complete information on the properties of atoms and molecules, and to investigate polarization of atoms or orientation of molecules.

1. INTRODUCTION

To perform a "complete" quantum-mechanical experiment from which one can extract all theoretical parameters describing the process under consideration, is an ultimate goal of all experiments. In the electric-dipole approximation one needs five parameters (or even three) to characterize the atomic photoionization process, and many more in the case of molecular photoionization (Kessler 1981, Cherepkov 1979, 1983). One way to obtain necessary parameters is to measure the spin polarization of photoelectrons, in addition to the usual partial photoionization cross section and the angular asymmetry parameter measurements.

It was shown at first by Fano (1969) that photoelectrons ejected by circularly polarized light from unpolarized alkali atoms should be highly spin polarized in the vicinity of the cross section minimum near ionization threshold. It was in contradiction with the previous belief that the spin polarization of photoelectrons is pure relativistic effect which appears at rather high photon energies (Nagel 1960). The Fano effect arises due to the spin-orbit interaction in the continuous spectrum, which converts the angular momentum of the circularly polarized photon into the spin polarization of photoelectrons.

Later it was shown (Cherepkov 1983, and references therein) that due to the spin-orbit splitting of atomic levels with orbital angular momentum $l \neq 0$ photoelectrons ejected at a definite angle are highly spin-polarized for all photon energies and for any photon polarization provided photoelectrons corresponding to one of the two spin-orbit components of either initial or final state are separated. In this case the spin polarization of photoelectrons appears due to the dipole selection rules while the spin-orbit interaction enters only indirectly through the spin-orbit splitting of atomic levels. This is why the degree of spin polarization in this case is always high, while in the case of Fano effect in alkali atoms the degree of spin polarization is high near the cross section minimum only.

Measurements of spin polarization of photoelectrons performed during the last two decades in Germany (Kessler 1985, Heinzmann 1987) fully confirmed the theoretical predictions and revealed many new and interesting features. In several cases the complete experiment has been performed (Heinzmann 1980, Müller *et al.* 1990). But, unfortunately, the spin polarization measurements are quite complicated and are connected with the loss of at least three orders of magnitude in intensity. Therefore, these very elegant measurements remain up to now unique.

On the other hand, the problem of complete quantum-mechanical experiment remains very attractive. To make the experiment much more simple, one had to exclude the electron spin polarization measurement. And one way to do it was found by Lubell and Raith (1969) in the very first observation of the Fano effect in alkalies. They measured not the spin polarization of photoelectrons, but a circular dichroism of polarized alkali atoms, that is a difference between absorption of left and right circularly polarized light. It was shown also by them that both phenomena can be expressed through the same parameter. After this pioneering work the idea of using the circular dichroism for photoionization studies was forgotten for many years. And though expressions for the angular distribution of photoelectrons ejected from polarized atoms (Klar and Kleinpoppen 1982) and oriented molecules (Dill 1976) have been derived, these authors have not made any remark on the circular dichroism or any other kind of optical activity of polarized atoms and oriented molecules. Only during the last few years it was fully realized that polarized atoms and oriented molecules are optically active (Parzyński 1980, Cherepkov 1982, Dubs *et al.* 1985, 1986, Cherepkov and Kuznetsov 1989, 1991, Schönhense 1990, Chandra 1989), and investigations of this optical activity enables one to perform the complete quantum-mechanical experiment for atoms or to make an essential step towards it for molecules without measuring the spin polarization of photoelectrons.

2. ANGULAR DISTRIBUTIONS

It is well known that if atoms are unpolarized or molecules are unoriented, the angular distribution of photoelectrons is defined by one parameter, the angular asymmetry parameter β, and has the same form for both atoms and molecules

$$I_{\vec{e}_z}^i (\vec{\kappa}) = \frac{\sigma_i(\omega)}{4\pi} \left[(1 + \beta \left(\tfrac{3}{2}\cos^2\vartheta - \tfrac{1}{2}\right)\right] \tag{1}$$

where $\vec{\kappa}$ is the unit vector in the direction of photoelectron momentum, $\sigma_i(\omega)$ is a partial photoionization cross section of the i-subshell, ω is the photon energy, and light is linearly polarized along the z axis. The process under consideration is described by two vectors, one of which is used to define a coordinate frame, while the other one remains free. Therefore, the angular distribution (1) depends on only one angle ϑ between these two vectors and possesses cylindrical symmetry. The coordinate frame is usually connected with the photon beam, so that its z axis coincides with the polarization vector $\vec{e}$ for linearly polarized light or with the direction of the photon beam (which will be characterized in the following by the unit vector $\vec{q}$) for circularly polarized and unpolarized light.

Consider now the cases when atoms or molecules in the initial state do not possess spherical symmetry and can be characterized by some vector. For atoms it means that there is some difference in population of substates with different projections M of the total angular momentum $J \neq 0$ on a given direction $\vec{n}$. In this case atoms are said to be polarized (aligned if states with projections M and −M are equally populated, and oriented otherwise). Atoms can be polarized by optical pumping, by photon excitation, by scattering processes. Atoms adsorbed at a surface are also polarized.

For molecules there are different ways to break the spherical symmetry of the initial state. Rotating molecules can be excited by polarized light and made aligned or oriented, like atoms, depending on light polarization. In this case there will be nonequal population of states with different projections of the total angular momentum including rotation. The other way is to stop rotation of molecules. For example, molecules adsorbed at a surface, or molecules in liquid crystals, are not rotating but are fixed in space. Molecules can be oriented by an external field (Harren *et al.* 1991). It is possible also to select the photoionization processes of oriented-in-space molecules in a gas phase by coincidence measurements of fragment ions and photoelectrons, provided the molecular ion decays rather fast after ionization (Golovin *et al.* 1990). Fixed-in-space molecules, depending on their structure, can be characterized by one, two or even three vectors. In the following, both fixed-in-space and aligned or oriented molecules will be called oriented for brevity.

So, polarized atoms or oriented molecules are characterized by (at least) one additional vector $\vec{n}$, and the angular distribution now will be defined by three independent vectors. Suppose that the laboratory frame is defined by the molecular orientation or by atomic polarization. Then the angular distribution of photoelectrons will depend on the spherical angles of two vectors, $\vec{e}$ (for linearly polarized light) or $\vec{q}$ (for circularly polarized light) and $\vec{\kappa} = \vec{p}/p$, where $\vec{p}$ is the photoelectron momentum, in this frame. It becomes essentially three-dimensional and can be presented as a double expansion over spherical functions (Dill 1976)

$$I_m^i (\vec{\kappa}, \vec{e}) = \sqrt{3}\, \sigma_i (\omega) (-1)^{1-m} \sum_{LM}^{2\ell_{max}} \sum_{JM_J}^{2} \begin{pmatrix} 1 & 1 & J \\ -m & m & 0 \end{pmatrix} A_{LM}^{JM_J} \cdot Y_{LM} (\hat{\kappa})\, Y_{JM_J} (\hat{e}) \qquad (2)$$

where m = 0 for linearly polarized light and m = ± 1 for circularly polarized light. In the last case the vector $\vec{e}$ should be substituted by $\vec{q}$. The parameters $A_{LM}^{JM_J}$ are expressed through the dipole matrix elements (Cherepkov and Kuznetsov 1987) and contain the dynamical information on the process, as does the angular asymmetry parameter β. They are normalized by the condition $A_{00}^{00} = 1$.

The number of terms in (2) for molecules is, in principle, infinite, and can be restricted by the number of partial waves (l_{max}) retained in the expansion of the photoelectron wave function. In the case of atoms, expansion (2) is restricted by the highest orbital angular momentum of photoelectron l_{max}. But even for p-subshells of atoms the number of terms in (2) can be as high as nine, and even more for molecules. To extract so many parameters from one measured curve is quite problematic.

Therefore, though the angular distributions of photoelectrons ejected from polarized atoms or oriented molecules contain rather exhaustive information on atoms or molecules, their practical use is restricted by the difficulty of extracting this information from experimental data.

3. QUALITATIVE CONSIDERATION OF CDAD AND LDAD

In classical optics (Barron 1982) Circular Dichroism (CD) is defined as a difference in absorption of the left and right circularly polarized light, and Linear Dichroism (LD) is defined as a difference in absorption of linearly polarized light of two mutually perpendicular polarizations. By analogy, a difference between photoelectron fluxes ejected at a definite angle by left and right circularly polarized light will be called Circular Dichroism in the Angular Distribution of photoelectrons (CDAD) (Ritchie 1975). An analogous difference for two orthogonal linear polarizations will be called Linear Dichroism in the Angular Distribution (LDAD).

Consider now qualitatively the origin of CDAD. Suppose for simplicity that the initial state of a polarized atom or oriented molecule is characterized by a wave function which can be expressed through the spherical function with a given l

$$\Psi_i = \Psi_{ilm}(\hat{\vec{r}}) \sim R_{il}(r) \, Y_{lm}(\hat{\vec{r}}) \tag{3}$$

The photoelectron wave function $\Psi_{\vec{p}}^{-}(\hat{\vec{r}})$, which contains in the asymptotic region the superposition of a plane wave propagating in the direction of the electron momentum $\vec{p}$, and a converging spherical wave, is presented as a partial wave expansion

$$\Psi_{\vec{p}}^{-}(\vec{r}) \sim \sum_{l'm'} Y_{l'm'}(\hat{\vec{r}}) \, Y_{l'm'}^{*}(\hat{\vec{\kappa}}) \, e^{-i\delta_{l'm'}} \tag{4}$$

Suppose also that the light beam is propagating along the x axis of our laboratory frame defined by atomic polarization or molecular orientation. Then the dipole operator for left and right circularly polarized light is given by

$$\hat{d}_{\pm 1} = (\vec{e}_z \pm i\vec{e}_y)\,\vec{r} = z \pm iy = \sqrt{\tfrac{4\pi}{3}} \left\{ Y_{10}(\hat{r}) \pm \tfrac{1}{\sqrt{2}}[Y_{11}(\hat{r}) + Y_{1-1}(\hat{r})]\right\} \quad (5)$$

With these assumptions, we will find for the dipole matrix element after integration over spherical angles

$$\langle \Psi_{\vec{p}}^- | \hat{d}_{\pm 1} | \Psi_{i\ell m}\rangle \sim \sum_{\ell' m'} (-1)^{m'} e^{i\delta \ell' m'} \left[\begin{pmatrix} \ell' & 1 & \ell \\ -m'0 & m \end{pmatrix} \pm \tfrac{1}{\sqrt{2}} \begin{pmatrix} \ell' & 1 & \ell \\ -m' & 1 & m \end{pmatrix} \pm \tfrac{1}{\sqrt{2}} \begin{pmatrix} \ell' & 1 & \ell \\ -m' & -1 & m \end{pmatrix} \right]$$

$$Y_{\ell' m'}(\hat{\kappa}) \langle p\ell' \| d \| i\ell\rangle \quad (6)$$

The angular distribution of photoelectrons is proportional to the square modulus of this matrix element. The key point is that in (6) there are both terms having the same sign for two circular polarizations, and terms having opposite signs. Due to that, the angular distributions for two circular polarizations will also contain a group of terms having the same sign and a group of terms having opposite signs. The latter terms will contribute to CDAD, while the former will cancel. In the case of unpolarized atoms or unoriented molecules the laboratory frame can be always taken to be coincident with the photon frame, in which the dipole operator for left and right circularly polarized light is proportional to only one spherical function, $Y_{11}(\hat{r})$ or $Y_{1-1}(\hat{r})$, respectively, and CDAD does not appear.

If states with different m are degenerate and equally populated, the angular distribution has to be averaged over m

$$I_{\pm 1}^i(\vec{\kappa},\vec{q}) \sim \frac{1}{(2\ell+1)} \sum_m |\langle \Psi_{\vec{p}}^- | \hat{d}_{\pm 1} | \Psi_{i\ell m}\rangle|^2 \quad (7)$$

Using equation (6) and the condition that the dependence on m is given solely by the spherical function $Y_{\ell m}(\hat{r})$, one can show that terms which have different signs for two polarizations disappear from (7). And vice versa, if there is any unequivalence between states with different projections m, these terms will give nonzero contribution and will cause appearance of CDAD. Unequivalence between states with different m can be connected with i) unequal population of states with different m (alignment and orientation of atoms and molecules) and with ii) a dependence of a radial part of wave function in our laboratory frame on the projection m, as it takes place when molecules are fixed in space, and states with different m are not degenerate.

Since in this consideration nothing has been used but the dipole selection rules (given by 3j-symbols in (6)), CDAD is of the order of unity, that is, of the same order of magnitude as the differential cross section itself for the same angle.

Consider now LDAD. Here it is convenient to start derivation in the photon frame with the z' axis directed along the photon beam. Then the dipole operators for two orthogonal linear polarizations along the x' and y' axes are given by

$$(\vec{e}_x \vec{r}') = \sqrt{\frac{2\pi}{3}}\, r'\, [Y_{1-1}(\hat{r}') - Y_{11}(\hat{r}')]\,, \quad (\vec{e}_y \vec{r}') = \sqrt{\frac{2\pi}{3}}\, r'\, [Y_{1-1}(\hat{r}') + Y_{11}(\hat{r}')] \quad (8)$$

Again, there are both terms of the same sign and terms of opposite signs for two polarizations. The latter lead to the appearance of LDAD. But contrary to CDAD, LDAD is different from zero even for unpolarized atoms and unoriented molecules. It follows from the fact that equations (8) are written in the photon frame not connected with atomic polarization or molecular orientation. It is also evident from comparison of the angular distributions (1) for two perpendicular polarizations. Equation (1) is written in the coordinate frame with the z axis directed along the polarization vector $\vec{e}$. Transforming it to the coordinate frame with the z' axis directed along the photon beam, and x' and y' axes directed along the polarization vectors $\vec{e}_x$ and $\vec{e}_y$, respectively, we find

$$I^i_{LDAD}(\vec{\kappa}) = I^i_{\vec{e}_y}(\vec{\kappa}) - I^i_{\vec{e}_x}(\vec{\kappa}) = -\frac{3\sigma_i(\omega)}{8\pi}\beta\,\sin^2\theta'\cos 2\phi \quad (9)$$

Corresponding angular distributions in the plane perpendicular to the photon beam are presented in Figure 1 for a particular case of $\beta = 2$.

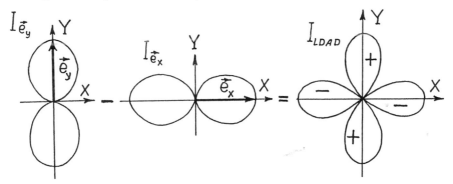

Fig. 1. The angular distributions (1) for two orthogonal linear polarizations, and LDAD angular dependence (9) for unpolarized atoms and unoriented molecules, $\beta = 2$, $\theta' = \pi/2$.

After integration over electron ejection angles one finds that polarized atoms and oriented molecules reveal also CD and LD, that is, they are optically active (Cherepkov and Kuznetsov 1987, 1989; Cherepkov and Schönhense 1991). To be more exact, CD is proportional to the orientation vector and LD is proportional to the alignment tensor (Blum 1981) of aligned or oriented atoms and molecules. All fixed-in-space molecules reveal LD, and only those with rather low symmetry reveal CD.

The usual optical activity of unoriented chiral molecules, and CD in particular, is caused by a dissymmetry in their structure. Namely, chiral molecules possess neither plane of symmetry nor centre of inversion. Like circularly polarized light, they can be presented as a helix. The absorption of light is different depending on whether the screw senses of these two helices are the same or the opposite. This leads to the appearance of CD in photoabsorption by chiral molecules.

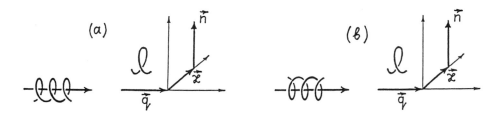

Fig. 2. Absorption of light depends on whether the screw senses of light and experimental arrangement are the same or opposite.

CDAD from oriented molecules appears due to dissymmetry in the geometry of the experiment. Three vectors describing the experiment, $\vec{n}$, $\vec{\kappa}$, and $\vec{q}$, can form a basis for a left or right coordinate system provided they are noncoplanar. Suppose that these vectors are mutually perpendicular. Then they can be arranged in a way shown in Figure 2, forming a part of a left-handed (as in Figure 2) or right-handed helix. Again, absorption of light will be different depending on whether the screw senses of light and of the experimental arrangement are the same or opposite. As a result, CDAD will appear.

It is also important to mention that the usual optical activity of unoriented chiral molecules is given by the electric dipole – magnetic dipole interference terms, which are α times smaller than the pure electric dipole terms, responsible for the effects considered here.

4. NUMERICAL EXAMPLES

To demonstrate the general properties of CDAD and LDAD, consider some model case of a diatomic molecule, for which we accept the following simplifying assumptions. i) An initial σ

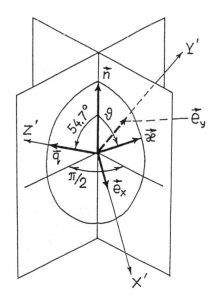

Fig.3. Geometry of experiments for which numerical results are presented in Figs.4-6.

state is ionized to only $s\sigma$, $p\sigma$, and $p\pi$ continuum states; ii) the dipole matrix elements for the $p\sigma$ and $p\pi$ transitions are equal to 1, and two times smaller for the $s\sigma$ transition; iii) the phase shifts are $\delta_{p\sigma} - \delta_{s\sigma} = \frac{\pi}{4}$, $\delta_{p\pi} - \delta_{p\sigma} = \frac{\pi}{18}$. Under these assumptions we have in (2) only terms with $L \leq 2$. For linear molecules the following conditions are fulfilled (Cherepkov and Kuznetsov 1987): $A_{LM}^{JM_J} = (-1)^J A_{L-M}^{J-M_J}$, and $M + M_J = 0$, so that the total number of parameters in (2) is reduced to ten (and only four of them are independent in our model case). To reduce further the number of parameters, consider the particular geometry shown in Figure 3. If we introduce the spherical angles ϑ, ϕ and ϑ_q, ϕ_q of the vectors $\vec{\kappa}$ and $\vec{q}$, respectively, in the laboratory frame, then this geometry will correspond to $\vartheta_q = 54.7°$ (so that $P_2 (\cos\vartheta_q) = 0$), $\phi_q = 0$, $0 \leq \vartheta \leq \pi$, and $\phi = 90°$ or $270°$. Under these conditions only two terms contribute to CDAD in heteronuclear molecules,

$$I_{CDAD}^i(\vec{\kappa}, \vec{q}) = I_{+1}^i(\vec{\kappa},\vec{q}) - I_{-1}^i(\vec{\kappa},\vec{q}) = \frac{\sigma_i(\omega)}{4\pi} 2i \left[\sqrt{2} A_{11}^{1-1}\sin\vartheta + \sqrt{15} A_{21}^{1-1} \sin\vartheta \cos\vartheta\right] \quad (10)$$

while the angular distribution for the same geometry is defined by five parameters

$$I_{\pm1}^i(\vec{\kappa},\vec{q}) = \frac{\sigma_i(\omega)}{4\pi} \left[1 + \sqrt{3} A_{10}^{00} \cos\vartheta + \sqrt{5} A_{20}^{00} P_2(\cos\vartheta) \pm \sqrt{2} i A_{11}^{1-1}\sin\vartheta \right.$$

$$\left. \pm \sqrt{15} i A_{21}^{1-1} \sin\vartheta\cos\vartheta - \frac{\sqrt{5}}{2\sqrt{2}} A_{22}^{2-2}\sin^2\vartheta\right]. \quad (11)$$

Figure 4 shows the results of our model calculations for this particular case. Evidently, it is much easier to extract two parameters from the CDAD curve in Figure 4b, than five parameters from the angular distribution curves in Figure 4a.

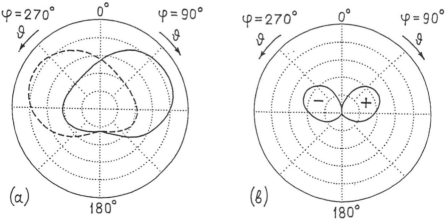

(a)

(b)

Fig. 4. Numerical results for the model case of an oriented heteronuclear diatomic molecule for the geometry of experiment presented in Fig. 3. (a) The angular distributions (11) $I_{+1}^i (\theta)$ (full curve) and $I_{-1}^i (\theta)$ (dashed curve). (b) CDAD (10) for the same case.

For homonuclear molecules the parameters $A_{LM}^{JM_J}$ with odd L for symmetry reasons are zero, and Equations (10) and (11) are simplified. Figure 5 shows the results of our model calculation for the case of a homonuclear molecule.

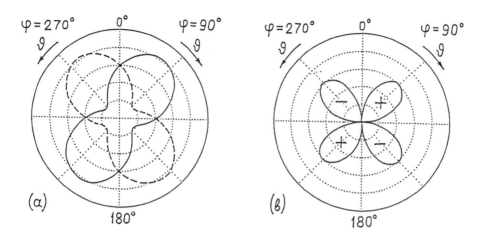

Fig. 5. The same as in Fig. 4, for a homonuclear diatomic molecule.

Consider now LDAD for the same geometry shown in Figure 3 when the y' axis is obtained by rotation on 45° around the z' axis from the initial position coincident with the y axis. In other words, primed coordinate system is obtained from unprimed one by rotation defined by the Euler angles $\alpha = 0$, $\beta = 54.7°$, $\gamma = 45°$. Figure 6 shows the results of our model calculations for LDAD (Cherepkov and Schönhense 1991) for the case of heteronuclear molecules.

5. WHAT CAN BE LEARNED FROM CDAD AND LDAD MEASUREMENTS

Compare now the kinds of information which can be extracted from the CDAD and LDAD measurements.

i) Neither CDAD nor LDAD depends essentially on the spin-orbit interaction (except for the case of atomic s-subshells), and if the fine-structure splitting of initial or final state is not resolved, both CDAD and LDAD remain at the same value, as does the angular asymmetry parameter β. This is in contrast to the spin polarization of photoelectrons, which is of the order $(\alpha Z)^2$ if the fine-structure splitting is not resolved (Cherepkov 1983). Therefore CDAD and LDAD can also be studied in light atoms and molecules where the photoelectron polarization measurements are practically impossible due to a very small spin-orbit splitting.

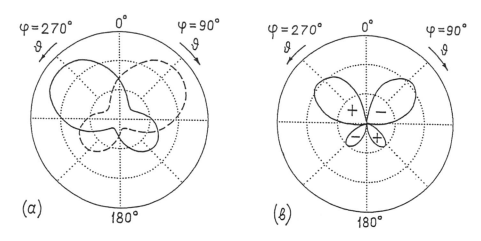

Fig. 6. LDAD for the model case of an oriented heteronuclear diatomic molecule and for the geometry of experiment presented in Fig. 3. (a) $I^i_{\vec{e}_x}(\vartheta)$ (dashed curve) and $I^i_{\vec{e}_y}(\vartheta)$ (full curve); (b) LDAD as a function of the angle ϑ.

ii) CDAD and LDAD are defined by the parameters $A^{JM_J}_{LM}$ with $J = 1$ and $J = 2$, respectively. Therefore CDAD and LDAD measurements supplement each other when one needs to extract the dipole matrix elements and phase shifts from the measured quantities.

iii) In principle, LDAD measurements can give more information than CDAD measurements since the general expression for LDAD contains more terms (Cherepkov and Schönhense 1991). CDAD is defined by the parameters which contain only interference terms between degenerate photoelectron continua differing by ±1 in the m values. For example, in photoionization from a σ orbital, CDAD will be due solely to the interference between the σ and π continua (Dubs *et al* 1985). LDAD is described by the parameters which contain interference terms between continua differing by $\Delta m = 0, \pm 1$ and ± 2.

Choosing a proper geometry, one can exclude the contribution of some parameters from LDAD, so that the number of parameters to be extracted from a particular experiment can be essentially reduced. Changing the geometry, one can select the parameters to be defined. Performing successively a series of measurements of CD, LD, CDAD, LDAD and of the angular distributions, one can extract all essential parameters characterizing the process provided only few parameters are extracted from each measurement. In the model case considered above, it is easy to find a sequence of measurements which enables one to extract all ten

Table 1. Possible set of experiments from which ten parameters $A^{JM_J}_{LM}$ with $L \leq 2$ can be extracted (marked by sign +).

PARAMETER	KIND OF MEASUREMENTS				
	CDAD eq. (10)	LD	LDAD case 1[a]	LDAD case 2[b]	$I^i_{\pm 1}$ eq. (11)
$A^{0\,0}_{1\,0}$	−	−	−	−	+
$A^{0\,0}_{2\,0}$	−	−	−	−	+
$A^{2\,0}_{0\,0}$	−	+	−	+	−
$A^{2\,0}_{1\,0}$	−	−	−	+	−
$A^{2\,0}_{2\,0}$	−	−	−	+	−
$A^{1-1}_{1\,1}$	+	−	−	−	+
$A^{1-1}_{2\,1}$	+	−	−	−	+
$A^{2-1}_{1\,1}$	−	−	+	−	−
$A^{2-1}_{2\,1}$	−	−	+	−	−
$A^{2-2}_{2\,2}$	−	−	−	+	+

[a] The geometry of the experiment is defined in Fig. 3.
[b] The geometry of the experiment is defined by the Euler angles $\alpha = 0$, $\beta = 45°$, $\gamma = 0$, and $0 \leq \vartheta \leq \pi$, $\phi = 90°$ or $270°$.

parameters provided no more than three parameters are extracted from each measured curve (Cherepkov and Schönhense 1991). Table 1 illustrates such a sequence for one particular set of experiments.

Another direction of possible use of CDAD and LDAD measurements is connected with determination of a direction and degree of atomic polarization and molecular orientation. Indeed, CDAD is different from zero only if atoms are polarized or molecules are oriented. Moreover, CDAD had characteristic zeroes, for example, when three vectors $\vec{\kappa}$, $\vec{n}$, and $\vec{q}$ are coplanar, from which one can define the direction of molecular orientation or atomic polarization. LDAD measurements can also be used for the same purpose, but here there is also the contribution given by equation (9) which appears even if atoms are unpolarized or molecules are unoriented. One practical way to exclude it is to perform an experiment with fixed relative orientation of three vectors $\vec{e}_x$, $\vec{e}_y$, and $\vec{\kappa}$, and varying only the direction of molecular orientation or atomic polarization $\vec{n}$ relative to them. In this case the contribution of (9) is constant, and every variation of LDAD is connected with the existence of molecular orientation or atomic polarization. And if $\vec{\kappa}$ is varied in the plane containing the vector $\vec{q}$ and forming an angle $\pm\frac{\pi}{4}$ with the vectors $\vec{e}_x$ and $\vec{e}_y$, the contribution of (9) turns out to be zero. Then any nonzero result for LDAD is caused by molecular orientation or atomic polarization, as in the case of CDAD.

Finally, it is worth while to note that CD and LD measurements in two-colour experiments with lasers can also be used to extract additional information on atoms and molecules. Here the first laser creates aligned or oriented atoms or molecules in a discrete excited state, while the second laser ionizes them and serves to probe CD or LD, by comparing, for example, the ionic currents for two orthogonal polarizations of light. The angular dependence of CD and LD is obtained by varying the angle between two laser beams or by varying the direction of linear polarization of laser beam(s). One or two parameters can be extracted for atoms even from this simple experiment. For experimentalists it is much easier to perform relative measurements; therefore, we are giving below the normalized expressions for LD and CD in atoms

$$A_{LD}^j = \frac{I_{\vec{e}_y}^j(\vec{n}) - I_{\vec{e}_x}^j(\vec{n})}{I_{\vec{e}_y}^j(\vec{n}) + I_{\vec{e}_x}^j(\vec{n})} = \frac{3}{2\sqrt{2}} \frac{\rho_{20}^n C_{202}^j [(\vec{n}\vec{e}_x)^2 - (\vec{n}\vec{e}_y)^2]}{\rho_{00}^n + \frac{1}{\sqrt{2}} C_{202}^j \rho_{20}^n P_2(\widehat{\vec{n}\vec{q}})} \tag{12}$$

$$A_{CD}^j = \frac{I_{+1}^j(\vec{n}) - I_{-1}^j(\vec{n})}{I_{+1}^j(\vec{n}) - I_{-1}^j(\vec{n})} = -\sqrt{\frac{3}{2}} \frac{\rho_{10}^n C_{101}^j (\vec{n}\vec{q})}{\rho_{00}^n + \frac{1}{\sqrt{2}} C_{202}^j \rho_{20}^n P_2(\widehat{\vec{n}\vec{q}})} \tag{13}$$

where the parameters C_{kLN}^j have been defined by Cherepkov and Kuznetsov (1989) and ρ_{N0}^n are the state multipoles (Blum 1981) characterizing the polarization of a resonantly excited state.

6. SURVEY OF EXPERIMENTAL RESULTS

The first theoretical prediction of CDAD for polarized alkali atoms in the ground state was made by Parzyński (1980). Like the spin polarization of photoelectrons, CDAD in this particular case is a consequence of the spin-orbit interaction. CDAD from aligned atoms was predicted by Dubs *et al.* (1986), and Cherepkov and Kuznetsov (1989) derived a general expression for CDAD for both aligned and oriented atoms. CDAD for space-fixed molecules in the electric-dipole approximation was predicted at first by Cherepkov (1982), and for the case of aligned molecules was derived by Dubs *et al.* (1986). Existence of CD for space-fixed molecules was mentioned by Cherepkov and Kuznetsov (1987). Since these theoretical predictions, several experiments have been performed which illustrate that CDAD measurements are indeed relatively simple and can serve for different purposes. Up to now, no experimental observations of LD or LDAD for polarized atoms or oriented molecules have been reported.

The first measurements of CDAD have been performed by Appling *et al.* (1986, 1987), who used CDAD to probe the alignment of NO molecules excited to the $A^2\Sigma^+$ state by linearly polarized light. In these experiments two counterpropagating laser beams crossed the molecular beam, which entered the interaction region at right angles to both the propagation direction of the lasers and the detector axis of the electron spectrometer. The angular dependence of CDAD was obtained by rotating the linear polarization vector $\vec{e}$ of the pump laser beam. These experiments for the first time demonstrated that CDAD is not a small effect, and that it can be successfully used to detect molecular alignment.

Winniczek *et al.* (1989) used this technique to probe the alignment of ground state NO fragments produced by the UV photodissociation of methyl nitrite, CH_3ONO. These measurements have shown the utility of the CDAD method for probing chemical processes in which alignment of fragments plays an important role.

CDAD for space-fixed molecules have been observed for the first time by Westphal *et al.* (1989, 1991). They investigated both diatomic (CO, NO) and polyatomic (benzene, CH_3I) molecules, oriented by adsorption on Pd(111) or graphite (0001) surface. Adsorption on a single-crystal surface yields a perfectly oriented ensemble of molecules with a rather high target density, about two orders of magnitude larger than in the gas phase, resulting in a huge photoelectron intensities. As a source of circularly polarized light they used synchrotron radiation emitted above or below the storage-ring plane. A disadvantage of these experiments is connected with the fact that photoelectrons are ejected into a hemisphere only, which is further restricted by the systematic effect of refraction of photoelectrons due to the electrostatic surface

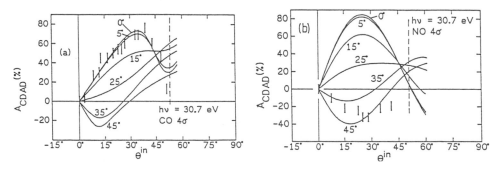

Fig. 7. Normalized CDAD spectra for the 4σ levels of CO (a) and NO (b) molecules, calculated for various bent orientations, and measured experimentally by Westphal *et al.* (1991) for the geometry of experiment shown in Fig. 3.

barrier. Therefore, the grazing emission parallel to the surface is equivalent to the emission angle $\theta^{in} < 90°$ (shown by dotted line in Figure 7) for the case of free molecules. For reasons of experiment, they measured normalized CDAD value A_{CDAD} which is a ratio of the difference to the sum of intensities, like it was used in (12), (13) for the case of CD and LD.

Westphal *et al.* (1991) compared experimental results for diatomic molecules CO and NO with calculations performed using Hartree-Fock orbitals for both the bound levels and the photoelectron. Generally the agreement between the calculated and measured CDAD values is quite encouraging, particularly in view of the very simple model of an isolated molecule assumed in their calculation. Westphal *et al.* (1991) investigated also a practical question of how sensitive are CDAD spectra to the tilt of the molecule. Figure 7 shows CDAD spectra for various angles of tilt calculated for 4σ photoemission of CO and NO molecules, as a function of the angle θ^{in} within the surface barrier, which corresponds to a free molecule fixed in space. For each angle of tilt the spectra correspond to an orientational average around the surface normal. The 4σ level in these molecules is only weakly influenced by the substrate, therefore calculations performed for free molecules are expected to be reliable. Results of calculations show that the CDAD spectra change significantly with increasing angle of tilt and even change sign. Comparison of these calculations with experimental data shows that CO molecules are adsorbed normal to the surface, while NO molecules have an average bent orientation between 35° and 45°. These results are consistent with earlier angle-resolved photoemission studies of Miazaki *et al.* (1987), and demonstrate the potential of CDAD measurements as a probe of adsorbate orientation.

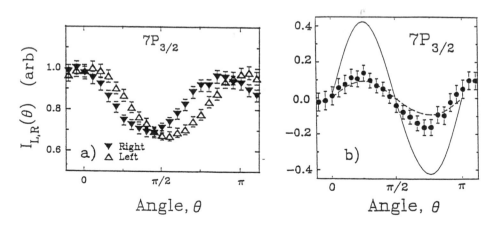

Fig. 8. CDAD for the $7P_{3/2}$ level of Cs (Cuéllar *et al.* 1991). (a) Photoelectron angular distributions for left (L) and right (R) circularly polarized light. (b) Experimental normalized CDAD signal (points) and theoretical calculations without (full curve) and with (dashed curve) hyperfine corrections.

Analogous measurements for benzene molecules have shown a substantial difference between CDAD spectra for monolayer and multilayer coverages. From these data Westphal *et al.* (1991) made a conclusion that for coverages above a monolayer not all molecules are lying flat on the surface, but some are strongly tilted or even standing upright.

And at last quite recently, the first measurements of CDAD in atoms have been performed by Cuéllar *et al.* (1991). They used CDAD spectra to examine the alignment of $7P_{3/2}$ and $7P_{1/2}$ levels of cesium, prepared by absorption of linearly polarized laser radiation. In accord with theory (Dubs *et al.* 1986), the $7P_{1/2}$ level could not be aligned and therefore exhibits no CDAD, whereas the $7P_{3/2}$ level shows a significant CDAD signal, although smaller than that predicted by theory (see Figure 8). The authors attribute this difference between experiment and theory to m_J mixing due to the hyperfine coupling during the finite ionization time of the laser pulse.

7. CONCLUSION

We have demonstrated that new observable quantities, CDAD and LDAD, can be measured with aligned or oriented atoms and molecules, and with space-fixed molecules. Investigations of CDAD and LDAD enable one to get rid of many terms giving contribution to the usual angular distributions, and to specify the kind of information extracted from the experiment. Changing the kind of measurement (CDAD or LDAD, CD or LD) and its geometry, it is possible to extract rather many parameters from a set of measurements and either to perform the complete quantum-mechanical experiment (in atoms), or to make an essential step towards it. CDAD measurements have been proposed to use in search of Cooper minima in molecules (Rudolph *et al.* 1990). The other purpose to use CDAD and LDAD measurements is to probe atomic polarization or molecular orientation, produced by any method, for example, by adsorption on a surface, by scattering processes, by chemical reactions, by external fields, by photon excitation, optical pumping and so on. Therefore CDAD and LDAD measurements have a vast variety of applications. With lasers or synchrotron radiation as a source of light, these experiments do not differ much from experiments currently performed in many laboratories. From a theoretical point of view, these phenomena are generally well understood, but there are only few numerical calculations for particular targets. Therefore, further investigations for different targets are desirable.

8. ACKNOWLEDGMENT

The author greatly appreciates the hospitality of the University of Nebraska, Lincoln, NE, where this talk has been prepared.

REFERENCES

Appling J R, White M G, Dubs R L, Dixit S N and McKoy V 1987 *J. Chem. Phys.* **87** 6927
Appling J R, White M G, Orlando T M and Anderson S L 1986 *J. Chem. Phys.* **85** 6803
Barron L D 1982 *Molecular Light Scattering and Optical Activity* (Cambridge: Cambridge Univ. Press)
Blum K 1981 *Density Matrix Theory and Applications* (New York: Plenum)
Chandra N 1989 *Phys. Rev. A* **39** 2256
Cherepkov N A 1979 *J. Phys. B* **12** 1279
Cherepkov N A 1982 *Chem. Phys. Lett.* **87** 344
Cherepkov N A 1983 *Adv. At. Mol. Phys.* **19** 395
Cherepkov N A and Kuznetsov V V 1987 *Z. Phys. D* **7** 271
Cherepkov N A and Kuznetsov V V 1989 *J. Phys. B* **22** L405
Cherepkov N A and Kuznetsov V V 1991 *J. Chem. Phys.* in press
Cherepkov N A and Schönhense G 1991 to be published
Cuéllar L E, Feigerle C S, Carman H S Jr and Compton R N 1991 *Phys. Rev. A* **43** 6437
Dill D 1976 *J. Chem. Phys.* **65** 1130
Dubs R L, Dixit S N and McKoy V 1985 *Phys. Rev. Lett.* **54** 1249
Dubs R L, Dixit S N and McKoy V 1986 *J. Chem. Phys.* **85** 656
Fano U 1969 *Phys. Rev.* **178** 131
Golovin A V, Kuznetsov V V and Cherepkov N A 1990 *Sov. Tech. Phys. Lett.* **16** 363
Harren F, Parker D H and Stolte S 1991 *Comments At. Mol. Phys.* **26** 109
Heinzmann U 1980 *J. Phys. B* **13** 4353, 4367
Heinzmann U 1987 *Phys. Scripta* **T17** 77
Kessler J 1981 *Comments At. Mol. Phys.* **10** 47
Kessler J 1985 *Polarized Electrons* (Berlin: Springer)
Klar H and Kleinpoppen H 1982 *J. Phys. B* **15** 933
Lubell M S and Raith W 1969 *Phys. Rev. Lett.* **23** 211
Miyazaki E, Kojima I, Orita M, Sawa K, Sanada N, Edamoto K, Miyahara T and Kato H 1987 *J. Electr. Spectr. Rel. Phen.* **43** 139
Müller M, Böwering N, Schäfers F and Heinzmann U 1990 *Phys. Scripta* **41** 42
Nagel B C H 1960 *Ark. Fys.* **18** 1
Parzyński R 1980 *Acta Phys. Polonica* **A57** 49
Ritchie B 1975 *Phys. Rev. A* **12** 567
Rudolph H, Dubs R L and McKoy V 1990 *J. Chem. Phys.* **93** 7513
Schönhense G 1990 *Phys. Scripta* **T31** 255
Westphal C, Bansmann J, Getzlaff M and Schönhense G 1989 *Phys. Rev. Lett.* **63** 151
Westphal C, Bansmann J, Getzlaff M, Schönhense G, Cherepkov N A, Braunstein M, McKoy V and Dubs R L 1991 *Surf. Science*, in press
Winniczek J W, Dubs R L, Appling J R, McKoy V and White M G 1989 *J. Chem. Phys.* **90** 949

Half-collisions in intense laser fields

Hanspeter Helm

Molecular Physics Laboratory, SRI International, Menlo Park, Ca 94025

ABSTRACT

Recent progress in the understanding of the response of atoms and molecules that are exposed to intense laser radiation is discussed. The modification of the ionization and dissociation dynamics by intense short laser pulses can readily be studied experimentally and provides tutorial insight into properties of collisions in strong electromagnetic fields.

1. INTRODUCTION

The term half-collision was coined by Durup (1979) to describe the intimate connection between a photodissociation process where two atomic particles are formed and a collision between the very same particles at an energy corresponding to that released in the dissociation process. The concept of a half-collision is also applicable to photoionization and photodetachment reactions. The location and shape of resonances in the photoexcitation cross sections for these three processes are related to resonances that appear in the scattering of the products. Hence if one knows the intermediate prepared in photoexcitation, when the products are still at short distance from each other, and determines the angular and energy distribution of the products, one in fact maps out the course of one-half of a collision. A collision of only a select set of partial waves because the intermediate is defined by photoexcitation in only a restricted set of quantum numbers.

In these experiments the photon is used as a gentle source of energy, employed merely to prepare the intermediate from which the half-collision proceeds. Precise definition of the intermediate is achieved with narrow-band radiation and the intensity is kept low so as not to modify events that pertain to the collision part. In other words the rate of photoexcitation is kept low compared to the rate at which the half-collision proceeds.

Over the past few years experiments have been carried out where these conditions have not been met, but rather intense lasers have been used to initiate photoionization and photodissociation using multiphoton excitation. Multiphoton processes can proceed at already low intensity when the wavelength employed excites multiphoton resonances. However, the subject of this review is the case of excitation in strong fields which is somewhat arbitrarily defined here as the range from 10^{13} to 10^{14} W/cm^2 at wavelengths which are at zero intensity not resonant for any specific excitation path.

A most important effect in the high intensity regime is the modification of excited state structure due to the ac-Stark shift which is induced by the electromagnetic field. This modification governs the path and the rates of photoabsorption and pertains to the dynamic part of dissociation into separated particles. As a consequence, results obtained in such studies are directly related to collisions in strong fields.

2. PONDEROMOTIVE EFFECTS

Various aspects of the modification of the electronic state structure in electromagnetic fields have been discussed by Avan *et al* (1974). One way to view this subject is to first consider the response of a free electron. Subjected to an electromagnetic field it performs an oscillatory (quiver) motion and its mean energy (ponderomotive energy) rises proportional to the intensity, I, and the square of the wavelength, λ

$$E_q = 9 \cdot 10^{-12} \, I \, \lambda^2 \, \text{(eV)} \tag{1}$$

where I is in W/cm^2 and λ in nm. At an intensity of 10^{13} W/cm^2 this energy is ~1 eV for a Nd-Yag laser and ~100 eV for a CO$_2$ laser. The variation of intensity over the focal volume of the laser implies a variation of the ponderomotive potential in space and the gradient of this potential accelerates electrons away from the focal volume (Kibble 1966). Boreham and Luther-Davis (1979) showed that the gradient caused by their powerful Nd-YAG laser accelerated electrons up to several hundred electron Volts. Bucksbaum *et al* (1987) showed that a slow electron beam could be scattered off a ponderomotive potential hill produced by a focused laser beam.

3. ATOMIC PHOTOIONIZATION

The existence of a ponderomotive potential of appreciable magnitude has several consequences for multiphoton ionization. For one if we wish to ionize an atom or molecule using an intense laser we not only have to provide sufficient photons to overcome the ionization threshold, but also provide the energy required for the electron to be a free electron in the field. In other words the effective ionization potential rises as the intensity rises by an amount given by equation (1). Secondly, the finite spatial distribution of the ponderomotive potential requires a finite amount of time for the electron to escape the focal volume.

The consequence of this is discussed in the context of Figure 1 which gives the time scale for an electron created with 3eV excess kinetic energy in the center of the focal volume of 20 μm diameter. The ponderomotive potential is 1 eV at the center and the time required by the electron to exit from the focal region is given along the x-axis. For a laser pulse that is "on"

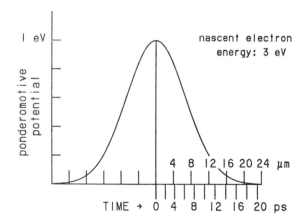

Fig. 1. Schematic of ponderomotive potential energy distribution across focal region. The time required for an electron to escape from the volume, including ponderomotive acceleration, is indicated at the x-axis.

for a long time the electron will be accelerated by the ponderomotive forces and an electron with 4 eV energy will be detected. On the other hand, if the duration of the laser pulse is short, of the order of 1 ps or less, the electron will not be able to experience this acceleration and an electron with ~3 eV energy will appear at the detector. Freeman *et al.* (1987) have shown that this case provides an opportunity for studying the nascent electron energy, as it is born in the presence of a strong field.

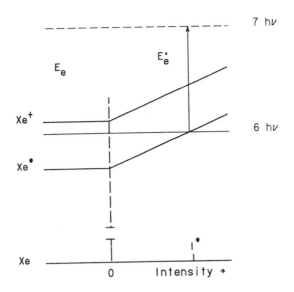

Fig. 2. Schematic energy diagram of xenon and its development with laser intensity.

We discuss this case at the example of photoionization of atomic xenon using visible radiation (see Figure 2). The wavelength in this experiment is 620 nm. Seven photons are required to reach the lowest ionization threshold, $Xe^+(^2P_{3/2})$, 12.13 eV. If xenon undergoes nonresonant photoionization we expect electrons to be produced with an energy

$$E_e = 7 \, h\nu - 12.13 \; eV \qquad (2)$$

Experimentally it is observed that this nonresonant process is an only weak channel. Rather, the dominant event is the shifting into resonance of excited states that are not resonant at zero field. This situation is indicated in Figure 2 where it is assumed that the excited state experiences an ac-Stark shift equal to the ponderomotive shift of the free electron. The excitation probability to the six-photon stark shifted intermediate maximizes at intensities near I^* and subsequent photoionization of the intermediate leads to an electron with drift energy

$$E'_e = 7 \, h\nu - 12.13 - E_q(I^*) \sim h\nu - IP^* \qquad (3)$$

where IP^* is the ionization potential of the excited state. When the laser is "on" for a long time the electron with energy (3) escapes from the focal volume and experiences acceleration equal to E_q, hence at the detector electrons will be detected with energies corresponding to the non-resonant seven photon process (equation 2). However, if the pulse is short, the electron will barely move prior to the end of the laser pulse and electrons with a reduced energy, near that given by equation (3) will be detected.

After its first demonstration by Freeman *et al* (1987) this phenomenon has been studied extensively. In Figure (3) we show a typical spectrum which reveals electron energy peak positions shifted down in energy from that expected in non-resonant 7-photon ionization. The observed shift is consistent with the assumption that the intermediate states, 4f and 5f experience an ac-Stark shift equal to the ponderomotive shift. The shift in photoelectron energy is also observed in the above-threshold ionization signal as apparent from Figure 4.

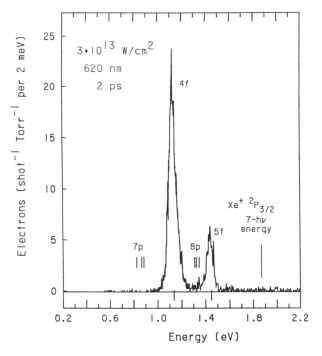

Fig. 3. Photoelectron spectrum of xenon. The laser parameters are indicated in the figure (Helm *et al* 1990).

Proof that indeed intermediate states with f-character contribute to the ionization process has come from measurement of the angular distribution of photoelectrons. Rottke *et al* (1989) have shown that the dominant character of the outgoing electrons reflects an $\ell = 4$ wave for electrons from the lowest order peaks in the photoelectron spectrum. These peaks, labelled S = 0 in Figure 4 correspond to one-photon ionization of the intermediate 4f or 5f state. On the other hand an $\ell = 5$ wave is indicated for the first above threshold ionization peak (S = 1 in Figure 4) which corresponds to two-photon ionization of the intermediate f-states. The particular states that can shift into resonance change with wavelength and with laser peak intensity as studied in detail by Agostini *et al* (1989).

The shifting of excited states into resonance is ubiquitous in strong field excitation, whenever direct ionization into the continuum is weak. Figure 5 gives another example in xenon, when excited at 328 nm. Four photons are required to ionize xenon at this wavelength as indicted by the energy diagram on the left. The experimental photoelectron spectrum recorded at a peak intensity of 80 TW/cm^2 and a pulse length of about 1.5 ps reveals three peaks, their energy location being consistent with the assumption that the intermediate states 8s, 6d and 7s are shifted into resonance at increasing intensity, and are subsequently one-photon ionized. This identification is supported by the appearance of only a single peak, corresponding to the position of the 6d states, when circular laser polarisation is used. In the latter case only the J = 3 components of the d states may serve as resonant intermediates, and the photoelectron spectrum reduces accordingly.

At the electron energy resolution employed in such studies to date the finding is that the ac-Stark shift of the excited states involved closely resembles the ponderomotive shift of the free electron. This is expected when the wavelength used is not near resonant with any particular bound-bound transition of the species studied (Avan *et al* 1976). The situation

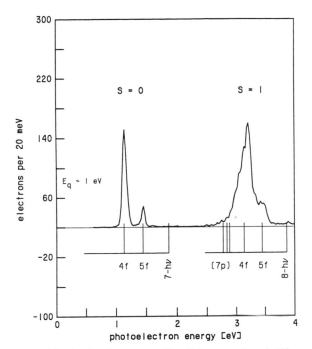

Fig. 4. Above threshold ionization in xenon. Conditions as given in Figure 3.

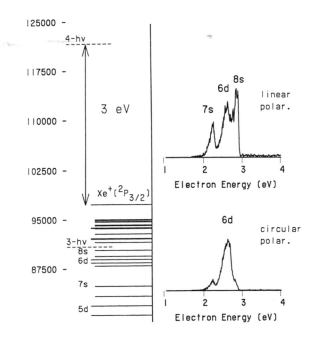

Fig. 5. Photoelectron spectra of xenon at 328 nm at 80 TW/cm^2 and a pulse length of 1.5 ps (Helm *et al*, 1990).

changes considerably when this is not the case as for example in atomic hydrogen at a wavelength of 616 nm, which is close to the Balmer-α line (656 nm). When driven by a strong electromagnetic field the $n = 3 \rightarrow n = 2$ transition energy is strongly modified.

This is apparent from Figure 6 which gives theoretical predictions of the development of excited state structure of some of the states of atomic hydrogen with intensity (Dörr *et al* 1991). The strongly driven 2p-3d transition leads to a lifting of the degeneracy of the various ℓ-states which propagates into the higher manifolds with $n = 4$ and 5. A direct confirmation of these effects has come from experiments by Rottke *et al* (1990) which display the predicted 4f-4p and 5f-5p splitting in the photoelectron spectra.

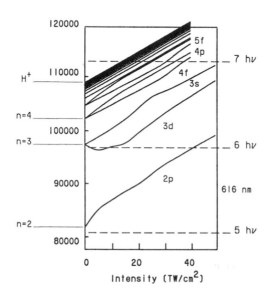

Fig. 6. Development of the excited state structure of atomic hydrogen with laser intensity for a wavelength of 616 nm (Dörr *et al* 1990).

4. MOLECULAR PHOTOIONIZATION

Experiments on molecular photoionization in intense fields are complicated by the additional degrees of freedom available in molecular species. Rotational and vibrational excitation provide a greater manifold of final states as do the dissociation of the product molecular ion or neutral molecule. We restrict the discussion here to experiments in molecular hydrogen and nitrogen at intensities below 10^{14} W/cm^2.

Studies on multiphoton ionization of molecules in intense laser fields have revealed a great diversity of photoelectron energy spectra that at times appear to defy assignment of specific excitation paths. In the light of the multitude of molecular excited states that can participate, this is not surprising when we consider that ac-Stark effects further modify the excited state structure.

In multiphoton ionization of H$_2$ and N$_2$ in intense fields very simple, atomic-like photoelectron features can appear when molecular Rydberg states act as intermediates in the photoionization event (Gibson *et al* 1991, Helm *et al* 1991). This finding is quite analogous to the atomic case discussed above. A key element for the appearance of simple spectra is to bypass excitation of valence states by proper choice of wavelength and intensity.

An example is shown in Figure 7 which gives photoelectron spectra of H_2 at 326 nm. The electron energies expected from one-photon ionization of excited atomic hydrogen,

$$H(n\ell) + h\nu \rightarrow H^+ + e \qquad (4)$$

are indicated in the figure and the peak positions suggest this process is a dominant ionization channel. This observation is made at high intensity for wavelengths between 308 and 330 nm. By contrast, at bluer wavelengths and lower intensity more complex spectra appear (Allendorf and Szöke 1991, Helm *et al* 1991).

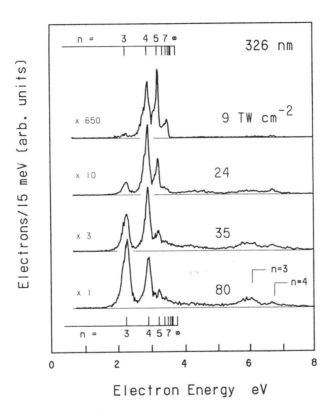

Fig. 7. Photoelectron spectra of H_2 recorded at various intensities. The labels refer to electron energies expected from one-photon ionization of excited atomic hydrogen.

Figure 8 shows the excited states of H_2 important in this wavelength range. Also indicated are the excitation energies at 326 nm, and the location of the vibrational wavefunction of ground state $H_2(v=0)$. At molecular distances most of the excited states of H_2 are well described by the Rydberg character derived from the core potential energy curve of ground state H_2^+. The most profound deviation from this pure Rydberg picture arises for the lowest $^1\Sigma$ states owing to their long-range ion pair character and at large internuclear separation for all states owing to the interaction with the Rydberg states belonging to the $2p\sigma$ core of the molecular ion (Lembo *et al* 1990).

We see from Figure 8 that the Franck-Condon overlap into the $B^1\Sigma_u^+$ state is poor at 326 nm and that the first "good" resonant states occur at the four-photon energy. This suggests that an important ionization channel may involve four-photon excited states of the molecule:

$$H_2 + 4\,h\nu \rightarrow H_2(n\ell\lambda,v) \qquad (5)$$

that are subsequently ionized by a fifth photon:

$$H_2(n\ell\lambda,v) + h\nu \rightarrow H_2^+(v^+) + e. \qquad (6)$$

In the light of studies of atoms in strong fields it appears natural to assume that at elevated intensity, specific intermediates are shifted into resonance due to the ac-Stark effect.

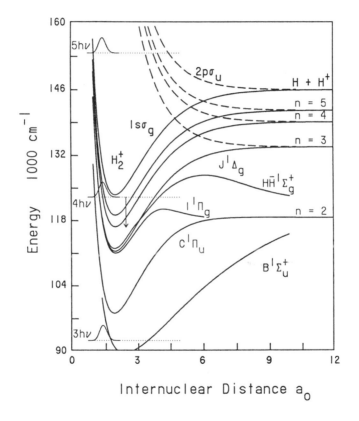

Fig. 8. Potential energy diagram of relevant states of H_2. The energy scale is relative to the lowest level of H_2 $X^1\Sigma_g^+$. The multiphoton energies for 326 nm are marked. The magnitude of the arrow indicates the ponderomotive shift at $8 \cdot 10^{13}$ W/cm^2.

    In a first approximation we may assume that the Rydberg states experience a shift equal to the ponderomotive energy of a free electron. At the highest intensity shown in Figure 7, equation (1) predicts a shift of 6400 cm^{-1}, about three vibrational spacings of H_2^+. The magnitude of this shift is indicated in Figure 7 by the downward arrow. Specifically under these conditions, the vibrational levels $v = 5,4,3$ of the states belonging to the $n=3$ manifold become sequentially resonant with the four-photon dressed ground state. In addition, the vibrational levels $v = 2,1,0$ for $n = 4$, the levels $v = 1,0$ for $n = 5$ and the levels $v = 0$ for $n = 6$ and 7 are swept through resonance for such a pulse. The simplicity of the electron energy spectra observed can be reconciled with excitation to such a wide variety of vibrational states from five Rydberg manifolds if we consider the following:

The postulate that the ac-Stark shift of the Rydberg electron is equal to the ponderomotive shift predicts that the photoelectron resulting from one-photon ionization of the Rydberg (equation 6) initially has the kinetic energy:

$$E_e = h\nu - E_{H_2^+(v^+)} + E_{H_2}(n\ell\lambda,v) \qquad (7)$$

where the molecular energies quoted are at zero intensity. Here we have made the implicit assumption that the ac-Stark shift of the ion core of the Rydberg is equal to that of H_2^+.

Relation (7) describes the electron's energy when it is born in the field, irrespective of the intensity at which four-photon excitation of the molecular Rydberg occurred. For a laser pulse of duration short compared to the time required by the electron to leave the focal region (a 3 eV electron moves 1 μm in 1 ps) this is also the final energy detected in the experiment. The quantum defects of gerade Rydberg states of molecular hydrogen are with few exceptions very small. Therefore, the molecular photoelectron energies (7) are practically indistinguishable from those from excited atomic hydrogen ionized in process (4), provided that the final vibrational state of the molecular ion, v^+, is equal to that of the molecular Rydberg, v. This last condition is fulfilled when the quantum defect of the Rydberg state is small at all internuclear distances sampled by the vibrational wavefunction. From the excitation sequence (5) and (6) we expect that H_2^+ will be produced in a variety of vibrational levels, depending on intensity and wavelength, but the electron's energy will merely reflect the quantum defect of the molecular Rydberg states being ionized, regardless of their vibrational quantum numbers. Similar conclusions were reported by Gibson *et al* (1991) in molecular nitrogen.

Since specific vibrational levels of the molecular ion are rapidly photodissociated at the wavelengths used here, we also need to consider the fate of the final ion state, and for that matter the fate of the molecular Rydberg core when subjected to intense fields. Photodissociation of the molecular ion proceeds through a transition from the $1s\sigma_g$ orbital to the repulsive $2p\sigma_u$ orbital. The molecular Rydberg states share this very same core. To estimate the importance of photodissociation of the Rydberg

$$H_2(n\ell\lambda,v) + h\nu \rightarrow H(1s) + H(n\ell). \qquad (8)$$

we approximate this cross section by the photodissociation cross section calculated by Dunn (1968) for H_2^+. The cross sections suggest that Rydberg states with high vibrational excitation will follow path (8) rather than path (6). However, owing to the small quantum defects, the electron signature from ionizing excited atoms formed in process (8) using a sixth photon is indistinguishable from that from molecular ionization. Hence both dissociation followed by ionization of the excited atom (4), as well as direct ionization into H_2^+ are consistent with the energy spectra observed at high intensities.

The variation of the photoelectron spectra with intensity in Figure 7 can be simulated (Helm *et al* 1991) using a model which treats the individual vibrational states in the Rydberg manifold as uncoupled. The simulations predict that at high intensity the photoelectron peak for $n = 3$ dominates over the contribution from $n = 4$ and $n = 5$ just as is observed in Figure 7 an that as the peak intensity is lowered, the contribution from $n = 4$ first overtakes that from $n = 3$ and at yet lower intensity the contribution from $n = 5$ dominates, consistent with what is found experimentally.

Why do more complex spectra appear at shorter wavelengths and at low intensities? At shorter wavelength higher vibrational levels in the B state and C state are accessed with three photons with better Franck-Condon factors (Figure 8). The non-Rydberg character of the B state well allows transitions that are not diagonal in vibration to four-photon states and to H_2^+. Photoelectron energies will then reflect the change in vibrational quantum number and the (unknown) ac-Stark shift of the B- and C-state.

5. MOLECULAR PHOTODISSOCIATION

Multiphoton phenomena in photodissociation have been predicted for the molecular hydrogen ion (Guisti-Suzor *et al* 1990) and recently first experiments on this subject have been reported (Bucksbaum *et al* 1990, Yang *et al* 1991). A tutorial picture for describing multiphoton dissociation is that of molecular dressed states which are eigenstates of the Born-Oppenheimer Hamiltonian and of the radiation field. In this basis set the optically allowed bound-free photodissociation process (Figure 9a) appears as predissociation of the bound state, when dressed by one photon more than the dissociative state (Figure 9b). The coupling that induces predissociation is the product of transition moment and electric field, $\mu \cdot E$. Thus in the dressed state picture the bound state acquires a field-dependent width owing to predissociation. In the low intensity limit this width is equivalent to the inverse of the rate of photodissociation of the bound state, obtained from the product of the photodissociation cross section times the photon flux.

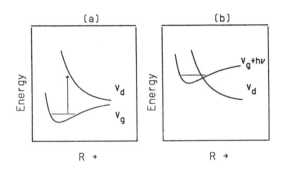

Fig. 9. Equivalence of photodissociation in the Born-Oppenheimer picture and predissociation of dressed states.

When extended to a frame of reference where the molecule is imbedded in a bath of many photons multiple crossings occur as shown in the example of Figure 10. The dotted curves give the diabatic dressed states where g stands for the ground state of H_2^+ and u for the dissociative state and N gives the photon number. Since the parity of "g" states dressed with (N+even) photons is equal to that of "u" states dressed with (N+odd) photons all crossings that appear in this picture are avoided. When the diabatic states are diagonalized in the $\mu \cdot E$ operator, adiabatic states are obtained which describe the situation for a specific intensity. These adiabatic states are shown by the full curves. One photon dissociation of the bound state appears in this picture as tunneling through the adiabatic barrier, giving rise to photofragments with an energy marked by W1. In addition, near its equilibrium position the bound state potential is crossed by the diabatic u(N-3) state. Predissociation to this curve corresponds to 3-photon absorption and if this predissociation follows the diabatic state, then photofragments with energy W3 will be formed. However, if the adiabatic path is chosen the products will tend to the final state g(N-2) with a photofragment energy corresponding to W2, as if only two photons had been absorbed. This balance comes about because choosing the adiabatic path as indicated in Figure 10 corresponds to stimulated emission at the avoided crossing between the g(N-2) and u(N-3) states.

Acknowledgements: This research was supported by NSF under Grant No. PHY-9024710 and by AFOSR under Contract No. F49620-88-K006.

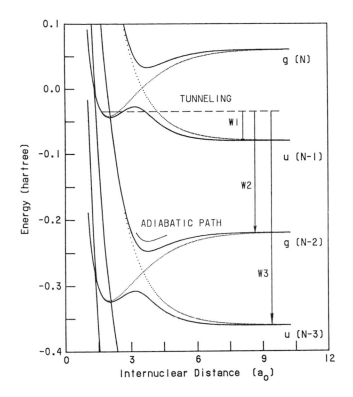

Fig. 10. Multiphoton dissociation of H_2^+ viewed in the framework of adiabatized dressed states (Guisti-Suzor *et al* 1990).

REFERENCES

Agostini P, Antonetti A, Breger P, Crance M, Migus A, Muller H G, Petite G 1989 J. Phys. B At. Mol. Opt. Phys. 22 1971.

Allendorf S and Szöke A 1991 Phys. Rev. A **44** 518.

Avan P, Cohen-Tannouddji C, Dupont-Roc J, and Fabre C, 1976 Le Journal de Physique **37** 993.

Boreham and Luther-Davis 1979 J. Appl. Phys. **50** 2533.

Bucksbaum P H, Bashkansky M, and McIlrath T J, 1987 Phys. Rev. Lett. 58 349.

Bucksbaum P H, Zariyev A, Muller H G, and Schumacher D W, 1990 Phys. Rev. Lett. **64** 1883

Dörr M, Potvliege R M, and Shakeshaft R, 1990 Phys. Rev. A **41** 558.

Dunn G H, 1968 Phys. Rev. **172** 1.

Durup J, 1979 J. Chim. Phys. Biol. **19** 779.

Freeman R R, Bucksbaum P H, Milchberg H, Darack S, Schumacher D, and Geusic M E, 1987 Phys. Rev. Lett. **59** 1092.

Gibson G N, Freeman R R, and McIlrath T J, 1991 NATO Workshop on Coherence Phenomena in Atoms and Molecules in Laser Fields, Hamilton, Ontario, and Phys. Rev. Lett. submitted for publication.

Guisti-Suzor A, He X, Attabeck O, and Mies F 1990 Phys. Rev. Lett. **64** 515.

Helm H, Dyer M J, and Bissantz H, 1990 Bull. Am. Phys. Soc. **35** 1151.

Helm H, Dyer M J, and Bissantz H, 1991 Phys. Rev. Lett. **67** 1234.

Kibble T W B, 1966 Phys. Rev. **150** 1060.

Lembo L J, Bjerre N, Huestis D L, and Helm H, 1990 J. Chem. Phys. **92** 2219.

Rottke H, Wolff B, Tapernon M, Feldmann D and Welge K G 1989 Lecture Notes in Physics 339, Fundametals of Laser Interactions II (Berlin: Springer Verlag) pp 25-36.

Rottke H, Wolff B, Brickwedde B, Feldmann D, and Welge K G, 1990 Phys. Rev. Lett. **64** 404.

Yang B, Saeed M, DiMauro L F, Zavriyev A, and Bucksbaum P H, 1991 Phys. Rev. Lett. A **44** R1458 .

Intermediate energy R-matrix theory

T T Scholz

ITAMP, Harvard–Smithsonian Center for Astrophysics, 60 Garden Street, Cambridge MA 02138, U.S.A.

Abstract. The theoretical difficulties which plague intermediate energy electron–atom scattering are described and an extension of the standard R–matrix theory designed to overcome such difficulties is presented. Justification for the method along with an application to electron scattering by atomic hydrogen is provided. The latter clearly indicates that theory and experiment have yet to be fully reconciled.

1. Introduction

The theoretical study of electron scattering by atoms and molecules is an enormous field of endeavour with a history dating back to the early 1930's. Despite this, there remain a great many unsolved problems. The Schrödinger equation for the electron–atom scattering interaction has evaded analytic solution for even the simplest systems. Theorists have therefore been forced to rely on a combination of approximation and numerical computation, the latter being greatly enhanced through the advent of modern computing facilities. Despite substantial progress, the fact remains that theoretical studies of electron scattering by even the simplest of all atomic targets, the hydrogen atom, have yet to be fully reconciled with experiment.

Theorists have enjoyed consistent success in calculating the results of scattering experiments under certain conditions. History has demonstrated that when the energy of the incident electron does not exceed the energy required to excite the target atom beyond its lower lying excited states, the so called close–coupling approximation is valid. In this approximation it is assumed that the scattering wavefunction, Ψ, may be written as a summation over the energetically accessible channels. That is,

$$\Psi(\mathbf{X}, r_{\mathrm{s}}) = \mathcal{A} \sum_{i=1}^{\mathrm{nchan}} \overline{\phi_i}(\mathbf{X}) r_{\mathrm{s}}^{-1} f_i(r_{\mathrm{s}}) \tag{1}$$

where $\mathbf{X}$ collectively denotes all electronic coordinates except the modulus of the position coordinate, $r_{\mathrm{s}} = |\mathbf{r_s}|$, of the scattered electron and the operator $\mathcal{A}$ ensures that the wavefunction is antisymmetrized. In this expression, the $\overline{\phi_i}$ channel functions represent the bound states of the atom coupled to the spin and angular dependence of the scattered electron and the total number of open channels, nchan, is determined by the set of possible final excited atomic states allowed through conservation of energy. Scattering amplitudes for electron impact induced transitions to the $\overline{\phi_i}$ channels are given by the asymptotic dependence of the f_i functions. From these amplitudes, all scattering observables associated with these transitions are easily derived. The equations which define the f_i functions are obtained by substituting the close–coupling expansion (1) into the Schrödinger equation for the scattering interaction and projecting onto the channels functions,

$$\langle \overline{\phi_j} | H_S - (E + E_0) | \Psi \rangle = 0 \qquad j = 1, \ldots, \text{nchan} \tag{2}$$

where E and E_0 denote of the kinetic energy of the incident electron and the ground state energy of the target atom respectively.

The form of the coupled equations may be simplified by subdividing the scattering Hamiltonian,

$$H_S = H_A - \frac{1}{2} \frac{\partial^2}{\partial \mathbf{r}_s^2} + V \tag{3}$$

where H_A denotes the Hamiltonian of the target atom and V is the interaction potential between the atom and scattered electron. Ignoring terms resulting from electron exchange, equation (2) reduces to the following set of coupled equations

$$\left[\frac{d^2}{dr_s^2} - \frac{\ell_j(\ell_j + 1)}{r_s^2} + k_j^2 \right] f_j(r_s) = 2 \sum_{i=1}^{\text{nchan}} V_{ji}(r_s) f_i(r_s) \tag{4}$$

where k_j^2 and ℓ_j denote the final kinetic energy of the scattered electron responsible for excitation into the $\overline{\phi_j}$ channel and its orbital angular momentum respectively. The coupling matrix is given by

$$V_{ji}(r_s) = \langle \overline{\phi_j} | V | \overline{\phi_i} \rangle . \tag{5}$$

Solution of this set of coupled equations with the appropriate boundary conditions yields the f_i functions and hence the scattering amplitudes.

As the scattering energy is increased, more channels become energetically accessible. These additional open channels must be included in the expansion of the scattering wavefunction for the close–coupling approximation to remain valid. In principle, a complete expansion involving a sum over all bound states and an integral over continuum states of the atom may be employed to yield an accurate solution at any scattering energy. Naturally this is impractical. If the number of channels is greater than about 30 then the numerical solution of the coupled equations (4) exceed the capabilities of the current generation of supercomputers. It is the method of solution of these equations which distinguishes the various theoretical methods based upon the close–coupling expansion. These include, but are not limited to, the Kohn variational (and its variants), Schwinger variational, R-matrix and linear algebraic theories, each of which have been employed to varying degrees of success. As the number of open channels which can be included into a close–coupling calculation is limited, these close–coupling theories have only been successfully implemented at scattering energies below the ionization threshold of the atom. They are therefore referred to as low energy theories.

The theories which have proven successful at the other end of the scattering energy spectrum take advantage of the fact that the scattered electron moves so quickly that neither it nor the target atom is strongly influenced by the electron–atom interaction potential V. The most common of these, the Born approximation, treats this potential as a perturbation. In its crudest form, V is set to zero in which case the scattering wavefunction is simply a product of a free electron wave and the ground state of the atom ϕ_0. Ignoring the spin of the scattered electron, this may be written as

$$\Psi = \phi_0 \, \exp(i\mathbf{k_i} \cdot \mathbf{r_s}) \tag{6}$$

where $\mathbf{k_i}$ denotes the wavevector of the scattered electron prior to the collision. Higher order versions of the Born approximation retain perturbative corrections to this initial approximation. Like all perturbative procedures, this approach yields accurate results when the interaction potential does not greatly distort the initial unperturbed wavefunction (6). When the scattering energy is high, this condition is satisfied. However at lower energies the potential plays an increasingly dominant role resulting in very slow convergence of the perturbative expansion of the scattering wavefunction. In fact, perturbative methods usually become impractical when the scattering energy[1] is less than several times the energy required to ionize the atom. There are more sophisticated versions of this perturbative procedure including the distorted wave Born approximation but all such methods are dogged by slow convergence except at high scattering energies.

Theorists have successfully applied both the low and high energy theories to their regimes of validity. However between these regimes lies an energy region where neither method is able to accurately model the scattering interaction. In this *intermediate* energy regime the number of open channels is too large for the equations (4) resulting from the close–coupling approximation to be solvable and the interaction between the scattered electron and the target atom creates too much distortion for the perturbative approaches to converge rapidly. While no well defined boundaries exist to separate the different regions, the intermediate energy regime may be loosely defined to range from just below the energy required to ionize the target atom to several times this scattering energy. In this paper we present an extension of the standard low energy R–matrix theory which includes a number of modifications designed to take account of the effect of the large or infinite number of channels which are open at intermediate energies. In the next section we provide a predominantly descriptive exposé of the theory. Mathematical detail has been avoided so as not to cloud the central features of the method. In section 3 the results of an application of the method to electron scattering by atomic hydrogen are presented. It is clear from section 3 that intermediate energy electron scattering by this, the simplest of all atomic targets, is still very much an unclosed chapter in atomic scattering physics. Future directions are presented in section 4.

2. Theory

The goal of this paper is not to present a mathematical derivation of intermediate energy R–matrix (IERM) theory or a detailed analysis of its strengths and weaknesses but rather to provide a comprehensible description of the most important facets of the theory and some justification for its use. The reader interested in the former is referred to Burke *et al* (1987), Scott *et al* (1989) and Scholz (1991). A comprehensive description of the standard R–matrix theory is to be found in the review articles by Burke (1987) and Burke and Robb (1975) along with articles referenced therein.

The philosophy behind R–matrix theory is very simple. Real three dimensional space is partitioned into two distinct regions. The *internal region* is a sphere centred on

---

[1] In this paper, the scattering energy is defined as the kinetic energy of the incident electron.

the nucleus of the atom which is just large enough to envelope the atomic states involved in any transitions of interest. The *external region* is all space outside of this internal region. The reason for such partitioning is clear. All of the complicated interactions between the scattered electron and electrons of the target atom, such as correlation and exchange, are confined to the region where the scattered electron overlaps those of the target atom. This region will extend over the space occupied by the most diffuse atomic state excited by the collision, that is, the region which we have defined as the internal region. So separating this part of space enables the most complicated part of the wavefunction to be confined to, and hence solved in, the smallest possible region. In the external region, both electron exchange and correlation are unimportant. The wavefunction in this region may therefore be accurately represented by an unsymmetrized close–coupling expansion over the channels of interest. Final solutions are obtained by matching the wavefunction of each region at the boundary. The R–matrix, after which the theory is named, defines the ratio of the wavefunction to its derivative at this boundary and therefore provides the ideal mechanism for the matching process.

The best way to demonstrate IERM theory is through the use of a simple example. The hydrogen atom provides the simplest of all atomic targets and is, therefore, an excellent choice. In particular, we shall consider 1s–2s–2p scattering, that is, scattering for which the initial and final atomic states include only the 1s, 2s and 2p eigenstates of the atom. These states are effectively enveloped in an internal region of radius $25a_o$ [2].

The IERM theory is a direct extension of the standard R–matrix approach. The main difference between the two methods is how the wavefunction of the internal region is constructed. In the standard theory, the scattering wavefunction within the internal region, Ψ_E^{INT}, is written as a sum over energy independent basis states, ψ_k, usually referred to as R–matrix basis states. As the case under consideration involves the interaction between just two electrons whose spin–orbit interactions produce negligible effects, the total spin of the system may be factored out *a priori*. We need therefore only consider the spatial part of the scattering wavefunction,

$$\Psi_E^{INT}(\mathbf{r}_1, \mathbf{r}_2) = \sum_k A_{Ek} \psi_k(\mathbf{r}_1, \mathbf{r}_2) \ . \tag{7}$$

In the standard approach, the R–matrix basis states are themselves expanded in terms of orbital products:

$$\psi_k^{SRM}(\mathbf{r}_1, \mathbf{r}_2) = \mathcal{A} \left[\sum_{i=1}^{3} \sum_{j=1}^{3} \phi_i(\mathbf{r}_1)\phi_j(\mathbf{r}_2)a_{ijk} + \sum_{i=1}^{3} \sum_{p=1}^{pmax} \phi_i(\mathbf{r}_1)u_p(\mathbf{r}_2)d_{ipk} \right] \ . \tag{8}$$

In this expression, the ϕ_i orbitals represent the 1s, 2s and 2p eigenstates of the hydrogen atom[3] and the u_p orbitals are members of a discrete set of continuum orbitals generated in a model potential and truncated at the edge of the internal region. The a and d coefficients are obtained by diagonalizing the scattering Hamiltonian within the internal region. It is not difficult to show that the expansion of the scattering wavefunction defined by (7) and (8) is equivalent to a close–coupling expansion (1) which includes only the 1s, 2s and 2p channels. We have already mentioned that such an expansion is

---

[2] $1a_o = 1$ Bohr radius.
[3] Expansions (8) and (1) may also include pseudostates of the atom which help to model electron correlation and flux losses into omitted channels.

inappropriate for intermediate energy scattering. Clearly then, some modification will be required.

The IERM theory rectifies the situation by extending the expansion on the R–matrix basis states to include a series of continuum–continuum orbital products. The new basis states become

$$\psi_k^{\text{IERM}}(\mathbf{r}_1, \mathbf{r}_2) = \mathcal{A}\left[\sum_{i=1}^{3}\sum_{j=1}^{3} \phi_i(\mathbf{r}_1)\phi_j(\mathbf{r}_2)a_{ijk} + \sum_{i=1}^{3}\sum_{p=1}^{\text{pmax}} \phi_i(\mathbf{r}_1)u_p(\mathbf{r}_2)d_{ipk} + \right.$$
$$\left. \sum_{q=1}^{\text{qmax}}\sum_{p=1}^{\text{pmax}} u_p(\mathbf{r}_1)u_q(\mathbf{r}_2)g_{qpk} \right]. \tag{9}$$

Once the a, d and g coefficients are obtained by diagonalizing the scattering Hamiltonian, the scattering wavefunction of the internal region may be constructed as in the standard method using (7).

The additional continuum–continuum orbital products are appended to the basis state expansion of the standard method to restore completeness. The basis states of the standard method, ψ_k^{SRM}, only contribute to the 1s–2s–2p close–coupling expansion and completely ignore the large or infinite number of channels which are open at intermediate energies. Such basis states are, therefore, unable to span the part of the exact intermediate energy scattering wavefunction, $\Psi_{E\ \text{exact}}^{\text{INT}}$, contained within the internal region. That is,

$$\Psi_{E\ \text{exact}}^{\text{INT}}(\mathbf{r}_1, \mathbf{r}_2) \neq \sum_k A_{Ek}\psi_k^{\text{SRM}}(\mathbf{r}_1, \mathbf{r}_2) \tag{10}$$

irrespective of how many continuum orbitals are included in expansion (8). However the additional continuum–continuum orbital products included in the expansion of the IERM basis states overcomes this limitation. Provided sufficient continuum orbitals are included in the expansion of ψ_k^{IERM}, this basis set will be effectively complete at intermediate energies. So, to a very good approximation, we may write

$$\Psi_{E\ \text{exact}}^{\text{INT}}(\mathbf{r}_1, \mathbf{r}_2) = \sum_k A_{Ek}\psi_k^{\text{IERM}}(\mathbf{r}_1, \mathbf{r}_2). \tag{11}$$

Before we go on to describe the treatment of the scattering wavefunction in the external region, we turn our attention to the nature of the coupling between open channels. Recall that, in the preceding section, the close–coupling expansion (1) led to a series of coupled equations (4) for the all important f_i radial functions. Although these equations ignore terms arising from electron exchange and are, as such, an oversimplification, they will serve to illustrate how IERM theory is able to accurately model electron scattering at intermediate energies. It is clear from (4) that the matrix V couples together the different f_i functions and therefore determines how they will influence each other. It can be shown that approximating the scattering wavefunction to a close–coupling expansion which includes only the 1s, 2s and 2p channels is equivalent to neglecting coupling between these channels and all other channels. As illustrated in the next section, calculations which neglect such couplings fail to predict the results of intermediate energy scattering experiments. Following the steps which lead to the coupled equations (4) from the close–coupling expansion (1), it is clear that only an infinite close–coupling expansion including all channels describing both bound state excitation and ionization is able to account for coupling between all channels. As such an expansion is complete, it may be

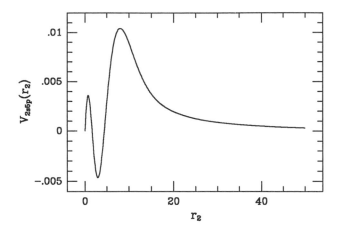

Figure 1. The matrix element coupling a 2s channel to a 5p channel.

used to describe the exact scattering wavefunction in any region of space, including the internal region. That is, we may write

$$\Psi_{E\ \text{exact}}^{\text{INT}}(\mathbf{r}_1, \mathbf{r}_2) = \mathcal{A}\sum_{i=1}^{\infty} \overline{\phi}_i(\mathbf{r}_1, \hat{\mathbf{r}}_2) r_2^{-1} f_i(r_2) \qquad r_2 < 25 a_o . \tag{12}$$

Comparing this expression with (11) it is evident that the scattering wavefunction of the IERM theory is equivalent to a complete close–coupling expansion within the internal region. With this we arrive at the very important conclusion that the IERM theory accurately models coupling between all channels within the internal region.

The validity of IERM theory is most clearly demonstrated through an analysis of the $V_{ij}(r_2)$ coupling matrix elements. In Figure 1 we plot the element which couples the 2s and 5p channels. Notice that the function falls off rapidly beyond $25 a_o$. The influence of the 5p channel on the 2s channel is therefore restricted almost exclusively to the region $r_2 < 25 a_o$ which defines our internal region for 1s–2s–2p scattering. Plots of other matrix elements which couple 1s, 2s or 2p channels to other channels show similar behaviour. This implies that the coupling between the 1s, 2s and 2p channels and all other channels is completely dominated inside the internal region. However we have already shown that this is precisely where IERM theory accurately models coupling between all channels.

The preceding analysis implies that coupling between the 1s, 2s and 2p channels and the other channels may be neglected in the external region without introducing significant errors. The IERM theory therefore assumes that the external region scattering wavefunction may be approximated to the following close–coupling expansion which includes only the 1s, 2s and 2p channels:

$$\Psi_E^{\text{EXT}}(\mathbf{r}_1, \mathbf{r}_2) = \sum_{i=1s,2s,2p} \overline{\phi}_i(\mathbf{r}_1, \hat{\mathbf{r}}_2) r_2^{-1} f_i(r_2) \qquad r_2 > 25 a_0 . \tag{13}$$

In this case we neglect the antisymmetrization operator because both electrons exist in different regions of space. This condition excludes coupling due to electron exchange which has also been demonstrated to be either zero or insignificant in the external region (Scholz 1991).

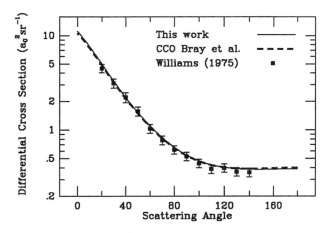

Figure 2. Elastic scattering cross section at 16.5eV.

Apart from treatment of the internal region wavefunction, an IERM calculation precedes as in the standard approach. The R–matrix is calculated by diagonalizing the scattering Hamiltonian within the internal region and then used to match to the external region wavefunction to yield the f_i functions and hence the scattering amplitudes.

3. Results

Most scattering calculations designed to reproduce experimental data involve a lot of technical detail. The present work is no exception. Scott *et al* (1989) gives a complete description of the calculation used to derive the scattering observables presented in this section. Briefly, IERM theory was used to evaluate contributions to the scattering amplitudes from partial waves in the range $0 \leq L \leq 4$ where L is the total orbital angular momentum of the electron–hydrogen atom system. In the range $5 \leq L \leq 16$, the more conventional pseudostate approach was adopted to account for the large or infinite number of channels which are open at intermediate energies. Corrections for angular momentum orbitals omitted from both the IERM and pseudostate basis sets were evaluated using the second Born plane wave approximation. This approximation was also implemented for partial waves $L > 16$ where interactions are expected to be weak.

The net result of the calculational procedure was the scattering amplitudes for electron impact induced transitions between the 1s, 2s and 2p states of the hydrogen atom. From these, a great many scattering observables may be derived. Those of most relevance may be found in the work of Scott *et al* (1989) and Scholz *et al* (1991). Here we present a sample of these results, concentrating on those for which experimental data exist.

The agreement between experiment and theory is excellent for elastic scattering. Figure 2 depicts a typical example of our results. Here we plot the experimental 16.5eV elastic scattering differential cross section against the theoretical results of both this work and the very recent unpublished coupled–channel optical (CCO) calculations of Bray and collaborators (see the article by Bray in this issue). The CCO theory uses an optical potential to account for the large or infinite number of channels which are

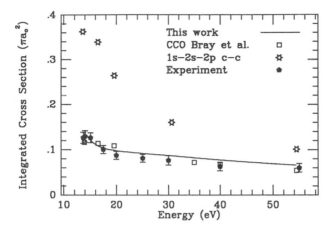

Figure 3. Integrated 1s–2s cross section.

open at intermediate energies. Our 1s–2s and 1s–2p integrated cross sections are also in accord with experiment. The former of these is plotted in Figure 3. Also included are results of calculations based upon the CCO and 1s–2s–2p close–coupling approximations. As expected, both intermediate energy theories are in good agreement with experiment but the 1s–2s–2p close–coupling calculations are more than 100% too large near the ionization threshold. Clearly the neglect of couplings between the 1s, 2s and 2p channels and all other channels produces inaccurate results at intermediate energies.

Despite this success, not all of our theoretical results are in agreement with experiment. Discrepancies exist in the inelastic differential data, an example of which is plotted in Figure 4. It is clear, from this figure, that the theoretical 54.4eV differential cross section for 1s–2p scattering resulting from both this work and the CCO calculations lie just below the experimental data in the 20°–40° angular range. While this discrepancy is not particularly noticeable the situation worsens when we turn our attention to the observables more closely related to the details of the scattering amplitudes – electron–photon coincidence parameters. These parameters depend on the phase differences between amplitudes for exciting different magnetic sublevels of the 2p eigenstate and are, therefore, a stringent test of the computational procedure. Unfortunately a long standing discrepancy between theory and experiment has not been resolved by either this work or the CCO calculations of Bray *et al* . In Figure 5 we plot the alignment angle γ of the excited 2p charge cloud. While experiment and theory are in reasonable agreement for scattering angles less than 90°, they diverge at larger angles. At present, it is not clear where the problem lies. The discrepancies depicted in Figure 5, which are also evident in the other electron–photon coincidence parameters, clearly demonstrate the electron scattering by atomic hydrogen is still very much an unclosed chapter in atomic scattering physics.

4. Future directions

The intermediate energy R–matrix theory has only been applied to electron scattering by atomic hydrogen and is, therefore, still very much in its infancy. There are several extensions of the theory which are now being pursued at The Queen's University of Belfast

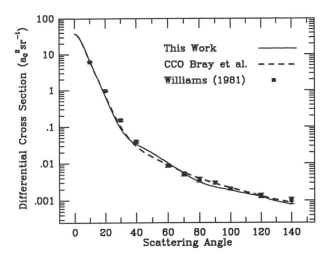

Figure 4. Differential 1s–2p cross section at 54.4eV.

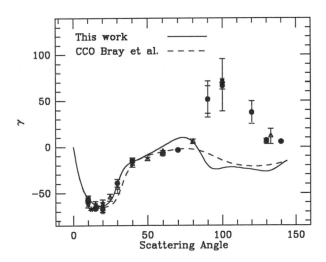

Figure 5. The electron–photon coincidence parameter γ at 54.4eV.

and ITAMP at the Harvard–Smithsonian Center for Astrophysics. The first and most obvious of these is the extension to electron scattering by the general atom. The algebra for this method is now in place (Burke *et al* 1987) but the arduous task of computer programming has yet to be completed. The next extension is the so–called propagation theory where a very large internal region is subdivided into manageable subregions. The scattering wavefunction is solved within each subregion and then propagated outwards to the edge of the region by matching at the subregion boundaries. This method has the advantage of enabling the scattering wavefunction to be spanned over a region much larger than is currently possible. The benefits are two fold: (i) channel couplings will be accurately modelled over a larger region of space and (ii) transitions to highly excited diffuse target states may be studied. The final extension of IERM theory currently on the agenda is to electron impact ionization. In this approach, instead of modelling transitions between bound states within the internal region, the scattering amplitude for an electron impact induced transition to a continuum orbital is evaluated. The excited continuum orbital is constructed to be identical to a continuum eigenstate within the internal region. Presumably the results of a such a calculation will become more accurate as the internal region is enlarged. Combining this with propagation theory may therefore yield an accurate procedure for evaluating electron impact ionization processes. Given the present lack of successful ionization theories except at high energies, this method is well worth pursuing.

Acknowledgments

The author acknowledges with thanks the substantial contributions to this work by Dr James Walters and Dr Penny Scott of Queen's University. This work was supported by The Queen's University of Belfast through a Postgraduate Scholarship, the Science and Engineering Research Council through a Research Assistantship and a grant for computing time and the National Science Foundation through a grant to the Institute of Atomic and Molecular Physics at Harvard University and the Smithsonian Astrophysical Laboratory.

References

Burke P G 1987 *Atomic Physics* **10** eds Narumi H and Shimamura I (Elsevier Science Publishers B.V.) 243

Burke P G, Noble C J and Scott M P 1987 *Proc. R. Soc.* A **410** 289

Burke P G and Robb W D 1975 *Ad. At. Mol. Phys.* **6** 320

Scholz T T, Walters H R J, Burke P G and Scott M P 1991 *J. Phys. B: At. Mol. Opt. Phys.* **24** 2097

Scholz T T 1991 *J. Phys. B: At. Mol. Opt. Phys.* **24** 2127

Scott M P, Scholz T T, Walters H R J, Burke P G 1989 *J. Phys. B: At. Mol. Opt. Phys.* **22** 3055

Williams J F 1975 *J. Phys. B: At. Mol. Phys.* **8** 2191

Williams J F 1981 *J. Phys. B: At. Mol. Phys.* **14** 1197

Electron–atom scattering calculations using the coupled-channel optical method

I Bray

Electronic Structure of Materials Centre, School of Physical Sciences, The Flinders University of South Australia, G.P.O. Box 2100, Adelaide 5001, Australia

Abstract. The coupled-channel optical method is applied to the calculation of elastic and inelastic phenomena for electron scattering on hydrogen and sodium atoms. The energy range for the projectile is taken to be the complete set for which experimental data is available. Excellent agreement with experiment is found for elastic differential cross sections across the entire energy range for both hydrogen and sodium. The agreement with experiment for inelastic differential cross sections is quite good, but agreement with the angular correlation parameters for the 2p channel of hydrogen is only fair.

1. Introduction

We use the coupled-channel optical (CCO) method for the calculation of electron-atom scattering phenomena. This is an *ab initio* approach to electron atom scattering. In the CCO method a finite number of channels (P space) are treated explicitly via the coupled-channel formalism. The rest of the channels (Q space) are treated indirectly via the complex non-local polarization potential, which together with the first order terms, forms the optical potential.

Our implementation of the CCO method Bray *et al* (1991) uses Feshbach projection operators P and Q which are symmetric in the space of the valence and projectile electrons. This has enabled our CCO approach to get very good agreement with experiment across the entire range of projectile energies for which data are available.

The aim of this paper is to present a broad range of CCO calculations for various elastic and inelastic phenomena, and to compare these results with available experimental data. One of the great advantages of the CCO method is that calculations may be done in stages, indicating what aspects have the most important contributions. For example, it is often very instructive to compare corresponding CC, CCO⁻ and CCO calculations. These show respectively the effect of leaving out the polarization potential, including just the Q-space bound states, and including the complete set of target states.

2. Outline of the CCO method

Before any calculation of scattering phenomena may be undertaken the target wave functions must be defined. For hydrogen this is simple as it is the only atom for which the target wave functions are known analytically. For alkali atoms we use the independent particle model of one valence electron above a frozen Hartree-Fock core.

The Lippman-Schwinger equation for the T matrix, which depends on the total spin S, of the nonrelativistic electron-atom scattering problem in the CCO approach is

$$\langle \mathbf{k}i \mid T^S \mid i_0 \mathbf{k}_0 \rangle = \langle \mathbf{k}i \mid V_Q^S \mid i_0 \mathbf{k}_0 \rangle$$
$$+ \sum_{i' \in P} \int d^3 k' \frac{\langle \mathbf{k}i \mid V_Q^S \mid i'\mathbf{k}' \rangle}{E^{(+)} - \epsilon_{i'} - k'^2/2} \langle \mathbf{k}'i' \mid T^S \mid i_0 \mathbf{k}_0 \rangle, \qquad (1)$$

where the projectile with momentum $\mathbf{k}_0$ is incident on the valence electron in state i_0 with energy ϵ_{i_0}, and where $E = \epsilon_{i_0} + k_0^2/2$ is the on-shell energy. This equation is solved in partial wave formalism. The complete description of the method of solution may be found in McCarthy and Stelbovics (1983). Writing the coordinate space-exchange operator as P_r, the matrix elements of V_Q^S are given by (McCarthy and Stelbovics 1983),

$$\langle \mathbf{k}i \mid V_Q^S \mid i'\mathbf{k}' \rangle = \langle \mathbf{k}i \mid v^{FC} + v_{12}(1 + (-1)^S P_r) \mid i'\mathbf{k}' \rangle$$
$$+ (-1)^S \langle \mathbf{k}i \mid (\epsilon_i + \epsilon_{i'} - E)P_r \mid i'\mathbf{k}' \rangle$$
$$- \delta_{ii'} \sum_{j \in C} \langle \mathbf{k}j \mid (2\epsilon_j - E)P_r \mid j\mathbf{k}' \rangle$$
$$+ \langle \mathbf{k}i \mid V_Q + (-1)^S V_Q P_r \mid i'\mathbf{k}' \rangle, \qquad (2)$$

where v^{FC} is the projectile-core potential, including core exchange, and v_{12} is the projectile-valence electron potential. For hydrogen v^{FC} is just the electron-proton potential and there is no sum over the core states C. The notation P in (1) indicates a finite set of discrete target states that are coupled explicitly. The remaining excited Q-space target states, are used to calculate the matrix elements of the polarization potential operator V_Q. This operator has a complicated structure. It is non-local, energy dependent, and for energies above first Q channel excitation threshold it is non-Hermitian. The detailed description of the calculation of V_Q matrix elements may be found in Bray *et al* (1991).

The approximation made is that of weak coupling in Q space. By this we mean that we leave out any direct coupling between distinct Q space channels. For example, the calculation that we denote by 1CCO for hydrogen has just the ground state in P space and has couplings of the form 1s-2s-1s, 1s-2p-1s, etc., but none of the form 1s-2s-2p-1s. This turns out to be generally a good approximation. It can be easily tested by doing a corresponding 3CCO and then a 6CCO calculation, which have 1s, 2s, 2p, and 1s, 2s, 2p, 3s, 3p, 3d in P space, respectively. These test the neglect of coupling between distinct $n = 2$, and $n = 3$ channels, respectively. It turns out that for hydrogen, the elastic differential cross sections calculated using the 1CCO and 6CCO models are only significantly different at a small energy range around the ionization threshold. Even so these differences are marginal, with the 6CCO results giving even better agreement with experiment. By virtue of the fact that our nCCO and mCCO calculations for $n \neq m$ give similar corresponding results, we say that our CCO approach to electron-atom scattering is internally consistent.

All Q-space target states are target eigenstates. If the target state is discrete the projectile is treated as a distorted wave orthogonalised to P space, with the distorting potential being due to this target state. If the target state is in the continuum the projectile is treated as a plane wave orthogonalised to P space. A full discussion may be found in Bray *et al* (1991). There the approximation for continuum states is tested by calculating the total ionization cross section as a function of the projectile energy, and is found to be in reasonable agreement with experiment.

All of the numerics are done to at least 1% convergence. In the partial wave expansion of (1) we take up to 100 partial waves, depending on the incident energy. In the expansion of the polarization operator V_Q in (2) we take target bound states with n and l up to 10 and 3 respectively. To get convergence in the continuum part of V_Q we take up to 30 quadrature points with l ranging up to 7 for each. A typical calculation would treat around 250 distinct excited target states. A 1CCO calculation takes about 10 minutes on an IBM RISC 6000 series CPU. A typical 6CCO calculation takes about 60 hours on this CPU.

3. Results and discussion

The philosophy of the CCO approach is to treat the most important channels via direct coupling in P space, and to treat the rest of the channels (Q space) indirectly via the non-local polarization potential. This, however, does not provide for a strong test of the treatment of Q space channels, as their total effect is then relatively small. The strongest test is to take the minimum number of P space channels.

The calculations of elastic scattering using the 1CCO model provides just such a test. In these, just the ground state is treated explicitly. All of the excited channels are put into Q space. It is well known that dipole excitation has a very large effect on elastic scattering, so a good test of the non-local polarization potential is to see how well it reproduces that effect.

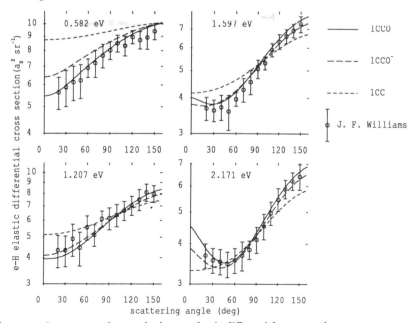

Figure 1. Low energy electron-hydrogen elastic differential cross sections.

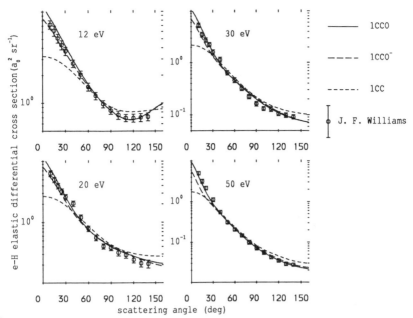

Figure 2. Intermediate energy electron-hydrogen elastic differential cross sections.

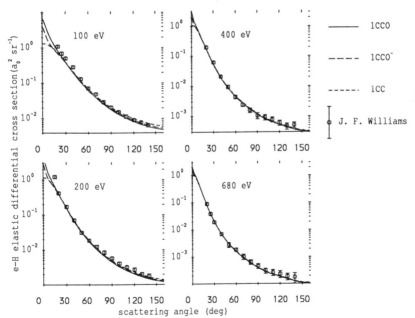

Figure 3. High energy electron-hydrogen elastic differential cross sections.

For elastic scattering there is a large quantity of data for the differential cross section across the "complete" energy range. This absolute data is mainly due to Williams (1975a, 1975b). In figures 1, 2 and 3 we present 1CCO calculations for elastic differential cross sections of electron-hydrogen scattering at low, intermediate and high energies, respectively. These are compared with the 1CC calculations, which are also known as "static-exchange", to show the effect of the non-local polarization potential. To show the effect of the continuum we present the calculations denoted by 1CCO⁻ which only have states with negative energy (bound states) in Q space.

From figure 1 we can see that the 1CCO calculations are in excellent agreement with experiment. The effect of the continuum is particularly interesting to note for the lowest considered energy. Other than for this energy the 1CC calculation is not too different from experiment. This is not so for intermediate energies.

In figure 2 we can see that the 1CC calculation is very much lower than experiment at forward angles. This problem is fixed by restoring the non-local polarization potential just for the bound states of Q space, which contain dipole excitation contributions. Subsequent adding of the continuum makes the result a little too high for the smaller energies. At these energies a 3CCO or a 6CCO calculation drops the differential cross section at forward angles to be in complete agreement with experiment. This indicates that at these energies direct coupling between the 2s and 2p states is more important.

Already at 50 eV the 1CCO, 3CCO and 6CCO give similar results for the elastic differential cross section. Thus there is a small discrepancy between theory and experiment at the forward angles at this energy. This is also a problem at the next considered energy of 100 eV in figure 3. Agreement with experiment is then very good for the higher energies. It is interesting to note how the effect of the polarization potential diminishes with increasing energy.

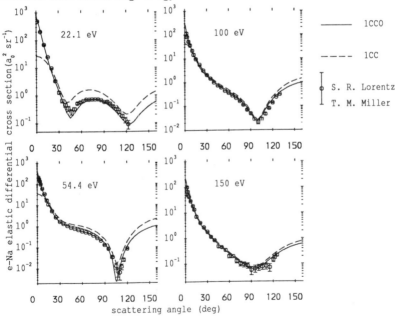

Figure 4. Elastic differential cross sections of electron-sodium scattering. A subset of experimental points has been selected for plotting.

In figure 4 we present 1CCO and 1CC calculations for elastic scattering on atomic sodium and compare them with the relative data of Lorentz and Miller (1991). Again, the effect of the polarization potential is extremely large, particularly at the lowest considered energy, and gives excellent agreement with experiment.

As agreement between our CCO theory and experiment is generally very good across the complete energy range considered, we believe that this theory is reliable in calculating elastic differential cross sections. The situation with inelastic phenomena is not quite so clear.

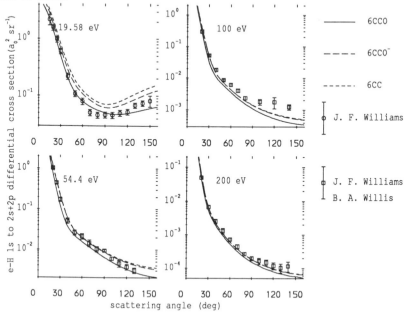

Figure 5. Summed 2s+2p differential cross sections for electron-hydrogen scattering.

In figure 5 we present 6CCO calculations for the electron-hydrogen differential cross section of 1s to 2s+2p excitation and compare with the measurements of Williams and Willis (1975) and Wiliams (1976). We also give the 6CC and the 6CCO⁻ calculations to show the effects of neglecting the polarization potential, and of neglecting just the continuum, respectively. By comparing the 6CC and the 6CCO⁻ calculations we see that the effect of the higher discrete states ($n = 4, ..., 10$) is only significant at the lowest considered energy, where it is necessary to get good agreement with experiment.

Generally, the 6CCO calculation is a little lower than experiment. It is interesting to note that unlike the difference between the 1CCO and 1CC calculations, the difference between the 6CCO and 6CC calculations is not very large. Thus, it would have been hoped that the polarization potential should work very well in this case. The results at 100 eV look a little anomalous. At all considered energies the effect of the polarization potential is to reduce the differential cross section. This effect is primarily due to the continuum. However, the simple 6CC calculation is already considerably below the experiment at 100 eV. Subsequent addition of the polarization potential only makes things worse. This is also true, but to a much lesser extent at 200 eV.

The results at 54.4 eV indicate the importance of a convergent treatment of the continuum. Taking fewer quadrature points or partial waves in the continuum may well result in even better agreement with experiment.

In figure 6 we present the 3p differential cross sections for sodium calculated with the 3CC and 3CCO models (P space $\in$ 3s,3p,3d). The effect of the polarization potential in this case is not large, but generally improves agreement with experiment.

Whilst the agreement with experiment for the inelastic differential cross sections is not as good as for the elastic differential cross sections, it can be described as generally good. This cannot be said for the calculations of hydrogen 2p angular correlation parameters.

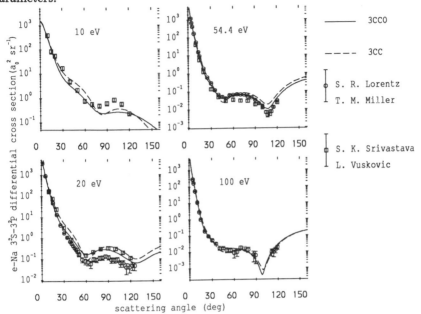

Figure 6. Differential cross sections of the 3^2p excitation of sodium by electrons. A subset of experimental points has been selected for plotting.

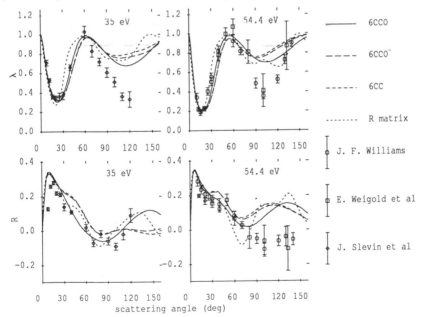

Figure 7. Angular correlation parameters λ_{2p} and R_{2p} for electron-hydrogen scattering.

In figure 7 we present 6CC, 6CCO$^-$ and 6CCO calculations for the hydrogen 2p angular correlation parameters λ and R. We compare these with various measurements and the R matrix calculations of Scholz *et al* . There are some definite discrepancies between theory and experiment at some angles. It is very interesting that unlike the effect of higher discrete states ($n = 4, ..., 10$), the effect of the continuum is very large and significantly improves agreement with experiment, particularly the R parameter at 35 eV. This suggests that the treatment of the continuum is crucial for these parameters. Whilst the R matrix and 6CCO calculations give different results, they are in better agreement with each other than with experiment at those angles where the discrepancy between theory and experiment is greatest.

4. Conclusions

We have found that our CCO approach yields very good agreement with the measured elastic differential cross sections, and good agreement with the inelastic differential cross sections considered above. We therefore conclude that we have a reasonably reliable theory for the prediction of differential, and hence integrated cross sections for the two atoms considered here, across the broad energy range of projectile energies.

We believe that our treatment of Q space bound states is satisfactory from internal consistency considerations. A good test of the treatment of the continuum is provided by the comparison of CCO$^-$ and CCO calculations with experiment. The effect of the continuum on differential cross sections is not very large. Up to 50% change could still leave the calculation in reasonable agreement with experiment. However for hydrogen, this effect is quite large in calculating the angular correlation parameters. It would be of great assistance, in testing the treatment of the continuum, if further measurements of these parameters at a wide range of energies and angles were available.

Acknowledgments

The author would like to express his thanks to fellow team members Ian E. McCarthy and Dmitry A. Konovalov. He is also very grateful to Don H. Madison who's early contribution helped to initiate this work.

References

Bray I, Konovalov D A and McCarthy I E 1991 *Phys. Rev.* A43
Lorentz S R and Miller T M 1991 submitted to *Phys. Rev. A*
McCarthy I E and Stelbovics A T 1983 *Phys. Rev. A* **28** 2693
Scholz T T, Walters H R J, Burke P G and Scott M P 1991 *J. Phys. B* **24** 2097–2126
Slevin J, Eminyan M, Woolsey J M, Vassilev G, Porter H Q, Back C G and Watkin S 1982 *J. Phys. B* **26** 1344
Srivastava S K and Vušković L 1980 *J. Phys. B* **13** 2633
Weigold E, Frost L and Nygaard K J 1979 *Phys. Rev. A* **21** 1950
Williams J F 1975a *J. Phys. B* **8** 1683–92
Williams J F 1975b *J. Phys. B* **8** 2191–99
Williams J F 1976 *J. Phys. B* **9** 1519
Williams J F and Willis B A 1975 *J. Phys. B* **8** 1641

Collisions of polarized electrons with atoms and molecules

G F Hanne
Physikalisches Institut, Universität Münster, 4400 Münster, Germany

Abstract: We report on recent collision experiments with polarized electrons that have been performed in Münster. Experimental data for left-right scattering asymmetries and electron-photon coincidences in collisions of polarized electrons with complex atoms are presented and discussed. Exchange collisions have been observed directly by measuring the change of the spin polarization after scattering of polarized electrons from atoms and molecules. We found that exchange in collisions of electrons with O_2 and NO molecules is much less important than it is for Na atoms.

1. Introduction

The exploration of spin effects in collisions of electrons with atoms and molecules has been a permanent topic of the ICPEAC conferences since they started in 1958. This report continues a series of invited contributions about this topic.

Why is this topic still alive and experiencing growing interest? The answer is two-fold. Firstly, experimental progress during the last ten years made feasible accurate measurements of the change of polarization and of spin-dependent cross sections by means of intense beams of polarized electrons ('GaAs' source) and polarized atoms (see e.g. Kessler 1985 and 1991, Raith 1988, Hanne 1988). Even (e,eγ) and (e,2e) coincidence experiments and superelastic scattering from laser-excited atoms involving polarized electrons were reported (Wolcke *et al* 1984, Goeke *et al* 1989, Baum *et al* 1991, McClelland *et al* 1985, Nickich *et al* 1990)! Spin effects can be explored on the most fundamental level by means of these types of experiments. Secondly, along with the experimental progress, a variety of theoretical calculations that include spin-dependent 'forces' (exchange, spin-orbit effects) were developed during the last ten years.

We report here on recent experiments involving polarized electrons that have been performed in Münster. For results from other groups presented at this conference, we refer to this book and the Book of Abstracts (XVII ICPEAC).

2. Spin up-down Asymmetries in Collisions of Polarized Electrons with Unpolarized Complex Atoms

What is it that is so interesting with spin up-down asymmetries in collisions of polarized electrons with unpolarized atoms? There are, in general, two mechanisms of very different nature involved in such collisions (Figure 1). The first mechanism is the spin-orbit interaction of the continuum electron in the field of the atomic nucleus. This effect is well-known as Mott scattering: In the rest frame of the continuum electron we observe a magnetic field $B_1 = -(v \times E)$ due to the electron's motion with velocity v in the electric field E of the atomic nucleus. The magnetic field is perpendicular to the scattering plane, and thus the spin-orbit energy $\mu_e \cdot B_1$ is different for electrons with magnetic moment μ_e up and down. As a result we obtain a spin up-down asymmetry:

$$A = \frac{\sigma(\uparrow)-\sigma(\downarrow)}{\sigma(\uparrow)+\sigma(\downarrow)} = S_A P = S_A \frac{N(\uparrow)-N(\downarrow)}{N(\uparrow)+N(\uparrow)} , \tag{1}$$

where $\sigma(\uparrow)$ and $\sigma(\downarrow)$ mean the differential cross sections for electrons with polarization ('spin') up and down, respectively. Equation (1) also defines the polarization P of the incident electrons, where $N(\uparrow)$ and $N(\downarrow)$ are the numbers of electrons with spin up and spin down, respectively. For elastic collisions the asymmetry function A/P is called the Sherman function S, i.e. $A/P = S_A = S$.

The second mechanism is the result of an interplay between the spin-orbit splitting of atomic fine-structure states, an orientation dependent electron-electron Coulomb interaction and the Pauli principle (electron exchange). A simple picture of this 'fine-structure effect' is shown in Figure 1. Electrons are scattered elastically from, e.g., $^2P_{1/2}$ atoms. We assume that the cross section for scattering from atoms with orbital orientation $<L>$ up is much larger than with $<L>$ down. In that case, electrons are preferentially scattered from atoms with spin orientation down, since in a $^2P_{1/2}$ state orbital and spin angular momentum are 'opposite'. If exchange processes are important, there is, evidently, a spin up-down asymmetry since this corresponds to a antiparallel-parallel asymmetry of the spins of the colliding electrons.

In elastic collisions from closed-shell atoms, such as Hg, only Mott scattering needs to be considered. In elastic collisions from open-shell atoms that have a fine-structure splitting in their ground state, and also in inelastic collisions, polarization phenomena may as well be generated by the 'fine-structure effect'. The quantitative understanding of polarization effects is sound for Mott scattering, but less satisfactory for the 'fine-structure effect', particularly for its role in elastic collisions.

In Figure 2, calculations of the asymmetry function S for elastic collisions of electrons from Hg and B atoms (Bartschat 1987, 1990b) are shown. Contrary to Hg ($Z = 80$) where only Mott scattering can apply, B is an open-shell atom with configuration $^2P_{1/2}$ in its ground state, and it is light ($Z = 5$). Hence, Mott scattering effects are negligible for B, but the 'fine-structure effect' may apply. Those of you that are not familiar with this topic may find it amazing that, at the low energies shown here ($E = 1.36$ eV), both mechanisms can cause asymmetries that are similar in shape and are of comparable size! It is therefore difficult to distinguish these effects alone by inspection of experimental data. Unfortunately, it is, in practice, not feasible to resolve the $^2P_{1/2}$ state in B from the $^2P_{3/2}$ state (splitting only 2 meV). One must use heavier targets, such as In, where the two mechanisms may work simultaneously.

In order to help to clarify this situation, a program has been set up in our laboratory to provide relevant experimental data that can be compared with theoretical calculations.

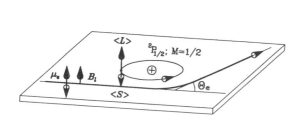

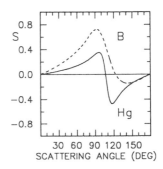

Fig. 1. Simple picture of spin up-down asymmetries caused by Mott scattering and 'fine-structure effect' in elastic collision of electrons from atoms.

Fig. 2. Numerical results for the Sherman function S for B (- - -) and Hg (——) at $E = 1.36$ eV (Bartschat 1987, 1990b).

Scattering asymmetries have been determined for a variety of targets (Geesmann *et al* 1991, Bartsch *et al* 1991, Dümmler *et al* 1990, 1991). From the results of these investigations we show here only some data for the targets Cd, In, Hg and Tl. Cd and In have neighboring atomic numbers ($Z = 48$ and 49) while their ground state configurations differ significantly, one being a closed-shell atom ($5s^2$ $^1S_0$) and the other being an open-shell atom ($5s^2$ $5p$ $^2P_{1/2}$). The situation is analogous for Hg and Tl ($Z = 80$ and 81, ground state configurations $6s^2$ $^1S_0$ and $6s^2$ $6p$ $^2P_{1/2}$, respectively). Comparison of the scattering asymmetries observed with the various targets should yield information on the role of the different spin effects in elastic and inelastic collisions.

A scheme of the experiment is shown in Figure 3. Linearly polarized light from a GaAlAs laser diode is converted to left or right circularly polarized light by means of a Pockels cell. This circularly polarized light creates longitudinally polarized photoelectrons from a GaAs Crystal. Typical emission currents range from 5...50 μA with a polarization $P \approx 0.24$ and an energy spread of about 0.15 eV. The electrons pass through a 180° monochromator and a 90° deflector which converts the longitudinal polarization into transverse polarization. The electron beam is focused onto the target by a lens system with energies down to 0.3 eV. The current at the target is > 0.15 μA. Electrons that are scattered to the left and to the right through the same scattering angle are detected by two electron analyzers, each consisting of an entrance lens system, a cylindrical mirror analyzer (CMA), and a position sensitive detector (PSD) at its exit. The PSD allows one to accumulate scattered electrons from different scattering channels simultaneously. The polarization of the incident electron beam is measured by means of a Mott detector that operates at 120 keV.

Experimental results are shown in Figures 4 and 5. For the inelastic np $^2P_{1/2} \to np$ $^2P_{3/2}$ and $(n+1)s^2$ $S_{1/2}$ transitions the asymmetries A/P of In and Tl show a nearly identical behavior at $E = 4$ eV (Figure 4). The theoretical calculations (Goerss *et al* 1991, Bartschat 1991) include the 'fine-structure effect', but only for Tl are Mott scattering effects included as well. The agreement with the experimental data is differing. Evidently, shape and size of the experimental asymmetry do not depend significantly on the atomic number Z. Hence the interpretation of these results is straightforward:

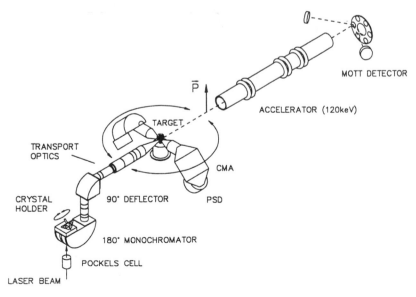

Fig. 3. Scheme of the left-right ('up-down') scattering experiment.

Exchange interaction between the scattered and atomic electrons ('fine-structure effect') is the dominant mechanism for producing scattering asymmetries in inelastic collisions at 4 eV. This is in accordance with previous results obtained with other atoms (Dümmler *et al* 1990, Geesmann *et al* 1991). Similarly, the results for elastic scattering at 2 eV from closed-shell atoms (Figure 5) are not difficult to interpret: The spin-orbit interaction of the scattered electron in the field of the atomic nucleus (Mott scattering) is the mechanism that causes the asymmetries in this case. As expected, the asymmetries are smaller for Cd ($Z = 48$) than for Hg ($Z = 80$). On the other hand, the interpretation of the elastic scattering asymmetries from open-shell atoms is somewhat involved. Bartschat's (1991) calculations for In at 2 eV illustrate that the dominant features are well described by the 'fine-structure effect' alone. On the other hand it is expected that Mott scattering also plays an important role in electron collisions with Tl ($Z = 81$). The R-matrix calculation of Bartschat (1990b) includes the Mott scattering as well as the fine-structure effect. The agreement of this calculation with the experimental data is fair. Surprisingly, the agreement between theory and experiment is better for a calculation of Haberland and Fritsche (1987) in which only the Mott scattering effect is included. It is difficult, from the experimental data alone, to disentangle the role of the 'fine-structure effect' and Mott scattering. This must be done by refined theoretical calculations for the atoms under study in conjunction with the experimental data (Bartschat 1990a).

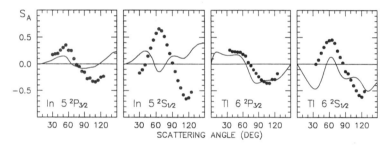

Fig. 4. Asymmetry function A/P for In and Tl for inelastic collisions at 4 eV.
• Experimental data, —— theory (In: Bartschat 1991, Tl: Goerss *et al* 1991).

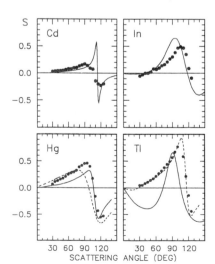

Fig. 5. Sherman function S for Cd, In, Hg and Tl for elastic collisions at 2 eV. • Experiment; —— , - - - theory (Cd: McEachran and Stauffer 1991, In: Bartschat 1991, Hg and Tl: —— Bartschat 1990b, - - - Haberland and Fritsche 1987).

3. Charge-Cloud Distribution of Hg*(6^3P_1) from Electron-Photon Coincidences with Polarized Electrons

The population of the magnetic sublevels of atoms excited by electron impact can be visualized by the corresponding charge-cloud (probability) distribution. Andersen *et al* (1988) have introduced a parameter set that characterizes shape and orientation of such a distribution. We generalize it for electron-photon coincidence experiments with polarized electrons. Figure 6 represents an example of the angular part $\rho(\theta,\phi)$ of a charge cloud. We assume an arbitrary transverse polarization vector for the incident electron beam. Therefore, reflection symmetry with respect to the scattering plane is, in general, broken, and the charge-cloud distribution may be tilted by an angle ϵ with respect to the scattering plane. Evidently, anisotropically populated atoms, such as shown in Figure 6, will emit linearly polarized photons. In fact, it is feasible to determine the parameters that characterize the anisotropic distribution of the radiating atom from measurements of the linear polarization of the emitted photons.

A scheme of the experiment is shown in Figure 7. A beam of electrons from a source similar to that shown in Figure 3 with spin polarization ($P_e = P_x$ or P_y) perpendicular to the beam direction is fired onto the Hg target. Electrons that are scattered inelastically (energy loss 4.9 eV) through an angle θ_e are detected in coincidence with photons (6^3P_1-6^1S_0 transition, $\lambda = 254$ nm) that are emitted perpendicular to the x-z scattering plane or, with a second photon analyzer not shown in Figure 7, into the -x direction, i.e., parallel to the scattering plane. Linear polarization filters (piles-of-quartz-plates) and $\lambda/4$ plates are used to determine the polarization state of the emitted photons.

The number of true coincidences is proportional to intensities $I(\alpha,P_e)$, where α is the angle by which the filter for linear polarization is inclined to the z axis. Intensities for unpolarized electrons are simulated by

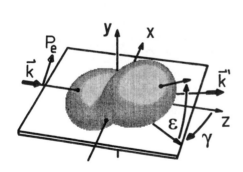

Fig. 6. Example of an anisotropic charge-cloud distribution.

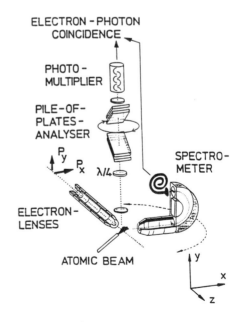

Fig. 7. Scheme of the electron-photon coincidence experiment with polarized electrons.

$$I(\alpha) = \tfrac{1}{2} I(\alpha, P_e) + \tfrac{1}{2} I(\alpha, -P_e) \ . \tag{2}$$

How can we determine the charge-cloud distribution of the excited $Hg^*(6^3P_1)$ state from the measured intensities such as in (2)? We can calculate polarization components (Stokes parameters) and spin up-down asymmetries, similar to the work of Goeke et al (1989). Bartschat et al (1981) have generalized the state multipole formalism of Blum (1981) by which these parameters can be expressed. State multipoles

$$<T(J)^{\dagger}_{KQ}> \tag{3}$$

characterize the angular momentum state of excited atoms with quantum number J ($-K \leq Q \leq K$). The state multipole with $K = 0$ is proportional to the overall population of the magnetic sublevels, whereas the state multipoles with $K = 2$ describe the anisotropic part of the population (alignment). The state multipoles with $K = 1$ describe the angular momentum orientation of the atom. The relationship between the Stokes parameters or the spin up-down asymmetries and the state multipoles are not shown here; they can be found in Bartschat et al (1981) and Goeke et al (1989). From measurements of such parameters all state multipoles (3) that are relevant for the construction of the charge cloud can be determined. Blum (1985) has shown how the state multipoles are related to the angular part $\rho(\theta,\phi)$ of the charge-cloud distribution. We give here the general expression for an arbitrary angular momentum J:

$$\rho(\theta,\phi) = \sum_{KQMM'} (-1)^{J-M}(JM'J-M|KQ)<T(J)_{KQ}> Y_{JM'}(\theta,\phi)Y^*_{JM}(\theta,\phi) \ , \tag{4}$$

where $(JM'J-M|KQ)$ are Clebsch-Gordan coefficients and the $Y(\theta,\phi)$ are the angular momentum functions of the atomic $|JM>$ substates. It turns out that for $J = 1$, only state multipoles with $K = 0$ and $K = 2$ and can contribute to $\rho(\theta,\phi)$.

Due to space limitation we cannot show here results of Stokes parameters and spin up-down asymmetries. Instead, we illustrate in Figure 8 the influence of the spin polarization on the shape of the charge-cloud of the excited atoms for scattering at 8 eV and a scattering of $\theta_e = 20°$. The relative size of the charge-cloud distribution is significantly different for $P_y = +1$ and $P_y = -1$ illustrating a spin up-down asymmetry. The alignment angle γ as well as the shape of the charge-cloud is very different for opposite spin polarizations $\pm P_y$, where the shape of the charge cloud for unpolarized electrons is just the average of those for $P_y = +1$ and -1. The alignment angle ϵ is clearly different from zero if $P_e = P_x = 1$. No such tilt out of the scattering plane would be observed, however, when the entire collision system could be described in LS coupling. This will

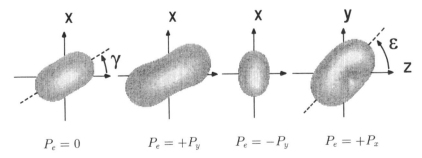

$$P_e = 0 \qquad\qquad P_e = +P_y \qquad\qquad P_e = -P_y \qquad\qquad P_e = +P_x$$

Fig. 8. Charge-cloud distribution of excited $Hg^*(6^3P_1)$ state after collisions with polarized electrons ($E = 8$ eV, $\theta_e = 20°$). For $P_e = 0$, $+P_y$ and $-P_y$ the view is perpendicular to the scattering plane (from the y direction), and for $P_e = P_x$ the view is parallel to the scattering plane (from the -x direction).

be shown in a forthcoming paper (Sohn and Hanne 1991), where detailed experimental results and a comparison with R-Matrix calculations by Bartschat (1991) will also be presented.

These experiments are currently extended in our group to excitation of noble gases by polarized electrons, where $np^6 \rightarrow np^5(n+1)s$ transitions are studied. This may result in shapes of the charge-cloud distribution which differ significantly from those of the $6s^2 \rightarrow 6s6p$ transitions in Hg.

4. Study of Exchange in Collisions of Polarized Electrons with Atoms and Molecules

While exchange effects in electron-atom collisions have been studied for more than 30 years (for a review see e.g. Kessler 1985), the first measurements for molecules were performed just recently by Ratliff *et al* (1989), and these were measurements of average spin-exchange cross sections at thermal energies for collisions of electrons with O_2 and NO molecules. These measurements correspond to spin-exchange cross sections integrated over all scattering angles and integrated over the spread of thermal energies of the electrons. Here we report the first measurement of relative differential spin-exchange cross sections for elastic scattering from molecules. When we started this project, it was anticipated that differential spin-exchange cross sections for elastic collisions of electrons from the open-shell molecules $O_2(X\,^3\Sigma_g^-)$ and NO ($X\,^2\Pi$) should be comparable to those for elastic scattering from Na (3^2S), where significant exchange effects were predicted up to an energy of 15 eV. However, we find no similarity between the corresponding atomic and molecular cases as described below.

We report here on investigations in which the change of the electron polarization has been measured to determine relative spin-exchange cross sections. A schematic diagram of the apparatus is shown in Figure 9. A lens system is used to focus the polarized electron beam from a source similar to that shown in Figure 3 onto the target, which is either a beam of Na atoms from an oven system, or a beam of O_2 or NO molecules from a gas inlet system. The energy of the electrons is varied from 4 to 15 eV. Some electrons are scattered through an angle θ and pass through a system of lenses and electrostatic deflectors. These can be rotated to select arbitrary scattering angles, where the 180° deflector serves as an energy analyzer. After that the electrons are accelerated to 100 keV, the energy at which the Mott analyzer operates to measure both the polarization P' after scattering as well as the initial polarization P (target off).

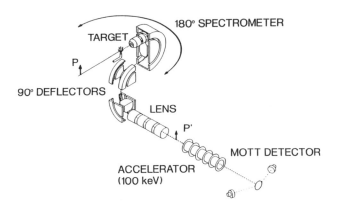

Fig. 9. Scheme of the P'/P experiment to determine relative spin-exchange cross sections.

For scattering from low-Z nuclei such as Na, O and N, spin-orbit effects can be neglected and 'spin flips' are only possible through spin-exchange collisions. With w_d denoting the probability for non spin-flip collisions and with w_{sf} denoting that for spin-flip collisions, the polarization P' after the collision is given by

$$P' = (w_d - w_{sf}) P = (1 - 2 w_{sf}) P \ , \qquad (5)$$

where P is the initial polarization. In the absence of spin-exchange collisions ($w_{sf} = 0$) we have $P'/P = 1$. Any deviation from $P'/P = 1$ is thus a direct observation of exchange collisions.

Some results of our measurements are shown in Figure 10. The error bars represent the statistical uncertainty (one-standard deviation) of the determination of P' and P. For elastic collisions from Na atoms, significant deviations from $P'/P = 1$ are observed which are, as expected, more pronounced at 4 eV than at 12.1eV. For elastic collisions from light one-electron atoms, we have $2w_{sf} = |g|^2/\sigma$, where g is the exchange amplitude and σ is the differential cross section. Thus, Eq. (5) yields (Kessler 1985, Hanne 1988)

$$P'/P = (1 - |g|^2/\sigma) \ . \qquad (6)$$

We have used the close-coupling results of Moores (1971) and Moores and Norcross (1972) to obtain P'/P for Na using Eq. (6) and these results are also shown in Figure 10. While experiment and theory are in reasonable agreement for elastic scattering, at 12.1 eV the minimum observed in P'/P is less pronounced and shifted towards smaller scattering angles than predicted.

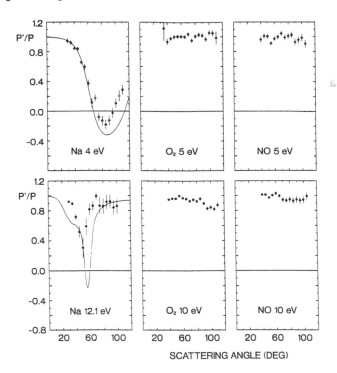

Fig. 10. P'/P plotted versus scattering angle for collisions of polarized electrons with Na atoms and O_2 and NO molecules. • Experiment, —— theory (12.1 eV; Moores 1971, 4 eV: Moores and Norcross 1972).

The measurements for Na demonstrate that the experimental method works. Another purpose of our investigation is the measurement of relative spin-exchange cross sections for scattering of electrons from open-shell molecules like O_2 and NO. To our surprise, their relative spin-flip cross sections are much smaller than those found for Na atoms at all energies (5-15 eV) and scattering angles studied so far. This is illustrated in Figure 10, which shows some typical results for O_2 and NO. In a recent measurement, Ratliff *et al* (1989) have determined the averaged spin-flip cross section for scattering of electrons from O_2 and NO molecules at thermal energies. They found averaged cross sections that are significantly smaller than those found for scattering from alkali or hydrogen atoms. While small average cross sections do not necessarily imply small differential spin-exchange cross sections, our results demonstrate that the differential spin-exchange cross sections are also small, even at large scattering angles where they are, in general, most likely to occur.

Evidently, spin exchange of continuum electrons with target electrons in the valence orbitals of O_2 and NO — which are π^* orbitals — is much less important than exchange with the s orbitals of alkali atoms. This would not have been anticipated, however, since it is known that exchange with the σ_g orbital in electron-H_2 collisions influences significantly the magnitude of cross sections (Lane 1980). Unfortunately, exchange cannot be observed directly for elastic scattering of electrons from a closed-shell molecule such as H_2. A possible explanation for the experimental findings is that the coupling of the electron spins to other angular momenta (spin-orbit coupling, rotational coupling) suppresses exchange of electrons with opposite spins from molecular π^* orbitals. Theoretical results investigating this phenomena should be available soon (Burke 1990, McKoy 1990). While a preliminary calculation (da Paixao 1991) confirmed that spin-exchange cross sections for these molecules are small, no explanation was provided. More experimental and, in particular, theoretical work is required to provide further insight into the strange behavior observed so far. Experimental work will be continued in our laboratory, and we hope that the publication of these first results (Hegemann *et al* 1991) will stimulate the discussion about the significance of exchange collisions in electron-molecule collisions.

Acknowledgements. The author gratefully acknowledges a fruitful cooperation with Prof. J. Kessler and the strong support that he gave to our work. The experimental results presented here have been obtained by several coworkers. I want to thank in particular Dr. H. Geesmann, Dr. M. Bartsch, M. Dümmler, M. Sohn and T. Hegemann. Furthermore I am indebted to my theoretical colleagues Dr. K. Bartschat and Prof. K. Blum for helpful discussions. This work has been supported by the Deutsche Forschungsgemeinschaft in Sonderforschungsbereich 216 'Polarization and Correlation in Atomic Collision Complexes'.

References

Andersen N, Gallagher JW and Hertel IV 1988 *Phys. Rep.* **165** 1
Bartsch M, Geesmann H, Hanne GF and Kessler J 1991 *Preprint*
Bartschat K 1987 *J. Phys. B* **20** L815
  — 1990a *J. Phys. B* **23** 2341 S
  — 1990b *private communication*
  — 1991 *private communication*
Bartschat K, Blum K, Hanne GF and Kessler J 1981 *J. Phys. B* **14** 3761
Baum G, Blask W, Freienstein P, Frost L, Hesse S and Raith W 1991 *Verh. DPG* **7** 677

Blum K 1981 *Density Matrix Theory and Applications* (Plenum, New York)
    — 1985   *Fundamental Processes in Atomic Collision Physics* (Plenum, New York) 103
Burke PG 1990 *private communication*
Dümmler M, Bartsch M, Geesmann H, Hanne GF and Kessler J 1990 *J. Phys. B* **23** 3407
    — 1991 *in preparation*
Geesmann H, Bartsch M, Hanne GF and Kessler J 1991 *J. Phys. B* **24**
Goeke J, Hanne GF and Kessler J 1989 *J. Phys. B* **22** 1075
Goerss H-J, Nordbeck R-P and Bartschat K 1991 *J. Phys. B* **24** *and private communication*
Haberland R and Fritsche L 1987 *J. Phys. B* **20** 121
Hanne GF 1988 *Coherence and Correlation in Atomic Collision Physics* (Plenum, New York) 41
Hegemann T, Oberste-Vorth M, Vogts R and Hanne GF 1991 Phys. Rev. Lett. **66** 2968
Kessler J 1985 *Polarized Electrons* (Springer, Berlin)
    — 1991 *Adv. Atom. Mol. Opt. Phys.* **27** 81
Lane NF 1980 *Rev. Mod. Phys.* **52** 29
McClelland JJ, Kelley MH and Celotta RJ 1985 *Phys. Rev. Lett.* **55** 688
McKoy BV 1990 *private communication*
McEachran RP and Stauffer AD 1991 *Preprint*
Moores DL 1971 *Comp. Phys. Commun.* **2** 360
Moores DL and Norcross DW 1972 *J. Phys. B* **5** 1482
Nickich V, Hegemann T, Bartsch M and Hanne GF 1990 *Z. Physik D* **16** 261
da Paixao LF 1991 *private communication*
Raith W 1988 *Fundamental Processes of Atomic Dynamics* (Plenum, New York) 429
Ratliff JM, Rutherford GH, Dunning FB and Walters GK 1989 *Phys. Rev.* **A39** 5584
Sohn M and Hanne GF *to be published*
Wolcke A, Goeke J, Hanne GF, Kessler J. Vollmer W, Bartschat K and Blum K 1984 *Phys. Rev. Lett.* **52** 1108

Double ionization of atoms by electron impact e,(3−1)e and (e,3e) techniques

A Lahmam-Bennani and A Duguet
*Laboratoire des Collisions Atomiques et Moléculaires (URAD0281), Université de Paris-Sud,
91405 ORSAY Cedex, FRANCE*

ABSTRACT : A preliminary investigation of the e,(3-1)e and (e,3e) double ionization processes in argon and krypton, at ~5.5 keV impact energy is presented. The relevance of the resulting four-fold and five-fold differential cross sections to the study of electron correlation and/or to the identification of the double ionization mechanism is emphasized. Experimental difficulties are outlined.

1. INTRODUCTION

Extensive experimental and theoretical information exists on *total* double ionization (DI) cross sections under various projectiles impact. Such studies most usually only determine the ratio σ^{2+}/σ^{+} of double to single ionization cross sections as a function of the incoming particle energy, and are therefore not sensitive to finer details of the ionization dynamics. *Differential* cross sections with respect to the energies and/or solid angles of emission of the final particles have seldom been reported. Measurements of final electron energies and momenta in the electron impact DI of an atomic target ((e,3e) process) allow one to determine the energy state in which the ion is left, i.e. to find the orbitals from which the outgoing electrons are knocked. Such completely differential experiments are extremely sensitive to the inter-electronic correlation.

A schematic diagram for such an experiment is shown in figure 1, for the case of a coplanar geometry. The fast incident electron, with an energy of several keV, is indexed 0. The three outgoing electrons, though indistinguishable, are indexed a for the fast, mostly forward "scattered" one and b and c for the two slow "ejected" ones. Conservation of energy and momentum requires

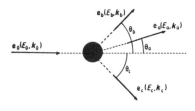

$$E_0 = E_a + E_b + E_c + E_i^{2+} \qquad (1a)$$

and $\qquad k_0 = k_a + k_b + k_c + q_r \qquad (1b)$

Fig.1. Schematic diagram of a coplanar (e,3e) experiment

where E_j and k_j (j=0, a, b or c) are the electron energies and momenta, E_i^{2+} is the DI energy of the target, and the atom and the residual ion are in their ground state. The ion recoil momentum is denoted by q_r. All electrons lie in the collision plane defined by k_0 and k_a, at angles θ_a, θ_b and θ_c with respect to the incident direction. Finally, $K^{2+}=k_0-k_a$ is the total momentum transfer to the target.

In an (e,3e) experiment, all emitted electrons are analysed both in direction and in energy and are detected in coincidence, yielding the most detailed information about the DI process. Hence, the quantity that is measured at a given impact energy is a fivefold differential cross section, 5DCS, or $d^5\sigma/dE_a\, dE_b\, d\Omega_a\, d\Omega_b\, d\Omega_c$, analogous to the triple differential cross section, TDCS or $d^3\sigma/dE_a\, d\Omega_a\, d\Omega_b$, measured in an (e,2e) single ionizing experiment. The e,(3-1)e experiments stand halfway between (e,2e) and (e,3e), as an arbitrary pair of electrons is detected in coincidence, irrespective of the direction of the third unobserved one. Therefore, integration is performed over the solid angle of emission of the undetected electron, say Ω_c, yielding a fourfold differential cross section, 4DCS or $d^4\sigma/dE_a\, dE_b\, d\Omega_a\, d\Omega_b$. Its energy E_c is in general also unknown since the ion final state is not determined.

2. ELECTRON CORRELATION

The key-word to understand why doing these fully differential experiments is electron correlation. Indeed, as in the (e,2e) case, depending on the kinematical parameters, one may distinguish between structure studies and ionization dynamics studies.

2.1 Structure studies : initial state correlations

These experiments have first been proposed by Neudatchin and coworkers (Neudatchin *et al* 1977, Yudin *et al* 1985 and references therein). The authors considered the case where both the ejected electrons have high energies and may therefore be described legitimately by plane waves. They showed, using the orthogonalized plane-wave first Born model that the cross section measured in an (e,3e) experiment is directly related to the square of the two-electron Fourier amplitude $\left|\psi(\mathbf{p}_1, \mathbf{p}_2)\right|^2$, similar to the impulsive (e,2e) TDCS which are proportional to the analogous one-electron quantity, $\left|\psi(\mathbf{p})\right|^2$. Therefore, one can study in detail the structure of a two-electron system, and in particular the momentum distribution of relative motion of electrons, which is particularly sensitive to electron-electron correlations. Typical recommended experimental conditions correspond to 5 keV incident electrons which are scattered strictly forward ($\theta_a \sim 0$) and symmetrical energy distribution between outgoing ejected electrons $E_b=E_c \sim 250$ eV. Under these conditions, the predicted (e,3e) 5DCS are of the order of 10^{-5} au or less, which unfortunately is appreciably below the sensitivity of the present experimental techniques.

2.2. Ionization mechanism

It is precisely for this reason that we have chosen to perform our (e,3e) experiments at lower ejected electron energies, 10 to 60 eV, where the cross sections are larger. Indeed, the measured absolute 5DCS are two orders of magnitude larger than given above.

Our work was motivated by several unanswered questions concerning the dynamics of the DI process : (a) How does the energy partitioning between the two "atomic" electrons depend on their respective directions of emission ? (b) What are the angular distributions of the electrons ? (c) How do these energy and angular distributions vary with the amount of momentum transferred to the target ? (d) what is the main process responsible for outer-shell DI under given kinematical conditions (i.e., is it a direct DI process or does it proceed via correlation between the electrons) ? And we are back again to this idea of correlation. To illustrate how sensitive is the 5DCS to electron correlations, let us consider one of the rare theoretical studies of this quantity (Dal Cappello and Le Rouzo 1991). In fact, it is well-known since the work of Byron and Joachain (1967) that it is the correlation between the two target electrons which is responsible for double photoionization. In the case of electron impact, outer-shell DI can be envisaged either as a first-order or as a second-order process. In the first case, the projectile electron interacts only once with a single target electron which is ejected without further interaction with other target electrons. Subsequent ejection of a second electron occurs, due to the electronic relaxation caused by the sudden change in the effective charge seen by this electron. This is the so-called shake-off mechanism (SO). The second case is a two-step mechanism (TS) in which the incident electron may either (i) interact with one target electron which subsequently collides with a second electron leading to ejection of the pair (TS1), or (ii) collide with two target electrons, resulting in their double ejection (TS2). In other words, DI may then be regarded as equivalent to a first ionization of the atom by the incident electron, followed by a second ionization of the resulting ion provoked by one of the intermediate electrons from the primary ionization event, either the slow "ejected" electron (TS1) or the fast "scattered" electron (TS2). Mc Guire (1982) has estimated the importance of each process in the case of a helium target and for total DI cross sections : at the energies in the present experiments, i.e. ~ 5 keV, the SO contribution should be dominant, about 3 times larger than the TS contribution. Based on this conclusion, DalCappello and Le Rouzo (1991) calculated (e,3e) angular distributions for He (figure 2) using a theory in which they take account only of the first term of the Born series and describe the incident and fast scattered electrons by plane waves. To investigate the sensitivity of the 5DCS to the electronic correlations, the authors used i) four different initial state wavefunctions including more or less of radial and angular correlations, and ii) two final state wavefunctions whose most sophisticated is the orthogonalized Coulomb wave (OCW) model (figure 2b) with angular dependent effective charges. This model takes into account the final state Coulomb interactions between each ejected electron and the nucleus, and includes in addition, but in an approximate way, the Coulomb interaction between the two ejected electrons. Such ee interaction is not present in the Coulomb wave (CW) model (figure 2a) with fixed effective charge, Z=2.

A sample of the results is shown in figure 2, for a symmetric energy-sharing $E_b=E_c=15$ eV. For CW calculations (figure 2(a)), the main facts to be underlined are i) the very high sensitivity of the 5DCS to the amount of correlation in the initial state wavefunctions, and ii) the physically astonishing result that the most probable event would be the ejection of both electrons in the direction of momentum transfer. In fact, this is the consequence of the complete neglect of electronic correlation in the double continuum state. By use of variable effective charges (OCW model, figure 2(b)), the final state electronic correlation is partly accounted for, and electrons actually depart in different directions, ~ 40° from each other. Clearly, electronic correlations have such a dramatic effect on the 5DCS for DI that one may speak about a "correlation explosion" (Dal Cappello), both in the initial and the final states.

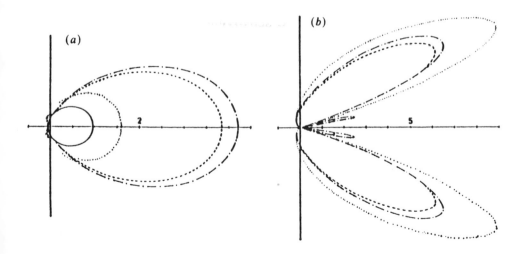

Fig. 2. (e,3e) cross sections for double ionization of He in coplanar geometry at $E_0=5.5$ keV, $E_b=E_c=15$ eV and $\theta_a=0.4°$. One electron is detected along the momentum transfer direction, represented by the horizontal axis), while the second one is detected at variable angle. The curves correspond to various initial state wavefunctions, from the non-correlated HF one (full curve, scaled by x0.1) to the highly correlated function of Tweed and Langlois (1987) (chain curve). (a) and (b) are the CW and OCW calculations, respectively.

3. EXPERIMENTAL TECHNIQUES

The measurement of the fourfold or fivefold differential cross section for an (e,3e) process requires the determination of all energies and directions of the three outgoing electrons. Coincidence techniques are also needed to guarantee that these electrons originate from one interaction event. A schematic diagram of the Orsay (e,3e) spectrometer is shown in figure 3.

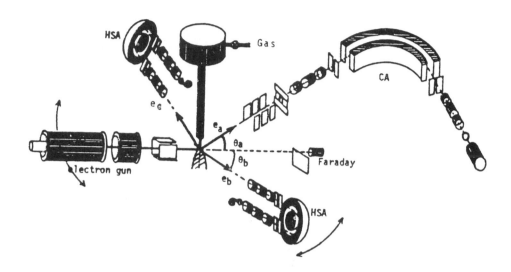

Fig.3 Schematic expanded view of the (e,3e) spectrometer, showing the cylindrical analyser (CA) and the two hemispherical ones (HSA), together with their associated electron optics.

A narrow electron beam is formed by an electrostatic gun at an energy of 1 to 10 keV. Its intensity is kept constant to better than 0.05% over several days by a feedback control on the Wehnelt voltage. The electron beam is scattered from a gas beam produced by a multichannel capillary. Each of the three outgoing electrons is analysed by an electrostatic analyser preceded by an electron optical lens system. The fast scattered electrons are decelerated to 1 keV prior to analysis in a fixed 127° cylindrical analyser. Then, they are reaccelerated to their nominal value and are detected with very high efficiency by a scintillator-photomultiplier arrangement. The slow "atomic" electrons are analysed in twin 180° hemispherical analysers, and are detected on channeltron detectors.

Both the electron gun and one of the hemi-spherical analysers are independently rotatable about the gas jet axis, while the cylindrical analyser remains stationary, thus defining the scattering angle θ_a and the ejection angle θ_b. In the present set-up, the second hemispherical analyser and hence θ_c is also fixed. As in the (e,2e) case, the system employed for triple coincidence detection in our (e,3e) experiments relies on the measurement of the difference in the arrival times of the pulses originated from the three detectors. The arrangement is essentially based on two identical time-to-amplitude converters (TAC), simultaneously started by the pulses from one detector, say a, and respectively stopped by the b and c-electron pulses. A logical AND gate allows the output signal of the TACs to be processed by two analogue-to-digital converters, and a triple coincidence to be registered *if and only if the TACs are both stopped* within a time window of 300 ns.

The data are displayed as a three-dimensional histogram of the arrival time coincidences, figure 4. The uniform background is caused by the arrival of fully uncorrelated electrons in the detectors, thereby giving rise to counts at any position of the spectra. A second type of background appears as edges (or walls) caused by one pair of correlated electrons, the third electron being random.

Finally, the peak superimposed on these background contributions is the triple coincidence peak due to fully correlated events. It is important to note that the AND gate allows most events due to a correlated electron pair to be rejected, so that the observed walls are a measure of the respective e,(3-1) e cross sections with a very low efficiency given by the probability of simultaneously finding a third electron at θ_c, within the 300 ns time interval corresponding to the TAC's ramps.

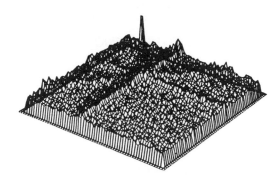

Fig. 4. 3D coincidence time spectrum.

The number N_T of true coincidences registered in the peak during the accumulation time t is given by (Dupré *et al* 1991)

$$N_T = N_{TA} - \sum_{\alpha=a}^{c} r_\alpha \, N_A^\alpha + 2r_u N_A^u \qquad (2)$$

where N_{TA} is the total number of counts, true plus accidental, in the n channels of the peak region, N_A^u is the total number of fully accidental counts registered in N_u channels of the uniform background, N_A^α is the number of accidental coincidences registered in N_α channels of the "α-wall" (α=a,b or c), and r_α and r_u are the ratios $r_\alpha = n/N_\alpha$ and $r_u = n/N_u$.

The two quantities that determine the "quality" of the experiment are the signal-to-background ratio, SBR, and the relative statistical uncertainty in N_T, σ_T/N_T, where σ_T is given by

$$\sigma_T = \left[N_{TA} + \sum_{\alpha=a}^{c} r_\alpha^2 \, N_A^\alpha + 2r_u^2 \, N_A^u \right]^{1/2} \qquad (3)$$

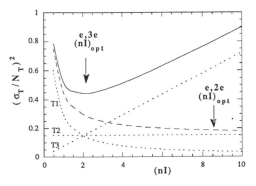

Fig. 5. Schematic variation of the square of the percentage statistical error, $(\sigma_T/N_T)^2$, for a given accumulation time, plotted against *nI*, the gas density-electron beam intensity product for an (e,2e) and an (e,3e) experiment, shown as a broken and a full curve, respectively. The arrows indicate the optimum choice for *nI* (from Dupré *et al* 1991).

Dupré *et al* (1991) investigated in detail the dependence of the SBR and σ_T/N_T upon nI, where n is the target gas density and I the incident electron beam current. Their results show that increasing nI i) reduces the signal-to-background ratio as $(nI)^{-1}$ and $(nI)^{-2}$ for the (e,2e) and (e,3e) cases respectively ; ii) improves, in the (e,2e) case, the relative statistical uncertainty down to the limit where it approaches a constant value (figure 5), while iii) in the (e,3e) case one observes a minimum in σ_T/N_T which determines the optimum choice for nI, nI_{opt}. Therefore, the major difference between (e,2e) and (e,3e) experiments is that, increasing the nI product, one does not gain in the e,2e case from the point of view of statistics, but at least one is not loosing. While this is not the case in (e,3e) case where an optimum choice does exist which minimizes the percentage statistical error.

4. RESULTS AND DISCUSSION

Three examples of results on the 4DCS, obtained on argon at ~ 5.6 keV incident energy, will be considered.

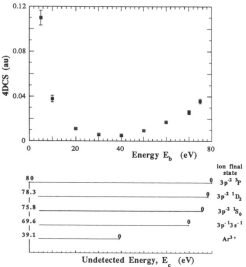

Fig. 6 : Absolute 4DCS showing the energy sharing between the two "atomic" electrons, see text. The horizontal lines represent the possible ranges of energy left for the undetected electron, corresponding to the main final ion states.

Figure 6 is an energy-sharing spectrum (Duguet *et al* 1991). Electron a is detected at θ_a =0.55° and fixed energy E_a=5500 eV. Electron b is observed at θ_b=50°, and its energy E_b is varied such that the excess energy above the DI threshold, $E_b+E_c=E_0-E_a-E_i^{2+}$ is kept fixed to 80 eV. That is to say that while E_b is varied, the unobserved c-electron carries out the complementary amount (80-E_b) when the ion is left in its ground state. The result is that the energy-sharing is far from being even. The probability is much higher that one electron carries out almost all energy and the second one almost nothing. The probability is very low to have two electrons coming out with equal energies. This distribution with a kind of a U-shape is well-known in single ionisation processes and has already been observed in integrated cross sections. But main difference here is that it is not symmetric. Indeed, if we were measuring total cross sections, integrated over all angles of emission of the b and c electrons, the result would necessarily have to be symmetric with respect to the mean energy because b and c are undistinguishable, and they would be playing the same role here. But in our case, we are measuring differential cross sections where electron b is observed only at one particular angle, 50°. The result would only be symmetric if both b and c electrons had the same angular distribution. Because this is not necessarily the case, our measured U-distribution has no reason of being symmetric. Of course, if we could measure similar U-distributions at all ejection angles, and add them together, the sum would have to be symmetric again.
The next examples are two angular distributions, both obtained for the same excess energy $E_b+E_c = 80$ eV.
The first one, shown in figure 7, (Lahmam-Bennani *et al* 1989) is an (a,c) coincidence between the scattered and the slow c-electron detected at 5 eV energy. The angular distributions for both double and single ionization events have been measured, for the same E_c value of 5 eV. As well-known, the binary and the recoil lobes of the single ionization distribution are found to

peak in the corresponding $\pm$ **K** directions. Whereas the DI distribution shows a very different angular behaviour. Moreover, the momentum transfer direction has lost its significance as a symmetry axis.

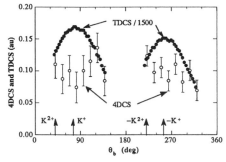

Fig. 7. Absolute fourfold (open symbols) and triple (closed symbols) differential cross sections for the coincidence detection of the pair $(e_a\, e_c)$ issued from double and single ionization of Ar $3p$, respectively, plotted against θ_c. $E_a=5500$ eV, $E_c=5$ eV, and $\theta_a= 0.55°$. Energy of the unobserved double-ionized b electron is $E_b=75$ eV. Single-ionization cross sections (i.e., TDCS) are divided by 1500.

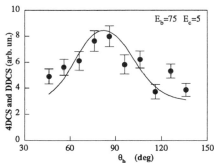

Fig. 8. Fourfold differential cross sections (full circles) for double ionization of argon in coplanar geometry at $E_0=5623$, $E_b=75$ and $E_c=5$ eV. The two "atomic" electrons are detected in coincidence with one angle fixed to $\theta_c=255°$, while θ_b is varied. The full curve is the corresponding double differential cross section measured under the same kinematical parameters.

The second angular distribution is a (bc) coincidence between the two slow ejected electrons, integrating over all scattering angles for the fast one, that is all momentum transfer directions.

The direction of emission of the slowest electron is fixed, and angle θ_b is scanned. The obtained angular distribution shows that the two electrons are rather strongly correlated. The similitude between the DDCS and the 4DCS distributions is consistent with an interpretation of the DI process in terms of a SO mechanism (Lahmam-Bennani *et al* 1991b).

Let us now turn to the triple coincidence measurements. Figures 9 and 10 show a sample of the measured (e,3e) 5DCS for double ejection from the 4p orbital of Kr (Lahmam-Bennani *et al* 1991a), at the indicated energies and angles. For each diagram all kinematical parameters are kept constant except θ_a in figure 9 and θ_b in figure 10 which are varied. The polar diagrams illustrate the kinematics employed in each case. Energies have been choosen to correspond (i) to a sum $(E_b+E_c) = 30$ eV, unevenly shared (figure 10e), or evenly shared (figure 10c-d) ; or (ii) to a sum $(E_b+E_c) = 90$ eV evenly shared (figure 9, 10a and b). In the even sharing case, electron b is detected either in the same half-plane as the a one, figure 10b and d, or in the opposite half-plane, figure 10a and c. (In (e,2e) studies, these are the so-called recoil and binary regions, respectively).

These data suffer from two major limitations. First, due to geometrical constraints, only a limited angular range could be covered. Second, due to the low triple coincidence rates, they were obtained at modest angular resolutions ($\Delta\Omega_a \sim 5\times10^{-5}$, $\Delta\Omega_b$ and $\Delta\Omega_c \sim 4\times10^{-2}$ sr) and modest energy resolutions ($\Delta E_a \sim 9$ to 15 eV, ΔE_b and $\Delta E_c \sim 5$ to 8 eV). However, the angular information is preserved, as shown by measurements made on well-known angular distributions

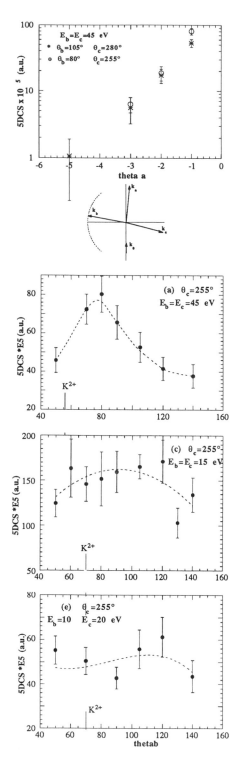

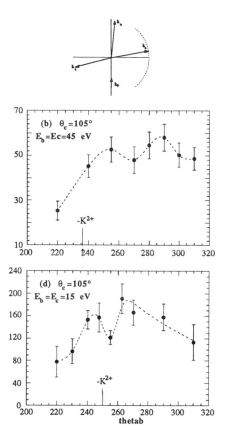

Fig. 9. Absolute (e,3e) five-fold differential cross sections for Kr $4p^{-2}$, versus θ_b. E_0= 5628.4 eV and E_a= 5500 eV.

Fig. 10. Absolute (e,3e) 5DCS for Kr $4p^{-2}$ versus θ_b. θ_a= -1°, E_a= 5500 eV. Left column : θ_c = 105°. Right column : θ_c = 255°.

(e.g. see figure 7). The modest energy resolution does not allow us to discriminate among various final Kr^{2+}-ion states, mainly the $4p^{-2}\ ^3P$ ground state ($E_i^{2+} = 38.4$ eV) and the $^1D_2$ and $^1S_0$ metastable states with excitation energies of 1.9 and 4.1 eV, but also the $4s^{-1}\ 4p^{-1}$ state whose threshold is $E_i^{2+} = 54$ eV. However, from observations made at low impact energy, we may reasonably expect the ^3P ground state to be the dominant contribution.

In spite of these modest energy and angular resolutions, the triple coincidence count rate is small, 25 counts/hour at maximum. And one should recall from the discussion in section 3 that it is useless to increase the beam current, as it can only result in degradation of the signal-to-noise ratio, and worse of the percentage statistical error. (The beam current used here ranges between 2 and 50 nA).

Parallel to these limitations, the data in figures 9 and 10 present a number of positive and very useful aspects. First, they have been assigned an absolute scale for the cross sections. This is a very important issue in future comparison with theory. Indeed, it is evident (see for instance figure 2) that data obtained only on a relative scale cannot distinguish between different theoretical models which predict about the same shape for the angular distributions but might differ in intensity by very large factors. Part of the sensitivity of the (e,3e) angular distributions to the electron correlations would then be lost. The absolute scale ($\sim 20\%$ accuracy) was determined by comparison with known elastic cross sections for Kr, and known (e,2e) triple differential cross sections, σ_{ab}^+ and σ_{ac}^+, for single ionization of He.

In the absence of any theoretical calculations to compare with, we can try to outline the essential features of the distributions in figures 9 and 10.

First, about the magnitude of the cross sections :
(i) from figure 9, the 5DCS is extremely rapidly decreasing with increasing scattering angle, much faster than the analogous TDCS for single ionization. The drop is about two orders of magnitude over an angular range of only few degrees. This certainly limits the experiments for the time being to very small θ_a-values, and explains the value $\theta_a = 1°$ used in figure 10 (or even 0.5° in the argon experiments (Lahmam-Bennani *et al* 1989)).
(ii) Also, from figure 10, the 5DCS decreases as the total ejected energy increases, from $\sim 2\times10^{-3}$ au at $E_b + E_c = 30$ eV to $\sim 5\times10^{-4}$ at 90 eV. These values are at the limit of what can be measured with the present experimental techniques, since they already necessitate ~ 24 hrs accumulation time for each single data point. Therefore, going to structure studies (section 2) where the quantity $|\psi(p_1, p_2)|^2$ would be aimed for, with predicted cross sections of less than 10^{-5} au, seems a rather difficult task.
(iii) At least at $E_b + E_c = 30$ eV, the cross section for an even energy-sharing is appreciably larger than the uneven case. This is in apparent contradiction with the well-known U-type distribution observed for total cross sections. However, as discussed above, the contradiction is entirely due to the different angular distributions of the differential cross sections whose sum over all angles must yield a symmetrical U.

Second, we can make some remarks about the shape of the angular distributions :
(i) When the b and a electrons are detected in opposite half-planes (i.e. $\theta_c = 255°$), the distribution is rather flat at $E_b = 10$ eV, and shows a more and more pronounced maximum as E_b increases.
(ii) Whereas, when b and a are in the same half-plane ($\theta_c = 105°$), the distribution seems to show a minimum between two maxima. The observed minimum is not very deep, but this might be due to the limited angular resolution which would tend to fill it. Why there is a minimum in one case and none in the other case is not understood.

(iii) Contrary to the TDCS distribution for single ionization where the ejected electron is known to preferentially depart in the momentum transfer direction or its opposite, $\pm K^{2+}$, these directions do not play a role anymore here.

4. CONCLUSION

In conclusion, it must be emphasized that the present experiments suffer from their modest energy and angular resolutions as well as from the limited θ_b-angular range which could be covered for each distribution. In spite of these limitations, they clearly demonstrate (i) the feasibility and the potential of (e,3e) and e,(3-1)e experiments, (ii) they help revealing all the technical problems involved in these experiments as well as mastering this new technique, (iii) they open up a new exciting domain especially for the study of electron correlations, and (iv) last but not least, they have prompted new theoretical efforts in this direction which should provide some guidance in planning the experiments and some support in interpreting the data.

However, even at the present energy and angular resolutions, the triple coincidence count rate is small, typically 7-8 counts/hr, and cannot be improved by increasing the beam current or the target gas density. Therefore, we are presently building a new experimental set-up which will incorporate a multiangle detection technique by using the cylindrical symmetry of two toroidal analysers. In this system, about 200 experiments such as the ones described here will be simultaneously performed.

REFERENCES

Byron F W Jr and Joachain C J 1967 *Phys.Rev.* A **164** 1

Dal Cappello C and Le Rouzo H 1991 *Phys.Rev.* A **43** 1395

Duguet A, Dupré C and Lahmam-Bennani A 1991 *J.Phys.B : At.Mol.Opt.Phys* **24** 675

Dupré C, Lahmam-Bennani A and Duguet A 1991 *Meas.Sci.Technol.* **2** 327

Lahmam-Bennani A, Duguet A and Grisogono A M 1991a to be published

Lahmam-Bennani A, Dupré C and Duguet A 1989 *Phys.Rev.Lett.* **63** 1582

Lahmam-Bennani A, Ehrhardt H, Dupré C and Duguet A 1991b *J.Phys.B : At.Mol.Opt.Phys.* **24** in press

McGuire H 1982 *Phys.Rev.Lett.* **49** 1153

Neudatchin V G, Smirnov Yu F, Pavlitchenkov A V and Levin V G 1977 *Phys.Lett.* **66A** 31

Tweed R J and Langlois J 1987 *J.Phys.B : At.Mol.Phys.* **20** 5213

Yudin N P, Pavlitchenkov A V and Neudatchim V G 1985 *Z.Phys.* A **320** 565

(e, 2e) on inner and outer shells at intermediate energies

L. Avaldi,R. Camilloni, E Fainelli and G. Stefani[+]

Istituto di Metodologie Avanzate Inorganiche del CNR
Area della Ricerca di Roma
CP10, 00016 Monterotondo Scalo, ITALY

[+] Dipartimento di Matematica e Fisica, Universita' di Camerino
62032 Camerino, ITALY

   Abstract : The possibility to use electron-electron coincidence techniques at intermediate energy and momentum transfer to study processes in which the direct ionisation competes with ionisation via resonant channels or in which ionisation and simultaneous excitation of the target occur is investigated. The experiments show that the contribution of resonant channels results in a shift of the peaks in the coincidence energy separation spectra and in unexpected distortions of the angular distributions. The measured triple differential cross section of the ionisation of helium to the n=2 states shows peculiarities typical of continuum final-state correlations, which appear to be enhanced when the residual ion is left in an excited state.

1. Introduction

Since the pioneering experiments by Amaldi et al. (1969) and Ehrhardt et al.(1969) electron-electron coincidence measurements have been successfully used either to investigate the ionisation mechanism upon several different kinematics or to study the electron momentum distribution of atomic and molecular orbitals.In the former case the simplest available systems, atomic hydrogen and helium, were used as targets and attention have been focussed to the interaction mechanism, i.e. the problem of three charged particles interacting through long range Coulomb forces. To the pourpose several kinematics were adopted from the threshold experiments of Fournier-Lagarde et al.(1984) and Jones et al.(1989) to the relativistic high energy one of Bonfert et al.(1991). In all these experiments the transition to the ground state of the ion has been investigated.In the latter case the dipolar (van der Wiel et al. 1976) and the impulsive kinematics (Leung and Brion 1983, McCarthy and Weigold 1988), in which the interaction mechanism is satisfactorily described by first order models, have been chosen and (e,2e) has been shown to be a useful tool to study the momentum density of systems as complex as polyatomic molecules (Barathi et al 1990) and solid target (Ritter et al.

1984, Gao et al. 1989, Lower et al. 1991).
In the present paper is described the possibility to use (e,2e)
experiments in order to investigate other processes in which
either direct ionisation competes with ionisation via resonant
channels or ionisation and simultaneous excitation of the
target occur.

2. Direct and resonant ionisation by (e,2e) experiments

Experimental and theoretical works on the photoionisation of
small molecules (Dehmer et al. 1987)have shown that whenever
the ionisation proceeds through shape resonances or
autoionising transitions the measured vibrational branching
ratios can be substantially different from those predicted by
the Franck-Condon (FC) principle and the β asymmetry parameters
display a strong dependence on the vibrational state of the
ion.
In a recent asymmetric (e,2e) experiment(Avaldi et al.
1990) the ionization of the C $\sigma 1s$ orbital in C_2H_2 has been
studied upon quasi-dipolar conditions. The measured angular
distributions are characterised by large recoil lobes, breaking
of the symmetry around the momentum transfer direction and
unusual bending of the recoil lobes towards smaller deflection
angles. Moreover the (e,2e) energy separation spectra showed
that the position of the C $\sigma 1s$ peak is shifted with respect to
the value measured by X-ray photoelectron spectroscopy . These
features are ascribable to several effects. Among them Post
Collision Interaction (PCI) between the slow ejected electron
and the decay products of the inner hole and the contribution
of the resonant channel $\sigma_g 1s \rightarrow Kb\sigma_u$ was supposed to play a
relevant role.
To better investigate the interplay between resonant and direct
ionisation channels without contributions of PCI effects an
investigation of molecular outer shells has been undertaken. To
the purpose the (e,2e) energy separation spectrum of the C_2H_2
$1\pi_u$ orbital and the angular distributions of the N_2 $3\sigma_g$ orbital
have been measured in asymmetric conditions at intermediate
incident energy and momentum transfer, K. The crossed beam
apparatus used in the present experiments as well as the
calibration procedure of the incident, scattered and ejected
electron energies and the test of the reliability of the
experimental set-up in measuring coincidence angular
distributions have been described in detail elsewhere (Avaldi
et al. 1987 and 1991).

2.1 Energy separation spectra of $1\pi_u$ orbital in C_2H_2

An extensive investigation (Langhoff et al. 1981,Machado et al.
1982 and Parr et al. 1982) of the photoionization of the C_2H_2
$1\pi_u$ orbital has shown that the relative intensities of the
different vibrational components suffer large departures from
the FC values for photon energies up to $h\nu \approx 16$ eV, i.e. about 5
eV above the $1\pi_u$ threshold.These effects are due to the
presence of autoionising states and shape resonances, although a

definite agreement on the assignment of the contributing
channels has not been reached yet. The experiments have been
performed at about 300 eV incident energy with a beam current of
a few hundred nA. Scattered and ejected electrons are collected
at fixed angles with respect to the incident beam. The scattered
electron energy, E_a, is always held fixed at 270 eV, while
several values of the ejected electron energy, E_b, between 2.2
and 10 eV are selected. In this way the selected energy loss,
$\Delta E=E_0-E_a$, in the ionising collision is tuned across the region
of excitation energies where resonant processes are dominant.
In measuring an energy spectrum E_a and E_b are held fixed, while
the energy E_0 of the incident beam is varied over the range of
interest. From the energy balance the "apparent" value of the
binding energy, $\varepsilon=E_0-(E_a+E_b)$, of the ionised orbital is
obtained.In the experiment ε is derived from the energy
position of the centroid of the vibrationally unresolved peak,
which corresponds, in the (e,2e) spectrum, to the ionisation of
the $1\pi_u$ orbital. Any change in the relative intensities of the
vibrational components of the transition will result in a
displacement of the centroid of the peak and, as a consequence,
in a different ε value.A series of measurements of the He 1s
(e,2e) energy separation spectra , at the same values of E_a and
E_b used in the C_2H_2 study, enabled the determination of the
energy resolution ($\cong$1eV FWHM) and response function of the
coincidence spectrometer, as well as the overall accuracy of the
ε measurements ($\pm$50 meV). The measured (e,2e) spectra of the $1\pi_u$
orbital are shown in fig. 1. The response function of the
spectrometer has been fitted to each spectrum to give the peak
positions and the ε values obtained by this procedure are given
in fig.2. Away from any resonance , a constant value for ε
(11.34$\pm$0.06 eV) is found, in good agreement with the one given
by photoionization (11.4 eV) (Machado et al. 1982) . Approaching
the resonance region, a

Fig. 1 : (e,2e) energy
separation spectra of
the $1\pi_u$ orbital. The
coincidence yield is
reported in arbitrary
units. The solid line
is the trial function
fitted to each spectrum.

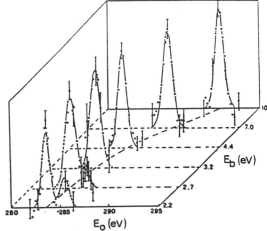

shift towards larger ε is observed. The measured shifts,
typically 250$\pm$50 meV, are in fair agreement with the
observations of Parr et al.(1982). Indeed the convolution of
the relative intensities of the vibrational transitions
measured by those authors at $h\nu\cong$14 eV with the response
function of the coincidence apparatus results in a peak which

is 170 meV shifted with respect to the one obtained by the convolution of the intensities measured, by the same authors, at hν=22 eV. Therefore the observed shifts are attributed to resonant processes that, perturbing the relative intensities of the vibrational components of the electronic transition ,result in a displacement of the centroid of the peak in the energy separation spectrum.

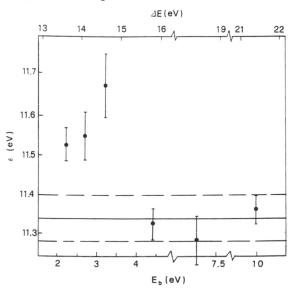

Fig. 2 : Measured ε values The solid and dashed lines show the ε value and its uncertainty obtained from the measurements away from any resonance.

At values of Eb of 2.7 and 2.2 eV an extra feature appears on the high energy side of the main peak (fig. 1). It can not be assigned to any stable $C_2H_2^+$ state (the $3\sigma_g^{-1}$ ionic state is at ε=16.36 eV (Parr et al. 1982)), and so it must result either from a satellite transition which ends up in an excited ionic state or from an autoionising transition. No information exists in the literature about ionic excited states 2 eV above the $1\pi_u^{-1}$ state. On the other hand, autoionising transitions from neutral states belonging to the npσu and npπu Rydberg series converging to the $3\sigma_g^{-1}$ ionic state have been invoked for the interpretation of the photoionization results (Machado et al. 1982, Parr et al. 1982).From the values of ΔE and Eb of the present experiment autoionisation from a Rydberg state to the $C_2H_2^+$ ($1\pi_u^{-1}$) ionic state can be excluded. A possible process is the dissociation of the excited molecule into two neutral fragments, one of which is in an excited state. Then the observed peak may correspond to an autoionising transition in this excited fragment. Little appears to be known about the excited states of the possible fragments. However similar processes have been observed in the photoionisation of diatomic molecules (Cafolla et al. 1989)

2.2 Angular distributions of the N_2 $3\sigma_g$ orbital

The   shape   resonance   occurring   at   hν≅34   eV   in   the photoionisation of molecular nitrogen leading to the $X^2\Sigma_g^+$ state of the ion is known to produce significant non FC effects in the final vibrational state distributions. These effects, predicted by Dehmer et al.(1979),have been observed experimentally by West et al.(1980) and Iga et al.(1989) both in the vibrationally resolved photoelectron cross sections and angular distributions. The ionisation of the $3\sigma_g$ orbital has been studied by a coplanar asymmetric (e,2e) measurement in which E_a has been kept fixed at 270 eV and two different E_b values have been selected, namely 18.4 and 10 eV. Being $\varepsilon=15.6$ eV the ionisation potential of the $3\sigma_g$ state the selected E_b's, which correspond to ΔE's of 34 and 25.6 eV, enable the investigation of the ionisation process in the shape resonance region and away from it.In measuring a coincidence angular distribution the scattered electrons are collected at a fixed angle ϑ_a, while the ejected electrons are collected at several angles ϑ_b by rotating one of the analysers. In this work the angular distributions have been measured at two different ϑ_a's in order to select K values which approach the dipolar limit.The measured angular distributions are shown in fig. 3(a-d). As expected the coincidence yields are distributed in two lobes, the binary lobe almost in the K direction and the recoil lobe in the opposite one. In order to extract quantitative informations on the symmetry of the lobes, their widths and the recoil to binary ratio a Legendre polynomial series has been separately fitted to the experimental data of the binary and recoil lobes (solid line in fig.3). At E_b= 10 eV a series truncated at the second order satisfactorily reproduces the experiments . When the ΔE in the collision is increased to 34 eV a polynomial truncated at the third order is needed to obtain an acceptable fit of the binary lobe at the smaller K and at larger K the trial function has to be extended up to the seventh order to reproduce the multilobe structure of the angular distribution. Any contribution to the observed shape from the $1\pi_u$ ionic state, whose ionisation potential is 1.3 eV apart from the $3\sigma_g$ one, can be excluded on the basis of the stated energy resolution . The non-coplanar symmetric experiments by Cook et al.(1990) have shown that the $3\sigma_g$ angular distribution displays a plateau for ion recoil momentum ($q = K_0-(K_a+K_b)$) values between 1 and 2 a.u., due to the shape of the molecular wave function in the momentum space. Despite the fact that in the present kinematics the impulse approximation does not hold, if the observed feature was due to the momentum distribution it should appear at the same q independently of the ϑ_a's.All the four measured angular distributions reach q values larger than 1 a.u. at the largest ϑ_b's, but only the binary lobes at E_b=18.4 eV display such a feature.At the same value of E_b the recoil lobes are well represented by a second order polynomial at both the scattering angles, but they are bent towards larger ϑ_b.Moreover all the lobes of the angular distribution become broader when E_b increases. The large body of the (e,2e) experimental and theoretical work on the ionisation of H and He (Ehrhardt et al. 1986) pointed out that interactions beyond the first order or final state continuum interactions result in the bending of the angular distribution with respect to the K direction. In

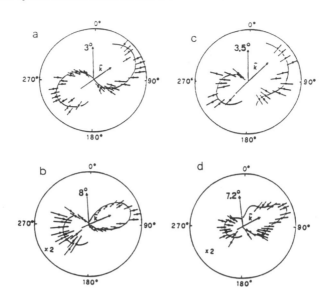

Fig. 3 : Angular distribution of the N2 3σg orbital at Ea=270 eV and Eb=10 eV (a-b) and 18.4 eV (c-d). The coincidence yields are reported in arbitrary units and the solid lines are the polynomial fits to the experiments.

particular it has been shown that the recoil lobe is bent toward smaller ϑ_b's. At intermediate K and E0 the effect decreases as K and/or E_b increase. Moreover all the previous experiments agree on a narrowing of the lobes as E_b increases. The present findings appear to be opposite. Therefore, although the extension of the results of the studies on the atomic species to more complex systems as molecules might not be straightforward, the effects observed in the angular distribution are likely to be ascribable to a contribution to the direct ionisation of a competitive process via the shape resonance $3\sigma_g\rightarrow\varepsilon\sigma_u, \varepsilon\pi_u$.

3. Ionisation and simultaneous excitation of helium

Processes which involve ionisation and simultaneous excitation of the target are of fundamental interest because they occurr only because of electron-electron correlation. Two classes of correlations are recognised to be relevant to such a process:
 i) internal correlation in the initial state of the target and/or in the final ionic state, which manifest as satellite lines in the energy separation spectrum; ii)dynamic correlation in the continuum final state that are switched on during the ionising collision. The latter ones involve short- and long-range interactions of the free electrons with each other and/or with the bound electrons.
(e,2e) experiments can be succesfully used to investigate electron correlations (Stefani et al 1990)and, by a suitable choice of the transition and the kinematics, they enable to

enhance either kind of correlation.The helium atom provides the simplest case for studying electron-electron correlations because internal correlations are present only in the initial state of the target. Their effect shows up in the (e,2e) energy separation spectrum with satellite peaks that correspond to the He II states with principal quantum numbers n=2,3,... The angular distribution relative to the more intense n=2 satellite at ε=65.4 eV was investigated at length with symmetric (e,2e) experiments and provides a rigorous test of the He ground state correlated wavefunction (Cook et al. 1984, and Smith et al. 1986).

The angular dependence of the same peak has been measured in asymmetric geometry. Three different experiments have been performed.The energies of the scattered and ejected electrons in the different experiments were 1500 and 20 eV,570 and 40 eV, 570 and 10 eV respectively. In all the experiments the scattered electrons were collected at $\vartheta_a=4°$ which correspond to K values of 0.8, 0.7 and 0.6 a.u.,respectively. To emphasize the relevance of the final state to the shape of the cross section the angular dependence of the He II n=1 peak at 24.5 eV has also been measured in the same experimental conditions as the ones used for the n=2 measurements. The angular distributions for the transition to the ground state are shown in figures 4 (a-c) while the ones for the transition to n=2 in figures 4 (d-f). The experiments for n=1 are compared with a plane wave first Born approximation, (solid line in figures 4 (a-c)) with an uncorrelated ground-state wavefunction (Popov and Shabalina 1986). At the highest incident energy theory and experiment are in fairly good agreement as far as the shape of the lobes is concerned, but the relative intensity of the recoil lobe is underestimated by a factor of about six. This finding can be explained by the interaction of the slow ejected electron with the nucleus not accounted for by the first order model. This interaction removes intensity from the forward (binary lobe) to the backward (recoil lobe)direction .At lower incident energy the experiments present a departure from the symmetry around K, which increases as K decreases and is not explained by the first order model. Also in these cases the relative intensity of the recoil lobes is underestimated by the model. These results show once more that in these kinematics the ionisation process has to be treated as a full three body problem in which both the interactions of the two escaping electrons with the residual ion and the non-central interaction between the two electrons play a relevant role. These findings are in general agreement with the ones of previous experimental investigation in analogous kinematics (Jung et al. 1985) . The angular distributions shown in fig. 4(d-f) share pratically the same kinematics with the experiments of figures 4 (a-c). The only difference is the ion core that is left either in its ground state or in the excited n=2 state. The larger width of the n=2 lobes can be partially explained by the wider n=2 momentum distribution measured by previous (e,2e) symmetric experiments (Cook et al. 1984, Smith et al. 1986). The binary versus recoil intensity ratio, which is never larger than 0.4 for the n=1 transition in the studied cases, approaches unity for n=2. Then a large deviation of the symmetry axis of the n=2 binary and recoil lobes from the K direction is observed. The break down of the symmetry might be

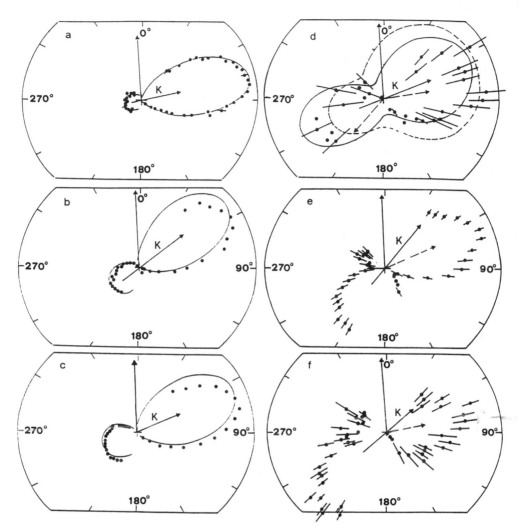

Fig. 4 : (e,2e) angular distributions of helium for the transition to the n=1 (a-c) and n=2 (d-f) states. The kinematics are : E_a=1500 eV and E_b=20 eV (a,d); E_a=570 eV and E_b=40 eV (b,e) and E_a=570 eV and E_b=10 eV (c,f). The scattering angle is always 4°. The solid lines in (a-c) are the predictions of the plane wave first Born approximation (Popov and Shabalina (1986).The solid and dashed lines in (d) are the predictions of the Brest group (Langlois et al. 1991) and Mota-Furtado and O'Mahony (1990), respectively. The broken arrows are the best symmetry axis of the individual lobes.

removed by the presence of a further feature in the angular
distributions at ϑ_b's smaller than $40°$, a region which can not
be investigated with our present experimental setup.
The deviation from the symmetry around **K**, mainly ascribed to the
interaction of the outgoing electrons with each other, are
dependent on K and on the relative velocity of the electrons, as
observed also in the ionisation to the n=1 state. However there
is a meaningless variation of the K value and the velocities of
the outgoing electrons are exactly the same in the experiments
on n=1 and n=2 . Hence an identical break down of the symmetry
should have been observed in the two sets of experiments. The
above observations drive to the conclusion that the observed
differences are related to the different final states in which
the target is left and rise the question whether the dynamical
effects observed are due to the fact that the slow electron was
initially coupled to a second bound electron or whether the
interaction of the ejected electron with the ion is heavily
dependent on the quantum state of the ion.
So far no theoretical predictions exist for the two experiments
at 600 eV, while two calculations for the high energy case have
been recently reported(Mota-Furtado and O'Mahony 1990, Langlois
et al. 1991). Both the models use highly correlated, although
different, wave functions for the target initial state(Tweed
and Langlois 1987, Mota-Furtado and O'Mahony 1988). They differ
in the treatment of the final state of the raection.
Mota-Furtado and O'Mahony (1990) use a 3-state close coupling
expansion to represent the wavefunction of the ejected electron,
where the 1s, 2s and 2p states are included. The scattering
amplitude is then evaluated in the plane wave first Born
approximation. The result of this calculation is represented by
the dashed line in fig. 4d . The predicted shape of the angular
distribution is no longer made by two separate lobes, in which
most of the cross section is concentrated, but a non-vanishing
cross section is predicted over a large ϑ_b range. This shape
result from the combination of the partial cross section to the
two quasi-degenerate $He^+(2p)$ and (2s) states, with the 2p
partial cross section dominating in the forward direction. As
far as the comparison with the experiment is concerned, the
model predicts a lobe in the forward direction too wide , but an
approximately correct relative intensity of the cross section in
the K and -K directions.
The second approach, developed by the Brest group (Langlois et
al.1991), uses a channel dependent central potential to
calculate the ejected electron wavefunction. This potential,
established on physical criteria independent of the collisional
process under study, has a parametric form and accounts for
radial and angular correlations, channel coupling and exchange.
The predictions of this model are also represented in fig. 4d
by the solid line.This latter model predicts an angular
distribution distributed in two lobes of comparable intensities
and its predictions are in better agreement with the
experiments then the ones of the previous model.Being the
target initial state described by highly correlated wave
functions in both the models, the differences between the two
models observed in fig.4d are to be ascribed to the description
of the ejected electron and its interaction with the residual
ion left in an excited state. However whether further

refinements in the description of the ejected electron wavefunctions would improve the agreement between theory and experiment or whether an extension of the calculation of the scattering amplitude beyond the first order is needed, is still matter of investigation.

4. Conclusions

The (e,2e) technique, succesfully used to investigate ionisation mechanism as well as momentum distributions, turns out to be a good way to investigate electron-electron correlations and processes dominated by resonant channels.
A resonant process was invoked to interpret the results of an (e,2e) experiment on the innermost orbital of acetylene (Avaldi et al. 1990) . However PCI among the slow ejected electron and the decay products of the inner hole might contribute to the observed phenomena. The study of the valence shell of C_2H_2 shows that whenever the excitation energy corresponds to a resonant channel the peak relative to that orbital is displaced in the energy spectra. An analogous study of the angular distribution of the N_2 $3\sigma_g$ orbital gives some evidences that resonant channels may also affect the coincidence angular distribution. This latter observation needs, however , further investigation. The experimental TDCS of the He^+ n=2 states display effects characteristic of continuum final-state correlations. These effects are enhanced when the ion is left in an excited state. So far a satisfactorily theoretical description of the phenomena has not been reached, but a comparison between theory and experiment suggests that the calculated angular distributions are strongly dependent on the description of the three body effects in the final state.

Acknowledgments

The authors are grateful to the colleagues R. Multari, S. Cvejanovic and G.C. King for their contribution to the experimental work in Rome and to R.J. Tweed for providing theoretical calculations for the He^+ (n=2) experiment.
Work partially supported by the EEC Contract :Science N.SCI*0175-C(EDB)

References
Amaldi U, Egidi A, Marconero R and Pizzella G 1969 Rev. Sci. Instrum. **40** 1001
Avaldi L,Camilloni R, Fainelli E,Stefani G,Franz A, Klar H and McCarthy I E 1987 J. Phys. B:At. Mol. Phys. **20** 5827
Avaldi L, Camilloni R and Stefani G 1990 Phys. Rev. **A41** 134
Avaldi L, Camilloni R, King G C and Stefani G 1991 Phys. Rev. A to be published
Barathi S M, Datta S K, Grisogono A M, Pascual R and Weigold E 1990 J. Electron Spectrosc. Relat. Phenom. **53** 51
Bonfert J, Graf H and Nakel W 1991 J. Phys. B:At. Mol. Opt. Phys. **24** 1423
Cafolla A A,Reddish T and Comer J 1989 J. Phys. B:At. Mol. Opt. Phys. **22,** L273
Cook J P D, McCarthy I E, Stelbovics A T and Weigold E 1984 J. Phys. B:At. Mol. Phys. **21** 2415

Dehmer J J,Parr A C and Soutworth S H 1987 *Handbook on Synchrotron Radiation* Vol.2, G.V. Marr Ed.(Amsterdam:North Holland), p.241 and references therein included

Dehmer J L, Dill D and Wallace S1979 Phys. Rev. Lett.**43** 1005

Ehrhardt H, Schulz M, Tekaat T and Willmann K 1969 Phys. Rev. Lett. **22** 89

Ehrhardt H, Knoth G, Schlemmer P and Jung K1986 Z. Phys. D1 3

Fournier-Lagarde P. Mazeau J and Huetz A 1984 J. Phys. B:At. Mol. Phys. **17** L5

Gao C, Wang Y Y, Ritter A L and Dennison J R 1989 Phys. Rev. Lett. **62** 945

Iga I, Svensson A and West J B 1989 J. Phys.B:At. Mol. Opt. Phys. **22** 2991

Jones T J, Cvejanovic S, Read F H and Woolf M B 1989 XVIth ICPEAC , New York, Book of Abstracts, p233

Jung K, Müller-Fiedler R, Schlemmer P, Ehrhardt H and Klar H 1985 J. Phys. B:At. Mol. Phys. **18** 2955

Mota-Furtado F and O'Mahony P F 1990 in *(e,2e) Collisions and Related Problems,*Kaiserlautern, Germany,p7

Mota-Furtado F and O'Mahony P F 1988 J. Phys. B: At. Mol. Opt. Phys. **21** 137

Langhoff P W, McKoy B V, Unwin R and Bradshaw A M 1981 Chem. Phys. Lett. **83** 270

Langlois J, Roubaix O and Tweed R J 1991 in *1990 (e,2e) Collisions and Related Problems,* Rome ,Italy,p234

Leung K T and Brion C E 1983 Chem. Phys. **82** 8711

Lower J, Barathi S M,Chen Y, Nygaard K J and Weigold E 1991 to be published

McCarthy I E and Weigold E 1988 Rep. Prog. Phys. **51** 299

Machado L E, Leal E P, Csanak G, McKoy B V and Langhoff P W 1982 J. Electron Spectrosc. Relat. Phenom. **25** 1

Parr A C,Ederer D L, West J B, Holland D M P and Dehmer J L 1982 J. Chem. Phys. **76** 4349

Popov Yu V and Shabalina E K 1986 J. Phys. B:At. Mol. Phys. **19** L855

Ritter A L,Dennison J R and Jones R 1984 Phys. Rev. Lett.**54** 2054

Smith A D, Coplan M A, Chorney D J, Moore J H, Tossell J A, Mrozek J, Smith V H Jr and Chant N S 1986 J. Phys. B:At. Mol. Phys. **19** 969

Stefani G, Avaldi L and Camilloni R 1990 in *Proceedings of the international symposium on correlation and polarization in electronic and atomic collisons*, NIST 789, P A Neill, K H Becker and M H. Kelley Ed.s, Gaithersburg , USA, p16

Tweed R J and Langlois J 1987 J. Phys. B:At. Mol. Phys. **20** 5213

van der Wiel M J , Wight G R and Tol R R 1976 J Phys. B:At. Mol. Phys. **9** 15

West J B,Parr A C, Cole B E,Ederer D L, Stockbauer R and Dehmer J L 1980 J. Phys.B:At. Mol. Phys. **13** L105

New applications of (e, 2e) techniques

E. Weigold

Electronic Structure of Materials Centre, The Flinders University of South Australia, GPO Box 2100, Adelaide 5001, Australia.

Abstract: The flexibility of the (e,2e) technique in obtaining information on both structure and collision dynamics is demonstrated. Examples of structure information are EMS studies of laser excited Na atoms and amorphous carbon films.   The role of post-collision effects and correlations is explored by measurements in the autoionising region of helium and in innershell ionisation of argon.

1. INTRODUCTION

The (e,2e) process, in which the momenta of the incident electron and two emitted electrons in an ionizing collision are completely determined, is capable of revealing a rich variety of information.   Depending on the kinematics employed, it is possible to investigate in detail either the dynamics of the ionizing collision or to use the reaction to elucidate the structure of the target and the ion.   When used for structure determination, high energies and high momentum transfers are normally employed to ensure the clean knockout of a target electron.   (e,2e) spectroscopy or electron momentum spectroscopy (EMS) has recently been extensively reviewed by McCarthy and Weigold (1988, 1991).

For most ionizing collisions, however, the kinematics is asymmetric, the two outgoing electrons having very different energies and the momentum transfer to the target is usually small.   Such asymmetric collisions have generally been studied using simple targets such as hydrogen (Weigold *et al.* 1979) or helium (e.g. Ehrhardt *et al.* 1969, 1972) whose structure is known or assumed to be known in order to test our understanding of the ionization mechanism.   As the momentum transfer approaches zero, the (e,2e) reaction simulates photo-ionization, and this kinematic region has been used to obtain much useful information on partical oscillator strengths and structure information (e.g. Hamnett *et al.* 1976).

The (e,2e) reaction has also been used to investigate final state correlation effects between the continuum electrons.   This has mainly focussed on post collision interaction effects (PCI) in the ionization of inner shells resulting in the emission of Auger electrons, particularly the ionization of 2p shell of argon (Sewell and Crowe 1982, 1984; Sandner and Völkel 1984; Stefani *et al.* 1986 and Weigold 1990).   The reaction has also been used to study the correlations between resonance and direct ionisation amplitudes in the autoionizing region of helium (Weigold *et al.* 1975, Pochat *et al.* 1982 and Weigold 1990).

In this communication I discuss briefly several (e,2e) studies recently carried out at Flinders University. These include the first measurements of the electron momentum distributions of an atom in an excited target state and of an oriented target, the application of the EMS technique to condensed matter targets, as well as the measurement of correlation effects in the autoionizing region of helium and in the Auger spectrum of argon.

2. NOTATION

The (e,2e) reaction can be written

$$e_0 + A \rightarrow A_f^+ + e_s + e_e, \tag{1}$$

where the subscripts 0,*s* and *e* denote the incident scattered and ejected electron. Although the two emitted electrons are indistinguishable, it is often convenient to call the fast outgoing electron the scattered one and the other the ejected one. Conservation of energy and momentum requires

$$\epsilon_f = E_0 - E_s - E_e, \text{and} \quad \mathbf{k}_0 = \mathbf{k}_s + \mathbf{k}_e - \mathbf{q}, \tag{2}$$

where ϵ, the separation energy of the electron, is equal to the energy difference between the initial target state A and final state $| f >$ of the ion, and $-\mathbf{q}$ is the ion recoil momentum of the ion. In the plane wave impulse approximation $\mathbf{q}$ is the momentum of the struck target electron. The ion recoil energy has been neglected. The momentum transfer to the target is given by

$$\mathbf{K} = \mathbf{k}_0 - \mathbf{k}_s. \tag{3}$$

In noncoplanar and coplanar symmetric (e,2e) experiments K is maximised by choosing $k_e = k_s$ and $\theta_e = \theta_s$.

At high enough energies and momentum transfer the (e,2e) differential cross section is given by

$$\sigma(e, 2e) = C f_{ee} \sum_{av} |< f \mid a(\mathbf{q}) \mid i >|^2, \tag{4}$$

where $\sum_{av}$ denotes the usual sum and average over final and initial degeneracies, C contains kinematical factors and f_{ee} is the half-off-shell Mott scattering cross section. In the noncoplanar symmetric geometry f_{ee} is essentially constant and the cross section is simply proportional to the square of the target-ion overlap amplitude resulting from the annihilation of an electron of momentum **q** in the target state $| i >$. This is the EMS region of the (e,2e) reaction.

The weak-coupling expansion of the target ion overlap is

$$< f \mid a(\mathbf{q}) \mid i > = \sum_{\ell} < f \mid \ell > < \ell \mid a(\mathbf{q}) \mid i >, \qquad (5)$$

where the orthonormal basis states $| \ell >$ are linear combinations of configurations formed by annihilating one electron in a target ion state. Generally only a single hole state $| j >$ contributes to the expansion in (6). We can define an experimental orbital $\psi_j(\mathbf{q})$ and the spectroscopic factor $S_f^{(j)}$ by

$$\psi_j(\mathbf{q}) \equiv < j \mid a(\mathbf{q}) \mid i > \quad \text{and} \quad S_f^{(j)} = |< f \mid j >|^2 \qquad (6)$$

3. ELECTRON MOMENTUM SPECTROSCOPY OF LASER EXCITED ATOMS

Excited target atoms can be prepared in well defined states by optically pumping atoms with a tuneable laser. By using polarised laser light it is possible to excite specific magnetic substates. This offers the possibility of measuring momentum distributions for excited targets as well as for atoms in aligned and oriented states. In the recent measurements of Zheng *et al.* (1990) and Bell *et al.* (1991a) sodium atoms are optically pumped by right hand circularly polarized laser light tuned to the $3^2S_{1/2}(F=2) \leftrightarrow 3^2P_{3/2}(F' = 3)$ transition. The excited pumped atoms are in the $m_{F'} = 3$ state, or in terms of orbital momentum in the $m_\ell = +1$ state.

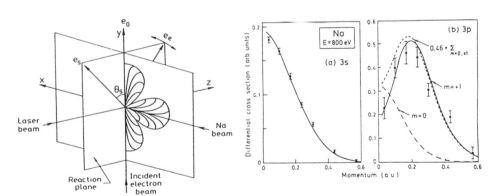

Fig.1 Schematic diagram of the experimental arrangement and the 3p($m_\ell = 1$) electron densities.

Fig.2 The momentum distributions for (a) the 3s ground state and (b) 3p excited state of Na compared with HF momentum distributions

The kinematics is shown in fig. 1, which also shows schematically the $3p_x$ and $3p_y$ electron density distribution of the excited sodium. The scattering plane is the zy plane and $\theta_s = \theta_e = 45°$, the azimuthal angle ϕ of the ejected electron is varied to probe the different momentum components of the target. Thus the experiment measures the (e,2e) triple differential cross section as a function of the x component of momentum, q_x. The z and y components of momenta are fixed and essentially zero for small binding energies. Due to the finite resolution of the spectrometer, the average values of q_y and q_z are not quite zero, being of the order of 0.06 a.u.

The momentum distributions obtained for the 3s ground state transition and for the $3p(m_\ell = 1)$ excited state are shown in figs. 2(a) and (b) respectively compared with calculated momentum distributions given by the Hartree-Fock 3s and 3p wavefunctions. The finite momentum resolution has been included in the calculations, its main effect is to fill in the 3p momentum distributions at momenta close to zero. The momentum distribution for the excited state peaks at very small momenta (~ 0.2 a.u.) because of the diffuse nature of the 3p orbital in coordinate space.

4. THE SPECTRAL MOMENTUM DENSITY OF AMORPHOUS CARBON

The (e,2e) technique can be used to obtain the spectral momentum density of electrons in solids, that is the probability of an electron possessing a particular value of binding energy ϵ and momentum $\mathbf{q}$. Ritter, Dennison and Jones (1984) reported the first spectral momentum density measurement of the valence band of a solid, the target material in this case being an amorphous carbon film. With an (e,2e) energy resolution of approximately 6eV, they were barely able to resolve two bands in the binding energy spectrum.

There has been considerable interest in determining the electronic properties of amorphous carbon (see Gao *et al.* 1989 and references therein). Amorphous carbon (a-C) films range from black, soft "graphite" films to hard, transparent "diamond-like" films depending on the method of preparation and the concentration of hydrogen in the sample (Ritter *et al.* 1984). The structure of a-C films is still not understood.

In order to obtain further information on the electronic structure of a-C, we have undertaken a series of (e,2e) measurements on evaporated films of a-C at much higher energy resolution than the above earlier work. In the present work a primary electron beam of 10keV plus binding energy intersects a free-standing 80Å amorphous carbon film. Scattered electrons leaving the collision region at angles of 45 degrees with respect to the incident beam direction pass through two identical hemispherical electrostatic analysers positioned on opposite sides of the beam. The analysers are used to determine the momenta and arrival times of each emitted electron by means of position sensitive detectors (Lower and Weigold 1989). Each analyser is adjusted to accept electrons in a 20eV band of energies centred around 5000eV. The combined energy resolution of each analyser and detector together is about 1eV. Different values of momenta q perpendicular to the incident direction and in the scattering plane are sampled by varying $\Delta E = E_s - E_e$ keeping $E_s + E_e = E_0 - \epsilon(\mathbf{q})$ constant. The values of q are given by

$$q = k_0 \frac{\Delta E}{2E_0} \left(1 + \frac{\epsilon}{E_0} \right) \qquad (7)$$

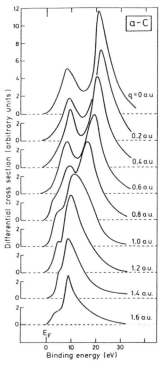

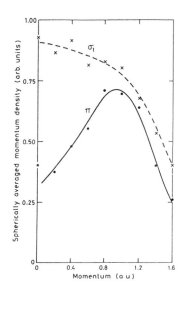

Fig.3 Spectral momentum density spectra for amorphous carbon (Lower *et al.* 1991).

Fig.4 Momentum densities for the valence π and σ bands of a-C (Lower *et al.* 1991).

Fig. 3 shows binding energy spectra obtained at different target electron momenta (Lower *et al.* 1991). This is a plot of the spectral momentum density. The low energy peak is attributed to the excitation of a graphitic π-band and the higher energy peak to the excitation of graphitic σ-bands. The data has been deconvoluted to allow for the effects of energy loss by incident and scattered electrons within the target material due to multiple scattering. The strongly dispersive nature of the higher energy σ band can be seen from the figure. The outer π band also shows some dispersion. In addition to the dispersion, there is for both bands a marked change in intensity as a function of q. This is displayed in fig. 4, which shows the momentum density as a function of q for the two bands.

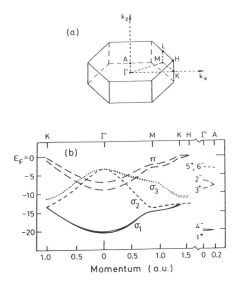

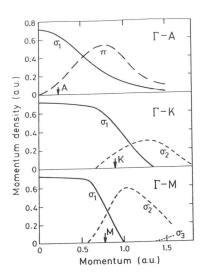

Fig.5   Brillouin zone and valence
band structure of graphite
(Goa *et al.* 1988).

Fig.6   Calculated spectral momentum
densities for the valence bands
of graphite in the different
crystal directions

The band structure of graphite as calculated by Gao *et al.* 1988 is shown in fig. 5, and the momentum densities for the π and three σ bands in fig. 6. The lowest σ band is σ_1 and its wavefunction is s-like, i.e. the momentum density is maximum at $q = 0$ and decreases monotonically as q increases. The wavefunctions for the upper two σ bands are p-like, the momentum density being zero at $q = 0$, and rising abruptly from zero near the Brillouin zone boundary to peak in the second and third zones, and then falling as q increases. Thus the σ_2 and σ_3 bands are essentially unoccupied at small q (≤ 0.5 a.u.). The momentum density for the π band is zero for **q** in the $\Gamma - K - M$ plane, since the π orbital has a node there. For **q** along the C axis perpendicular to the basal plane the momentum density of the π band displays p-wave character, it is zero at the origin and peaks at $q \approx 0.7$ a.u. In the present work we expect random orientations in the disordered phase of the evaporated films. This would still preserve the p-wave and s-wave character of the π and σ_1 bands, respectively.

Fig. 4 shows that the σ_1 band does indeed have s-wave character, having greatest momentum density at smaller q. The momentum density for the lower-energy π band does increase somewhat with q, as expected from the predicted p-wave character. There is some evidence that for diamond-like structures the momentum density for the outer-valence band does not go to zero (Gao *et al.* 1989). The spectral density for the π band is rather larger than expected for graphite. The data are therefore not consistent with the film being completely graphitic.

5. CORRELATIONS IN INNER-SHELL IONISATION : AUGER PROCESSES IN ARGON

The (e,2e) technique has been applied to study the ionisation of the 2p core level of argon by a number of groups. These experiments, proposed by Berezhko *et al.* (1978), detect the decay Auger electron in coincidence with the scattered electron, thus selecting momentum transfer and energy loss. The reaction can be written as

$$e^- + Ar \rightarrow Ar^+(J, M) + e_s^- + e_e^-$$
$$\downarrow$$
$$Ar^{2+}(J', M') + e_A^-. \tag{8}$$

Sewell and Crowe (1982,1984), Stefani *et al* (1986) and Sandner and Völkel (1984) measured angular correlations between the Auger and scattered electrons for the $L_3 M_{23} M_{23}(^1S_0)$ transition. Sandner and Völkel (1984) and Stefani *et al.* (1986) observed post collision interaction (PCI) effects close to threshold, the slow ejected electron transferring energy to the decay Auger electron. These experiments were all based on single parameter detection techniques, and the measurements could only be carried out by using very large energy windows on the scattered electron detector with a corresponding large spread in the energy of the near threshold ejected electron. In the present experiments we use a multiparameter coincidence spectrometer to examine simultaneously with improved energy resolution the whole argon Auger spectrum in the region 200-208eV in coincidence with scattered electrons as a function of the energy and angle of emission of the Auger electrons at several incident energies close to threshold. The spread in energy of the scattered and ejected electrons was only 3eV in these measurements, compared with about 10eV in previous measurements.

In the large distance eikonal approximation (Kuchiev and Sheinerman 1989), the energy shift ΔE_A in the Auger electron spectrum is given by

$$\Delta E_A = -\frac{\xi \Gamma}{2}, \tag{9}$$

where Γ is the width of the intermediate state and

$$\xi = \frac{1}{v_{sA}} - \frac{1}{v_s} + \frac{1}{v_{Ae}} - \frac{1}{v_e}. \tag{10}$$

In all measurements $v_e << v_s, v_A$ and $\xi \sim -1/v_e$. Thus the shift in energy $\Delta E_A = \Gamma/2v_e$ is positive and decreases as $E_e^{-1/2}$. For $v_e \sim v_S \sim v_A$ the post collision effects are a strong function of the angle of emission of the three electrons.

Fig. 7 shows a coincidence and noncoincidence Auger spectrum taken at just above threshold (Bell *et al.* 1991b). The coincidence spectrum shows an observable shift to higher energy, showing the significance of post collision interactions. The shift as a function of energy above threshold is shown in fig. 8 for the $L_3M_{23}M_{23}(^1D_2)$ transition compared with $\Delta E = \Gamma/2v_e$. The shape of the Auger line also depends sensitively on the parameter ξ (Kuchiev and Sheinerman 1989).

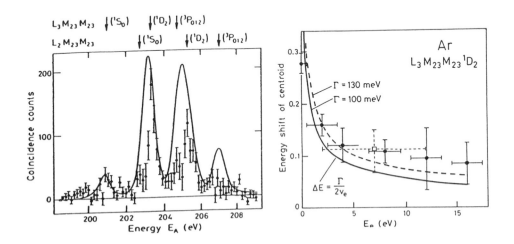

Fig.7 Coincidence spectra and non coincidence spectra (solid line) taken simultaneously with $E_0 = 750$eV, $E_s = 496$eV, $\theta_A = 135°$, $\theta_s = -18°$.

Fig.8 The PCI shift ΔE_A of Auger peak centroids as a function of E_e compared with eq.9 for the $L_3M_{23}M_{23}{}^1D_2$ transition ($\bullet$). The square is the data point of Stefani *et al.* (1986).

Berezhko *et al.* (1978) showed in the two-step plane-wave Born approximation that the probability of Auger electron emission in the scattering plane is given by

$$I(\theta_A)\alpha 1 + \beta\cos 2(\theta_A - \psi), \tag{11}$$

where the anisotropy parameter β and the angle ψ are related to alignment and statistical tensors. In the Born approximation $\psi = \theta_K$, the direction of momentum transfer. Thus angular correlations can lead to the determination of electron correlation effects.

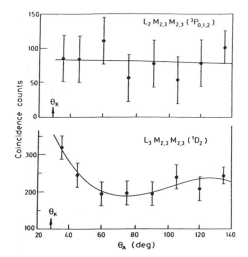

Fig.9 Angular correlations for argon Auger electrons in coincidence with scattered electrons $E_s = 496\text{eV}$, $\theta_s = -16°$, $E_0 = 753\text{eV}$.

Fig. 9(a) and (b) show respectively the angular correlations observed by Bell *et al.* (1991c) for the $L_2M_{23}M_{23}(^3P)$ and $L_3M_{23}M_{23}(^1D_2)$ transitions. The former must be isotropic since $J = \frac{1}{2}$ for the $2p_{\frac{1}{2}}$ shell and the initial state cannot be aligned.

6. (e,2e) COINCIDENCE MEASUREMENTS IN THE AUTO-IONISING REGION OF HELIUM

The autoionisation of atoms by electron impact involves in general the interference between the direct and the resonance ionisation amplitudes. This interference depends on the momenta of the scattered and ejected electrons and on the momentum transfer . Therefore (e,2e) cross section measurements in the autoionising region can provide very sensitive information on details of the excitation process of the resonance as well as on the interference of the resonance process with direct ionization.

Tweed (1976) showed that the triple differential (e,2e) cross section in the vicinity of the rth autoionising resonance can quite generally be written in the parameterized form originally due to Shore (1967),

$$\frac{d^5\sigma}{d\Omega_e \, d\Omega_s \, dE_e} = f(\mathbf{k}_e, \mathbf{K}) + \sum_\mu \frac{a_\mu(\mathbf{k}_{e\mu}, \mathbf{K}_\mu)\epsilon_\mu + b_\mu(\mathbf{k}_{e\mu}, \mathbf{K}_\mu)}{1 + \epsilon_\mu^2}, \tag{12}$$

$$\text{where} \quad \epsilon_\mu = 2(E_e - \bar{E}_\mu)\Gamma_\mu^{-1}, \tag{13}$$

and $\bar{E}_\mu$ and E_e are respectively the energies of the rth autoionising resonance and the energy of the ion plus continuum emitted electron (relative to the energy of the residual ion) with total angular momentum and spin quantum numbers denoted by

$\mu = \{r; L, M, S\}$. f is simply the cross section for direct ionization, a_μ is a measure of the asymmetry of the resonance profile and b_μ of its contributions to the cross section.

Fig. (10) provides some examples of the observed coincidence ejected electron spectra (Lower and Weigold 1990). We see a series of resonance profiles superimposed upon a background of direct ionization events. The final fitted function is represented by the solid curve, whilst the fitted direct ionization background f and the individual fitted resonance profiles $(a\epsilon + b)/(\epsilon^2 + 1)$ convoluted with the instrumental response, are indicated by dashed lines. Both spectra were obtained in the binary region with the only difference being 18° in the angle of the ejected electron. This is a clear demonstration of how sensitive the cross sections are to the ejected electron momentum.

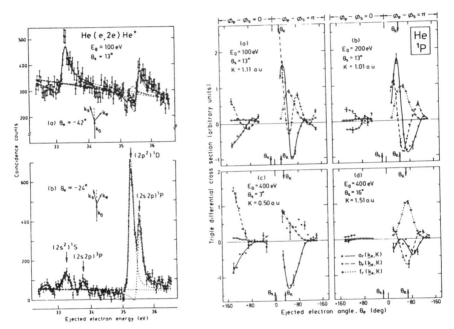

Fig.10 Coincidence ejected electron spectra for He(e,2e)He$^+$.

Fig.11 The direct (e,2e) cross section f and parameters a_r and b_r for the $(2s2p)^1P$ resonance in He (Lower and Weigold 1990).

Values for a_r, b_r and f_r deduced from the fitting of individual ejected electron spectra for the $(2s^2)^1S$, $(2s2p)^1P$ and $(2p^2)^1D$ resonances show rapid variations as a function of θ_e. Fig. 11 shows the results for the $(2s2p)^1P$ state for four different combinations of incident energies and scattering angles. Since absolute cross sections were not obtained the results have been normalized to the maximum value of f under the $(2s^2)^1S$ resonance profile at each particular energy and scattering angle, and this maximum has been set to 1 unity. The direct ionization cross section f for each incident energy and scattered electron angle appears in essentially two lobes. The major lobe (the binary lobe) occurs approximately in the momentum transfer direction. The second occurs roughly 180° away and can be associated with recoil collisions involving momentum transfer to the residual ion. For each resonance and at each energy and scattering angle, the a_r and b_r parameters show smooth oscillation between positive and negative values as the ejection angle varies. Positive values of b_r correspond to constructive interference between direct and resonant ionization amplitudes, resulting in localized increases in the (e,2e) cross section. Negative values of b_r are related to a decrease in the cross section due to the effects of destructive interference. Positive values of a_r correspond to a profile showing a minimum followed by a maximum in the sense of increasing ejection energy, whilst the reverse occurs when a_r is negative.

The results for all three resonances show strong interference, both constructive and destructive, between the direct and resonance amplitudes. The resonance parameters show even more rapid variation in magnitude as a function of the ejected electron angles than does the direct cross section. Although the peak structures in a_r and b_r are correlated with the direction of the momentum transfer θ_K, as is that for the direct cross section f_r, the correlation is not simple. Considerable amount of theoretical effort is needed to explain the data. First order theories are inadequate since they would give the **K** direction as a symmetry axis, which is clearly in violation of the measurements.

7. SUMMARY

The richness of information that can be obtained by the application of the (e,2e) technique has been demonstrated by discussing four different types of experiments. Two examples involve the determination of structure information in kinematic regions where the (e,2e) collision process is well understood. The other two examples involve the determination of subtle effects in the collision dynamics.

ACKNOWLEDGEMENTS

I am grateful to the ARC for financial support of this work and would like to acknowledge the essential role played by the other members of the Flinders (e,2e) group.

References

Bell S, Shen Y, Weigold E and Zheng Y 1991a Proc. 17th ICPEAC p177
Bell S, Hall D, Samardzic O and Weigold E 1991b Proc. 17th ICPEAC p200
Bell S, Shen Y, Weigold E and Zheng Y 1991c Proc. 17th ICPEAC p201
Berezhko E G, Kabachnik N M and Sizov V V 1978 J Phys B: At Mol Phys **11** 1819
Ehrhardt H, Hesselbacher K H, Jung K, Schultz M, and Willmann K 1972 J Phys B At Mol Phys **5** 2107
Ehrhardt H, Schultz M, Tekaat T and Willmann K 1969 Phys Rev Lett **22** 89
Gao C, Ritter A L, Dennison J R and Holzworth N A W 1988 Phys Rev B **37** 3914

Gao C, Wang Y Y, Ritter A L and Dennison J R 1989 Phys Rev Lett **62** 165
Hamnett A, Stoll W, Branton G, Brion C E and van der Wiel M J 1976 J Phys B: At Mol Phys **9** 945
Kuchiev M Yu and Sheinerman S A 1989 Sov Phys Usp **32** 569
Lower J and Weigold E 1989 J Phys E:Sci Instrum **22** 421
Lower J and Weigold E 1990 J Phys B **23** 2819
Lower J, Bharathi S M, Chen Y, Nygaard K J and Weigold E 1990 (to be published)
McCarthy I E and Weigold E 1988 Rep Prog Phys **51** 299
McCarthy I E and Weigold E 1991 Rep Prog Phys **54** (June issue)
Ritter A L, Dennison J R and Jones R 1984 Phys Rev Lett **53** 2054
Sandner W and Völkel M 1984 J Phys B: At Mol Phys **17** L597
Sewell E C and Crowe A 1982 J Phys B: At Mol Phys **15** L357
Sewell E C and Crowe A 1984 J Phys B: At Mol Phys **17** 2913
Shore B W 1967 J Opt Soc Am **57** 881
Stefani G, Avaldi L, Lahmam-Bennani and Duguet A 1986 J Phys B **19** 3787
Tweed R J 1976 J Phys B **9** 1725
Weigold E, Noble C J Hood S T and Fuss I 1979 J Phys B: At Mol Phys **12** 291
Weigold E 1990 Aust J Phys **44** 277
Weigold E, Ugbabe A and Teubner P J O 1975 Phys Rev Lett **35** 209
Zheng Y, McCarthy I E, Weigold E and Zhang D 1990 Phys Rev Lett **64** 1358

Resonant electron scattering from adsorbed molecules

J.P. Gauyacq, V. Djamo and D.Teillet-Billy

Laboratoire des Collisions Atomiques et Moléculaires (URA0281), Université de Paris-Sud, 91405 ORSAY, Cedex, FRANCE

ABSTRACT : A theoretical study of resonant electron scattering by static molecules adsorbed on a metal surface is presented. Using a new non perturbative method, the Coupled Angular Mode method, the characteristics of the quasistationary negative ion formed during the collision (resonance position and width, angular distribution) can be determined and analyzed. Specific results on the N_2-Ag and CO-Ni systems are presented.

1 INTRODUCTION

In the course of a gas phase electron-molecule collision, the transfer of energy between the electronic motion and the molecular vibration is often dominated by resonant processes. Indeed, a direct transfer from the electron to the nuclear motion is not efficient, due to the large mass difference. In contrast, resonant scattering which implies the capture of the incident electron to form a quasi stationary temporary anion considerably increases the collision time and the energy transfer. If a molecule is adsorbed on a metal surface, one can expect similar effects, i.e. energy transfer from the electron to the vibration will be favoured by the temporary formation of a negative ion. Furthermore, in the case of a molecule physisorbed on a metal surface, the molecule is only very slightly modified by the adsorption process, its scattering properties should be similar to those of the same molecule in the gas phase and one can then expect to observe for physisorbed molecules the same resonances as those observed for free molecules. These processes have been studied by a few groups over the past years (see the review by Sanche, 1990). The first experimental studies of vibrational excitation of physisorbed molecules by electron impact appeared in 1981 (Demuth et al 1981, Sanche and Michaud 1981). They revealed that gas phase resonances appear at significantly lower energy than in the gas phase, and that the resonance lifetime is decreased by the adsorption. For example in the case of N_2 adsorbed on Ag (Demuth et al 1981), which will be discussed below, the so called "boomerang structure" visible in gas phase collisions in the region of the $^2\Pi_g$ resonance disappears in the adsorbed case indicating a shortening of the resonance lifetime. More recently, Jacobi et al (1990) studied resonance scattering by adsorbed molecules ; they showed that the resonance formation was also inducing a vibrational excitation of the relative molecule-surface movement. This process is very similar to the "usual" intra molecular vibrational excitation and was originally suggested by Gadzuk (1985). Angular distributions for resonant scattering have also been investigated experimentally (Palmer et al 1988, Jones et al 1989) ; as in gas phase collisions, they are associated with the symmetry of the intermediate negative ion. The theoretical studies on this subject are much more scarse : Gerber and Herzenberg (1985) studied the case ot N_2 molecules with a spherical metallic environment. They also performed as well as Gadzuk (1983) a semi empirical analysis of the experimental results of Demuth et al on N_2-Ag. Rous et al (1989) studied the effects of multiple scattering on the resonant angular distributions.

Recently we developped a new non perturbative method to study negative ion states of atoms and molecules in front of a metal surface (Teillet-Billy and Gauyacq 1990, 1991). The first study concerned C^- ions in front of a Cs/W surface, it allowed the discussion of the electron capture process in collisions of C atoms with a Cs/W surface. In a second study, a preliminary approach of the resonant angular distributions was presented for the case of CO adsorbed on Ni, it is briefly discussed in part 5. Below, we present in parts 2-4 a theoretical study of the resonant scattering of electrons by N_2 molecules adsorbed on Ag. The N_2 molecule presents in the gas phase the well know $^2\Pi_g$ resonance around 2.3 eV, associated with a l=2 angular mode ; the aim of the present work is to determine the modifications of the N_2^- ($^2\Pi_g$) resonance introduced by the adsorption.

Let us first consider qualitatively the effects introduced by the metal surface, on the resonant scattering. First of all, when an electron approaches a metal surface, it induces in the metal an image charge, that results in an attractive potential (-1/4z) where z is the electron-surface distance. As a consequence, the energy of a negative ion should decrease, when the molecule approaches the surface. Since the electron is attracted toward the surface, one can also expect that the lifetime of the resonance is decreased. The e^--surface interaction has a symmetry completely different from the molecular scattering symmetry and then, the surface will mix the various molecular symmetries thus modifying the scattered electron angular distributions. A few other effects can also influence the resonant scattering : multiple scattering where an electron is successively scattered by a few adsorbed molecules (see e.g. Rous et al 1989), interactions between neighbours, modifications of the neutral molecule caused by the adsorption process. In the work presented below, only one molecule is considered, that is supposed to have not undergone any change due to the adsorption, so the latter effects are not taken into account.

2. THE COUPLED ANGULAR MODE (CAM) METHOD

The method has been presented with our study of the $C^-(^4S)$ in Teillet-Billy and Gauyacq (1990) and will be only briefly outlined here. The basic idea of the method is to study electron scattering by the compound molecule + surface system. The two interactions, e^--molecule and e^--surface interactions, are supposed to be additive. This approximation limits the applicability of the method to large molecule-surface distances where molecules are not significantly perturbed (physisorption). The e^--molecule interaction is described in the effective range approximation (ERT) (Gauyacq, 1987). This approximation was developped to treat low energy electron-molecule scattering : vibrational excitation and dissociative attachment (Gauyacq 1983, Teillet-Billy and Gauyacq 1984) as well as electronic excitation (Teillet Billy et al 1987). Within the ERT one only considers the region outside of the molecule ($r > r_c$, r is the e^--molecule distance) where uncoupled angular modes can be defined (spherical harmonics Y_{lm} in the present case) and where the e^--molecule interaction is described by a local potential V_{ext}^{lm} (r). The short range forces are represented by a boundary condition on the electron wavefunction at $r=r_c$

$$\frac{1}{\psi}\frac{\partial \psi}{\partial r}\bigg|_{r=r_c} = f_{lm}(R) \qquad (1)$$

f is energy independent for low energy scattering, it depends on the molecular internuclear distance. *A priori*, both V_{ext}^{lm} and f_{lm} depend on the considered (l,m) angular mode. For the N_2 study, the external potential was taken from the work of Le Dourneuf et al (1982), and the boundary condition is taken such that the ($^2\Pi_g$) resonance (l=2, m=1) lies at 2.3 eV.

The e^--Ag surface interaction is represented by a local potential V(z) only function of the e^--surface distance, taken from the work of Jennings et al 1988. It smoothly joins a constant potential inside the metal ($-V_0$) to an image charge potential outside the metal.

The e⁻-molecule and e⁻-surface interactions have a different symmetry. In the CAM method the scattered electron is described by an expansion over spherical harmonics Y_{lm} centered around the molecule :

$$\Psi_m = \sum_l^{l_M} Y_{lm} \frac{1}{r} F_{lm}(r) \tag{2}$$

In the present study, the N_2 molecular axis is supposed to be perpendicular to the surface. The system is then invariant by rotation around the molecular axis and m the projection of the electron angular momentum on this axis is a good quantum number (m=1 for N_2^- $(^2\Pi_g)$). When (2) is brought into the Schroedinger equation, it yields a set of coupled equations :

$$-\frac{1}{2} \frac{\partial^2 F_{lm}}{dr^2} + V_{ext}^{lm} F_{lm} + \sum_{l'} <lm|V|l'm> F_{l'm} = E F_{lm} \tag{3}$$

with the following ERT boundary conditions :

$$\frac{1}{F_{lm}} \frac{dF_{1m}}{dr}\bigg|_{r=r_C} = f_{lm}$$

The expansion (2) is well suited to treat the e⁻-molecule interactions, it is not for the e⁻-surface interaction V(r) which introduces couplings in the coupled equations (3). The potential and couplings in the equation (3) are illustrated on figures 1 and 2. Figure 1 presents the diagonal potentials (termed diabatic) in the equation (3) : at small r, the curves are well separated due to the centrifugal energies. In contrast, at large r, all the curves tend to $-V_0/2$ i.e. half the potential inside the metal ; this value corresponds to the average of V(r), the e⁻ surface interaction, over spherical harmonics. At large r, the problem is dominated by the e⁻-surface interaction, and the coupling terms do not vanish in the equations (3). This is illustrated on the figure 2 which presents adiabatic potentials obtained by diagonalizing the matrix $V_{ext}^{lm} \delta_{ll'} + <lm|V|l'm>$. At small r, the diabatic and adiabatic potentials are almost identical, showing that the scattering is dominated by the e⁻- molecule interaction. At large r, the two sets are quite different : the adiabatic potentials split into two groups : a first group goes to zero at infinity and corresponds to adiabatic angular modes on the vacuum side, whereas the second group going to $-V_0$ at infinity corresponds to adiabatic angular modes on the metal side. The diagonalization provides adiabatic angular modes adapted to the symmetry of the problem i.e. at infinity, modes for which the electron is either in the vacuum *or* in the metal.

One can recognize on figures 1 and 2 the trapping mechanism for the formation of the $^2\Pi_g$ resonance. At small r, on figure 1, the second potential energy curve corresponds to the d wave and exhibits a rotational barrier that traps the electron. The shape of this barrier is due to the superposition of the gas phase barrier and of the attractive e⁻-metal potential. In addition, the various modes are coupled together ; in particular, the l=2 mode is coupled with the l=1 mode, since the latter does not exhibit a significant rotational barrier, this coupling will lead to an enhancement of the decay rate of the resonance together with a change in the resonant angular mode. The equations (3) are solved for a collision problem and the scattering S matrix is extracted for a large r value : r_M. Indeed the S matrix is extracted in the adiabatic basis defined above, where the couplings vanish at large r. From the S matrix, one defines the time delay matrix Q (Smith 1960)

$$Q = \frac{i}{2} S \frac{dS^+}{dE} \tag{4}$$

One of the eigenvalues of Q,q_1, has a resonance behaviour, with a maximum q_{max}. q_1 corresponds to the N_2^- resonance. In a first approximation the position of this maximum yields the resonance energy and the resonance width is given by

$$\Gamma' = 2/q_{max} \tag{5}$$

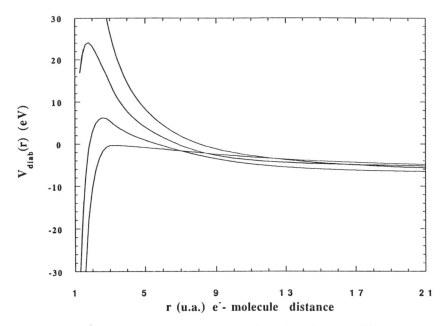

FIg. 1 : Diabatic potential energy curves for a molecule-surface distance of $5a_0$

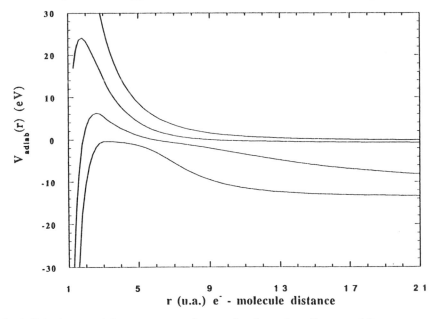

Fig. 2 : Adiabatic potential energy curves for a molecule-surface distance of 5 a_0.

In fact, this method suffers from stability problems when the number of waves l_{Max}, or the integration domain (r_M value) are varied. The stability and convergence problems will be discussed elsewhere (D. Teillet Billy et al 1991), and only an outline is presented here. The e⁻-surface interaction contains long range terms (the image potential for example) which are not explicitly taken into account in the S matrix extraction. In addition, the non resonant e⁻-surface scattering included in the treatment is not well described by a finite expansion over spherical harmonics. As a consequence, the S matrix contains spurious and not converged terms ; since they depend on the energy and on r_M and l_M, they yield contributions to the time delays that depend on r_M and l_M. These effects are rather limited at small r, and become more and more important as r increases. These problems can be solved by removing the resonant part of the scattering which is the non converged one. Then, a second calculations is performed in which the resonance is moved away to negative energies by changing the boundary condition f_{21}. This provides the non resonant contributions to the scattering (time delay matrix Q_{nr}) and the resonance width is now defined from :

$$\Gamma = \text{Min}\left(\frac{2}{T_r(Q) - T_r(Q_{nr})}\right) \tag{6}$$

The position of this minimum gives the position of the resonance. One can note, that the trace of Q is equal to the derivative of the eigenphase sum with respect to the energy. With the definition (6), a good convergence of Γ and E_R can be achieved with a rather limited calculation ; for example, the expansion over l can be limited to 4 terms (see in Teillet-Billy et al 1991).

3. POSITION AND WIDTH OF THE RESONANCE IN N_2-Ag

Figure 3 presents the position of the resonance E_R as a function of Z the molecule-surface distance. Indeed, in the real N_2-Ag system, the molecule-surface distance has a well defined value. Here, for illustrative purpose, we vary this distance, in order to vary the importance of the coupling due to the surface. The resonance energy is found to decrease when the molecule approaches the surface, roughly following an image charge variation.

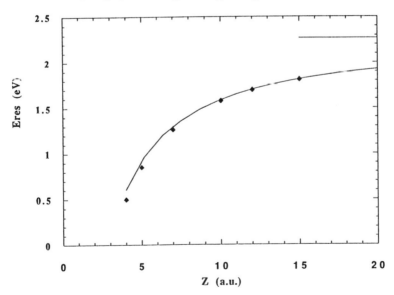

Fig. 3 : Energy of the N_2^- resonance as a function of the N_2-surface distance ($\blacklozenge$: CAM; Full line : E_∞ -1/4Z)

Figure 4 presents the resonance width as a function of Z. It is almost constant at large Z and rapidly increases below Z ~ 8 a_0. These results can be compared with those of Gerber and Herzenberg (1985) who considered a N_2 molecule with the metal wrapped around it. As expected, in the real 3D problem, the effect of the surface on the width is much weaker (a factor

3 in the width around $Z=5a_0$). Qualitatively the E_R and Γ behaviours correspond to the experimentally observed resonances that appear broader and at lower energy in the case of adsorbed molecules. It is difficult to make a more precise comparison, since the experiment do not directly determine the width Γ. However, from the resonance position experimentally observed (Demuth et al, 1981) and from figure 3, one can deduce that the molecule-surface distance corresponding to Demuth experiment lies in the range 5-$7a_0$ for our model where the molecule-surface distance is measured from the electrical image reference plane.

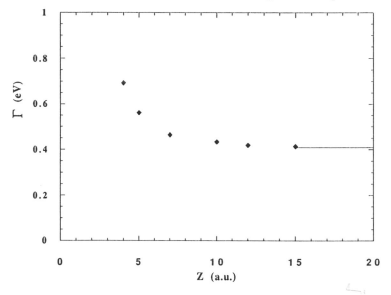

Fig. 4 : Width of the N_2^- resonance as a function of the N_2-surface distance.

The width increase between infinite Z value and Z ~ 5-7 a_0 may seem rather moderate. However, it is noteworthy that the Γ increase is associated with a substancial E_R decrease so that the ratio Γ/E_R drastically increase : at $Z=5a_0$, the width is around 0.56 eV for a resonance at 0.85 eV. This aspect is further illustrated on figure 5 which presents the width as a function of the resonance position for various molecule-surface distances. These curves were obtained by varying the f_{21} value which governs the resonance position in the gas phase. The various resonance positions (or f_{21} values) correspond to different internuclear distances of the N_2 molecule. Figure 5 shows a rather substantial increase of the width for a constante E_R. It also shows that the resonance width in the adsorbed molecule does not vanish for vanishing energy as it does in the gas phase. This feature is due to the possibility of decay for the N_2^- resonance by emitting an electron inside the metal ; this decay channel is open for negative energies above $-V_0$.

4. DISCUSSION OF THE N_2-Ag RESULTS

The experimental results of Demuth et al (1981) were analysed in a semi empirical study by Gerber and Herzenberg (1985) and by Gadzuk (1983). They chose a functional form for the width Γ (E), and using different approximate vibrational excitation treatments, they adjusted the $\Gamma(E)$ function to reproduce the experimental ratio between the various overtone excitations. Gadzuk (1983) used a constant Γ difficult to compare with the present one. On the other hand

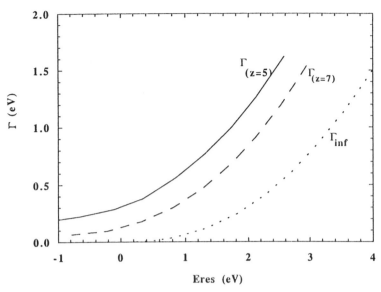

Fig. 5 : $\Gamma(E)$ function. Full line : CAM, $Z=5a_0$; dashed line : CAM, $Z=7a_0$; dotted line : gas phase molecule.

Gerber and Herzenberg (1985) used $\Gamma(E)$ functions suitable for gas phase collisions for $l=1$ and $l=2$ angular waves. (The $l=2$ wave corresponds to an unperturbed N_2^- resonance, whereas the N_2^- resonance perturbed by the surface is expected to exhibit a significant $l=1$ component).

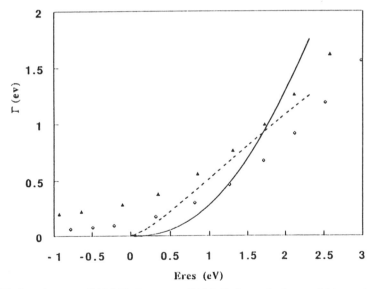

Fig. 6 : $\Gamma(E)$ function : ▲ : CAM,$Z=5a_0$; ◇ : CAM, $Z=7a_0$; Gerber and Herzenberg results : Full line ($l=2$), dashed line ($l=1$).

Figure 6 presents the result of their adjustment for the l=1 and l=2 case together with the CAM results at Z=5 and Z=7a₀. The two sets are found to be quite similar in the energy region important for vibrational excitation and they are significantly larger than the gas phase $\Gamma(E)$. This gives confidence in the capability of the CAM method associated with a dynamical treatment of the vibrational excitation process to reproduce the experimental findings of Demuth et al. It should be noted that the $\Gamma(E)$ width adjusted by Gerber and Herzenberg unrealistically vanish at zero energy.

5. ANGULAR DISTRIBUTION OF RESONANTLY SCATTERED ELECTRONS

The CAM method can also be used to discuss the angular distributions for resonant scattering. They are different for free and adsorbed molecules. For adsorbed molecules, the molecular axis is fixed in space, if one excepts the frustrated rotation movement, and so the scattered electron angular distribution directly reflects the resonance wavefunction (Davenport et al 1978). The e⁻-surface interaction also induces modifications in the resonant angular distribution which can be split into two components. The first effect is classical : when the electron is ejected from the molecular ion into the vacuum, it feels the long range image charge potential that deflects it. This effect can be evaluated in classical mechanics, it results in scattering angles larger than expected from the resonant wavefunction. The second effect is quantal : it corresponds to the coupling between the angular modes induced by the e⁻-surface interaction in the region of the rotational barrier holding the resonance. As explained in part 2, this coupling results in a modification of the resonance decay scheme and consequently in a decrease of the angular momentum associated with the resonance. These two effects of the e⁻- surface interaction were studied with the CAM method (Teillet-Billy and Gauyacq 1991) for the case of CO adsorbed on Ni (110) which was experimentally studied by Jones et al (1989). CO presents a $^2\Sigma$ shape resonance around 19 eV, associated with an f wave (l=3). For this particular case, the modifications introduced by the surface are very weak. Indeed, the classical effect is very limited due to the large energy of the emitted electron. The second effect connected with the modification of the resonance decay scheme can be studied with the CAM method. For this, the coupled equations (3), are solved up to a rather small r_M value (6a₀) where the S matrix is extracted to yield the time delay matrix Q. The calculation with a small r_M value emphasizes the effect of mode mixing at small r i.e. in the resonance barrier region.

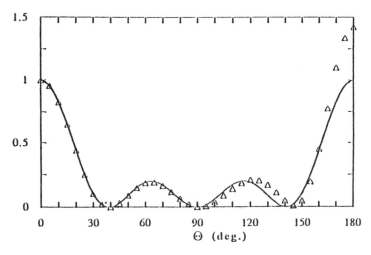

Fig. 7 : Resonant angular dependance for the CO-Ni system. Full line : pure l=3 wave ;   : resonance angular behaviour. θ is measured with respect to the surface normal.

Figure 7 presents the angular wavefunction of the eigenvector of Q associated with the largest eigen time delay, together with a pure l=3 wave corresponding to the unperturbed molecule. For the case of this model CO study, the effect of mode mixing is a slight asymmetry of the angular distribution : due to the metal attraction, the resonance decay is favored toward the metal. The angular distributions experimentally observed by Jones et al (1989) are substancially broader than a pure (l=3) distributions. Other effects can be invoked to account for this broadening : the frustrated rotation movement of the CO molecular axis, modifications of the CO molecule itself introduced by the adsorption process that result in a modification of its scattering properties, or multiple scattering effects. The modifications of the angular distribution introduced by the e⁻-surface coupling is small in this CO case, because the rotational barrier for a l=3 wave is located at small e⁻-molecule distance where the coupling introduced by the surface is small and the energy distance between the potentials very large (see figure 1). Indeed, these modifications should be much larger in the case of resonances associated with a lower angular momentum. For lower l values, considerable mode mixing can occur ; this is illustrated by our study on the C⁻ ion decay in front of metal surface which revealed a substancial mixing between s and p waves (Teillet-Billy and Gauyacq 1990).

CONCLUSIONS

Results on resonant electron scattering by molecules adsorbed on a metal surface were obtained with a non perturbative approach, the CAM method. It allows to determine the position and width of the resonances, as well as the resonant angular distributions. As a main result, the resonance is lower in energy and broader for adsorbed molecules than for gas phase molecules. The interaction between the collisional electron and the surface is also found to destroy the molecular symmetry and induce angular mode mixing.

These results concern molecules with a fixed internuclear distance. The next step will be to take into account the vibration of the molecule during the collision, i.e. the quantitative study of the vibrational excitation process.

These studies should be of interest for a variety of other processes which invoke negative ions as collision intermediates for example for inducing substancial energy transfer between electronic and nuclear motions : charge exchange in atom-surface collisions, reactive scattering (harpoon), desorption of ions or neutral.

REFERENCES

Davenport JW, Ho W and Schrieffer JR 1978 *Phys.Rev.* **17** 3115
Demuth JE, Schmeisser D and Avouris P 1981 *Phys.Rev.Let* **47** 1166
Gadzuk JW 1983 *J.Chem.Phys.* **79** 3982
Gadzuk JW 1985 *Phys.Rev.B* **13** 6789
Gauyacq JP 1987 *Dynamics of negative ions* (World Scientific, Singapore)
Gauyacq JP 1983 *J.Phys.B* **16** 4049
Gerber A and Herzenberg A 1985 *Phys.Rev.B* **31** 6219
Jacobi K, Bertolo M and Hansen W 1990 *J.El.Spect.Rel.Phen.* **54** 529
Jennings PJ, Bertolo M and Hansen W 1988 *Phys.Rev.* **37** 6113
Jones TS, Ashton MR, Ding MQ and Richardson NV 1989 *Chem.Phys.Let.* **161** 467
Ledourneuf M, Vo Ky Lan and Launay JM 1982 *J.Phys.B* **15** L685
Palmer RE, Rous PJ, Wilkes JL and Willis RF 1988 *Phys Rev.Let.* **60** 329
Rous PJ, Palmer RE and Willis RF 1989, *Phys.Rev.B* **39** 7552
Sanche L and Michaud M 1981 *Phys.Rev.Let.* **47** 1008
Sanche L 1990 *J.Phys.B* **23** 1597
Smith FT 1960 *Phys.Rev.* **118** 349

Teillet-Billy D and Gauyacq JP 1984 *J.Phys.B* **17** 4041
Teillet-Billy D, Malegat L and Gauyacq JP 1987 *J.Phys.B* **20** 3201
Teillet-Billy D and Gauyacq JP 1990 *Surf.Sci.* **239** 343
Teillet-Billy D and Gauyacq JP 1991 *Nucl.Inst.Meth.B* in press
Teillet-Billy D, Djamo V and Gauyacq JP 1991 *to be published*

Low energy electron–molecule collision cross sections

Stephen J. Buckman, Michael J. Brunger, Michael J. Brennan, Dean J. Alle, Robert J. Gulley and Stan J. Newman

Electron Physics Group, R.S.Phys.S.E., Australian National University, Canberra, Australia

ABSTRACT: Electron-molecule collisions at low incident energies present an exciting challenge for both theory and experiment. The energy regime below 10 eV is of special importance for many reasons - it is here that scattering calculations must pay special attention, amongst other things, to the effects of correlation/polarization and exchange; the experiments must provide accurate absolute measurements with reactive species; and both are severely tested in describing the sharp structures which dominate many near-threshold excitation processes. At the Australian National University we are engaged in a program of absolute cross section determinations on relatively simple diatomic and polyatomic molecules at very low energies. We present results for H_2, N_2 and NH_3 at incident energies between 1.0 and 15 eV which are critically compared with available calculations and other experimental results.

1. INTRODUCTION

Interest in low energy electron-molecule collisions has grown rapidly over the past two decades, fuelled both by fundamental interest in the nature of the scattering process and by the need for accurate collision cross sections and reaction rates in models of a wide and ever increasing range of technological devices and processes based on gas discharges. The increasing sophistication of the scattering calculations applied to electron molecule collisions and the sheer brute force offered by modern computers has served as somewhat of an inspiration to both theoretical and experimental investigators such that there are now several groups working on the absolute determination of cross sections for electron-induced reactions in a wide variety of gases. Many of these pose major technical problems for the experimentalist and, by their very nature molecules, particularly polyatomics, present problems for the scattering theorist that can only be solved in the majority of cases by extensive approximations and/or model calculations.

Notwithstanding this sizeable effort, there remain many mysteries and long-standing discrepancies between experiments and between experiment and theory, even for relatively simple diatomic and polyatomic molecules. Specific examples include the magnitude and energy dependence of the vibrational excitation cross section in H_2 (v=0-1) and the nature of the sharp structures that have been observed at threshold in vibrational excitation of the hydrogen halides and some simple polyatomics such as CO_2.

In the Electron Physics Group at the ANU we have recently embarked on a program of absolute, low energy differential cross section measurements on a range of molecular targets. The initial motivation was to investigate vibrational excitation in the simplest molecular target, H_2, where there were unacceptable discrepancies between experiment and theory. These studies have been followed by a series of measurements on N_2 at energies away from the Π_g resonance, and on NH_3, a relatively simple, polar polyatomic for which there is a body of theoretical cross section calculations. This article will attempt to describe our recent experiments, to give an overview of the theory, and to discuss how well they agree.

2. EXPERIMENTAL CONSIDERATIONS

Experiments such as the one to be described here lie at the heart of the subject matter of ICPEAC and are no doubt familiar to many. We shall therefore not dwell too heavily on the techniques employed in the present studies but rather give an overview of the central components and some further detail concerning the experimental techniques.

The differential cross sections (DCS) are measured with a crossed electron-molecule beam apparatus which has been described in some detail previously (Brunger *et al.* 1990,1991). The monochromatic ($\Delta E\sim50\text{-}100$ meV FWHM), variable-energy electron beam is produced by a combination of electrostatic electron optics and a hemispherical deflector and is incident upon a molecular beam formed by effusive flow through a multicapillary array. Scattered electrons are transported and energy analysed by a similar arrangement of electron optics and hemispherical analyser and detected by a channeltron. A schematic of the experimental apparatus is shown in Figure 1.

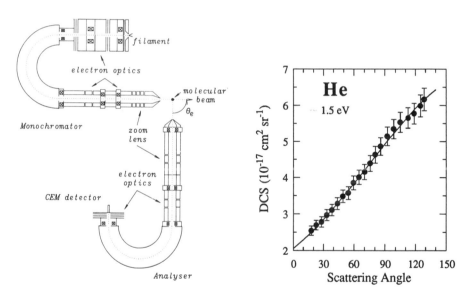

Figure 1: Schematic of the experimental apparatus

Figure 2: Elastic differential cross section for helium at 1.5 eV. (●) Present data: (—) Nesbet

The differential cross section for the scattering of an electron of energy E through an angle θ is given by

$$\frac{d\sigma}{d\Omega}(E,\theta) = \frac{I_s(E,\theta)}{I_o(E)}\left(Nld\Omega\varepsilon\tau\right)^{-1}$$

where I_o and I_s are the incident and scattered electron intensities, N the number density in the molecular beam, $ld\Omega$ the effective interaction length, and ε and τ the transmission and detection efficiencies of the analyser and channeltron respectively. A true absolute determination of the scattering cross section involves the accurate, absolute measurement of each of these quantities and, in practice, this is not done. Generally, "absolute" values are placed on the relative angular distributions by the use of a normalisation technique. For potential exactly, and the rotational motion adiabatically. The static and polarization terms in the Hamiltonian are treated at the Hartree-Fock level. Local polarization effects (long-range) and non-local (short-range) correlation effects are treated by a potential that includes adiabatic polarization effects exactly, models non-local and non-adiabatic correlation effects, and does not include any adjustable parameters. This work, in particular, has highlighted the importance of the inclusion and correct treatment of exchange and polarization for vibrational excitation (see Buckman *et al.* (1990) and references therein for further details).

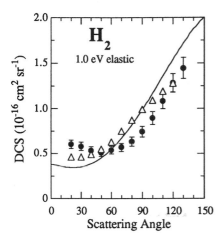

Figure 3(a): Elastic differential cross section for H$_2$ at 1.0 eV. (●) Present data; (Δ) Linder and Schmidt; (—) Snitchler *et al.*

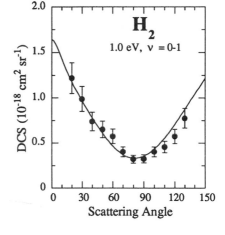

Figure 3(b): Differential cross section for vibrational excitation of H$_2$ v=0-1 at 1.0 eV. (●) Present data; (—) Snitchler *et al.*

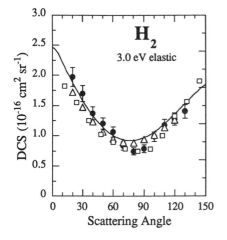

Figure 4(a): Elastic differential cross section for H$_2$ at 3.0 eV. (●) Present data; (Δ) Linder and Schmidt; (□) Shyn and Sharp; (—) Snitchler *et al.*

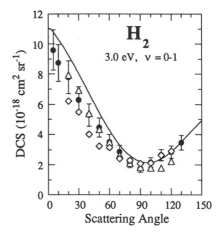

Figure 4(b): Differential cross section for vibrational excitation of H$_2$ v=0-1 at 3.0 eV. (●) Present data; (Δ) Linder & Schmidt; (◊) Nishimura *et al.*; (—) Snitchler *et al.*

targets which are gaseous at room temperature the most commonly used technique is the relative flow method (Srivastava *et al.* 1975) and that is the method which has been applied to all of the data presented here.

The relative flow technique is based on comparative measurements of scattering intensities for the gas of interest and a known or "standard" gas. It is now widely accepted that the elastic differential scattering cross section for helium is known to within a few percent at energies below the first excited state threshold ($\approx$19.8 eV). The general acceptance of this cross section as the standard cross section arises from the excellent agreement which exists between experimental determinations of it, based on phase shift analysis techniques, and the *ab initio* variational calculations of Nesbet (1978). Our own absolute experimental investigations of this cross section at 1.5 eV further confirm it as an appropriate choice (see Figure 2).

The relative flow technique has been described in great detail previously so we will not add further to it here other than to emphasise that extreme care is required to ensure that the scattering volume for each target gas is identical. In the present measurements we further check the scattering volume at each energy by measuring the corresponding angular distribution for elastic scattering from helium under identical electron optical conditions to those used for the gas of interest.

Having ensured that the above conditions have been satisfied the elastic scattering cross section for the target gas of interest can be calculated from the known elastic differential cross section for helium. Given the agreement that exists between our own experimental determinations of this cross section and the calculation of Nesbet (1978), of which Figure 2 is an example, we choose the "standard" values of the helium cross section to be those given by the theory. We estimate that they are accurate to within a few percent.

Absolute cross sections for vibrationally inelastic events are then established relative to the elastic cross section by measuring ratios of the various inelastic scattering intensities at each scattering angle to that for elastic scattering. For these measurements, not only is extreme care taken to optimise the transmission of the scattered electron analyser for each scattered electron energy, but this relative transmission is actually measured. This is done by measuring the yield of ionization electrons following near-threshold ionization of helium. This technique has also been discussed in detail elsewhere (Brunger *et al.* 1990).

3. RESULTS AND DISCUSSION

3.1   H$_2$

For many years there has been a serious discrepancy between various experiments, both single-collision and swarm, and between swarm experiments and theory, over the magnitude and energy dependence of the absolute total cross section for near-threshold excitation of H$_2$ v=0-1. Given the intrinsic importance of this molecule as the simplest diatomic structure for molecular scattering theory, we have undertaken a number of low energy absolute differential scattering experiments to provide a specific test for the theory. We have also derived from these measurements, absolute integral cross sections which can be compared directly with the cross section which results from the analysis of swarm experiments.

The present data, both for vibrationally elastic scattering and ro-vibrational excitation were measured at seven discrete energies between 1.0 and 5.0 eV. In Figures 3 and 4, we show representative examples of the present DCS measurements for (a) elastic scattering and (b) ro-vibrational excitation at energies of 1.0 and 3.0 eV respectively. Comparison can also be made in these figures with other experimental results and, in particular, with recent theoretical calculations by Morrison and co-workers at the University of Oklahoma (Snitchler *et al.* 1990). These calculations have evolved over the past decade to the point where they clearly represent the most accurate calculations to date for electronically elastic scattering from H$_2$. Snitchler *et al.* solve the Schrödinger equation using body-frame vibrational close-coupling theory which treats the effects of nuclear motion and the non-spherical nature of the interaction

In the near-threshold region (≤1.5 eV), whilst there are some differences in shape and magnitude between the present data and the theory, the overall level of agreement for both elastic and ro-vibrational scattering is very good.

The differential cross sections for ro-vibrational excitation have been extrapolated and integrated to yield discrete values of the total cross section which are compared with other experimental and theoretical determinations in Figure 5. In particular we compare with the swarm-derived result and find, that in the near-threshold region, the present results are in excellent agreement with both the earlier beam measurements and the theory of Morrison and co-workers but are incompatible with the swarm cross section. Indeed at 1.5 eV the present result is in excellent agreement with the theory but some 60% higher than the swarm cross section. This, of course, has major ramifications for either the transport analysis used to derive the swarm cross section or indeed for the beam experiments and theory.

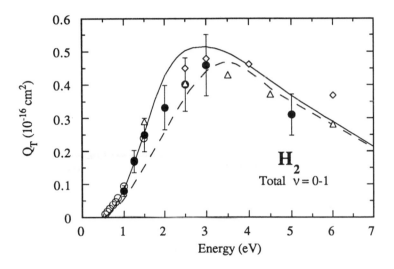

Figure 5: Total scattering cross section for vibrational excitation of H_2 v=0-1. (●) present data; (Δ) Linder and Schmidt; (◊) Nishimura *et al.*; (o) Ehrhardt *et al.*; (— —) England *et al.*; (—) Snitchler *et al.*

3.2 N_2

Whilst from a structural point of view N_2 is a considerably more complicated molecular target than H_2, there appear to have been many more theoretical studies made of this molecule and, in some sense, it appears to be a favoured choice for the application of molecular scattering theory. This is in a large part due to the presence, at around 2.5 eV, of the large shape resonance which dominates vibrational excitation and plays such an important role in many gas discharge devices containing N_2. In addition to some similar calculations by the Oklahoma group (Morrison *et al.* 1987) to those outlined above for H_2, there have been hybrid calculations (eg. Chandra and Temkin 1976) which attempt to combine elements of the adiabatic-nuclei and close-coupling methods in an effort to simplify the calculation for scattering in the vicinity of a resonance. In this approximation nuclear effects are treated adiabatically off-resonance and by vibrational close coupling in the resonance region. There have also been R-Matrix calculations (Gillan *et al.* 1987) and calculations based on the Schwinger multichannel method (Huo *et al.* 1987).

On the experimental side there is a wealth of integral cross section information for this molecule but very few absolute differential cross section results which cover a wide range of (low) energy and scattering angle away from the resonance. The present results for elastic scattering and vibrational excitation (v=0-1,2,3) at energies of 1.5, 2.1, 3.0 and 5.0 eV fill a few of these gaps and enable a comprehensive comparison with the available theory.

Figures 6 and 7 illustrate the situation for (a) elastic scattering and (b) vibrational excitation (v=0-1) at incident energies of 1.5 and 3.0 eV respectively. The striking feature which emerges from the data at 1.5 eV is the discrepancy between the present measurements and those of both Sohn *et al.* (1986) and Shyn and Carignan (1980). In particular, the present differential cross sections for elastic scattering are almost uniformly a factor of two higher than those of Sohn *et al.*, although the shapes are remarkably similar. Given that both the present measurements and those of Sohn *et al.* are normalised in a similar fashion relative to helium, this is alarming (the data of Shyn and Carignan are also normalised relative to helium at 10 eV but by a significantly more complicated procedure). Some further insight into this discrepancy can be obtained from a comparison of the total elastic cross sections obtained from these differential measurements to the grand total cross section measured by Kennerly (1980) using time-of-flight (TOF) techniques (Note that this comparison is a valid one for the present purposes as at this energy, over 98% of the grand total cross section comes from vibrationally elastic scattering). The present measurements yield a total elastic cross section of 12.6±2.5 Å^2 in reasonable agreement with the TOF value of 11.24±0.34 Å^2, whilst the grand total cross section quoted by Sohn *et al.* is 7.48 Å^2. Whilst one must be careful in making such comparisons for integral cross sections derived from differential measurements as differences in shape can sometimes lead to erroneous conclusions, the present case indicates that a normalisation error may be present in the elastic data of both Sohn *et al.*and Shyn and Carignan.

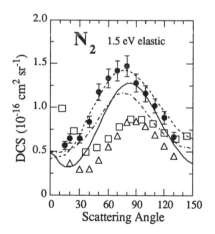

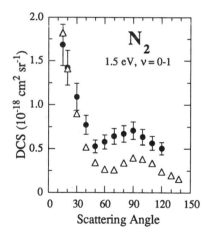

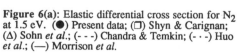

Figure 6(a): Elastic differential cross section for N$_2$ at 1.5 eV. (●) Present data; (□) Shyn & Carignan; (△) Sohn *et al.*; (- - -) Chandra & Temkin; (- · -) Huo *et al.*; (——) Morrison *et al.*

Figure 6(b): Differential cross section for vibrational excitation of N$_2$ at 1.5 eV. (●) Present data; (△) Sohn *et al.*

At both energies, where a comparison with theory is possible, the result is quite heartening and is indicative of the general level of agreement that is found at all energies we have studied. In particular we find fair agreement with the calculations of both Chandra and Temkin (1976) and Huo *et al.* (1987), and at 1.5 eV, with the calculation of Morrison *et al.* (1987). For

vibrational excitation, the overall level of agreement with the earlier data of Brunger *et al.*
(1989), performed on an entirely different apparatus, at 2.1 and 3.0 eV is generally fair,
although there are differences in some details at middle angles.

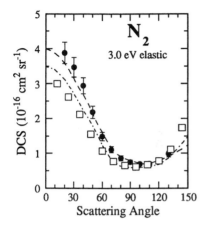

Figure 7(a): Elastic differential cross section for
N_2 at 3.0 eV. (●) Present data; (□) Shyn &
Carignan; (- - -) Chandra & Temkin; (- · -) Huo *et al.*

Figure 7(b): Differential cross section for
vibrational excitation of N_2 at 3.0 eV. (●) Present
data; (♦) Brunger *et al.*(1989); (◊) Tanaka *et al.*; (- - -)
Chandra & Temkin; (— · —) Gillan *et al.*

3.3 NH_3

Ammonia is one of the simpler, polar polyatomic molecules for which a reasonable amount of
scattering theory exists. These include Schwinger variational calculations by Pritchard *et al.*
(1989) and *ab-initio* model calculations by Gianturco (1990) and Jain and Baluja (1991). The
first of these is calculated in the fixed-nuclei, static exchange approximation and as a result only
provides information for intermediate and backward scattering angles as the fixed-nuclei
approximation leads to divergent cross sections for elastic scattering from polar molecules at
forward angles. On the other hand, Gianturco's calculations circumvent this effect for high
partial waves by a partitioning of the problem into one region where the pure dipole interactions
are calculated in the First Born Approximation and another region where the stronger
interaction with the full target potential is calculated in a close-coupling approximation. This
calculation includes exchange as an energy-dependent local potential, correlation/polarization
effects via a density function approach, and is carried out using the adiabatic nuclei
approximation. The calculation of Jain and Baluja is similar but treats exchange exactly and
uses a different approach to polarization effects.

To date we have measured differential vibrationally elastic collisions for ammonia at incident
energies of 2.0, 5.0, 7.5 and 15.0 eV. Figures 8 and 9 show the present experimental results
at 2.0 and 7.5 eV respectively. In both cases the differential cross sections indicate the strong
forward scattering characteristic of polar molecules. At 2.0 eV, the present absolute cross
section is compared with the calculations of Gianturco and Jain and Baluja. It bears little
resemblence to either, although it is in better accord with the absolute magnitude of the
calculation of Gianturco.

At 7.5 eV, in addition to comparing the present data with the calculations of Gianturco (at 7 eV)
and Jain and Baluja, we also show a comparison with what is, to our knowledge, the only
other absolute determination of a low energy differential elastic scattering cross section in NH_3,

that by Ben Arfa and Tronc (1987). The agreement with this measurement, which was placed on an absolute scale by use of the relative flow technique in combination with the elastic cross section for N_2 measured by Srivastava *et al.* (1976), is excellent at all angles except for a few in the forward direction. Given the different normalising standards used, this level of agreement is extremely heartening. There is also fair agreement with both theoretical calculations at this energy.

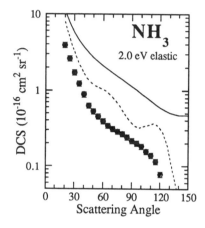

Figure 8: Elastic differential cross section for NH_3 at 2.0 eV. (●) Present data; (—) Jain and Baluja; (- - -) Gianturco

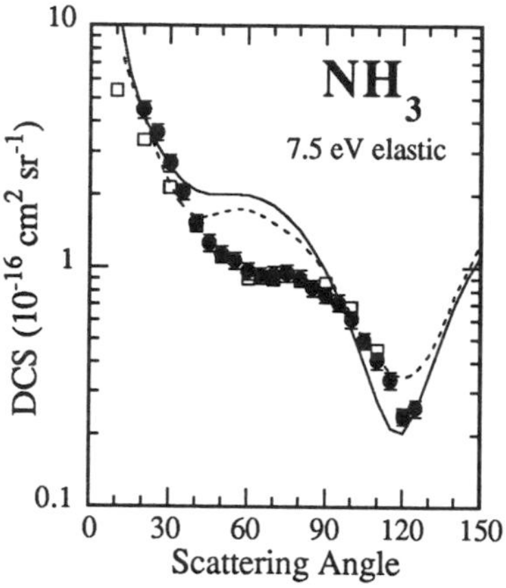

Figure 9: Elastic differential cross section for NH_3 at 7.5 eV. (●) Present data; (□) Ben Arfa & Tronc; (—) Jain and Baluja; (- - -) Gianturco (at 7.0 eV)

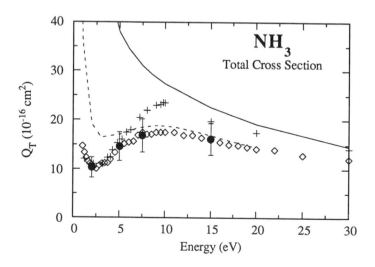

Figure 10: Total scattering cross section for NH_3. (●) Present total elastic cross section; (◊) Sueoka *et al.*; (+) Szmytkowski *et al.*; (—) Jain and Baluja; (- - -) Gianturco

We have also integrated the measured differential cross sections to yield both total elastic and elastic momentum transfer cross sections at the four energies studied. To accomplish this we have used the theory of Jain and Baluja as a guide for the *shape* of the cross section at forward angles and that of Gianturco for the backward angles, to extrapolate the present measurements to those areas we cannot access experimentally. The resultant cross sections are shown and compared with other available experimental and theoretical values in Figures 10 and 11. In the case of the total cross section the agreement is excellent between the present values, which have an uncertainty of ±20%, and the TOF measurements of Sueoka *et al.* (1987). In this case it should be realised that the present result represents a lower bound for the grand total cross section as it does not include inelastic cross sections, although we expect these to be a small contribution. For the momentum transfer cross section there is fair agreement between the present values and those of Hayashi (1981) and the theoretical calculation of Pritchard *et al.*, although at around 2 eV it appears that the minimum in the momentum transfer cross section is substantially deeper than has been previously published.

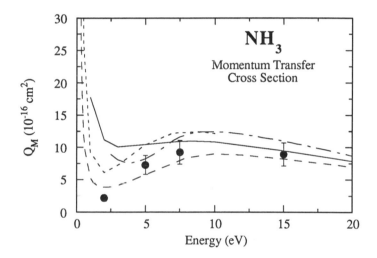

Figure 11: Momentum transfer cross section for NH_3. (●) Present data; (—) Jain and Baluja; (— —) Hayashi; (- - -) Gianturco; (— - —) Pritchard *et al.*

5. FINAL REMARKS

We have presented the results of an ongoing program to provide accurate, absolute differential and integral cross sections for molecular targets of both fundamental and technological interest. In H_2, we find by a direct comparison of the differential cross sections that the close coupling calculations of Morrison and co-workers are in good accord with experiment for both elastic scattering and vibrational excitation. We have also integrated our differential vibrational cross sections to yield a total vibrational cross section at seven discrete energies between 1.0 and 5.0 eV, which does not compare well with that derived from swarm experiments in the near-threshold region (≤ 1.5 eV). For N_2 we find generally good agreement with other available experimental data and theory with the exception of the elastic scattering measurements at 1.5 eV of Sohn *et al.* and Shyn and Carignan. This we believe is indicative of an error in the absolute normalisation of these latter measurements.

For NH_3, the only polyatomic molecule in the present study, we were genuinely surprised at the level of agreement with various theoretical calculations and heartened by the excellent agreement with the only other absolute experimental differential cross section at 7.5 eV. At energies above 2 eV, there is fair agreement between the present measurements and theory. At 2 eV there are substantial discrepancies, both in shape and absolute magnitude, between the present cross sections and the theory of Gianturco and that of Jain and Baluja. In contrast there is good agreement between the total cross section derived from the present results and the TOF measurements of Sueoka *et al.* at all energies studied here.

ACKNOWLEDGEMENTS

It is a pleasure to acknowledge Michael Morrison and his colleagues at the University of Oklahoma with whom we have had a most fruitful collaboration on studies of diatomic molecules. We also wish to thank Professors Gianturco and Jain for providing us with their calculated cross sections prior to publication.

REFERENCES

Ben Arfa M and Tronc M 1987 *J. Chim. Physique* **85** 889
Brunger M J, Buckman S J and Newman D S 1990 *Aust. J. Phys.* **43** 665
Brunger M J, Buckman S J, Newman D S and Alle D T 1991 *J. Phys. B: Atom. Molec. Opt. Phys.* **24** 1435
Brunger M J, Teubner P J O, Weigold A M and Buckman S J 1989 *J. Phys. B: Atom. Molec. Opt. Phys.* **22** 1443
Buckman S J, Brunger M J, Newman D S, Snitchler G, Alston S, Norcross D W, Saha B, Danby G, Trail W and Morrison M A 1990 *Phys. Rev. Lett.* **65** 3253
Chandra N and Temkin A 1976 *Phys. Rev.* A **13** 188
Ehrhardt H, Langhans L, Linder F and Taylor H S 1968 *Phys. Rev.* **173** 222
England J P, Elford M T and Crompton R W 1988 *Aust. J. Phys.* **41** 573
Gianturco F A 1990 *Private Communication*
Gillan C J, Nagy O, Burke P G, Morgan L A and Noble C J 1987 *J. Phys. B: Atom. Molec. Phys.* **20** 4585
Huo W, Lima M A P, Gibson T L and McKoy V 1987 *Phys. Rev.* A **36** 1642
Hayashi M 1981 *Inst. Plasma Phys. Report, Nagoya Univ., Japan* IPPJ-AM-19
Jain A and Baluja K L 1991 *Private Communication*
Kennerly R E 1980 *Phys. Rev.* A **21** 1876
Linder F and Schmidt H 1971 *Z. Naturforsch.* **26a** 1603
Morrison M A, Saha B C and Gibson T L 1987 *Phys. Rev.* A **36** 3682
Nesbet R K 1978 *Phys. Rev.* A **20** 58
Nishimura H, Danjo A and Sugahara H 1985 *J. Phys. Soc. Japan* **54** 1757
Pritchard H P, Lima M A P and McKoy V 1989 *Phys. Rev.* A **39** 2392
Shyn T W and Carignan G R 1980 *Phys. Rev.* A **22** 923
Shyn T W and Sharp W E 1981 *Phys. Rev.* A **24** 1734
Szmytkowski C, Maciag K, Karwasz G and Filipovic D 1989 *J. Phys. B: Atom. Molec. Opt. Phys.* **22** 525
Snitchler G, Alston S, Norcross D, Saha B, Danby G, Trail W and Morrison M A 1990 *Private Communication*
Sohn W, Kochem K-H, Scheuerlein K-M, Jung K and Ehrhardt H 1986 *J. Phys. B: Atom. Molec. Phys.* **19** 4017
Srivastava S K, Chutjian A and Trajmar S 1975 *J. Chem. Phys.* **63** 2659
Sueoka O, Mori S and Katayama Y 1987 *J. Phys. B: Atom. Molec. Phys.* **20** 3237
Tanaka H, Yamamoto T and Okada T 1981 *J. Phys. B: Atom. Molec. Phys.* **14** 2081

R-matrix calculations of electron scattering from neutral and ionized molecules

Charles J Gillan

Department of Applied Mathematics and Theoretical Physics, Queen's University of Belfast, N. Ireland BT7 1NN, UK

Abstract. A number of new calculations on low energy electron scattering by diatomic molecules, carried out within the R-matrix formalism, are briefly reviewed. Elastic scattering studies for HeH^+, He_2^+ and CO are reported along with electronic excitation calculations on H_2, N_2 and O_2. Finally, future directions of this work are discussed.

1. Introduction

One of the most potent methods used in the calculation of low energy electron scattering from atomic or molecular targets is the close coupling expansion of Massey and Mohr (1932) which is based upon an expansion of the collision complex in terms of the wavefunctions of the target eigenstates. The R-matrix method is a powerful derivative from standard close coupling theory and has been developed over the past decade by our group in the UK to yield accurate data on electron and positron collisions with diatomic molecules. At the XVIth ICPEAC Morgan (1990) reported on a number of calculations in progress at that time; this article follows on from that previous one and discusses the developments which have occurred in the intervening years. Much of these have been in the area of vibrational and electronic excitation and so form the bulk of this report, however, elastic scattering is also mentioned as is the highly successful application of a new bound state evaluation technique to obtain Rydberg states of the molecule HeH. The reader should remember that the many results reported here are the work of several scientists and not just the author.

2. Method

It is not appropriate within the context of this review to give a detailed discussion of the R-matrix theory. Instead the salient features of the fixed nuclei and nuclear motion

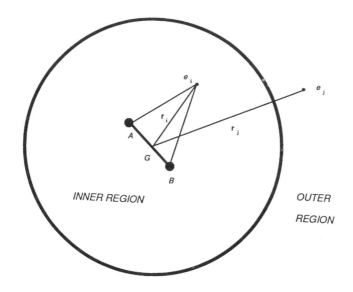

Figure 1. Illustration of the division of configuration space for fixed nuclei scattering.

methods are summarized. A more detailed discussion of the theory can be found in Gillan *et al* (1987) while the computational aspects are presented in Gillan (1991).

The basic philosophy behind the R-matrix approach is that the interaction of the target and projectile exhibits fundamentally different physical characteristics in different regions of configuration space. One takes advantage of this fact and formulates different equations for the wavefunction appropriate to the region in question. Continuity is then ensured by joining the solutions at the boundaries by an R-matrix.

In the fixed nuclei approximation there are just two regions of configuration space separated by a sphere of radius $r = a$, where r is the distance of the scattering electron from the centre of mass of the target; this division of configuration space is illustrated in figure 1. A multi-centre configuration interaction type expansion of the scattering wavefunction is chosen in the inner region, with the scattering wavefunction expanded in a basis ψ_k with

$$\psi_k = \sum_{ij} \mathcal{A}\bar{\Phi}_i(x_1, x_2, \ldots, x_n, \sigma_{n+1})\eta_j(\hat{\mathbf{r}}_{n+1})\alpha_{ijk} + \sum_j \chi_j(x_1, \ldots, x_{n+1})\beta_{jk} \qquad (1)$$

where $\mathcal{A}$ is the anti-symmetrization operator. The functions $\bar{\Phi}_i$ are eigenfunctions of the total spin operator, S^2, and its z-component, S_z, formed by coupling the products of the spin function of the continuum electron with the N electron target states Φ_i. The η_j functions are continuum molecular orbitals on the center of gravity of the molecule which are non-zero on the boundary of the inner region. The second summation is over square integrable L^2 functions, χ_j, descriptive of states with all of the electrons in bound orbitals. The coefficients α_{ijk} and β_{jk} are obtained by diagonalizing the operator $(H_{N+1} + L)$ in the basis ψ_k in the inner region,

$$< \psi_{k'} \mid H_{N+1} + L \mid \psi_k >= E_k \delta_{kk'}. \tag{2}$$

H_{N+1} being the Schrödinger Hamiltonian and L a surface projection operator introduced by Bloch (1957), so that $(H_{N+1} + L)$ is Hermitian over the finite inner region.

The formal solution for the scattering wavefunction, Ψ, at any energy E may be written as

$$\Psi = \sum_k \frac{\mid \psi_k >< \psi_k \mid L \mid \Psi >}{E_k - E} \tag{3}$$

The R-matrix which joins the inner and outer regions is then defined by projecting equation (3) onto the channel function $\bar{\Phi}_i Y_{l_i m_i}$ and evaluating the expression on the R-matrix sphere. At fixed geometry, the collision problem can then be solved by adopting a single centre, no exchange, close coupling expansion of the wavefunction in the outer region using the R-matrix to match the inner and outer region solutions at the boundary. We note that for any given scattering system the choice of target states and L^2 terms in equation (1) leads to the different scattering approximations (Gillan 1991). In particular, retaining only the ground state in (1) yields either the static exchange, SE, or static exchange plus polarization, SEP, models depending upon the associated choice of L^2 functions.

When nuclear motion is introduced into the theory, configuration space is again partioned into different regions, but the inner region is now taken to be a hypersphere as described by Gillan *et al* (1987). The scattering wavefunction is now expanded in a basis θ_k where

$$\theta_i = \sum_{jk} \psi_k(R) u_j(R) c_{ijk} \tag{4}$$

with the ψ_k being the fixed nuclei functions defined above and the u_j orthogonal polynomials. Clearly, to construct this expansion it is first necessary to obtain the ψ_k at several internuclear separations. The coefficients c_{ijk} are obtained by diagonalizing a Hamiltonian matrix following the method proposed by Schneider *et al* (1979).

To obtain vibrational excitation cross sections it is necessary to solve the collision problem in the external region. We use a single centre, no exchange, vibrational close coupling expansion of the scattering wavefunction which yields a set of coupled second order ordinary differential equations identical to those in the hybrid close coupling method of Chandra and Temkin (1976).

3. Electronic Excitation in e-H$_2$ Scattering

The previous review by Morgan (1990) gave a preliminary account of research conducted at the University of London on R-matrix calculations in which up to seven low lying electronic states of H$_2$ are coupled. In this work a full CI representation, within a small orbital set, was being used for each target state yielding excellent threshold energies and correspondingly every possible L^2 term that can be constructed in the basis. The total cross sections obtained in these calculations have now been reported

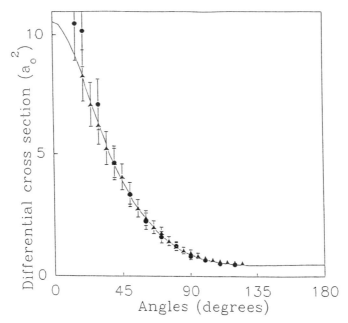

Figure 2. Differential cross section for elastic scattering in e-H_2 at 15 eV. ——, R-matrix results (Branchett *et al* 1991); $\triangle$, experiment (Khakoo and Trajmar 1986); $\bullet$, experiment (Nishimura *et al* 1986)

(Branchett and Tennyson 1990, Branchett *et al* 1990) and are a considerable improvement on the two state work of Baluja *et al* (1985). These new results are in accord with experiment.

In order to obtain a better understanding of the observed resonances Branchett *et al* (1991) have calculated differential cross sections from their T-matrices. Additionally, these differential cross sections are a much more rigorous test of the theory as the available experimental data is more accurate than for total cross sections. Figure 2 shows the excellent agreement, for the case of elastic scattering, between the R-matrix data and the experimental results of Khakoo and Trajmar and Nishimura *et al* at an incident energy of 15 eV.

4. Electronically elastic scattering by HeH$^+$ and He$_2^+$

At the time of writing there have been no experimental studies of e-He$_2^+$ scattering nor, as far as we are aware, have there been any there any scattering calculations. This is despite the fact that a knowledge of this system is important since it opens the avenue to studying a host of related theoretical problems (Gillan *et al* 1991). Accordingly, a detailed study was undertaken at three internuclear geometries in which the R-matrix method was employed in a two state approximation. SCF, and later full CI, representations of the X $^2\Sigma_u^+$ and A $^2\Sigma_g^+$ states of He$_2^+$ were used in this work in which cross sections were obtained for elastic scattering when the total symmetry of the collision complex is $^3\Sigma_u^+$.

Resonance	Γ_{CI} (Ryd)	Γ_{SCF} (Ryd)
$3p\sigma$	0.296^{-2}	0.175^{-2}
$4p\sigma$	0.199^{-2}	0.802^{-3}
$5p\sigma$	0.460^{-3}	0.544^{-4}

Table 1. Comparison of autoionizing widths Γ obtained using SCF and CI representations for the target states in the R-matrix calculations on e-He_2^+ at $R = 2.0625$ a_0.

As predicted theoretically for scattering from an ion the cross section below the first excitation threhold, is dominated by series of Rydberg resonances converging on the first excited state. During the scattering process there is a rearrangement of the He_2^+ core with the continuum electron attaching itself to the unstable excited A $^3\Sigma_g^+$ state. We observed two overlapping resonance series in our calculations corresponding to the continuum electron being in a diffuse $np\sigma$ or $nf\sigma$ orbital. We analysed the first three members of the $np\sigma$ series fitting the resonances to a Breit Wigner form to obtain their positions and widths. Table 1 presents a comparison of the autoionizing widths, at the X $^2\Sigma_u^+$ state equilibrium geometry, calculated with the two different target state representations. The substantial difference between the two sets of values demonstrates the importance of using suitably correlated wavefunctions in any model used to obtain these widths. In short caution must be exercised in using results for any system obtained within the framework of an SCF representation. At present this study is being pursued along similar lines to the very detailed study of e-H_2^+ collisions undertaken by Shimamura *et al* 1990 and at the next stage nuclear motion will be introduced into the work.

A similar study has been carried out for electron scattering from the HeH$^+$ molecular ion by Sarpal *et al* 1991a. These workers have already incorporated non-adiabatic nuclear motion in their calculations finding that it produces a vibrational excitation cross section rich in resonant structures. These features can only be explained by the fact that the electron is captured into a vibronically excited state of HeH. The fixed nuclei cross sections do not exhibit these resonant peaks indicating that they are vibrational levels associated with electronic Rydberg states of HeH converging onto the HeH$^+$ ground state. The mechanism for their formation is thus similar to that observed in non-abiabatic e-HF and e-HCl calculations (Burke *et al* 1990). Sarpal *et al* 1991b have applied the bound state method of Seaton (1985) to evaluate these Rydberg states at the internuclear geometry $R = 1.455$ a_0 calculating transition energies which agree to better than 50 cm^{-1} with spectroscopic determinations. A code is now being developed to determine the transition moments among these diffuse levels.

5. Electronic excitation in e-N_2 and e-O_2

For many years the favourite proving ground for any new theory of electron molecule scattering has been a study of the well known $^2\Pi_g$ shape resonance in elastic e-N_2 scattering. There has been comparatively little work on the electronic excitation process however, a statement which is also true for the e-O_2 system. On the one hand this is somewhat surprising given the importance of electron scattering from N_2 and O_2 in atmospheric physics while on the other hand it is understandable due to the difficulty in modelling accurately the electronic excitation process.

In an attempt to elucidate the physics of the excitation process in these systems, a set of R-matrix calculations has been in progress over the past three years. In the case of N_2 an initial calculation was performed in which the X $^2\Sigma_g^+$ ground state and the three low lying valence excited states A $^3\Sigma_u^+$, B $^3\Pi_g$ and W $^3\Delta_u$ were retained in the expansion of ψ_k (Gillan *et al* 1990). This has now been extended to include the higher lying B' Σ_u^-, a' $^1\Sigma_u^-$, a $^1\Pi_g$ and w $^1\Delta_u$ states. For O_2, a three state calculation was reported by Noble and Burke (1986) and this has now been extended to the include in addition to those three lowest states X $^3\Sigma_g^-$, a $^1\Delta_g$, b $^1\Sigma_g^+$, the next six states: C $^1\Sigma_g^+$, A' $^3\Delta_u$, A $^3\Sigma_u^+$, B $^3\Sigma_u^+$, $^1\Delta_u$ and $^1\Sigma_u^+$.

A particularly difficult aspect of the computational process is obtaining a compact representation of the electronic states in a common orbital basis, a feature for which there is no ab-initio prescription. For these molecules complete CI is not feasible due to the very large number of configurations involved and for scattering work it is not yet practical to use the type of CI expansions which are commonplace in quantum chemical evaluations. In scattering it is, however, more important to obtain the correct relative spacing of the target potential energy curves than to have correct absolute eigenenergies. Thus, for both targets the eigenstates are represented by small CI wavefunctions in which all single and double valence excitations are allowed from the relevant dominant electronic configuration. This level of CI accounts for a measure of target correlation without unduly increasing the magnitude of the computational task. For O_2 the target states are generated in a set of SCF orbitals obtained from the work of Saxon and Liu (1977), while for N_2 the orbital set is a hybrid obtained from MCSCF calculations on the X $^1\Sigma_g^+$ ground state and some of the excited states.

In figure 3 differential cross sections obtained at 10 eV impact energy, for the transition $X^3\Sigma_g^+ \rightarrow a^1\Delta_g$ in e-O_2 scattering , are compared with experimental values from Trajmar *et al* al (1971). The experimental points beyond 90° are extrapolations from the measurements at lower angles and are thus not as reliable as the measured data. This is probably one source of the disagreement between theory and experiment. A careful analysis of the calculated R-matrix data shows that in this nine state model there is an insufficient representation of the long range dipole polarizibility of the $X^3\Sigma_g^+$ ground state. This results in a larger than expected elastic scattering cross section at impact energies from threshold to beyond 10 eV. This is the source of large disagreement between theory and experiment at forward scattering angles. It is clear, therefore that pseudo states must be incorporated in the calculation to provide the correct polarizibility and this work is already underway.

Turning to e-N_2, we note first of all that the experimental differential cross sec-

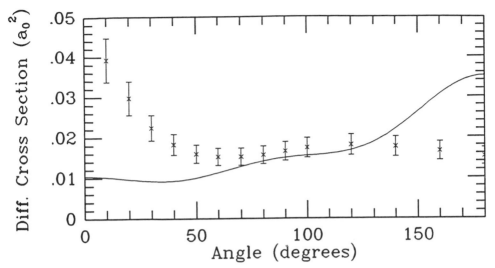

Figure 3. Differential cross section for the transition $X^3\Sigma_g^+ \to a^1\Delta_g$ in e-O_2 scattering at 10 eV. ———, nine state R-matrix calculations; × experiment (Trajmar *et al* 1971).

tions results are debatable. The measurements by Brunger and Teubner (1990) are at variance with the older data as revised Trajmar and co-workers (1983). Unfortunately, we have not yet evaluated differential cross sections in the R-matrix work but show in figure ?? the total cross sections for the $X^1\Sigma_g^+ \to A^3\Sigma_u^+$ transition at impact energies up to 25 eV. In general, there is good agreement between the calculated values and the experimental total cross sections. The structure near 19 eV is due to a pseudoresonance effect since it lies above the thresholds of all the states in our calcuation. These investigations are continuing and we hope to report differential cross sections in the near future.

6. Resonant e-CO scattering

Elastic e-CO scattering displays similar physical features to elastic e-N_2 scattering at low energy. In particular, there exists a low lying shape resonance in the $^2\Pi_g$ symmetry whose lifetime is long enough to produce the characteristic peak structure seen in the corresponding e-N_2 cross section. Motivated by the work of Gillan *et al* 1987 who successfully studied low energy e-N_2 scattering in the non-adiabatic R-matrix formalism, Morgan 1991 has recently completed a very detailed study of the corresponding situation in e-CO.

In this work only the lowest X $^1\Sigma^+$ target state was ultiately retained in the expansion of the R-matrix basis functions ψ_k since this approach has been found to give excellent results for low energy scattering from the isoelectronic N_2 target. Within this approximation a detailed set of calculations was performed, at fixed geometry, to check the sensitivity of the scattering cross section to different choices of target state

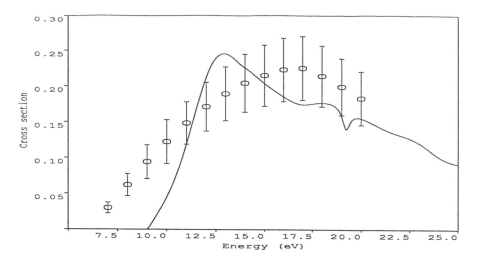

Figure 4. Differential cross section for the transition $X^1\Sigma_g^+ \to A^1\Sigma_u^+$ in e-N_2 scattering. ——, nine state R-matrix calculations; $\bigcirc$ experiment from Trajmar *et al* (1983).

representation. Both SCF and CI representations of the ground state were used, in two different bases of atomic centred Slater type functions (STFs), along with SE, SEP. It should be noted that in order to check the single state approximation some coupled calculations were performed. The best results for the position and width of the $^2\Pi$ resonance, when compared to experimental estimates, were obtained by using an SCF target representation and a large set of L^2 correlation terms in a diffuse STF basis.

The optimum fixed geometry approximation was then used in a non-adiabatic R-matrix calculation to obtain vibrationally inelastic cross sections; three scattering symmetries $^2\Sigma^+$, $^2\Pi$ and $^2\Delta$ were studied. In figure 5 we see that there is excellent agreement between calculated cross sections, from the $^2\Pi$ symmetry, for the transition $v = 0 \to v = 1$ and the recent experimental results of Allan (1989). There remains a small discrepancy in the theoretical and experimental values on either side of the main peak, something which is not remedied by the inclusion of the non-resonant symmetries. This feature could be due however to the truncations inherent in the calculation. Obviously, the next stage in performing calculations is to incorporate electronic excitation channels in a similar fashion to the previous section.

7. Future Directions

The purpose of this paper has been to provide a brief review of recent work on electron collisions with molecular targets carried out within the framework of the R-matrix method. It has been shown that these R-matrix studies are being vigorously pursued at present and this situation should continue for the foreseeable future. Much of this article has been devoted to excitation, either vibrationally or electronically, of

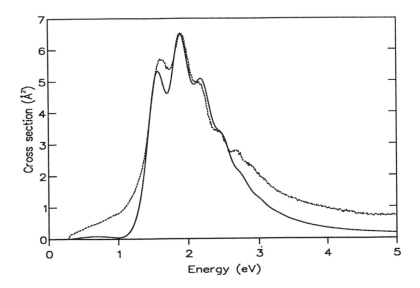

Figure 5. Cross section for excitation of the v=1 level in e-CO. ———, $^2\Pi$ contribution from the SEP model (Morgan 1991); · · · · ·, experiment (Allan 1989).

the target molecule during the scattering interaction. The results obtained in e-H_2 scattering indicate the importance of using many coupled states while that on e-He_2^+ displays the necessity to use accurate target state representations. This knowledge is assisting in our present work on e-O_2 and e-N_2 scattering where we are attempting to resolve the discrepancies that exist between experimental results. All this is in sharp contrast to the situation ten years ago when single state static exchange calculations were a state of the art achievement.

For each system, we have indicated planned enhancements to our calculations. Additionally, our group is developing new codes to study such processes as dissociation and photoionization for diatomic targets and at the same time implementing new and more efficient algorithms within our present suite. We note here also that calculations are being carried out on positron scattering (Tennyson and Danby 1990) but these are beyond the scope of this review. An exciting development in our group recently has been the modification of the R-matrix codes to study scattering by polyatomic molecules. As well as continuing with a programme of detailed work on diatomics we plan to begin a study of electron scattering by small polyatomics in the near future. In the field of electron molecule scattering there remain many new challenges to be met and so it is appropriate to conclude with the words of Robert Frost

and we have miles to go before we sleep,
and we have miles to go before we sleep.

Acknowledgements

The author wishes to express sincere thanks to Miss C Gillan and Mrs E Kirkpatrick for their generous help with the preparation of the manuscript in electronic form. The target states for He_2^+ and N_2 were generated using ALCHEMY II from the IBM MOTECC package.

References

Allan M 1989 {\em J. Electron. Spectrosc. Rel. Phenomena} {\bf 14} 3517
Baluja K, Noble C J and Tennyson J 1985 \JPB {\bf 18} L851
Bloch C 1957 {\em Nucl. Phys} {\bf 4} 503
Branchett S E and Tennyson J 1990 {\em Phys. Rev. Lett.} {bf 64} 2889
Branchett S E, Tennyson J and Morgan L A 1990 \jpb {\bf 23} 4625
Branchett S E, Tennyson J and Morgan L A 1991 submitted
Brunger M J and Teubner T J O 1990 \PR A. {\bf 41} 1413
Burke P G, Gillan C J and Morgan L A 1989 ed Herzenberg A (pub: AIP) pp 15?
Chandra N and Temkin A 1976 \PR A {\bf 13} 188
Gillan C J, Nagy O, Burke P G, Morgan L A and Noble C J 1987 \JPB {\bf 20} 4585\\
Gillan C J, McLaughlin B M, Burke P G and Dahler J S 1991 \PR A. submitted
Gillan C J 1991 {\em Comput. Phys. Commun} submitted for publication
Khakoo L and Trajmar S 1986 \PR A {\bf 34} 146
Massey H S W and Mohr C B O 1932 {\em Proc. Roy. Soc (London)} {\bf A 136} 289
Morgan L A 1990 {\em Proc XVI ICPEAC Conference} ed. Dalgarno A, Freund R S,
        Koch P M, Lubell M S and Lucatorto T B (pub: AIP NY) pp 96ff.
Morgan L A 1991 \jpb submitted for publication
Nishimura K and Danjo L 1986 {\em J Phys Soc Japan} {\bf 55} 3031
Noble C J and Burke P G 1986 \JPB {\bf 19} L35
Sarpal B K, Tennyson J and Morgan L A \jpb 1991a {\bf 24} 1851
Sarpal B K, Branchett S E, Tennyson J and Morgan L A 1991b \jpb submitted
Saxon R P and Liu B 1977 \jcp {\bf 67} 5432
Schneider B I, LeDourneuf M and Burke P G 1979 \JPB {\bf 12} L35
Seaton M J 1985 \JPB {\bf 18} 211
Shimamura I, Noble C J and Burke P G 1990 \PR A. {\bf 41} 3545
Sohn W, Kochem K-H, Erhardt H and Chang E S 1985 \JPB {\bf 18} 2049
Tennyson J and Danby G 1990 \JPB {\bf 23} 1005
Trajmar S, Cartwright D C and Williams W 1971 \PR A {\bf 4} 1482
Trajmar S, Register D F and Chutjian 1983 {\em Phys. Reports} {\bf 97} 219

Optical potentials for electron molecule collisions

Hans—Dieter Meyer

Theoretische Chemie, Physikalisch—Chemisches Institut, Universität Heidelberg, Im Neuenheimer Feld 253, W—6900 Heidelberg, Federal Republic of Germany

Abstract: The optical potential for elastic electron—molecule scattering is investigated. This optical potential is usually defined either by performing a Feshbach projection of the Schrödinger—equation onto the elastic channel or by identifying it with the self—energy of the one—particle Green's function. The two approaches are analyzed in comparison and it is shown that the Green's function approach is superior because it accounts for the correlation of the ionic and ground state in a balanced way. Numerical calculations on the N_2 $^2\Pi_g$—shape resonance are performed to demonstrate the accuracy and stability of the Green's function method.

1. Introduction

The electronically elastic scattering of low—energy electrons from small molecules is a well studied problem in molecular physics (Lane 1980, Rescigno et al 1979, Csanak et al 1984, Lucchese et al 1986). *Ab initio* calculations going beyond the static—exchange level, however, are rather rare as compared to the large number of high—quality *ab initio* calculations on molecular bound states. The reason for this lack is the simultaneous appearance of two difficulties, the many body problem (correlation) and the scattering problem (continuum).

The introduction of an optical potential allows for a very convenient separation of the many–body problem from the scattering problem. The optical potential itself can be computed by bound–state quantum–chemistry methods. The scattering problem is reduced to single–particle scattering. This procedure is similar to the route usually taken in heavy particle scattering. There one first computes the adiabatic Born–Oppenheimer potential–energy surfaces and then performs a scattering calculation by employing these surfaces. A calculation where the coordinates of the nuclei and electrons are simultaneously treated as dynamical variables is avoided.

Returning to the electron–molecule scattering problem one realizes that an adiabatic approximation is not possible. However, an optical potential can always be defined which (formally exact) reduces the many–body scattering problem to a one–particle scattering problem. There are two established approaches to the optical potential. One may define the optical potential via a Feshbach projection (Feshbach 1958, 1962) of the configuration interaction (CI) matrix onto the elastic channel (Schneider and Collins 1982, 1983) or one identifies it with the self–energy of the one–particle Green's function (Bell and Squires 1956, Csanak et al 1971).

It is shown that it is difficult within the CI–approach to account for the ground state correlation. In most applications only the correlation within the temporarily formed anionic state is accounted for. As a consequence of this unbalanced treatment the computed energy of a shape resonance may become too low. In the Green's function approach, on the other hand, the ground state correlation is included in a *balanced way*, i.e. the ground state correlation is evaluated at the same level as the correlation in the anionic state.

To demonstrate the accuracy and stability of the Green's function approach we have performed numerical calculations on the N_2 ${}^2\Pi_g$ shape resonance. The energy and width of this resonance are deduced from a phase shift analysis.

2. The CI–approach to the optical potential

We designate the exact $(N+1)$–electron continuum state as $|\Psi_{\vec{k}}^{N+1}\rangle$. The superscript indicates the number of electrons; the neutral target is assumed to have N electrons. The subscript $\vec{k}$ denotes the wave vector of the incoming electron. To avoid a too clumsy notation the superscript + which indicates the outgoing boundary conditions is omitted. The exact continuum wavefunction converges asymptotically to the antisymmetrized product of the ground state $|\Psi_0^N\rangle$ and a one particle scattered wave $f_{\vec{k}}$. Hence $|\Psi_{\vec{k}}^{N+1}\rangle$ can be written as

$$|\Psi_{\vec{k}}^{N+1}\rangle = a_f^\dagger|\Psi_0^N\rangle + |\chi^{N+1}\rangle \qquad (2.1)$$

where $a_f^\dagger$ denotes the creation operator of the scattered wave $f_{\vec{k}}$

$$a_f^\dagger = \int dr\ \hat{\psi}^\dagger(r)\ f_{\vec{k}}(r) \qquad (2.2)$$

and where $\hat{\psi}^\dagger(r)$ denotes the field–operator (Fetter and Walecka 1971). The function $|\chi^{N+1}\rangle$ accounts for correlation effects. It is an L^2–function when the energy of the incoming electron is below the first excitation threshold of the target. To make the decomposition (2.1) unique we require

$$\langle\Psi_0^N|\hat{\psi}(r)|\chi^{N+1}\rangle = 0 \qquad (2.3)$$

for all r. It is clear that all the scattering information is contained in the asymptotic part of $f_{\vec{k}}$ since $|\chi^{N+1}\rangle$ is L^2. We hence seek for a way to compute the optical wavefunction $f_{\vec{k}}$ without explicitly computing $|\chi^{N+1}\rangle$ or $|\Psi_{\vec{k}}^{N+1}\rangle$. The Hamiltonian which determines $f_{\vec{k}}$ is called the optical Hamiltonian. To derive it, we use a variational principle (Miller 1969) and require that the functional

$$F = <\Psi_{\mathbf{k}}^{N+1} | E + E_0 - H | \Psi_{\mathbf{k}}^{N+1}> \tag{2.4}$$

is stationary. Here E_0 denotes the energy of the ground state and $E = k^2/2$ is the energy of the scattered electron. (Atomic unites are used throughout). Alternatively one may project the (N+1)–particle Schrödinger equation onto the elastic channel (Schneider and Collins 1982, 1983, Rescigno et al 1989). The resulting equation for $f_{\mathbf{k}}$ can be more easily stated when introducing basis sets $\{\varphi_n\}$ and $\{|\chi_i^{N+1}>\}$ to expand $f_{\mathbf{k}}$ and $|\chi^{N+1}>$, respectively. The function $f_{\mathbf{k}}(r)$ is now represented by the coefficient vector $\underline{f}$ and the desired equation reads

$$(\underline{A} - \underline{D}^\dagger \, \underline{B}^{-1} \, \underline{D}) \cdot \underline{f} = 0 \tag{2.5}$$

where

$$A_{pq} = <\Psi_0^N | a_p (E + E_0 - H) \, a_q^\dagger | \Psi_0^N> \tag{2.6}$$

$$B_{ij} = <\chi_i^{N+1} | E + E_0 - H | \chi_j^{N+1}> \tag{2.7}$$

$$D_{iq} = <\chi_i^{N+1} | (E + E_0 - H) \, a_q^\dagger | \Psi_0^N> \tag{2.8}$$

and where a_p denotes the annihilation operator for the orbital φ_p. Employing the second–quantized form for the Hamiltonian the above equations may be further evaluated. In particular A can be expressed by the static–exchange Hamiltonian, coulomb matrix elements and density matrices.

3. The Green's function approach to the optical potential

Let us start this section by introducing a second optical wavefunction $g_{\vec{k}}(r)$ defined as

$$g_{\vec{k}}(r) = <\Psi_0^N | \hat{\psi}(r) | \Psi_{\vec{k}}^{N+1}> \tag{3.1}$$

This function is related to the previously defined function $f_{\vec{k}}$ by

$$g_{\vec{k}}(r) = \int <\Psi_0^N | \hat{\psi}(r) \; \hat{\psi}^\dagger(r) \; f_{\vec{k}}(r')| \Psi_0^N > dr' = \int \sigma(r,r') \; f_{\vec{k}}(r')dr' \tag{3.2}$$

or

$$g = \underline{\sigma} \cdot \underline{f} \tag{3.3}$$

where $\sigma_{pq} = <\Psi_0^N | a_p \; a_q^\dagger | \Psi_0^N >$. Because $\sigma(r,r') \rightarrow \delta(r-r')$ for r or $r' \rightarrow \infty$ one finds that $g_{\vec{k}}$ and $f_{\vec{k}}$ obey the same asymptotic behavior. Moreover, because of eqs. (3.2, 3.3) one can easily express one optical function by the other.

The optical wavefunction $g_{\vec{k}}$ can be evaluated by employing the one–particle Green's function (Csanak et al 1971, Brenig 1967, Bell and Squires 1956)

$$g_{\vec{k}}(r) = \lim_{\eta \to +0} i\eta \int G(r,r', E + i\eta) \; \varphi_{\vec{k}}(r')dr' \tag{3.4}$$

where $\varphi_{\vec{k}}$ denotes a plane wave with wave vector $\vec{k}$.

The Green's function G obeys the well–known Dyson equation (Fetter and Walecka 1971) which we formally write as

$$G(E) = G^{HF}(E) + G^{HF}(E) \; \Sigma(E) \; G(E) \tag{3.5}$$

Here Σ denotes the self–energy and G^{HF} the Hartree–Fock Green's function. The Dyson equation is equivalent to

$$G^{-1}(E) \;=\; (G^{HF}(E))^{-1} - \Sigma(E) \;=\; E - H_0^{SE} - \Sigma(E)$$

(3.6)

where H_0^{SE} denotes the static–exchange Hamiltonian defined with respect to the charge distribution of the Hartree–Fock ground state. The self–energy Σ can be split into three terms (Schirmer et al 1983)

$$\Sigma(E) = \Sigma(\infty) + M^{I}(E) + M^{II}(E)$$

(3.7)

The dynamical parts, M^{I} and M^{II}, are associated with excitations of $N{+}1$ and $N{-}1$ particles, respectively. The static part, $\Sigma(\infty)$, accounts for the difference in the charge distributions of the Hartree–Fock and the exact ground state, i.e.

$$H^{SE} = H_0^{SE} + \Sigma(\infty)$$

(3.8)

where H^{SE} denotes the static–exchange Hamiltonian defined with respect to the charge distribution of the exact ground state. Eq. (3.4) thus becomes equivalent to the Schrödinger equation

$$(E - H^{SE} - M^{I}(E) - M^{II}(E))\, g_{\vec{k}} = 0$$

(3.9)

subject to scattering boundary conditions.

We emphasize that the Green's function approach has lead to an optical Hamiltonian (cf. (3.9)) which is structurally different from that of the CI–approach (cf. (2.5)). The M^{I} term is quite similar to the $D^{\dagger}B^{-1}D$ term, however, there appears nothing like the

M^{II} term in the CI–approach. Both approaches are formally exact and yield the same optical wavefunction $g_{\vec{k}}$. When approximations are introduced, however, the two approaches perform quite differently.

The evaluation of the optical potential via the CI–approach requires that two approximations have to be introduced (except for the choice of the one–particle basis set). The first one is the specification of the space $\{|\chi_i^{N+1}>\}$ and the second is the choice of an approximation to the exact ground state $|\Psi_0^N>$. The evaluation of the optical potential significantly simplifies if one chooses the Hartree–Fock determinant as an approximation to $|\Psi_0^N>$. Doing so one neglects the ground state correlation whereas the correlation in the (N+1)–particle anionic state is permitted via the functions $\{|\chi_i^{N+1}>\}$. This leads to an unbalanced treatment and it has been found (Schneider and Collins 1984) that the computed resonance energy of the N_2 $^2\Pi_g$ shape resonance becomes the lower, the larger is the space $\{|\chi_i^{N+1}>\}$. For large spaces the computed resonance energy is below its correct value by more than 1/2 eV.

Turning to the Green's function approach we remark that we have used the algebraic diagrammatic construction (ADC) (Schirmer et al 1983) to compute the self–energy. Within the ADC, the ground state $|\Psi_0^N>$ is evaluated by Rayleigh–Schrödinger perturbation theory. The order of the perturbation expansion uniquely determines the spaces $\{|\chi_i^{N+1}>\}$ and $\{|\chi_i^{N-1}>\}$ used to evaluate M^I and M^{II}, respectively. The procedure is hence balanced. The correlation of the ground state and the correlation of the anionic state are treated on the same footing. It has been shown by perturbation theoretical analysis of ADC(2) (Schirmer et al 1983, Meyer 1989) that the (N+1)–particle contribution M^I accounts for the relaxation in the anionic state, whereas the (N–1)–particle term M^{II} accounts for the ground state correlation. Both parts contribute to the correlation in the anionic state.

4. The N_2 $^2\Pi_g$ shape resonance

The shape resonance of the nitrogen molecule serves as an excellent test for the optical potential because the resonance energy E_r and width Γ depend sensitively on the interaction potential. These resonance parameters are deduced from the scattering calculation by a phase shift analysis.

When computing the self—energy by ADC one has to adopt a single—particle basis set and decide on the order n of the ADC scheme. In the calculations reported below we have used three different basis sets consisting of cartesian gaussians. The first two sets are extensions of the one given by Salez and Veillard (1968). They consist of (5s,4p, 2d) and (5s,7p,3d) contracted functions yielding 58 and 88 MO's, respectively (see Meyer 1989). The third basis set is due to Partridge (1989) and consist of (14s,12p, 3d) contracted functions yielding 130 MO's. These sets have been used to compute the self—energy by ADC(2), by the two—particle—hole Tamm—Dancoff approximation (2ph—TDA) (Schirmer and Cederbaum 1978) and by ADC(3) (von Niessen et al 1984). ADC(2) is identical to the well known second order expression of the self—energy. It has been used previously by Klonover and Kaldor (1977) to study the electron scattering off H_2. The 2ph—TDA and ADC(3) have previously been used for computing optical potentials by Berman et al (1983) and by Meyer (1989), respectively.

Table I shows the computed resonance parameters. The three different basis sets are designated by their number of MO's (see above). The optical potential was evaluated by ADC(2), i.e. the well—known second order expression for the self—energy, by 2ph—TDA and by ADC(3). To discuss the results we note that the resonance parameters in the fixed nuclei limit at equilibrium geometry $(R = 2.0693\ a_o)$ are expected to assume the values $E_r = 2.31$ eV and $\Gamma = 0.41$ eV (Berman et al 1983).

The column designated SE depicts the static exchange results. The static exchange resonance energy is about 1.5 eV too high, indicating the importance of relaxation and correlation. Already the simple second order approximation to the self–energy, ADC(2), is able to incorporate a large fraction these effects and yields reasonable results. The next step in the hierarchy, 2ph–TDA, further lowers resonance energies and widths and brings them in exellent agreement with the expected values (see above). The most elaborate method investigated, ADC(3), on the other hand, increases the values of the resonance parameters as compared to 2ph–TDA but they remain lower than the ADC(2) values. The 2ph–TDA seemingly gives better results then the more elaborate ADC(3). This is due to error compensation. The 2ph–TDA is expected to yield optical potentials which are too attractive. The introduction of an incomplete basis set, on the other hand, makes the computed optical potential less attractive which (partially) compensates the error introduced by 2ph–TDA.

The dependence of the resonance parameters on the basis set is larger as one might have expected. It seems that one needs quite a large basis set in order to converge the results to better than 0.05 eV. In conclusion we would like to say that the optical potential defined as the self–energy of the one–particle Green's function is a very promising tool for ab–initio calculations on low energy electron–molecule collisions.

Table I   Resonance energy (upper entry) and width (lower entry) in eV for different basis sets and evaluation schemes

Basis set	SE	ADC(2)	2ph–TDA	ADC(3)
58	3.800	2.740	2.411	2.653
	1.230	0.503	0.365	0.487
88	3.808	2.615	2.268	2.545
	1.230	0.584	0.419	0.541
130	3.804	2.617	2.277	2.586
	1.236	0.532	0.380	0.423

References

Bell S J and Squires E 1956 Phys. Rev. Lett. $\underline{3}$ 96

Berman M, Walter O and Cederbaum LS 1983 Phys. Rev. Lett. $\underline{50}$ 1979

Berman M, Estrada H, Cederbaum LS and Domcke W 1983 Phys. Rev. A $\underline{28}$ 1363

Brenig W 1967 Z. Phys. $\underline{202}$ 340

Csanak G. Cartwright D C, Srivwastava S K and Trajmar S 1984 "Electron–Molecule
Interactions and their Applications" Vol. 1 edited by Christophorou L G (New York:
Academic Press) pp 1–153

Csanak G, Taylor H S and Yaris R 1971 Adv. Mol. Phys. $\underline{7}$ 287

Feshbach H 1958 Ann. Phys. (N.Y.) $\underline{5}$ 357

Feshbach H 1962 Ann. Phys. (N.Y.) $\underline{19}$ 287

Fetter A L and Walecka J D 1972 "Quantum Theory of Many Particle Systems"
(New York: Mc Graw Hill)

Klonover A and Kaldor U 1977 Chem. Phys. Lett. $\underline{51}$ 321

Lane N F 1980 Rev. Mod. Phys. $\underline{52}$ 29

Lucchese R R, Takatsuka K and McKoy V 1986 Physics Reports $\underline{131}$ 147

Meyer H D 1989 Phys. Rev. A $\underline{40}$ 5605

Miller W H 1969 J. Chem. Phys. $\underline{50}$ 407

Partridge H 1989, NASA Technical Memorandum 101044, Ames Research Center,
Moffett Field, California

Rescigno T N, McCurdy C W and Schneider B I 1989 Phys. Rev. Lett. $\underline{63}$ 248

Salez C and Veillard A 1968 Theor. Chim. Acta $\underline{11}$ 441

Schirmer J Cederbaum L S and Walter O 1983 Phys. Rev. A $\underline{28}$ 1237

Schirmer J and Cederbaum L S 1978 J. Phys. B $\underline{11}$ 1889

Schneider B I and Collins L A 1982 J. Phys. B $\underline{15}$ L 335

Schneider B I and Collins L A 1983 Phys. Rev. A $\underline{27}$ 2847

Schneider B I and Collins L A 1984 Phys. Rev. A $\underline{30}$ 95

Ab Initio calculations on collisions of low energy electrons with polyatomic molecules

Thomas N. Rescigno
Lawrence Livermore National Laboratory
P. O. Box 808
Livermore, CA 94550

ABSTRACT: The Kohn variational method is one of simplest, and oldest, techniques for performing scattering calculations. Nevertheless, a number of formal problems, as well as practical difficulties associated with the computation of certain required matrix elements, delayed its application to electron—molecule scattering problems for many years. This paper will describe the recent theoretical and computational developments that have made the "complex" Kohn variational method a practical tool for carrying out calculations of low energy electron—molecule scattering. Recent calculations on a number of target molecules will also be summarized.

1. INTRODUCTION

In the study of electron—molecule collision processes, increased attention is being paid, by both experimentalists and theoreticians, to small polyatomic targets. With these systems, the many vibrational degrees of freedom and the opportunities for interaction between excited electronic surfaces in several dimensions present a challenging opportunity to study a body of physics far richer, and more complicated, than that found in diatomic systems. The study of electron—polyatomic molecule scattering, apart from its fundamental theoretical importance, is also driven by the practical need to develop a better understanding of the physics and chemistry of the key processes that drive low temperature plasmas (Ensslen and Veprek 1987, Capezzuto and Bruno 1988, Veprek and Heintze 1990). For example, the detailed modelling of a chemical—vapor—deposition process may require many reaction rates involving neutral radicals and ionic fragments, but the rate determining reactions which initiate this chemistry frequently involve the electron impact dissociation of a polyatomic molecule, such as SiH_4 or CH_4, into neutral fragments. Our understanding of these electron collision processes has not kept pace with their importance and there is a critical need for more information about the basic physics underlying processes that are crucial to the development of emerging technologies.

The ability to confidently predict cross sections for complex electron—molecule processes requires a formalism that is robust enough to deal with the various complexities of the problem— the non—spherical nature of the interaction, electron exchange, target correlations, electron induced polarization — without resorting to parametrization or the use of model potentials. Our approach is based on a completely *ab initio* treatment of the $(N+1)$— electron time—independent Schroedinger equation. The treatment of processes such as electronic excitation demands such a fully *ab initio* approach. We use an algebraic variational principle, namely the Kohn variational method (Kohn 1948), in a formulation that includes complex outgoing—wave boundary conditions. The use of complex boundary conditions in the trial function has the profound effect of eliminating a classic pathology of the traditional Kohn method — the so—called Kohn "anomalies" or spurious resonances (Nesbet 1968, 1969). This idea was originally used in nuclear physics by Mito and Kamimura (1976), but went largely unnoticed until it was introduced into atomic and molecular physics several years ago by Miller and co—workers in the context of reactive heavy—particle scattering (Miller and Jansen op de Haar 1987, Zhang and Miller 1989).

The development of the complex Kohn method for electron—molecule scattering is the result of a collaborative effort that has taken place over the past few years. The connection between Miller's original work, the Kohn method and Kapur—Perils theory was pointed out by McCurdy, Rescigno and Schneider (1987). The essential formulation of the complex Kohn method for electron scattering — with the elimination of continuum exchange matrix elements through the use of separable expansions — was done in collaboration with B. Schneider at Los Alamos (Rescigno and Schneider 1988, Schneider and Rescigno 1988). Our initial treatment relied on numerical single—center expansions and was, for all practical purposes, limited to linear molecular targets. The use of adaptive three—dimensional quadrature and the development of a formalism capable of handling nonlinear polyatomic targets derived from a collaborative effort between C. W. McCurdy and myself (McCurdy and Rescigno 1989). The modifications of the complex Kohn formalism that were needed to treat photoionization and electron scattering by ionic molecular targets were developed by A.E. Orel and myself (Orel and Rescigno 1990, Rescigno and Orel 1991). Finally, the coupling of the complex Kohn formalism to modern electronic methods — a development that allows us to take full advantage of the heavy arsenal of computational machinery that has been developed to study the molecular electronic structure problem — was carried out by Byron Lengsfield at Livermore (Orel et al. 1990, Lengsfield et al. 1991, Lengsfield and Rescigno 1991). This development has allowed us to explore the use of elaborate correlated target wave functions and to incorporate the effects of target polarization into our calculations by the efficient construction of optical potentials from first principles.

2.   THEORY

To describe the scattering of low energy electrons, incident in a channel denoted by the label Γ_0, by an N—electron target molecule, we formulate the problem in body—frame coordinates within the framework of the fixed—nuclei approximation using an antisymmetrized trial wave function of the form,

$$\Psi_{\Gamma_0} = \sum_\Gamma A(\chi_\Gamma \, F_{\Gamma\Gamma_0}) + \sum_\mu d_\mu^{\Gamma_0} \otimes_\mu . \tag{1}$$

where the first sum runs over the energetically open N—electron target states, denoted by the normalized functions χ, and the operator A antisymmetrizes the orbital functions F into the functions χ. The functions F are, by construction, orthogonal to all the orbitals used to build the target functions χ. The set $\{\otimes\}$ are square—integrable (N+1)—electron terms used to relax any constraints imposed by the orthogonality requirement placed on F and to represent polarization and correlation effects not included in the first summation. The symbol Γ is being used to label all the quantum numbers needed to represent a physical state of the composite system, ie. the internal state of the target molecule as well as the energy and orbital angular momentum of the scattered electron.

In the complex Kohn method, the channel continuum functions, F, are further expanded as:

$$F_{\Gamma\Gamma_0}(\vec{r}) = \sum_{\ell m} [\, f_\ell^\Gamma(r)\delta_{\ell\ell_0}\delta_{mm_0}\delta_{\Gamma\Gamma_0} + T^{\Gamma}{}_{\ell m}{}^{\Gamma_0}{}_{\ell_0 m_0}\, g_\ell^\Gamma(r)]\, Y_{\ell m}(\hat{r})/r + \sum_k c_k^{\Gamma\Gamma_0}\varphi_k^\Gamma(\vec{r}) \quad (2)$$

where $Y_{\ell m}$ is a normalized spherical harmonic, the φ_k are a set of square–integrable functions, and the functions {f} and {g} are linearly independent continuum orbitals which are regular at the origin and, in the case of neutral targets, behave asymptotically as regular and *complex, outgoing* Riccati–Bessel functions, respectively. To treat ionic targets, the functions {f} and {g} may be modified to incorporate Coulomb boundary conditions (Orel et al. 1990). The coefficients [T] are elements of the T–matrix and are the fundamental dynamical quantities that determine scattering amplitudes and cross sections.

Electron scattering cross sections can be expressed solely in terms of the T–matrix elements which, in turn, are extracted from the open–channel part of the wave function. The correlation functions ⊗ can be formally incorporated into an effective optical potential. This is accomplished in standard fashion by using Feshbach (1958, 1962) partitioning with operators P and Q that project onto the open– and closed–channel subspaces, respectively. $P\Psi$ satisfies a Schroedinger equation with the effective Hamiltonian:

$$H_{eff} = H_{PP} + H_{PQ}\,(E - H_{QQ})^{-1}\,H_{QP} \quad (3)$$

In calculations which include the effect of closed channels, the set of Q–space configurations can become quite large. The optical potential formulation allows one to use bound–state molecular structure methods to treat the Q–space portion of the problem very efficiently and to divorce this part of the calculation from the rest of the problem.

The T–matrix can be characterized as the stationary value of the Kohn functional:

$$T_{stat}^{\Gamma\Gamma_0} = T^{\Gamma\Gamma_0} - 2 \int P\Psi_\Gamma\,(\,H_{eff} - E\,)\,P\Psi_{\Gamma_0} . \quad (4)$$

The variation is carried out by substituting the trial wave function into Eq. (4) and requiring the derivatives with respect to the linear trial coefficients to be zero. The result is a set of linear equations for the stationary T–matrix elements. In a condensed matrix notation, in which channel indices are supressed, the result can be written:

$$[T_{stat}] = -2\,(\,M_{00} - M_{0q}\,M_{qq}^{-1}\,M_{q0}\,) , \quad (5)$$

where $M = H_{eff}-E$, 0 refers to $A(\chi_\Gamma\, f_\ell^\Gamma\, Y_{\ell m})$ and q refers to the space spanned by $A(\chi_\Gamma\, g_\ell^\Gamma\, Y_{\ell m})$ and $A(\chi_\Gamma\, \varphi_k^\Gamma)$. Because of the outgoing–wave behavior of the functions {g}, the matrix M_{qq} is complex symmetric and thus its inverse will generally be non–singular for real energies. This result is in marked contrast to the traditional Kohn method where a real Hermitian matrix is inverted.

The efficacy of this approach hinges on the ease with which the required Hamiltonian matrix elements can be assembled. Matrix elements involving only bound functions can be evaluated using the standard techniques of bound–state molecular electronic structure theory. The matrix elements involving continuum (Bessel or Coulomb) functions are more problematic, and the principal difficulty of any molecular scattering calculation is their evaluation. There are two critical steps in making this problem tractable for molecules. The first practical step is to reformulate the entire procedure in terms of mutually orthogonal bound and continuum functions. This can be done quite rigorously and follows from a property of the Kohn principle called *transfer invariance* (Nesbet 1980), which is just a fancy way of saying that the T–matrix is unchanged by any unitary transformation between the functions $\{\varphi\}$ and the continuum functions $\{f\}$ and $\{g\}$ of Eq. (2). This reformulation is especially useful since it allows us to eliminate the entire class of 'bound–free' and 'free–free' exchange matrix elements (Rescigno and Schneider 1988). This is accomplished by invoking a separable represention of all exchange operators in terms of the same square–intergable basis to which the continuum functions have been orthogonalized. If an optical potential is used, it too is treated in separable form. I will give no further details about the use of separable approximations here because it has been discussed in detail elsewhere (Rescigno and Orel 1981a, 1981b, Schneider and Collins 1981, Rescigno and Schneider 1988). The only remaining matrix elements involving continuum functions are "direct". Thus the second step in rendering the molecular problem tractable is to devise an efficient numerical scheme for evaluating these intergals.

All "direct" bound–free and free–free matrix elements can be reduced to single three–dimensional quadrature after analytic integration over the target coordinates. We do not use single–center expansions to further reduce these 3–D integrations because single–center expansions can converge quite slowly for polyatomic targets. We evaluate the three–dimensional integrals directly using a novel quadrature scheme (McCurdy and Rescigno 1989). The key to making this quadrature practical is to make it adaptive so that points are clustered with nearly spherical symmetry around each nucleus while the points at large distances from the molecule are arranged with spherical symmetry around its center. We accomplish this by starting with a quadrature consisting of shells of points around the center of the molecule which we can construct from standard Gauss quadrature points and weights. We then transform these points under a mapping that rearranges the points anound each nucleus but leaves the distant points unchanged. The mapping is analytic, so that knowing the Jacobian gives us the appropriate quadrature weights and lets us evaluate the integrals in our new coordinates.

The formulation of the variational method I have just outlined allows for considerable flexibility in the choice of continuum basis functions, because it does not rely on any particular analytic techniques for evaluating the required bound–free and free–free integrals. For example, the "traditional" choice for the continuum basis functions is to use Riccati–Bessel (or, for ionic targets, Coulomb) functions regularized at r=0 by a simple exponential cutoff. We have recently shown (Rescigno and Orel 1991) that a better choice for the outgoing wave function is (for neutral targets, with an obvious generalization for ionic targets):

$$g_\ell(r) = \int G_\ell^+(r,r')\, w(r')\, dr' \Big/ \int j_\ell(kr')\, w(r')\, dr' \tag{6}$$

where w is an arbitrary square–integrable function, j_ℓ is the regular Riccati–Bessel function and G_ℓ^+ is the free–particle Green's function, defined here as,

$$G_\ell^+ = j_\ell(kr_<)\,[-n_\ell(kr_>) + i\,j_\ell(kr_>)]\ . \tag{7}$$

It is simple to show that the function defined in Eq. (6) is a pure outgoing wave at large r. These dynamically prepared functions, which can be thought of as the first Born approximation to the wave function for some model problem, turn out to be quite easy to generate and remove the sensitivity to the cutoff function that can be a problem with the traditional choice.

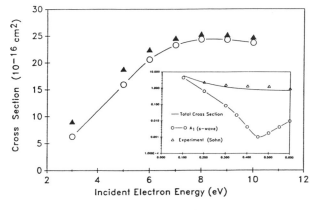

Fig.1 Total cross sections for e +CH$_4$

Because the complex Kohn method is based on a Hamiltonian formulation of the scattering problem, most of the matrix elements needed can be assembled by using bound–state structure methodology. For example, the formalism does not become particularly difficult to apply even when elaborate correlated target states are incorporated into the Kohn trial wave function. Moreover, there are cases, as I will show below, where it is essential to use multi–configuration target states to obtain accurate results (Orel et al. 1990). The use of

correlated target states does, however, introduce a new set of *formal* difficulties when one tries to extend the calculations into the intermediate energy range. The problem, which is common to most close–coupling methods, is that in improving the description of the target states explicitly included, one begins to indirectly describe other excited states which have not been explicitly included and this can lead to pseudo–resonances. We have recently devised an approximation scheme (Lengsfield and Rescigno 1991) which allows us to systematically isolate and eliminate the terms that give rise to the spurious behavior. I will also illustrate this below.

3. EXAMPLES

3.1 Low Energy Electron Scattering by CH$_4$, SiH$_4$ and C$_2$H$_4$

Complex Kohn calculations have been carried out on a number of closed–shell polyatomic molecules. Our goal was to develop a type of trial wave function that gives accurate elastic, differential and momentum transfer cross sections over an energy range from a few tenths of an electron volt – the vicinity of the Ramsauer–Townsend (RT) minimum – through the shape resonance region below the first excited electronic threshold. To achieve this, we need a scattering function that balances short and long range interactions, since the RT effect is the result of a cancellation of a long–range polarization interaction and a short–range repulsion. It is well known that polarization of the electron cloud leads to a breakdown of the static–exchange approximation below a few electron volts even for a molecule as simple as H$_2$. The type of trial function we have used is based on a polarized–SCF target wave function. In this approach, the target ground state is described by a SCF wave function and closed–channel functions are generated by singly exciting the occupied target orbitals into a set of virtual orbitals. The P–space portion of the (N+1)–electron wave function consists of the static–exchange terms formed from the direct product of the SCF target function and orbitals used to describe the scattered electron. The complementary Q–space consists of configuration state functions formed from the direct product of a scattering orbital and a closed–channel target function. In order to achieve a compact representation of the total wave function, only a subset of the virtual orbital space, the polarized–virtual orbitals, is used to construct the closed–channel target states. These polarized orbitals, as we have shown elsewhere (Lengsfield et al. 1991), can be constructed by diagonalizing the matrix

$$\rho_{ij}^{nk} = \frac{<\phi_i \mid \mu_k \mid \phi_n><\phi_n \mid \mu_k \mid \phi_j>}{(\epsilon_i - \epsilon_n)(\epsilon_j - \epsilon_n)} \tag{8}$$

where μ is the dipole operator, n refers to the SCF orbital that is being polarized, the virtual orbitals denoted by i and j are eigenvectors of a $V(N-1)$ Fock operator, and ϵ is an orbital energy. The use of polarized orbitals provides a very compact representation of the total wave function (there are at most three polarized orbitals for each occupied orbital); we have also found this prescription to provide a good balance between the long–range and short–range interactions in these types of problems.

Figure 1 compares total cross sections we have recently obtained for methane using the complex Kohn method (Lengsfield et al. 1991) with experiment. The experimental points in the 3–10 eV range are those of Lohmann and Buckman (1986). The inset to Fig. 1 shows the cross section in the region of the

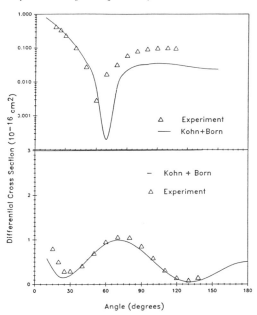

Fig. 2 Differential cross sections for CH_4 at .5 (upper) and 3.0 eV (lower).

Ramsauer–Townsend minimum. The low–energy measurements are those of Sohn et al. (1986). The agreement is seen to be quite good. Fig. 2 compares the differential cross sections we obtained at .5 and 3.0 eV with the experimental results of Sohn et al. (1986). These are the first *ab initio* calculations on this system to achieve this kind of accuracy.

The same treatment has been applied to the e + SiH_4 system with comparable success (Sun et al. 1991). Fig. 3 shows the results obtained for the total cross section, along with the experimental results of Wan et al. (1989) and the model

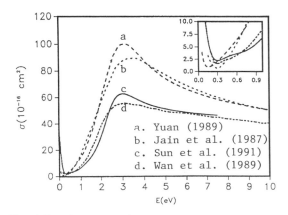

a. Yuan (1989)
b. Jain et al. (1987)
c. Sun et al. (1991)
d. Wan et al. (1989)

Fig. 3 Total cross sections for e+SiH_4

potential results of Jain et al. (1987) and Yuan (1989). Although the complex Kohn results are, of the theoretical results, in closest agreement with experiment, they evidently over estimate the cross section at the resonance maximum. Discrepancies in this energy range are even more pronounced in the momentum transfer cross section. For a large target like silane, which has low–lying unoccupied d–orbitals, the use of a SCF target wave function may not be adequate to achieve quantitative cross sections in the vicinity of the d–wave shape resonance.

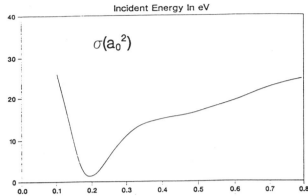

FIG4 ^2A_g partial integrated cross section for e^--C_2H_4 scattering.

Recent complex Kohn calculations by Schneider et al. (1991) on low–energy electron– ethylene (C_2H_4) scattering were significantly the first *ab initio* calculations to confirm the existence of a Ramsauer–Townsend minimum in a molecule that possesses a permanent quadrupole moment. Although there was some experimental evidence to suggest this finding (Boness et al. 1967), there was considerable reluctance among experimentalists to accept this data as fact. Fig. 3 shows the ^2A_g (s–wave) component of the total cross section we have obtained from our variational calculations including polarization. The deep minimum near 200 meV is characteristic of the Ramsauer–Townsend effect seen in other atoms and molecules.

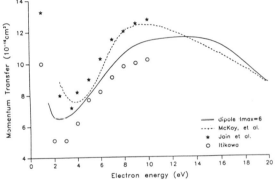

Fig. 5 e + NH_3 momentum transfer cross sections

3.2   Elastic Scattering from Polar Polatomics: e^- + NH_3

Low energy electron scattering by a molecule that possesses a permanent dipole moment is dominated by the long–range dipolar field of the target. It is well known that many partial waves must be included to obtain converged cross sections. However, it not necessary to carry out elaborate calculations to obtain all the necessary contributions, since the weakly scattered, high partial wave components of the T–matrix are well represented in the first Born approximation with the asymptotic potential

$$V(r) \approx \frac{\vec{\mu} \cdot \hat{r}}{r^2} \ . \tag{9}$$

Another problem with polar targets is related to the frame in which the calculations are performed. Body–frame, fixed–nuclei calculations which ignore the rotational motion of the target molecule give differential cross sections that diverge in the forward direction and, consequently, infinite total cross sections . This is again related to the slow fall off of the large ℓ–components of the T–matrix for a fixed dipole potential. Norcross and Padial (1982) have suggested a hybrid treatment which they call the Multipole Extracted Adiabatic Nuclei (MEAN) approximation. The idea is to add and subtract the Born approximation to the expression for the cross section:

$$\sigma = \sigma^{\text{Born}} + \Delta\sigma$$

$$= \sigma^{\text{Born}} + \frac{4\pi}{k^2} \sum_{\ell,\ell'} |\, T_{\ell\ell'} |^2 - |\, T_{\ell\ell'}^{\text{Born}} |^2 \; . \qquad (10)$$

The first term is the full Born cross section and is evaluated in a framework that includes the rotational motion (Crawford et al. 1967). The second sum, which contains the dynamically important low ℓ terms, converges very rapidly and is evaluated using the fixed–nuclei approximation.

Implementations of the hybrid treatment outlined above have largely been limited to linear rotors. We have carried out complex Kohn calculations on $e^- + NH_3$ using a modification of the above formalism (Parker et al. 1991). For a symmetric top molecule like ammonia, it is still not possible to obtain finite total cross sections since the Born approximation in this case diverges, even when proper allowance for rotation is included (Crawford 1967). The momentum transfer cross section is, however, well defined. We used the complex Kohn method to obtain T–matrix elements for values of $\ell,\ell' \leq 6$ in the static–exchange approximation. One does not expect polarization effects to be terribly important in a system like this where many partial waves contribute, even at low collision energies. In computing differential and momentum transfer cross sections, we did use Born closure to include the high ℓ contributions. However, unlike other investigators who have applied the analogue of Eq. (10) to the differential cross section, we found that this procedure is more properly applied to the scattering amplitude itself. Fig. 5 shows our results for the momentum transfer cross section, along with the theoretical results of Pritchard,Lima and McKoy (1989), Jain and Thompson (1983), and the suggested values of Itikawa (1974). The theoretical results are in reasonably good accord, although the previous theoretical studies did not account for the high partial wave contribution. This is most evident at energies below 4 eV.

3.3 $e^- + H_2CO$: Resonant Vibrational and Electronic Excitation

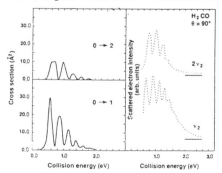

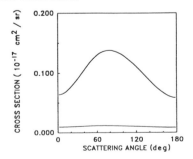

FIG. 6 Comparison of calculated vibrational excitation cross sections (left) with experimental excitation functions of Benoit and Abouaf

FIG. 7 Differential cross sections for electron impact excitation of formaldehyde at 5 eV. Upper curve, $^1A_1 \rightarrow \, ^3A_2$; lower curve $^1A_1 \rightarrow \, ^1A_2$.

Formaldehyde was one of the first polyatomic molecules to be studied using the complex Kohn technique (Rescigno et al. 1989). Calculations carrid out at the equilibrium geometry of the target revealed a shape resonance in $^2B_1$ symmetry near 1 eV. These calculations included an optical potential to account for the effects of target distortion. Static–exchange calculations placed the resonance too high in energy and with too broad a width. These results have recently been extended to account for the effects of nuclear motion (Schneider et al. 1990). Because the experimental energy loss spectrum shows that scattering near 1 eV strongly favors excitation of the ν_2 mode (CO stretch), we carried out additional calculations for different values of the CO bond distance. The resonance parameters extracted from these calculations were used in a simple one–dimensional boomering model to compute resonant vibrational excitation cross sections. The results of this study are shown in Fig. 6 and are found to display vibrational substructure in reasonable agreement with the measurements of Benoit and Abouaf (1986).

We have also calculated cross sections for electronic excitation of H_2CO (Rescigno et al. 1990), the first *ab initio* treatment of electronic excitation in a polyatomic molecule to be reported. Three state close–coupling calculations were carried out which included the ground $^1A_1$ state and the two lowest $^{1,3}A_2$ states. Simple, single–configuration wave functions were employed for the target states. The differential excitation cross sections we obtained at 5 eV are shown in Fig. 7. As of this writing, there are no experimental measurements available for comparison. Our calculated $^1A_1 \rightarrow {}^3A_2$ cross sections were found to be small near 0^0 and 180^0 over the entire energy range we considered (5–20 eV). This finding is suggestive, as formaldehyde is isoelectronic with O_2, of a selection rule that explains this behavior for the analogous type of transition in a linear molecule (Cartwright et al. 1971).

3.4 Correlated Target Wave functions and Intermediate Collision Energies

In many of the studies I have described, it was important to include correlation in the *(N+1)–electron system* to account for target distortion and polarization. This was achieved by constructing optical potentials. The inclusion of correlation in the *N–electron* target states themselves is a more difficult problem and requires special consideration. Our initial efforts in this area have so far been restricted to a few diatomic molecules, but additional studies are underway.

There are instances were we have found it absolutely essential to use multi–configuration target states. The $e^- + N_2^+$ system was such a case (Orel et al. 1990). An accurate description of excitation from the ground state to the optically allowed B–state requires target wave functions that give the correct value for the dipole transition moment We found that we could not abtain a reasonable value for the X→B transition moment with single–configuration wave functions, no matter how the molecular orbitals were chosen. By using state–averaged natural orbitals, we were able to construct compact, accurate target states and obtain excitation cross sections in excellent agreement with the most recent experiments on this system.

There are formal difficulties that arise when one attempts to carry out close–coupling calculations with correlated target states at intermediate energies. To simplify the formulation of the collision problem, the scattering orbitals are generally kept orthogonal to the set of orbitals from which the target states are built. To relax any unphysical orthogonality constraints, appropriate (N+1)–electron correlation terms, which are constructed from the target orbitals, are introduced into the trial wave function and their coefficients determined variationally. As more elaborate multi–configuration target states are used, the number of (N+1)–electron correlation terms grows correspondingly. Because these terms are allowed to freely vary, they can begin to represent excited states not explicitly included in the open–channel space.

The signature of this problem is a plethora of resonances near the thresholds of these pseudo—states. We have recently shown how to alleviate this problem by contracting the correlation terms in a way consistent with the target states being used and still relax the orthogonality constraints properly (Lengsfield and Rescigno 1991). The technique was illustrated with extensive calculations on the e^- + F_2 system.

ACKNOWLEDGEMENT

This work was performed under the auspices of the U.S. Department of Energy by the Lawrence Livermore National Laboratory under contract No. W—7405—ENG—48.

REFERENCES

Benoit C and Abouaf R 1986 *Chem. Phys. Letts.* **123**, 134
Capezzuto P and Bruno G 1988 *Pure and Appl. Chem* **60** 633
Crawford O H 1967 *J. Chem. Phys.* **47** 1100
Crawford O H, Dalgarno A and Hays P B 1967 *Molec. Phys.* **13** 181
Ensslen K and Veprek S 1987 *Plasma Chem. and Plasma Proc.* **7** 139
Feshbach H 1958 *Ann. Phys. (New York)* **5** 357
Feshbach H 1962 *Ann. Phys. (New York)* **19** 287
Itikawa Y 1974 *At. Data Nucl. Data Tables* **14**
Jain A K and Thompson D G 1983 *J. Chem Phys. B* **16** L629
Jain A K, Tripathi A N and Jain A 1987 *J. Phys. B* **20** L389
Kohn W 1948 *Phys. Rev.* **74** 1763
Lengsfield B H and Rescigno T N 1991 *Phys. Rev. A* **44** xxxx
Lengsfield B H, Rescigno T N and McCurdy C W 1991 *Phys. Rev. A* **XX** xxxx
Lohmann B and Buckman S 1986 *J. Phys. B* **19** 2565
Miller W H and Jansen op de Haar B M D D 1987 *J. Chem. Phys.* **86** 6213
McCurdy C W, Rescigno T N and Schneider B I 1987 *Phys. Rev. A* **36** 2061
McCurdy C W and Rescigno T N 1989 *Phys. Rev. A* **39** 4487
Mito Y and Kamimura M 1976 *Prog. Theor. Phys.* **56** 583
Nesbet R K 1968 *Phys. Rev.* **175** 134
Nesbet R K 1969 *Phys. Rev.* **179** 60
Nesbet R K 1980 *Variational Methods in Electron—Atom Scattering Theory* (New York: Plenum) pp 69—78
Norcross D W and Padial N T 1982 *Phys. Rev. A* **25** 226
Orel A E and Rescigno T N 1990 *Phys. Rev. A* **41** 1695
Orel A E , Rescigno T N and Lengsfield B H 1990 *Phys. Rev. A* **42** 5292
Parker S D, McCurdy C W and Rescigno T N 1991 (to be published)
Rescigno T N, Lengsfield BH and McCurdy C W 1990 *Phys. Rev. A* **41** 2462
Rescigno T N, McCurdy C W and Schneider B I 1989 *Phys. Rev. Letts.* **63** 248
Rescigno T N and Orel A E 1981a *Phys. Rev. A* **23** 1134
Rescigno T N and Orel A E 1981b *Phys. Rev. A* **24** 1267
Rescigno T N and Orel A E 1991 *Phys. Rev. A* **43** 1625
Rescigno T N and Schneider B I 1988 *Phys. Rev. A* **37** 1044
Schneider B I and Collins L A 1981 *Phys. Rev. A* **24** 1264
Schneider B I and Rescigno T N 1988 *Phys. Rev. A* **37** 3749
Schneider B I, Rescigno T N, Lengsfield B H and McCurdy C W 1991 *Phys. Rev. Letts.* **66** 2728
Schneider B I, Rescigno T N and McCurdy C W 1990 *Phys. Rev. A* **42** 3132
Sohn W, Kochem K H, Scheuerlein K M, Jung K and Eherhardt 1986 *J. Phys. B* **19** 3625
Sun W, McCurdy C W and Lengsfield B H 1991 *Phys. Rev. A* (submitted for publication)
Veprek S and Heintze M 1990 *Plasma Chem. and Plasma Proc.* **10** 3
Wan H X, Moore and J H Tossel J A 1989 *J. Chem. Phys.* **91** 7340
Yuan J M 1989 *J. Phys. B* **22** 2589
Zhang J Z H and Miller W H 1989 *J. Chem. Phys.* **91** 1528

Electron scattering from polyatomic molecules using exact exchange plus parameter-free polarisation potential

Ashok Jain

Department of Physics, Florida A & M University, Tallahassee, Florida 32307 (USA)

ABSTRACT: The low–energy electron scattering (electronically elastic) from closed–shell polyatomic molecules is reviewed. We discuss our recent *ab initio* calculations on the CH_4, SiH_4, GeH_4, NH_3, H_2O and H_2S molecules in which the exchange correlation is treated <u>exactly</u> by solving non–local coupled integro-differential equations in an *optimized* iterative scheme, while polarisation effects are included approximately under the perturbation theory. This method does not require any prior information, experimental or theoretical, on the scattering parameters and is free from any spurious effects. A comparison with experimental data on the differential and total cross sections in the minima and maxima regions is presented for the above molecules.

1.   INTRODUCTION

The first *ab initio* calculations on low energy electron (e^-) scattering from non–linear molecules (CH_4, H_2O and H_2S) were performed by Gianturco and Thompson (1976a, 1976b, 1980). They employed fixed–nuclei adiabatic close–coupling technique under the single–center–expansion (SCE) formalism; the static interaction was included exactly at the Hartree–Fock level, however, the non–local exchange force was approximated by a local free–electron–gas–exchange (FEGE) (Hara 1967) model along with the orthogo-nalization of bound and continuum functions, while the target polarisation was included phenomenologically. These earlier calculations of Gianturco and Thompson were lim-ited to rotationally summed differential (DCS), integral (σ_t) and momentum transfer (σ_m) cross sections only. Later their work was extended by Jain and Thompson to more molecules (CH_4, SiH_4, NH_3, H_2O, H_2S) for ro–vibrationally elastic and inelastic processes (Jain and Thompson 1982; 1983a, b, c; 1984a, b; 1987). Jain and Thompson employed the same local model exchange potential (i.e. FEGE plus orthogonalization) alongwith a parameter–free polarisation potential based on the polarised orbital type approach (Jain and Thompson 1982). Thus the approach of Gianturco, Thompson and Jain, to be denoted in the following as SMEP (static–model–exchange plus polar-isation) gave satisfactory description of ro–vibrationally elastic, inelastic and summed processes. However, scattering parameters such as the DCS, rotationally inelastic cross sections, cross sections near a shape–resonance or a minima (Ramsauer–Townsend (RT) effect) etc. require a more sophisticated theoretical treatment in which exchange and polarisation effects are included from *first principles*.

Recently, McNaughten and Thompson (1988) have included non–local exchange interaction exactly by solving integro–differential coupled equations iteratively for the e^-–CH_4 system. More recently this *iterative* program is employed to study e^- scattering with several polyatomic gases such as the CH_4 (McNaughten et al 1990, Jain et al 1989), SiH_4 (Jain and Thompson 1991a), NH_3 (Jain and Thompson 1991b), H_2O (Jain et al 1991a, b) and GeH_4 (Jain et al 1991c) molecules. The calculations on H_2S, CF_4, SF_6, PH_3 etc. targets are also being undertaken by our group. This model, to be denoted as SEEP (static–exact–exchange plus polarisation) is free from any tuning procedure and is capable of producing reliable physical parameters below 1 eV, rotationally inelastic DCS, σ_t near the resonance and the DCS at all energies below 10 eV.

Electron scattering from polyatomic molecules has also been studied at the static–exact–exchange (SEE) level by employing Schwinger multichannel (SMC) (Lima et al 1985) and complex Kohn variational (CKV) (Rescigno et al 1989, McCurdy et el 1989) methods. The two methods (i.e. SMC and CKV) do not seem to give identical results at the SEE level for the e^-–CH_4 case (see McCurdy et al 1989). The SMC and CKV theories have been extended to include target polarisation (Lima et al 1989; see Schneider et al 1989 for a review on variational calculations on e^-–polyatomics collisions). The SMC and CKV methods have been extended for other polyatomics such as H_2O, NH_3, SiH_4, CF_4, CH_2O etc. (see Schneider et al 1989). One big advantage of the SMC and CKV methods is the direct evaluation of scattering parameters without adopting any SCE scheme and in addition electronic excitation can also be investigated in a coupled–state theory. However, an accurately polarized variational calculation for any of the above molecular systems has yet to be published. Recently Gianturco et al (1987) have developed a local semi–classical exchange approximation which works quite well below 1 eV impact energy.

2. STATIC EXACT EXCHANGE PLUS POLARISATION MODEL

At the static–exchange level (Gianturco and Jain 1986), the Schrödinger equation for the continuum e^- function $F(\mathbf{r})$ in the body–fixed (BF) coordinate system can be written as (neglecting vibrational and electronic excitations),

$$[-\frac{1}{2}\vec{\nabla}_r^2 + V_s(\mathbf{r}) - \frac{1}{2}k^2]F(\mathbf{r}) = \sum_\alpha \int \phi_\alpha^*(\mathbf{r}')|\mathbf{r} - \mathbf{r}'|^{-1}F(\mathbf{r}')d\mathbf{r}'\phi_\alpha(\mathbf{r}), \qquad (1)$$

where k is the incident e^- wavevector and the static potential V_s is given by,

$$V_s(\mathbf{r}) = \int |\Phi_0|^2 \{\sum_{j=1}^N |\mathbf{r} - \mathbf{r_j}|^{-1}d\mathbf{r_1}d\mathbf{r_2}\cdots d\mathbf{r_N}\} - \sum_{i=1}^M Z_i|\mathbf{r} - \mathbf{R_i}|^{-1}. \qquad (2)$$

Here Φ_0 is the target ground state wavefunction given as a single Slater determinant of one–electron N spin orbitals $\phi_i(\mathbf{r})$ and M is the number of nuclei in the molecule.

In the static–exchange approximation (Eq. 1), not all the short–range correlation is included and long–range polarisation of the target charge cloud is totally neglected. However, these effects (particularly the long–range polarisation) must be included for a proper description of low–energy e^- scattering and also for a comparison with experiment. The charge distortion effects are included approximately by introducing a local $V_{pol}(\mathbf{r})$ in (1). In brief, the V_{pol} is parameter–free and calculated by the method of Pople and Schofield (1957) (PS) and the non–penetration criterion of Temkin (1957). In the PS method (Jain and Thompson 1982), the second–order energy of the target is determined from first–order target wavefunction Φ_1 which is expanded in terms of Φ_0 and the expansion coefficients are evaluated variationally. The polarisibility tensor (α_{ij}) of various molecules determined from the PS method is within 20% error relative to exact value. The V_{pol}^{JT} is normalized to yield the correct asymptotic form. The V_{pol}^{CP} denotes the correlation polarisation (CP) potential (O'Connel and Lane 1983).

Finally, after projecting the Eq. (1) onto the symmetry–adapted angular basis functions of $F(\mathbf{r})$, we obtain a set of coupled integro–differential equations which can be written in a convenient matrix form, $\mathbf{LF} = \mathbf{WF}$, where $\mathbf{WF}$ is the exchange term. The iterative scheme is $\mathbf{LF}^i = \mathbf{WF}^{i-1}$, where $i = 1, 2, \ldots$. In order to start the solution, we chose $\mathbf{F}^0$ to be the solutions obtained from the FEGE potential with orthogonalization. Eq. (1) is solved for each irreducible representation (IR) of the molecular point group. The convergence of the reaction matrix with respect to number of iterations is tested for each state and also at lower and the upper ends of the present energy region.

Recently we (Jain et el 1991a) suggested an economic and practical way to use the iterative scheme efficiently without sacrificing any numerical accuracy and presented test calculations on the e–H_2O system at the SEE level. In brief, we demonstrated that (1), it is sufficient to solve Eq. (1) only for a few lower partial waves; (2), the SEE scattered function, calculated at any energy, is a very good guess to start solutions of the coupled equations at the next or other energies; this reduces the number of iterations (n) appreciably to make the calculations possible on a supercomputer.

First, we show that one needs to solve the full integro–differentinal Eq. (1) only for a few low partial waves; for higher partial waves, a model exchange (SME) calculation is adequate enough. Jain et al (1991a) have shown the e–H_2O K–matrix elements for $\ell \geq 4$ to be identical in the SEE and SME models. It is also interesting that the first–Born–approximation (FBA) dipole $K_{\ell,\ell\pm1}$ elements are almost equal to the close–coupling (SEE or SME) values for $\ell \geq 3$. The FBA quadrupole $K_{\ell,\ell\pm2}$ elements are within 10% of the exact results. This means that for higher values of angular momentum the FBA is a good approximation. We have illustrated this scheme of low–, intermediate– and high–ℓ values in Fig. 1 which implies that for low partial waves (say $\ell_{max} = L_e$), we need the full SEE scheme, for $\ell \geq L_e$ (say $\ell_{max} = L_m$), we can employ the SME model and finally for $\ell \geq L_m$ (say $\ell_{max} = L_B$), the Unitarised–Born–Approximatuion (UBA) (Norcross and Collins 1982) will be a good approximation. The UBA matrix is calculated up to a final $\ell_{max} = L_B$ value and finally the MEAN approximation can be used to get the converged cross sections. (Norcross and Padial 1982).

TABLE 1.   K–matrix elements, $K^{A_1}_{\ell m,\ell'm'}$, eigenphase sums and partial total cross section for e–H_2O A_1 symmetry collisions at 1 eV. The n is the number of iterations when F^{SME}_0 (at 1 eV) is used, while $\bar{n}$ is the corresponding number of iterations when $\tilde{F}^{SEE}_n$ (at E=0.5 eV) is employed to start the iteration at 1 eV. The number of channels is 9, i.e., $\ell_{max} = 4$.

Model for F_0	Iterations	CPU Units	$K_{00,00}$	$K_{00,10}$	$\delta^{A_1}_s$	$\sigma^{A_1}_t$
SME	-	2	1.20657	0.37415	0.44386	48.7096
F^{SME}_0(1 eV)	$n = 24$	100	−0.38338	0.39166	−0.71237	27.2794
F^{SEE}_{24}(0.5 eV)	$\bar{n} = 3$	10	−0.38438	0.39131	−0.71349	27.2818
F^{SEE}_{24}(0.5 eV)	$\bar{n} = 5$	20	−0.38333	0.39190	−0.71209	27.2818
F^{SEE}_{24}(0.5 eV)	$\bar{n} = 7$	25	−0.38353	0.39176	−0.71239	27.2805
F^{SME}_0(1 eV)*	$n = 34$	265	−0.40834	0.40489	−0.71585	26.76

*Using 16($\ell = 6$) channels

We now discuss second aspect of the *optimized iterative* scheme. Suppose one wants to determine the SEE K–matrix at energy E_1. When $F^{SME}_0(E_1)$ is used to start the iterative scheme, it leads to the final $\tilde{F}^{SEE}_n(E_1)$ converged continuum function after n iterations. However, if we want to apply the same iterative scheme at another energy, say E_2, we can use either $F^{SME}_0(E_2)$ (requiring approximately the same number of n iterations) or $\tilde{F}^{SEE}_n(E_1)$ (requiring only $\bar{n}$ iterations) to obtain the final $\tilde{F}^{SEE}_{\bar{n}}(E_2)$ at energy E_2.

Recently Jain et al (1991c) found that $\bar{n}$ is much less than n and the final scattering matrix converge to the same values, i.e., $\tilde{F}^{SME}_n(E_2) = \tilde{F}^{SEE}_{\bar{n}}(E_2)$. If $\tilde{F}(E)$ is a slowly varying function with respect to the impact energy E, one can use a well converged scattering function at an energy E_1 to start the iteration at the next energy E_2. First, in Fig. 2, the scattering component $f_{00}(r)$ is plotted at 1 eV for the e–H_2O A_1 symmetry. The top solid curve (increasing rapidly with r) is the F^{SME}_0 (at 1 eV) (using the FEGE potential) case, while the lower solid curve is the corresponding converged result with 24 iterations which is denoted as $\tilde{F}^{SEE}_{24}$ (at 1 eV). The dash curve represents the $\tilde{F}^{SEE}_{24}$(at 0.5 eV) function at 0.5 eV. We clearly see that the two SEE curves at 0.5 and 1 eV are very close to each other as compared to SME curve at 1 eV.

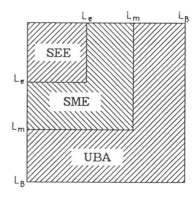

FIG. 1: All notations are explained in the text.

We see that the two SEE functions at 0.5 and 1 eV are very close to each other as compared to the SME curve at 1 eV. In the optimized scheme, however, if one uses the dash curve of Fig. 2 to start the iterations at 1 eV, one ends up with the same continumm function (lower solid curve in Fig. 2) with a much smaller number of iterations, $\bar{n}$. In table 1, we have shown various scattering parameters with respect to SME and SEE (with different values of n and $\bar{n}$) models. Thus, with the criterion,

$$F_0^{SEE}(E_p) = \tilde{F}_n^{SEE}(E_{p-1}), \qquad (3)$$

to start iteration at a given energy E_p, one can save a significant amount of computer time without loosing any accuracy (see table 1). It is also worth mentioning that the number of iterations depends almost linearly on size of the K–matrix.

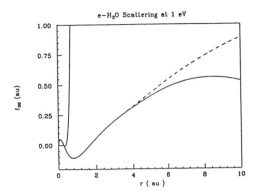

FIG. 2: *Scattered radial spherical function of the e–H_2O A_1 symmetry collision at 1 eV. Upper solid curve: FEGE model; Lower solid curve: the SEE function using upper solid curve as starting point for iteration of Eq. 3. The dash curve is the continuum radial function at 0.5 eV.*

For the cross section formulae, we refer to a recent review by Gianturco and Jain (1986). Here, we provide only a summary of various cross section quantities. As mentioned earlier, we discuss here vibrationally elastic, rotationally elastic, inelastic and summed channels for differential, integral and momentum–transfer cross sections. In order to obtain collisional parameters, we transform the BF scattering amplitude $f(\hat{k}\cdot\hat{r}; Q_n)$ ($\hat{k}$ and $\hat{r}$ are respectively the initial and final directions of the projectile and Q_n represents normal coordinates) into space–fixed (SF) amplitude $f(\hat{k}\cdot\hat{r}'; Q_n; \alpha\beta\gamma)$ (where $\hat{r}'$ now refers the direction of scattered electron with respect to SF coordinate system and $(\alpha\beta\gamma)$ are the three Euler angles) by making use of rotation matrices. Even at the lower bound of the present energy region, the collision time is much less than the molecular rotational time period. hence,the adiabatic–nuclei ro–vibrational approximation is valid and the scattering amplitude for a particular nuclear transition $VJKM \rightarrow V'J'K'M'$ can be written as (Chase 1957),

$$f_{VV'}(JKM \rightarrow J'K'M') = \langle\phi_V(Q_n)\psi_{JKM}|f(\hat{k}.\hat{r}'; Q_n; \alpha\beta\gamma)|\phi_{V'}(Q_n)\psi_{J'K'M'}\rangle, \qquad (4)$$

where ϕ_V is initial vibrational wave function with V vibrational quantum numbers and the $|\psi_{JKM}\rangle$ represents rotational eigenfunctions of a spherical top defined as,

$$\psi_{JKM}(\alpha\beta\gamma) = \left[\frac{2J+1}{8\pi^2}\right]^{\frac{1}{2}} D_{KM}^{J^*}(\alpha\beta\gamma), \qquad (5)$$

Here K and M are the projections of J along the BF and SF frame principle axes respectively. The DCS for a particular $VJK \rightarrow VJ'K'$ transition is obtained by summing over all final magnetic substates M' and averaging over all initial substates M, i.e.,

$$\frac{d\sigma}{d\Omega}(VJK, V'J'K') = \frac{k'}{k}\frac{1}{(2J+1)}\sum_{MM'}|f_{VV'}(JKM \to J'K'M')|^2, \qquad (6)$$

where the final wave vector, k' is calculated from the energy conservation relation. For spherical top molecules, the summation over K' and averaging over K can also be done. Finally, It is convenient to expand the DCS in terms of Legendre polynomials,

$$\frac{d\sigma}{d\Omega}(VJK, V'J'K') = \frac{k'}{k}\sum_{L} A_L(VJK, V'J'K')P_L(\cos\theta). \qquad (7)$$

Here θ is the scattering angle between the vectors $\hat{\mathbf{k}}$ and $\hat{\mathbf{r}}'$. For the special case of a spherical top molecule and for vibrationally elastic scattering ($V = V'$), the expansion A_L coefficients are given as (Jain 1983),

$$A_L(JJ') = \frac{(2J'+1)(2L+1)(-1)^L}{4k^2(2J+1)}\sum_{ll'\bar{l}\bar{l}'}[(2l+1)(2l'+1)(2\bar{l}+1)(2\bar{l}'+1)]^{\frac{1}{2}}i^{l-l'}(-i)^{\bar{l}-\bar{l}'}$$

$$\begin{pmatrix} \bar{l} & l & L \\ 0 & 0 & 0 \end{pmatrix}\begin{pmatrix} l' & \bar{l}' & L \\ 0 & 0 & 0 \end{pmatrix}\sum_{j=|J-J'|}^{J+J'}(-1)^j W(ll'\bar{l}\bar{l}'; jL)M_{ll'}^{jm_j}M_{\bar{l}\bar{l}'}^{jm_j*} \qquad (8)$$

where the M–matrix is constructed as

$$M_{ll'}^{jm_j} = \sum_{mm'hh'p\mu}(-1)^m \bar{b}_{lhm}^{p\mu}\begin{pmatrix} l & l' & j \\ m & m' & -m_j \end{pmatrix}\bar{b}_{l'h'm'}^{p\mu}\, \mathbf{T}_{lh,l'h'}^{p\mu}\,. \qquad (9)$$

Here, as usual, the $\mathbf{T}$–matrix is defined in terms of the $\mathbf{S}$–matrix as $\mathbf{T} = (\mathbf{S} - \mathbf{1})$ where $\mathbf{S} = (1 + i\mathbf{K})(1 - i\mathbf{K})^{-1}$ is written in terms of scattering $\mathbf{K}$ matrix for each symmetry $(p\mu)$. In eq. (8), $\begin{pmatrix} a & b & c \\ d & e & f \end{pmatrix}$ is a $3 - j$ symbol, $W(abcd; ef)$ is the Racah coefficient and the $b_{lhm}^{p\mu}$ coefficients are expansion terms in the definition of symmetry–adapted basis functions in terms of real spherical harmonics (see Gianturco and Jain 1986). A computer program (EROTVIB) is available due to Jain and Thompson (1984b) in order to evaluate the A_L coefficients (Eq. 7). The scattering matrix can be obtained as a function of internal or normal coordinates of the molecule for vibrational transitions in the adiabatic–nuclei theory. However, a proper vibrational close–coupling theory for e^-–polyatomics is still a computational challenge. A different approach such as the partial differential equation technique may be required for this purpose (Weatherford 1991). From equation (6), it is easy to derive simple forms for the σ_t and σ_m cross sections in terms of A_L coefficients, namely,

$$\sigma_t^{JJ'} = 4\pi A_0(JJ') \quad ; \quad \sigma_m^{JJ'} = 4\pi[A_0(JJ') - \frac{1}{3}A_1(JJ')]. \qquad (10)$$

3. RESULTS AND DISCUSSION

We classify the discussion of the SEEP results in terms of (a) DCS, (b) σ_t below 1 eV and (c) σ_t around a shape–resonance region. These are the physical situations where the approximations involved in the inclusion of exchange and polarization effects are important. First we would like to mention that the SEE level e–CH_4 iterative calculations (McNaughten and Thompson 1988, McNaughten et al 1990) are in excellent agreement with the corresponding CKV results of McCurdy et el (1989).

In Fig. 3, 4, 5, 6 and 7 we have shown DCS for e–CH_4 (at 0.6 eV), e–SiH_4 (at 3 eV), e–GeH_4 (2 eV), e–NH_3 (8.5 eV) and e–H_2O (6 eV) respectively. The SEEP (JT) results in Fig. 3 are compared with experimental data of Sohn et al (1986) (crosses). The other *ab initio* SMC or CKV results are not available for comparison with the results of FIG. 3, where it is clear that any structure in the DCS is created by target polarisation. The high sensitivity of DCS with respect to polarisation effect in this low energy region is discussed by Jain et al (1989). In Fig. 4 the e–SiH_4 DCS are shown at 3 eV along with recent experiment of Tanaka et al (1990). No other multichannel calculation at the exact exchange level with polarisation is available to compare with Fig. 4. Next in Fig. 5 we illustrate low energy DCS for another spherical top but heavier GeH_4 molecule at 2 eV along with measurements of Boesten and Tanaka (1991). For all three spherical tops the SEEP(JT) model gives reliable differential and integral cross sections.

In the present fixed–nuclei approximation, the forward angle DCS and thus the σ_t for polar molecules are not defined due to long–range dipole interaction. We have employed the MEAN prescription (Norcross and Padial 1983) to get converged DCS for NH_3 (Fig. 6) and H_2O (Fig. 7) at 8.5 and 6 eV shown respectievly along with recent relative measurements and the the SMC calculations (NH_3 only).

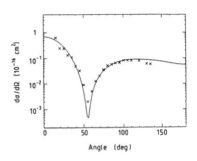

FIG. 3: *Elastic DCS for the e–CH_4 scattering at 0.6 eV. Theory: solid curve, present SEEP (JT) theory. Expt. , ×, Sohn et al (1985).*

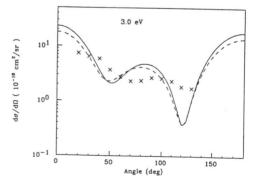

FIG. 4: *DCS for the e–SiH_4 scattering in SEEP(JT) model at 3 eV (solid line). Expt. , ×, Tanaka et el (1989).*

The RT minimum occurs in low energy e^-–molecule (atom) collisions mainly due to polarisation of the target. The CH_4 is known to have RT effect around 0.25 eV. Previous calculations on the e–CH_4 system do not reproduce experimental shape of the σ_t or σ_m curve in the 0–1 eV region. The swarm analysis predicts quite accurate momentum transfer cross sections at such low energies. We have shown in Fig. 8 the SEEP(JT) σ_m calculations of McNaughten *et al* (1990) alongwith swarm data. The agreement between SEEP(JT) theory and experiment at all energies between 0.1 and 1 eV is rather too good. To the best of our knowledge, no other *ab initio* calculation is available for comparison with data of Fig. 8. In addition, the scattering length for CH_4 and SiH_4 molecules has been found to be in excellent agreement with measured values (McNaughten et al 1990; Jain and Thompson 1991a).

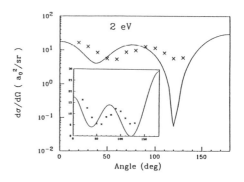

FIG. 5: DCS for the e–GeH_4 system at 2 eV in the present SEEP (JT) model (full line). Expt., ×, Boesten and Tanaka (1991). Note that we have shown both the linear and log scales.

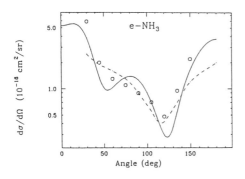

FIG. 6: DCS for the e–NH_3 system at 8.5 eV in the present SEEP(JT) (solid line) and SMC (dashed) (Pritchard et al 1989) theories. Expt. , ×, Shyn (1991).

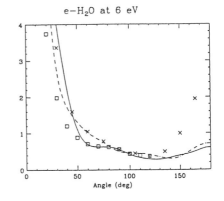

FIG. 7: DCS for the e–H_2O system at 6 eV in our SEEP (JT) (solid line) and SEE (dashed curve) models. Expt. , ×, Shyn and Cho (1987); □, Danjo and Nishimura (1985)

4. CONCLUSION

We conclude that the low energy electron polyatomic scattering can be accurately described in a model where exchange interaction is treated exactly while polarisation is included approximately in a local but parameter–free theory. Except the near–threshold collision, the fixed–nuclei SCE scheme under the $\tau_c \leq \tau_r$ situation (τ_c and τ_r are respectively the collision and nuclear time periods) is adequate to get reliable information on ro–vibrational elastic, inelastic and rotationally summed processes. We presented here only few examples of success of the SEEP(JT) model. Currently, we are in the process of completing detailed results on NH_3, H_2O, H_2S and GeH_4 molecules. Some preliminary test calculations have been carried out on the heavy CF_4 molecule by using multicenter target wavefunctions in our SCE approach (Gianturco et al 1991).

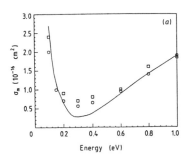

FIG. 8: The e–CH_4 σ_m cross sections below 1 eV in the SEEP (JT) model (full curve). Expt. data: □, Hadded (1985); ○, Ohmari et al (1987).

ACKNOWLEDGEMENT

This research is funded by the US Department of Airforce, Wright-Patterson Air Force Base, under the contract No. FY33615-90-C-2032. This work is partially supported by the US Army Scientific Office of Research under the contract no DAAL03–89-9-0111. A travel grant from NATO under contract No CRG890470 is thankfully acknowledged. The computational facilities were provided by the FSU Supercomputer Research Institute (SCRI). We thank Professor Charles Weatherford for fruitful discussions. This research was carried out in collaboration with Professor F A Gianturco (Rome) and D G Thompson (Belfast). We appreciate some computational help from Drs. Kasturi Baluja and S N Rao.

REFERENCES

Boesten L and Tanaka T 1991 (private communication)
Chase D M 1956, Phys. Rev. **104**, 838.
Danjo A and Nishimura H 1985, J. Phys. Soc. Jpn. **54**, 1224.
Gianturco F A and Thompson D G 1976a, Chem. Phys. Lett. **14**, 110.
Gianturco F A and Thompson D G 1976b, J. Phys. **B9**, L383.
Gianturco F A and Thompson D G 1980, J. Phys. **B13**, 613.

Gianturco F A and Jain A 1986, Phy. Rep. **143**, 348.

Gianturco F A, Pantano L C and Scialla S 1987 Phys. Rev. **A36**, 557;
     Gianturco F A and Scialla S 1987, J. Phys. **B20**, 3171.

Gianturco F A, Jain A and Martino Vi D 1991, (submitted for publication).

Haddad G N 1985, Aust. J. Phys. **38**, 677.

Hara S 1967, J. Phys. Soc. Japan **22**, 710.

Jain A 1983, Ph.D. thesis (Queen's University, Belfast).

Jain A and Thompson D G 1982, J. Phys. **B15**, L631.

Jain A and Thompson D G 1983a, J. Phys. **B16**, 3077.

Jain A and Thompson D G 1983b, J. Phys. **B16**, 443

Jain A and Thompson D G 1983c, J. Phys. **B16**, 2593

Jain A and Thompson D G 1984a, Phys. Rev. **A30**, 1098.

Jain A and Thompson D G 1984b, Comp. Phys. Commun. **32**, 367.

Jain A and Thompson D G 1987, J. Phys, **B20**, 2861.

Jain A and Thompson D G 1991a, J. Phys. **B24**,xxxx

Jain A and Thompson D G 1991b, Phys. Rev. (to be submitted).

Jain A, Weatherford C A, McNaughten P and Thompson D G 1989,
     Phys. Rev. **A40**,6730.

Jain A, Gianturco F A and Thompson D G 1991a, J. Phys. **B24**, xxxx

Jain A, Gianturco F A and Thompson D G 1991b, (in preparation)

Jain A, Baluja K L, Martino Vi D and Gianturco F A 1991c, Chem. Phys. Lett.
     (submitted for publication)

Lima M A P, Gibson T L, Huo W M and McKoy V 1985, Phys. Rev. **A40**, 6730

Lima M A P, Watari K and McLoy V 1989, Phys. Rev. **A39**, 4312

McCurdy C W and T N Rescigno 1989, Phys. Rev. **A39**, 4487.

McNaughten P and Thompson D G 1988 J. Phys. **B21**, L703.

McNaughten P, Thompson D G and Jain A 1990, J. Phys. **B23**, 2405.

Norcross D W and Collins L A 1982, Adv. At. Mol. Phys. **18**, 341.

Norcross D W and Padial N T 1982, Phys. Rev. **A25**, 226.

O'Connel J K and Lane N F 1983, Phys. Rev. **A27**, 1893.

Ohmori Y, Kitamori K, Shimozuma M and Tagashira H 1986, J. Phys. **D19**, 437.

Prichard H P, Lima M A P and McKoy V 1989, Phys. Rev. **A39**, 2392.

Pople J A and Schofield P 1957, Phil. Mag. **2**, 591.

Rescigno T N, McCurdy C W and Schneider B I 1989, Phys. Rev. Lett. **248**, 248.

Schneider B I 1989, in *Aspects of Electron–Molecule Scattering and Photoionization*,
     ed. A Herzenberg, (American Institute of Physics, New York).

Shyn T W 1991 (private communication)

Shyn T W and Cho S Y 1987, Phys. Rev. **A36**, 5138.

Sohn W, Kochem K H, Scheuerlein K M, Jung K and Ehrhardt H 1986,
     J. Phys. **B19**, 3625

Tanaka H, Boesten L, Sato H, Kimura M, Dillon M A and Spence D 1990,
     J. Phys., **B23**, 577.

Temkin A 1957 Phys. Rev. **107**, 1004.

Weatherford C A 1991 (private communication).

Electron-impact excitation of ions

D C Griffin[*], M S Pindzola[+], and N R Badnell[+]

[*]Department of Physics, Rollins College, Winter Park, Florida 32789 USA
[+]Department of Physics, Auburn University, Auburn, Alabama 36849 USA

ABSTRACT: Methods for performing theoretical calculations of electron-impact excitation of positive ions are reviewed and results using various levels of approximation are compared with experimental measurements. The important contributions of recombination resonances to the excitation cross sections are discussed and various theoretical methods of treating these resonances are considered. In addition to excitation of valence electrons, we also consider the contributions of inner-shell excitation followed by autoionization to the ionization cross section for cases where these indirect processes dominate the total cross section.

1. INTRODUCTION

Electron-impact excitation of multi-electron ions continues to present a formidable challenge to atomic theory. Excitation cross sections are quite sensitive to the accuracy of the description of the N-electron target states. Furthermore, at short range, the incident electron becomes part of an (N+1)-electron system which requires a detailed description of the electron-electron interactions. Finally, the long-range attractive Coulomb interaction causes many partial waves to contribute to the cross sections.

In this paper, we shall consider two methods for the calculation of electron-impact excitation cross sections; one is based on the close-coupling approximation and the other is based on the distorted-wave (DW) approximation. In the close-coupling approximation, one first expands the wavefunction for the (N+1)-electron scattering problem in the form

$$\psi^{\Gamma}(N+1) = \mathcal{A} \sum_{i} \theta_i(x_1, \ldots x_N, \hat{r}_{N+1}, \sigma_{N+1}) \frac{1}{r_{N+1}} F_i(r_{N+1}) + \sum_{j} c_j \phi_j(x_1, \ldots x_{N+1}), \quad (1)$$

where Γ represents the quantum numbers needed to describe the (N+1)-electron system; $\mathcal{A}$ is an operator that antisymmetrizes the scattered electron coordinate with the N-electron coordinates; θ_i are channel functions formed by coupling a target wavefunction with the angular and spin parts of the (N+1)-electron wavefunction, which has a radial function $F_i(r_{N+1})$; $x_i = \vec{r}_i \sigma_i$ represents the space and spin coordinates of the ith electron; and the functions ϕ_j are (N+1)-electron bound-state wavefunctions, which must be included in order to make the expansion complete when the radial wavefunctions for the continuum orbitals are forced to be orthogonal to radial wavefunctions for the bound orbitals, with the same angular momentum. The target wavefunctions are determined using configuration-interaction theory and a common set of radial wavefunctions. For atomic systems with low Z, we can ignore relativistic effects and employ the LS coupling approximation; then $\Gamma = LSM_LM_S\Pi$, where Π is the parity of the (N+1)-electron system.

CC calculations of excitation cross sections have been performed using two approaches. In the first method, one applies the variational principle to derive a set of coupled differential equations (the CC equations), which are solved using finite difference formulae. This is the technique employed in the University College London CC program, Impact (Crees et al 1978). The second approach is to employ the R-matrix method (Burke and Robb 1975) in which the configuration space is divided into two regions: an inner region where electron exchange is important, and an outer region where the bound radial wavefunctions have negligible amplitude and exchange can be ignored. In the inner region, the continuum orbitals for each channel are expanded in terms of a discrete set of basis orbitals which satisfy logarithmic boundary conditions on the surface of a sphere dividing the two regions. The total wavefunction given by Eq. 1 then becomes an expansion in terms of discrete functions in the inner region and the continuum problem in this region reduces to a diagonalization of the (N+1)-electron problem, which only has to be solved once to determine the so called R-matrix at all energies. In the outer region, one solves the coupled differential equations corresponding to a long-range approximation of the CC equations without exchange at each energy of interest. These solutions then allow one to relate the R-matrix on the boundary of the inner sphere to the scattering matrix at infinity. The first method then involves the solution of coupled differential equations in all regions of space at all energies, while the R-matrix method only requires one solution in the more complex inner region, which is good for all energies; therefore, the R-matrix method is especially advantageous when a fine energy mesh must be employed to describe a rapidly varying cross section.

In the DW approximation, all terms in the CC equations corresponding to coupling between scattering channels are treated as perturbations. The problem then reduces to the solution of a single uncoupled differential equation for each scattering channel. This method is especially advantageous for complex highly ionized species where relativistic effects are large and the number of N-electron states, which should be included in the CC equations, is determined by the number of intermediate-coupled levels rather than the smaller number of LS terms. Furthermore, the DW approximation grows more accurate in highly ionized atoms since the effects of coupling between channels grows smaller as the net ionic charge grows larger (Pindzola and Griffin 1989).

In many cases, excitation cross sections are dominated by strong resonance features. These occur when an electron in the continuum of the N-electron ion coincides in energy with a compound doubly excited state of the (N+1)-electron ion. In such a case, the continuum electron can be captured into the compound state, a process which we shall refer to as resonant recombination (RR). The resulting doubly excited state can autoionize back to the initial state of the N-electron ion, resulting in a resonance in the elastic cross section; it can autoionize to an excited state of the N-electron ion, resulting in a resonance in the excitation cross section; or finally, it can radiatively decay to a bound state of the (N+1)-

electron ion, resulting in dielectronic recombination. Resonances result in a natural way within the CC approximation when both open and closed channels are included in the wavefunction expansion. Furthemore, the CC approximation includes interference effects with the nonresonant background and between individual resonances.

One can most easily include these resonance features within the DW framework by invoking the independent-processes (IP) approximation, in which one ignores interference effects and writes the total excitation cross section σ_E as a simple sum of the non-resonant excitation cross section σ_{NR} and the resonant excitation cross section σ_R,

$$\sigma_E(i \rightarrow f) = \sigma_{NR}(i \rightarrow f) + \sigma_R(i \rightarrow f). \tag{2}$$

The resonant excitation cross section in atomic units is given by the equation

$$\sigma_R(i \rightarrow f) = \frac{2\pi^2}{k_i^2} \sum_j \frac{q_j}{2q_i} \frac{A_a(j \rightarrow i) A_a(j \rightarrow f)}{\sum_k A_a(j \rightarrow k) + \sum_n A_r(j \rightarrow n)} L_j(k_i^2), \tag{3}$$

where k_i is the linear momentum of the incident electron; q_i and q_j are the statistical weights of the initial state i of the N-electron ion and the intermediate doubly excited state j of the (N+1)-electron ion, respectively; $A_a(j \rightarrow k)$ is the autoionizing rate from the state j to a lower level k of the N-electron ion; $A_r(j \rightarrow n)$ is the radiative rate from state j to a bound state n of the (N+1)-electron ion; and the Lorentz profile $L_j(k_i^2)$ is given by

$$L_j(k_i^2) = \frac{\Gamma_j/2\pi}{(k_i^2/2 + E_i - E_j)^2 + \Gamma_j^2/4}. \tag{4}$$

In the above equation, $k_i^2/2$ is the energy of the incident electron; E_i and E_j are the energies of the initial state i and the intermediate doubly excited state j, respectively; and Γ_j is the total width of state j and is given by

$$\Gamma_j = \sum_k A_a(j \rightarrow k) + \sum_n A_r(j \rightarrow n). \tag{5}$$

Thus, in the IP approximation, we include the effects of radiation damping on the resonant contribution through the use of radiative rates. In the CC approximation, the inclusion of these effects is much more difficult. A doubly excited state can radiatively decay through transitions involving either one of the inner electrons or the outer Rydberg electron. The first of these can be included within a CC formalism using quantum-defect theory as was first done by Bell and Seaton (1985) for dielectronic recombination, or in theory, by including radiation damping within a complex Hamiltonian in the CC equations; this latter approach has not been tried to date. However, it is not possible to incorporate transitions involving the outer Rydberg electron within either formalism. Both types of transitions can be important in highly ionized atoms where radiative rates become comparable to autoionizing rates; however, then interference effects on the non-resonant background as well as on the resonant contributions become less important and the accuracy of the IP approximation improves.

2. EXCITATION CROSS SECTION RESULTS

We now consider calculations of excitation cross sections, with a focus on those performed using the R-matrix method. However, in several cases, we shall compare R-matrix results with those obtained from the IP approximation using distorted waves. In such cases, the same radial wavefunctions for the N-electron bound orbitals were employed for both calculations; thus, any differences are due to the effects of interference incorporated in the CC approximation and not in the IP approximation. For the IP approximation, DW autoionizing rates and all radiative rates were determined using the Autostructure program (Badnell 1986).

We performed R-matrix and IP calculations of the 2s→2p excitation in the Li-like ions C^{3+}, O^{5+}, and Ne^{7+} and the 3s→3p and 3s→3d excitations in the Na-like ions Si^{3+}, Ar^{7+}, and Ti^{11+} (Badnell et al 1991a). The bound-state orbitals were generated using a single-configuration, Hartree-Fock approximation. For the Li-like ions, we made 5-state (2s,2p,3s,3p,3d) CC calculations and included resonances of the form $3l3l'$ in the IP calculation. For the Na-like ions, we made 7-state (3s,3p,3d,4s,4p,4d,4f) CC calculations and, in our IP calculations, we included resonances of the form $3dnl'$ and $4lnl'$ for the 3s→3p excitation and $4lnl'$ for the 3s→3d excitation. These calculations were performed in LS coupling; effects of intermediate coupling and radiation damping were investigated with the IP approximation and were negligible for these Δn = 0 transitions in the ions considered. The results of the two types of calculations for the 3s→3p excitation are shown in Figure 1 for Si^{3+} and in Figure 2 for Ti^{11+}.

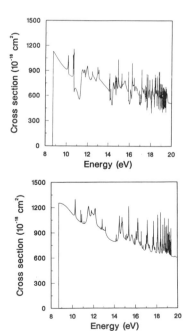
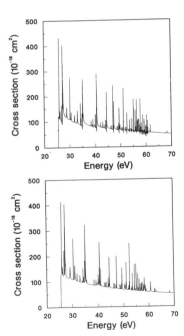

Fig. 1 Excitation cross section for the 3s→3p excitation in Si^{3+}. Upper, 7-state CC approximation; lower, IP approximation

Fig. 2. Excitation cross section for the 3s→3p excitation in Ti^{11+}. Upper, 7-state CC approximation; lower, IP approximation.

For Si^{3+}, there are some significant differences between the CC and IP results; for example, there is a large window resonance in the CC results at 11 eV which is apparently due to interference between the background cross section and the 3d4d ^3G resonance. Similar features are repeated in the CC results at higher energies. As can be seen in Figure 2, the agreement between the CC and IP results is much better for Ti^{11+} in terms of both the background and resonance contributions. Improvement in agreement between the two approximations with ionization stage is also observed in the 3s→3d excitation for these ions and the 2s→2p excitation in the Li-like ions. The 3s→3p excitation cross section has recently been measured in the threshold region using a merged-beams electron energy loss technique by Wahlin et al (1991). As shown in Figure 3, the agreement between these measurements and our CC calculations is very good after the theory has been convoluted with a 0.2 eV Gaussian, to simulate the experimental electron energy distribution.

We have also made R-matrix calculations of electron-impact excitation for Zn^+, with a focus on the 4s→4p resonance transition (Pindzola et al 1991). In order to check on the completeness of our calculations, we performed 6-state, 15-state, and 31-state CC calculations for this ion. The 6-state calculation includes all the LS terms arising from the $3d^{10}4s$, $3d^{10}4p$, $3d^9 4s^2$, $3d^{10}5s$, $3d^{10}4d$, and $3d^{10}5p$ configurations. In the 15-state calculation, we added to these, the 9 terms arising from the $3d^9 4s4p$ configuration, while for the 31-state calculation, we also added the 16 terms arising from the $3d^9 4s4d$ and the $3d^{10}4f$ configurations. The radial wavefunctions for the bound-state orbitals were generated by solving for them in a

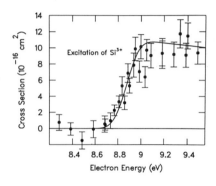

Fig. 3. The 3s→3p excitation cross section for Si^{3+} in the threshold region. Experimental points from Wahlin et al (1991). Solid curve: the 7-state R-matrix calculation convoluted with a 0.2 eV Gaussian.

local model potential, generated from Slater-type orbitals using a modified version of the program SUPERSTRUCTURE (Eissner et al 1974). In the 6-state and 15-state calculations, the configuration-interaction basis set included all configurations listed above for the 15-state R-matrix calculation, and was optimized by varying a radial parameter on the model potential to minimize a weighted sum of energies. For the 31-state calculation, the additional configurations $3d^9 4s4d$ and $3d^{10}4f$ were, of course, included in the basis set, but no further optimization was performed.

We have made detailed comparisons of the results of these three calculations with respect to angular differential cross sections at selected energies and total cross sections over a wide range of energies. Even though angular differential cross sections are much more sensitive to coupling effects than total cross sections (Griffin and Pindzola 1990), the agreement between the differential 4s→4p cross sections calculated from the 6-state, 15-state, and 31-state calculations was surprisingly good even at energies just above threshold. In addition, including more states had little effect on the 4s→4p total non-resonant background cross section; however, there were some effects on the resonance structures, which are due both to the new resonances which appear as the CC coupling expansion is increased, as well as differences between these calculations with respect to the energies of the (N+1)-electron structure.

The 4s→4p excitation cross section in Zn^+ was first measured by detecting fluorescence radiation resulting from collisions in a crossed-beams arrangement by Rogers et al (1982) and more recently using a merged-beams electron energy loss technique by Smith et al (1991). The crossed-beams measurements include the effects of radiative cascading from higher lying states, while this particular merged-beams experiment measures the cascade-free cross section at forward scattering angles from 0^0 to 90^0. In order to provide a meaningful comparison with the merged-beams experiment, we performed angular differential cross section calculations of the 4s→4p excitation at a large number of energies and then numerically integrated the results from 0^0 to 90^0 to obtain a partial cross section. Our 15-state results for the partial and total cross sections are shown in Figure 4, in comparison to the merged-beams experiment. This figure illustrates nicely how the cross section is dominated by forward scattering at higher electron energies. However, near threshold, there are significant differences between the partial and total cross section, especially with respect to the resonance structure. This indicates that the Auger electrons emitted from the doubly excited resonance states are focused primarily at angles greater than 90^0. The agreement between the measured and calculated partial cross sections is reasonably good near threshold; however, the experimental results fall noticeably below the theoretical cross section at higher energies.

In Figure 5, we show our 15-state results for the total cross section over a wide range of energies in comparison to the crossed-beams measurements

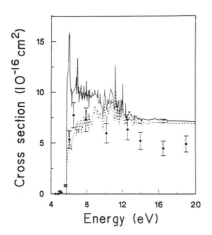

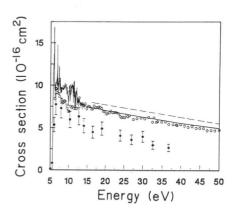

Fig. 4. Cross section for 4s→4p Zn^+ transition. Solid curve: total cross section, 15-state calculation; dashed curve: 0^0–90^0 partial cross section, 15-state calculation; experimental points: merged-beams measurements, Smith et al (1991).

Fig. 5. Cross section for 4s→4p Zn^+ transition. Solid curve: total cross section, 15-state calculation; dashed curve: plus cascade enhancement. Solid circles: merged-beams measurements, Smith et al (1991); open circles: crossed-beams measurements, Rogers et al (1982).

of the total cross section and the merged-beams measurements of the partial cross section. The excellent agreement between the calculated cross section and the crossed-beams measurements is somewhat fortuitous; this experiment includes the effects of cascading from higher lying levels, which start at an energy of about 11 eV and these are not included in the calculation. The dashed curve in Figure 5 shows that the cascade-corrected theoretical cross section is above the measured cross section in the energy range from 15 to 50 eV. We might have expected the theoretical results to be somewhat high, since the oscillator strength calculated with the basis set used in our 15-state calculation is about 17% higher than that obtained from a multiconfiguration Hartree-Fock calculation performed by Froese-Fischer (1977). Thus, we conclude that there is reasonably good agreement between theory and the crossed-beams measurements; however, the merged-beams measurements appear to be low at higher electron energies.

3. EXCITATION-AUTOIONIZATION CONTRIBUTIONS TO IONIZATION CROSS SECTIONS

The number of experimental measurements of excitation cross sections for ionized species is relatively small. However, there exists a large number of crossed-beams measurements of electron-impact ionization where indirect processes dominate the total ionization cross section. That is, in addition to direct electron-impact ionization of an electron, we can have inner-shell excitation followed by autoionization or resonant recombination to a doubly excited autoionizing state of the (N+1)-electron ion followed by sequential autoionization of two electrons. Higher order processes are also possible; however, we will concentrate here on these two indirect processes since they have been found to contribute significantly to the total ionization cross section, even in highly ionized species (see, for example, Griffin et al 1987 and Chen et al 1990), and they provide information on inner-shell excitation cross sections and their resonant contributions.

If we invoke the IP approximation for ionization, the total ionization cross section σ_I from an initial state i is given by

$$\sigma_I(i) = \sigma_{DI}(i \to f) + \sum_d \sigma_{NR}(i \to d) B^a(d) + \sum_d \sigma_R(i \to d) B^a(d), \tag{6}$$

where the first term represents the direct ionization cross section; the second term represents the non-resonant contribution to the cross section through inner-shell excitation to all possible doubly excited states d, followed by autoionization; and the third term represents the contribution from resonant recombination autoionization, given by Equation 3, to all possible doubly excited states d, followed by a second autoionization. In this expression, $B^a(d)$ is the branching ratio for autoionization from the doubly excited state d of the N-electron ion to the lower states of the (N-1)-electron ion and is given by

$$B^a(d) = \frac{\sum_k A_a(d \to k)}{\sum_k A_a(d \to k) + \sum_n A_r(d \to n)}. \tag{7}$$

For highly ionized species in which the non-resonant, inner-shell excitation cross sections dominate the indirect cross section, the IP approximation, using distorted waves, has been very successful in predicting the total ionization cross sections, and the agreement between theory and experiment has been quite good (see, for example, Griffin and Pindzola 1988). However, for complex highly ionized ions in which the resonant part is also important, the calculation of the indirect contributions to ionization using the IP approximation is difficult because of the very large number of levels which are involved in the sequential double autoionization process, and the first complete calculation of this process in such an ion has only recently been completed (Chen et al 1990).

As already discussed, the CC approximation incorporates the contributions of resonant recombination followed by autoionization in a natural way; therefore, in low stages of ionization, where radiative rates are normally small compared to autoionizing rates, it can be used to calculate the contributions of inner-shell excitation followed by autoionization and resonant recombination followed by sequential double autoionization to the ionization cross section. Furthermore, it includes the effects of interference between the single electron continuum states and the recombination resonance states, but not with the two-electron continuum states associated with the (N-1)-electron bound states.

One of the first ionization experiments which demonstrated the importance of excitation autoionization to the ionization cross section was performed by Peart and Dolder (1975) on Ca^+. The experiment showed a very large excitation-autoionization contribution attributed to the inner-shell excitation $3p^6 4s \rightarrow 3p^5 3d4s$, with no major resonant features. More recently, this experiment was repeated at much higher resolution by Peart et al (1989), and they obtained essentially the same results. However, until recently, the shape and magnitude of theoretical cross sections did not agree with the measured cross section in the threshold region.

Burke et al (1983) carried out an R-matrix calculation in which the autoionizing terms $3p^5 4s^2$ ^2P, $3p^5 3d(^3P)4s$ $^{2,4}P$, and $3p^5 3d(^1P)4s$ ^2P were included in the CC expansion. They obtained a cross section dominated by a large broad resonance feature, which they attributed to resonance recombination into the $3p^5 3d^2 4s$ $^{1,3}F$ terms of Ca followed by the sequential emission of two Auger electrons. Subsequently, Griffin et al (1984) carried out a DW calculation and showed that significant contributions to ionization come from all nine terms of the $3p^5 3d4s$ configuration, that the branching ratios for autoionization from Ca^+ to Ca^{2+} differ significantly from unity for several of these terms, and that there is a large term dependent effect due to the 3d orbital in $3p^5 3d(^1P)4s$ ^2P being different from that in the remaining terms of $3p^5 3d4s$. Nevertheless, their calculated cross section was well above the measured cross section in the threshold region, even though they did not include resonant recombination double autoionization in their calculation. Pindzola et al (1987) carried out a CC calculation using a modified version of the program Impact (Crees et al 1978). In this calculation, only the $3p^5 3d(^1P)$ ^2P and $3p^5 3d(^3P)$ $^{2,4}P$ autoionizing terms were included in the calculation but a different 3d orbital was used for the first term than that employed for the other two terms to account for the term-dependent effect. However, like Burke et al (1983), they obtained a large resonance feature in the cross section, not observed in the experiment.

We recently carried out a 13-state R-matrix calculation of the contributions of the $3p^6 4s \rightarrow 3p^5 3d4s$ excitation to the ionization cross section of Ca^+ in which we included the bound terms from $3p^6 4s$, $3p^6 3d$, and $3p^6 4p$ and all 10 autoionizing terms from $3p^5 4s^2$ and $3p^5 3d4s$ in our CC expansion (Badnell et al 1991b). The bound-state orbitals were determined using a local model potential generated from Slater-type orbitals, as discussed in the last section for Zn^+. The effect of term dependence in the 3d orbital was partially taken into account by including configuration interaction with the $3p^5 4d4s$ configuration (see Griffin et al 1984). We multiplied each excitation cross section by the branching ratios for autoionization from Ca^+ to Ca^{2+}, as determined by Griffin et al (1984); summed them; convoluted the result with a 0.20 eV Gaussian to simulate the experimental electron energy distribution; and added them to a direct ionization cross section determined from a single parameter formula by Lotz (1969), scaled to agree with experiment before the excitation-autoionization threshold. The result, shown by the solid curve in Figure 6, reproduces the general shape and size of the measured cross section by Peart et al (1989), although it is still somewhat high.

We also performed an R-matrix calculation which included only those three terms of $3p^5 3d4s$ used in the earlier CC calculations. The result multiplied by the branching ratios, convoluted with a 0.20 eV Gaussian, and added to our estimate for direct ionization, is shown by the dashed curve in Figure 6. It is similar to the results obtained by Burke et al (1983) and Pindzola et al (1987). The significant difference between the two theoretical cross sections shown in Figure 6 is primarily due to the limited number of Auger channels that are available for a given autoionizing state when only a few N-electron terms arising from a given configuration are included in the continuum part of the CC expansion. However, when all possible terms are included, the collision strength of the resonance is distributed over many more channels with interference between each. Therefore, inclusion of more channels increases the non-resonant cross section but decreases the size of the resonance structure.

It was shown experimentally and also theoretically by Falk et al (1981) that ionization from the $3p^6 3d$ ground state of Ti^{3+} is another case where inner-shell excitation followed by auto-ionization dominates the total cross section. This was found to be due to the $3p^6 3d \rightarrow 3p^5 3d^2$ transition, even though only three of the 19 terms of $3p^5 3d^2$ are autoionizing. Bottcher et al (1983) performed a calculation of the ionization cross section for this ion using the IP approximation and distorted waves. However, their results were approximately a factor of 1.5 higher than experiment. Burke et al (1984) performed an R-matrix calculation in which they included nine terms of $3p^5 3d^2$ in the CC expansion and obtained results which were improved over those of Bottcher et al (1983), but but still well above experiment in the threshold region.

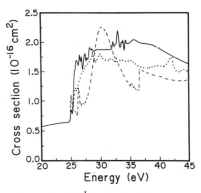

Fig. 6. Ca^+ ionization cross section. —, 13-state and − −, 4-state CC calculations of the $3p^6 4s \rightarrow 3p^5 3d4s$ excitation both convoluted with 0.20 Gaussian and added to a Lotz estimate of the direct ionization cross section; ···· Experiment due to Peart et al (1989).

We have completed an R-matrix calculation for Ti^{3+}, which included all 19 terms of $3p^5 3d^2$ in the CC expansion. The calculated ionization cross section is in excellent agreement with experiment. Furthermore, the non-resonant part of the excitation cross section to the 16 bound terms of $3p^5 3d^2$ is much lower than that obtained from the earlier DW calculation, although resonance features significantly enhance the cross section. This is then a case where coupling between the terms of the doubly excited configuration strongly affects the non-resonant excitation cross section.

4. CONCLUSIONS

It is clear from the examples presented here that the interference effects incorporated within the CC approximation are very important in ions in relatively low stages of ionization. However, the IP approximation shows great promise for highly ionized species where the effects of relativity and radiation damping often render the CC approximation impractical. The most challenging electron-impact excitation calculations are those associated with heavy atoms in low stages of ionization where relativistic and continuum coupling effects are both important and the size of the required CC expansion can be extremely large.

5. ACKNOWLEDGEMENTS

This work was supported by the Office of Fusion Energy, U. S. Department of Energy under Contract No. DE-AC05-84OR21400 with Martin Marietta Energy Systems Inc. and Contract No. DE-FG05-86ER53217 with Auburn University.

6. REFERENCES

Badnell N R 1986 J. Phys. B**19** 3827
Badnell N R, Pindzola M S, and Griffin D C 1991a Phys. Rev. A**43** 2250
Badnell N R, Griffin D C, and Pindzola M S 1991b J. Phys. B**24**
Bell R H and Seaton M J 1985 J. Phys. B**18** 1589
Bottcher C, Griffin D C, and Pindzola M S 1983 J. Phys. B**16** L65
Burke P G and Robb W D 1975 Adv. At. Mol. Phys. **11** 143
Burke P G, Kingston A E, and Thompson A 1983 J. Phys. B**16** L385
Burke P G, Fon W C, and Kingston A E 1984 J. Phys. B**17** L733
Chen M H, Reed K J, and Moores D L 1990 Phys. Rev. Lett. **64** 1350
Crees M A, Seaton M J, and Wilson P M H 1978 Comput. Phys. Commun. **15** 23
Eissner W, Jones M, and Nussbaumer H 1974 Comput. Phys. Commun. **8** 270
Falk R A, Dunn G H, Griffin D C, Bottcher C, Gregory D C, Crandall D H, and Pindzola M S 1981 Phys. Rev. Lett. **47** 494
Froese-Fischer C 1977 J. Phys. B**10** 1241
Griffin D C, Pindzola M S, and Bottcher C 1984 J. Phys. B**17** 3183
Griffin D C, Pindzola M S, and Bottcher C 1987 Phys. Rev. A**36** 3642
Griffin D C and Pindzola M S 1988 J. Phys. B**21** 3253
Griffin D C and Pindzola M S 1990 Phys. Rev. A**42** 248
Lotz W 1969 Z. Phys. **220** 466
Peart B and Dolder K 1975 J. Phys. B**8** 56
Peart B, Underwood J R A, and Dolder K 1989 J. Phys. B**22** 2789.
Pindzola M S, Bottcher C, and Griffin D C 1987 J. Phys. B**20** 3535
Pindzola M S and Griffin D C 1989 Phys. Rev. A**39** 2385
Pindzola M S, Badnell N R, Henry R J W, Griffin D C, and van Wyngaarden W L 1991 submitted Phys. Rev. A
Rogers W T, Dunn G H, Olsen J O, Reading M, and Stefani G 1982 Phys. Rev. A**25** 681
Smith S J, Man K F, Mawhorter R J, Williams I D, and Chutjian A 1991 Phys. Rev. Lett **67** 30
Wahlin E K, Thompson J S, Dunn G H, Phaneuf R A, Gregory D C, and Smith A C H 1991 Phys. Rev. Lett **66** 157

Electron–ion interaction cross sections determined by x-ray spectroscopy on EBIT

P. Beiersdorfer, R. Cauble, S. Chantrenne,[a] M. Chen, D. Knapp, R. Marrs, T. Phillips, K. Reed, M. Schneider, J. Scofield, K. Wong, D. Vogel, R. Zasadzinski

University of California, Lawrence Livermore National Laboratory, Livermore, CA 94550

B. Wargelin

Space Sciences Laboratory, University of California, Berkeley, CA 94720

M. Bitter and S. von Goeler

Princeton Plasma Physics Laboratory, Princeton University, Princeton, NJ 08543

ABSTRACT

The Livermore electron beam ion trap (EBIT) is used to measure electron-ion interactions with high-resolution x-ray spectroscopy. Measurements are presented of the Kα x-ray emission of heliumlike Fe^{24+} that demonstrate the effect of various processes on the spectrum of highly charged heliumlike ions. In particular, we have studied how dielectronic recombination into high-n Rydberg levels and resonance excitation processes contribute to the x-ray emission near threshold for direct electron-impact excitation. From these and other measurements we infer the cross sections for impact excitation of heliumlike titanium, chromium, manganese, and iron. Comparing the results with theoretical cross sections from distorted-wave calculations we find excellent agreement for all transitions but the heliumlike resonance transition from $1s2p\ ^1P_1$ to ground, whose excitation cross section is measured to be 10%-20% smaller than calculated.

1. INTRODUCTION

Many novel devices for the study of highly charged ions have been developed in the past few years, such as storage rings and electron coolers coupled with heavy-ion accelerators, and electron-beam ion sources (Hvelplund 1989, Habs et al. 1989, Schuch et al. 1989, Schmieder 1990). A common feature of these devices is that cross sections of electron-ion interactions are measured by detecting particles and analyzing the charge-state distribution. The processes studied, therefore, involve electron-ion interactions that change the charge state of the ion, such as dielectronic recombination or electron-impact ionization (Andersen et al. 1989, Ali et al. 1990, Kilgus et al. 1990).

The electron beam ion trap (EBIT) at Livermore is another novel device for the study of highly charged ions. It uses a ≤ 240 mA electron beam, squeezed to 60 μm by a 3-T magnetic field, magnetic field, to ionize, trap, and excite highly charged, heavy ions in a 2 cm-long trap region (Levine et al. 1988). Like in other facilities, ionization and recombination can be studied by analyzing the charge of extracted ions (Schneider et al. 1990, Penetrante et al. 1991). Unlike other heavy-ion facilities, however, it can also be used to study electron-ion interactions by monitoring and analyzing the x-ray flux both with solid-state x-ray detectors and with high-resolution crystal spectrometers. X-ray techniques have allowed us to measure the cross sections of dielectronic recombination for heliumlike ions up to Mo^{42+} (Knapp et al. 1989, 1990) and for neonlike ions up to U^{82+} (Schneider et al. 1991). In addition, x-ray techniques allow us to measure the cross sections of excitation processes, such as electron-impact excitation, radiative cascades, or resonance excitation, which for highly charged ions have not or cannot be studied with particle-detection techniques. High-resolution crystal spectroscopy was used, for example, to determine the electron-impact excitation cross sections of several 2-3 transitions in neonlike Ba^{46+} (Marrs et al. 1988), and it established the autoionization resonance contribution to the excitation of the magnetic quadrupole transition in Ba^{46+} (Beiersdorfer et al. 1990). By contrast, electron-impact excitation cross sections, for example, have so far been studied with particle-detection techniques only for low-charge ions, such as Si^{3+}, with the merged-beam technique (Wåhlin et al. 1991).

In the following we present measurements of the Kα line emission of heliumlike ions that illustrate our techniques for determining electron-ion interaction cross sections from x-ray observations. We place special emphasis on measurements of electron-ion interaction processes near threshold for electron-impact excitation, since accurate calculations exist with which we can compare our measurements, yet the processes in this energy region are complex and involve a variety of resonance phenomena that illustrate the need for high-resolution measurements. Our measurements can also be compared to observations from high-temperature plasma sources, where high-resolution measurements of the x-ray line intensities provide information on ion and electron temperatures, plasma motion, ion abundances, and transport phenomena (DeMichelis and Mattioli 1982, Källne et al. 1985). Several recent measurements of the Kα emission spectra of heliumlike ions from tokamaks (Bitter et al. 1985, Zastrow et al. 1990) and the sun (Doschek 1990) have shown puzzling discrepancies with model calculations, as no self-consistent set of plasma parameters could be found to explain the observed intensities of the heliumlike x-ray lines. So far it has not been resolved whether these discrepancies are the result of inaccurate atomic data, inadequacies of the atomic model, or are consequences of processes in the plasma that are not understood. The measurements we present in the following assess the accuracy of the theoretical cross sections employed in these model calculations, and represent a step toward identifying all atomic processes that affect the Kα x-ray emission of heliumlike ions in low-collisional, high-temperature plasmas.

2. X-RAY INSTRUMENTATION

The EBIT electron beam is aligned vertically. This configuration provides easy, direct spectroscopic access to the trap through six radial ports, as shown schematically in Fig. 1. Two types of instrumentation are used to analyze the x-rays produced in electron-ion interactions: Solid-state detectors, which provide full spectral coverage in the x-ray regime with near 100% quantum efficiency, but with only moderate energy resolution, and Bragg crystal spectrometers, which provide high spectral resolution. The counting efficiencies of crystal spectrometers, however, are greatly reduced over those of solid-state detectors because of low crystal reflectivities.

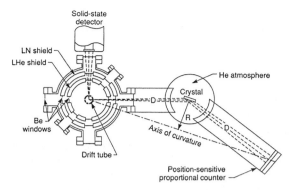

Fig. 1. Schematic diagram of the Livermore EBIT showing the layout of the high-resolution von Hámos spectrometer. The electron beam direction is out of the page.

EBIT represents a line source. Its width determined by the 60-μm diameter electron beam (Levine et al. 1989) is almost what one would have chosen for an entrance slit for a flat-crystal or von Hámos spectrometer, and both types of spectrometers are used on EBIT (Beiersdorfer et al. 1990a). However, since the von Hámos geometry (von Hámos 1933), shown in Fig. 2, provides higher throughput than a flat-crystal arrangement, EBIT's von Hámos spectrometer described by Beiersdorfer et al. (1990a) has become the instrument of choice for many high-resolution measurements. By choosing an analyzing crystal with a 75-cm radius of curvature, spectral resolving powers of up to $\lambda/\Delta\lambda = 4500$ have been achieved (Beiersdorfer 1991). Crystal spectrometers can thus be used to resolve transitions from individual atomic levels. As a result, we have been able to perform measurements of level-specific electron-ion interaction cross sections with 1-2 eV resolution when studying dielectronic recombination (Beiersdorfer 1991) despite the fact that the energy spread of the electron beam is about 50 eV FWHM.

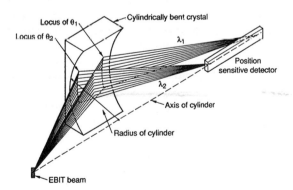

Fig. 2. Focusing properties of the von Hámos geometry. Monochromatic x rays from a line source (EBIT) are imaged by a cylindrically bent crystal onto a position-sensitive detector. The layout is such that the distance source to crystal equals the distance crystal to detector.

3. EXCITATION PROCESSES IN HELIUMLIKE IRON Fe^{24+}

In many plasma environments electron-impact excitation is the most important line formation process. In heliumlike ions it is the dominant process for exciting the 1s2p $^1P_1$ level, whose decay results in the resonance line commonly referred to as w. To a lesser extent direct electron collisions also excite the 1s2p $^3P_{1,2}$ and 1s2s $^3S_1$ levels, resulting in the intercombination lines y and x and the forbidden line z, as shown schematically in Fig. 3. Calculations show that the electron-impact excitation cross section of the $^1P_1$ level increases with electron energy, while those of the triplet levels decrease (Bely-Dubau et al. 1982).

Consequently, the intensity ratio of the triplet lines to the single line, $(x + y + z)/w$, depends on electron temperature and suggests itself as a temperature diagnostic (Pradhan and Shull 1981, Pradhan 1982).

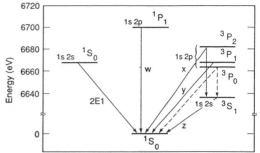

Fig. 3. Energy level diagram of heliumlike Fe^{24+}. Radiative transitions to ground are labeled in the notation of Gabriel (1972). Level 1s2s $^1S_0$ decays via two photons. Level 1s2p $^3P_0$ decays to ground state via the hyperfine interaction. Level 1s2p $^3P_2$ decays 18.3% of the time to level 1s2s $^3S_1$.

In many observations the triplet lines are noted to be more intense relative to w than predicted by model calculations (Bitter et al. 1985, Zastrow et al. 1990, Doschek 1990). This intimates the possibility that the theoretical electron-impact excitation cross sections are inaccurate. However, the heliumlike lines may also be excited by a variety of other processes such as electron capture by the hydrogenic ion, inner-shell ionization of the lithiumlike ion, or resonance excitation. In a plasma virtually all excitation channels are open concurrently. These processes, therefore, must be included in model calculations, which complicates the analysis and precludes a definite test of theoretical cross sections in plasma experiments.

In EBIT we employ an electron beam whose energy spread is about 50-eV FWHM (Knapp et al. 1989) to excite ions confined in the trap. This is narrow enough to select and study a particular excitation process, as illustrated in Fig. 4. Here we have plotted the $K\alpha$ emission of heliumlike Fe^{24+} measured with a Ge detector as a function of electron beam energy between 5.5 keV and 8.0 keV. The figure clearly shows the contribution from different dielectronic resonances to the $n=2\rightarrow1$ x-ray emission. Resonances are labeled in the usual Auger notation; the KLM resonances, for example, involve intermediate states in which an electron from the K shell is excited to the L shell as a free electron is captured into the M shell.

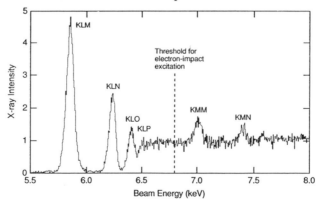

Fig. 4. Excitation function of $K\alpha$ x rays from heliumlike Fe^{24+} iron ions. X rays below threshold for electron-impact excitation are produced in the radiative stabilization of the L-shell electron of doubly excited levels in lithiumlike ions populated by dielectronic recombination resonances. Resonance excitation enhances the $K\alpha$ x-ray intensity above threshold. The data are obtained with a Ge solid-state detector viewing normal to the electron-beam direction.

Radiative stabilization of the L shell electron generates the Kα x-rays observed at beam energies below threshold for direct electron-impact excitation. Dielectronic resonances also contribute strongly to the Kα emission above threshold; the KMM resonances at 7.0 keV, for example, nearly double the observed emission. The processes with which they do so are, however, different from those below threshold, as resonances above threshold populate doubly excited states that can decay via autoionization to an excited level in the heliumlike ion, thus resonantly exciting the heliumlike transitions *w*, *x*, *y*, and *z*. For the KMM resonances this process can be schematically expressed as

$$1s^2 + e- \rightarrow 1s\,3\ell3\ell' \rightarrow 1s2\ell + e^- \ .$$

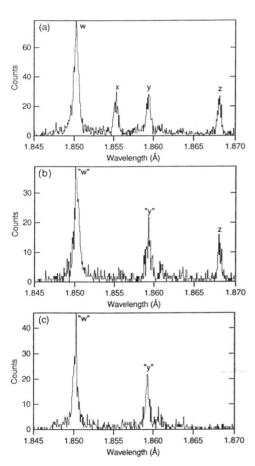

The data in Fig. 4 are obtained by rapidly ramping the beam energy from 4.5 keV to 8.5 keV and back down again and using acquisition techniques described by Knapp (1990). The cycle time is 2 ms. Thus the time spent on any particular beam energy, especially on any dielectronic recombination resonance, is very short, guaranteeing that the ionization balance is unchanged throughout the measurement. As a result, the relative emission intensities, normalized to the intensity at threshold for direct electron-impact excitation, reflect the relative excitation cross sections of the processes occurring at each beam energy.

The excitation cross sections of dielectronic resonances involving the L shell appear to smoothly match that of electron-impact excitation. In order to study the progression from dielectronic resonance excitation to electron-impact excitation, we have measured the Ka x-ray emission near threshold for electron-impact excitation as a function of electron-beam energy with a high resolution crystal spectrometer. Representative spectra taken with the von Hámos Bragg-crystal spectrometer are shown in Fig. 5 for three different beam energies.

Fig. 5. Kα x-ray spectra of heliumlike Fe^{24+} measured with a high-resolution von Hámos spectrometer on EBIT. The spectra are obtained with the electron beam energy set to (a) 6.74 keV, (b) 6.66 keV, and (c) 6.61 keV. For comparison, the threshold energies to excite lines *w*, *x*, *y*, and *z* by direct electron impact are 6.701 keV, 6.682 keV, 6.668 keV, and 6.637 keV, respectively. Lines labeled in quotes are excited by dielectronic capture of beam electrons into high-n levels.

The measurements in Fig. 5 clearly resolve lines *w*, *x*, *y*, and *z*. For electron-beam energies below threshold, a particular heliumlike line cannot be excited. Indeed, lines *x* and *z* are not seen below their respective thresholds. By contrast, lines *w* and *y* appear to be present even when the electron-beam energy is set to a value well below their thresholds. What is observed, however, are not the apparent heliumlike lines but dielectronic satellite lines produced by dielectronic recombination of beam electrons into Rydberg levels, i.e., the process

$$1s^2 + e^- \rightarrow 1s2pn\ell, \ n \gg 2 \ .$$

The Rydberg levels decay radiatively,

$$1s2pn\ell \rightarrow 1s^2 \, n\ell + h\nu \ ,$$

by emitting a photon whose energy differs only imperceptibly from those of their heliumlike parent lines w and y, as the high-n spectator electron interacts only weakly with the 1s2p-core electrons. Only w and y have these high-n satellite lines, because of their dipole-allowed nature. Lines x and z are electric-dipole forbidden and do not have appreciable high-n satellite lines, e.g., the n=2→1 fluorescence yields of their high-n satellites are on the order of 10^{-3} or less. The smooth matching of the excitation function of the dielectronic resonances with the excitation function of the heliumlike transitions implied by Fig. 4 is, therefore, only true for the dipole-allowed transitions. This behavior is demonstrated in Fig. 6, where we have plotted the observed intensities of the apparent lines x, y, and z relative to the apparent line w as a function of beam energy. Only the excitation function of y (and implicitly that of w) matches smoothly with its associated high-n satellites. High-n satellites associated with y and w play a significant role in the proper interpretation of spectra observed on tokamaks or the sun (Bely-Dubau et al. 1979, Bitter et al 1981) and have been studied on EBIT earlier in the case of heliumlike Ti^{20+} and V^{21+} (Beiersdorfer et al. 1990c, 1991).

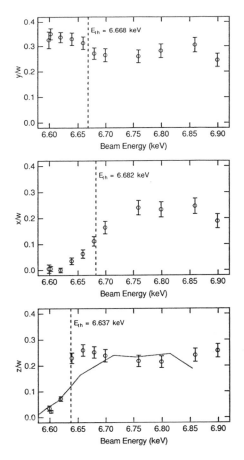

Fig. 6. Relative intensities of the triplet lines in Fe^{24+} as a function of electron beam energy. The intensities are normalized to the intensity of w. E_{th} denotes the threshold energy for direct electron-impact excitation. The dashed line represents the measured excitation function of x shifted by -45 eV to illustrate the differences in the onset of the lines at threshold.

Comparing the intensity of z with that of x as their respective electron-impact excitation threshold energies are crossed, we note that z rises faster than x. To illustrate this effect, we have plotted the rise of x, normalized to the threshold energy of z, as a dashed curve in Fig. 6. The enhancement of z at threshold suggested by the comparison may be attributed to resonance processes involving high-n autoionizing levels associated with line x populated by dielectronic capture:

$$1s^2 + e^- \rightarrow 1s2p \, (^3P_2) \, n\ell \ .$$

We have already said that x does not have associated high-n satellite lines because these levels decay by autoionization instead of radiative stabilization. However, those doubly excited levels produced in the resonant capture of beam electrons into sufficiently high n levels, i.e., those with energies between the electron-impact excitation thresholds of z and x, are energetically allowed to autoionize to the $^3S_1$ level with a probability as high as 0.9:

$$1s2p \ (^3P_2) \ n\ell \rightarrow 1s2s \ (^3S_1) + e^- \ .$$

Thus dielectronic capture into high-n levels with a $^3P_2$ core may result in an enhanced excitation of line z at threshold and could account for the faster rise in z. Evidently, this process cannot take place if the beam energy is above the threshold for direct excitation of x. Similarly, the $^3S_1$ level may be populated by resonance processes involving the high-n doubly autoionizing levels associated with y and w, but with much less probability because of the high fluorescence yield of these satellites. Moreover, y may receive contributions from the decay of autoionizing high-n levels associated with x and w, and x may receive contributions from the decay of high-n levels associated with w. Our measurements are, however, not sensitive enough to indicate the presence of such processes.

4. ELECTRON-IMPACT EXCITATION CROSS SECTIONS

Because of possible contributions to the line intensities near threshold from resonances we have chosen to determine the cross section for electron-impact excitation of the four x-ray lines at a beam energy that is centered between the threshold energy for w at 6.7 keV and the KMM excitation resonances at 7.0 keV. The values of the electron-impact excitation cross sections are determined in a procedure similar to the one described by Marrs et al. (1988). In particular, the summed intensity of the four heliumlike lines measured with the crystal spectrometer is related to their combined intensity observed with a solid-state detector, which in turn is normalized to the intensity of the x rays given off in the radiative capture of a beam electron to the n = 2 levels in the heliumlike ion, i.e.,

$$1s^2 + e^- \rightarrow 1s^2 2\ell + h\nu \ .$$

This procedure allows us to determine the electron-impact excitation cross section relative to that of radiative recombination, i.e., the inverse of photoionization, which is theoretically well known. Results of the measurement are listed in Table I. Similar measurements were carried out for heliumlike Ti^{20+}, Cr^{22+}, and Mn^{23+}, the results of which are also listed in Table I.

Because of the directionality introduced by the electron beam the x-ray emission from EBIT is non-isotropic. Calculations of the magnetic sublevel populations, including cascades, with the relativistic distorted-wave code developed by Zhang, Sampson, and Clark (1990) show that the polarization of x, y, z, and w are -52%, -19%, -7%, and +60%, respectively. This is taken into account in inferring the electron-impact cross sections of the heliumlike lines from our measurements, not only because it modulates the angular distribution of the x-ray emission (i.e., the intensity observed at 90° to the beam direction differs from its 4π average), but also because crystal spectrometers act as polarimeters and preferentially reflect x rays polarized perpendicular to the plane of dispersion.

For comparison, Table I lists the electron-impact excitation cross sections calculated with the code of Zhang, Sampson, and Clark (1990). The values have been adjusted to take into account radiative cascades and the hyperfine decay of the $^3P_0$ level. Radiative cascades strongly affect the $^3S_1$ level. Based on branching ratios calculated by Lin, Johnson, and Dalgarno (1977) 18.3% of the total population of the $^3P_2$ level decays to the $^3S_1$ level in iron. In addition, radiative decay of the $^3P_0$ level to the $^1S_0$ ground state is forbidden in iron, chromium, and titanium, whose most abundant isotopes have no nuclear magnetic moment. It decays instead to the $^3S_1$ level, further increasing the intensity of z (cf. Fig. 3). In contrast, the manganese nucleus has a finite magnetic moment, and the $^3P_0$ level decays directly to the ground state as a result of the hyperfine interaction. Because its energy is within 1.8 eV of that of the $^3P_1$ level, the resulting x ray appears to enhance the intensity of y. The effect of the hyperfine interaction on the x-ray line intensity of heliumlike Mn^{23+} is seen in the effective excitation cross sections of y and z, which clearly differ from the trend set by those of iron, chromium, and titanium.

TABLE I.   Electron-impact excitation cross sections of K x-ray transitions in heliumlike Ti^{20+}, Cr^{22+}, Mn^{23+}, and Fe^{24+} measured on EBIT. Theoretical values have been calculated with the relativistic distorted-wave code of Zhang, Sampson, and Clark (1990) and take into account radiative cascade contributions discussed in the text. All values are in units of 10^{-22} cm^2. The experimental values are accurate to within about 10%. The electron-ion interaction energies are known to within 100 eV.

Element	Energy keV	w expt	w theo	x expt	x theo	y expt	y theo	z expt	z theo
Ti^{20+}	4.8	6.05	7.50	2.30	2.59	2.20	2.49	2.95	3.11
Cr^{22+}	5.9	4.40	5.34	1.81	1.86	1.46	1.74		1.89
Mn^{23+}	6.5	3.59	4.55	1.56	1.58	1.87	1.89	1.04	1.08
Fe^{24+}	6.8	2.98	3.72	1.65	1.51	1.24	1.38	1.31	1.25

A comparison of the theoretical with the measured values shows excellent agreement for the triplet lines. By contrast, the measured electron-impact excitation cross section of w is about 20% lower than calculated. This is an unexpected result. In plasma observations, where the intensity of the triple lines relative to that of w in many cases is found to be larger than predicted, the suggestion has been made that this is due to errors in the atomic data of the triplet lines. The present result suggests instead that the the theoretical data for w is slightly off. The 20% discrepancy, however, is small and could account for only part of the discrepancies noted in plasma observations. It certainly cannot explain the discrepancies noted by Bitter et al. (1985) in the case of heliumlike Ti^{20+} on the TFTR tokamak that differed from theory by one order of magnitude. We have considered instrumental effects that could account for the difference between our measurement and calculations, such as the relative detection efficiency of the $K\alpha$ emission and radiative-recombination photons. The difference, however, persists for different elements, and thus for different experiments with different systematics, and we have not been able to find a plausible source of error that might change the results. Cross section measurements at beam energies other than threshold will be performed in the future to check whether the discrepancy persists at higher beam energies. Such measurements are important not only as a check on the present results, but also because plasmas average over the entire excitation cross section.

5. SUMMARY AND CONCLUSION

The Livermore EBIT allows not only the production of very highly charged ions (at present up to carbonlike uranium U^{86+}), but because of easy line-of-sight access to the trap also allows investigation of electron-ion interaction processes not presently possible at other heavy-ion facilities. We have started to perform such measurements using high-resolution x-ray techniques following the development of efficiently high-resolution bent-crystal spectrometers and novel data-acquisition techniques.

We have presented data that include measurements of the electron-impact excitation cross sections of heliumlike ions just above threshold. Other measurements based on high-resolution x-ray techniques are presently in progress. These include measurements of innershell and outershell ionization cross sections, level-specific dielectronic recombination cross sections and radiative rates, as well as excitation cross sections for beam energies as high as three times threshold. Our initial results provide tantalizing glimpses of the individual atomic processes that affect the x-ray spectra of highly charged ions. Comprehensive studies as a function of beam energy along isoelectronic sequences will allow us to test model calculations in detail and thus will greatly enhance the precision of x-ray spectroscopy for the diagnostic of high-temperature laboratory and astrophysical plasmas.

ACKNOWLEDGMENT

The expert technical support of Mr. E. Magee and Mr. D. Nelson are greatly appreciated. We thank Dr. A. Hazi and Dr. M. Eckart for their support and encouragement. Work performed under the auspices of the U.S. Department of Energy by Lawrence Livermore National Laboratory under contract W-7405-Eng-48.

REFERENCES

[a]present address: Hewlett Packard Company, Silicon Process Laboratory, Circuit Technology Research and Development, Box 10350, Palo Alto, CA 94303

Ali R, Bhalla, C P, Cocke C L, and Stöckli M 1990 Phys. Rev. Lett . **64** 633

Andersen L H, Hvelplund P, Knudsen H, and Kvistgaard P 1989 Phys. Rev. Lett. **62** 2656

Beiersdorfer P, Marrs R E, Henderson J R, Knapp D, Levine M A, Platt D B, Schneider M B, Vogel D A, and Wong K L 1990a Rev. Sci. Instrum. **61** 2338

Beiersdorfer P, Osterheld A L, Chen M H, Henderson J R, Knapp D A, Levine M A, Marrs R E, Reed K J, Schneider M B, and Vogel D A 1990b Phys. Rev. Lett. **65** 1995

Beiersdorfer P, Chantrenne S, Chen M H, Marrs R, Vogel D, Wong K, and Zasadzinki, in *Proceedings of the Vth International Conference on the Physics of Highly Charged Ions, Giessen, Sept 10-14 1990* to be published in Z. Phys. D

Beiersdorfer P 1991 Nucl. Instrum. Methods **B56/57** 1144

Beiersdorfer P, Chen M H, Marrs R E, Schneider M B, and Walling R S, 1991 Phys. Rev. A (in press)

Bely-Dubau F, Gabriel A H, and Volonté S, 1979 Not. R. Atron. Soc. **189** 801

Bely-Dubau F, Faucher P, Steenman-Clark L, Bitter M, von Goeler S, Hill K W, Cmahy-Val C, and Dubau J, 1982 Phys. Rev. A **26** 3459

Bitter M, von Goeler S, Hill K W, Horton R, Johnson D, Roney W, Sauthoff N, Silver E, and Stodiek W, 1981 Phys. Rev. Lett. **47** 921

Bitter M, Hill K. W, Zarnstorff M, von Goeler S, Hulse R, Johnson L C, Sauthoff N R, Sesnic S, and Young K M, 1985 Phys. Rev. A **32** 3022

De Michelis C and Mattioli M, 1981 Nucl. Fusion **21** 677

Doscheck G A in *X-ray and Inner-Shell Processes Knoxville TN 1990,* Aip Conference Proceedings No 215, ed. by Carlson T A, Krause M O, Manson S T (AIP, New York 1990) 603

Gabriel A H, 1972 Mon. Not. R. Astron. Soc **160** 99

Habs D, et al. 1989 Nucl. Instrum. Methods B **43** 390

Hámos L V 1933 Ann. der Physik **17** 716

Hvelplund P, 1989 J. Phys. (Paris) **50** C1-459

Källne E, Källne J, Marmar E S, and Rice J E 1985 Phys. Scripta **31** 551

Kilgus G et al. 1990 Phys. Rev. Lett. **64** 737

Knapp D A, Marrs R E, Levine M A, Bennett C L, Chen M H, Henderson J R, Schneider M B, and Scofield J H, 1989 Phys. Rev. Lett. **62** 2104

Knapp D, in *Proceedings of the Vth International Conference on the Physics of Highly Charged Ions, Giessen, September 10-14, 1990* to be published in Z. Phys. D

Levine M A, Marrs R E, Henderson J R, Knapp D A, and Schneider M B, 1988 Phys. Scripta **T22**, 157

Levine M A, Marrs R E, Bardsley J N, Beiersdorfer P, Bennett C L, Chen M H, Cowan T, Dietrich D, Henderson J R, Knapp D A, Osterheld A, Penetrante B M, Schneider M B, and Scofield J. H, 1989 Nucl. Instrum. Methods **B43** 431

Lin C D, Johnson W R, and Dalgarno A 1977 Phys. Rev. A **15** 154

Marrs R E, Levine M A, Knapp D A, and Henderson J R, 1988 Phys. Rev. Lett . **60** 1715

Pradhan A K and Shull J M, 1981 Astrophys. J. **249** 821

Pradhan A K, 1982 Astrophys. J. **263** 477

Schmieder R W in *Proceedings of the NATO Workshop on the Physics of Highly Ionized Atoms, Cargese, Corsica June 1988,* ed by Marrus R (Plenum Press: New York 1990) 231

Schneider D, DeWitt D, Clark M W, Schuch R, Cocke C L, Schmieder R, Reed K J, Chen M H, Marrs R, Levine M, and Fortner R 1990 Phys. Rev. A **421** 3889

Schneider M B, Knapp D A, Chen M H, Scofield J H, Beiersdorfer P, Bennett C L, DeWitt D, Henderson J R, and Marrs R E, 1991 Bull. Am. Phys. Soc. **36** 1265

Schuch R, Bárány A, Danared H, Elander N, and Mannervik S, 1989 Nucl. Instrum. Methods B **43**, 411

Wåhlin E K, Thompson J S, Dunn G H, Phaneuf A, Gregory D C, and Smith A C H 1991 Phys. Rev. Lett. **66** 157

Zhang H L, Sampson D H, and Clark R E H, 1990 Phys. Rev. A **41** 198

Zastrow K-D, Källne E, and Summers H P, 1990 Phys. Rev. A **41** 1427

Electron impact double ionization of ions

P. Defrance and Yu De Jiang
Universite Catholique de Louvain, Departement de Physique, Chemin du
cyclotron 2, B1348 Louvain-la-Neuve (Belgium)

Electron impact double ionization of atoms or ions was not extensively studied in the past. A few cases of atoms and singly charged ions only were subject to investigation until rare gas ions singly and multiply charged were studied by the Giessen group: Argon (Tinschert et al, 1989), Krypton (Tinschert et al, 1987) and Xenon, for charge states up to 7. The main feature of these measurements is the role of inner-shell single ionisation followed by autoionization. This indirect process is seen to dominate the ejection of two electrons from the concerned ions.

Although direct double ionization has a low probability to play a role – usually more than two orders of magnitude less than direct single ionization – it is interesting to study this process, because it has been seen by Joachain and Byron (1967) that it is very sensitive to the correlation between target electrons. This early conclusion was obtained for the helium atom and it holds also for other particles, atoms and negative or positive ions.

In addition, let us note that, up to now, no definite scaling law could be settled for this particular process.

In our laboratory some cases were recently analysed for which direct double ionization cross-sections can be precisely estimated, that is where indirect processes do not take a dominant role in the reaction. These are singly charged ions of Carbon, Nitrogen, Oxygen and Neon (Zambra et al, 1991 a, b).
This series allows us to observe the evolution of the cross-sections (σ_D) with increasing number of electrons (n) present in the L-shell, while the double ionization threshold $(I_{1,3})$ does not change drastically with increasing nuclear charge.

Cross-sections were obtained with a crossed electron-ion beam experimental set-up (Defrance et al, 1990). In this set-up, the animated crossed beams method has been employed (Brouillard and Defrance, 1983). The electron beam is swept electrostatically across the ion beam in a linear see-saw motion at constant velocity u which is measured by means of two thin wires located on both sides of the ion beam. The ion beam is extracted from a Colutron Ion Source. The usual acceleration voltage is a few keV. The electron energy E_e ranges from a few eV to 2.5 keV. The electron impact ionization cross-section is related to the measured quantities by:

$$\sigma = \frac{v_e v_i}{(v_e^2 + v_i^2)^{1/2}} \frac{uK}{(I_i/q_e)(I_e/e)} \tag{1}$$

where u is the scanning velocity, K is the total number of events produced during one passage of electrons across the ion beam, v_e and v_i, I_e and I_i, e and q_e are the velocities, currents and charges of the electrons and ions, respectively.

In table 1, measured cross-sections for direct double ionization at the maximum (σ_D) are reported together with relevant data for C^+, N^+, O^+ and Ne^+ ions.

Let us first note that the cross-section for inner-shell ionization followed by autoionization (σ_{i-a}) can be estimated from the K-shell ionization cross-section (σ_K) of the parent atoms for which these data exist (Xianguan et al, 1990):

$$\sigma_{i-a} = \sigma_K (1 - f) \tag{2}$$

In this expression, f express the fluorescence proportion of excited states that reduces the triply-charged ion production. As expected the fluorescence is of low importance, that is of order of 10^{-2} (see table 1).

Ionization potentials for the K-shell I_K are taken from the Clementi and Roetti calculations (1974). The ratio between I_K and $I_{1,3}$ ranges from 4.5 for C^+ to 8.7 for Ne^+ that is above the energy where the direct double ionization cross-section reaches its maximum value, usually around 3-4 times the ionization threshold. This indicates that the K-shell ionization contribution is well separated from the direct one for these ions.

In addition, σ_K is reduced by an order of magnitude when passing from C^+ to Ne^+. This is confirmed by the prediction of the classical scaling law that is included in the semi-empirical Lotz formula (Lotz, 1968)

$$\sigma_K = \frac{2 \times 4.5 \times 10^{-14} \ln E_e}{I_K E_e} \tag{3}$$

The maximum cross-section obtained with this formula is in good agreement with data reported in Table 1.

TABLE 1

Ion	n_e	$I_{1,3}(eV)$	$I_K(eV)$	$\sigma_K(10^{-19} cm^2)$	$f(10^{-2})$	$\sigma_D(10^{-19} cm^2)$
C^+	3	72.3	323.8	3.28	0.27	2.5
N^+	4	74.3	443.1	1.79	0.47	5.2
O^+	5	90	580.8	0.95	0.65	8.5
Ne^+	7	104.6	914.6	0.4	1.55	11.1

The variation of $_D$ clearly disagrees with the classical scaling law that predicts a linear variation with the number of electrons and an I^{-2} dependence on the ionization energy.

The direct double ionization reactions under consideration suggest that the process is related to the number of different electron pairs that can be formed in the shell. This number is simply

$$C_2^{ne} = \frac{n_e(n_e-1)}{2} \tag{4}$$

According to this model a simple scaling law can be drawn, in which the classical ionization energy dependence is also introduced. This leads to

$$\sigma_{1,3} = \frac{a\,n_e(n_e-1)}{2\,I_{1,3}^2} \tag{5}$$

From our measurements the average value of the constant a is estimated to be 5.5×10^{-16} cm^2 eV^2 25%.

The test of the scaling law passes through the comparison of predictions with experimental data for other systems. For example, atomic Neon (**Lebiars** et al, 1988), N$_e^{2+}$ ion (**Saner** et al, 1987) and Ar^{7+} (Rachafi et al, 1991) have been studied. Let us note that for the sodium-like ion Ar^{7+} it has been seen that the dominant contribution to double ionization comes from the closed L-shell.

Experimental data and predictions are summarized in table 2.

Table 2.

Ion	Scaling	Measured
Ne	3.9×10^{-18} cm^2	3.2×10^{-18} cm^2
Ne^{2+}	3.2×10^{-19} cm^2	3.4×10^{-19} cm^2
Ar^{7+}	2.1×10^{-20} cm^2	1.8×10^{-20} cm^2

In spite of the simplicity of the proposed law the agreement between predicted and measured cross-sections is remarkable.

The application of this law to the higher shells M, N, ... is not immediate, because of the increasing role of indirect processes. The prediction for the Argon isonuclear sequence, from neutral to charge state 5 overestimates the measured cross-sections (Tinschert et al 1989) by a factor of 8. From these data a value of 7×10^{-16} cm^2 50% can be assigned to the constant a present in the formula. This seems to indicate that the constant a depends on the shell that is concerned.

Acknowledgments

Authors co-workers have participated to the production of the experimental
results on which this paper rely.
In particular, authors are grateful to M. Zambra, J. Jureta and D. Belic
for their collaboration.

References

Brouillard F and Defrance P, Physica Scripta T3, 68 (1983).
Byron F W and Joachain C J 1967 PRA 164 1.
Clementi and Roetti 1974 At. Data and Nucl. Dat. Tables 14 177.
Defrance P, Chantrenne S , Rachafi S , Belic S D, Jureta J, Gregory D
    and Brouillard F 1990 J. Phys B 23 2333.
Lebins H, Binder J, Koslowski H R, Wiesemann K and Huber B A 1988
    J. Phys B
Lotz W 1968 Z. Phys 216 241
Müller A, Achenbach C, Salzborn E and Becker R 1984 J. Phys B 17 1427
Rachafi S, Belic D S, Duponchelle M, Jureta J, Zambra M, Zhang Hui and
    Defrance P 1991, J Phys B 24 1037
Saner R, Tinschert K, Hofmann G, Müller A, Becker R and Salzborn E,
    XVth ICPEAC Brighton 1987 Book of abstracts 377
Tinschert K, Müller A, Phaneuf R A, Hofmann G and Salzborn E 1989,
    J. Phys B 22 1241
Tinschert K, Müller A, Becker R and Salzborn E 1987, J. Phys. B 20 1823
Xianguan Long, Mantion Liu, Fuquing Ho and Xin Feng Peng 1990, At. Data
    and Nucl. Data Tables 45 353
Zambra M, Belic D, Defrance P, Yu De Jiang, XVIIth ICPEAC, Brisbane 1991
    Book of abstracts p298
Zambra M, Belic D, Defrance P, Yu De Jiang, XVIIth ICPEAC, Brisbane 1991
    Book of abstracts p296

Electron–atom collisions in strong laser fields

Alfred Maquet

Laboratoire de Chimie Physique, Université Pierre et Marie Curie,
11, Rue Pierre et Marie Curie. F 75 231, Paris Cedex 05, France.

Abstract: We present and discuss some recent advances made in the theory of electron-atom collisions in the presence of a strong, non-resonant, laser field. We show, in particular, that recent experiments can be interpreted with the help of a simplified version of the theory, relevant in the low-frequency regime.

1. INTRODUCTION

Free-Free Transitions (FFT's) experienced by electrons scattered, in the presence of a CO_2 laser field, by an atomic target remaining in its ground state, have been observed already more than fifteen years ago (Andrick and Langhans 1976, Weingartshofer *et al.* 1977). In these experiments, the electron-target system has been shown to be able to exchange up to twelve photons with the field (Weingartshofer *et al.* 1983). By contrast, it is only very recently that one has demonstrated the possibility of observing laser-assisted electron-impact atomic excitation in the presence of a (relatively) strong field (Mason and Newell 1987, Wallbank *et al.* 1988, Hippler *et al.* 1991), a process much more difficult to evidence. In the course of this type of laser-assisted collisions, which have been dubbed as "Simultaneous Electron-Photon Excitation", (SEPE), the electron-target system can absorb (emit) one or several photons from the laser field, the atom ending in an excited state. So far, only the electron-helium system in the presence of a low frequency field has been considered and SEPE has been demonstrated by experiments in which relatively slow electrons, with incoming energies below the excitation threshold of the metastable 2^3S state, collide with atoms in their ground state 1^1S, the laser supplying the needed energy to achieve excitation. Note that SEPE of higher excited states has been also observed recently by Wallbank *et al.* (1990). These results raise several questions related to the theory of this class of processes and we shall report on some recent advances made in the field.

A convenient way to think of the systems under study is to consider that they contain basically three components, namely: the projectile, the target and the field. The physics of each of the subsystems composed of the couples: (electron-atom), (atom-field) and (electron-field) is now quite well understood, at least in its broader features. Essentials of the theory of electron-atom scattering processes can be found in text-books (Joachain 1983) and a comprehensive treatment of atom-laser interactions, at least in the moderate intensity range which is of interest here, is given in the books by Mittleman (1982) and Faisal (1987). As regards to the electron-laser system, the corresponding Schrödinger equation can be solved exactly, the solutions being the so-called Volkov waves (Volkov 1935).

The situation is much less favourable when considering the whole "3-body" system encompassing the electron-atom-laser system. No exact solution of the corresponding Schrödinger equation can be found and one has to resort to somewhat simplified models in order to account for the experimental observations. We shall discuss below some of the simplifying features of the problem which help to keep the computational effort at a tractable level.

One of the main influence of the laser field on the dynamics of the collision is to act as a reservoir of energy, coupled to the projectile-target system via radiative coupling operators, which allow the system to gain (loose) energy via absorption (stimulated emission) of photons. Another, more subtle, influence of the laser field, is to modify the symmetry properties of the (otherwise cylindrically symmetric) electron-atom system, via the dynamic polarization of the target along the polarization vector of the field. Energetic aspects are nevertheless dominant as they provide a convenient way to classify the various laser-assisted processes, depending on the final state reached by the atomic system when the collision is completed.

Indeed, one can distinguish between:
i) Free-Free Transitions (FFTs):

$$e^- + A + N\gamma \longrightarrow e^- + A + (N \pm n)\gamma, \qquad (1)$$

where A represents a ground state atom, γ symbolizes the laser photons, $N \gg 1$ being the occupation number of the relevant laser mode and n is the net number of photons exchanged during the course of the collision, + standing for stimulated emission and - for absorption. The energy conservation relation reads here (atomic units shall be used throughout this paper unless otherwise mentioned):

$$k_i^2/2 = k_f^2/2 \pm n\,\omega, \qquad (2)$$

where k_i (k_f) is the momentum of the incoming (outgoing) electron and ω is the laser frequency;

ii) Laser-assisted electron-impact atomic excitation or "Simultaneous Electron-Photon Excitation" (SEPE):

$$e^- + A + N\gamma \longrightarrow e^- + A^* + (N \pm n)\gamma. \qquad (3)$$

Here the atom ends in an excited state A* and conservation of energy requires that:

$$k_i^2/2 + E_g = k_f^2/2 + E_e \pm n\,\omega, \qquad (4)$$

where E_g and E_e are the energies of the ground and excited atomic states, respectively.

iii) Laser-assisted electron-impact atomic ionization or "Simultaneous Electron-Photon Ionization" (SEPI):

$$e^- + A + N\gamma ----> 2\,e^- + A^+ + (N \pm n)\gamma. \qquad (5)$$

Here the atom is ionized and two electrons are present in the continuum of the remaining ionic core. The energy conservation relation is:

$$k_i^2/2 + E_g = k_f^2/2 + k_b^2/2 \pm n\,\omega, \qquad (6)$$

where k_b is the momentum of the second "ejected" electron. Note that, in order to characterize the process, one has to detect the two electrons in coincidence, which amounts to realize an (e,2e) experiment in the presence of a laser, which, by no means, is a simple task.

As already mentioned, up to now, experiments have only been reported for FFTs on the one hand and for excitation of He 2^3S on the other hand. Moreover, in both cases the lasers used were operated in a low-frequency domain, namely in the infrared range. As regards to the excitation process, the laser used in the pioneering work by Mason and Newell (1987) was a cw CO_2 one (λ = 10.6 μm, ω = .117 eV) with an average power around 10^4 W/cm^2, see also Mason (1989) and Mason and Newell (1989 and 1991). At such a relatively low intensity, they have been able to observe the exchange of only one photon between the field and the projectile-target system. It is in fact necessary to go to higher intensities in order to observe the exchange of several photons: this has been indeed verified by Wallbank *et al.* (1988, 1990) who used a pulsed CO_2 laser with a peak intensity around 10^8 W/cm^2 and have been able to observe the absorption of up to four photons during the collision process. Even more recently, Hippler *et al.* (1991) have used a pulsed Nd laser (λ = 1.06 μm, ω = 1.17 eV) with a peak intensity around 10^{12} W/cm^2 which, in spite of the much higher intensity, allowed them to observe the exchange of only one photon with the projectile-target system. This behavior shows explicitely that the dynamics of laser-assisted collisions depends, not only on the laser intensity but can also critically depend on its frequency. This point, as well as other general features of this class of processes, taking place in a low-frequency regime, will be discussed in the next section.

2. Low-Frequency Approximation

As already mentioned, the experiments reported so far have been realized in a low-frequency regime, such that the laser frequency (photon energy) ω is much smaller than any excitation frequency (energy) E_{exc} of the target, i.e. it fulfills the condition $\omega \ll E_{exc}$. This is clearly the case in the above reported experiments in He, with E_{exc} = 19.8 eV and $\omega \le 1.$ eV.

Amongst the main features of the low-frequency regime are the facts that the cross sections are larger and, as another favourable by-product, the theory is somewhat simpler. These facts are direct consequences of the infrared divergence of Quantum Electrodynamics which, when specialized to the case of electron-atom scattering, states that the bremsstrahlung cross section becomes divergent in the low frequency (soft-photon) limit (Low 1958):

$$d\sigma_{Brems} \underset{\omega \to 0}{\approx} \omega^{-1} [d\sigma_{elast} + a\omega + ...] \tag{7}$$

where $d\sigma_{Brems}$ is the cross section for the (spontaneous) emission of one photon, with frequency ω, taking place during the electron-atom collision, as a result of the coupling between the system and the empty modes of the (vacuum) field. Here $d\sigma_{elast}$ is the field-free elastic cross section and a stands for an approximately constant term, related in fact to the derivative $d(d\sigma_{elast})/dE$, which brings no contribution to the overall cross section in the limit of vanishing ω. This result shows that the cross sections for radiative processes accompanying a collision become proportional to the field-free ones and are larger at low frequencies.

The generalization to the case of stimulated radiative processes (FFTs), involving the exchange of several photons, in the course of an electron-atom collision in the presence of a classical field, $\mathbf{F} = \mathbf{F_0} \sin \omega t$, has been given by Kroll and Watson (1973), in the following form:

$$d\sigma^{(n)} \approx | J_n (\alpha. \mathbf{q}) |^2 d\sigma_{elast} . \tag{8}$$

Here n is the number of photons exchanged, $J_n (x)$ is a Bessel function, $\alpha = \mathbf{F_0}/\omega^2$, with the dimension of a length, is the (classical) excursion vector of a free electron embedded in the field, $\mathbf{q} = \mathbf{k_i} - \mathbf{k_f}$ is the momentum transfer and the "exact" elastic scattering cross section $d\sigma_{elast}$ is evaluated at a shifted energy $E_i = k_i^2/2 \pm n\omega$, compatible with the energy conservation Eq. (2). This formula, as it stands, is valid *to lowest order in* ω. However, its domain of validity can be extended to the next order in ω by using another shifted energy $E^*_i = (k_i - n\omega \mathbf{F_0}/F_0. \mathbf{q})^2/2$, as prescribed by Kroll and Watson (1973).

Such a formula, which has been established for potential scattering by a short-range potential, correctly describes FFTs as in the experiments by Weingartshofer *et al.* (1977, 1983). No equivalent general result exists for the case of atomic excitation, in which the role of the atomic structure can no longer be neglected. However, if one considers a first Born treatment of the collisional stage, similar to the one proposed by Bunkin and Fedorov (1966) for FFTs, one can straightforwardly derive a formula similar to Eq. (8), in which the exact elastic scattering cross section $d\sigma_{elast}$ is replaced by the first Born approximation $d\sigma_{excit}^{(B1)}$ for the field-free atomic excitation process (Beigman and Chichkov 1987). The extension to higher order Born contributions in $d\sigma_{excit}$ has been recently discussed (Geltman and Maquet 1989,

Maquet and Cooper 1990, see also Chichkov 1990). It can be shown in particular that, still in the low frequency domain and *provided the laser intensity is kept at a moderate level,* a result similar to Kroll and Watson's holds also for atomic excitation, i.e. one has, to lowest order in ω and for the exchange of n photons:

$$d\sigma^{(n)}_{excit} \approx | J_n (\alpha. q) |^2 d\sigma_{excit}, \qquad (9)$$

where $d\sigma_{excit}$ stands for the *exact* field-free excitation cross section, evaluated at the shifted energy $E_i = k_i^2/2 \pm n\omega$. It should be mentioned that higher order corrections in ω *are not* included in this simple model.

One of the interesting result of the above mentioned analysis is related to the range of validity of the formula Eq. (9). It appears in fact that an important criterion is the magnitude of the argument of the Bessel function $J_n (\alpha. q)$ entering Eq. (9). Indeed, a detailed study of the low-frequency limit of the Born series, as generalized to the inelastic scattering of Volkov waves by an atomic system, shows that, if the argument of the Bessel function is such that $|\alpha. q| \gg 1$, it is no longer valid to neglect the possible exchange of large numbers of (soft-) photons, which makes questionable the validity of the low-frequency approximation (Maquet and Cooper 1990). This shows that an important parameter, which governs the dynamics of this class of laser-assisted collisions, is the magnitude of the excursion length $\alpha = F_0/\omega^2$. More precisely, if α is larger than one (as expressed in units of the Bohr radius $a_0 = .53 \ 10^{-10}$ m), i.e. is larger than the typical size of an atomic target, one expects that the influence of the laser field on the collision can become dominant. And the important point to notice is that α can be large not only if the field amplitude F_0 itself is large, but also if the frequency ω is small. This basically explains why it has been much easier to observe laser-assisted atomic excitation with the help of infrared lasers than with the ones operated at higher frequencies in the optical and *a fortiori* in the ultra-violet ranges.

As regards to the applicability of the formula Eq. (9), it is of interest to compare its predictions with the results of the experiments (Mason and Newell 1989, Wallbank *et al.* 1989). This is done in the Figures [1] - [4], in which we display both the results of the experiments and of the calculations performed for the total cross section:

$$[\sigma_{excit}]_{assist} = \Sigma_n \ \sigma^{(n)}_{excit} \qquad (10)$$

where $\sigma^{(n)}$ stands for the total cross sections obtained from the differential ones, Eq. (9), by integrating over the scattering angles and the sum runs over the number of exchanged photons. The partial cross sections for the absorption (stimulated emission) of n photons have been calculated by inserting in Eq. (9), the "exact" values of the corresponding field-free cross sections computed and tabulated by Fon *et al.* (1981). In Figs. [1] and [3] are shown the results of such a calculation (Geltman and Maquet 1989) for the field parameters correponding to, or

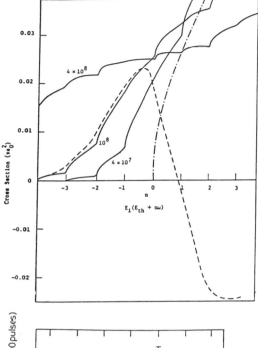

Fig. [1]. Calculated laser-assisted cross sections (from Eq. 10) as a function of electron energy (in units of the photon energy from threshold) for several laser intensities as indicated (solid line). The dashed line is the difference between the laser-assisted and the field-free cross section for $I \approx 10^8$ W/cm^2, convoluted with a Maxwell distribution of incoming electron energies. (from Geltman and Maquet 1989).

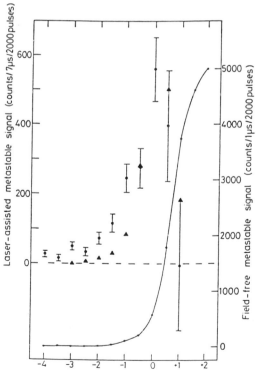

Fig. [2]. Experimental results by Wallbank et al. (1988), obtained at a peak laser intensity $I \approx 10^8$ W/cm^2. Solid curve, left scale: field-free signal. The laser-assisted signal corresponds to the dots with the error bars. It should be compared with the dashed line in Fig.[1].

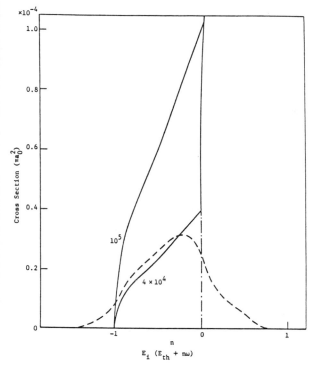

Fig. [3]. Calculated laser-assisted cross sections (from Eq. 10) as a function of electron energy (in units of the photon energy from threshold) for several laser intensities as indicated (solid line). The dashed line is the difference between the laser-assisted and the field-free cross section for $I \approx 4.10^4$ W/cm^2 , convoluted with a Maxwell distribution of incoming electron energies, (from Geltman and Maquet 1989).

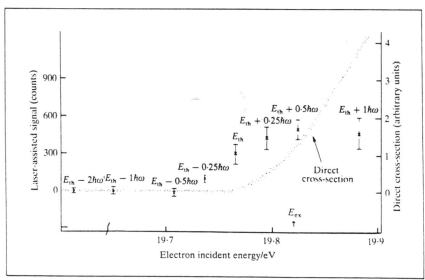

Fig. [4]. Experimental results by Mason and Newell (1987), obtained at a cw laser intensity $I \approx 10^4$ W/cm^2. Dots, left scale: field-free signal. The laser-assisted signal corresponds to the crosses with the error bars. It should be compared with the dashed line in Fig.[3]. Note the change of energy scale.

close to, the ones relevant for the experiments by Wallbank *et al.* 1988 and Mason and Newell (1987,1989), respectively.

As a first remark, this simple model correctly accounts for the number of exchanged photons as observed in the experiments. In the low intensity regime prevailing in the case of the Mason and Newell's experiment ($I \approx 10^4$ W/cm^2; $\alpha \approx 0.03$ a_0), one observes the absorption of only one photon, which means that we are likely to be in a perturbative regime in which the polarization of the target by the field is weak. At higher field intensities, as in the case of the experiment by Wallbank *et al.* ($I \approx 10^8$ W/cm^2; $\alpha \approx 3$ a_0), one observes the absorption of up to four photons, in agreement with the theory, see Figs [1], [2].

The situation is more involved in the case of the experiment by Hippler *et al.* (1991), who used a pulsed neodymium laser with a much higher peak intensity around $I \approx 10^{12}$ W/cm^2. However, as $\omega = 1.17$ eV, which is ten times larger than for the CO_2 laser, the overall effect of the field on the collision is in fact somewhat reduced ($\alpha \approx 0.3$ a_0) and they have observed the absorption of only one photon. Another complication arises from the presence of a $He^-(1s2s^2\ ^2S)$ resonance located at 19.4 eV, i.e. at less than the energy of one photon below the excitation threshold (19.8 eV), and it is likely that the model described here is no longer valid, at least in its simpler form.

Another remarkable feature of this set of results is that the laser-assisted total cross section becomes smaller than the field-free one at incoming electron energies larger than the excitation threshold. This fact results from the overall conservation of the probability, regardless of the presence of the laser: more channels are open to the projectile in its final state when the laser is on and the new channels are populated at the expenses of the ones already present when the laser is off. A similar effect, which can be traced back to the properties of the Bessel functions, holds also for FFTs.

3. PERSPECTIVES

As we have shown, a quite simple theoretical model allows to account, at least qualitatively, for the essential features of laser-assisted electron-atom collisions, in the low frequency regime. However, more detailed theoretical studies, conducted on the electron-H atom system, have shown that this is no longer true when considering *differential cross sections* and specially for small-angle scattering (Byron and Joachain 1984, Dubois *et al.* 1986, Jetzke *et al.* 1987, Francken and Joachain 1987, Byron *et al.* 1988, Francken *et al.* 1988). Dressing effects of the target become important and must be consistently included in the theory. We have shown that these effects are particularly important in the case of ionisation, namely for SEPI (Joachain *et al.* 1988, Martin *et al.* 1989).

This set of studies has also shown that such dressing effects can be particularly important at higher laser frequencies, i.e. in the optical or uv range. Then, depending on the dynamic polarizability of the target, strong modifications of the dynamics of electron-atom collision are expected.

The laser intensity plays also an important role on these dressing effects. However, it should be kept in mind that, if it becomes comparable to the atomic unit of intensity, namely $I_0 = 3.5 \; 10^{16}$ W/cm^2, the atom is ionized via multiphoton ionization. It has been observed, in fact, that ionization can become dominant at intensities of the order of $I \approx 10^{13}$ W/cm^2 . This will set a limitation to the possibility to observe strong field effects in laser-assisted electron-atom collisions.

References:

Andrick D and Langhans L 1976 *J. Phys. B: At. Mol. Phys.* **9** L459.

Beigman I L and Chichkov B N 1987 *Sov. Phys. JETP Lett.* **46** 395.

Bunkin F V and Fedorov M V 1966 *Sov. Phys. JETP* **22** 844.

Byron Jr. F W and Joachain C J 1984 *J. Phys. B: At. Mol. Phys.* **17** L295.

Byron Jr. F W, Francken P and Joachain C J 1987 *J. Phys. B: At. Mol. Phys.* **20** 5487.

Chichkov B N 1990 *J. Phys. B: At. Mol. Phys.* **23** L333.

Dubois A, Maquet A and Jetzke S 1986 *Phys. Rev. A* **34** 1888.

Francken P and Joachain C J 1987 *Phys. Rev. A* **35** 1590.

Francken P, Attaourti Y and Joachain C J 1988 *Phys. Rev. A* **38** 1785

Faisal F H M 1987 *Theory of Multiphoton Processes* (Plenum, New York).

Fon W C, Berrington K A, Burke P G and Kingston A E 1981 *J. Phys. B: At. Mol. Phys.* **14** 2921.

Geltman S and Maquet A 1989 *J. Phys. B: At. Mol. Phys.* **22** L419.

Hippler R, Luan S and Lutz H O 1991; This Conference.

Jetzke S, Broad J and Maquet A 1987 *J. Phys. B: At. Mol. Phys.* **20** 2887.

Joachain C J 1983 *Quantum Collision Theory* (North Holland, Amsterdam).

Joachain C J, Francken P, Maquet A, Martin P and Véniard V 1988 *Phys. Rev. Lett.* **61** 165.

Kroll N M and Watson K M 1973 *Phys. Rev. A* **8** 804.

Low F E 1958 *Phys. Rev.* **110** 974.

Maquet A and Cooper J 1990 *Phys. Rev. A* **41** 1724.

Martin P, Véniard V, Maquet A, Francken P and Joachain C J 1989 *Phys. Rev. A* **39** 1724.

Mason N J 1989 *Contemporary Physics* **30** 449.

Mason N J and Newell W R 1987 *J. Phys. B: At. Mol. Phys.* **20** L323.

-------- 1989 *J. Phys. B: At. Mol. Phys.* **22** 777.

-------- 1991; This Conference.

Mittleman M H 1982 *Theory of Laser Atom Interactions* (Plenum, New York).

Volkov D M 1935 *Z. Phys.* **94** 250.

Wallbank B, Holmes J K, LeBlanc L and Weingartshofer A 1988 *Z. Phys. D* **10** 467.

Wallbank B, Holmes J K and Weingartshofer A 1989 *Phys. Rev. A* **40** 5461.

-------- 1990 *J. Phys. B: At. Mol. Phys.* **23** 2997.

Weingartshofer A, Holmes J K, Caudle G, Clarke E M and Kruger H 1977 *Phys. Rev. Lett.* **39** 269.

Weingartshofer A, Holmes J K, Sabbagh J and Chin S L 1983 *J. Phys. B: At. Mol. Phys.* **16** 1805.

Collisional excitation transfer in a magnetic field

Takashi Fujimoto

Department of Engineering Science, Kyoto University, Kyoto 606, Japan

ABSTRACT: Experimental and theoretical studies of collisional excitation transfer between the magnetic substates are reviewed. By using the pulsed excitation laser-induced-fluorescence spectroscopy, we have measured the disalignment rate of neon $2p^5$ $3p$ atoms due to neon collisions, and found the rate coefficient which depends on the magnetic field strength. We have also observed the excitation transfer from the helium singlet states to the triplet state due to helium collisions. The rate coefficients show magnetic field dependences which are proportional to the degree of mixing of the singlet and triplet wavefunctions.

1. INTRODUCTION

A magnetic field may affect the collision processes in various ways. Specifically for low-energy atom-atom collisions, effects of the field have been investigated on excitation transfer between the magnetic substates of excited atoms. Gay and Schneider (1979a) present a theory which is based on the impact parameter method. They treat the cases in which the interaction is an electrostatic long-range force; the first of these cases is the resonance collisions between atoms of the same species, and the second is the collisions between atoms of different species in which the van der Waals force is responsible (foreign gas collisions). In both the cases, the effect of a magnetic field on collisions is appreciable only when the condition ω $T_c > 1$ is satisfied, where ω is the Larmor frequency and T_c is the collision time.

Experiment on the resonance collision is reported for $Hg(6\ ^3P_1)$ + $Hg(6\ ^3S_0)$ and $Na(3\ ^2P)$ + $Na(3\ ^2P)$ systems (Gay and Schneider 1979b,c), and the observed decrease in the excitation transfer rates between the magnetic substates with an increase in the field strength (B) is well reproduced by the theory. The foreign gas collision has been studied on $Hg(6\ ^3P_1)$ atoms with the collision partner of rare gas atoms or several species of molecules (Gay and Omont 1976, Fuchs et al 1979). The agreement between the experimental results and the theoretical predictions is rather poor: only the observed B-dependence of $g^{(11)(10)}/g^{(1-1)(10)}$ for xenon and krypton perturbers is approximately reproduced by the theory, where $g^{(11)(10)}$ stands for the excitation transfer rate from the J = 1, m_J = 0 state to the J = 1, m_J = 1 state. For other cases, the experimental results are inconsistent with the theoretical predictions. It is suggested that interactions other than the van der Waals force may be dominant in these collisions. It is noted that the criterion ω $T_c \gtrsim 1$ is still valid in these cases.

In the following, the two experimental studies performed in Applied Spectroscopy Laboratory, Kyoto University, are presented.

2. DISALIGNMENT COLLISIONS OF NEON $2p^5$ $3p$ ATOMS

The experimental setup is described in detail in the forthcoming paper
(Matsumoto, in press). The superconducting magnet, capable of producing
a magnetic field up to 10 T, was installed in a cryostat having the
cylindrical-shape free space of 60mm diameter. A discharge tube of 10 mm
inner diameter made from pyrex glass was placed on the axis of the
solenoid. Neon gas of 1 - 7 torr was introduced to the discharge tube and
a dc mild glow discharge was produced with a current of 0.8 - 2.2 mA. By
careful alignment of the discharge tube axis along the magnetic-field, we
obtained a stable discharge.

A nitrogen-laser pumped dye laser produced a laser pulse of a 5 ns
duration and a repetition rate of 26 pulses per second. This laser light
was transmitted through an optical fiber into the free space of the
cryostat, reflected by a prism, focused by a lens and illuminated the
plasma from the direction perpendicular to the magnetic field.

Some of the neon $1s_3$ (Paschen notation, $2p^5$ $3s$ configuration, J = 0)
metastable atoms in the discharge plasma were excited by the 616.4 nm
light to the $2p_2$ level ($2p^5$ $3p$ configuration, J=1), and we observed the
fluorescence light of the 659.8 nm ($1s_2$ (J=1) - $2p_2$) line. We thus used
the (J=0) $\rightarrow$ (J=1) $\rightarrow$ (J=1) excitation-observation scheme. The
absorption line for the laser excitation was Zeeman split, and for the
field strength of $B >$ 2 T, we excited only the central π-component,
producing a population only in the m_J = 0 magnetic substate of the upper
level.

The fluorescence light emanating from the plasma in the direction at right
angles to the directions of the magnetic field and the laser beam was
reflected by two mirrors and focused by a lens onto the entrance slit of
the monochromator. Its focal length was 250 mm and the linear reciprocal
dispersion was 3.4 nm/mm. Owing to the low intensity of the fluorescence
signal we did not attempt to resolve each component of the Zeeman split
line. Rather, by using a polarizer we observed the π- and σ-components
of the fluorescence, separately. The temporal development of the output
from the photomultiplier was sampled with a 1.6 ns time resolution,
averaged and processed by a boxcar averager system.

From the observed intensities of the π- and σ-components of the
fluorescence, I_p and I_s , we calculated the relative total population of
the upper level (I_p + $2I_s$), and the longitudinal alignment
(I_p - I_s)/(I_p + $2I_s$), as functions of time. The longitudinal alignment
was plotted on a semi-logarithmic scale, and its slope gives the
disalignment rate. This rate is three times the average excitation
transfer rate ($g^{(11)(10)}$ + $g^{(1-1)(10)}$)/2. The disalignment rate was
measured for varying atom densities, and the slope of the best fitted line
against the atom density gives the disalignment rate coefficient. This
whole procedure was repeated for varying magnetic fields.

The disalignment rate coefficient thus obtained showed a dependence on the
magnetic field strength; it starts out at the zero field from the value of
1.7 (x 10^{-16} m^3 s^{-1}), decreases to a minimum of 1.2 at B = 4 T,
increases to reach a maximum of 1.9 at B = 7 T and then continues to

decrease, reaching a value of 1.2 at 10 T. The critical field strength is estimated to be about 35 T from the excitation transfer rate coefficient. Thus the present Ne* + Ne system shows an obvious violation of the criterion.

3. EXCITATION TRANSFER OF HELIUM SINGLET-TRIPLET STATES

Fujimoto and Fukuda (1981) reported the processes, which apparently violate Wigner's spin conservation rule

$$He(3\ ^1P) + He(1\ ^1S) \rightarrow [He(3\ ^3D) + He(1\ ^1S)] \quad \Delta E = 104\ cm^{-1} \quad (1)$$
$$He(3\ ^1D) + He(1\ ^1S) \rightarrow [He(3\ ^3D) + He(1\ ^1S)] \quad \Delta E = \ 4\ cm^{-1} \quad (2)$$

However, for helium the LS coupling scheme is only approximately valid, and owing mainly to the spin orbit interaction the $3\ ^3D$ wavefunction, for example, contains a small amount of the pure singlet wavefunction. The degree of the mixing is known to be $\omega^2 = 2.4 \times 10^{-4}$ (Derouard et al 1976, Fujimoto et al 1986). When we compare the rate coefficient 2.5×10^{-19} ($m^3\ s^{-1}$) for process (1) with the rate coefficient 3.7×10^{-16} for process

$$He(3\ ^1P) + He(1\ ^1S) \rightarrow [He(3\ ^1D) + He(1\ ^1S)] \quad (3)$$

it is seen that one third of the former coefficient is accounted for within the framework of the spin conservation rule. When the rate coefficient 2.2×10^{-18} for process (2) is compared with the rate coefficient 3.2×10^{-15} for disalignment of $3\ ^1D$ atoms due to atom collisions (Fujimoto and Fukuda 1981) we again arrive at a similar conclusion.

When helium is placed in a magnetic field the helium $3\ ^1D$ and $3\ ^3D$ levels are Zeeman split, and undergoes crossings as shown in Fig. 1(a). Owing to the decrease in the energy separation between the magnetic sublevels having the same m_J value the mixing coefficient increases as shown in Fig. 1(b). At about B = 7.3 T the four pairs of the magnetic sublevels intermix completely, leading to level anticrossing as shown in Fig. 1(b) and the mixing coefficient at this magnetic field is $\omega^2 = 1/2$. Hirabayashi et al (1986) and Matsumoto et al (1989) performed measurements of the rate coefficients for processes (1) and (2) under the magnetic field of up to 10 T.

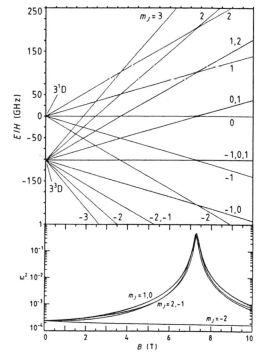

Fig. 1

The experimental setup and the procedure are basically the same as described in the preceding section except that we used a helium discharge and we excited the metastable 2 ^{1}S atoms to the 3 ^{1}P level, and that we observed the direct fluorescence light together with the 667.8 nm (mainly 2 ^{1}P -3 ^{1}D) line and the 587.6 nm (mainly 2 ^{3}P -3 ^{3}D) line. We deduced the population densities of the three upper levels as functions of time, and determined the excitation transfer rates for processes (1) and (2). The excitation rate coefficients are shown in Fig. 2 (Matsumoto et al 1989). The points ● are for process (1) and ○ for process (2). In this figure the averaged mixing coefficient over the five pairs of the magnetic substates is shown with the solid curve. It is seen that both the rate coefficients follow the increase and decrease in the degree of the wave-function mixing closely.

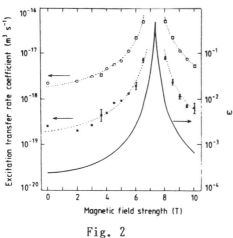

Fig. 2

Derouard A, Jost R, Lombardi M, Miller A T and Freund R S 1976 Phys. Rev. A 14 1025
Fuchs B, Hanle W and Scharmann A 1979 Phys. Letters 75A 50
Fujimoto T and Fukuda K 1981 J. Phys. Soc. Jpn. 50 3476
Fujimoto T, Hirabayashi A, Okuda S, Shimizu K and Takuma H 1986 J. Phys. B: At. Mol. Phys. 19 571
Gay J C and Schneider W B 1979a Phys. Rev. A 20 879
Gay J C and Schneider W B 1979b Phys. Rev. A 20 894
Gay J C and Schneider W B 1979c Phys. Rev. A 20 905
Gay J C and Omont A 1976 J. Phys. (Paris) 37 L69
Hirabayashi A, Okuda S and Fujimoto T 1988 J. Phys. B: At. Mol. Phys. 19 581
Matsumoto S, Shiozawa K and Fujimoto T 1989 J. Phys. B: At. Mol. Opt. Phys. 22 L373
Matsumoto S, Shiozawa K, Ishitani Y, Hirabayashi A and Fujimoto T (in press) Phys. Rev. A

Collisional and radiative processes in a quantizing magnetic field

Michelangelo Zarcone

Istituto di Fisica dell'Università,
Via Archirafi 36, 90123 Palermo, Italy

ABSTRACT: The potential scattering total cross section in the presence of a strong magnetic field is derived in the Higher Order Modified Born Approximation, obtaining an expression that does not diverge at the Landau thresholds. Stimulated and spontaneous bremsstrahlung cross sections and absorption coefficients in a magnetic field such that $\omega_c \gg \omega$ and $\hbar\omega_c \gg kT_e$ are also derived.

1. INTRODUCTION

The study of various collisional and radiative processes in the presence of strong magnetic fields has attracted considerable interest in recent years (for a complete list of references see McDowell and Zarcone 1985). In this paper we show how these processes are modified by the presence of a quantizing magnetic field. Quantum effects become important when the de Broglie wavelength associated with the motion of a charged particle in the presence of a magnetic field is comparable or higher than the radius of its classical helicoidal orbit, i.e. $\lambda_B \geq R$ or $\hbar/mv \geq (mc/eB)v$, that gives

$$B \geq B_q \left(\frac{v}{c} \right)^2 \qquad (1)$$

with $$B_q = \frac{m^2 c^3}{e\,\hbar} = 4.414\ 10^9\ \text{T} \ .$$

This unit of magnetic field plays a fundamental role. In a magnetic field of such intensity the difference of energy between the spin up and the spin down levels for a particle with spin 1/2 is $\Delta E = mc^2$, the splitting is then equal to the rest mass of the electron. A magnetic field $B = B_q$ is a huge magnetic field thought to exist only on astrophysical objects like white dwarfs or neutron stars. Under laboratory conditions we deal usually with fields $B \ll B_q$ and the scattering processes are only slightly affected by the presence of the field. However, in experiments with highly excited atoms, in solid state physics and for low energy incident particles, situations are encountered where the effect of the field on the collision process is not negligible (Canuto and Ventura 1977). In all these cases, the presence of the magnetic field changes, significantly, the conditions of the scattering process. The field introduces a new axis of symmetry changing the spherical geometry into a cylindrical one. The electron motion remains free along the direction of the magnetic field but becomes confined to discrete Landau orbitals in the transverse direction. This one-dimensional character of the scattering particle's free motion along the field gives rise to a density of final states whose discontinuous and oscillatory behavior is responsible of quite all the new effects arising in magnetic field assisted collision processes. In fact the quantization of the energy relative to the particle's motion perpendicular to the magnetic field gives rise to a discretization of the total cross section in a sum over the final Landau quantum numbers, while its free motion along the direction of the field gives rise to a one-dimensional density of states $\eta(E)=1/k_{zf}$. When the initial kinetic energy along B matches exactly the energy difference between the initial and final Landau levels, the final momentum k_{zf} goes to zero and the cross section undergoes an infinite growth. To remove these infinities the collisional process must be treated in a more precise way. This can be achieved going over the FBA in the calculation of the T matrix elements. In section 2, to establish notations we derive the potential scattering cross section in the presence of a strong magnetic field. In section 3 we analyze the Born series of the T-matrix for potential scattering in a strong magnetic field in cylindrical coordinates. In section 4 we derive the Higher Order Modified Born Approximation (HOMBA) and we show that in this approximation it is possible to sum up all the Born series getting a closed form for the total cross section that does not diverge at Landau thresholds. Stimulated and spontaneous bremsstrahlung cross sections and absorption coefficients for a magnetic field intensity such that $\omega \gg \omega_c$ and $\hbar\omega \gg kT$ are derived in section 5 as a typical case where the effect of a strong magnetic field on the electron ion interaction modifies, significantly, a radiative process. Concluding remarks are given in section 6.

2. THE TOTAL CROSS SECTION

In the presence of a quantizing magnetic field the electron motion is free along the direction of the field and becomes confined in the transverse direction. In cylindrical coordinates and in the Landau gauge the wave function for an electron in the presence of a magnetic field is given by (Ferrante et al 1980)

$$\psi_{nsk_z}(\vartheta,\rho,z) = e^{ik_z z} \Phi_{ns}(\vartheta,\rho) \tag{2}$$

with

$$\Phi_{ns}(\vartheta,\rho) = \left[\frac{\pi}{\gamma}\right]^{1/2} e^{i(n-s)\vartheta} I_{ns}(\gamma\rho^2) \tag{3}$$

where

$$I_{ns}(t) = (n!s!)^{-1/2} e^{-t/2} t^{|n-s|} \begin{cases} Q_n^{s-n}(t) & \text{for } s \geq n \\ \\ Q_s^{n-s}(t) & \text{for } s < n \end{cases} \tag{4}$$

where $Q_n^\alpha(t)$ are associated Laguerre polynomials defined as

$$Q_n^\alpha(t) = t^{-\alpha} e^t \frac{d^n}{dt^n} \left[e^{-t} t^{n+\alpha} \right], \tag{5}$$

$t = \gamma \rho^2$ with $\hat{\rho} = \hat{x}x + \hat{y}y$ and $\gamma = eB/2\hbar c$; in atomic units $\gamma(a.u.) = B/B_o$ with $B_o = 4.7 \ 10^5$ T; n,s=0,1,2....; n indicates the Landau quantum number; s is related to the distance of the center of the orbit of the spiraling electron from the origin, and $(n-s)\hbar = m\hbar$ is the particle angular momentum. The corresponding energy eigenvalues are given by

$$E_n = \frac{k_z^2}{2} + \left(n + \frac{1}{2} \right) 2\gamma \tag{6}$$

where 2γ (a.u.)=$\hbar\omega_c$ is the cyclotron energy in atomic units. The energy levels are degenerate with respect to the quantum number s. In the S-matrix approach the cross section is obtained dividing the transition probability per unit time P_{if} by the incident current density $\tilde{J}_z$ and summing over all the final state quantum numbers (n_f, s_f, k_{zf}) and over the degenerate initial states s_i:

$$\sigma_T = \sum_{n_f s_f s_i} \sum_{k_{zf}} \frac{P_{if}}{\tilde{J}_z} \tag{7}$$

where

$$P_{if} = 2\pi |\langle \psi_f | T | \psi_i \rangle|^2 \ \delta(E_f - E_i) \tag{8}$$

with $|\langle \psi_f | T | \psi_i \rangle|$ indicating the T-matrix between the initial and final Landau levels. The current density of the incident particle summed over the degenerate initial quantum number s_i is given by

$$\tilde{J}_z = \frac{\gamma}{\pi} k_{zi} . \tag{9}$$

Transforming in Eq. (7) the sum over k_{zf} into an integral as

$$\sum_{k_{zf}} \longrightarrow \int \eta(E_f) \ dE_f$$

we obtain the one-dimensional density of final states

$$\eta(E_f) = \frac{1}{2\pi} \frac{1}{|k_{zf}|} \tag{10}$$

Using Eq. (9) and Eq. (10), the total cross section results

$$\sigma_T(\pm) = \frac{\pi}{\gamma} \sum_{n_f s_f s_i} \frac{1}{k_{zi}|k_{zf}|} |<\psi_f|T|\psi_i>|_\pm^2 \tag{11}$$

In the above expressions

$$k_{zf} = \mp \left[k_{zi}^2 + 4\gamma \ (n_i - n_f) \right]^{1/2} \tag{12}$$

the positive values corresponding to forward (+) and the   negative ones to backward (−) scattering. The total cross section (11)   shows two important features: i) the dependence over $k_{zf}^{-1}(n_f)$, introduced through the one-dimensional final state density $\eta(E_f)$; ii) the dependence over γ^{-1}, due to the bound electron motion in the plane perpendicular to the field. The problem of the divergence of the total cross section for a vanishing magnetic field ($\gamma \to 0$) has already been studied in a previous paper (Nuzzo and Zarcone 1989a). In the next sections we show how it is possible to remove the unphysical infinity of the total cross section arising when the initial kinetic energy (along z) matches exactly the energy difference between the initial and final Landau levels.

3. THE T-MATRIX

The T-matrix series for potential scattering in the presence of a strong magnetic field may be written as

$$<f|T|i>_\pm = <f|V|i>_\pm + \sum_a \frac{<f|V|a>_\pm <a|V|i>}{(\epsilon_i - \epsilon_a + 2\gamma n_{ia} + i\eta)} +$$

$$+ \sum_{a,b} \frac{<f|V|a>_\pm <a|V|b> <b|V|i>}{(\epsilon_i - \epsilon_a + 2\gamma n_{ia} + i\eta)(\epsilon_i - \epsilon_b + 2\gamma n_{ib} + i\eta)} + \dots . \tag{13}$$

in the above equation the state vector $\psi = |t> = |n_t s_t k_t> = |n_t s_t>|k_t>$, is given by Eq.(2)−(4); $<f|V|a> = <n_f s_f \pm k_f |V| n_a s_a k_a>$ and the sum over the intermediate states a,b, etc. must be understood as a summation over the discrete values of (n_a, s_a), (n_b, s_b) etc. and an integration in $dk_a \ dk_b$ etc. as

$$\sum_{k_a} \longrightarrow \frac{1}{2\pi} \int dk_a \ , \tag{14}$$

finally $n_{ba} = n_b - n_a$ and $\epsilon_a = k_a^2/2$ is the energy of the scattering particle along B. The matrix elements for a screened Coulomb potential

$$V(r) = V_0 \frac{e^{-\lambda/r}}{r} \tag{15}$$

are given by (Ventura 1973)

$$<f|v|i>_{\pm} = V_0 \left[\frac{s_i! \, s_f!}{n_i! \, n_f!} \right]^{1/2} U(s_i + 1, \, s_i - s_f + 1, \, \xi) \, Q_{n_i}^{n_f - n_i}(-\xi) \tag{16}$$

for n≤s and

$$<f|v|i>_{\pm} = V_0 \left[\frac{n_i! \, n_f!}{s_i! \, s_f!} \right]^{1/2} U(n_i + 1, \, n_i - n_f + 1, \, \xi) \, Q_{s_i}^{s_f - s_i}(-\xi) \tag{17}$$

for n>s

where $U(a,b,z)$ are second kind hypergeometric functions and $\xi_{\pm} = [(k_z \pm k_z)^2 + \lambda^2]/4\gamma$. The T-matrix, Eq.(13) has the structure of the scattering series which is obtained when collisions of particles with internal discrete structure are considered. In our case the free particle entering a region where a magnetic field is present undergoes a change in its motion; in particular, now the particle is spiraling in the xy plane. As a result of this bound motion a discrete internal structure, given by the Landau levels, is added to the free motion along B and transitions can occur also between two of these discrete levels.

4. HIGHER ORDER MODIFIED BORN APPROXIMATION

From an inspection to the N-th term of the Born series for the T-matrix derived in the previous section we see that near a resonance, i.e. for an incident energy of the scattering electron along the direction of the magnetic field equal to the difference between the final and initial Landau levels, the leading terms are those relative to scattering events consisting of N-1 elastic transitions between the same Landau level and only one inelastic transition from the initial level n_i to the final level n_f. Moreover, for a spherical potential the conservation of the angular momentum imposes for any transition: $n_f - s_f = n_a - s_a = \ldots\ldots = n_i - s_i$. The above considerations permit to reduce the number of terms in the series and to perform the integrations over dk_a, dk_b, etc., obtaining for the N-th term of the T-matrix the expression (Ohsaki 1983)

$$<f|T|i>_{\pm}^N = \sum_{M=0}^{N-1} \left[\frac{i <f|V|f>}{k_f} \right]^M <f|V|i>_{\pm} \left[\frac{i <i|V|i>}{k_i} \right]^{N-M-1}. \tag{18}$$

Moreover, assuming that $<a|V|a>/k < 1$, (Nuzzo and Zarcone 1989b) the series can be summed obtaining a closed form for the HOMBA cross section given by

$$\sigma_{n_i n_f} = \frac{\pi}{\gamma} \sum_{s_i} \frac{k_i \, k_f \, |<f|V|i>|^2}{[k_f^2 + |<f|V|f>|^2] \, [k_i^2 + |<i|V|i>|^2]} \tag{19}$$

if $n_f \neq n_i$, and

$$\sigma_{n_i n_i} = \frac{\pi}{\gamma} \sum_{s_i} \frac{1}{k_i} \frac{|<i|V|i>|_{\pm}^2}{k_i^2 + |<i|V|i>|^2} \tag{20}$$

if $n_f = n_i$.

In the above equations $<f|V|i>_{\pm} = <n_f s_f \pm k_f |V| n_i s_i k_i>$, $<f|V|f> = <n_f s_f |k_f||V| n_f s_f |k_f|>$, $<i|V|i> = <n_i s_i k_i |V| n_i s_i k_i>$ and $s_i = s_f + n_f - n_i$. Since the HOMBÁ T-matrix is proportional to k_{zf}, in the vicinity of the Landau thresholds when k_{zf} goes to zero, the total cross sections do not present any infinity.

5. THE BREMSSTRAHLUNG PROCESS

In this section we give a concise treatment of the bremsstrahlung problem as a typical radiative process modified by the presence of a strong magnetic field. An electron collides with an ion in the presence of a strong magnetic field and makes a radiative transition from the Landau initial state n_i to the final Landau state n_f with the emission or the absorption (inverse bremsstrahlung) of ν photons of frequency ω

$$e^-_{n_i} + (Z,A) \;\rightarrow\; e^-_{n_f} + (Z,A) + \nu \; (\hbar\omega) \; . \tag{21}$$

The total cross section for the above process, in the nonrelativistic case, is evaluated using as unperturbed electron wavefunctions the exact solutions of the Schroedinger equation for a particle embedded in a static magnetic field and a radiation field of arbitrary polarization (Seely 1974). The emitted radiation is taken in the dipole approximation. In view of the applications to the emission mechanisms of pulsars our emphasis is on electrons populating only the low lying Landau levels $n_i = n_f = 0$. For a magnetic field so strong that $\omega_c \gg \omega$ and $\hbar\omega_c \gg kT$ where T is the electron temperature, for photons of energy $\hbar\omega = 1$ KeV at $T = 10^7$ $^\circ$K, these conditions imply $B \gg 10^{11}$ G. Following the same procedures outlined in section 2 the total bremsstrahlung cross section with emission or absorption of ν photons results (Ferrante et al 1982)

$$\sigma_T(\pm) = \frac{2\pi \, m \, Z^2 \, e^4}{\omega_c \, \hbar^3 \, k_{zi} \, k_{zf}} \int_0^\infty d\rho \, \frac{e^{-\rho}}{(\rho + \xi_\pm)^2} \, J_\nu^2 \, [\lambda(\rho)] \tag{22}$$

where J_ν are Bessel functions of order ν. Their argument $\lambda(\rho)$ assumes different values according to the different polarization of the emitted or absorbed radiation:

a) for a linear polarization along B we have $\lambda(\rho)=\lambda_z$

$$\lambda_z = \frac{e\ E}{m\ \hbar\ \omega^2}\ \left[2\ m\ \hbar\ \omega_c\ \xi\right]^{1/2} \tag{23}$$

b) for right hand circular polarization in the xy plane (extraordinary wave) we have

$$\lambda(\rho) = t_x\ \rho^{1/2} \tag{24}$$

with

$$t_x = \frac{e\ E\ 2^{1/2}}{m\ \omega\ (\omega - \omega_c)\ \rho_o} \tag{25}$$

c) for left hand circular polarization in the xy plane (ordinary wave) we have

$$\lambda(\rho) = t_o\ \rho^{1/2} \tag{26}$$

with

$$t_o = \frac{e\ E\ 2^{1/2}}{m\ \omega\ (\omega + \omega_c)\ \rho_o}. \tag{27}$$

If we consider only the emission of one weak photon ($\nu=1$ and $\lambda(\rho)\ll1$), using $J_1(\lambda)\approx\lambda/2$ in eq. (22), we can easily obtain the total cross section for spontaneous bremsstrahlung:

For case a) we have

$$\sigma_z(\pm) = \frac{\pi\ Z^2\ e^6\ E^2}{\hbar^4\ k_{zi}\ k_{zf}\ \omega^4}\ \xi_\pm\ \Sigma_o(\xi_\pm) \tag{28}$$

with

$$\Sigma_o(\xi_\pm) = \int_0^\infty d\rho\ \frac{e^{-\rho}}{(\rho + \xi_\pm)^2} \tag{29}$$

for case b) and c) since $\omega_c\gg\omega$ we have

$$\sigma_\perp(\pm) = \frac{\pi\ Z^2\ e^6\ E^2}{\hbar^4\ k_{zi}\ k_{zf}\ \omega^2\ \omega_c^2}\ \Sigma_1(\xi_\pm) \tag{30}$$

with

$$\Sigma_1(\xi_\pm) = \int_0^\infty d\rho \; \frac{\rho \; e^{-\rho}}{(\rho + \xi_\pm)^2} \tag{31}$$

where (+) is for forward and (−) for backward scattering. Finally, using the above cross sections we calculate the absorption coefficient for photons with linear (case a) and circular polarization (case b and c)

$$A_z = \frac{8 \; \pi^2 \; Z^2 \; e^6 \; n_e \; n_i}{m \; c \; \hbar \; \omega^3 \; (p_{zi}^2 + 2\nu m\hbar\omega)^{1/2}} \left[\xi_+ \; \Sigma_o(\xi_+) + \xi_- \; \Sigma_o(\xi_-) \right] \tag{32}$$

and

$$A_\perp = \frac{8 \; \pi^2 \; Z^2 \; e^6 \; n_e \; n_i}{m \; c \; \hbar \; \omega \; \omega_c^2 \; (p_{zi}^2 + 2\nu m\hbar\omega)^{1/2}} \left[\Sigma_1(\xi_+) + \Sigma_1(\xi_-) \right]. \tag{33}$$

Assuming that $\hbar\omega \gg p^2/2m$ it results $\xi \ll 1$ and we can approximate $\Sigma_o \approx 1/\xi_\pm$ and $\Sigma_1 \approx \log(\xi^{-1}) - \gamma - 1$, where $\gamma = 0.577$ is the Euler's constant. Considering only the absorption of $\nu = \pm 1$ photons, if the initial and final particle's energies along B are much greater than the photon energy we get a simple form for the absorption coefficient for linear polarization along B

$$A_z = - \frac{32 \; \pi^2 \; Z^2 \; e^6 \; n_e \; n_i}{c \; \omega^2 \; p_{zi}^3} \tag{34}$$

The negative sign indicates that the emission process is dominant. For circular polarization we get instead

$$A_\perp = A_z \left[\frac{\omega}{\omega_c} \right]^2 \left[\log \frac{\omega_c}{\omega} - 1 - \gamma \right] \tag{35}$$

where we have assumed $\xi = \xi_+ \approx \omega/\omega_c$. Since $\omega_c \gg \omega$ also in this case the emission of one photon is the dominant process. Moreover if the electrons have a definite temperature, averanging the absorption coefficient with a Maxwellian distribution function in p_z

$$f(p_z) = \frac{1}{(2 \; \pi \; m \; k \; T_e)^{1/2}} \exp \left[- \frac{p_z^2}{2 \; m \; k \; T_e} \right] \tag{36}$$

we obtain the averaged coefficients for the absorption of $\nu = \pm 1$ photons

$$\tilde{A}_z = \frac{16 \, \pi^2 \, z^2 \, e^6 \, n_e \, n_i}{m \, c \, \hbar \, \omega^3 \, (2 \, \pi \, m \, kT_e)^{1/2}} \left[1 - e^{-(\hbar\omega/kT_e)} \right] e^{(\hbar\omega/2kT_e)} K_0 \left[\frac{\hbar \, \omega}{2kT_e} \right] \quad (37)$$

where K_0 is the modified Bessel function of zero order, and

$$\tilde{A}_\perp = \tilde{A}_z \left[\frac{\omega}{\omega_c} \right]^2 \left[\log \frac{\omega_c}{\omega} - 1 - \gamma \right]. \quad (38)$$

In the limit $\hbar\omega \ll kT_e$ the longitudinal absorption coefficient eq.(37) may be approximated as

$$\tilde{A}_z = \frac{2^{7/2} \, \pi^{3/2} \, z^2 \, e^6 \, n_e \, n_i}{c \, \omega^2 \, (m \, k \, T_e)^{3/2}} \, \ln \left[\frac{4 \, k \, T_e}{\hbar \, \omega} \right] \quad (39)$$

This equation can be compared with the expression for the absorption coefficient in the absence of a magnetic field given by Beilin et al 1990

$$\tilde{A}_0 = \frac{2^{9/2} \, \pi^{3/2} \, z^2 \, e^6 \, n_e \, n_i}{3 \, c \, \omega^2 \, (m \, k \, T_e)^{3/2}} \, \ln \left[\frac{2^{3/2} \, kT_e}{\hbar \, \omega} \right] \quad (40)$$

obtaining, in the limit $kT_e \gg \hbar\omega$,

$$\tilde{A}_z = \frac{3}{2} \, \tilde{A}_0 \quad (41)$$

and

$$\tilde{A}_\perp = \frac{3}{2} \left[\frac{\omega}{\omega_c} \right]^2 \left[\log \frac{\omega_c}{\omega} - 1 - \gamma \right] \tilde{A}_0 \quad (42)$$

From eqs. (41) and (42) we note that the presence of a very strong magnetic field affects only sligthly the absorption coefficient for a linearly polarized radiation along the field, which is instead strongly reduced for a circular radiation polarization in the plane perpendicular to it. Moreover, due to the fact that the absorption coefficient has been averaged over a one-dimensional distribution function the dominant one photon process becomes now absorption instead of emission.

6. CONCLUDING REMARKS

We have derived the T-matrix and the total cross sections for potential scattering in the presence of a magnetic field. The resulting analytic structure of the Born series resembles closely that of scattering between systems having discrete structure. In our case, the discrete structure is bound to the confined spiraling particle motion due to the presence of the field. Near the Landau resonances the total cross section in the Born approximation undergoes a resonant growth. As shown in section 4, an effective way to remove the infinities is to sum all the Born series terms in the T-matrix, at least for small k_f. At the Landau thresholds, for vanishing k_f, first order treatments like the first Born approximation are no longer valid and therefore the information obtained is not reliable. In fact for the first Born approximation to be valid, both k_i and k_f are required to be large. The HOMBA permits to sum all the orders of the scattering series and gives a closed form for the total cross section not divergent at the Landau thresholds. To evaluate the N-th term of the Born series we have considered only the contributions given by N-1 elastic transitions between the same Landau level and only one inelastic transition from n_i to n_f and we have neglected other contributions. This approximation is good at the Landau thresholds where the coupling of the energy of the incident particle to the two Landau levels n_i and n_f becomes strong. The problem of potential scattering in the presence of a magnetic field is one of the few cases where, with some approximations, it is possible to sum the scattering series at all orders. The close expressions of the total cross section Eq. (24) and (25) are useful for calculating of a number of measurable quantities in astrophysics, plasma physics and solid state physics. The bremsstrahlung process has been analized in the last section as a typical radiative process strongly affected by the presence of a magnetic field. We have shown that the absorption coefficient is greatly reduced over the field free value if the polarization of the radiation is perpendicular to B. This occurs because the presence of a strong magnetic field tends to confine the motion of the electron and to suppress their response to any external perpendicular force.

REFERENCES

Beilin E L, Fedorov M V and Zon B A 1990 J.Phys.B: At. Mol. Phys. 23 4181

Canuto V and Ventura J 1977 Fundamentals of Cosmic Physics 2 203

Ferrante G, Nuzzo S, Zarcone M and Bivona S 1980 J. Phys. B: At. Mol. Phys. 13 731

McDowell M R C and Zarcone M 1985 Adv. At. Mol. Phys. 21 255

Nuzzo S and Zarcone M 1989a Il Nuovo Cimento D 11 419

Nuzzo S and Zarcone M 1989b J. Phys. B: At. Mol. Opt. Phys. 22 L627

Nuzzo S and Zarcone M 1991 Il Nuovo Cimento D (in press)

Seely J F 1974 Phys. Rev. A10 1863

Ventura J 1973 Phys. Rev. A9 3021

Ohsaki A 1983 J. Phys. Soc. Japan 52 431

Recoil ion momentum spectroscopy in fast ion atom collisions

R.Dörner,[1] J.Ullrich[2], O.Jagutzki[1], S. Lencinas,[1] A. Gensmantel[1] and
H.Schmidt-Böcking[1]

[1] Institut für Kernphysik der Universität Frankfurt
   Frankfurt, 6000 Frankfurt FRG
[2] GSI Darmstadt, 6100 Darmstadt FRG

Abstract:

*The kinematics of the recoil ion in fast ionising ion atom collisions is discussed.
It is shown that the recoil ion longitudinal momentum yields direct information on
the Q-value and electron mass transfer in the collision, whereas its transverse momentum is a good measure for the impact parameter. With a new very cold target
system momentum resolution in the sub a.u. level can be obtained yielding excellent Q-value resolution (~ 100 eV) even in GeV collisions. For p on He highly
differential cross sections of the ionisation process are presented.*

I. Intoduction

Fast collisions between ions and atoms have been studied in numerous experimental
and theoretical investigations over the last few decades. Using electron momentum and
photon spectroscopy in particular or measuring the energy loss or gain of the outgoing projectile, detailed information has been obtained on the different reaction
channels like ionisation, excitation, electron transfer or superpositions of these processes. It is well established that highly differential cross sections yield detailed information on the reaction mechanism and in general are more valuable for the understanding of the reaction mechanism of the processes involved than total cross sections.
Thus multiparameter coincidence experiments, where several collision parameters have
been measured simultaneously, have yielded very detailed information on such processes. So far one important parameter for ion atom collisions, the recoil ion momentum,
was measured only in few experiments to study ionisation or electron capture reaction processes (Ullrich et al. 1989; Dörner et al. 1989; Cocke et al. 1991a,b).
The reason is mainly an experimentally one, since the interesting recoil ion momenta
are rather small, comparable with the thermal motion at room temperature. As discussed by Schmidt-Böcking et al. (1991) and as it will be shown below, the momentum
of the recoiling ion contains important information on the ion-atom reaction like the
inelastic Q-value and the collision trajectory. Furthermore if recoil ion and projectile
final momenta are measured in coincidence, the sum momentum of the ejected electrons in multiple ionising collisions can be determined too.

This electron sum momentum contains the information on possible collective behavior of the emitted electrons, on multielectron correlations in the initial and final channel and on dynamical e-e-correlations during the collision (Gonzales et al. 90, Forberich et al. 91).

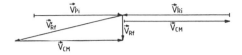

Fig 1. Velocity vectors for a pure elastic collision (no electron transfer)

Since the fast moving projectile is providing the force to induce electronic reactions, the detailed knowledge of this force is very important for a comparison between experimental and theoretical data. In order to know this force the trajectory of the ion nuclear motion has to be measured. Up to now this has been done by determining the projectile scattering angle. For fast ion atom collisions however these angles of interest are in the sub milli rad regime typically. Such angles are very difficult to measure.

Furthermore for distant collisions, where ionisation or excitation mainly takes place, the projectile transverse momentum, i.e. its deflection, is comparable to the sum momentum of the ejected electrons. Thus the projectile deflection is strongly influenced by the scattering on theses electrons (so called binary collisions) and no unique correspondence between projectile scattering angle and projectile trajectory i.e. the nuclear impact parameter (b) exists (see Dörner et al. 1989, Salin 1989, Kristensen et al. 1990, Kamber et al. 88). The recoiling ion however cannot be scattered on its "own" electrons. Its final transverse momentum reflects indeed the nuclear-nuclear momentum exchange, i.e. the classical nuclear trajectory over a much wider impact parameter region.

According to nCTMC calculations (Olson 1991) for p on He at such large b the momentum of the recoiling ion is less influenced by the ejected electron sum momentum. In this paper (section II) the kinematics of the recoiling ion will be discussed. It will be shown that the longitudinal momentum of the recoil ions is a sensitive measure of the Q-value and of the electron mass transfer between the collision partners. The recoil ion transverse momentum, which is independent of the Q-value, yields direct information on the impact parameter of the collision even in GeV collisions. The resolving power of the recoil ion momentum spectroscopy is superior enough to determine reaction Q-values in the sub 100 eV regime.

In section III the experimental technique for measuring recoil momenta below 1.a.u. is discussed. In section IV experimental and theoretical data for the p on He system are presented. Section V gives an outlook on near future experiments using the recoil ion momentum spectroscopy with very cold targets.

II. Recoil Ion Kinematics in fast Ion Atom Collisions

All projectile scattering angles of interest, i.e. the ratio of the maximal final transverse momenta of the projectile to its initial momentum, are very small compared to one. E.g. in 1 MeV p on He singly ionising collisions the mean impact parameter is in the order of 0.5 a.u. thus yielding scattering angles of below 10^{-4} rad. In 10 MeV/u U^{q+} on He electron capture collisions the typical laboratory system scattering angles are even in the 10^{-6} regime.

For small deflection angles and a complete elastic collision the kinematics of the recoil ion is displayed in figure 1. The center of mass velocity vectors are indicated by the prime sign. In a fully elastic collision the length of the initial and final recoil velocity vectors in the CM-system are identical. The slightly different direction between both vectors represent the CM-deflection angle $\vartheta_r{}'$. Since the initial vector $\vec{v}_{ri}$ is just $(-\vec{v}_{cm})$ the final recoil momentum velocity in the Lab system $\vec{v}_{rf} = (\vec{v}^{\perp}_{rf}, v^{\parallel}_{rf})$ is the vector sum $\vec{v}_{rf} - \vec{v}_{ri}$ where $\vec{v}^{\perp}_{rf}$ are the transverse and $v^{\parallel}_{rf}$ the longitudinal components respectively. For a two-body collision the recoil ion transverse momentum is identical with the projectile transverse momentum $P^{\perp}_{pf} = \vartheta_p \cdot P^o_p$, where P^o_p is the initial projectile momentum. Thus the transverse recoil momentum represents exactly the projectile scattering angle. It is to notice that $P^{\perp}_{rf}$ is not affected by the Q-value or possible mass transfer. For a pure elastic collision the longitudinal component of the recoil ion $\vec{P}^{\parallel}_{rf} = M_r \cdot \vec{v}^{\parallel}_{rf}$ is equal to $\vartheta_p P^{\perp}_{rf}$ and thus very small compared to P_{rf}. This means that all recoil ions in an elastic two body collision are emitted at nearly $\vartheta_r = 90^o$, where ϑ_r is the recoil emission angle. If the collision is inelastic in the CM–system i.e. a Q-value (Q < 0 energy gain; Q > 0 energy loss. See figure 2) is involved, and if furthermore n_e electrons are transfered from the target to the fast moving projectile and $q_r - n_e$ electrons are finally emitted to the continuum (assuming the final sum momentum of the $q_r - n_e$ electrons in the Lab system is zero), the longitudinal component of the recoil ion is then :

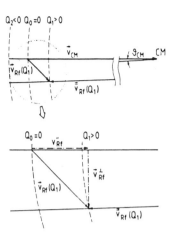

Fig 2 *Velocity vectors in the Lab system for different Q-values (no electron transfer)*

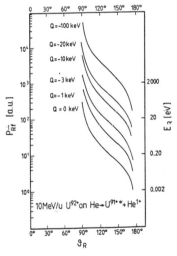

Fig 3 *Relation between recoil ion emission angle and transverse momentum for different Q-values in 10 MeV/u $U^q + He \rightarrow U^{q-1} + He$ collisions*

$$P^{\parallel}_{rf} \approx \frac{Q}{v_p} - \frac{n_e + q}{4} m_e v_p \qquad (1)$$

($P_{rf} \lesssim 0$ corresponds to $\vartheta \gtrsim 90$ respectively). For more details see reference Schmidt-Böcking et al. 1991. Equation (1) means that in a purely elastic collision but with n_e electrons transfered from the target to the projectile the recoil ion is emitted into backward direction. This is easy to understand, since even for Q=0 a small mass transfer from the target to the projectile enlarges $\vec{v}_{rf}$ and reduces $\vec{v}_{rf}$ compared to the initial values because of momentum conservation in the CM system. Thus the backward directed increased velocity vector $\vec{v}_{rf}'$ results in a backward directed recoil emission in the Lab system. The influence of non-zero Q-value and a finite mass transfer between ion and atom on the recoil emission angle will be illustrated below for the 10 MeV/u $U^{92+} + He \rightarrow U^{91+} + He^{1+}$ transfer reaction. For $n_e=1$ the mass transfer contribution to the longitudinal recoil momentum is about 10 a.u., and it increases linearly with the projectile velocity. A Q-value of Q=-1 keV yields an additional contribution of 25 a.u. and increases

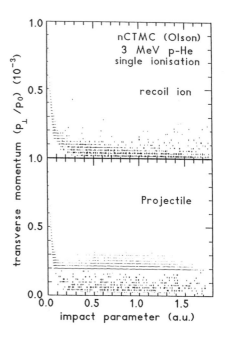

Fig. 4 nCTMC- calculations for the correspondence between the impact parameter (b) and transverse momentum of the projectile (fig b) and recoil ion (fig a) in 3 MeV p on He collisions. In fig. 4b for $P_{pf}^{\perp}/P_0 < 0.2 \; 10^{-3}$ less events were calculated within the nCTMC-approach.

linearly with Q too. Thus a Q-value of -1 keV would shift $P_{rf}^{\parallel}$ by about 25 a.u. into backward direction yielding for a total $P_{rf}^{\parallel}$ of -35 a.u. n_e =1. If the transverse momentum is 35 a.u. , a recoil ion is emitted under $\vartheta_r=135°$, whereas for Q=0 the recoil ion is detected at $\vartheta_r=110°$. In figure 3 the recoil ion emission angles as a function of the recoil ion transverse momenta for different Q-values are given. Furthermore on the right ordinate the recoil energies are shown. The final Q-value resolution depends strongly on the recoil transverse momentum transferred to the recoil ion at this particular impact parameter. More details on the estimated Q-value resolution for 10 MeV/u U^{92+} on He collision will be presented in section IV.

The recoil ion transverse momentum with zero electron momentum in the continuum is completly independent of the Q-value. It dependends only on the nuclear-nuclear-trajectory, in the screened two center potential.

So far we have neglected that the q_e-n_e electrons emitted to the continuum always carry momentum. Since in fast ionisation collisions the transfered nuclear momenta are typically in the order of 1 to 1000 a.u. the influence of the final continuum electron momenta on the final nuclear momenta must be taken into account too.

With P_{ef}^{λ} being the final momentum of each of the q_r-n_e emitted electrons with:

$$\vec{P}^s_{ef} = \sum_\lambda^{q-n_e} \vec{P}^\lambda_{ef}$$

where $\vec{P}^s_{ef}$ is the sum momentum of all ejected electrons, we obtain from momentum conservation (the recoil is completly at rest before the collision)

$$\vec{P}_p = \vec{P}_{pf} + \vec{P}_{rf} + \vec{P}^s_{ef} \quad (2)$$

Since at present no complete multi-parameter coincidence experiment can be performed, where all emitted electrons are detected in coincidence with at least one of the nuclear partners, it is not possible to measure all vector components of equation (2) simultaneously.

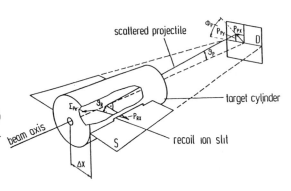

Fig. 5 Principle of recoil ion momentum spectroscopy

Depending on the collision system however the recoil momentum should be less affected by the final electron momenta than the projectile momentum $\vec{P}^\perp_{rf}$. In binary collisions the projectile can transfer much larger momenta to the emitted electrons. The critical value of recoil momentum P_c, when the electron momentum influence becomes important, depends strongly on the target mass. For He as a target P_c ist about 2 a.u., whereas for Ne P_c might be close to 10 a.u. . Thus the recoil transverse momentum is over a very large b regime still a good measure for the nuclear- nuclear interaction, whereas the projectile final momentum is over a wider b regime influenced by the projectile-electron interactions. In figure 4 for 3 MeV p on He nCTMC calculations for the b-$P^\perp_f$ relation including electron momenta are shown (figure 4b for the projectile, figure 4a for the recoil ions). It is obvious that for the projectiles only for $\vartheta_p > m_e/M_p$ a unique relation between b and ϑ_p exists, whereas for the recoil ions the relationship is preserved even to much larger b values. Only for very small recoil transverse momenta the influence of $\sum \vec{P}^\lambda_{ef}$ on the recoil momenta becomes important. (It is to notice that in figure 4a the statistics below $P^\perp_{pf}/P^p_0 = 2 \cdot 10^{-4}$ has to be multiplied by factor $5 \cdot 10^{-3}$ yielding then a smooth density distribution).

By measuring with high precision both nuclear transverse momenta in coincidence the influence of the electron momemta on $|P_{rf}|$ gives us vice versa the unique chance to obtain direct experimental information on the electron sum momentum.

To use recoil momentum spectroscopy as a powerful tool for measuring highly differential cross sections in

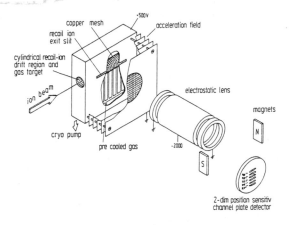

Fig. 6 Present cooled recoil ion spectrometer system

fast ion atom collisions, we need an experimental set up, which allows recoil ion momentum spectroscopy (RIMS) on the 1.a.u. level or below. He gas atoms at room temperature have a mean momentum of 3 a.u. , thus it is obvious that only cooled target systems can provide sufficient resolution.

III Principle of RIMS and its Experimental Realisation

As mentioned above recoil momenta on the 1 a.u. level are of interest and have to be measured. For He as the typical multi-electron target this corresponds to about 3 meV recoil energy in comparison to 25 meV of thermal motion. Thus the He target atoms have to be cooled well below 50 K. Furthermore in order to measure these energies the gas target region must be completely free of any electrostatic field (e.g. contact potential

between the material in the target region). Also the product of target density times beam current must be sufficiently small to avoid any mikro-plasma along the beam path . If this is ensured for a sufficiently thin gas target, the recoil momentum is measured by time-of-flight techniques. In figure 5 the principle of RIMS is shown for an extended target region. The well collimated projectile beam penetrates the target cell (approximately 4cm long). The produced recoil ions are moving in the field free target cylinder with the constant recoiling velocity towards the wall of the target cylinder. A small fraction of the recoil ions can leave the cell in the horizontal plane through a narrow horizontal slit covered with a high transmission grid. The ions are then post-accelerated, charge state analyzed and detected by a position sensitive channelplate detector.

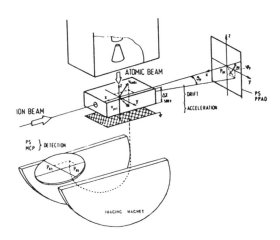

Fig. 7 Future gas jet recoil ion spectrometer system

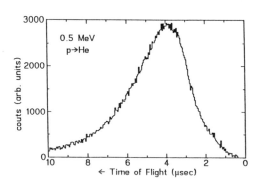

Fig. 8 Typical recoil ion time of flight spectrum

The projectiles are detected in a two-dimensional position sensitive particle detector mounted several meters away from the target. Thus scattering angles below 10^{-4} rad can be resolved. The time difference of both the projectile and recoil ion detection signal yields the recoil ion flight time in the target cylinder. Since all recoil ions beween $\vartheta_r = 20°$ and $160°$ are detected with equal efficiency the measured time difference represents the transverse recoil velocity component only. This extended gas cell does not allow ϑ_r-angular resolved measurements. Such experiments can be done however with the new generation target set up described below. In figure 6 the existing and working RIMS system is shown. Details of this spectrometer system are given in reference Dörner (1991) , Forberich et al (1991) .

The whole target system is mounted on the cold finger of a cryopump thus a target gas temperature of 25 to 30 Kelvin can be achieved.

With this system recoil velocities above $3 \cdot 10^4$ cm/sec can be measured. Since the time resolution is in the 20 nsec range, the serious limitations of the recoil momentum resolution result from non zero target temperature and possible contact potentials at the recoil ion exit slits. However, varying the drift length in the target cell by moving the primary beam position it can be verified that the experimental results are not influenced by non linear flight time effects. Varying several times this target geometry the results could be reproduced within the estimated experimental uncertainty of ± 5 meV for the He target. This gives confidence that the results presented below are accurate within the estimated error bars.

In figure 7 the new generation RIMS system is displayed. An ultrasonic cold jet with an internal temperature of about 1 K in all three dimensions is crossed by the well collimated projectile beam. Since the atoms are moving with about 2000 m/sec all recoil ions are projected on a two dimensional position sensitive recoil ion detector. From the recoil detection position and its flight time the complete recoil ion momentum vector can be determined with high precision.

According to test measurements of the jet thermal distribution the recoil momentum resolution in transverse direction can be below 1 meV, even down to 100 μeV. Due to the extension of the atomic jet

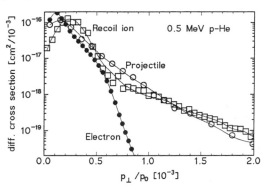

Fig.9 *Doubly differential He^{1+} ionisation cross-sections as function of the different transverse momenta of the projectile, the electrons and the recoil ions for 0.5 MeV p on He*

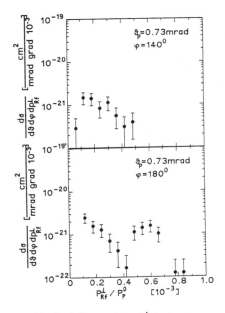

Fig. 10 *Triple differential He^{1+} ionization cross sections as function the recoil ion transverse momentum for fixed projectile transverse momentum $P^{\perp}_{pf} = 0.73 \cdot P^0_p$ (azimuthal 180° and 140°) for 3 MeV p on He (Gensmantel et al. 1991)*

in longitudinal direction only about 1 meV can be obtained. The important advantage of this new system compared to the used standard system is its 4π detection efficiency. This high efficiency allows even low rate coincidence studies, where the complete recoil momentum vector is measured.

IV Data and Discussion

This paper will only present multi differential ionisation cross section for p on He. These data are part of the thesis of Dörner (1991). In figure 8 a typical He-recoil ion time-of-flight spectrum is shown. This spectrum is obtained in coincidence with projectiles scattered in the whole ϑ,φ-region, where ϑ and φ are the polar and the azimuthal projectile scattering angles, respectively (see fig.5).

In figure 9 double differential cross section for single ionisation for 0.5 MeV p on He as a function of the different transverse momenta of the projectile, the electron and the recoil ion are displayed. The He$^+$ cross sections as a function of the electron transverse momenta are derived from the data of Rudd et al. 1976. The data clearly show that there is a classical cut-off for the electron transverse momentum at $P_e^\perp = m_e \cdot v_p \approx 4.6$ a.u., which corresponds to 0.54 mrad projectile scattering angle independent of the proton impact energie. The new data on He$^+$ ionisation obtained by RIMS show that there is no shoulder or cut-off near 4.6 a.u in dependence of the recoil transverse momentum. In such binary collisions of projectiles with electrons the recoil ions are indeed spectators and their transverse momentum is comparable to that expected from the pure ion-ion inter-action. The He$^+$-data as a function of $P_{rf}^\perp$ are smooth and reflect indeed more or less the impact paramter dependence of the ionisation process. Only at very small transverse momenta the recoil curve exceeds the projectile curve. This intersection must occur since the integral over both differential cross sections namely the He$^+$ total ionisation cross section is identical.

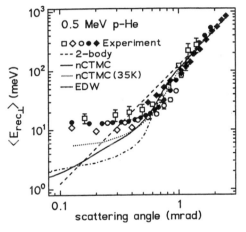

Fig 11 Mean recoil energies as function of the projectile scattering angles in 3 MeV p-He collisions. The dashed line represents the pure two body interaction (0 K target temperature). the solid lines represents nCTMC calculations for a target temperatur of 30 K. The dotted line shows quantummechanical calculations of Fukuda et al. 1991.

In figure 10 the triple differential He$^+$ cross sections for $\vartheta_p = P_{pf}^\perp/P_p^O = 0.73$ and $\varphi = 180^O$ and 140° are shown as a function of the recoil ion transverse momentum. The large contribution at low recoil momenta represents collisions, where the recoil ion is spectator and the projectile is deflected off the electron. The peak around $6\cdot10^{-4}$rad represents the "close" nucleus-nucleus collisions, where the electron transverse momentum is small. When the recoil momenta distribution is plotted for $\varphi=140^O$, i.e. out of the two body nuclear scattering plane only the low recoil momenta remain, indicating that again the projectile is scattered at the electron to an angle of $\varphi = 140^0$.

In figure 11 the mean transverse recoil energies $\langle E_r \rangle = 1/N \sum_i^N P_{rf}^2/2M_r$ are plotted as function of the transverse projectile momentum. The dashed line represents the relation of a pure two body nuclear collision. The deviations from this line account for the influence of the electron momenta on the heavy particle momenta.

At $P_p^\perp / P_p^o < 0.54 \cdot 10^{-3}$ the projectile is scattered predominantly at the electron, thus the recoil ion is partially spectator yielding lower transverse momenta. At very low $P_p^\perp < 0.3 \cdot 10^{-4} P_p^o$ the mean recoil ion energy saturates at about 12 meV. Even taking into account the target temperature of 30 K this value does not agree with the theoretical predictions (Dörner et al. 89). The nCTMC and also the quantummechanical calculation by Fukuda et al. 91 predict mean recoil ion energies which are a factor of 3 below the experimental values. The experimental value corrected for the target thermal motion would nicely correspond to the total electron Compton profile, assuming that the He^+ momentum distribution is always inverse to that of the ionized electron.

In present theoretical approaches, the projectile only interacts with electrons at the lower part of the momentum tail of the Compton profile at these large b, when the projectile is scattered with very small $P_p^\perp$. This yields lower $\langle E_r \rangle$ value for $P_p^\perp \to 0$.

Since the theory does not include any e-e-correlation this discrepancy is probably a very interesting one and may be taken as an indication that our understanding of the ionisation process at large b collisions is still far from completeness. On the other side the discrepancies are close to the resolution limit and more precise data are needed to etablish possible e-e correlations.

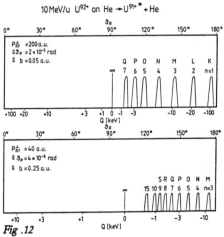

Fig .12

Recoil ion angular distribution for 10 MeV/u
$U^{q+} + He \to U^{q-1} + He^{1+}$ *collisions for two different fixed transverse recoil ion momenta*

V Outlook:

The recoil momentum spectroscopy in fast ion atom collision using the new target jet system (figure 7) yields a very efficient method to study highly differential ionisation and charge transfer cross sections even in GeV collisions with nearly 4π solid angle. The experimental detection limit of the recoil momentum (e.g. for He as target) will be in the order 0.1 a.u. .

For 10 MeV/u U on He collisions such a small recoil transverse momentum corresponds to a projectil scattering angle in the order of 10^{-8} rad. Thus RIMS used as inverse kinematical technique allows to measure differential cross sections over the whole interesting b regime from b inside the Uranium-K-shell radius to several Ångstrom.

It is to notice that the requirements for such a measurement are only a precise determination of the beam location at the intersection with the atomic gas jet. The divergence of the beam, however, can exceed several degrees. An error of one degree in the beam direction would mean a rotation of the target system by 1° and therefore would have "no" practical influence on the momentum kinematics. Thus RIMS is in particular of interest in situations where no beam collimation is possible, e.g. internal experiments in storage rings, antiproton or relativistic heavy ion beam experiments.

Since the new jet target allows the determination of ϑ_r within a few degree, the longitudinal momentum component can be measured with good resolution too. In figure 12 the expected kinematical recoil angles folded with about 5° angular resolution (FWHM) are presented for two different impact parameters in 10 MeV/u U^{92+} + He --> U^{91+} + He^{+} electron transfer reactions into the different Uranium n-states.

The different capture probabilities are all set equal. It can be seen that RIMS should provide enough resolution to resolve the final n-state in the Uranium ion by Q-value measurements. In the two-dimensional plot of recoil angle ϑ_r versus recoil transverse momentum the reaction can be investigated within details. The dependence on $P_r^{\perp}$ yields the whole information about the impact parameter.

Depending on the transverse momenta, i.e. on the impact parameter, relative longitudinal energy transfers of below 10^{-8} or momentum transfers below 10^{-6} can be resolved.

We believe that RIMS will open a new window to study the multi electron processes in fast ion atom collisions with such a precision (recoil momentum of 0.1 a.u. corresponds to electron energy resolution in the order of 1 eV) that dynamical e-e-correlations might be measurable in the electron sum momentum.

Acknowledgement:

we want to thank our friends and collegues R.E. Olson, C.L. Cocke, R. Dreizler, S. Hagmann, J. McGuire, J. Reading, V. Schmidt, R. Rivarola and A. Salin for many helpful and supporting discussions.

R.Buck made important contributions to the new atomic jet target. The work was supported by **DFG, BMFT** and the **Humboldt Foundation.**

References:

C.L. Cocke, R.E. Olson Phy. Rep. 205(4) 153 (1991), C.L. Cocke in High-Energy Ion-Atom Collisions ed. D. Berenyi, G. Hock, Springer-Verlag 1990

R. Dörner, J. Ullrich, H. Schmidt-Böcking, R.E. Olson Phys.Rev.Lett 63, 147 (1989)

R. Dörner Thesis, Univ. Frankfurt/M and to be published

E. Forberich, R. Dörner, J. Ullrich R.E. Olson, K. Ullmann, A. Gensmantel, S. Lencinas, O. Jagutzki, H. Schmidt-Böcking, accepted for publication in J. Phys B 1991

H. Fukuda, I. Shimamura, L. Vegh, T. Watanabe, to appear in Phys. Rev. A. 44 1991

A. Gensmatel, J. Ullrich, R. Dörner, R.E. Olson H. Schmidt-Böcking, submitted to Phys. Rev.

A. Gonzales, S. Hagmann, T.B. Quinteros, B. Krässig, R. Koch, A. Skutlartz, H. Schmidt-Böcking, J. Phys. B23 L303 1990

E.Y. Kamber, C.L. Cocke, S. Cheng, S.L. Varghese, Phys Rev. Lett. 60, 2026 (1988)

F.G. Kristensen, E. Horsdal-Pedersen, J. Phys. B23, 4129 (1990)

R.E. Olson priv. comm. and to be published

M.E. Rudd, L.H. Toburen, N. Stolterfoth, At. Data Nucl. Data Tables 18, 413 (1976)

A. Salin J. Phys. B22, 3901 (1989)

H. Schmidt-Böcking, R. Dörner, J. Ullrich, J. Euler, H. Berg, E. Forberich, S. Lencinas, O. Jagutzki, A. Gensmantel, K. Ullmann, R.D. DuBois, Feng Jiazhen, R.E. Olson, A. Gonzales, S. Hagmann in High-Energy Ion-Atom Collisions ed. D. Berenyi, G. Hock, Springer-Verlag 1990

J. Ullrich, R.E. Olson, R. Dörner, V. Dangendorf, S. Kelbch, H. Berg, H. Schmidt-Böcking J.Phys. B22, 627 (1989)

Inner shell alignment in ion—atom collisions

J. Pálinkás

Institute of Nuclear Research of the Hungarian Academy of Sciences
(ATOMKI), H-4001 Debrecen, Pf. 51, Hungary

ABSTRACT: Recent experimental investigations of the alignment of L_3 subshell
vacancies produced by ionisation and by electron capture is reviewed.   The
experiments were based on the measurements of the angular distribution of L x-
rays and L-MM Auger electrons. The results justify the use of coupled channel
calculations to estimate the alignment in multiple ionisation by heavy projectiles.
The experimental findings for capture induced alignment at lower energies indicate
the failure of the the existing theories to describe the capture-induced alignment.

1. INTRODUCTION

In the ionization of atoms by a directed beam of charged particles, the ions with
vacancies in subshell with total angular momentum $j > 1/2$ can be aligned (Mehlhorn
1968). The alignment, which reflects the spatial anisotropy of the ionized state, has
been extensively studied in the last decade (Kabachnik 1988). The workhorse of these
investigations is the L_3 $(2p_{3/2})$ subshell, the lowest state, which can be aligned.

In a fast ion-atom collision, the L_3 subshell vacancy can be created either by the
direct ionization or by the capture of the L_3 electron by the projectile. Furthermore,
for light projectiles the vacancy creating process can be a single ionization (capture),
but in the case of heavy projectiles other (outer shell) electrons from the target can be
removed simultaneously, leaving the target in a multiply ionized state. The alignment
produced in single ionization by light projectiles has been studied by many laboratories
(Jitschin *et al* 1979, Pálinkás *et al* 1980, Richter *et al* 1981, Barros Leite *et al* 1982,
Konrad *et al* 1984) and the experimental results are fairly well described by first order
perturbation theories (Sizov and Kabachnik 1980, Rösel *et al* 1982). The capture-
produced alignment and the alignment of multiply ionized atoms, however, continues
to present puzzling feature for the theories. The alignment depends on the dynamics
of the collision and presents a sensitive testing ground for collision theories. In this
paper we give a short summary of the main features of recent studies of the L_3 subshell
alignment and present results for the alignment of multiply ionized heavy elements
and the alignment produced by electron capture.

2. BACKGROUND OF THE ALIGNMENT MEASUREMENTS

The alignment of the L_3 ($2p_{3/2}$) subshell can be quantitatively described by the alignment parameter

$$\mathcal{A}_{20} = \frac{\sigma_{2p_{3/2}}(|m_j| = 3/2) - \sigma_{2p_{3/2}}(|m_j| = 1/2)}{\sigma_{2p_{3/2}}(|m_j| = 3/2) + \sigma_{2p_{3/2}}(|m_j| = 1/2)}, \tag{1}$$

where $\sigma_{2p_{3/2}}(|m_j| = 3/2)$ and $\sigma_{2p_{3/2}}(|m_j| = 1/2)$ are the vacancy production cross sections of the magnetic substates with total angular momentum component $|m_j| = 3/2$ and $|m_j| = 1/2$, respectively. The alignment parameter measures the relative contribution of the differently asymmetric $|m_j| = 3/2$ and $|m_j| = 1/2$ magnetic substates to the vacancy production and therefore measures the charge cloud asymmetry created in the collision.

The L_3 subshell alignment can be revealed by studying the angular distribution of x-radiation and Auger-electrons arising from the decay of collisionally aligned L_3 subshell vacancies, which is given by (Cleff and Mehlhorn 1974, Berezhko and Kabachnik 1977)

$$I(\theta) = (I_0/4\pi)(1 + \beta P_2(\cos\theta)), \tag{2}$$

where $I(\theta)$ is the intensity at angle θ with respect to the beam direction, I_0 is the total intensity emitted into the 4π solid angle, $P_2(\cos\theta)$ is the second-order Legendre polynomial, and β is the anisotropy parameter, which for independent ionization and decay processes and can be expressed as

$$\beta_2 = \kappa\alpha\mathcal{A}_{20}, \tag{3}$$

where $\mathcal{A}_{20}$ is the alignment parameter, α charecterizes the anisotropy property of a given transition, and depends on the total angular momentum of the initial and final states. The correction factor κ takes into account that L_3 holes can also be created by Coster-Kronig transitions (Jitschin *et al* 1979), and its value for Auger electrons is unity. To determine the L_3 subshell alignment for heavy elements from the measured L x-ray spectrum the best candidate is the L_l line, which is a single (M_1-L_3) transition and the corresponding α parameter is high (0.5). In the Auger electron spectra the L_3-$M_{2,3}^2(^1S_0)$ transition is the most suitable line since its α parameter is minus unity.

In collisions of fast light projectiles with heavy targets the dominant vacancy production process for the L_3 subshell is direct ionization. In these collisions electron capture contributes a few percent at most to the vacancy production and the experimental procedure to determine the ionization-produced alignment simply requires the precise measurement of the angular distribution of the above x-ray or Auger lines, fitting the measured angular distribution with equation (2) and use equation (3) to obtain $\mathcal{A}_{20}$, supposed α and κ is known for the transition. For capture induced alignment, however, the experimental procedure is considerably more complicated, since one has to verify that the L_3 vacancy has been created by capture. This can be done by measuring the angular distribution of x-rays and Auger electrons in coincidence with the charge changed projectiles.

3. L₃ SUBSHELL ALIGNMENT IN MULTIPLY IONISED ATOMS

The L_3 subshell alignment in heavy elements ionised by heavy ion impact has been studied by measuring the angular distribution of the L x-ray spectrum (Pálinkás *et al* 1983, Jitschin *et al* 1983, Stachura *et al* 1984, Berinde *et al* 1984). For heavy projectile impact, simultaneously with the L_3 electron outer shell electron(s) may be removed. The L_3 subshell vacancy in the multiply ionized atom decays with the emission of many satellite lines which are transition between different angular momentum states. If these satellites are not resolved equation (3) has to be modified to give the anisotropy parameter for this diagram and satellite complex

$$\beta = \sum_{if} \epsilon_{if}\alpha_{if}\kappa_{if}\mathcal{A}_{20,i}, \tag{4}$$

where $\mathcal{A}_{20,i}$ is the alignment parameter of the ion with vacancies in a given angularmomentum state, α_{ij} accounts for the angular momentum coupling of the initial and final states of the given transition, κ_i takes into account the reduction of the alignment due to rearrangement processes prior to the x-ray emission, and ϵ_{if} are weight factors for the different satellite transitions contributing to the investigated x-ray line. The complicated satellite structure practically prevents the determination of the alignment of the initial state of the multiply ionized atom since the satellites can not be resolved even with a crystal spectrometer.

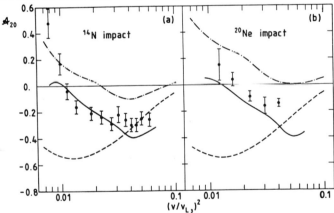

Figure 1. L_3 subshell alignment of gold ionised by (a) nitrogen (Pálinkás *et al* 1983) and (b) neon (Jitschin *et al* 1983) as the function of the relative collision velocity. The curves display first-order Born approximation - - - (Rösel *et al* 1982); second-order Born approximation — · — · — (Sarkadi 1986a); and coupled states calculation —— (Sarkadi 1986b).

The problem of unfolding the collision induced L_3 subshell alignment from the measured x-ray angular distribution presented a serious problem in the L x-ray angular distribution measurements on heavy targets using C and N (Pálinkás *et al* 1983), O and Ne (Jitschin *et al* 1983) projectiles. The the anisotropy parameters measured in those experiments were in strong disagreement with the predictions of first order perturbation theories. Sarkadi (1986ab) made and attempt to explain the discrepancy on the basis of the second order Born and a coupled channel calculations, and got better agreement with the theories as can be seen in figure 1. The agreement, however, may be regarded somewhat fortuitous since these higher order theories calculate the single ionization cross sections, and consider the decay of the ion with a single vacancy. The measurements, however, include different degrees of multiple ionization. This problem becomes more serious if one uses even heavier projectiles.

To circumvent this problem a new experimental method has been devised to separate the diagram and satellite lines and measure their angular distribution (Papp *et al* 1991). The crucial point in the experiment is the separation of the single ionization diagram lines from the multiple ionization satellites, which can be done by combining a Si(Li) detector with an appropriate absorber with its K-absorbtion edge falling between the x-ray lines to be resolved (Papp and Pálinkás 1989). In this way one can measure the angular distribution of the single ionisation diagram lines in case of heavy ion impact. In the L x-ray spectrum of gold and thallium one can use zinc and copper absorbers, respectively, to separate the satellites of the $L_{\alpha 1}$ line (Au) and that of the L_l line (Tl) from their diagram lines.

The experimental setup in the measurement was similar as used in our previous angular distribution studies (Pálinkás et al 1983) and will be described in details elsewhere (Papp *et al* 1991). 1 MeV/amu N and Cl beams were collimated on thin Au and Tl targets and the L x-rays were detected with a Si(Li) detector, which could be rotated around a small scattering chamber. The x-ray exit slit on the scattering chamber was covered with 4 mgcm^{-2} Al and 107 mgcm^{-2} PET foils and the angular distribution of the L x-ray spectra of Au and Tl were measured at 5 different angles in the range of 15° -105° using 25 mgcm^{-2} Zn and Cu absorber in front of the Si(Li) detector to absorb the satellites of the $L_{\alpha 1}$ and L_l lines of Au and Tl, respectively. A typical Tl spectrum measured with Cu absorber can be seen in figure 2.

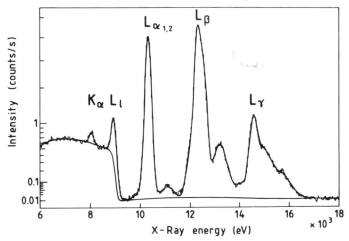

Figure 2. L x-ray spectrum of Tl induced by 1 MeV/amu N ions and measured with 25 μgcm^{-2} Cu absorber in front of a Si(Li) detector.

The x-ray spectra, measured without absorber, were analyzed with a Gaussian fitting program, which fitted a quadratic polynomial background and appropriate number of Gaussian peaks simultaneously. Linear function plus a step function convolved with a Gaussian was used to fit the background in spectra measured with absorbers. The intensities of the L_3 x-ray lines were normalized to the intensity of the isotropic $L\gamma$ line. These normalized intensities were fitted with function (1) using the anisotropy parameter β as fitting parameter.

The attenuation of the x-ray lines by the absorber does not affect the determination of the anisotropy parameter, since the absorption is the same at each angle. The Coster-Kronig satellites are emitted from the decay of L_3M double hole states created when single L_1 vacancies decay into these states by Coster-Kronig transitions. These L_3M Coster-Kronig satellites are isotropic and the usually applied Coster-Kronig correction takes into account this effect, but the absorber cuts some of these satellites and therefore reduces this correction. This reduction has been estimated previously (Papp *et al* 1990) and found to be only a few percent.

The type of satellites concern us here are the L_3M^n satellites emitted when M-shell holes have been created by the projectile simultaneously with the ionization of the L_3 subshell. In the case of Tl L x-rays the extent of the L_3M^n satellite contribution can be estimated from the L_l/L_α intensity ratio measured using Cu absorber (Papp et al 1991). The Cu absorber attenuates weakly the L_l diagram line but does attenuate strongly the satellites. Therefore, as the relative contribution of the satellites increases, the L_l/L_α ratio decreases since the attenuation of the L_α line depends only very weakly on the L_α satellite structure. The single ionization value of the L_l/L_α ratio for Tl measured with Cu absorber can be obtained from data by fast proton impact, where collision induced multiple ionization is small. It has been found (Papp et al 1991) that the value of the measured L_l/L_α ratio changes from 0.81 ± 0.07 for 2.5 MeV proton impact to 0.180 ± 0.014 for 1 MeV/amu Cl impact on Tl, i.e. in the latter case approximately 80 percent of the L_l intensity is emitted as satellites.

The anisotropy parameter of the L_l line of Tl ionized by 1 MeV/amu N ions changes from 0.10 ± 0.02 to 0.07 ± 0.02 if one measures the angular distribution with and without Cu absorber. For 1 MeV/amu Cl impact the corresponding changes are from 0.12 ± 0.03 to 0.10 ± 0.01. This tendency is in accord with the expectation that the L_l radiation emitted from the decay of single vacancy states is more anisotropic than in the case of multiply ionized atoms. The extent of the change, however, indicates that the effect of the multiple ionization satellites does not modify drastically the angular distribution, and one may expect that the anisotropy parameter as function of the projectile velocity will have the same form for atoms singly ionized by heavy projectiles as it was found previously (Jitschin *et al* 1983, Pálinkás *et al* 1983) and the comparison with the single ionization theoretical calculation (Sarkadi 1986ab) can be justified.

4. L_3 SUBSHELL ALIGNMENT INDUCED BY ELECTRON CAPTURE

For heavier target elements the dominant vacancy production mechanism is the direct Coulomb ionisation. For light target atoms the contribution of the electron capture to the vacancy production may not be negligible, especially at low collision velocities.

There exists only very limited experimental information for the L_3 subshell alignment induced by electron capture owing to the inherent difficulty that one has to apply coincidence technique to separate the capture and the ionisation channels.

Until recently only noncoincidence experiments (Rødbro *et al* 1978, DuBois *et al* 1981, Menzel and Mehlhorn 1987) indicated that capture induced alignment of the $2p_{3/2}$ vacancy may be important when light targets are ionised by slow projectiles. Including the capture channel into the ionisation theory with use of the Oppenheimer–Brinkmann–Kramers (OBK) approximation (Sizov and Kabachnik 1983), a considerable improvement has been achieved in the description of the experimental data. In a different type of experiment, the alignment of He-like 2P states populated mainly by electron capture when 32–105 MeV S ions are passed through a thin carbon foil (Pálinkás and Watson 1985) has been found in gross contradiction with the OBK, but in agreement with the 'one and a half centered atomic orbital expansion (OHCE)' approximation of Reading *et al* (1982).

The aim of our recent work (Sarkadi *et al* 1990) was to measure directly the alignment parameter for L_3 subshell vacancies created by electron capture in 0.3–1 MeV p+Ar collisions. The alignment parameter have been determined by measuring the non-isotropic angular distribution of the L_3-$M_{2,3}^2$ Auger electrons resulting from the decay of the excited state. The electron capture channel has been separated from the ionisation channel by coincident detection of the Auger electrons with the outgoing neutralised H atoms. The scheme of the experimental set-up is shown in figure 3, and is very similar to that used in our previous coincidence investigations of the forward electron 'cusp' peak (Kövér *et al* 1989).

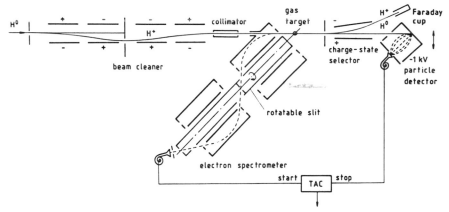

Figure 3. Sketch of the experimental arrangement used to study the capture induced alignment.

The proton beam obtained from the Van de Graaff accelerator of ATOMKI were 'cleaned' from H^0 projectiles by an electrostatic 'cleaner' in front of the beam collimator and passed through an effusion gas target mounted in the focal point of a double-stage cylindrical mirror electron spectrometer. The spectrometer makes the energy (0.36% relative resolution) and angle selection of the electrons counted by a channel electron multiplier. The proton beam leaving the interaction zone was deflected electrostatically into a Faraday-cup and the outgoing H^0 atoms were counted by a particle detector. Standard coincidence electronics were applied to record simultaneously the single spectrum, the spectrum belonging to the true+random, and to the random coincidences. The time resolution was about 17 ns (FWHM) and the random to true ratio was kept below 20%. The pressure of the target gas was kept constant during the long measurements and the single collision condition was verifed by the linear pressure dependence of the yield of events. The angular distribution of the Auger electrons given by equation (2) has only I_0 and β as fitting parameters, and one can determine $\mathcal{A}_{20}$ from the angular distribution measured at two different angles. Choosing the two angles close to $90°$ and $180°$ the coincidence yield can be ineased by using large acceptance angle because $P_2(\cos\theta)$ varies slowly around $90°$ and $180°$.

The recorded electron spectra have been evaluated with special care in order to reduce the error caused by the limited counting statistics. The spectra have been computer fitted by a curve consisting of known Auger transitions and a linear function for the background. The shape of the peaks was approximated by linear combination of a Gaussian and a Lorentzian peak. The angular distribution of the anisotropic L_3-$M_{2,3}^2(^1S_0)$ transition has been normalized to the sum of the intensities of all the isotropic lines originating from the decay of the L_2 vacancy. Figure 4 shows the $\mathcal{A}_{20}$ values obtained from our coincidence measurements performed at 0.3, 0.5, 0.7 and 1 MeV proton energies as a function of the relative collision velocity v/v_{L_3}. In the figure we have also plotted the results of non-coincidence measurements made at several other impact energies. The latter $\mathcal{A}_{20}$ values have been determined taking electron spectra at five angles between $90°$ and $180°$.

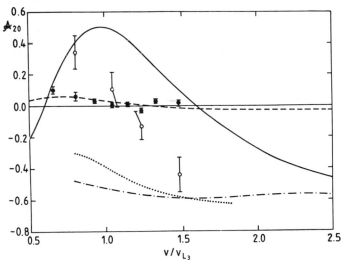

Figure 4. L_3-subshell alignment parameters of Ar for proton impact as function of the relative collision velocity. Experimental data: o, Auger electrons detected in coincidence with the outgoing H^0; •, results of the non-coincidence measurements. Theories: ——, OBK (Berezhko *et al* 1980, 1981); ······, CDW (Belkić *et al* 1984); — · — · —, SPB (Jakubassa–Amundsen 1987); - - -, effective alignment calculated as the weighted avarage of ionisation and capture induced alignment.

As is seen from the figure the alignment produced by electron capture depends strongly on the collision velocity taking large negative and positive values at the limits of the velocity range of the measurements. This is in a strong contrast with the small non-coincidence alignment values. The A_{20} values observed for the capture channel have been compared with the OBK calculations of Berezhko *et al* (1980, 1981) applying screened hydrogen-like wave functions, the SPB calculations of Jakubassa–Amundsen (1987) made in the 'transverse peaking' approximation, and the predictions of the CDW theory calculated with the computer code of Belkić *et al* (1984), taking into account capture into the 1s, 2s and 2p states of the projectile. The results of the non-coincidence measurements have been compared with a curve representing the effective alignment, the weighted avarage of the ionisation and capture induced alignment, using the plane-wave Born-approximation (PWBA) (Sizov and Kabachnik 1980) and the OBK alignment values Berezhko *et al* (1980, 1981), respectively. For the capture to ionisation cross section ratios the model of Brandt and Lapicki (1979) has been applied. The small (1.5 – 9 %) contribution of the capture channel in the energy range of our measurements makes the effective alignment insensitive to the alignment due to the capture, and explains that in spite of the apparent failure of the OBK for capture, a reasonable agreement can be observed between the combined PWBA and OBK theory and the results of the non-coincidence experiments.

Before comparing the experimental and theoretical alignment parameters, an effect which may cause an increasing error in the measured data with decreasing bombarding energy has to be considered. Due to the finite acceptance angle of the particle detector, the H$^{\circ}$ atoms emerged from the collision with scattering angles larger than 0.27° were not detected in this experiment. We have estimated the resulting error calculating theoretical A_{20} values with the exclusion of the contribution of the larger scattering angles. The limitation in the scattering angle means that the integration excludes a certain range of small impact parameters and this range decreases with increasing collision energy. Using the results of the IA calculations of Jakubassa–Amundsen (1981, 1987) performed in the semiclassical approximation (Briggs 1977), the IA curve corrected for the finite acceptance angle of the particle detector has been calculated. To convert scattering angles to impact parameters we have solved the Kepler problem for scattering of protons in the electric field of the target nucleus screened by its electrons. This calculation did not take into account that the projectile emerging from the collision is neutral, consequently the obtained b_c values are probably overestimated, but the correction procedure cannot lead to a large positive A_{20} value observed at 0.3 MeV, even if we calculate the b_c values assuming Rutherford scattering of bare nuclei. Although the correction shifts the IA curve closer to the experimental points, it cannot resolve the discrepancy between the theory and experiment. A further indication that the limitation of the capture events at small impact parameters was not strong in our measurement is that at the lowest proton energy the experimental capture to ionisation cross section ratio (the ratio of the coincidence to single Auger-electron yield corrected for the counting efficiency) was found to be 0.072±0.11. This value is in good accordance with the value 0.06 obtained from the phenomenological model of Brandt and Lapicki (1979), which has been tested on a large experimental data base.

The disagreement between the data and the OBK (see figure 4) is not surprising, although it is remarkable, that this approach cannot give account even most reliable experimental point at 1 MeV. At the same time, the acceptable agreement between the experimental result at 1 MeV and the SPB and CDW curves in figure 4 indicates that the higher-order capture theories provide a correct description at high collision velocities. Concerning the unsatisfactory performance of the latter theories in the low-velocity range, one has to take into account that the higher-order approaches have been developed mainly to understand the features of the electron capture at asymptotically high collision velocities. Belkić *et al* (1984) give the low-velocity limit of the validity of their CDW theory as a function of the initial and final energies of the captured electron. For our case the energy limit of the CDW is 0.72 MeV, i.e. the comparison with the experimental results at lower energies is not justified. However, the rather large deviation between the observed and calculated value at 0.7 MeV can hardly be explained by the restricted validity of this theory at low impact energies.

5. CONCLUSIONS

The measurement of the angular distribution of Tl L_l x-rays using the combination of a Cu absorber and a Si(Li) detector (Papp *et al* 1991) made possible the separation of the LM^n multiple ionisation satellites from the diagram line. The experimental results at 1 MeV/amu indicate that the effect of the multiple ionization satellites does not modify drastically the angular distribution. Therefore one may expect that the anisotropy parameter as function of the projectile velocity will have the same form for atoms single ionized by heavy projectiles as it was found previously (Jitschin *et al* 1983, Pálinkás *et al* 1983) and the comparison with the single ionization theoretical calculations (Sarkadi 1986ab) can be justified.

The study of the capture induced alignment (Sarkadi *et al* 1990) demonstrated that the coincidence measurement of the L_3-subshell alignment provides a sensitive test of the charge transfer theories. The reasonable agreement between the predictions of the SPB, CDW, IA theories and the large negative alignment observed at 1 MeV proves that the higher-order theories provide a better description of the high-velocity charge transfer processes than the first-order approach. The higher-order theories predicting negative values in the intermediate range are in contradiction with the experiment. The uncertainty of our data due to the limited detection of the scattered projectiles at lower energies, however, does not allow to draw firm conclusions. The reliability of the obtained alignment values could be checked repeating the experiment using a particle detector with a larger acceptance angle.

ACKNOWLEDGEMENT: Financial support was provided by the Hungarian Scientific Research Foundation (OTKA).

REFERENCES

Barros Leite C V, de Castro Faria N V, Horowicz R J, Montenegro E C and de Pinho A G 1982 *Phys. Rev.* A **25** 1880

Belkić Dž, Gayet R and Salin A 1984 *Comput. Phys. Commun.* **32** 385

Berényi D, Cserny I, Kádár I, Kövér Á, Ricz S, Sarkadi L, Varga D and Végh J 1984 *J. Phys. B: At. Mol. Phys.* **17** 829

Berezhko E G and Kabachnik N M 1977 *J. Phys. B: At. Mol. Phys.* **10** 2467

Berezhko E G, Kabachnik N M and Sizov V V 1980 *Phys. Lett.* **77A** 231
Berezhko E G, Sizov V V and Kabachnik N M 1981 *J. Phys. B: At. Mol. Phys.* **14** 1981
Berinde A, Ciortea C, Enulescu Al, Flucraşu D, Piticu I, Zoran V and Trautmann D 1984 *Nucl. Instrum. Methods* B **4** 283
Brandt W and Lapicki G 1979 *Phys. Rev.* A **20** 465
Briggs J S 1977 *J. Phys. B: At. Mol. Phys.* **10** 3075
Cleff B and Mehlhorn W 1974 *J. Phys. B: At. Mol. Phys.* **7** 593
DuBois R, Mortensen L and Rødbro M 1981 *J. Phys. B: At. Mol. Phys.* **14** 1613
Folkmann F and Dahl P, unpublished data, see Sizov and Kabachnik (1983)
Jakubassa–Amundsen D H 1981 *J. Phys. B: At. Mol. Phys.* **14** 2647
——1987, private communication
Jitschin W, Kleinpoppen H, Hippler R and Lutz H O 1979 *J. Phys. B: At. Mol. Phys.* **12** 4077
Jitschin W, Hippler R, Shanker R, Kleinpoppen H, Schuch R and Lutz H O 1983 *J. Phys. B: At. Mol. Phys.* **16** 1417
Kabachnik N M 1988 *Electronic and Atomic Collisions* ed. H B Gilbody *et al* (Elsevier Science Publishers B. V.) p 221
Konrad J, Schuch R, Hoffmann R and Schmidt-Böcking H 1984 *Phys. Rev. Lett.* **52** 188
Kövér Á, Sarkadi L, Pálinkás J, Berényi D, Szabó Gy, Vajnai T, Heil O, Groeneveld K O, Gibbons J and Sellin I A 1989 *J. Phys. B: At. Mol. Opt. Phys.* **22** 1595
Mehlhorn W 1968 *Phys. Lett.* **26A** 166
Menzel W and Mehlhorn W 1987 *J. Phys. B: At. Mol. Phys.* **20** L277
Pálinkás J, Sarkadi L and Schlenk B 1980 *J. Phys. B: At. Mol. Phys.* **13** 3829
Pálinkás J, Sarkadi L, Schlenk B, Török I, Kálmán Gy, Bauer C, Brankoff K, Grambole D, Heiser C, Rudolph W and Thomas H J 1983 *J. Phys. B: At. Mol. Phys.* **17** 131
Pálinkás J and Watson R L 1985 *Phys. Lett.* **110A** 298
Papp T, Pálinkás J, 1989 *Acta Physica Hungarica* **65** 209
Papp T, Pálinkás J, Sarkadi L 1990 *Phys. Rev.* A **42** 5452
Papp T, Pálinkás J, Sarkadi L, Neelmejer C, Brankoff K, Grambole D, Heiser C, Rudolph W, Herrmann F and Thomas H J 1991 to be published
Reading J F, Ford A L and Becker R L 1982 *J. Phys. B: At. Mol. Phys.* **15** 625
Richter G, Brüssermann M, Ost S, Wiggert J, Cleff B and Santo R 1981 *Phys. Lett.* **82A** 412
Rødbro M, DuBois R and Schimdt V 1978 *J. Phys. B: At. Mol. Phys.* **11** L551
Rösel F, Trautmann D and Baur G 1982 *Z. Phys.* A **304** 75
Sarkadi L 1986a *J. Phys. B: At. Mol. Phys.* **19** 2519
——1986b *J. Phys. B: At. Mol. Phys.* **19** L755
Sarkadi L, Vajnai T, Pálinkás J, Kövér Á, Végh J and Mukoyama T 1990 *J. Phys. B: At. Mol. Phys.* **23** 3643
Sizov V V and Kabachnik N M 1980 *J. Phys. B: At. Mol. Phys.* **13** 1601
——1983 *J. Phys. B: At. Mol. Phys.* **16** 1565
Stachura Z, Bosch F, Hambsch F J, Liu B, Maor D, Mokler P H, Schönfeldt W A, Wahl H, Cleff B, Brüssermann M and Wigger J 1984 *J. Phys. B: At. Mol. Phys.* **17** 835

Ultrarelativistic atomic collisions

C. Bottcher and M. R. Strayer

Physics Division, Oak Ridge National Laboratory,
Oak Ridge TN 37831-6373, U.S.A.

ABSTRACT: Calculations of the coherent production of free pairs and of pair production with electron capture from ultrarelativistic ion-ion collisions are discussed. Theory and experiment are contrasted, with some conjectures on the possibility of new phenomena.

1. INTRODUCTION

The proposed new colliders designed to investigate the collisions of heavy ions at very high velocities have led to extensive theoretical speculation concerning possible new phenomena. In particular, the large and essentially classical electromagnetic fields associated with such collisions mean that the electromagnetic coupling to ordinary matter is greatly enhanced (Bottcher 1988). The production of multiple electron-positron pairs, muon pairs, vector bosons, and possibly the yet-unconfirmed Higgs boson all occur at nearly atomic distance scales (Bottcher 1989, 1990).

The process of electron-positron production with electron capture by one of the ions is the principal beam loss mechanism for highly charged ions in a storage ring and thus plays a central role in the design and the operation of these machines (Rhoades-Brown 1990). At present, there are no measurements of this process, and the various theoretical calculations seem to differ by as much as two orders of magnitude. These calculations encompass many different approaches – two-photon perturbation theory (Rhoades-Brown 1989), coupled-channel expansions (Rumrich 1991), distorted-wave techniques (Becker 1987), numerical lattice calculations (Strayer 1990), as well as use of the Weizsäcker-Williams approach (Bertulani 1988, Nikishov 1982, Rhoades-Brown 1991). Thus there is a pressing need for experiments to aid in resolving this ambiguity.

The production of electron pairs observed in the collisions of cosmic rays with nuclei was studied by Racah (Racah 1937) using the Weizsäcker-Williams, or equivalent photon method (Williams 1934, Weizsäcker 1934, Landau 1934). In more modern studies, this technique has been extensively used to study free pair production in different circumstances: electron pairs (Brodsky 1971, Bertulani 1988, Rau 1990), muon pairs, heavy vector bosons (Grabiak 1989), and possibly the Higgs boson (Papageorgiu 1989, Grabiak 1989, Drees 1989, Cahn 1990, Müller 1990). In contrast, direct Monte Carlo integration of perturbation theory has also been employed to study these systems. Although the total cross sections are in reasonable agreement between different approaches, details of differential cross sections, spectra, and impact parameter dependence are notably different (Bottcher 1989, 1990, 1990a, 1990b). A critical review of these different approaches will be given elsewhere (Bottcher 1992). A comparison of three different equivalent photon calculations with the Monte Carlo integration of the exact two-photon terms, as shown in Fig. 1, illustrates the point that the total cross sections obtained from the different approaches are nearly the same. The Lorentz factor, γ, is nearly equal to the energy per nucleon of the beam in GeV.

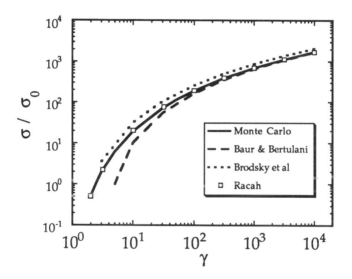

Fig. 1. The cross section for the production of electron pairs from the symmetric collision of Au nuclei as a function of the Lorentz γ of each beam. The constant σ_0 is given in the text.

The relation between this Lorentz factor in the center-of-mass system and the factor in an equivalent fixed-target frame is

$$\gamma_T = 2\gamma^2 - 1. \tag{1}$$

Equation (1) expresses the important result that collisions in colliders correspond to fixed-target experiments at very high energies. The curves in Fig. 1 are taken, respectively, from the references Brodsky (1971), Baur (1987), Racah (1937), and the Monte Carlo curve from Bottcher (1989) is given by the formula

$$\sigma = \sigma_0 C_\infty \frac{(\ln \gamma)^3}{1 + a_1 y + a_2 y^2 + \cdots} \quad ; \quad y = \ln[1 + (g_0 / \gamma)^2],$$

(2)

where the constants in Eq. (2) are given below

$$C_\infty = 3.40, \quad a_1 = 0.4570, \quad a_2 = 0.0222, \quad g_0 = 10^3.$$

The scale for the cross sections, assuming point nuclear charge densities, is expressed in terms of the charge of the ion, and the Compton wavelength of the electron, and for Au ions is

$$\sigma_0 = (Z\alpha)^4 \lambdabar^2 \approx 165 \, b.$$

(3)

A brief discussion of the classical field theory and the Monte Carlo method is discussed in the next section. Details of the method are given elsewhere (Bottcher 1989, Wu 1990).

2. TWO-PHOTON PAIR PRODUCTION

Most modern approaches to pair production begin with quantum field theory (Berestetskii 1979). However, for the collisions of highly charged ions it is possible to treat the electromagnetic field classically. We begin with the S-matrix

$$S = \text{T} \exp[-i\int d^4x \, \mathcal{H}_{\text{int}}(x)] ,$$

(4)

expressed in terms of the time-ordered interaction Hamiltonian which governs the coupling of electrons to the electromagnetic field

$$\mathcal{H}_{\text{int}}(x) = e : \overline{\Psi}(x)\gamma^\mu \Psi(x) A_\mu(x) : + e : J^\mu(x) A_\mu(x) : .$$

(5)

In Eq. (5) Ψ denotes the electron field operator and the electromagnetic current and field operators are, respectively, J^μ and A^μ. The classical field model is then obtained by replacing J^μ and A^μ with their classical counterparts. Under these assumptions, the second term in Eq. (5) can be ignored since it only contributes a c-number phase to the S-matrix in Eq. (1).

The electromagnetic potentials come from the motion of the two ions

$$A^\mu(x) = A_a^\mu + A_b^\mu ,$$

(6)

where a,b label the two ions. These potentials can be calculated from the nuclear charge density, ρ, of each ion expressed as a function of the four-momentum transfer $-q^2 > 0$

$$f(q^2) = \frac{F(-q^2)}{-q^2} = \frac{4\pi}{-q^2}\int_0^\infty \frac{rdr}{q}\sin(qr)\rho(r).$$

(7)

In the rest frame of each ion only the time component is nonzero, yielding solutions to Maxwell's equations as

$$A_{a,b}^0(q) = 2\pi e Z_{a,b}\,\delta(q^0)f_{a,b}(q^2)\exp(\pm i\vec{q}\cdot\vec{b}/2),$$

(8)

where b is the impact parameter separating the two ions. Lorentz boosting these potentials into the center-of-velocity frame of the a+b system gives the result

$$A_{a,b}^\mu(q) = 2\pi e Z_{a,b}\,u_{a,b}^\mu\,\delta(q^0 \mp \beta q^z)f_{a,b}(q^2)\exp(\pm i\vec{q}\cdot\vec{b}/2),$$

(9)

as a function of the "boost" four-velocities

$$u_{a,b}^\mu = (1,0,0,\pm\beta).$$

(10)

The total cross section is obtained from Eq. (1) by integrating over all impact parameters and summing over all final states

$$\sigma = \int d^2b \sum_f |\langle f|S|0\rangle|^2.$$

(11)

The lowest-order contributions to the S-matrix which create pairs out of the vacuum are the second-order terms given by

$$\left\langle \vec{k}_1 s_1 \vec{k}_2 s_2 \middle| S^{(2)} \middle| 0 \right\rangle = -ie^2\int\frac{d^4k}{(2\pi)^4}\,\bar{u}(k_1 s_1)\{A_a'(q_a)\frac{1}{\slashed{k}-m}A_b'(q_b)$$
$$+ A_b'(q_b)\frac{1}{\slashed{k}'-m}A_a'(q_a)\}v(k_2 s_2).$$

(12)

The four momenta q_a, q_b are the momenta carried by the fields of each ion, and k, k' are the corresponding intermediate momenta for the direct and crossed terms

$$k = k_1 - q_a = -k_2 + q_b; \quad k' = k + q_a - q_b.$$

The slashes denote the usual Feynman contractions with the γ matrices (Berestetskii 1979). After a modest amount of analysis, the resulting total pair production cross section can be reduced to an eight-dimensional integration as follows

$$\sigma = \frac{Z_a^2 Z_b^2 (4\pi\alpha)^4}{4\beta^2} \int \frac{d^3k_1 d^3k_2 d^2q_{a\perp}}{(2\pi)^8 2E_1 2E_2} (f_a(q_a) f_b(q_b))^2$$

$$\times \sum_{s_1 s_2} \left| \bar{u}(k_1 s_1)(u_a \frac{1}{\not{k} - m} u_b + u_b \frac{1}{\not{k}' - m} u_a) v(k_2 s_2) \right|^2 .$$

$$(13)$$

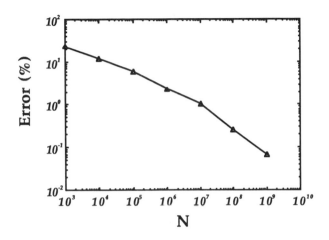

Fig. 2. Error in the Monte Carlo calculation for the case studied in Fig. 1 at an energy per nucleon of about 100 GeV as a function of the number of Monte Carlo points used in the integration.

In the above equation the longitudinal and time-like components of q_a and q_b are fixed by the delta function in Eq. (9) for the classical fields, while the transverse parts must be integrated over.

The eight-dimension integration needed to evaluate the above cross section is carried out using the Monte Carlo method. The integrand is nonsingular and positive definite so that simple mappings can be devised to bias the Monte Carlo sampling near the peaks in the integrand. This results in rapid convergence of the integration. The error in evaluating the cross section in Fig. 1 is shown in Fig. 2 as a function of the number of Monte Carlo points. We observe that the behavior of the errors goes as $N^{-1/2}$ for the larger values of N. In evaluating the curve in Fig. 1, we typically have used more than 10^7 points with a corresponding error of better than about one percent.

The algorithms that are needed to evaluate these expressions can easily be optimized on supercomputers with either vector instructions like the Cray 2, or massively parallel architectures like the Intel 860 hypercube. In Table 1 we compare timing measurements for a typical Monte Carlo calculation on these two machines. The calculation on the hypercube is repeated with a different number of nodes, demonstrating that the effects of message passing and communication among the nodes is not important. Note that the speed on the Cray 2 is about 150 MFLOPS, or about three times faster than most large scientific calculations. The overall speed on the Intel hypercube of about 1.6 GFLOPS is more than ten times faster than the Cray 2.

Nodes	16	32	64	128	Cray 2
Time (Sec)	65.7	29.8	14.2	6.9	74.7
MFLOPS	170	375	787	1620	150

Table 1. Comparison of the speed of Monte Carlo integration for a typical case having 2×10^6 points on the Cray 2 and on the Intel 860 hypercube with a different number of nodes.

Differential distributions can easily be calculated with this method by simply binning the integrand in terms of the requisite variables, thereby avoiding the expensive calculation of Jacobians. The doubly differential cross section in terms of the opening angle of the pair, Θ, and the angle specifying the total momentum of the pair with respect to the beam axis, θ, are shown in Fig. 3 for the case of symmetric Au collisions at a beam energy per nucleon of 100 GeV. In the figure the different curves correspond to histograms with very closely spaced angular bins, and thus appear as smooth curves.

Only recently has it become possible to measure this type of pair production at relativistic energies. Presently, the highest energy heavy-ion experiments are those of the WA90 collaboration at CERN (Vane 1991) using the 200 GeV per nucleon beams on fixed targets. Measurements of electron-positron pair production have been reported for collisions of S+Au, and future experiments with Pb beams are scheduled for 1993.

The experiment detects both electrons and positrons within an angular region of 20 degrees with respect to the beam and for a range of momenta between 0.9 and 7 MeV/c. In Fig. 4 the positrons have been summed over all angles, and the corresponding electrons have been summed over all angles and momenta in the range. The solid curve is the two-photon Monte Carlo calculations, and the dashed curve are the results of an equivalent photon calculation (Eby 1989). The experimental cross section is reported as 75±25 b as compared to the Monte Carlo result of 51 b. While these results are preliminary, there is an indication that theory and experiment may disagree for the higher momentum events.

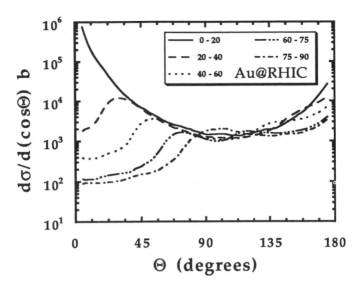

Fig. 3. Electron pair differential cross sections with respect to the opening angle of the pair, Θ, for various cuts on the pair orientation angle, θ, given in degrees. The case studied is the same as in Fig. 2.

3. PAIR PRODUCTION WITH CAPTURE

As previously stated, electron pair production with electron capture is an important process for designing heavy-ion colliders. The mechanism is a second-order one where the pair is first formed and the electron subsequently captured into an orbit of one of the ions. This formulation has been followed using the two-photon Monte Carlo method (Rhoades-Brown 1989). From Eq. (11) and Eq. (12), the cross section for capture into the bound state, Φ_T, can be written as

$$\sigma_c = \int d^2b \sum_s \int \frac{d^3k}{(2\pi)^3} \left| \left\langle \Phi_T \, \vec{k}s \, \middle| \, S \, \middle| \, 0 \right\rangle \right|^2 \; .$$

(14)

The S-matrix in the equal velocity frame is given by Rhoades-Brown (1989) and is not repeated here. The results of evaluating this cross section using the Monte Carlo method for a series of target and projectile combinations is shown in Fig. 5 as a function of beam energy. For the case of an energy of 100 GeV per nucleon, the cross section for Au beams is about 80 b and would yield a beam lifetime of about 10 hours in the RHIC collider. These results are within ten percent of those of Lee and Weneser (Lee 1986).

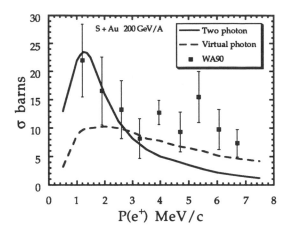

Fig. 4. The distribution of positrons as a function of momentum for collisions of S+Au at a laboratory energy per nucleon of 200 GeV.

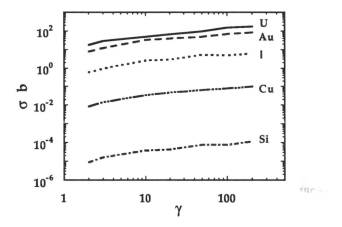

Fig. 5. The cross section for pair production with electron capture for symmetric collisions as a function of beam energy.

At lower energies, a variety of methods have been employed and are shown in Table 2 for comparison. Table 2 shows a variation in the capture cross section by more than a factor of ten and underscores the need for reliable experiments to help clarify the situation.

Method	Reference	σ (b)
Equivalent Photon	Baur 1987	2.8
Two Photon	Rhoades-Brown 1989	4.6
Distorted Wave	Becker 1987	0.3
Coupled Channel	Rumrich 1991	1.5

Table 2. Different capture calculations for collisions of Pb+Pb at an energy of 1.2 GeV per nucleon.

3. SUMMARY

This paper has addressed the phenomena of electron pair production and pair production with electron capture from the coherent fields associated with relativistic collisions of highly charged ions. We have discussed calculations of these phenomena using methods of quantum field theory and modern Monte Carlo algorithms for numerical integration. In discussing the production of electron pairs, we have compared our results with the new measurements of the WA90 group at CERN. Generally these measurements and the theory are in reasonable agreement. However, experiments with heavier beams are needed in order to test higher-order effects such as multiple pair production. Pair production with electron capture to one of the ions has not been measured at these energies, and theoretical calculations at lower energies differ by an order of magnitude. This disagreement could be more than a factor of 100 for the highest RHIC energies. Thus there is a clear need to obtain definite capture measurements as soon as possible.

This research was sponsored by the Office of Basic Energy Sciences, and the Division of Nuclear Physics of the U.S. Department of Energy under contract No. DE-AC05-84OR21400 with Martin Marietta Energy Systems, Inc.

REFERENCES

Baur G and Bertulani C A 1987 *Phys. Rev.* **C35** 836
Berestetskii V B, Lifshitz E M and Pitaevskii L P 1979 *Quantum Electrodynamics* translated by J B Sykes and J S Bell (New York: Pergamon)
Bertulani C A and Baur G 1988 *Phys. Rep.* **163** 299
Becker U, Grün N and Scheid W 1987 *J. Phys.* **B20** 2075

Bottcher C and Strayer M R 1988 *Nucl. Inst. Meth.* **B31** 122
Bottcher C and Strayer M R 1989 *Phys. Rev.* **D39** 1330
Bottcher C and Strayer M R 1990 *J. Phys.* **G16** 975
Bottcher C, Strayer M R, Albert C and Ernst D J 1990a *Phys. Lett.* **B237** 175
Bottcher C, Kerman A K, Strayer M R and Wu J S 1990b *Part. World* **1** 174
Bottcher C and Strayer M R 1992 *Rev. Mod. Phys.*, to be published
Brodsky S J, Kinoshita T and Terazawa H 1971 *Phys. Rev.* **D4** 1532
Cahn R and Jackson J D 1990 *Phys. Rev.* **D42** 3690
Drees M, Ellis J and Zeppenfeld D 1989 *Phys. Lett.* **223** 454
Eby P B 1989 *Phys. Rev.* **A39** 5451
Grabiak B, Müller B, Greiner W, Soff G and Koch P 1989 *J. Phys.* **G15** L25
Landau L D and Lifshitz 1934 *Phys. Z. Sowjetunion* **6** 244
Müller B and Schramm A J. 1990 *Phys. Rev.* **D42** 3699
Nikishov A I and Kurov N V 1982 *Sov. J. Nucl. Phys.* **35** 561
Papageorgiu E 1989 *Phys. Rev.* **D40** 92
Racah G 1937 *Nuovo Cimento* **14** 93
Rau J, Müller B, Greiner W and Soff G 1990 *J. Phys.* **G16** 211
Rhoades-Brown M J, Bottcher C and Strayer M R 1989 *Phys. Rev.* **A40** 2831
Rhoades-Brown M J 1990 in *"Can RHIC be used to test QED?"* BNL–52247
Rhoades-Brown M J and Weneser J 1991 *Phys. Rev.* **A44** 330
Rumrich K, Momberger K, Soff G, Greiner W, Grün N and Scheid W 1991
    Phys. Rev. Lett. **66** 2613
Strayer M R, Bottcher C, Oberacker V E and Umar A S 1990 *Phys. Rev.* **A41**, 1399
Vane C R, Datz S, Dittner P F, Krause H K, Belkacem A, Feinberg B, Gould H,
    Schuch R, Knudsen H and Hvelplund P 1991 in *Abstracts of XVII ICPEAC*, this
    conference, p. 497
Williams E J 1934 *Phys. Rev.* **45** 729
Wu J S, Bottcher C and Strayer M R 1990 *Phys. Lett.* **252B**, 3

Electron capture in very low energy collisions of multicharged ions with H and D in merged beams

C. C. Havener, F. W. Meyer, and R. A. Phaneuf

Oak Ridge National Laboratory, Oak Ridge, Tennessee 37831-6372 USA

ABSTRACT: An ion-atom merged-beams technique is being used to measure total absolute electron-capture cross sections for multicharged ions in collisions with H (or D) in the energy range between 0.1 and 1000 eV/amu. Comparison between experiment and theory over such a large energy range constitutes a critical test for both experiment and theory. Total capture cross-section measurements for O^{3+} + H(D) and O^{5+} + H(D) are presented and compared to state selective and differential cross section calculations. Landau-Zener calculations show that for O^{5+} the sharp increase in the measured cross section below 1 eV/amu is partly due to trajectory effects arising from the ion-induced dipole interaction between the reactants.

1. INTRODUCTION

With the merged-beams technique it is now possible to measure with a single apparatus total electron capture cross sections for multicharged ions in collisions with H or D over an energy range covering four orders of magnitude: 0.1 eV/amu to over 1000 eV/amu. Comparison of experiment with theory over such a wide energy range provides a stringent test of our understanding of such processes which are fundamental to modelling of both laboratory and astrophysical plasmas. Simple scaling laws (e.g., see Gilbody 1986) which have been used to parametrize the behavior of the total capture cross section in the keV/amu energy range, are not appropriate in this range. At these low energies, theoretical calculations must take into account that the nuclear motion between collision partners is slow compared to the orbital motion of the active electrons of the system. Electrons of the temporary quasi-molecule formed in the collision have sufficient time to adjust to the changing interatomic field as the nuclei approach and separate. State-of-the-art molecular orbital coupled-state calculations are rather complex and, at these low energies, treat the collision dynamics quantum-mechanically.

Comparison between theory and experiment over this large energy range permits evaluation of a number of the theoretical parameters and methods used in these calculations. For example, for the O^{5+} + H(D) system, such comparison (Andersson et al. 1991) at the higher energies permits evaluation of theoretical parameters (e.g., adiabatic potential energies and non-adiabatic coupling matrix elements) which are used to calculate the collision dynamics throughout the whole energy range. As will be illustrated for both O^{3+} and O^{5+}, electron capture for these systems is extremely state-selective at low energies, with the dominant (nl) final state of the multicharged ion changing as a function of energy.

In addition, there are several aspects of the theoretical calculation of electron capture cross sections whose importance is energy dependent. For example, electron translation effects are of decreasing importance below 1 keV/amu (Gargaud and McCarroll 1985). Calculations have predicted that rotational coupling mechanisms, while important for some systems at energies around 1 keV/amu (Gargaud et al. 1988), are of negligible importance below 10 eV/amu. As collision energies decrease below 1 eV/amu, the ion-induced dipole attraction between the ion and neutral may become strong enough to significantly affect the reactant trajectories. These trajectory effects can lead to an enhancement in the cross section that increases with decreasing collision energy. In addition, at these low collision energies several quantal calculations (Rittby et al. 1984, Shimakura and Kimura 1991) have predicted "orbiting" resonances. These resonances are due to the excitation of ro-vibrational states in the shallow potential well formed by the attractive induced dipole in combination with the repulsive angular momentum barrier.

The ion-atom merged-beams apparatus (Havener et al. 1989) at Oak Ridge National Laboratory has recently been improved (Havener and Phaneuf, in prep.), allowing measurements to access collision energies as low as 0.1 eV/amu. Previous measurements (Havener et al. 1989, Huq et al. 1989, Havener et al. 1991) have been restricted to the 1 to 1000 eV/amu collision energy range. The first measurements to extend significantly below 1 eV/amu were made for the O^{3+} and O^{5+} + D systems and are presented here. Comparison with previous measurements and with state-selective (Gargaud 1987) and differential (Andersson et al. 1991) cross-section calculations illustrates the state selective nature of electron capture and accompanying, potentially large angular scattering of the products. To provide some physical insight into the collision dynamics below 1 eV/amu, Landau-Zener calculations are used to determine the energy dependence of the capture cross section for different reactant trajectories. Resultant estimates of the cross section show that the observed sharp increase in the cross section for O^{5+} + D below 1 eV/amu can be attributed in part to the ion-induced dipole attraction modifying the reactant trajectories. This enhancement in the cross section can be further amplified by considering collisions with H atoms rather than the heavier isotope D.

2. EXPERIMENTAL

The merged-beams method (Havener et al. 1989, and ref. within) is well suited for these measurements. In this technique, beams of neutral atoms and multicharged ions each having energies in the keV range are merged, resulting in a relative velocity of the two beams that can be "tuned" over a very large range. Figure 1 is a simplified schematic of the apparatus. The multicharged ion beam X^{q+} is merged electrostatically with a neutral H or D beam. The merged beams interact in a field-free region for a distance of 47 cm, after which the primary beams are magnetically separated from each other and from the product or "signal" $H(D)^+$ ions. The $X^{(q-1)+}$ product of the reaction is not measured separately, but is collected together with the primary X^{q+} in a large Faraday cup. The neutral beam intensity is measured by secondary-electron emission from a stainless steel plate, and the signal H^+ or D^+ ions are recorded by a channel electron multiplier operated in pulse-counting mode. A 99.98% pure ground-state beam of H or D atoms is produced by passing a 6- to 9-keV beam of H^- or D^- ions through the optical cavity of a 1.06-μm Nd:YAG laser, where up to 600 W of continuous power circulates and typically 0.5% of the negative ions undergo photodetachment. A nearly parallel beam of H(D) atoms is produced having a diameter of 2 to 4 mm FWHM and an equivalent intensity of 10 to 20 (particle) na. The divergence of this beam is typically less than 0.2°. A 50- to 90-keV, 2- to 5-μA beam of X^{q+} ions is produced by the ORNL-ECR source with a typical diameter of 6 to 8 mm FWHM in the merge path and a divergence of less than 0.5°. The finite divergence of the primary beams results in a distribution of merging angles, creating an energy spread of about 0.1 eV/amu at collision energies near 0.1 eV/amu.

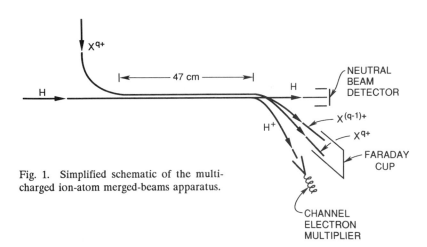

Fig. 1. Simplified schematic of the multi-charged ion-atom merged-beams apparatus.

Electron-capture cross sections are determined absolutely by measuring the rate of $H^+(D^+)$ ions produced by the beam-beam interaction over the merged path. The cross-section value is determined at each velocity from directly measurable parameters which include the signal count rate, intensities of the two beams, and the form factor which is a measure of the spatial overlap of the two beams. The integrated three-dimensional form factor is estimated from two-dimensional measurements of the overlap at three different positions along the merge path.

There have been some recent changes that have led to significant improvements to the technique. These improvements and their benefits are discussed elsewhere (Havener and Phaneuf, in prep.) in more detail, and are only briefly outlined here. While most of the beam-beam signal is generated at the beginning of the merge path where the overlap of the beams is typically better, the background due to H stripping on background gas is created uniformly along the merge-path. For this reason, improvements in vacuum and reduction of the merge-path from 80 to 47 cm has resulted in an increase in the signal-to-noise by at least a factor of 3, thereby extending the range of measurements downward to almost a factor of 10 lower in energy. Reduction of the small excited state component of the H(D) beam formed by stripping of H^- on background gas has resulted in a 99.98% pure ground-state beam. Such beam purities are essential, since even a small residual fraction of excited states can contribute significantly to the beam-beam signal because of the significantly larger cross sections for electron capture from excited atoms by multicharged ions. Enlargement of the physical aperture of the signal collection optics has resulted in a significant increase in the angular collection. Ion-trajectory modelling of the product D^+ trajectories has indicated that for a typical 8 keV D neutral beam, the angular collection of the apparatus was increased from an average angle of $1.8°$ to a minimum of $2.3°$.

3. ANGULAR COLLECTION

An important advantage of the merged-beams technique in low-energy measurements is the potentially large angular collection of the reaction products. The low-energy electron capture collisions under study are exoergic and both products are positively charged, so that significant angular scattering can occur in the center-of-mass frame (Olson and Kimura 1982). However, due to the kinematic frame transformation, this angular scattering is significantly compressed

Fig. 2. Angular scattering and collection in the center-of-mass frame for the O^{5+} + (8 keV) D collision system. The angular collection estimates correspond to the original ($\theta_c^{lab} = 1.8°$) and to the present ($\theta_c^{lab} = 2.3°$) apparatus. The angular scattering estimate corresponds to a simple "half-Coulomb" Rutherford scattering calculation (see text).

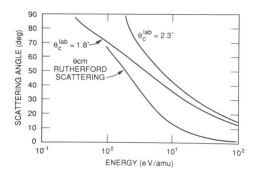

in the laboratory frame, the frame in which the products are collected. As an example, Fig. 2 shows the angular collection for the O^{5+} + (8 keV) D system in the center-of-mass frame for a lab frame angular collection of 1.8 and 2.3°. Note that with the present apparatus, which has a 2.3° minimum angular collection, all products will be collected below 2 eV/amu for this system. Also shown in Fig. 2 is the Rutherford minimum scattering estimate (Olson and Kimura 1982). Recent detailed differential cross-section calculations (Andersson et al. 1991) predict significantly larger scattering for the O^{5+} + D system, requiring an angular collection in the lab of at least 2.3° for a collision energy between 1 and 10 eV/amu. Further discussion specific to this system is presented in the next section.

4. CROSS-SECTION MEASUREMENTS

The ion-atom merged-beams apparatus has been used to measure total capture cross sections for O^{3+} and O^{5+} + D collisions; these measurements along with other experimental and theoretical results are shown in Figs. 3-5. The error bars shown correspond to an uncertainty in the reproducibility of the measurements estimated with a 90% confidence level. The absolute uncertainty in the measurements corresponds to about 12% and must be added in quadrature to the relative uncertainty. For both the O^{3+} + H(D) and O^{5+} + H(D) collision systems, the merged-beams data join smoothly with other measurements (Phaneuf et al. 1982, Meyer et al. unpublished) at the higher energies based upon ion beam-gas target methods, and verify the normalization methods used for the latter.

For the O^{3+} + H(D) system the present measurements lie significantly below the calculations of Bienstock et al. (1983) which predict a large contribution from capture to the $(1s^22s^22p3s)$ and $(1s^22s^22p3p)$ configurations at these energies (see Fig. 3). Capture to the 3d state is not expected to contribute significantly in this energy range. Both the old and more recent merged-beams measurements are more consistent with the calculations of Gargaud et al. (1989) which predict a significantly smaller contribution from capture to the 3p state at the lower energies. Both calculations predict that capture to the 3s state dominates at the higher energies (see Fig. 4). This prediction agrees with the state-specific translational spectroscopy measurements of Wilson et al. (1988) which were performed at collision energies between 200 and 700 eV/amu.

Our previous O^{3+} measurements, which could only be extended down to 1 eV/amu, were not able to verify the cross section rise due to the predicted capture to the 3p state at low energies. However, the present merged-beam measurements, which extend down to 0.1 eV/amu, follow closely the predicted energy dependence of the 3p capture cross section. Nevertheless, as can be seen in Fig. 4, there remains a discrepancy between the measurements and theory throughout

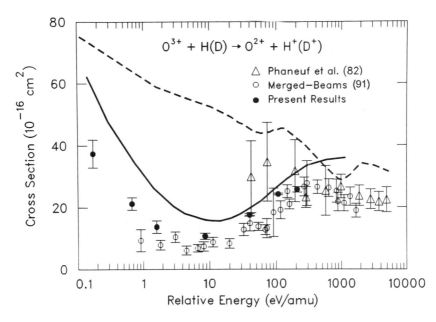

Fig. 3. Comparison of merged-beams data for O^{3+} + H(D) with other measurements (Phaneuf et al. 1982) and theoretical calculations (dashed curve, Bienstock et al. 1983; solid curve, Gargaud et al. 1989).

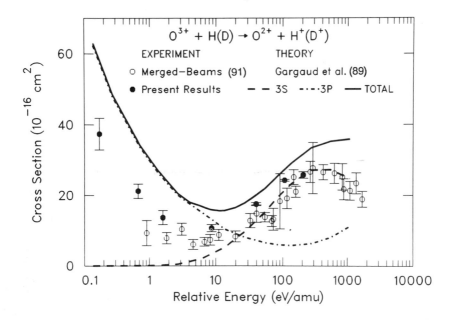

Fig. 4. Comparison of merged-beams data for O^{3+} + H(D) with theoretical calculations (Gargaud et al. 1989) for capture to the 3s and 3p states.

the whole energy range. A possibility exists that the calculations may overestimate the contribution due to capture to the 3p state. However, it should be noted that a finite fraction of the B-like O^{3+} ion beam is in the $(1s^22s2p^2)^4P$ metastable state, which may also contribute at least in part to this discrepancy. This fraction has been previously estimated (Phaneuf et al. 1982) to be on the order of 16% for beams from a Penning multicharged ion source.

In Fig. 5, the merged-beam measurements for O^{5+} + H(D) are presented along with other measurements and theoretical calculations (Gargaud 1987, Bottcher unpub.). Unlike the O^{3+} case, the present measurements for O^{5+} show as much as a 30% deviation from previous merged-beams results (Havener et al. 1989) in some energy regions. This discrepancy is now understood as having originated from two sources: the small fraction of excited states in the H(D) beam and the insufficient angular collection of the $H^+(D^+)$ products in the previous apparatus between 1 and 10 eV/amu. It has been observed experimentally that for collision energies less than about 50 eV/amu, the collision-energy-dependent beam-beam signal due to the estimated 0.1% of the H beam which was in excited states can account for a significant fraction of the measured signal. In the first merged-beams measurements, this contribution was not subtracted from the measured signal and therefore resulted in artificially high cross section values below 50 eV/amu.

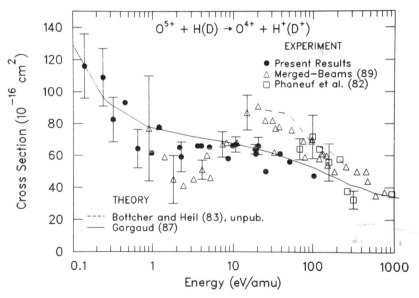

Fig. 5. Comparison of merged-beams data for O^{5+} + H(D) with other measurements and theoretical calculations.

The previous O^{5+} measurements also show a distinct minimum in the cross section between 1 and 10 eV/amu. For this energy range, the calculations predict that the total capture cross section consisted of a decreasing contribution from capture to the 4s state and an approximately equally increasing contribution from 4p capture. As reported recently (Andersson et al. 1991), no reasonable variation of theoretical parameters was successful in bringing theory into agreement at a few eV/amu without destroying the good agreement with the measurements at energies greater than 20 eV/amu. However, differential-cross-section calculations, as shown in Fig. 6, show that while the angular scattering of the 4s capture channel is forwardly peaked, the capture to the 4p state results in large scattering angles in the center-of-mass frame. A detailed

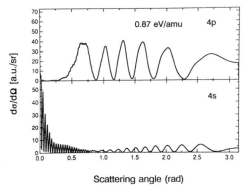

Fig. 6. Differential cross sections (Andersson et al. 1991) for the 4s and 4p electron capture channels in the center-of-mass frame for the O^{5+} + H system at 0.87 eV/amu collision energy.

analysis of the angular collection of the experiment versus angular scattering has been performed by Andersson et al. (1991). The angular collection of the present apparatus which is at least 2.3° in the laboratory frame, is sufficient to guarantee no significant loss of signal. Indeed, as can be seen in Fig. 5, the present O^{5+} measurements agree well with the calculations of Gargaud. The sharp increase in the cross section below 1 eV/amu is attributed to a trajectory-enhanced capture into the 4p state. The degree of this enhancement is estimated using simple Landau-Zener calculations and is discussed in the next section.

5. TRAJECTORY EFFECTS AT LOW ENERGIES

As the ion and neutral approach, a dipole is induced in the neutral atom, causing an attractive force between them. The resultant interaction potential is given by

$$V(r) = -\alpha \ q^2/2r^4 \tag{1}$$

where α is the polarizability of H, q is the charge of the ion, and r is the internuclear separation. For sufficiently large q and for collision energies below 1 eV/amu the attraction is strong enough to significantly modify the trajectories of the reactants and thereby possibly affect the total capture cross section. Indeed, the simple classical orbiting model (Gioumousis and Stevenson 1958), which uses a straightforward geometrical interpretation of orbits that decay into a reaction sphere, predicts a strong $1/v$ increase in the cross section at these low energies.

To obtain an estimate of such trajectory effects, Landau-Zener calculations were used to determine the energy dependence of the cross section with and without the ion-induced dipole attraction taken into account. The capture cross section is calculated by integrating the total transition probability, $2p(1-p)$, over the impact parameter b,

$$\sigma = 2\pi \int_0^{b_{max}} 2p(1-p) \ bdb \tag{2}$$

where p is the Landau-Zener probability given by

$$p = \exp\left(-\frac{2\pi\Delta^2}{v_r |H'_{11} - H'_{22}|}\right) \tag{3}$$

Δ is the energy splitting at the avoided crossing, R_c; v_r is the radial component of the collision velocity; and $|H'_{11} - H'_{22}|$ is the difference in slopes of the diabatic potential energy curves at the crossing.

Since the crossings occur at relatively large internuclear separations, the potential energy, V, of the initial state can be approximated by that due to the ion-induced dipole interaction [Eq. (1)] and the final-state potential energy by $(q-1)/r$ (a.u.). Δ is estimated by matching, for a specific state, the maximum in the Landau-Zener cross section to the corresponding maximum in the full quantal calculations.   Having obtained Δ by normalization in this way, Landau-Zener calculations for other final states which cross the initial state at different internuclear separations R'_e could be performed by estimating the appropriate energy splitting using an $R_e^2 e^{-R_c}$ scaling (Butler and Dalgarno 1980). The Landau-Zener cross sections could be evaluated with straight-line or with trajectories modified by the ion-induced dipole potential (polarization interaction) by straightforward modifications to H_{11}, b_{max} and v_r. The polarization interaction results in the reactant trajectories with impact parameters larger than R_c being able to access the avoided crossing where capture occurs. In Eq. (2) the integrand thus becomes

$$b_{max} = R_c \sqrt{1 - V/E} \tag{4}$$

where E is the collision energy. Also the radial velocity, v_r, used in Eq. (3), is modified along the incident trajectory and is given by

$$v_r = v_0 \sqrt{1 - b^2/r^2} \tag{5}$$

where v_0 is the initial collision velocity.

Figure 7a shows the results for $O^{3+} + D$. The Landau-Zener parameters were estimated by normalization at the calculated (Gargaud et al. 1989) cross section maximum for capture to 3s. As may be seen in the figure, the energy dependence for capture to 3s and for capture to 3p agree qualitatively with the full quantal calculations of Gargaud et al. (1989) (see Fig. 4). The Landau-Zener estimates are plotted with and without the effect of the polarization interaction. Note that only for capture to 3p at the lowest energies is there any enhancement due to the interaction.   Figure 7b shows similar calculations for $O^{5+} + D$, where the ion-induced polarization attraction is stronger due to the increased charge. For O^{5+}, capture to the 4s shows only a small enhancement at the low energies, while capture to the dominant 4p channel shows a significant enhancement that increases toward lower energies. Capture to the 4d (not shown) shows a similar enhancement but only accounts for less than 20% of the total cross section. At an energy of 0.1 eV, the enhancement in the 4p accounts for 30% of the 4p capture cross section. The sharp increase at low energies in the measured cross section can be interpreted as evidence of trajectory effects due to the ion-induced dipole attraction for this system.

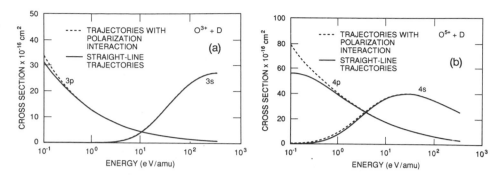

Fig. 7. Landau-Zener calculations for the O^{3+} + D system(a) and O^{5+} + D system (b) for capture to specific states (as labelled). Both straight-line (solid line) and trajectories with the polarization interaction (dashed line) are shown.

Until now we have not made a distinction between measurements with atomic hydrogen or the heavier isotope deuterium. Generally in our past measurements the heavier isotope was used both to decrease the angular scattering after the reaction, and to increase the angular collection of the apparatus in the center-of-mass frame. Detailed angular scattering estimates (Andersson et al. 1991) show that for measurements with a 6-keV-H or 8-keV beam, the angular collection of the present apparatus is sufficient below 1 eV/amu to collect all the signal. To obtain direct evidence for the enhancement in the cross section due to trajectory effects, we are in the process of performing measurements with both H and D below 1 eV/amu for O^{5+}. The expected enhancement for these different systems has been estimated using Landau-Zener calculations for capture to the 4p state and is shown in Fig. 8. In the figure the cross section is calculated for an H or D projectile, with and without trajectory effects. The difference in enhancement at 0.1 eV/amu between H and D is estimated to be on the order of 20×10^{-16} cm^{-2}, or 20% of the total cross section with H. The full quantal calculations (Gargaud 1987) for the O^{5+} + H system have also been performed for O^{5+} + D for energies as low as 0.87 eV/amu. At this energy, though, the enhancement amounts only to a few percent in the total capture cross section, in agreement with the Landau-Zener estimate.

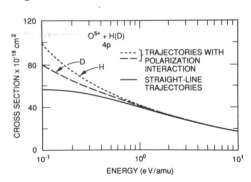

Fig. 8. Landau-Zener 4p cross section calculations for the O^{5+} + H (short-dashed line) and D (long-dashed line) collision systems with trajectories modified by polarization effects. Also shown is the calculation due to straight-line trajectories for H or D (solid line).

6. CONCLUSIONS

The merged-beams method now makes possible absolute total electron-capture cross section measurements in the energy range between 0.1 and 1000 eV/amu for multicharged ions colliding with ground-state H or D atoms. Comparison with state-of-the-art theory over such a large energy range permits a systematic evaluation of both experiment and theory and can provide further insights into collision mechanisms and dynamics. New measurements for O^{5+} have observed the predicted rise in the cross section due to capture to the 3p state below 1 eV/amu. Detailed differential cross section calculations have predicted surprisingly large angular scattering for the O^{5+} system. An upgrade of the merged-beams apparatus to accommodate such scattering has resolved the discrepancy between measurement and theory for this system. Landau-Zener estimates for the cross section indicate that trajectory effects may be significant for the O^{5+} + D system at low energies. Comparison with measurements planned for O^{5+} + H should be capable of directly probing such enhancements.

7. ACKNOWLEDGEMENTS

The research was sponsored by the Division of Chemical Sciences, U.S. Department of Energy, under Contract No. DE-AC05-84OR21400 with Martin Marietta Energy Systems, Inc. The authors are grateful to Lars Andersson, M. Gargaud, and R. McCarroll for fruitful discussions.

8. REFERENCES

Andersson L R, Gargaud M and McCarroll R 1991 *J. Phys. B: At. Mol. Opt. Phys.* **24** 2073
Bienstock S, Heil T G and Dalgarno A 1983 *Phys. Rev. A* **27** 2741
Bottcher C and Heil T G, unpublished; for method, see 1982 *Chem. Phys. Lett.* **86** 506
Butler S E and Dalgarno A 1980 *Astrophys. J.* **241** 838
Gargaud M and McCarroll R 1985 *J. Phys. B: At. Mol. Opt. Phys.* **18** 463
Gargaud M 1987 Doctoral Thesis, L'Université de Bordeau I
Gargaud M, McCarroll R and Opradolce L 1988 *J. Phys. B: At. Mol. Opt. Phys.* **21** 521
Gargaud M, McCarroll R and Opradolce L 1989 *Astron. Astrophys.* **208** 251
Gioumousis G and Stevenson D P 1958 *J. Chem. Phys.* **29** 294
Gilbody H B 1986 Adv. At. Mol. Phys. **22** 143
Havener C C, Huq M S, Krause H F, Schulz P A and Phaneuf R A 1989 *Phys. Rev. A* **39** 1725
Havener C C, Nesnidal M P, Porter M R and Phaneuf R A 1991 *Nucl. Instrum. Methods B* **56/57** 95
Havener C C and Phaneuf R A, in preparation
Huq M S, Havener C C and Phaneuf R A 1989 *Phys. Rev. A* **40** 1811
Meyer F W, Howald A M, Havener C C and Phaneuf R A, unpublished
Olson R E and Kimura M 1982 *J. Phys. B: At. Mol. Opt. Phys.* **15** 4231
Phaneuf R A, Alvarez I, Meyer F W and Crandall D H 1982 **26** 1892
Rittby M, Elander N, Brändas E and Bárany A 1984 *J. Phys. B: At. Mol. Opt. Phys.* **17** L677
Shimakura N and Kimura M, to be published in *Physical Review A*
Wilson S M, McCullough R W and Gilbody H B 1988 *J. Phys. B: At. Mol. Opt. Phys.* **21** 1027

Slow collisions between ions of very high charge and neutral atoms

H Cederquist, H Andersson, E Beebe, C Biedermann, L Broström, Å Engström, H Gao, R Hutton[+], J C Levin[++], L Liljeby, M Pajek, T Quinteros, N Selberg, and P Sigray

Manne Siegbahn Institute of Physics, S-104 05 Stockholm, Sweden

[+]Present address: University of Lund, S-223 62 Lund, Sweden

[++]University of Tennessee, Knoxville, TN 37996-1200, U.S.A.

ABSTRACT: We have measured cross sections for single-electron capture, true double-electron capture, and transfer ionisation in slow (0.1-0.2 a.u.) Xe^{q+}-Xe, He ($15 \leq q \leq 42$) collisions. The ratio of two- to one-electron removal from Xe is twice of that of He for high q, although no difference is expected from the extended classical over-the-barrier model. The probabilities for radiative stabilisation of double Rydberg states populated after transfer of two electrons to the projectile rise sharply in the $q=25$ to $q=35$ region for Xe and He targets. Finally, we report on pure ionisation in slow Xe^{q+}-Xe collisions.

1. INTRODUCTION

In this invited paper we will discuss three different aspects of slow collisions between highly charged ions and atoms. First we will consider the transfer-excitation mechanism (Cederquist 1991), in which one electron is transferred from the target (B) to the highly charged projectile (A^{q+}), while another electron is excited:

$$A^{q+} + B \rightarrow A^{(q-1)+} + (B^+)^*. \tag{1}$$

Then we will turn our attention to the problem of radiative stabilisation of double Rydberg states, formed after transfer of two electrons from the target to the projectile:

$$A^{q+} + B \rightarrow A^{(q-2)+}(nl, n'l') + B^{2+} \rightarrow$$

$$\xrightarrow{P_{rad}} A^{(q-2)+} + B^{2+} \tag{2}$$

$$\xrightarrow{(1-P_{rad})} A^{(q-1)+} + B^{2+} + e^- \tag{3}$$

We define P_{rad} as the probability that both electrons, initially captured to the double Rydberg state $(nl, n'l')$ ((n, n') and (l, l') are the principal and angular momentum quantum numbers, respectively), remain bound to the projectile at large internuclear separations.

Finally, we will discuss experimental indications that pure ionisation:

$$A^{q+} + B \rightarrow A^{q+} + B^+ + e^- \tag{4}$$

takes place for high q even in these *slow* collisions, where orbital velocities of active electrons are much larger than projectile velocities.

For slow collisions it is meaningful to discuss charge-transfer processes as well-localized transitions between adiabatic quasi-molecular states of the same spin and symmetry. It has been demonstrated repeatedly (see e.g. Cederquist *et al* 1985, Biedermann *et al* 1990, or Andersson *et al* 1991) that this approach is fruitful when all the relevant quasi-molecular states are taken into consideration. However, it becomes prohibitively difficult to include all possible reaction channels for the high densities of projectile capture states, which occur at high projectile charge. Then the much more simplified approach of the extended classical over-the-barrier model (Bárány *et al* 1985 and Niehaus 1986) becomes useful since it assumes that a projectile capture state always is available once the outermost target electron is classically allowed to pass the internuclear potential barrier between the target and the projectile. This assumption becomes more justified for higher q, since the density of projectile capture states increases as

$$n^2 = q^{3/2}/I_1, \tag{5}$$

for hydrogen-like capture states in the high-q limit (I_1 is the first ionisation potential of the target). Thus experimental investigations of charge transfer to projectiles of very high q are extremely interesting since we may then discuss electron-transfer mechanisms in very simple terms without having to bother with the details of complicated energy-level structures.

We will give a brief account of the experimental technique in the next section. In the last section we will present measurements of the ratio of the cross sections for two- and one-electron removal, σ_2/σ_1, for Xe^{q+}-Xe and Xe^{q+}-He ($15 \leq q \leq 42$) collisions. We find that σ_2/σ_1 approaches 0.64 for the Xe target, while the corresponding value is 0.33 for He. This difference is not expected from the original extended classical over-the-barrier (ECB) model, but it can be accounted for quantitatively by associating a range of impact parameters with a specific three-step transfer excitation process. We will further present results for the probability for radiative stabilisation P_{rad} as function of q for double Rydberg states populated after transfer of two electrons from either Xe or He. Surprisingly enough, $P_{rad}(q)$ is quite similar for the two targets although more asymmetric and lower excited states are formed in the case of the He target. Finally, we will report absolute cross sections for pure ionisation in slow Xe^{q+}-Xe collisions for $q = 15$, 20, 25, 30, and 35.

2. THE EXPERIMENTAL TECHNIQUE

For the present set of experiments we injected a 90% enriched ^{136}Xe gas into the cryogenic electron beam ion source (CRYSIS) at the Manne Siegbahn Institute of Physics in Stockholm. Xe ions were extracted at about 4.5 kV by raising the bottom of the potential well until the ions no longer were trapped axially by the potential distribution of the drift tubes (see e.g. Andersson *et al* 1990). Radial trapping was provided by a 2 T axial magnetic field, furnished by the superconducting solenoid. The current of the 10 keV electron beam was 200 mA, the current density ~200 A/cm^2, and the total loss between cathode and collector was less than 1.5 permille. The source output was diagnosed by deflecting a small fraction of the 50 μs pulse into a ~3 m long drift tube terminated by an open electron multiplier. The charge state distributions peaked at $q = 26$, 32, 36, and 42 for ion-confinement times of 150, 250, 500, and 2000 ms.

The beam was transported without any essential intensity loss over a distance of about 6 m (at a pressure of ~1·10^{-9} Torr) to the entrance slits of a 90° double-focusing magnet with 0.5 m radius. At this point the current was 1-5 μA (in the 50 μs pulse) and the focused beam had a diameter of about 20 mm. After selection of the mass-to-charge ratio at the exit slits of the magnet, the beam was focused by a quadrupole triplet and directed by two pairs of vertical and horisontal deflectors to the 0.5 mm entrance aperture of the 10 mm long collision gas cell.

During data acquisition the beam pulses were extended to about 50 ms. This was done in order to avoid saturation in the position-sensitive channelplate detector used to register the highly charged Xe ions after their passage through the collision cell and a 160° electrostatic hemispherical energy analyzer. Recoil ions were extracted from a 3 mm central region in the gas cell by a static potential of 100 V providing an electric field perpendicular to the projectile beam of ~30 V/mm. A second potential (V_B c.f. Figure 1) was set to 254 V in order to optimize the resolution in the time-of-flight spectrum. The recoil ions detected by a Ceratron at an energy of 4.5q kV provided the start pulses of a time-to-amplitude converter (TAC), which was stopped by a suitably delayed fast signal from the back of the second channelplate in the projectile detector. Several charge states of the outgoing projectile could be analyzed simultaneously for a fixed analyzer voltage setting due to the large separation (30 mm) between the analyzer spheres. Data was collected in list mode by means of a PDP 11/55 computer during the time the beam pulse went through the apparatus in order to minimize the influence from the detector background.

Figure 1: Schematic of the experimental set up. Highly charged Xe^{q+} ions ($15 \leq q \leq 42$) enter the collision gas cell. The voltage between the analyzer spheres, $V_+ - V_-$, is constant during a measurement and we indicate a situation where the direct beam, Xe^{q+}, and a charge reduced beam, $Xe^{(q-1)+}$, hit the detector. Recoil ions, produced in the gas cell, are extracted by a static electric field provided by a positive pusher voltage, V_P. The recoils are further accelerated by the 'buncher potential', V_B, and the potential ,V_C of the Ceratron detector (c.f. text).

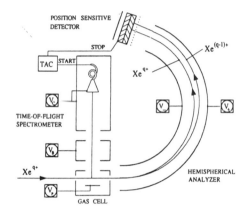

The position-sensitive detector was placed at the position where the center of the collision cell was imaged by the 100 mm radius energy analyzer. An example of the projection of the two-dimensional detector image on the axis of energy dispersion is shown in Figure 2 for the case of Xe^{25+}-Xe collisions. The largest peak is due to the primary beam (Xe^{25+}), while the two smaller peaks are due to the net transfer of one and two electrons, respectively. The middle and the lower spectra of Figure 2 are the time-of-flight distributions recorded in coincidence with Xe^{24+} and Xe^{23+}, respectively. In the lower spectrum the occurence of Xe^{1+} recoils are entirely due to double collisions. This provides a very good handle on the fraction of double collisions. Figure 3 displays an example for Xe^{37+}-Xe collisions. In Figure 3a, we show the projected position spectrum in coincidence with Xe^+ recoils. The large peak is due to the single-capture process, while the smaller structure (labelled '35+') is due to double collisions. In Figure 3b coincidences with Xe^{2+} are shown. The (q-2)-peak (35+) results mainly from true double-capture events (c.f. equ. (2)), with a minor contribution from double collisions. We estimate the double-collision fraction by multiplying the intensity of the transfer ionisation peak (coincidences between Xe^{36+} and Xe^{2+} in Figure 3b) with the ratio of (q-2) to (q-1) events in the Xe^+ coincidence spectrum (Figure 3a). Target gas pressures were set between 0.04 and 0.20 mTorr, depending on the charge and the target, in order to keep rates of double collisions within reasonable limits.

Primary Xe-ion intensities were typically limited to ~100-200 s^{-1} for each 50 ms long CRYSIS pulse independent of charge state in order to avoid saturation of the detector. The repetition rate of the ion-production cycle ranged from about 15 Hz to 0.5 Hz for the lowest (q=15) to the highest (q=42) projectile charge, respectively.

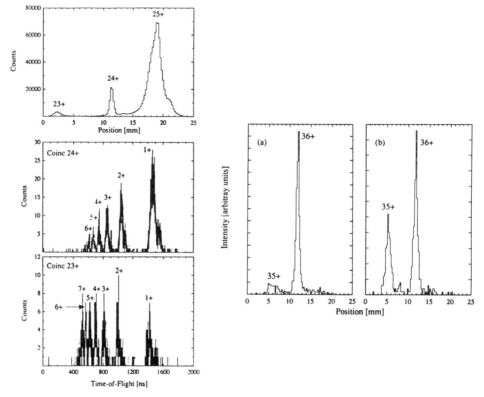

Figure 2: The intensity on the two-dimensional position-sensitive detector projected on the axis of energy dispersion for Xe^{25+}-Xe collisions (upper spectrum). The active diameter of the detector is 25 mm. The differences in peak widths are due to the different focusing properties for the different paths through the analyzer. In the middle spectrum we show a time-of-flight spectrum coincident with Xe^{24+}, while the lower spectrum exhibits coincidences with Xe^{23+}. The 1+ peak in the latter spectrum is exclusively due to double collisions (c.f. text).

Figure 3: The intensities on the two-dimensional position-sensitive detector projected on the axis of energy dispersion in coincidence with Xe^+ (a) and Xe^{2+} (b) recoils for Xe^{37+}-Xe collisions. In both spectra, the peaks around 12 mm are due to Xe^{36+}, while the peaks at about 6 mm are due to Xe^{35+}. Clearly the small peak produced by the Xe^{35+}-Xe^+ coincidences in the left spectrum is entirely due to double collisions. Through the relation between this peak and the single capture peak we obtain the fraction of the double-collisions in the peak for true double-electron capture (Xe^{35+}-Xe^{2+} coincidences).

3. RESULTS AND DISCUSSIONS

A major part of the discussion to follow will be based on the extended classical over-the-barrier (ECB) model. Therefore we start with a brief account of its experimental justifications at high q.

The ECB model gives the absolute cross sections

$$\sigma_m = \pi(R_m^2 - R_{m+1}^2) \qquad (6)$$

for removing m electrons from the target. R_m is the internuclear distance at which the m:th target electron is classically allowed to enter the projectile phase space. Furthermore projectile scattering angles (see e.g. Hvelplund *et al* 1987), collision inelasticities (see e.g. Hvelplund *et al* 1985), and thereby projectile capture state populations follow from the assumptions of the ECB model. Danared *et al* (1987) have shown that the projectile scattering angle essentially is a function of the recoil-ion charge state, which is in accordance with model predictions, while e.g. Hvelplund *et al* (1987) have shown that the projectile energy gain is well accounted for by the model. The expected n-state population has been demonstrated by Tawara *et al* (1985), while Martin *et al* (1989) have measured photon-emission from double Rydberg states populated in Ar^{9+}-Cs collisions. In Figure 4 we show a comparison between ECB and experimental cross sections for removing one (Figure 4a) and two (Figure 4b) target electrons from Xe in slow Xe^{q+}-Xe collisions for projectile charge states ranging between $q=15$ and $q=35$.

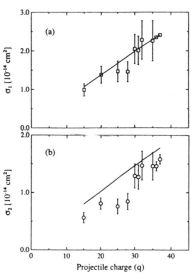

Figure 4: A comparison between experimental and extended classical over-the-barrier model cross sections for removal of one (a), σ_1, and two (b), σ_2, electrons from the target. The cross sections for $q = 36$ and $q = 37$ in (b) are obtained by normalizing experimental σ_1 to model σ_1 (c.f. (a)).

3.1 Transfer excitation

We have measured the ratio σ_2/σ_1 for Xe^{q+} ($15 \leq q \leq 42$) colliding on Xe and He. σ_1 is simply the cross section for single-electron capture

$$Xe^{q+} + B \rightarrow Xe^{(q-1)+} + B^+, \tag{7}$$

while σ_2 is the sum of the cross sections for true double-electron capture (2), σ_{DC}, and transfer ionisation (3), σ_{TI}. The ECB cross section for σ_1 is

$$\sigma_1 = \pi(R_1^2 - R_2^2) \tag{8}$$

for both Xe and He. The cross section for removing two electrons from Xe is:

$$\sigma_2 = \pi(R_2^2 - R_3^2), \tag{9}$$

whereas it is

$$\sigma_2 = \pi R_2^2 \tag{10}$$

for the He target. Incidentally σ_2/σ_1 should approach 0.71 for both Xe and He targets in the high q limit according to the ECB model. However, we measure σ_2/σ_1 to be 0.64 for Xe^{q+}-Xe collisions at the higher q investigated, while the ratio for Xe^{q+}-He collisions decreases slowly to a level of about 0.33. This difference was predicted by an earlier explanation (Cederquist 1991) for unexpectedly low σ_2/σ_1 for Xe^{q+}-He at somewhat lower q (<30). A three-step transfer excitation mechanism

$$Xe^{q+} + He \rightarrow Xe^{(q-1)+}(n) + He^+ \rightarrow$$

$$\rightarrow Xe^{(q-2)+}(n,n') + He^{2+} \rightarrow Xe^{(q-1)+}(n) + (He^+)^* \tag{11}$$

which rapidly becomes more important with increasing q may account for low σ_2/σ_1 in the case of the He target. For the Xe target, the distance R_3 where a third electron becomes quasi molecular is reached before a process similar to (11) may occur. The second of the two active electrons in (11) is recaptured to an excited state of the target at internuclear separations just inside R_2, where the second electron becomes quasi molecular. In Figure 5 we show for Xe^{30+}-He collisions that all quasi-molecular levels dissociating to transfer-excitation channels cross the double-capture channels just inside R_2. The prediction that transfer-excitation processes are much less important for multi-electron targets is corroborated by the present measurement of σ_2/σ_1 for Xe^{q+}-Xe, shown in Figure 6. In Figure 7 we show that the inclusion of transfer excitation (11) quantitatively explains the σ_2/σ_1 data for the He target. The large spread at lower q in Figure 7 is presumably due to the breakdown of the assumption about a quasi-continuum of projectile capture states (c.f. Andersson *et al* 1990).

Figure 5: Total Coulombic potential energies as functions of the internuclear distance R (a.u.) for different charge changing processes. The potential denoted 'IN' is the incident channel Xe^{30+}-He, the one labeled 'SC' is Xe^{29+}-He$^+$, while 'DC' is the Xe^{28+}-He^{2+} channel. The dashed lines show potentials for channels dissociating to transfer excitation ('TE') processes. The upper dashed curve corresponds to He$^+$-excitations at the ionisation limit, while the lower dashed curve gives the lowest excitation. Note that all these channels cross within a very narrow region of internuclear distances just inside R_2.

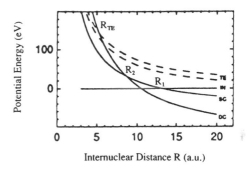

Figure 6: The ratio (σ_2/σ_1) between cross sections for removing two (σ_2) and one (σ_1) electrons from Xe in slow Xe^{q+}-Xe collisions as a function of the projectile charge state q. The line is from the extended classical over-the-barrier model $\sigma_2/\sigma_1 = (R_2^2 - R_3^2)/(R_1^2 - R_2^2)$.

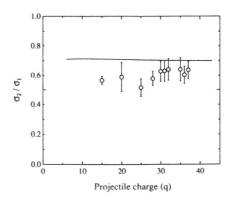

Figure 7: The ratio (σ_2/σ_1) between cross sections for removing two (σ_2) and one (σ_1) electrons from He in slow Xe^{q+}-He and Ar^{q+}-He collisions as a function of q (present measurement Xe^{q+} projectiles: $\triangle$, Ar^{q+} projectiles: $\bigcirc$ and Andersson *et al* 1988 Xe^{q+} projectiles:). The dashed line shows the results of the original classical over-the-barrier model $\sigma_2/\sigma_1 = R_2^2/(R_1^2 - R_2^2)$, while the full line represents the same model but with the inclusion of the transfer excitation mechanism (c.f. eq. (11) in the text). In the latter case $\sigma_2/\sigma_1 = (R_2^2 - R_{TE}^2)/(R_1^2 - R_2^2 + R_{TE}^2)$, where $R_{TE} = R_2/(1 + 2R_2/(q-3))$.

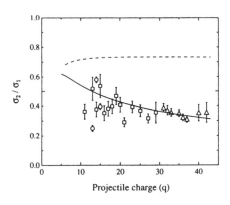

3.2 Radiative stabilisation

In earlier investigations at more moderate q it has been found that true double-electron capture dominates transfer ionisation at the lowest q, while the opposite relation prevails at somewhat higher q (see e.g. Justiniano *et al* 1984). This behaviour is well accounted for by the fact that doubly excited projectile states of excitations high enough for autoionisation are created only at sufficiently high q. An assumption about unit probability for electron emission (i.e. $P_{rad}=0$ in (2) and (3)) has in general been sufficient to explain the experimental data in the q-range between ~ 8 and ~ 20. It has , however, also been anticipated that the strong q scaling of the radiative decay may favour the charge-stabilisation process (2) at still somewhat higher q (Janev and Presnyakov 1981). For a specific state, $\Delta n=1$ electric dipole transition rates scale as q^4, while Coulombic autoionisation is independent of q. Against this it has been argued (Hansen 1989) that double Rydberg states formed after transfer of two electrons through

$$Xe^{q+} + B \to Xe^{(q-1)+}(nl) + B^+ \to Xe^{(q-2)+}(nl, n'l') + B^{2+} \qquad (12)$$

should autoionise rapidly due to the energy proximity to highly excited states of the $Xe^{(q-1)+}$ ions even at very high q. In Figure 8, we show absolute cross sections for transfer ionisation and double-electron capture for Xe^{q+}-Xe collisions. The latter cross sections increase by about one order of magnitude between $q=25$ and $q=35$, while the former ones are essentially constant throughout the same q-region. The probability for radiative stabilisation, as defined by (2) and (3), is given by:

$$P_{rad} = \sigma_{DC}/(\sigma_{DC} + \sigma_{TI}). \qquad (13)$$

Figure 8: Absolute cross sections for transfer ionisation ($\bigcirc$) and true double-electron capture ($\triangle$) for Xe^{q+}-Xe collisions as a function of the projectile charge state q.

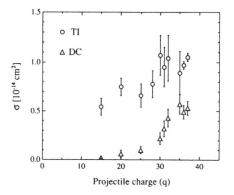

P_{rad} is then the product of probabilities for surviving from autoionisation for the intermediate states with excitation energies above the lowest $A^{(q-1)+}$ ionisation limit, averaged over all cascade paths and the distribution of initial capture states $(nl, n'l')$.

It is generally expected (Luc-Koenig and Bauche 1990 and Poirier 1988) that branching ratios for autoionisation should become smaller for more asymmetric (larger $n - n'$) doubly excited states. The ECB model predicts that this asymmetry is given by the first two ionisation potentials of the target. Thus since I_1/I_2 is much larger for He than for Xe a difference in $P_{rad}(q)$ for the two targets was expected. Surprisingly enough, however, $P_{rad}(q)$ for $Xe^{(q-2)+}$ double Rydberg states populated after capture of two electrons from Xe and He are very similar as shown in Figure 9. This indicates that the electronic structure of the incoming projectile rather than the exact nature of the initial capture state determines the subsequent stabilisation process. We note that the increase in P_{rad} occurs for Xe^{q+} projectiles with vacancies in the $3d$-shell, which allows a direct photon decay cascade along the yrast line down to the singly excited $Xe^{(q-2)+}$ system.

Figure 9: The probabilities for radiative sta-bilisation, P_{rad}, as functions of the projectile charge state q for double Rydberg states of $Xe^{(q-2)+}$ after transfer of two electrons from Xe (●) and from He (□ : Andersson *et al* 1988; △ : present measurement). The error bars are due to statistical uncertainties and uncertainties in the corrections for double col-lisions.

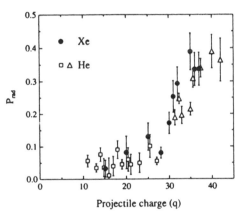

While the preferred n, n' population seems to be well accounted for by the ECB model, some uncertainty still remains about a possible l-state selectivity in the present (v=0.1-0.2 a.u.) velocity regime. A preference for high l has been demonstrated for collision velocities approaching 1 a.u. (Dijkkamp *et al* 1986), while collisions in the velocity range below 0.01 a.u. show a complete lack of l-selectivity (Nielsen *et al* 1985). The l, l' population of the initial capture process is very important for the discussion of the relaxation process since it has been argued that autoionisation rates decrease drastically with increasing l (Luc-Koenig and Bauche 1990 and Poirier 1988). Thus the tentative explanation given here for the $P_{rad}(q)$ behaviour for states populated through capture from Xe and He can certainly still be debated: For instance, are high l, l' states preferentially populated in the present collisions and how do the probabilities for radiative decay and autoionization scale with l and l'?

3.3 Pure ionisation in slow collisions

Pure ionisation (4) is known to dominate over electron capture at high collision velocities. In the low velocity regime, however, electron capture strongly dominates pure ionisation (see e.g. Hvelplund *et al* 1980). The reason for this is that electronic rearrangements in the low velocity regime in general are governed by transitions between adiabatic quasi-molecular potential curves of the same spin and symmetry. Therefore, since ionisation without accompanying electron capture by necessity is an endothermic process, there is no simple reaction path by which radial couplings can connect to the ionisation channel. One should also bear in mind that direct impact ionisation, which dominates pure ionisation at high velocities, is prohibited by a low energy transfer to the electron in the low velocity region (Hvelplund *et al* 1980).

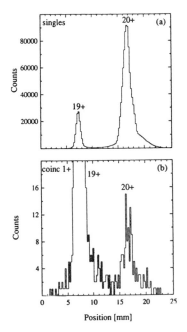

Figure 10: Projections of the intensities on the two-dimensional position-sensitive detector on the axis of energy (charge-state) dispersion for Xe^{20+}-Xe collisions. The singles spectrum (a) is shown together with the spectrum recorded in coincidence with Xe^+ recoil ions (b). The intensity at the position for Xe^{20+} projectile in (b) is assigned to pure ionisation.

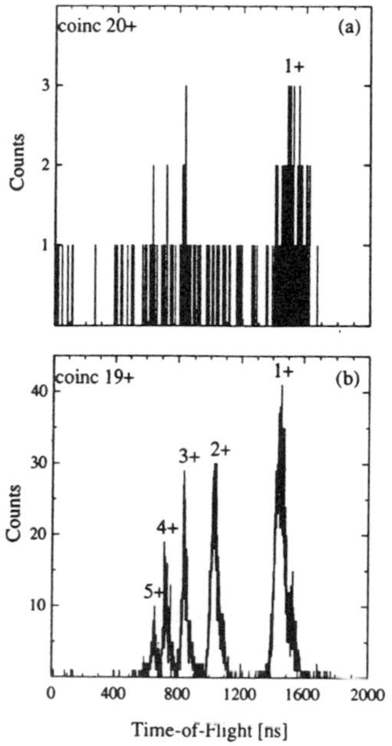

Figure 11: Time-of-flight spectra recorded in coincidence with Xe^{20+} (a) and with Xe^{19+} (b) for Xe^{20+}-Xe collisions. The structure to the left in (a) is possibly CO_2^+.

In figure 10 we display a comparison between a singles projectile charge state spectrum for Xe^{20+}-Xe and a projectile charge spectrum recorded in coincidence with Xe^{1+} recoils. Indeed there is a small peak at the position for the primary beam (Xe^{20+}). The comparison between time-of-flight spectra in coincidence with Xe^{19+} and Xe^{20+} shows the Xe^{1+} recoils due to pure ionisation events (Figure 11). We have measured absolute cross sections for pure ionisation in Xe^{q+}-Xe collisions for q=15, 20, 25, 30 and q=35. We find a rather strong increase with q (Figure 12). At this point it is necessary to stress that a metastable contamination in the beam in principle can mimic the pure ionisation process. It has been observed earlier for a few cases that a metastable incoming projectile A^{q+} captures one electron and creates a doubly excited $A^{(q-1)+}$ state which autoionises promptly giving again A^{q+} (c.f. Nielsen *et al* 1985 and Justiniano *et al* 1984). However, the fact that we observe the pure ionisation process for a variety of q makes us believe that our interpretation is correct. In figure 12 we show absolute cross sections σ_{DI}, reaching the $1\cdot10^{-15}$ cm^2 magnitude at q=30. It should be stressed that although these cross sections are fairly large on an absolute scale they are truly minor in comparison with the capture cross sections (typically a few percent c.f. Figure 4). The mechanism for pure ionisation in slow collisions is not yet understood, but several mechanisms have recently been discussed in connection with this problem (Bárány and Ovchinnikov 1991).

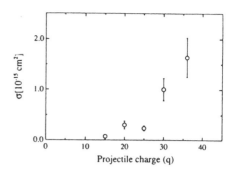

Figure 12: Absolute cross sections for pure ionisation in slow Xe^{q+}-Xe collisions as a function of the projectile charge state q.

Acknowledgements

The authors are grateful to Anders Bárány for reading the manuscript carefully. This work was supported by the swedish national research council (NFR).

References

Andersson H *et al* 1988 *J. Phys. B* **21** L187

Andersson H *et al* 1990 *Phys. Scr.* **42** 150

Andersson L R *et al* 1991 *Phys. Rev. A* **43** 4075

Bárány A *et al* 1985 *Nucl. Instrum. Methods, Phys. Res. B* **9** 397

Bárány A and Ovchinnikov S 1991 *to be published*

Biedermann C *et al* 1990 *Phys. Rev. A* **41** 5889

Cederquist H 1991 *Phys. Rev. A* **43** 2306

Cederquist H *et al* 1985 *J. Phys. B* **18** 3951

Danared H *et al* 1987 *J. Phys. B* **20** L165

Dijkkamp D *et al* 1995 *J. Phys. B* **18** 4763

Hansen J E 1989 *Journal de Physique* **50** C1-603

Hvelplund P, Bárány A, Cederquist H, and Pedersen J O K 1987 *J. Phys. B* **20** 2515

Hvelplund P *et al* 1980 *Phys. Rev. A* **22** 1930

Hvelplund P *et al* 1985 *Nucl. Instrum. Methods, Phys. Res. B* **9** 421

Janev R K and Presnyakov L P 1981 *Phys. Rep.* **70** 1

Justiniano *et al* 1984 *Phys. Rev. A* **29** 1088

Luc-Koenig E and Bauche J 1990 *J. Phys. B* **23** 1763

Martin S *et al* 1989 *Phys. Rev. Lett.* **62** 2112

Niehaus A 1986 *J. Phys. B* **19** 2925

Nielsen E *et al* 1985 *J. Phys. B* **18** 1789

Poirier M 1988 *Phys. Rev. A* **38** 3484

Tawara H *et al* 1985 *J. Phys. B* **18** 337

State-selective electron capture in collisions between ions and alkali–metal atoms

J. Pascale

CEA/DSM/DRECAM/Service des Photons, Atomes et Molécules, Centre d'Etudes de Saclay, Bâtiment 522, 91191 Gif-sur-Yvette Cédex, France

ABSTRACT : In this paper our interest is focused in the theoretical determination of total cross sections for state-selective single-electron capture at medium-energy collisions between ions and alkali-metal atoms, in the case where the electron is captured into high Rydberg states. We mainly report our recent three-body classical trajectory Monte Carlo calculations of final state distributions for $Na^+ + Na(28d)$ collisions, and N^{5+} and $Ar^{8+} + Cs(6s)$ collisions. The results are discussed along with available experimental data and related theoretical results.

1. INTRODUCTION

For collisions between multiply-charged ions (MCI) and neutral atoms, the electron capture processes occur generally with the production of a large density of final states, which makes both experiment and theory difficult. In particular, much efforts have been devoted in the past year for the understanding of collision mechanisms for single or multiple electron capture (see, for example, Harel and Salin 1988, Barany 1989, Fritsch and Lin 1991).

In the low to intermediate energy range, the efficiency of the electron capture can be very large for collisions between MCI projectiles and ground-state alkali-metal (AM) atoms. Then two electrons may be captured by the MCI core in doubly excited states (Martin et al 1989). When the AM atom is initially in an high Rydberg state, the efficiency of the single electron capture may be so large (Olson 1980) that one could also produce doubly excited Rydberg states in an experiment where two electrons would be captured by the MCI projectile, in two successive collisions with excited Rydberg AM atoms. In such an experiment, it is very important to have information in the n, l, m distribution of the captured electrons. This has been the main motivation of our recent work on this problem (Pascale et al 1990).

A theoretical approach of the problem by means of quantum mechanical or semiclassical methods becomes rapidly difficult in view of the large number of channels which are coupled when the electron is captured into high Rydberg states. In particular, the molecular or atomic states expansions used in the close-coupling calculations must include both ionization and excitation channels, and also the electronic-translation factors. The complexity of the problem is clearly shown by the recent discussion of Harel and Salin 1988 on the mechanisms responsible for the final n, l, m distribution of the electron capture in slow collisions between bare or non-bare MCI projectiles and H(1s) atoms. Another example of the difficulty of the problem is given by electron capture in H^+ + $Na(3s)$ collisions, for which there has been a long-standing controversy between theoretical (and also experimental) studies (see, Shingall and Bransden 1987, Fritsch 1987, Courbin et al 1990, and references therein). Also, an approach of the problem based on perturbative methods is not appropriate in the intermediate energy range. However, when the electron is captured into high-Rydberg states, the electron capture process is expected to be well described by a classical model. In particular, the three-body classical trajectory Monte Carlo (CTMC) method, first proposed by Abrines and Percival 1966 for H^+ + $H(1s)$ collisions, appears to be well appropriate to describe our collision problem (Banks et al 1976, Olson 1980, Becker and Mackellar 1984, Isler and Olson 1988). For H^+ + $H(1s)$ collisions, the domain of validity of the CTMC has been found to be $1 \lesssim v \lesssim 4$, where the reduced velocity v is defined as the ratio of the velocity of the projectile in the laboratory frame to the initial orbital velocity of the electron bound to the target core. For capture into excited Rydberg states, this range of validity is found to be wider, but its definition rests by comparisons with experimental or quantum-mechanical calculations when available.

In order to test the CTMC method, we have considered the two reactions :

$$Na^+ + Na\ (n_i l_i) \rightarrow Na\ (n\ l\ m) + Na^+ \tag{1}$$

$$X^+ + Cs\ (6s) \rightarrow X\ (n\ l\ m) + Cs^+ \tag{2}$$

with $X^+ \equiv N^{5+}$, and Ar^{8+}, for which two sets of experimental data are available for the n distribution of the captured electrons. For these two collisions, the electron is captured into highly excited states. However, the two collisions differ much in the initial state of the target and also in the charge of the ion projectile. Therefore, we have used in the CTMC calculations either pure Coulomb interactions, as usually done, or more realistic interactions between the valence electron and the ionic cores (Peach et al 1985, Reinhold and Falcon 1986).

Concerning Reaction (1), there has been an extensive experimental study by Rolfes and MacAdam 1982 for Na targets in excited Rydberg states ($n_i \simeq 24-34$, $l_i = 0, 2$), and for Ne^+, Ar^+ and Na^+ projectiles, in the range $v \simeq 0.6 - 1.8$. No significant core effect has been observed. These authors used a field ionisation technique to determine the n distribution of the captured electrons. They assumed that the electron are preferentially captured by the projectile into states with m = 0, 1, and 2, and therefore the captured electrons field-ionize adiabatically rather than diabatically in the detector. Becker and MacKellar 1984 have performed CTMC calculations for Na^+ + $Na(28d)$ collisions and used pure Coulomb interactions. Their results, for $v = 1.658$, have shown a good agreement with the relative measurements (Rolfes and MacAdam 1982). However, for $v \simeq 1.0$, there was a discrepancy between the experimental and theoretical data, for both the position of the peak and the width of the n distribution. Because MacKellar and Becker 1984 used a small number of trajectories (1.8 10^4) in their CTMC calculations, it was not possible to verify the assumption of Rolfes and MacAdam concerning the m population of final states, making doubtful the validity of the CTMC calculations for $v \lesssim 1.0$. In our CTMC calculations we have used up to 10^6 trajectories, and the final n, l, m distributions have been obtained with small statistical errors. Recently, MacAdam et al 1989 have observed that diabatic field ionization takes place also in the detector after electron capture into final states with m>2, and a new analysis of the data is now made (MacAdam et al 1990). Concerning Reaction (2), Martin et al (1989) have reported optical spectroscopic studies of the electron capture in N^{5+} and $Ar^{8+} + Cs(6s)$ collisions, at $v = 0.894$ and $v = 0.529$ respectively, and have determined unambiguously the predominant n population of the captured electron.

The CTMC method for calculating state-selective electron-capture distributions is outlined in Section 2. The results are then presented and discussed in Section 3. A few concluding remarks are given in Section 4.

2. THEORETICAL APPROACH

We reduce the collision between the ion and the AM atom to that of three particles (ion projectile, AM core and the valence electron e^-). Then a local model potential is well appropriate to describe the effective interaction between e^- and each of the two closed-shell cores. We have used a model potential of the general form.

$$V(r) = -\frac{Z(r)}{r} + V_p(r) \tag{3}$$

where $Z(r)$ is an effective charge and $V_p(r)$ represents polarization terms. For large r, $Z(r) \to z$, where z is the charge of the ionic core ; and for $r \to 0, Z(r)$ goes to the charge of the nucleus. Thus, the model potential has the correct behaviour at small and large distances. The parameters of the potential model are determined to fit the energy levels of the $e^- -$ core system to a good accuracy (see, for example, Pascale 1986). A Coulomb potential is assumed for the core-core interaction.

The CTMC method is based on solving the Hamilton's equations of motion for the three classical particles, given a set of initial conditions for the target and the projectile. Usually (see Abrine and Percival 1966), given the electronic energy corresponding to the initial state $n_i l_i$, the shape and the orientation of the Kepler orbit for e^- about the target core is determined by five parameters randomly selected from a microcanonical distribution. The method of Reinhold and Falcon 1986, which avoids the integration of the Kepler's equations, is used in our calculations. The initial orbital quantum number l_i is specified from the classical angular momentum $\vec{l_c} = \vec{r} \times \vec{k}$ and the condition

$$l_i \leq l_c < l_i + 1 \tag{4}$$

where $\vec{r}$ and $\vec{k}$ are, respectively, the position and momentum vectors of e^- relative to the target core. This condition allows to reproduce the quantal statistical weights for a n level (Becker and MacKellar 1984). The initial conditions for the projectile are given by its position relative to the target core, its velocity, and the impact parameter b randomly selected from b^2 in the interval $(0, b_{max}^2)$; b_{max} is defined as the maximum impact parameter which contributes significantly to the electron capture and ionization processes.

The final states values n, l, and m of the captured electron are determined in the following manner. A classical number n_c is defined from the calculated binding energy E of the electron relative to the ionic core projectile as $E = -\frac{z^2}{2n_c^2}$, where z is the charge of the projectile, and the principal quantum number n of the final state is defined by the condition (Becker and MacKellar 1984)

$$[(n - \frac{1}{2}) (n - 1) n]^{1/3} \leq n_c < [n (n + \frac{1}{2}) (n + 1)]^{1/3} \tag{5}$$

As for the initial state, the value l of the final state is obtained from condition (4) where $\vec{l_c}$ is replaced by the normalized classical momentum $\vec{L_c} = \frac{n}{n_c} \vec{l_c}$, relative to the projectile.

Similarly, we define an absolute value m of the magnetic quantum number of the final state, from the projection L_c^z along the initial velocity of the projectile (taken as the quantization axis) and the condition

$$m \leq |L_c^z| < m + 1 \tag{6}$$

The cross sections for state-selective electron capture are then calculated from the total number of trajectories which are sampled, and from counting the number of those trajectories which result in electron capture into a specific final state. A statistical error is also calculated.

3. RESULTS

A severe test of the CTMC method is provided by the calculation of total cross sections for electron capture into H(2s) and H(2p) in collisions between H^+ and ground-state AM atoms, in the keV-energy range. For all the AM-atoms, we have found an overall agreement between the CTMC results (Pascale, unpublished) and available theoretical and experimental data. The agreement in the absolute values of the cross sections is generally better than a factor 2-3 for energies larger than 2 keV. It has only been obtained by using effective interactions in the CTMC calculations.

These comparisons gives more credibility to the absolute values of the cross sections which are now reported in the case where the electron is captured into high Rydberg states of the projectile, and for which the CTMC method is more appropriate.

3.1. N^{5+}, Ar^{8+} − $Cs(6s)$ collisions

State-selective electron capture cross sections have been calculated at $v = 0.894$ for N^{5+} + $Cs(6s)$ collisions, and at $v = 0.529$ for Ar^{8+} + $Cs(6s)$ collisions. As expected, we have verified that the projectile core effect is unimportant in view of the large values of n and l which are populated, but that the target core effect is quite significant for both the magnitude of the cross sections and the position of the peak of the n distribution.

The CTMC results obtained from e^-− core effective interactions have been found in good agreement with the experimental findings of Martin et al (1989). These authors were able to determine unambiguously, from the relative intensities of the observed lines, the n values of the final levels which are predominantly populated in the electron capture process.

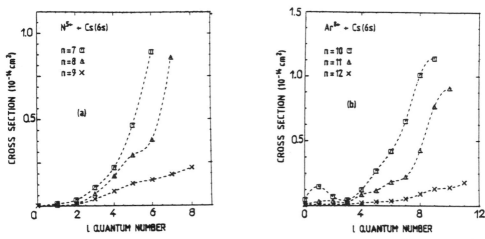

Figure 1 : The calculated final n, l distribution for electron capture in (a) N^{5+} + $Cs(6s)$ collisions at v = 0.894, and (b) Ar^{8+} + $Cs(6s)$ collisions, at v = 0.529, for some selected values of n. The values of the $\sigma(nl)$ cross sections, erroneously reported with a multiplicative factor 2 by Pascale et al 1990, have been corrected.

The calculated total cross sections $\sigma(nl)$ versus l, for n fixed, are reported for some values of n in Figs. 1(a) and 1(b), respectively, for the N^{5+} + $Cs(6s)$ and Ar^{8+} + $Cs(6s)$ collisions. For the two collisions systems, the largest possible values of l are preferentially populated, and the increase with l of the n, l distribution is very depending on n. Near the peak of the n distribution, the n, l distribution increases more rapidly than statistical, but becomes statistical for n far of the peak. Some structure is observed in the $\sigma(nl)$ cross sections, which is more pronounced in the case of Ar^{8+}. Similar structure has also been observed in the final n, l distributions calculated by Harel and Jouin 1988 for electron capture in slow collisions ($0.2 \lesssim v \lesssim 0.8$) between N^{5+} and Ar^{8+} projectiles and $H(1s)$. They used a semiclassical close-coupling method, with a molecular expansion in OEDM orbitals including electronic translation factors. These authors have found that the $\sigma(n, l)$ cross sections behave differently with the reduced velocity v. In the case of N^{5+} projectiles, their results are in very good agreement with the experimental results of Dijkkamp et al 1985, except for $\sigma(4f)$, for all the range of reduced velocited explored ($v \simeq 0.2 - 0.5$). For both N^{5+} and Ar^{8+} projectiles and for the largest values of v considered, their cross sections $\sigma(n, l)$ for $l = n - 1$ take large values. In particular, for Ar^{8+} projectiles, $\sigma(5g)$ increases more rapidly than statistical, as found in our calculations for N^{5+}, and Ar^{8+} + $Cs(6s)$ collisions. Moreover, similar semiclassical calculations performed by Salin 1984, for electron capture in slow collisions ($v \simeq 0.2 - 0.8$) between bare ions and H(1s), have show a very different variation of $\sigma(nl)$ with l at all values of v. In that case, the increase of $\sigma(nl)$ with l is nearly statistical. Therefore, one may conclude that the core projectile effect is important on the final n, l distribution of the captured electron (in agreement with the discussion of Harel and Salin 1988). Our results for N^{5+}, and Ar^{8+} + $Cs(6s)$ collisions indicate that there is also a significant target effect, probably because more channels are involved in the electron capture process since the target atom is initially excited.

The calculated total cross sections $\sigma(n\ l\ m)$ for electron-capture into $N^{4+}(n = 7,\ l = 6)$ and Ar^{7+} $(n = 10,\ l = 9)$ in collisions between N^{5+}, and Ar^{8+} + $Cs(6s)$ collisions are reported versus m in Fig. 2. These values of n, and l correspond to n l levels which are predominantly populated. The n, l, m distribution for N^{5+} projectiles is peaked at m = 0 and then decreases continuously with m. For Ar^{8+} projectiles, there is no peak in the n, l, m distribution at m = 0. The n, l, m distribution is found wider than in the case of N^{5+} projectiles. We have found that for the total electron capture, 82 % of final states with m=0,1 are populated for N^{5+} projectiles, against 64 % for Ar^{8+} projectiles.

This result indicates clearly a projectile core effect on the n, l, m distribution. Then, it is interesting to notice the experimental findings of Lembo et al 1985, and also Lin et al 1991, for the electron capture in Ne^{8+} + $Na(3s)$ collisions (note that both Ne^{8+} and N^{5+} are He-like core projectiles). From the measurement of the strong degree of polarization of the n = 8 to 9 line observed in Ne^{7+} after electron capture in collisions between 4 - keV Ne^{8+} ions and Na(3s), Lembo et al have concluded to a preferential population of final states with m = 0. The same result has been also found by Lin et al 1991, who have extended the experiment of Lembo et al to higher energies. As noted by Lembo et al 1985, their conclusion is in agreement with the semiclassical calculations of Salin 1984 for slow collisions between bare ions and H(1s). From our results for both the final electron-capture n, l and n, l, m distributions we may conclude that the target core effect is more significant for the n, l distribution than for the n, l, m distribution. And, the target core effect is comparatively less important than the projectile core effect on both final n, l and n, l, m distributions.

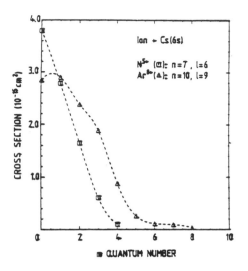

Figure 2 : The calculated n, l, m distribution for electron capture in N^{5+} + $Cs(6s)$ collisions, for n = 7, l = 6 at v = 0.894, and in Ar^{8+} + $Cs(6s)$ collisions, for n = 10, l = 9 at v = 0.529.

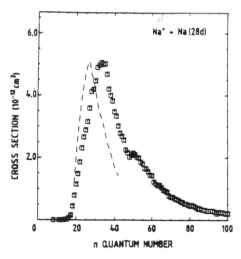

Figure 3 : The n distribution for electron capture in Na^+ + $Na(28d)$ collisions at v = 1.0. Open squares : CTMC results. Dashed line : relative experimental data of Rolfes and MacAdam 1982 normalized at the peak of the calculated n distribution.

3.2. $Na^+ + Na(28d)$ collision

For the $Na^+ + Na(28d)$ collisions, because final levels with large values of n are populated, we have verified that there is no significant change in both the shape of the n distribution and in the cross section values by using either pure Coulomb or model potential interactions in the CTMC calculations. We have also verified that, for the final n distribution, the low scale CTMC calculations of Becker and MacKellar 1984 and our calculations agree within less than 35 % for the two values of the reduced velocity $v = 1$, and 1.658 where the CTM calculations have been performed.

The calculated n state distribution for the electron capture in $Na^+ + Na(28d)$ collisions is shown in Fig. 3 for v = 1, along with the relative experimental data of Rolfes and MacAdam (1982) (we have assumed for these comparisons the final Stark number, determined from the field-ionization analysis, to be equal to n). As previously found by Becker and MacKellar 1984, the peak of the calculated n distribution is shifted to large values of n (n = 33) relative to the peak in the experimental distribution (n $\simeq$ 27).

Also, the calculated width of the n-distribution is $\Delta n \simeq 20$ compared with the experimental result $\Delta n \simeq 15$. Then it has to be noticed that in the experimental determination of the n-distribution, Rolfes and MacAdam 1982 have used a field-ionization detection technique. They have assumed a preferential population of final states with $m \lesssim 2$ in the electron capture process, which field ionize adiabatically in the detector. This assumption was based on theoretical predictions using the Brinkman-Kramers approximation). The calculated n, m distribution shown in Fig. 4, for n=33 corresponding to the peak of the n distribution, does not agree with this assumption. The calculated n, m distribution is peaked at m=0, but is relatively wide. We have found that less than 28 % of final states with m=0,1 are populated in the total electron-capture process.

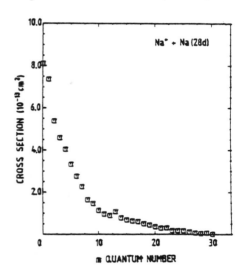

Figure 4 : The calculated n, m distribution for electron capture in $Na^+ + Na(28d)$ collisions, for n = 33 and at v = 1.0.

Similar results have also been obtained for other values of n and also for $v = 1.658$ (in that case, there is an apparent good agreement between the calculated and experimental n distributions). Our results indicate clearly that both adiabatic and diabatic processes have to be considered in the experimental determination of the n distribution of the captured electron by means of the field-ionization technique, not only for $v \lesssim 1$ but also for $v > 1$. It is clear from our results that some guidance has to be provided for the analysis of the experimental data, since an observed field-ionization peak may correspond to many n levels.

In the previous experiment of Rolfes and MacAdam 1982 for Reaction(1), the energies of the Ar^+ and Na^+ projectiles were in the range 550-2000 eV. More recently, MacAdam et al 1990 have extended this energy range to 60-2100 eV but for Ar^+ projectile.

In the new analysis of the measurements, MacAdam et al 1990 have assumed that the Stark levels field ionize adiabatically for $m = 0, 1$ and 2, and diabatically for $3 \le m \le n-1$. They have also assumed that each sublevel l m is equally populated for a given n (however, our results do not agree with this assumption). The analysis of final-state distributions is based on overlapping contributions of field ionization from adjacent n levels and the fitting of data by a parametric formula. The new set of experimental data is found consistent with the previous one (MacAdam and Rolfes 1982). Recently, MacAdam 1991 has improved again this analysis and a final n distribution has been obtained in the case of $Ar^+ + Na(28d)$ at the reduced velocity v = 0.462. This result is shown in Fig. 5 along with the calculated n distribution for $Na^+ + Na(28d)$ at the same reduced velocity (Pascale, unpublished). This comparison is justified since it has been verified both experimentally and theoretically that the core effect is insignificant for the n distribution.

The calculated n distribution shows a structure which seems to be related to the structure in the experimental velocity dependence of the cross section for the total electron capture in $Na^+ + Na(28d)$ collisions. This is shown in Fig. 6 along with our CTMC calculations (Pascale, unpublished). The agreement between the CTMC results and the experimental data normalized at $v = 1.0$ is found very good in the range $v \simeq 0.4 - 1.7$.

From the large range of reduced velocities which has been explored, $v \simeq 0.2 - 2$, and from the large choice of initially prepared Na states ($n_i \simeq 24 - 34, l_i = 0, 2$) MacAdam et al 1990 have been able to summarize their observations concerning the position of the peak and the width of the n distributions as a function of v.

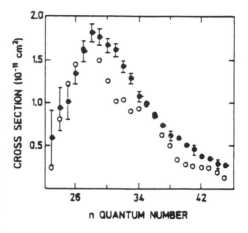

Figure 5 : The n distribution for electron capture in $X + Na(28d)$ collisions at v = 0.462. Open circles : CTMC result for $X \equiv Na^+$. Full circles : relative experimental data for $X \equiv Ar^+$ (MacAdam 1991) normalized at n = 28 to the calculated n distribution.

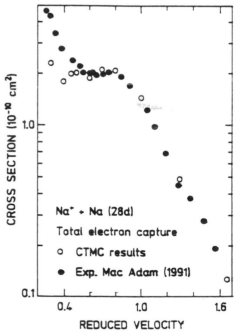

Figure 6 : Total cross section versus the reduced velocity for total electron capture in $Na^+ + Na(28d)$ collisions. The relative experimental data (MacAdam 1991) have been normalized to the calculated cross section at v = 1.0.

The peak of the n distribution shifts upward from a value less than the initial value n_i to a value larger than n_i ($\Delta n \approx 3$) at $v \simeq 0.7 - 0.9$ and then shifts downward below n_i as v is further increased. Also, it is found experimentally that the n distributions become sharper where the maximum upward shift occurs. These experimental findings are found in qualitative agreement with our results.

The calculated final n, l distribution is shown in Fig. 7, for $v = 1$ and for $n = 33$ corresponding to the peak in the n distribution. In general, we have found that the n, l distributions increase with l nearly statistically. This result is consistent with the semiclassical calculations of Salin 1984 for collisions between bare-ion projectiles and H(1s) targets, since we have observed no change by using either Coulomb or model potential interactions. For $n \gtrsim 33$, the n, l distributions go through a maximum for $l \simeq 18 - 22$ and then decreases abruptly. In contrast, for $n \lesssim 33$ the n, l distributions increase with l up to a value close to the maximum allowed value n-1. These results may be explained by the large l mixing which takes place in the target during the collision (see MacAdam et al 1987). Then, because the captured electron tends to preserve its orbital size and binding energy, final states with large values of l, but with $l \leq n_i - 1$ should be preferentially populated. To verify this point, we have recently used (Pascale, unpublished) the CTMC method to calculate the cross section σ_d for the total depopulation of the 28d level of Na by collision with Na^+, at $v = 1.0$. We have found the value $3.9 \, 10^{-7} cm^2$, which is three orders of magnitude larger than the total electron capture cross section, in agreement with the experimental findings (MacAdam et al 1987).

The final n, l, m distributions for electron capture in $Na^+ + Na(28d)$ collisions have been calculated at $v = 1.0$, for some values of n, and for values of l corresponding to the maximum of the n, l distribution. As an example, we show in Fig. 8 the n, l, m distribution for $n = 33$, $l = 21$.

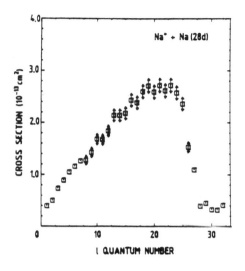

Figure 7 : The calculated n, l distribution for electron capture in $Na^+ + Na(28d)$ collisions at v = 1.0, for n = 33.

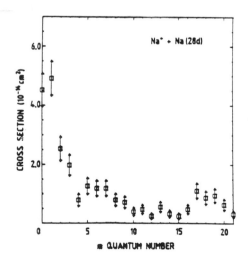

Figure 8 : The calculated n, l, m distribution for electron capture in $Na^+ + Na(28d)$ collisions at v = 1.0, for n = 33 and l = 21.

In general, we have found that for $n \gtrsim 33$, the n, l, m distribution are peaked at small m values and become wider as n increase from $n = 33$. In contrast, the n, l, m distributions for $n < 33$ become much broader than for $n \gtrsim 33$. However, we have presently not enough results, for other values of the reduced velocity to discuss further the n, l, m distributions.

4. CONCLUSION

In conclusion, we have shown that the three-body CTMC method is quite reliable for the calculation of state-selective electron-capture cross sections for collisions between ion and alkali-metal atoms, in the range of reduced velocities from 0.5 to 2.0 and when the electron is captured by the core projectile into high Rydberg states. Our results are consistent with recent available experimental data, and also with previous theoretical works on similar collision systems. However, an experimental determination of absolute cross sections is needed to assess definitively the accuracy of the calculated cross sections. This is certainly possible for total electron capture cross sections, but this appears quite difficult already in the case of the n-state distributions. Indeed, in view of the large density of final states produced in the electron capture process, for the collision systems that we have studied, it is clear that measurements of final n, l, m distributions are impossible by conventional methods. Then, other observables such as alignment and orientation parameters (see, Fano and Macek 1973) would be more appropriate to study, in order to bring experiment and theory into valuable comparisons (see, for example, Lin and Macek 1987, Andersen et al 1988).

For the two collisions systems that we have considered, we have found that the final electron-capture n, m distributions are peaked for m = 0, but are relatively broad. This result indicates that both diabatic and adiabatic ionization processes must be considered in the analysis of experimental data by means of a field ionization technique. Also, we have found that final state with large l-values are easily produced in the electron capture process ; in particular, in the case of MCI + Cs(6s) collisions we have found that the electron is preferentially captured by the projectile into nearly circular states.

AKNOWLEDGEMENTS

The collaboration with R.E. Olson and C.O. Reinhold is greatly aknowledged. I would also like to thank R.E. Olson and A. Pesnelle for useful discussions.

REFERENCES

Abrines R. and Percival I.C. 1966 Proc. Phys. Soc. <u>88</u> 861

Andersen N., Gallagher J.W. and Hertel I.V. 1988, Phys. Rep. <u>165</u> 1

Banks D., Barnes R.S. and Wilson J. 1976 J. Phys. B <u>9</u> L141 (1976)

Barany A. 1989 The Physics of Electronic and Atomic Collisions ed. A. Dalgarno et al (AIP, New York) p. 246-257

Becker R.C. and MacKellar A.D. 1984 J. Phys. B <u>17</u> 3923

Courbin C., Allan R.J., Salas P. and Wahnon P. 1990 J. Phys. B <u>23</u> 3909

Dijkkamp D. Ciric D., Vlieg E., de Boer A. and de Heer F.J. 1985 J. Phys. B <u>18</u> 4763

Fano V. and Macek J. 1973 Rev. Mod. Phys. <u>45</u> 553

Fritsch W. 1987 Phys. Rev. A <u>35</u> 2342

Fritsch W. and Lin C.D. 1991 Phys. Rep. 202 1

Harel C. and Jouin H 1988 J. Phys. B 21 859

Harel C. and Salin A. 1988 Electronic and Atomic Collisions ed H.B. Gilbody et al (Elsevier, Amsterdam) p. 631-642

Isler R.C. and Olson R.E. 1988 Phys. Rev. A 37 3399

Lembo L.J., Danzmann K., Stoller Ch. Meyerhof W.E. and Hänsch T.W. 1985 Phys. Rev. Lett. 55 1985

Lin C.D. and J.H. Macek 1987 Phys. Rev. A 35 5005

Liu C.J., Dinford R.W. Berry H.G. and Church D.A. 1991 Phys. Rev. A 43 572

MacAdam 1991 Nucl. Instrum. Methods B 56/57 253

MacAdam K.B., Gray L.G. and Rolfes R.G. 1989 Proc. 16th Int. Conf. on the Physics of Electronic and Atomic Collisions ed. A. Dalgarno et al (Elsevier, New York) p. 728

MacAdam K.B., Gray L.G. and Rolfes R.G. 1990 Phys. Rev. A 42 5269

MacAdam K.B., Rolfes R.G., Sien X., Singh J., Fuqua III W.L. and Smith D.B. 1987 Phys. Rev. A 36 4234

Martin S., Salmoun A., Do Cao G., Bouchama T., Andrä J. and Druetta M. 1989 J. Phys. (Paris) Colloq. 50 C1-357

Martin S., Salmoun A., Ouardane Y., Druetta M., Desesquelles J. and Denis A. 1989 Phys. Rev. Lett. 62 2112

Olson R.E. 1980 J. Phys. B 13 843

Pascale J. 1986 Quantum Optics ed. J. Fiutak and J. Mizerski (World Scientific) p. 38-93

Pascale J., Olson R.E. and Reinhold C.O. 1990 Phys. Rev. A 42 5305

Peach G., Willis S.L. and McDowell M.R.C. 1985 J. Phys. B 18 3921

Reinhold C.O, and Falcon C.A. 1986 Phys. Rev. A 33 3859

Rolfes R.G. and MacAdam K.B. 1982 J. Phys. B. 15 4591

Salin A. 1984 J. Phys. (Paris) 45 671

Shingall R. and Bransden B.H. 1987 J. Phys. B 20 4815

Theory of ionization in ion–atom collisions

D S F Crothers

Department of Applied Mathematics and Theoretical Physics, Theoretical and
Computational Physics Research Division, The Queen's University of Belfast, Belfast BT7
1NN, N.Ireland, UK

ABSTRACT: A progress report is given on the continuum distorted wave (CDW)
independent-event model and its extension at intermediate energies to double
ionisation and transfer ionisation. We use explicitly correlated CDW helium wave
functions and consider a variety of projectile ions. Both events and electrons are
correlated. The CDW eikonal initial state method of Crothers and McCann (1983,
J.Phys.B$\underline{16}$, 3229) for single ionisation is reviewed as is electron capture to the
continuum. The report is placed in the context of ionisation work published since the
last ICPEAC.

1.   INTRODUCTION

The continuum-distorted-wave eikonal-initial state (CDWEIS) model for single ionisation
of atoms by ion impact was originally proposed and utilised by Crothers and McCann
(1983) and well reviewed at ICPEAC XVI by Rivarola et al (1989), and independently by
Crothers (1989) and Gilbody (1989, 1986) who pointed out that for total cross sections for
ionisation of H by bare nuclei it is seen to provide the best overall agreement with
experiment at intermediate velocities. The principal advantages of the CDWEIS model are,
firstly, that the long range boundary conditions are satisfied exactly: in the final state by the
CDW description (Belkić 1978) of the three body Coulomb problem and in the initial state
by the EIS approximation; secondly, both initial- and final-state wave functions are
normalised; thirdly, the final state comprises regular Coulomb functions which are well-
behaved at short range. In addition, the original CDWEIS model was shown to predict the
electron capture to the continuum (ECC) cusp in the doubly-differential cross sections and
good scaling properties of total cross sections as a function of projectile charge (Zp): at 200
keV/u agreement between experiment and CDWEIS was perfect for Zp = 3 and 4, both
lying between Glauber and classical-trajectory Monte Carlo (CTMC) theories. At ICPEAC
XVI, Salzborn (1989) and Schiwietz et al (1989) also commented favourably on the
CDWEIS model of H^+–He^+ and H^+–He collisions, respectively. The model had of course
already been generalised to include fully stripped ions (Fainstein et al 1988a), antiprotons
(Fainstein et al 1988b) and carbon and oxygen ions (Rivarola et al 1989), albeit in a
Hartree-Fock independent-electron model of the helium target. Subsequently, Fainstein et
al (1989) extended their CDWEIS antiproton (as well as proton) work, substituting carbon
for helium, and including L as well as K shells at 1 MeV. It would be interesting to extend
such studies to gold (Sarkadi and Mukoyama 1990). Fainstein and Rivarola (1990) have
applied CDWEIS to single ionisation of neon by protons, while Bernardi et al (1989) have
applied our CDWEIS ansatz to total and singly- and doubly-differential cross sections
(DDCS) for the ionisation of helium by protons and by $^3He^{2+}$. Their lack of quantitative
agreement between CDWEIS and their experimental DDCS may or may not be due to a
lack of correlated helium bound state wave function. On the other hand, Andersen et al

(1990) show in their figure 4 that CDWEIS, even with an uncorrelated helium target, is the best theory for predicting the ratio between total single-ionisation cross sections of helium for antiprotons and for protons, as a function of energy. However, their statement in their abstract is misleading: for CDW, read CDWEIS. Schneider et al (1989) have considered strong continuum-continuum couplings in the direct ionisation of Ar and He atoms by 6 MeV/u U^{38+} and Th^{38+} projectiles. Fortunately, these authors, W G Graham included, have considered only plane-wave Born approximations and CTMC theories, since CDWEIS appears to be misunderstood ('Continuum-distorted-wave-eikonal approximation approximation'). Datz et al (1990) find that, as in the experiments of Bernardi et al, CDWEIS affords only qualitative agreement for single ionisation of helium by I^{q+} and U^{q+}. Once again this may or may not be the result of an uncorrelated helium wave function.

The above comment about the difference between CDW and CDWEIS is an important one. At high energies when normalisation ceases to be a major concern and charge transfer is dominated by small impact parameters then of course a CDW initial state is infinitely preferable to an eikonal initial state, since it is regular at short distances and facilitates Thomas double scattering. Actually, Crothers (1982) proposed the explicit use of CDW to describe the final state in the ionisation of H by electron impact. This has since been successfully taken up by Brauner et al (1989a,b). Indeed, we ourselves (Deb et al 1990) have applied CDW, albeit in target form to the process of electron capture by positrons from helium into the ground state of positronium. Above 160 eV we obtain excellent agreement with experiment for the total cross section and its fall-off with respect to energy (Fromme et al 1986), in contrast with the CTMC result of Schultz (1989).

In my own view one of the main reasons why CDW and CDWEIS are misunderstood is that the calculation of $H - i d/dt$ operating on a CDW function presents unnecessary barriers. I write this, since a former colleague tells me that one of his referees finds it obscure. Actually as pointed out by Crothers (1982) it is necessary only to introduce generalised non-orthogonal coordinates r_T, r_P and R, the position vectors of Te^-, Pe^- and TP respectively where T and P stand for target and projectile. Regarding them as independent for book-keeping purposes and by this I mean making sure one has differentiated every last function of a function (and hands up those who have never made this type of error in their career), one may write in Cartesian-tensor notation with Hamiltonian H and time t,

$$H - i\frac{d}{dt} = -\frac{1}{2}\partial_{r_P}^2 - \partial_{r_P} \cdot \partial_{r_T} - \frac{1}{2}\partial_{r_T}^2 + V - i\partial_t - \frac{1}{2}i\underline{v}\cdot\partial_{r_T} + \frac{1}{2}i\underline{v}\cdot\partial_{r_P} \quad (1).$$

Notice that this is the simplest way to obtain the infamous electron translation factors, from the last two terms, and a simple calculation shows that the perturbation is the second term, the so called non-orthogonal kinetic energy which connects a bound state function on one nucleus to the CDW function on the other. Eqn (1) naturally generalises to a fully wave time-independent treatment (Crothers 1982, McCann 1991).

2.   TWO-ELECTRON TRANSITIONS

In the introduction we have reviewed the CDWEIS model and its relative success in describing single ionisation, our principal caveat being the use of the independent-electron model in catering for multi-electron targets and in particular the treatment of K shell electrons as equivalent electrons. That Coulomb potentials can rarely be treated perturbatively was the raison d'être for CDW. Why should the electron-electron repulsion be any different? In 1987 Crothers and McCarroll studied the double and single capture in He^{2+} - He collisions and they took for their helium target a highly and explicitly correlated wave function of Pluvinage (1950). Indeed the latter comprises a CDW description of a bound state in which the two electrons are diametrically opposed. Another caveat concerning CDWEIS is that the EIS makes the theory less suitable for small impact parameters and therefore large angles, in contrast with a pure CDW treatment. This is seen particularly clearly in fig 7 of the very recent paper (Miraglia and Macek 1991) which presents an exact quantal impulse approximation of the initial state, with a CDW final state. Crothers and McCarroll therefore adopted a pure CDW initial state in deference to this latter caveat. The price to be paid is that the Pauli exclusion principle has to be obeyed and so we adopt the pragmatic independent-event model, with its own binomial distribution, and within the semiclassical impact parameter treatment. This has the advantage that the wave transition amplitude does not explicitly involve the electronic Coulomb repulsion potential, but rather the integrations over the two electronic coordinates may be done analytically, at the price of a numerical inverse Laplace transform. The impact parameter amplitude may then be obtained by the one-dimensional eikonal-like Hankel transform. This is sometimes disputed but may be justified by identifying the collision plane as the azimuthal plane in both wave and impact parameter treatments. Dunseath and Crothers (1991), Deb and Crothers (1990,1991) and Deb et al (1991) have extended the work to cover single capture with probability $2P_{c1}$, double resonant capture with probability P_{c1}^2, single ionisation with probability $2P_{i1}$, double ionisation with probability $P_{i1}P_{ii2}$ and transfer ionisation with probability $P_{c1}P_{ci2} + P_{i1}P_{ic2}$. Here P_{Rn} is the single probability for process R in the nth event, while P_{Srn} is the single probability for process r in the nth event *after* process S. Even as seven or eight numerical quadratures are required, depending on the process, for total cross sections, so optimisation of computer time is crucial with attention to numerical methods and problems, vectorisation and parallelism. Our results for the five processes are presented in figures 1 to 5. In the case of the multiple events of transfer ionisation (fig 4) and double ionisation (fig 5), a dichotomy arises regarding the correct target charge to be taken for the second event. In the case of fig 5 which shows double ionisation, we have actually used CDWEIS for the second event with Z_T replaced by (Z_T-1), on the grounds that the first ionised electron is in the low continuum and so partially screens the target nucleus from the outer remaining electron. Afterall, this is not unlike the intermediate event in Thomas double scattering. The results are particularly good down to 200 keV/u and might well be improved below this energy by adopting CDWEIS, rather that CDW, for the first event. This involves some minor changes in the Nordsieck parameters, provided we use the post form. We extend our results to proton impact in figure 6 and to Li^{3+} in table 1, and compare with Shingal and Lin (1991a,b) and Shingal (1991), who use different effective charges in the independent electron model. For one-electron processes they use the effective charge which gives the correct single ionisation energy, while for two-electron processes they use the effective charge which gives one half of the double ionisation energy. Like our own strategy this appears to be somewhat empirical and perhaps is difficult to generalise (Fritsch and Lin 1991). Indeed a not dissimilar strategy by Gayet et al (1991), in which the outer electron sees a charge of 0.69 and the inner electron sees a charge of 1.69, gives better results for resonant double capture, using CDW and CDWEIS, than our own results of figure 3. Equally well and returning to our dichotomy, our results in figure 4 for transfer ionisation are disappointingly high, being based on $Z_T = 2$ in the second event. This is understandable because in the dominating probability $P_{i1}P_{ic2}$ and assuming that the first electron is ionised and removed from the interaction zone, then actually the second event is dominated by the exact resonance charge transfer between He^{2+} and $He^+(1s)$. If on the other hand we supose as in double ionisation that the first electron screens the target nucleus before escaping, then the second event is more like the asymmetric process $He^{2+} + H$ giving principally $He^+(1s) + H^+$, and on an *accidental* resonance basis $He^+(2l) + H^+$. Our results for transfer ionisation, on the basis of replacing Z_T by (Z_T-1) in the second event, and neglecting $P_{c1}P_{ci2}$, are given in table 2. The agreement with Shah and Gilbody is immeasurably better. It would be interesting to extend these transfer ionisation calculations to much higher energies where Thomas double scattering between the two electrons may be a factor (Palinkas et al 1990, Knudsen et al 1987).

In figure 7 we present results for the ratio of double to single ionisation for a variety of projectiles. In particular for $N^{7+} + He$, the results of Deb et al (1991) using CDW, CDWEIS and Pluvinage are close to the forced impulse approximation of Ford and Reading (1988) and to the semiempirical fit of Knudsen et al (1984). Despite the disagreement with

Heber et al (1990) concerning the magnitude of R, nevertheless the agreement between the two ab initio theories is encouraging, given their utterly disparate approach to the question of correlation. The Born result of Salin (1987) for $He^+ + He$ at 1.4 MeV/u is R = 3.5×10^{-3} and agrees well with the lowest triangles in figure 7. Ford and Reading (1990) have established scaling laws in the first Born approximation for two electron ions. They find Zp^{-4}, Zp^{-6} and hence Zp^{-2} for single and double ionisation and their ratio, respectively. In figure 8 we present results for the same ratio as in figure 7, but for antiproton impact. Here the upgraded forced impulse method does particularly well, in relation to experiment below 1 MeV/u, but it is interesting that in this range the CTMC method of Olson (1987) and the CDW-CDWEIS method of ourselves show similar trends, given that the Bohr and Pluvinage models of the helium atom both keep the electrons apart angularly. Further details of figure 8 are discussed in a topical review by Schultz et al (1991). The Bohr CTMC model has also been applied successfully by Olson et al (1989) who find that the above ratio for 1 MeV proton impact on helium peaks as a function of deflection angle at c.1 mrad, is in close agreement with the experiment of Giese and Horsdal (1988) regarding shape. In similar collisions, Fukuda et al (1991) have used an eikonal theory to find singly ionised recoil ions with significantly lower average momenta than in the experiment of Dorner et al (1989). Other recent experiments include the single and double ionisation of helium by hydrogen atom impact (DuBois and Kover, 1989) in the 25 to 1000 keV/u range, the single and double ionisation of hydrogen molecules by alpha particles (Edwards et al 1990) in the 125 to 750 keV/u range and mutual ionisation in the collision of two negative hydrogen ions (Schulze et al 1991) also in the keV range. Clearly the number of active electrons is limitless and already the last mentioned includes comparison with CTMC for both double and triple ionisation. Certainly the triple ionisation of say a lithium atom by an alpha particle should be well within the capability of the CDWEIS model, perhaps combining the independent-electron and independent-event models.

A few words are in order concerning correlation (Pedersen 1990, McGuire 1987, McGuire and Straton 1989). Static correlation can be both angular and radial. Screening can be both static and dynamic. In our Pluvinage CDW model of helium we have essentially angular correlation. In our choice of net target charge for the second event in our independent-event model, we have dynamic screening. But in our model we do not have shake-off, because in the first event the spectator electron remains in the hydrogenlike He^+ K shell orbital. Of course single ionisation may include the contribution $P_{i1} P_{iex2}$ where in the second event the second electron is excited. It is just a question of point of view. Time ordering is important for the correlation of events, just as spatial ordering is important for the correlation of particles and their energies.

As mentioned in the introduction, a particular spatial correlation is electron capture to the continuum (ECC), for which CDW and, to a lesser extent, CDWEIS are notably well suited (Crothers and McCann 1983). The reason for the caveat is that small impact parameters are likely to be important for which CDWEIS is less well suited. The difference is well illustrated for proton-hydrogen atom collisions at 50 keV in figure 1 of Crothers and McCann (1983), in which CDW is seen to give a broader cusp with a greater negative discontinuity. Of course, normalisation problems at this impact energy must be a problem, but this does not affect the shape. In figure 6 of Miraglia and Macek (1991), the disagreement, between the impulse approximation (IA) and CDWEIS for He^{2+} and H^+ impact on helium at 100 keV/u, for the doubly differential cross sections may be understood in this context: CDW would be much better. Their positive value of the dipole parameter β for He^{2+} and for CDWEIS reinforces this point. In any case CDW gives good results for the monopole coefficient for helium targets for impact energies above 400 keV/u and for fully stripped projectile charges in the range 1 to 8 (McCann and Crothers 1987, Crothers and McCann 1987), that is, compared to experiment (Andersen et al 1986, Knudsen et al 1986). Figure 2 of Moiseiwitsch (1991) shows results for the dipole parameter β as a function of impact energy. Suffice it to say that the picture is more complicated than figure 5.9 of Fritsch and Lin (1991) and so β for H^+ and He^{2+} impact remains an open question. Note, however, that for C^{6+}, not shown by Moiseiwitsch, the CDW results of McCann and

Crothers (1987) and the experiment of Knudsen et al (1986) are in close agreement, as reviewed by Crothers (1989). Also, not shown are the results of Miraglia and Macek (1991), namely -0.55 for H^+ and -0.47 for He^{2+} at 100 keV/u. The former is closer to the data of Dahl (1985), the latter is closer to the -0.35 of McCann and Crothers (1987). The relativistic Oppenheimer-Brinkman-Kramers second order ((OBK2) approximation of Moiseiwitsch (1991) agrees well with the data of Dahl (1985) and Andersen et al (1986) for energies > 50 keV/u/Zp but not with the data of Meckbach et al (1981) and Gulyas et al (1986). However, OBK2 suffers from well known defects concerning boundary conditions and elastic divergences, and is unlikely to be valid at lower energies. It is intersting to note that Moiseiwitsch finds positive values for β at the highest energies. My own calculation confirms this: namely in the non-relativistic high energy limit we have in the joint OBK2/CDW2 theory

$$\beta = 24 \cos \pi/3 \; [(1+2/\gamma) \exp(-\gamma) +1 - 2/\gamma]/[1 - \exp(-2\gamma) J(\gamma)] \tag{2}$$

where γ is Z_p / Z_T and where using modified Bessel functions I_n

$$J(\gamma) = I_0(2\gamma) + I_1(2\gamma). \tag{3}$$

In the limit as $\gamma \to 0$, this β tends to $4\gamma \cos \pi/3$ and therefore 0. Expression (2) is never negative for non-negative γ. On the other hand, using a halfway house phase integral wave version of CDW (Crothers and Dubé, 1989), which satisfies all known short and long range boundary conditions, we obtain in the high-energy limit

$$\beta = - 2 \cos \pi/3 \; (1-\exp(-\gamma))/[1-\exp(-2\gamma) J(\gamma)]. \tag{4}$$

This is interesting because it is always negative for $\gamma > 0$, which makes sense physically, in tems of a parachute effect. That is, one would expect backward scattering to predominate. Another point of interest is that we have

$$\lim_{\gamma \to 0} \beta = - 2 \cos \pi/3 < 0, \tag{5}$$

and indeed we note that the minimum value of β is (-1) when $Z_p = 0$. This is in accord with the observation of electron capture into continuum states of neutral atoms (Sarkardi et al, 1989), characterised by a considerably smaller width than that for positively charged ions. Some attempt has also been made by Gulyas and Szabo (1991) with partial success to account for long range boundary conditions for ECC in He^{2+} + He collisions, with simultaneous target excitation or ionisation. However, to end this section on two-electron transitions, we should stress that most of this ECC work, including our own, uses the independent-electron approximation. The use of explicitly correlated wave functions in the independent-event model appears desirable.

3. OTHER RECENT IONISATION STUDIES

Shah et al (1987) have extended their data on H^+ - H ionisation down to 9.4 keV, which may reasonably be considered to be too low for distorted travelling atomic orbital theories, such as CDWEIS, to be successful. Similar remarks apply to Shah et al (1988) who have extended their data on He^{2+} + H ionisation down to 18.6 keV/u. The principal defect of CDWEIS at these energies is the as yet omitted intermediate charge transfer channels, although of course some intermediate continuum coupling is implicitly included. Nevertheless the rate of fall-off with decreasing impact energy appears to be reasonably well predicted by CDWEIS.

These experimental studies have in turn stimulated new theoretical approaches to the question of slow-adiabatic ionisation. Presnyakov and Uskov (1989), following Presnyakov and Uskov (1984), have applied the Keldysh method and obtained good agreement with

Shah et al (1988) for He^{2+} + H. Bandarage and Parson (1990) have reinvestigated H^+ + H using CTMC and paying particular attention to Wannier saddle-point electrons. Although showing improvement on previous CTMC studies, their cross sections are still lower than Shah et al (1987). Semiclassical, rather than classical, treatments of saddle-point electrons and hidden crossings have been developed by Solovev (1990), Ovchinnikov (1990) and Abramov et al (1990), and are well suited to barrier tunneling. Ovchinnikov (1990) has obtained good agreement with Shah et al (1987) for H^+ - H ionisation, though with not quite the right fall-off. The essence of these papers is the two transition point phase integral problem, modelled by the comparison parabolic cylinder function which is expanded asymptotically. The Wannier saddle-point electrons are also investigated by Wang et al (1991) and Miraglia and Macek (1991), while the role of barrier models in general has been reviewed by Barany (1990), who draws particular attention to the contribution at low energies to transfer ionisation from double capture into a doubly excited state of the projectile followed by autoionisation. A similar idea applies to the recent measurements (Hopkins et al 1991) of single ionisation of Ba^+ and Sr^+ ions by protons in the 50 to 500 keV centre of mass energy range; it is believed that the primary mechanism is that of excitation followed by autoionisation. Other experimental measurements will be covered by C L Cocke in his preceding talk and may include differential cross sections for single and multiple ionisation of Ne and Ar by fast protons (Kamber et al, 1990), total cross sections for single and double ionisation, transfer ionisation and ionisation excitation of D_2 by fast 0^{8+} (Cheng et al, 1990) and K-shell ionisation probabilities for proton impact on thin-foil targets of Ti, Ni and Cu (Simons et al, 1990). Other K-shell ionisation investigations include the CTMC calculations on 47 MeV Ca^{17+} collisions with Ar (Heber 1990), the measurement of total cross sections for 1 to 6.4 MeV proton and carbon ion collisions with Si, P, K, Ca, Zn and Ga (Geretschlager et al 1990) and the relativistic-Dirac-equation calculations of probabilities in proton collisions with Pb and Pt (Mehler et al 1989). Total ionisation cross sections have also been measured for L-shell ionisation of the rare earth elements (La, Pr, Nd, Eu, Tb, Ho and Er) by   H   and   He ions in the 0.1 to 3.9 MeV range (Braziewicz et al 1991). In the last two experiments, reasonable accord was obtained with perturbed-stationary-state theories which include the usual corrections such as relativity, recoil, Coulomb deflection and screening.

4.   CONCLUSIONS

Ionisation, single, double or multiple and their hybrids with capture, transfer and excitation, are amenable to theory making judicious use of CDW and CDWEIS on the one hand and of independent events and independent electrons on the other. Impact parameter treatments are transparent and pragmatic but correlation of events and/or particles is a continuing issue, as is ECC. Expertise in experimental technique has thrown down the gauntlet at low, high and relativistic energies. At intermediate energies it would appear that the independent-event model lends itself to judicious dynamical screening.

ACKNOWLEDGEMENTS

This work was supported in part by the Science and Engineering Research Council, United Kingdom, through grant GR/G 06244. The author is indebted to Dr Narayan Deb (SERC), Dr Kevin Dunseath (DENI) and Mr David Marshall (DENI). The author is grateful to Dr J F McCann, Professor B L Moiseiwitsch, Professor E Salzborn, Dr R Shingal and Dr I Shimamura for their kind communication of results.

REFERENCES

Abramov D I, Ovchinnikov S Y and Solovev E A 1990 Phys.Rev.A.42 6366.
Andersen L H, Hvelplund P, Knudsen H, Møller S P, Pedersen J O P, Sørensen A H,
   Uggerhoi E, Elsener K and Morenzoni E 1990 Phys.Rev.Lett.65 1687.
Andersen L H, Hvelplund P, Knudsen H, Møller S P, Pedersen J O P, Tang-Petersen S,
   Uggerhoj E, Elsener K and Morenzoni E 1990 Phys.Rev.A.41 6536.

Andersen L H, Hvelplund P, Knudsen H, Møller S P, Sørensen A H, Elsener K, Rensfelt K-G and Uggerhoj E 1987 Phys.Rev.A$\underline{36}$ 3612.

Andersen L H, Jensen K E and Knudsen H 1986 J.Phys.B.$\underline{19}$ L161.

Bandarage G and Parson R 1990 Phys.Rev.A$\underline{41}$ 5878.

Bárány A 1990 Phys.Scripta. $\underline{42}$ 280.

Barnett C F, Ray J A, Ricci E, Wilker M I, McDaniel E W, Thomas E W and Gilbody H B 1977 Oak Ridge National Laboratory Report No. ORNL-5206, table a.4.3.8, unpublished.

Belkić D Z 1978 J.Phys.B.$\underline{11}$ 3529.

Bernardi G C, Suarez S, Fainstein P D, Garibotti C R, Meckbach W and Focke P 1989 Phys.Rev.A.$\underline{40}$ 6863.

Brauner M, Briggs J S and Klar H 1989 J.Phys.B$\underline{22}$ 2265.

Brauner M, Briggs J S and Klar H 1989 Z.Phys.D.$\underline{11}$ 257.

Braziewicz E, Braziewicz J, Gzyzewski T, Glowacka L, Jaskola M J, Kauer T, Kobzev A P, Pajek M and Trautmann D 1991 J.Phys.B$\underline{24}$ 1669.

Cheng S, Cocke C L, Kamber E Y, Hsu C C and Varghese S L 1990 Phys.Rev.A$\underline{42}$ 214.

Crothers D S F 1982 J.Phys.B$\underline{15}$ 2061.

Crothers D S F 1989 Phys.Scripta $\underline{40}$ 634.

Crothers D S F and Dubé L J 1989 J.Phys.B.$\underline{22}$ L609.

Crothers D S F and Marshall D P 1991 in preparation.

Crothers D S F and McCann J F 1983 J.Phys.B$\underline{16}$ 3229.

Crothers D S F and McCann J F 1987 J.Phys.B$\underline{20}$ L19.

Crothers D S F and McCarroll R 1987 J.Phys.B$\underline{20}$ 2835.

Datz S, Hippler R, Andersen L H, Dittner P F, Knudsen H, Krause H F, Miller P D, Pepmiller P L, Rosseel T, Schuch R, Stolterfoht N, Yamazaki Y and Vane C R 1990 Phys.Rev.A.$\underline{41}$ 3559.

Deb N C and Crothers D S F 1990 J.Phys.B$\underline{23}$ L799.

Deb N C and Crothers D S F 1991 J.Phys.B$\underline{24}$ 2359.

Deb N C, Crothers D S F and Bhattacharjee B 1991 submitted.

Deb N C, Crothers D S F and Fromme D 1990 J.Phys.B$\underline{23}$ L483.

Dubois R D and Köver A 1989 Phys.Rev.A$\underline{40}$ 3605.

Dunseath K M and Crothers D S F 1991 J.Phys.B$\underline{24}$ in press.

Edwards A K, Wood R M and Ezell R L 1990 Phys.Rev.A$\underline{42}$ 1799.

Fainstein P D and Rivarola R D 1990 Phys.Lett.A.$\underline{150}$ 23.

Fainstein P D, Ponce V H and Rivarola R D 1988a J.Phys.B$\underline{21}$ 287.

Fainstein P D, Ponce V H and Rivarola R D 1988b J.Phys.B$\underline{21}$ 2989.

Fainstein P D, Ponce V H and Rivarola R D 1989 Phys.Rev.A$\underline{40}$ 2828.

Ford A L and Reading J F 1988 J.Phys.B$\underline{21}$ L685.

Ford A L and Reading J F 1990 J.Phys.B$\underline{23}$ 3131.

Fritsch W and Lin C D 1991 Phys.Rep. $\underline{202}$ 1.

Fromme D, Kruse G, Raith W and Sinapius G 1986 Phys.Rev.Lett. $\underline{57}$ 3031.

Fukuda H, Shimamura I, Vegh L and Watanabe T 1991 Phys.Rev.A$\underline{44}$ in press.

Gayet R and Salin A 1987 J.Phys.B$\underline{20}$ L571.

Gayet R, Rivarola R D and Salin A 1981 J.Phys.B.$\underline{14}$ 2421.

Gayet R, Hanssen J, Martinez A and Rivarola R 1991 Z.Phys.D$\underline{18}$ 345.

Geretschläger M, Smit Z and Benka O 1990 Phys.Rev.A$\underline{41}$ 123.

Giese J P and Horsdal E 1988 Phys.Rev.Lett.$\underline{60}$ 2018.

Gilbody H B 1986 Adv.At.Mol.Phys. $\underline{22}$ 143.

Gilbody H B 1989 Phys.Scripta. T$\underline{28}$ 49.

Gulyas L and Szabo G 1991 Phys.Rev.A$\underline{43}$ 5133.

Gulyas L, Szabo Gy, Berenyi D, Köver A, Groenfeld K O, Hoffmann D and Burkhard M 1986 Phys.Rev.A.$\underline{34}$ 2751.

Heber O, Bandong B B, Sampoll G and Watson R L 1990 Phys.Rev.Lett.$\underline{64}$ 851.

Heber O 1990 Phys.Rev.A$\underline{42}$ 1795.

Hopkins C J, Dunn K F and Gilbody H B 1991 J.Phys.B$\underline{24}$ 2379.

Kamber E Y, Cocke C L, Cheng S and Varghese S L 1990 Phys.Rev.A$\underline{41}$ 150.

Knudsen H, Andersen L H and Jensen K E 1986 J.Phys.B.$\underline{19}$ 3341.

Knusden H, Andersen L H, Hvelplund P, Astner G, Cederquist H, Danared H, Liljeby L and
    Rensfelt K G, 1984 J.Phys.B17 3545.
Knusden H, Andersen L H, Hvelplund P, Sørensen J and Ciric D 1987 J.Phys.B20 L253.
McCann J F 1991 J.Phys.B24 in press.
McCann J F and Crothers D S F 1987 Nucl.Instr.Meth.Phys.Res.B23 164.
McGuire J H and Straton J C 1989 The Physics of Electronic and Atomic Collisions XVI
    International Conference, New York, NY (AIP: New York, ed A Dalgarno et al) p 280-
    289.
Meckbach W, Nemirowsky I B and Garibotti C R 1981 Phys.Rev.A.24 1793.
Mehler G, Soff G, Rumrich K and Greiner W 1989 Z.Phys.D.13 193.
Miraglia J E and Macek J 1991 Phys.Rev.A43 5919.
Moiseiwitsch B L 1991 J.Phys.B.24 983.
Muller A, Schuch B, Groh W and Salzborn E 1987 Z.Phys.D.7 251.
Olson R E 1987 Phys.Rev.A.36 1519.
Olson R E, Ullrich J, Dörner R and Schmidt-Bocking H 1989 Phys.Rev.A40 2843.
Ovchinnikov S Y 1990 Phys.Rev.A42 3865.
Palinkas J, Schuch R, Cederquist H and Gustafsson O 1990 Phys.Scripta. 42 175.
Pedersen J O P 1990 Phys.Scripta. 42 180.
Pivovar L I, Novikov M T and Tubaev V M 1962 Sov.Phys.JETP 15 1035.
Pluvinage P 1980 Ann.Phys.5 145.
Presnyakov L P and Uskov D B 1984 Sovt.Phys.JETP. 59 515.
Presnyakov L P and Uskov D B 1989 Abstracts of Contributed Papers XVI ICPEAC, New
    York (ed. A Dalgarno et al) 602.
Reading J F and Ford A L 1987 J.Phys.B20 3747.
Rivarola R D, Fainstein P D and Ponce H 1989 Phys.Scripta T28 101.
Rivarola R D, Fainstein P D and Ponce H 1989 The Physics of Electronic and Atomic
    Collisions XVI International Conference New York NY (AIP: New York, ed. A Dalgarno
    et al) pp 264-72.
Salin A 1987 Phys.Rev.A36 5471.
Salzborn E 1989 The Physics of Electronic and Atomic Collisions XVI International
    Conference, New York, NY (AIP: New York, ed A.Dalgarno et al) pp 290-8.
Sarkardi L and Mukoyama T 1990 Phys.Rev A42 3878.
Sarkardi L, Palinkas J. Kover A, Berenyi D and Vajnai T 1989 Phys.Rev.Lett.62 527.
Schiwietz G, Skogvall B, Stolterfoht N, Schneider D and Montemayor V 1989 The Physics
    of Electronic and Atomic Collisions XVI International Conference, New York, Ny (AIP:
    New York, ed A.Dalgarno et al) pp 299-308.
Schneider D, De Witt D, Schlachter A S, Olson R E, Graham W G, Mowat J R, DuBois R
    D, Lloyd D H, Montemayor V and Schiwietz G 1989 Phys.Rev.A.40 2971.
Schram B L, Boreboom A H J and Kistemaker 1966 Physica 32 185.
Schultz D R 1989 Phys.Rev.A40 2330.
Schultz D R, Olson R E and Reinhold C O 1991 J.Phys.B24 521.
Schulz R, Melchert F, Hagmann M, Krüdener S, Krüger J, Salzborn E, Reinhold C O and
    Olson R E 1991 J.Phys.B24 L7.
Shah M B and Gilbody H B 1985 J.Phys.B18 899.
Shah M B, Elliot D S and Gilbody H B 1987 J.Phys.B20 2481.
Shah M B, Elliott D S, McCallion P and Gilbody H B 1988 J.Phys.B.21 2455.
Shingal R 1991 Nucl.Instrum.Meth.Phys.Res. in press.
Shingal R and Lin C D 1991a J.Phys.B24 251.
Shingal R and Lin C D 1991b J.Phys.B24 963.
Simons D G, Land D J, Brown M T and Cocke C L 1990 Phys.Rev.A42 1324.
Solovev E A 1990 Phys.Rev.A42 1331.
Wang J, Burgdörfer J and Bárány A 1991 Phys.Rev.A43 4036.

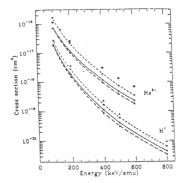

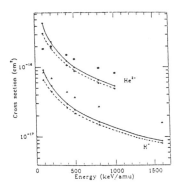

Figure 1: Single capture cross sections for H⁺ + He, He²⁺ + He collisions. Theoretical values: PCDW, HCDW refer to our CDW calculations using Pluvinage and Hydrogenic wave functions respectively, with the corresponding theoretical binding energies employed throughout. Small open circles represent points at which cross sections have actually been calculated (Dunseath and Crothers 1991) while the curves have been drawn as guide the eye. ———, PCDW (post form); — — —, PCDW (prior form); - - - -, HCDW. Experimental data: Shah and Gilbody (1985): ●, He²⁺ projectiles; ▲, H⁺ projectiles.

Figure 2: Single ionisation cross sections for H⁺ + He, He²⁺ + He collisions. Theoretical values: Dunseath and Crothers 1991 ———, PCDW (post form); - - - -, HCDW (post form). Experimental data: Shah and Gilbody (1985), as for Figure 1.

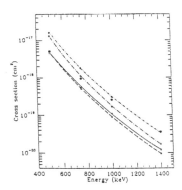

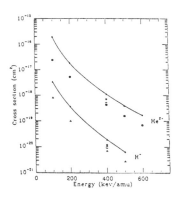

Figure 3: Resonant double capture cross sections for He²⁺ + He collisions. Theoretical values: Dunseath and Crothers (1991) ———, PCDW (post form); — — — — — , CDW (I); — — — — , (li) (Gayet et al 1981). Experimental data: Barnett et al (1977), Pivovar et al (1962).

Figure 4: Transfer ionisation cross sections for H⁺ + He, He²⁺ + He collisions. Theoretical values: Dunseath and Crothers (1991) ———, PCDW (post form): ■, CDW+Born; □, CDW+MEDOC (Gayet and Salin 1987). Experimental data: Shah and Gilbody (1985), as for Figure 1.

E (keV/amu)	DC	SL	SG	KD
200.0	236.2	303.0	192.0± 8.0	—
313.0	131.8	374.0	153.0± 8.0	—
390.0	99.73	260.0	126.0± 4.0	—
640.0	46.9	—	—	74.0± 8.1
1440.0	12.5	—	—	18.0± 2.0
2310.0	5.53	—	—	8.5 ± 0.9

Table 1: Total double ionisation cross sections in units of 10^{-19} cm² for Li²⁺ impact on helium atoms. DC: Deb and Crothers (1991); SL: Shingal and Lin (1991); SG: Shah and Gilbody (1985); KD: Knudsen et al (1984).

Energy (cm²)	Expt (cm²)	PCDW(+) (cm²)	PCDW(+)+20% (cm²)	CDW+B (cm²)	CDW+MEDOC (cm²)
100	2.4^{-12}	3.8^{-17}			
200	5.2^{-18}	5.0^{-18}			
400	4.3^{-19}	3.3^{-19}	$\left[4.0^{-19}\right.$	7.1^{-19}	4.9^{-19}
500	1.6^{-19}	1.2^{-19}	1.4^{-19}		
600	7.0^{-20}	5.2^{-20}	$\left.6.2^{-20}\right]$		

Table 2.

Transfer ionisation cross sections for He²⁺ + He collisions. Experimental data: Shah and Gilbody (1985). Theoretical data: PCDW(+), Crothers and Dunseath (unpublished, see text); CDW+Born, CDW+MEDOC, Gayet and Salin (1987). Bracketed terms are PCDW(+) increased by 20%.

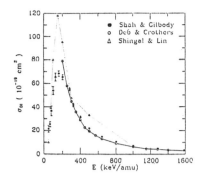

Figure 5: Total cross sections for double ionisation of He by He^{++} impact as a function of incident energy E (keV/amu). Solid circles: measurement of Shah and Gilbody (1985); open circles: Deb and Crothers (1990); open triangles: Shingal and Lin (1991). Solid and dotted lines are drawn through the theoretical results to guide the eye.

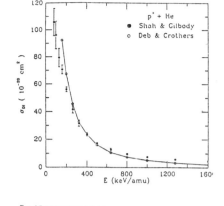

Figure 6: Total cross sections for double ionisation of He by H$^+$ impact as a function of incident energy E(keV/u). Full circles: measurement of Shah and Gilbody (1985); open circles: theory of Deb and Crothers (1991). A full curve is drawn through the latter results to guide the eye.

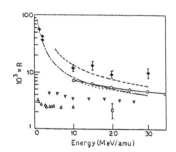

Figure 7: Double-to-single ionisation ratio multiplied by 10^3 for various projectiles: Measurements: (a) ▽, electron data (Schram et al, 1966) (b) △, proton data (Knudsen et al 1984; Shah and Gilbody 1985) (c) □ alphaparticle data (Heber et al 1990) and (d) ◆ N^{7+} data (Knudsen et al 1984, Heber et al 1990, Müller et al 1987) (e) the analysis of Andersen et al (1990) based on their data for H$^+$, p$^+$ and He^{++}. Theory for N^{7+} (a) ── ──, semiempirical fit of Knudsen et al (1984), (b) ...., forced impulse approximation of Reading and Ford (1987) and of Ford and Reading (1988) and (c) σ, Deb et al (1991). Solid line is drawn through the present results to guide the eye.

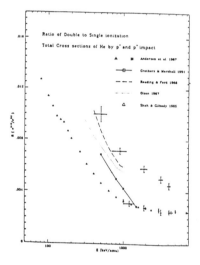

Figure 8: The graph shows the double ionisation to single ionisation cross section ratios (R) for both proton and anti-proton impact on helium. Anti-proton. Andersen et al (1987) dark squares, Crothers & Marshall (1991) solid line. Olson (1987) dotted line. Ford & Reading (1988) dashed line. Proton. Andersen et al (1987) dark triangles, Shah & Gilbody (1985) open triangles.

Theory of ion–atom excitation

J. E. Miraglia and V. D. Rodríguez

Instituto de Astronomía y Física del Espacio
C.C. 67, Suc. 28 - 1428 Buenos Aires - Argentina

ABSTRACT: A detailed study of the eikonal impulse approximation as applied to hydrogen excitation by bare projectiles and antiprotons impact is reported. Total and differential cross sections, sublevel population, scaling laws with the projectile charge, off-diagonal density matrix elements and related parameters are presented for the $n = 2$ and 3 hydrogen levels. The present theory proves to be a good approximation for the intermediate and high energy range.

1. INTRODUCTION

Ion-atom excitation has been the object of a series of experimental studies in the last decade (for a review see the work by Park 1983, more recent works are: Wohrer *et al.* 1986, Reymann *et al* 1988, Schartner *el al* 1989). The experiments on atomic hydrogen targets are very difficult to perform and the data reported are very few. At high impact energies, say larger than 200 keV for proton impact, data should tend to the first Born approximation which is known to give the correct nonrelativistic limit. On the other hand, at low energy, say less than 10 keV, the close-coupling of molecular states is known to be the appropiate theoretical technique to solve the problem. Theoretical difficulties arise in the intermediate energy range, say between 15 and 200 keV for proton impact, that is where the projectile has a velocity comparable with to that of the electron in the first Bohr orbit. The situation become worse for multicharged projectile impact; in this case, the Coulomb parameter Z_P/v, where Z_P is the projectile charge and v the impinging velocity, may be larger than unity even though $v > 1$. Therefore, the theoretical work is directed to solve the problem in the intermediate energy range, which is here defined as: $25 < E/Z_P < 200$ keV/amu where E is the energy per atomic unit mass.

In the intermediate energy range, most of the efforts were addressed to solve the time-dependent Schrödinger equation for the electron motion, for which a close-coupling of atomic states procedure is generally performed (see for example Shakeshaft 1978, Jain *et al.* 1987 and 1988, and references therein). On the other hand, for the intermediate and high energies, some distorted wave methods have been developed (Coleman 1969 and references therein, Franco and Thomas 1971, Theodosiou 1980), and they have been quite successful to explain some experiments. Common features of them are their relative simplicity and the standardization of the numerical calculation for any projectile charge Z_P.

In this context, we have developed the symmetric eikonal (SE) distorted wave method (Reinhold and Miraglia 1987, Rodríguez and Miraglia 1989a and b, Reinhold *et al* 1990) which preserves the Coulomb asymptotic conditions on both the initial and final channels. We have observed that the SE method provides good description for allowed transitions, but it seems to underestimate the forbidden ones when compared with sophisticated close-coupling calculations and the experiments in the intermediate energy range. A permitted (forbidden) process is defined when the leading term in the first Born approximation is a dipole (quadrupole) matrix element; $1s - np$ are permitted and $1s - ns$ and $1s - nd$ are forbidden. The reason for the deficiency of the SE method could be understood if we consider that in forbidden transitions second orders processes can be competitive with the direct one, that is: a forbidden transition can be achieved by two successive allowed transitions, as discussed by Inokuti (1971) and Vainshtein and Presnyakov (1969). This two-step transition can be nearly as efficient as the weak direct (optically-forbidden) one at intermediate energies. A way to improve the theory is to use a better wave function that considers, albeit approximately, these higher order terms, as the impulse wave function does (Coleman 1969). In the so-called impulse approximation (IA), the multiple scattering of the electron in the field of the projectile is built up into the corresponding wave function, though in this theory, when the internuclear interaction is switched-off, the Coulomb condition on the non-distorted channel is missing. Improvement on both, the SE and the IA has been done by introducing the eikonal-impulse approximation (EIA) in the context of excitation (Rodríguez and Miraglia 1990). This work intends to be a detailed report on the EIA. Comparison with the SE, IA and other theoretical calculations, as well as with experiments when available will be made throughout.

2. THE EIKONAL IMPULSE THEORY

The impulse hypothesis considers that the action of the projectile over the electron is a force of a very large magnitude which acts during a small period of time. From the quantum mechanical point of view, it is assumed that during such a small period the binding target-electron interaction may be neglected so that the electron evolves in the Coulomb continuum of the projectile. The formalism based on the impulse hypothesis leads to the so-called impulse wave functions ψ_i^I and ψ_f^I, satisfying the incoming and outgoing conditions, respectively (see for example, Coleman 1969). Within the distorted wave formalism, ψ_i^I and ψ_f^I approximate the exact scattering states Ψ_i^+ and Ψ_f^-, respectively. As these exact states, the impulse wave functions satisfy two important properties; first, they are properly orthonormalized, i.e., $\langle \psi_i^I | \psi_{i'}^I \rangle = \delta_{i,i'}$ and $\langle \psi_f^I | \psi_{f'}^I \rangle = \delta_{f,f'}$; and second, they have the correct asymptotic Coulomb distortions. Both ψ_i^I and ψ_f^I are represented by three-dimensional integrals on the momentum space, which has to be solved numerically; any attempt to approximate them, to obtain analytical forms, generally leads to the loss of orthonormalization property or to the deterioration of the asymptotic conditions. Such is the case of the peaking impulse approximation (Bransden and Cheshire 1963, Coleman 1969) which has been largely used to obtain high energy behaviors. We can construct two T-matrix elements, the standard post (IA^+) and prior (IA^-) impulse approximations, given by $T^{IA+} = \langle \psi_f^B | V_i^I | \psi_i^I \rangle$ and $T^{IA-} = \langle \psi_f^I | V_f^I{}^\dagger | \psi_i^B \rangle$ respectively, where $V_{i,f}^I$ are the corresponding perturbations and $\psi_{i,f}^B$ are the unperturbed (Born) wave functions. Total cross sections for excitation of hydrogen to n=2 by proton impact are displayed in fig.1 along with the experiments by Morgan *et al.* (1973) and by Park *et al* (1980) measured with the energy loss technique. After a considerable numerical task, the performance of the IA is frustrating, not only an inadmissible prior-post discrepancy is observed, but the simple first Born approximation produces much better results. This poor performance reported by Coleman (1968) in the sixties has probably been the reason of why the IA has been forgotten, as far as excitation is concerned.

One can rightly blame the unperturbed wave functions $\psi_{i,f}^B$ for the failure: they satisfy the orthonormalization but not the Coulomb asymptotic conditions. To solve this problem, we may employ the well known eikonal wave functions $\psi_{i,f}^E$, satisfying both, orthonormalization and asymptotic conditions (see for example Rodríguez and Miraglia 1990). However, the virtue of the impulse approximation is lost in the sense that the full electron-projectile continuum interaction is now approximated by its asymptotic phase factor. We can then construct a symmetric T-matrix element defining $T^{SE} = \langle \psi_f^E | V_{i,f}^E | \psi_i^E \rangle$. This approximation, called symmetric eikonal (SE), presents a good performance when compared with the data in fig. 1. One way of testing further the SE approximation is to compare it with data for $2p$ and $2s$ subshell excitation, as shown in fig. 2. We observe that the SE model produces good results only for allowed transition and it underestimates the $2s$ forbidden excitation. The explanation to this situation is simple: allowed transitions are known to take place at relatively large distances in a single step, so just the Coulomb conditions included in the SE method are enough to explain the data. On the other side, the $2s$-excitation takes place, as a forbidden transition, at smaller distances (than permitted ones) in one and two steps at high and intermediate energies, respectively.

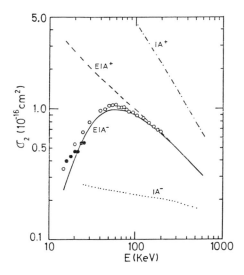

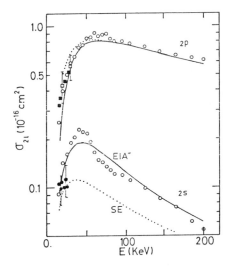

Fig. 1: Excitation cross sections of hydrogen to n=2 by impact of protons. Experiments: o by Park *et al.* (1976), • Morgan *et al.* (1973). Theory: IA$^+$ and IA$^-$ post and prior exact impulse approximations; EIA$^+$ and EIA$^-$ post and prior eikonal impulse approximations as indicated in the text.

Fig. 2: Excitation to 2p and 2s-states of hydrogen by impact of protons as a function of the impinging energy. Experiments: o ,extended data from Park (1983); • , Morgan *et al* (1973); □ , Young *et al.* (1968); ■ , Kondow *et al* (1974). Theory: solid line: prior eikonal impulse approximations; dotted line, symmetric eikonal method.

Therefore, the SE method can be used with some degree of confidence if we are interested in optically-allowed transitions. The T^{SE} is expressed in closed-forms in terms of the hypergeometric function $_2F_1$, and a total cross section is obtained in a few seconds in a personal computer. Since optically forbidden transitions are considerably weaker than the allowed ones, this theory accounts rather well for the excitation total cross sections to a given level. Thus, for example, total X-ray emission cross section of Ca^{19+} and Ca^{18+} by neutral atoms impact at fixed velocity $v = 18.5$ (a.u.) measured by Xu *et al.* (1988) were explained by theoretical calculations obtained with the SE model (Rodríguez and Miraglia 1989a). Also, helium excitation of multicharged ions were successfully explained with this simple model (Rodríguez and Miraglia 1989c).

To treat also forbidden transitions, we should improve the description at short distances by using the impulse wave functions on both channels. The theory so-constructed, called generalized impulse approximation (Miraglia 1982), requires a six-dimensional integral of very complicated functions to calculate the corresponding T-matrix element. This task is too hard even for today standards. We can considerably reduce the calculation by introducing ψ^E in one channel. In this way, we define the prior and post eikonal impulse approximation (EIA$^-$ and EIA$^+$) as follows: $T^{EIA+} = \langle \psi_f^E | V_i^I | \psi_i^I \rangle$, and $T^{EIA-} = \langle \psi_f^I | V_f^{I\,\dagger} | \psi_i^E \rangle$ (see Rodríguez and Miraglia 1990 for the mathematical details). We can consider the EIA approximation as an improvement on the SE where $\psi_{i,f}^E$ is replaced by $\psi_{i,f}^I$ to account for the multiple scattering of the electron in the field of the projectile, or an improvement on the IA where $\psi_{i,f}^B$ is replaced by $\psi_{i,f}^E$ to account for the proper Coulomb conditions. Total cross section to n=2 excitation is plotted in fig. 1, and we come to two conclusions: prior-post discrepancy vanishes for 100 keV onwards, and the prior version has the best performance. This last feature has a simple explanation: the mean velocity of the electron in the final excited state is smaller than the initial one, so ψ_f^I is in a better condition than ψ_i^I to describe the collision process in accordance with the impulse hypothesis, i.e., the collision time should be short compared with the corresponding orbital times. Examining fig. 2, we conclude that the EIA$^-$ describes quite well the 2s-forbidden transition.

3. RESULTS

In this section, we will restrict ourselves to the study of the EIA^- performance in the treatment of hydrogen excitation by multicharged ions, i.e.

$$P^{Z_P+} + H(1s) \rightarrow P^{Z_P+} + H(nlm).$$

The EIA$^-$ calculation involves a computing time about 10^5 longer than the SE method, so it could be considered as a relatively-heavy task, but still faster than a detailed atomic close-coupling calculation. The computing time lightly increases with the final quantum number (n) as well as with the projectile charge Z_P, but it does not require a previous knowledge of the collision system as required by the atomic or molecular close-coupling method. After producing the numerical code, the calculation reduces to a standard number crunching to solve three-dimensional integrals to obtain the corresponding T-matrix. Another numerical integral is needed to obtain the total cross section σ_{nlm} for a given transition. Numerical errors at the level of cross sections are estimated to be less tahn 3 percent, for nondiagonal elements of the density matrix elements ($\sigma_{nlm,nl'm'}$), the numerical uncertainties are larger.

3.1 Total cross sections

Total cross sections $\sigma_n = \Sigma_{lm}\sigma_{nlm}$ for hydrogen excitation to $n = 2$ and 3 by impact of multicharged ions are plotted in fig. 3 along with the experiments for proton impact (Park *et al* 1976). These data were normalized to the first Born approximation for 200 keV proton impact excitation of atomic hydrogen to its $n = 2$ level, *i.e.* $6.64 \pm 0.35 \ 10^{-17} \ cm^2$ as obtained by Bates and Griffing (1953); a more exact calculation gives $6.85 \ 10^{-17} \ cm^2$ (Mandal *et al* 1990) instead. Our result is $\sigma_2(200 \ keV) = 6.38 \ 10^{-17} \ cm^2$, which lies 4 percent below the experiment. This difference is of the order of our numerical uncertainty and experimental error (about 5 percent). For $n = 2$ excitation our theory presents a very good performance, the worst case is at the lowest energy plotted 25 keV proton impact where our theory lies 13 percent below the data; it is a quite good performance for a distorted wave method (if we normalized the data to our theory at 200 keV proton impact, the lowest result would be only 9 percent below). Disagreements are found for $n = 3$ excitation by proton impact. At 200 keV proton impact, the experimental data is $1.41 \ 10^{-17} \ cm^2$, and the Born approximation gives 1.25 in the same units (Mandal *et al* 1990). A point to take into consideration is that the first order is here smaller than the data, and it is known by experience that generally it is the other way around. The EIA$^-$ results run consistently around 25 percent below the experiment through all the energy range. The efficacy of the EIA^- theory to explain the $n = 2$ data seems to suggest that the difference may be due to the normalization procedure of the experiments.

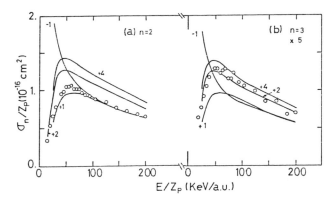

Fig. 3: Total cross section of hydrogen excitation to $n = 2$ (a) and $n = 3$ (b), by impact of multicharged ions calculated with the EIA^- theoretical method. The cross section and the impact energy are divided by the projectile charge. Experiments for proton impact from Park *et al* (1980).

Unfortunately, there are no experiments for p^- and multicharged projectiles; however, three features are worthy of mention:

- for $Z_P = 1, 2$, and 4, the total cross sections can be enclosed within a certain band if we plot $\sigma_n/|Z_P|$ versus $E/|Z_P|$. This parametrization is related to the Janev and Presnyakov scaling (1980), and it will be discussed in section 3.3, in connection with the optically-allowed transitions.

- at high impact energies, the excitation by protons and antiprotons tends to the same value, as expected,

- antiproton cross sections do not show the typical maximum as those of the positive-charged projectiles.

This last feature can be understood if we take into account that no electron capture is feasible by antiprotons and therefore there is no depletion of probability due to exchange for the lower energies.

3.2 Differential cross sections

Angular distributions are shown in fig. 4 and compared with the experiments of Park *et al* (1980). At 50 keV, the theory seems to underestimate the data at intermediate angles, while at 100 keV it overestimates the same experiments in the forward direction, but in general, the theory produces a satisfactory performance.

3.3 Scaling laws for allowed transition

Janev and Presnyakov (JP, 1980) have found that the np-excitation cross section (σ_{np}) divided by Z_P lies approximately on a universal curve in the impact energy range $E/Z_P = 25 - 100$ keV/amu. That is: $\sigma_{np}(Z_P, v) \cong Z_p \, U_{np}(v/\sqrt{Z_P})$, where with U we denote a universal curve and v is the projectile velocity. The JP scaling has been the object of a large number of theoretical and experimental studies, not only in excitation (Fritsch and Shartner 1987, Rodríguez and Miraglia 1989b, Reinhold *et al.* 1987), but also in ionization (Gilbody 1986, Rodríguez and Falcón 1990), and it has become a useful tool to plot the data in a consistent way. The JP scaling property is based on a picture of distant collision where it can be proved that the probability P_{np} as a function of the impact parameter ρ satisfies: $P_{np}(Z_P, v, \rho) \cong U'_{np}(v/\sqrt{Z_P}, \rho/\sqrt{Z_P})$. In order to examine this scaling, EIA^- probabilities for 2p-excitation are plotted in fig. 5 as a function of $\rho' = \rho/\sqrt{|Z_P|}$ for $Z_P = -1$ to 4.

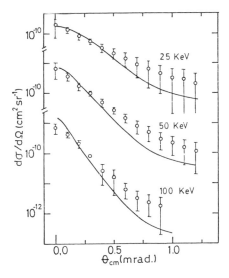

Fig. 4: Differential cross section for excitation of hydrogen to $n = 2$ by impact of protons as a function of the center-of-mass scattering angle at different energies. Experiments of Park *et al* (1980); and the solid line, the present EIA^- approximation.

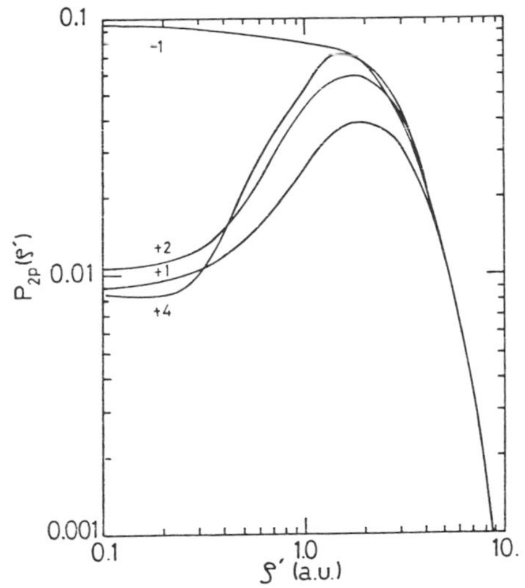

Fig. 5: 2p-excitation probabilities of hydrogen excitation by impact of multicharged ions as a function of the reduced impact parameter: $\rho' = \rho/\sqrt{|\bar{Z}_P|}$, for impinging energy of $E/|Z_P| = 50$ keV/amu, calculated with the present EIA^- method.

Two features should be noted: first, for large ρ', the scaling is satisfied quite well, and second, for small ρ', antiprotons do not yield probability. It is an indication that capture is here not feasible, and the EIA^- model rightly predicts this situation. It should be mentioned that, at the level of cross sections , the SE theoretical method follows quite well the JP scaling for permitted and forbidden transitions (Rodríguez and Miraglia 1989b), but it does not distinguish between positive and negative charges; this is a source of difference with the present EIA^- method which stems from the small impact parameter range. It indicates again the importance of the multiple scattering contribution included in the impulse wave function at short distances.

The scaling of the probability amplitude leads to a scaling of the T-matrix elements so that the differential cross section can also be expressed with a universal function: $d\sigma_{np}(Z_P, v, \theta)/d\Omega \cong Z_P^3 U''_{np}(v/\sqrt{Z_P}, \theta Z_P)$ for a fixed projectile mass and θ is the center-of-mass scattering angle.

According to the JP work, the universal function U_{np} is linearly proportional to the oscillator strength f_{np}. The ratio of total cross section then satisfies: $\sigma_{np}/\sigma_{n'p} = f_{np}/f_{n'p}$ or $\sigma_{3p}/\sigma_{2p} = 0.4162/0.0791 = 5.26$. Our EIA results satisfy:$\sigma_{3p}/\sigma_{2p} = 5.5 \pm 0.5$ for $Z_P = -1, 1, 2$, and 4 in the range $25 - 200$ keV/$|Z_P|$, indicating that the prediction of the JP scaling holds for the EI approximation within a band of 20 percent, as shown in Figure 6a. In this figure we plot the *reduced* cross section $\sigma_{np}/(f_{np}|Z_P|)$ as a function of $E/|Z_P|$.

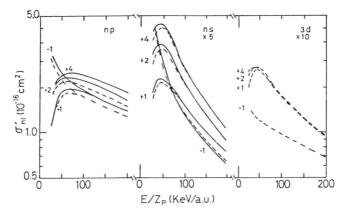

Fig. 6: Reduced cross section $\sigma'_{nl} = \sigma_{nl}/(|Z_P|f_{np})$ for hydrogen excitation to np (a), ns (b) and nd (c) by impact of multicharged ions calculated with the EIA^- theoretical method as a function of $E/|Z_P|$ ($f_{2p} = 0.4162$ and $f_{3p} = 0.0791$). The solid ($n = 2$) and dashed ($n = 3$) curves are labelled with the corresponding projectile charge Z_P.

A very sensitive parameter to test further the universality of the JP scaling is the so-called alignment factor defined as: $A_{20} = (\sigma_{2p1} - \sigma_{2p0})/\sigma_{2p}$. Results for different charges are plotted in fig.7 in the intermediate energy range. Even though 2p excitation follows quite well the JP scaling, A_{20} presents a much more disperse band in this frame. Although the error bars of the experiments for proton impact are very large, the EIA^- seems to follow the experiments. As it is well known the dipole interaction produces transitions between the sublevels 2s and $2p_0$ even at relatively long distances (Stark mixing). The agreement of the EIA^- model with the experiments may indicate that Stark mixing is perhaps included, at least, in part. More experiments are needed to confirm this.

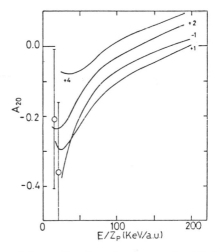

Fig. 7: Alignment factor A_{20} as a function of $E/|Z_P|$ for different projectile charges. The circles denote A_{20} related to the experimental polarization of Kauppila *et al* (1970) through equation (5) of Hippler *et al* (1988).

3.4 Forbidden transitions

Excitation to 2s by proton impact was discussed in section 2 and compared with the data in Fig. 2. Figures 6b and 6c show reduced cross sections for ns and $3d$ hydrogen excitations by impact of multicharged projectiles. These cross sections are between half and one order of magnitude smaller than the np allowed ones. Few points are worthy of mention. For ns excitation, the JP scaling is not observed for $Z_P < 4$ (we have observed this scaling starts to hold for $Z_P > 8$, indicating that forbidden transitions may be achieved by two successive allowed ones). The reduced cross section is independent on

n within a band of 15 percent. For antiproton impact, $3d$-excitation presents a peculiar behavior; we have no explanation for this feature. It should be mentioned that our results for proton impact for ns- and $3d$-excitations are larger than those of the first Born approximation in the range 100-200 keV (this is not the case for np-allowed transitions, where the theory runs below the first order). It suggests that our theory may include part of the intermediate dipole transitions (Vainshtein and Presnyakov 1969). Further discussions will be possible when more experiments are available.

3.5 Density matrix elements

The collisional population of a given level can be fully described by the so-called density matrix. The interest of this kind of studies rests on the different pictures of the collisional dynamics that can be deduced from them. There are three ways of presenting the results, apart from the simple tabulation of all the matrix elements, viz: (i) the state multipoles or statistical tensors $< T(l', l)_{k,q} >$ (see for example Blum 1981); (ii) the z-components of the dipole moment $\langle D_z \rangle$ and the perihelion velocity $\langle \mathbf{L} \times \mathbf{A} \rangle_z$, where $\mathbf{L}$ and $\mathbf{A}$ are the angular momentum and Runge-Lenz vectors (Burgdörfer 1986, Jain *et al* 1988); and (iii) the coefficients of the Legendre polynomial expansion β_K (Briggs 1986, Burgdörfer 1986). We here deal with the parameters $\langle D_z \rangle$ and $\langle \mathbf{L} \times \mathbf{A} \rangle_z$ which have classical interpretations, and the coefficients β_K which are closely related to the soft electron peak for ionization (Meckbach *et al.* 1981), and therefore suitable to be measured.

The z-components of the dipole $\langle D_z \rangle$ and perihelion velocity $\langle \mathbf{L} \times \mathbf{A} \rangle_z$ for hydrogen excitation to $n =2$ and 3 are plotted in Figs. 8a and 8b, respectively. Since there are no experiments available, we display the more sophisticated close-coupling calculation of Jain *et al* (1988) for proton impact. In general, our theory gives the same signs, but different structure: the close-coupling results present sharp shapes in contrast with our smooth behavior. For the dipole moment to n=2, our results give the same order as those obtained by Jain *et al* (1988), but large differences are found for n=3 excitation. Note, the dipole moment increases in magnitude with the principal quantum number n. In relation to Z_P, we find the following systematics: p^- produces results opposite in sign but with the same order as H^+ impact, and the JP scaling rule holds quite well.

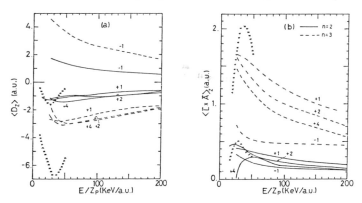

Fig. 8: z-components of the dipole $< D_z >$ (a) and the perihelion velocity $< \mathbf{L} \times \mathbf{A} >_z$ (b) of $n = 2$ (solid lines) and $n = 3$ (dashed lines) of hydrogen excitation as a function of the impact energy for different projectiles Z_p, as indicated in the figure. The crosses are the results of Jain *et al* (1988) for proton impact.

In relation to the perihelion velocity we find that for proton impact, our theory agrees quite well with the close-coupling calculation of Jain *et al* (1988). Note that the sign is always positive no matter the charge and it increases with the principal quantum number

n. For $n = 2$, and accepting the scaling in terms of the specific energy $(E/|Z_P|)$, we find a rough equality for all the systems. But this is not the case for $n = 3$

Based on an extrapolation below threshold describing the angular distribution of low energy ionized electrons, Schöller *et al* (1986) and Burgdörfer (1986) propose the following expansion:

$$\frac{d\sigma_n}{d\Omega_e} = \sum_K \beta_K^{(n)} P_K(cos\theta_e)$$

where Ω_e is the angle of the electron relative to the incident beam. The $\beta_K^{(n)}$ parameters are found to be related to the state multipoles and the density matrix elements, see for details the works by Schöller *et al* (1986) and Burgdörfer (1986) (there is a difference of sign of $(-1)^K$ in the definition of β_K between these authors). The parameters β_K are real and they vanish for odd K within the first Born approximation. We have considered it convenient to plot the ratio $\beta_{K0}^{(n)} = \beta_K^{(n)}/\beta_0^{(n)}$ in order to normalize the coefficients.

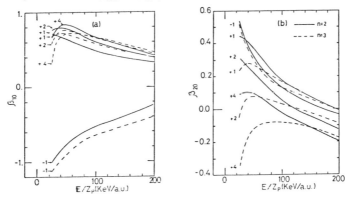

Fig. 9: Parameters $\beta_{10}^{(n)}$ (a) and $\beta_{20}^{(n)}$ (b) for hydrogen excitation to its level $n = 2$ (solid lines) and $n = 3$ (dashed lines) by impact of antiprotons, protons, He^{++} and Be^{4+} as a function of $E/|Z_P|$. The curves are labelled with the corresponding projectile charge Z_P.

Figure 9a displays $\beta_{10}^{(n)}$ calculated with the present EIA^- method. In the first Born approximation this parameter is null, so the values obtained represent the second and higher order contributions. For 200 keV proton impact, Schöller *et al* (1986) found roughly 0.2 and 0.6 (considering the appropriate change of sign), while our results are 0.32 and 0.43, for $\beta_{10}^{(2)}$ and $\beta_{10}^{(3)}$, respectively. The systematic is similar to the one we find for the dipole moment, *i.e.*, $\beta_{10}^{(n)}$ lightly increases with n, p^- produces results opposite in sign, but with the same magnitude as H^+ impact, and the Janev and Presnyakov scaling rule holds quite well. Figure 9b displays $\beta_{20}^{(n)}$. For 200 keV proton impact, Schöller *et al* (1986) find that their results along with the first Born approximation crosses the zero with negative slope, as our findings. It is difficult to find a systematic, but in general we could say that they follow the same tendency, at higher energies $\beta_{20}^{(3)}$ tends to $\beta_{20}^{(2)}$, but no firm routine is observed when varying Z_P. Following with higher order parameters we find that $\beta_{30}^{(3)}$ has a finite no-zero value and $\beta_{40}^{(3)}$ is quite small.

The fact that odd-order legendre polynomials contribute to the angular distribution has important implications in describing the soft electron peak (low energy with respect to the target nucleus) in ionization processes. Since the first Born approximation predicts that odd β cancels (Drepper and Briggs 1976), the angular distribution of the low-energy electrons has been considered to be symmetric. Based on the present calculation, it can be inferred that $\beta_1^{(n)}$ tends to a certain positive value as n tends to ∞, and so the soft energy cusp should be asymmetric enhanced in the direction of the projectile. Experiments would be welcome to test the present prediction.

4. CLASSICAL PICTURES

With the z-components of the Runge-lenz vector (which is parallel to the dipole vector) and the perihelion velocity, two possible classical 'trajectories' of the electron in the final state can be sketched, depending on the sign of the angular momentum perpendicular to the plane of the electron orbit, namely: $L_y(\theta) = -2\sqrt{2}\ Im\ \sigma_{n10,n11}/\sigma_n(\theta)$. Angular distributions of L_y for different projectiles for $n=2$ and $n=3$ hydrogen excitation are shown in Figure 10. Since the first Born approximation produces $L_y = 0$, the values shown in the figure are strongly related to the second- and higher-order perturbation contributions. An important conclusion can be drawn: L_y is negative for antiproton impact, and positive for impact of positive charges at grazing collision (forward direction). The sign of L_y is related to the attractive or repulsive feature of the projectile-electron interaction (Blum 1981), as studied by Fano and Kohmoto (1977). An interesting point to mention is that within the SE theoretical methods, it can be proved that $T^{SE}(Z_p) = T^{SE}\ ^*(-Z_p)$, and then: $L_y^{SE}(+Z_p, \theta) = -L_y^{SE}(-Z_p, \theta) > 0$, so the long distance conditions included in this method produce different signs according with the repulsive and attractive nature of the projectile-electron interaction at long distances, even though there is no difference at the level of total and differential cross sections. For positive-charged projectile, our EIA^- method produces some structures indicating that a deeper penetration of the projectile could find repulsive parts of the potential.

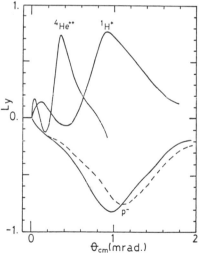

Fig. 10: Angular distributions of L_y for p^-, $^1H^+$, and $^9Be^{4+}$ impact as a function of the center of mass scattering at $E/|Z_P| = 50$ keV/amu for hydrogen excitation to its level $n = 2$ (solid lines) and $n = 3$ (dashed lines). The value of the azimuth is zero, i.e., along the x-axes, and the internuclear interaction is included in the calculation.

Figure 11 schematically shows the classical orbits of the electron in the final state when excited by negative and positive projectiles at grazing collision. The orbits follow the common sense; the Runge-Lenz vector is oriented along the projectile way out. In the case of positive projectile impact, for example, the electron orbit keeps the memory of a trajectory which encloses both heavy nuclei. We have found that the sign of L_y for small angles is not necessary the same to that obtained at large impact parameter in the semiclassical frame. For example L_y is negative for antiprotons in the forward direction and positive for large ρ. In this respect, we obtain the same pictures of distance collisions as Jain *et al* (1988).

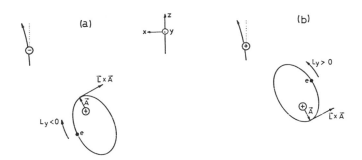

Fig. 11: Classical orbital pictures of the excited electron for impact of negative- (a) and positive-charged (b) projectiles at grazing collisions.

5. CONCLUDING REMARKS

In this work, we have studied in detail the eikonal impulse theoretical method for hydrogen excitation, using the eikonal and the exact impulse wave functions in the entrance and exit channels, respectively. Both wave functions have proper normalizations and long distance asymptotic conditions.

From section 2 we may conclude that: first, the correct treatment of the asymptotic conditions on both channels is necessary to get a good description of excitation, as other studies revealed in the context of charge exchange (Dewangan and Eichler 1985, 1986, 1987, Gravielle and Miraglia 1988, Belkić *et al.* 1987); and second, the multiple scattering built up into the impulse wave function is required to get a good description of forbidden excitations in the intermediate energy range. The first point is the reason for the failure of the standard impulse approximation, and the second point explains the short-coming of the SE method to treat the forbidden transitions.

In section 3, we have presented a detailed study of the eikonal impulse model for different projectile charges and final quantum numbers. Unfortunately, very few experiments are available to test our results. We have then compared our findings with the more sophisticated close-coupling theoretical results. We conclude that the EIA^- theoretical method presents a reliable alternative for excitation at intermediate and high energies, specially for high projectile charge and large final quantum numbers where the atomic close-coupling becomes very tedious.

REFERENCES
Bates D R and Griffing G 1953 *Proc. Pys. Soc.* **67** 961
Belkić Dz, Saini S and Taylor H S 1987 *Phys. Rev. A* **36** 1601
Blum K 1981 *Density Matrix Theory and Applications* (New York: Plenum)
Bransden B H and Cheshire I M 1963 *Proc. Pys. Soc.* **81** 820
Burgdörfer J 1986 *Phys. Rev. A* **33** 1578
Coleman J P 1968 *J. Phys. B: At. Mol. Phys.* **2** 567
——1969 *Case Studies in Atomic Collision Theory* vol 1 ed. E W McDaniel and M R C McDowell (Amsterdam: North Holland) p 149
Fano U and Kohmoto M 1977 *XV International Conference on the Physics of Electronic and Atomic Collisions* X ICPEAC, Paris, Abstract of Papers, p. 516
Dewangan D P and Eichler J. 1985 *J. Phys. B: At. Mol. Phys.* **18** L65
——1986 *J. Phys. B: At. Mol. Phys.* **19** 2939
——1987 *Nucl. Instrum. Methods B* **23** 160
Drepper F and Briggs J S 1976 *J. Phys. B: At. Mol. Phys.* **9** 2063

Fritsch W and Schartner K H 1987 *Phys. Lett.* **126A** 17
Gilbody H B 1986 *Adv. At. Mol. Phys.* **22** 143
Gravielle M S and Miraglia J E 1988 *Phys. Rev. A* **38** 504
Hippler R, Madehein H, Harbich W, Kleinpoppen H and Lutz H O 1988 *Phys. Rev. A* **38** 1662
Inokuti M 1971 *Rev. Mod. Phys.* **43** 340
Jain A, Lin C D and Fritsch W 1987 *Phys. Rev. A* **36**
——1988 *J. Phys. B: At. Mol. Opt. Phys.* **21**
Janev R K and Presnyakov L P 1980 *J. Phys. B: At. Mol. Phys.* **13** 4233
Kauppila W E, Teubner P J O, Fite W L and Girnius R J 1970 *Phys. Rev. A* **2** 1759
Kondow T Grinius R J, Chong T P , and Fite W L 1974 *Phys. Rev. A* **10** 1167
Mandal C R , Mandal Mita, and Mukherjee S C 1990 *Phys. Rev. A* **33** 1787
Meckbach W, Nemirovsky I B, and C R Garibotti 1981 *Phys. Rev. A* **24** 1793
Miraglia J E 1982 *J. Phys. B: At. Mol. Opt. Phys.* **15** 4205
Morgan T J, Geddes J, and Gilbody H B 1973 *J. Phys. B: At. Mol. Phys.* **6** 2118
Park J T, Aldag J E, George J M and Peacher J L 1976 *Phys. Rev. A* **14** 608
Park J T, Aldag J E, Peacher J L and George J M 1980 *Phys. Rev. A* **21** 751
Park J T 1983 *Advances in At. and Molec. Phys.* **19** 67
Reinhold C O and Miraglia J E 1987 *J. Phys. B: At. Mol. Phys.* **120** 1069
Reinhold C O, Olson R E, and Fritsch P R A 1990 *Phys. Rev. A* **41** 4837
Reymann K, Schartner K H, Sommer B and Trabert E 1988 *Phys. Rev. A* **38** 2290
Rodríguez V D and Falcón C A 1990 *J. Phys. B: At. Mol. Phys.* **23** L547
Rodríguez V D and Miraglia J E 1989a *Phys. Rev. A* **39** 6594
——1989b *Phys. Lett.* **137A** 123
——1989c *Electronic and Atomic Collisions. Abstracts of contributed papers of the XVI International Conference on the Physics of Electronic and Atomic Collisions XVI ICPEAC* (New York), p 490
Rodríguez V D and Miraglia J E 1990 *J. Phys. B: At. Mol. Phys.* **23** 3629
Schöller O, Briggs J S and Dreizler D M 1986 *J. Phys. B: At. Mol. Phys.* **19** 2505
Schartner K H, Detleffsen D, and Sommer B 1989 *Phys. Lett. A* **12** 55
Theodosiou C E 1980 *Phys. Rev. A* **22** 2556
Vainshtein L A and Presnyakov 1969 *Sov. Phys. JETP* **28** 156
Wohrer K, Chetioui A, Rozet J P, Jolly A, Fernández F, Stephan C, Brendlé B and Gayet R 1986 *J. Phys. B: At. Mol. Phys.* **19** 1997
Xu Xiang-Yuang, Montenegro E C, Anholt R, Danzmann, Meyerhof W E, Schlachter A S, Rude B S and McDonald R J 1988 *Phys. Rev. A* **38** 1848
Young R A, Stebbing R F, and McGowan J W 1968 *Phys. Rev.* **171** 2118

Two-electron transitions in non-resonant charge transfer and related processes

V N Ostrovsky

Institute of Physics, Leningrad State University, 198904 Leningrad, USSR

Abstract. Asymptotic theory valid for large internuclear separations is applied to non-resonant charge exchange and related processes. The non-stationary approach allows inclusion of the electron momentum transfer effects and transitions induced by the rotational coupling (that is important for the description of the electron orbital polarisation produced in the final state). The stationary theory of two-electron transitions is developed for charge exchange with excitation of the ion, transfer ionisation, Penning process, Penning process with excitation of the ion. For the collisions He^+-Hg, He^+-Cd the existence of alternative mechanism of successive one-electron transitions is revealed.

1. Introduction

In the framework of the quasimolecular approach the transition cross sections are governed by the interaction of the quasimolecular potential curves. In the basis of adiabatic states the interaction appears for instance as a splitting of the potential curves in the region of avoided crossing. In the case of two-electron transitions such an interaction generally is of a correlated nature.

The theoretical analysis below is based on the asymptoptic approach which is valid for large internuclear separations. We start from the general characteristic of the asymptotic theory first for the one-electron transitions (charge exchange). Some new developments are presented here related which take into account electron momentum transfer. Then we turn to the two-electron transitions in the charge exchange with excitation of the residual ion. The interpretation of the existing experimental data is discussed. As another type of two-electron transition, the Penning process is considered.

2. One-electron charge exchange: account for electron momentum transfer

Asymptotic theory treats the interaction between the *diabatic states* φ_a, φ_b in the case of large internuclear distance R. The straightforward expression for this interaction is

$$H_{ab} = \langle \varphi_a \mid H \mid \varphi_b \rangle, \tag{1}$$

where H is the electronic Hamiltonian of the system. In the case of one-electron charge exchange the diabatic orbitals φ_a, φ_b for large R are close to the atomic orbitals $\psi^{(a)}$, $\psi^{(b)}$ centered on different nuclei. The simple analysis shows that the principal contribution to the integral H_{ab} is given in the vicinity of the internuclear axis. This situation is inconvenient for accurate asymptotic calculations since at the extreme parts of the internuclear interval the atomic orbital $\psi^{(a)}$ is strongly perturbed by the ionic core b and vice versa. Therefore the specific representation was deduced for the matrix element H_{ab} as an integral over the surface S which divides location regions of the electron in the initial and final states. This surface passes through the center of the internuclear axis and perpendicularly to it. The expression for H_{ab} reads:

$$H_{ab} = \frac{1}{2} \int_S dS \left[\varphi_a \nabla_n \varphi_b - \varphi_b \nabla_n \varphi_a \right]. \tag{2}$$

where ∇_n is the gradient projection on the normal to S. The surface integral is essentially one-dimensional (due to the axial symmetry) and for large R can be calculated using the steepest descent method (the principal contribution is given in the vicinity of the internuclear axis). On the surface S the perturbation of the atomic orbitals is small and can be taken into account by some version of the eikonal method. The resulting expression takes a general form (for atomic s-states):

$$H_{ab} = \frac{1}{8} A^{(a)} A^{(b)} D_1 R \left[\frac{R}{2} \right]^\nu \exp \left(-\frac{\alpha + \beta}{2} R \right), \tag{3}$$

$$D_1 = \left(\frac{2}{\sqrt{e}} \right)^\mu, \qquad \mu = \frac{Z_b}{\alpha} + \frac{Z_a}{\beta}, \qquad \nu = \frac{Z_a}{\alpha} + \frac{Z_b}{\beta} - 2$$

(see e.g. the recent review by Chibisov and Janev (1988); note, however, that the expression (3) somewhat differs from that recomended by these authors). Here $\frac{1}{2}\alpha^2$ and $\frac{1}{2}\beta^2$ are the atomic ionisation potentials in the initial and final state respectively, Z_a and Z_b are the charges of the ionic cores, $A^{(a)}$ and $A^{(b)}$ are the coefficients which effectively characterise the deviation of the core potential from the pure Coulomb field. The typical exponential decrease of the matrix element for large R appears also in semiempirical expressions for H_{ab} (Olson *et al* 1971).

It should be noted that the expression (3) is valid in the case when the binding energies in the initial and final states are not very different $((\alpha - \beta)R < 1)$. In the opposite case $\alpha \gg \beta$ the region close to the nucleus a becomes most important in the integral (1) and the special asymptotic technique is required (see below, Section 3). However the quasiresonance case $\alpha \approx \beta$ seems to be the most important one in the one-electron exchange processes since it implies appreciable interaction of the potential curves at large internuclear separations where the asymptotic approach is applicable.

The other noticable feature is that the notion of diabatic states does not easily lend itself to an accurate definition. Thus precise specification of the interaction between the diabatic states requires some care. From this point of view the other approach in the asymptotic theory seems to be more natural: direct calculation of the charge exchange amplitude f for the collision with large impact parameter ρ (see Demkov and Ostrovsky (1975), Ostrovsky (1991) and bibliography therein). The charge exchange amplitude f can be presented *exactly* as a surface integral with the additional integration in time t:

$$f = \frac{i}{2} \int_{-\infty}^{\infty} dt \int_S dS [\Psi_2^* \nabla_n \Psi_1 - \Psi_1 \nabla_n \Psi_2^*]. \tag{4}$$

Here Ψ_1, Ψ_2 are the *exact* solutions of the non-stationary Schrodinger equation for the collision system. The simplest choice of *approximation* for the wave functions Ψ_1, Ψ_2 is to put them equal to their asymptotic expressions for $t \to -\infty$ and $t \to \infty$ respectively:

$$\Psi_1(\vec{r}, t) = \psi^{(a)} \exp\left(\frac{i}{2}\vec{v}\vec{r} + i\left(\frac{1}{2}\alpha^2 - \frac{1}{8}v^2\right)t\right), \qquad t \to -\infty; \tag{5}$$

$$\Psi_2(\vec{r}, t) = \psi^{(b)} \exp\left(-\frac{i}{2}\vec{v}\vec{r} + i\left(\frac{1}{2}\beta^2 - \frac{1}{8}v^2\right)t\right), \qquad t \to \infty; \tag{6}$$

where $\vec{v}$ is the collision velocity and the inertial coordinate frame is chosen so that the center of the internuclear axis is at rest.

It is seen from the expressions (5), (6) that the *electron momentum transfer* (EMT) is included in the asymptotic theory in this approach. The role of this effect was analysed recently in the Landau-Zener type interaction of the potential curves located at $R = R_c$ (Ostrovsky 1991). For instance, the amplitude for $s\sigma \to s'\sigma$ charge exchange is expressed as

$$f = \frac{1}{16\kappa}\left(\frac{2\pi}{\Delta F v}\right)^{\frac{1}{2}} A^{(a)}A^{(b)} D R_c(\alpha + \beta)\left[\frac{(\alpha + \beta)}{4\kappa}R_c\right]^\nu \exp(-\kappa R_c) \tag{7}$$

with the typical parameter κ

$$\kappa = \frac{1}{2}\sqrt{(\alpha + \beta)^2 + v^2}. \tag{8}$$

and ΔF being the difference of the potential curve slopes at $R = R_c$. Thus the asymptotic theory for one-electron charge exchange gives *explicit expressions* for the interaction matrix element H_{ab} or the charge exchange amplitude f via *the separated atoms parameters* α, β, $A^{(a)}$, $A^{(b)}$. One advantage of the stationary (adiabatic) approach as compared with the non-stationary asymptotic theory should be mentioned: the results of the former apply both in the case of adiabatic and diabatic passage of the interaction region whereas the latter works only when the probability of the charge exchange is small.

The EMT effect in (7) is characterised by the parameter $\zeta = v/(\alpha + \beta)$, i.e. the ratio of the collision velocity to the doubled characteristic electron velocity in the initial (or final) state. The contribution of EMT is proportional to ζ^2, i.e. small in many typical situations.

The non-stationary approach allows the evaluation of the amplitude of charge exchange generated by rotation of the internuclear axis, e.g. for $s\sigma \to p\pi$ transitions (this mechanism was not treated before in the framework of the asymptotic theory). Such an amplitude contains an additional factor ζ as compared with the amplitude of charge exchange $s\sigma \to p\sigma$ generated by the radial coupling. It becomes particularly important when the orbital polarisation of the final state (with non-zero orbital momentum) is analysed. Consider for simplicity $s \to p$ charge exchange. Transitions to various magnetic sublevels of the final atomic p-state proceed via *radial* ($s\sigma \to p\sigma$) or *rotational* ($s\sigma \to p\pi$) coupling. *Interference* of these contributions creates orbital polarisation which generally oscillates with the impact parameter ρ. The polarisation created in one passage of the interaction region proves to be equal to 2ζ (Ostrovsky 1991). Thus the effect of rotation-induced charge transfer is more important than the manifestations of EMT. It seems that this mechanism could explain the recent experimental findings for H^+ - $Na(3p)$ collisions (Dowek *et al* 1990) (the detailed calculations are in progress now).

Significant polarisation appears also when the electron orbital momentum in the final p-state is weakly coupled with the quasimolecule axis. In this case the polarisation is created even for small values of the parameter ζ when the rotation-induced charge exchange is negligible. In each of two passages of the interaction (avoided crossing) region some specific linear combination of the magnetic substates of the atomic p-state is populated via $s\sigma \rightarrow p\sigma$ radial coupling. If the electron orbital momentum is decoupled from the quasimolecular axis then this combinations of substates correspond to *different orientations in space of the electron cloud with the same form*. The population amplitudes for these combinations *differ in phase* (the difference is Stueckelberg phase). The joint action of this *two factors* gives rise to the preferential circular polarisation of the final atomic p-state which oscillates with the impact parameter (or scattering angle). This mechanism explains (Ostrovsky *et al* 1991) strong polarisation observed recently for small angle scattering in $B^{3+} + He \longrightarrow B^{++}(2p) + He^{+}$ charge exchange (Roncin *et al* 1990).

3. Charge exchange with excitation of the ion

Turning to the two-electron transitions we note that they can be subdivided in two types depending on whether one or both electrons change the center of localisation. Below, we consider only transitions of the first type (including the processes when one of the electrons is transferred and the other is ejected into continuum). Namely, we consider the following processes:

charge exchange with the excitation of the residual ion:

$$A^{+} + B \longrightarrow A + B^{+*}; \tag{9}$$

charge exchange with formation of the doubly excited state:

$$A^{+} + B \longrightarrow A^{**} + B^{+}; \tag{10}$$

transfer ionisation:

$$A^{+} + B \longrightarrow A + B^{++} + e; \tag{11}$$

Penning process:

$$A^{*} + B \longrightarrow A + B^{+} + e; \tag{12}$$

Penning process with excitation of the ion:

$$A^{*} + B \longrightarrow A + B^{+*} + e. \tag{13}$$

We consider in more detail the first of this processes. If the transferred electron does not change its energy significantly, then the asymptotic theory can be constructed employing the dividing surface technique mentioned above (Ostrovsky 1983, Ostrovsky and Tolmachev 1983). However, it seems that such a situation is not typical since the change of the energy of the transferred electron should be large enough to enable occurence of the second acitive electron excitation on the center B. Due to this reason we consider the opposite case when the ionisation potential of the atom A exceeds significantly that of the atom B. Then in the matrix element H_{ab} the principal contribution to the integral over the coordinates of the transferred electron comes from the configuration when this electron is close to the core a (the second electron performs transition near the core b). In this region of configuration space the wave function of the excited electron can be well approximated by the atomic orbitals centered on the core b. As the final state of the transferred electron, the atomic orbital for the atom A can be used. However the initial state of the transferred electron (correlated with the atomic orbital $\psi^{(b)}$) should be considered in the vicinity of the core a where it is strongly perturbed by the potential of the A^+ ion.

The continuation of the atomic orbital $\psi^{(b)}$ to the region of the core a was developed by Karbovanets *et al* (1984) in the framework of the asymptotic (and semiclassical) theory. Close to the core a the continued wave function is presented as an expansion over spherical harmonics:

$$\psi_{ba} = \frac{1}{r_{1a}} \sum_l c_l f_l(r_{1a}) P_l(\cos\theta_{1a}), \tag{14}$$

where r_{1a}, θ_{1a} are the spherical coordinates of the electron in the frame with the center a and the axis $\vec{R}$, $P_l(z)$ arc the Legendre polinomials (we consider here Σ-states of the quasimolecule). The radial functions satisfy the differential equation:

$$\left[\frac{d^2}{dr_{1a}^2} - 2V_a + \frac{l(l+1)}{r_{1a}^2} - \beta^2 \right] f_l(r_{1a}) = 0, \tag{15}$$

where $V_a(r_{1a})$ is the effective potential of the ion A^+. The solution $f_l(r_{1a})$ should be regular at the origin ($f_l(0) = 0$). The energy entering the equation (15) is equal to the energy of the electron in the initial state on the atom B. It does not correspond to the energy of any bound state in the field of the ion A^+. Therefore the solution $f_l(r_{1a})$ is divergent exponentially for $r_{1a} \to \infty$ and can be specified by the coefficient in the corresponding asymptote:

$$f_l(r_{1a}) \approx r^{-1/\beta} \exp(\beta r_{1a}), \qquad r_{1a} \to \infty. \tag{16}$$

The coefficients c_l are found by matching ψ_{ba} (14) and $\psi^{(b)}$ in the intermediate region between the atomic cores. The final result is:

$$c_l = D_2(R)\sqrt{2l+1}, \tag{17}$$

$$D_2(R) = \frac{1}{2} A^{(b)} n_b Q(R) \left(\frac{2}{n_b^2 Z_a} \right)^{n_b Z_a} \left(\frac{n_b Z_a}{2e} \right)^{n_b Z_b}, \tag{18}$$

$$Q(R) = R^{2n_b Z_a - 1} \exp\left(-\frac{R}{n_b}\right), \qquad n_b = \frac{Z_b}{\beta}.$$

The two-electron matrix element H_{ab} is governed essentially by the correlation term $1/r_{12}$ in the electronic Hamiltonian H. As it was discussed above, the integration over the coordinate of the transferred electron is localised primarily near the core a, whereas the other electron udergoes transition (excitation) in the vicinity of the core b. Therefore the *two-center* multipole expansion can be employed for $1/r_{12}$. The leading contributing term represents interaction dipole-multipole (the order of multipole is defined by the selection rules for the transition $B^+ \to B^{+*}$). The final result is:

$$H_{ab} = \frac{l'+1}{\sqrt{2l'+1}} D_2(R) \, R^{-(l'+2)} \left(R_{ns} \mid r_{2b}^{l'} \mid R_{n'l'} \right) \left(R_A \mid r_{1a} \mid \frac{f_1}{r_{1a}} \right) \tag{19}$$

(Duman *et al* 1983, Ivakin *et al* 1987). For simplicity we assume here that the initial configuration of the atom B is ns^2 and in the final state B^{+*} the outer electron is in $n'l'$-state. The related radial wave functions are R_{ns} and $R_{n'l'}$. The radial function of the transferred electron in the final state is denoted as R_A. The round brackets refer to the radial matrix element. It should be stressed that the radial integration in the matrix element $(R_A \mid r_{1a} \mid f_1/r_{1a})$ is convergent. Although the function $f_1(r_{1a})$ diverges (see Eq.(16)), the function $R_A(r_{1a})$ decreases much faster since $\alpha \gg \beta$.

Thus for two-electron transitions the asymptotic theory expresses the interaction H_{ab} via one-electron one-center matrix elements. The latter can be calculated by various methods developed in the theory of atoms (Hartree-Fock method, model potential approach *etc*).

The factor $R^{-l'}$ in (18) is decreasing with the increase of the order of multipole l'. However, simple analysis shows that the matrix element $\left(R_{ns} \mid r_{2b}^{l'} \mid R_{n'l'}\right)$ increases with l' (Ivakin *et al* 1987). Due to compensation of two factors the cross sections of the process (10) proves to be of the same order for various l' (that agrees with the experimental observations).

The theory of transfer ionisation (11) and charge exchange with formation of the doubly excited states (10) is constructed along the same lines. Below we discuss in more detail the charge exchange with excitation of the ion which is supposed to be important, for instance, in creation of the inversed population of ionic levels. The experiments of various groups exist for such processes in collisions of He^+ ions with Hg, Cd and Zn atoms (Turner-Smith *et al* 1973, Collins 1973, Baltayan *et al* 1985a,b). Extensive experimental study was performed by Prof.Tolmachev and coworkers in Leningrad University (Bochkova *et al* 1987, 1988, 1989, 1990, 1991, Alexandrov *et al* 1990, Kuligin 1991). Almost all experiments (with the exception of the beam experiment by Soskida and Shevera (1975)) were performed in the modulated gas discharge, i.e. for thermal collisions. The difficulties encountered in such experiments are well known: some lines emitted by the excited ion lie in a spectral interval which is difficult for observation; the competing population mechanism distinct from the charge exchange (such as direct excitation by electron impact, Penning process) should be separated; the possibility of the radiative cascades should be taken into account when the results are interpreted.

The atoms Hg, Cd, Zn have the outer shell configuration s^2. The two outer electrons are active in two-electron mechanism of charge exchange with excitation of the ion. Behaviour of the potential curves in the asymptotic (large R) region is governed by the polarisation interaction. The polarisability α_M of the atoms Hg, Cd and Zn exceeds significantly that of the helium atom. Therefore the quasimolecular potential curves correlated with the initial and final atomic states exhibit an avoided crossing at large $R = R_c$ for exothermic processes with small energy defect ε, namely

$$R_c = \left(\frac{\alpha_M}{2\varepsilon}\right)^{1/4}. \tag{20}$$

The first theoretical estimates (Turner-Smith *et al* 1973) treated the avoided crossing region $R \approx R_c$ in the Landau-Zener approximation choosing interaction of the potential curves in the form deduced empirically (Olson *et al* 1991) for the one-electron process (i.e. the charge exchange without excitation of the ion). Comparison with the present theory shows that the one-electron model overestimates significantly the curve splitting at the avoided crossing (this is manifestation of the correlated nature of two-electron transition).

This distinction is of qualitative importance since for thermal collisions the interaction region is passed adiabatically in the one-electron approximation and diabatically in the present two-electron theory.

In order to verify what is the real situation one can analyse the velocity dependence of the experimental cross section (which differs significantly for the diabatic and adiabatic situation). It is difficult to vary appreciably the temperature of the gas discharge. Therefore the experiments with the isotopic substitution of ^{4}He by ^{3}He were performed (Bochkova *et al* 1987, 1988). Such substitution is equivalent to enhancement of the collision velocity by 15% for the fixed temperature. The trajectories of low energy colliding atoms are influenced significantly by the orbiting in the attractive polarisation potential. As a result the rate constants for the process (9) are virtually independent of the temperature and the masses of the colliding atoms. Weak temperature dependence of the total rate constant for quenching of He$^+$ ions in He-Hg discharge was noted by Johnson *et al* (1973). The experiments by Bochkova *et al* (1988) demonstrate a similar situation for the rate constants describing excitation of 7p $^2P_{1/2}$ and 7p $^2P_{3/2}$ levels of Hg$^+$. These results substantiate the diabatic regime, i.e. small probability of transition between diabatic states in each passage of the avoided crossing. In particular, this regime allows independent treatment of the interaction regions corresponding to each final state of B^{+*} ion.

The calculations of the cross sections and rates were performed for the thermal collisions of He$^+$ ions with Hg, Cd and Zn atoms (Ivakin *et al* 1987). The population of 7p $^2P_{1/2}$ and 7p $^2P_{3/2}$ states of Hg$^+$ ion and of the large number of final states of Cd$^+$ ion was considered in detail. The typical energy defects ε are enclosed in the interval $0.07 \div 0.7$ eV and the crossing radii R_c vary from 10 to 6 a_0. The effect of orbiting was taken into account. The comparison with the experiments cited above reveals the general feature: the experimental rate constants exceed the theoretical data by one or two orders of magnitude for small energy defect ε. For larger ϵ (and smaller R_c) the theoretical results approach the empirical data and exceed them. However this correlation seems to be spurious since no trend for better agreement can be observed for the processes with the smallest energy defect ε (i.e. the largest values of R_c) where the asymptotic theory could be expected to give better results (note also that for the highest ionic states the role of radiative cascades decreases). The inability of the theory to reproduce large enough partial charge transfer cross sections is a manifestation of the fact that the interaction of the diabatic potential curves proves to be too small at large internuclear distances. It is interesting that the similar situation appears for the processes with the multicharged ions observed experimentally by Woerlee *et al* 1979 (for the asymptotic calculations see Ivakin (1989)).

As it was discussed above, the basic assumptions of the asymptotic theory are (i) the large value of the internuclear separation (as compared with the typical dimensions of the electron orbits) and (ii) large ratio of the ionisation potential of the atom A to that of the atom B. We have introduced in the theory some semiempirical modifications in order to relax the related applicability conditions (Bochkova *et al* 1988, Ivakin 1991). The first of these modifications somewhat enhances matrix element H_{ab} whereas the second diminishes it. However significant improvement was not achieved. The same situation was met in the asymptotic calculations of other authors who also have added some modifications to the asymptotic theory (Belyaev *et al* 1987, 1989, Ovchinnikova and Shalashilin 1988, Belyaev and Tzerkovnyi 1989) (as it was stressed above, only the theory reproducing the partial cross sections can be considered as reliable).

We interpret this disagreement between the theory and experiment as an evidence that some other efficient mechanism exists which is not related to the potential curves interaction at large internuclear distances. The arguments which follow tentatively specify the nature of this mechanism.

4. Alternative mechanism of charge transfer with excitation of the ion

First of all we list some important conclusions which follow from the experimental data.

(i) Observation of He^+ ion quenching in He-Hg afterglow allows an to estimate to be made of the total cross section of the charge transfer summed over the final states of the Hg^+ ion. It proves to be close to the orbiting cross section (Piotrovsky *et al* 1982). However the cross sections for the individual 7p $^2P_{1/2}$ ($\varepsilon = 0.72$ eV, $R_c = 6.2a_0$) and 7p $^2P_{3/2}$ ($\varepsilon = 0.27$ eV, $R_c = 8.0a_0$) levels of the ion are order of magnitude less ($0.9 \cdot 10^{-10}$ cm^3/s and $1.4 \cdot 10^{-10}$ cm^3/s respectively (Alexandrov *et al* 1990, Kuligin 1991)).

(ii) Analysis of radiation over a spectral range shows that the probability fluxes over radiative cascades are not more than $20 \div 30$ % of the total cross sections (Alexandrov *et al* 1990). It is of the same order for the principal and Beutler system of levels (the latter states have vacancy in 5d shell). These systems are effectively separated, the intersystem transition probabilities being an order of magnitude less than the typical probabilities of transition in each of the systems.

(iii) Efficient population of $5d^9 6s^2$ $^2D_{5/2}$ Hg^+ is observed (the energy defect is about 10 eV). Study of the afterglow kinetics shows that it can not be attributed to the direct excitation by the electron impact. In order to suppress the contribution of the Penning process some neon was added to He-Hg mixture. It is known that Ne atoms effectively quench helium metastables. However the observed population of $5d^9 6s^2$ $^2D_{5/2}$ Hg^+ state was not influenced. The rate constant for the charge transfer with formation of the $5d^9 6s^2$ $^2D_{5/2}$ Hg^+ state was estimated from experiments as $3.5 \cdot 10^{-10}$ cm^3/s (Alexandrov *et al* 1990) that is $0.15 \div 0.20$ of the total Hg^+ quenching rate.

The additional experimental verification of direct production of $5d^9 6s^2$ $^2D_{5/2}$ Hg^+ states in He^+-Hg charge exchange could be related to the fact that the process is strongly exothermic, giving Hg^{+*} ions with kinetic energy about 10 eV. This should result in substantial broadening of Hg^{+*} lines.

Charge exchange with excitation of $5d^9 6s^2$ $^2D_{5/2}$ Hg^+ states implies a transition of one electron from the inner $5d^{10}$ shell. The energy defect for this process is about 10 eV. Therefore the corresponding potential curve interacts with the initial state at small internuclear separations where the asymptotic approach is invalid. However some general observations can be made.

(i) In the final state $He(1s^2) + Hg^+(5d^9 6s^2)$ both partners have closed outer electron shells. The strong repulsion is typical for such situation resulting in the promotion of the potential curve. This implies that the region of interaction with the initial state should exist for small R (Ivakin *et al* 1987). This picture is confirmed by the more detailed study of the correlation diagrams (Ivakin *et al* 1990a).

(ii) If the internuclear separation is not large then the electrostatic interaction produces comparatively large potential curve splittings at the avoided crossing region. This circumstance could be in apparent contradiction to the diabatic regime established above. However the splitting due to relativistic effects (fine structure) could be small. The fine structure effects seems to be important in this process. Indeed, the radiation from $5d^96s^2 \, ^2D_{3/2}$ Hg$^+$ was not registered in experiments by Alexandrov *et al* (1990) although the related radiative transition probability exceeds that of $5d^96s^2 \, ^2D_{5/2}$ Hg$^+$ state by an order of magnitude (Ostrovsky and Scheinerman 1989).

As it was demonstrated above, two-electron transitions at large internuclear separations are unable to explain the magnitude of the cross sections observed in experiments. The other feasible mechanism implies the sequence of one-electron transitions at medium (or small) R. Thus some intermediate state should be involved. As a possible candidate we suggest the potential curve correlated with the He($1s^2$) + Hg$^+$($5d^96s^2$) separated atoms limit. Some general analysis was performed (Ivakin *et al* 1990a) which showed that this explanation is consistent with the current experimental data.

A similar mechanism can be proposed for the interpretation of the experimental data in He$^+$-Cd charge exchange. The tentative intermediate state here is correlated with the separated atoms He($1s^2$) + Cd$^+$($4d^95s^2$). In particular, this mechanism allows interpretation of the endothermic charge exchange with excitation of the ion observed recently in experiments (Bochkova *et al* 1991, Kuligin 1991). Since these processes correspond to negative energy defects ε, the avoided crossing at large R is absent here and the two-electron mechanism does not contribute. However population via an intermediate state is equally feasible for endothermic and exothermic processes.

A particularly interesting situation emerges when the states with negative ε belong to the same Rydberg series (i.e. have the same quantum numbers except the principal one) as the ionic states populated in the exothermic process. Based on the model of intermediate state and using simple arguments one can show that within the same Rydberg series the charge exchange cross sections are inversely proportional to the cube of the principal quantum number. In the case of $^2F_{5/2}$ Cd$^+$ states, the ratio of measured cross sections for population of the ionic levels with the principal quantum number $n = 6$ and $n = 5$ equals to 0.33. The same ratio for $^2F_{7/2}$ Cd$^+$ states is 0.47. The inverse ratio of the cubes of the principal quantum numbers is 0.58. Bearing in mind the uncertainty in the experimental data and the crudeness of the theoretical estimate it can be said that the agreement is reasonably good (Bochkova *et al* 1991, Ivakin 1989).

As it was indicated above, the asymptotic theory of the processes (10) and (11) is similar to that of (9). The process (10) is essentially the inverse of (9). It can be expected to be particularly important in collisions with the multicharged ions. Some sample calculations were performed by Ivakin (1989).

For the process of transfer ionisation (11) the function $R_{n'l'}$ in Eq.(19) should be substituted with the continuum wave function. Then the matrix element H_{ab} defines the autoionisation width of the quasimolecular state. In this respect the situation is similar to that for the Penning process considered in the next Section.

5. Asymptotic theory of Penning process

Description of the Penning process (12) (where the atom A^* is metastable) in the framework of the quasimolecular approach requires information about the autoionisation width Γ of the initial quasimolecular state as a function of the internuclear separation R. *Ab initio* calculation of this magnitude is a difficult problem. In most cases the empirical approach is employed when some analytical $\Gamma(R)$ dependence is assumed (usually in the form $\Gamma(R) = a\exp(-bR)$). The parameters (a and b) are restored by fitting the calculated cross sections with the experimental data. The asymptotic approach does not use empirical information being applicable for large R.

We construct the theory based on the important particular case when the atom A^* is the helium metastable $He(2^3S)$. Since the relativistic effects are small, the total spin of the quasimolecule is integral of motion (Wigner rule). This constraint dictates the principle mechanism of the process: the active electron from the atom B comes to the ground (1s) orbital of He atom and the outer (2s) electron from $He(2^3S)$ is ejected into continuum. The other feasible mechanisms imply transitions of more than two electrons. Therefore, they are less important. The energy exchange between the electrons is induced by their Coulomb interaction. The radius of the 1s orbital is small since the ionisation potential of He atom is high. Due to this the magnitude of the transition matrix element is defined primarily by the configuration when the transferred electron is close to the He atom. This argument shows that the Penning process is governed by an exchange mechanism to which the asymptotic approach is applicable.

The autoionisation width is evaluated in the first order perturbation theory:

$$\Gamma(R) = 2\pi \mid \langle \Psi_f \mid H - E \mid \Psi_i \rangle \mid^2 . \tag{21}$$

The (nonsymmetrised) wave functions for the initial and final states are presented as the products:

$$\Psi_i(\vec{r}_1, \vec{r}_2, \vec{r}_3) = \psi_{ba}(\vec{r}_1)\varphi_{2s}(\vec{r}_2)\varphi_{1s^*}(\vec{r}_3), \tag{22}$$

$$\Psi_f(\vec{r}_1, \vec{r}_2, \vec{r}_3) = \varphi_{1s}(\vec{r}_1)\varphi_{el}(\vec{r}_2)\varphi_{1s}(\vec{r}_3). \tag{23}$$

Here φ_{1s} is teh 1s orbital in the ground state of He atom. The similar orbital in the excited $He(2^3S)$ state is φ_{1s^*}. This distinction allows core relaxation to be accounted for. The excited electron (2s) orbital in $He(2^3S)$ is φ_{2s}. The ejected electron wave function φ_{el} should be calculated in the vicinity of the He atom. We employ for it the one-center approximation which accounts for the static potential of the ground state helium atom. The energy of ejected electron ϵ is defined by the energy conservation. Summation over its orbital momentum l is performed in the calculation of the width.

The function ψ_{ba} is the orbital $\psi^{(b)}$ describing the active electron in the atom B continued to the vicinity of the excited He atom. It is constructed in the form (14) with the functions $f_l(r_{1a})$ satisfying the equation (15). In the latter the potential $V_a(r_{1a})$ describes now the interaction of the electron with the $He(2^3S)$ atom. We account for this interaction in the static-field approximation.

The essential (correlation) part of the operator in the expression (21) for $\Gamma(R)$ is the electron-electron interaction $1/r_{12}$ of the first (transferred) and the second (ejected) electrons. The conventional *one-center* multipole expansion is used for it (compare with Section 3 where *two-center* expansion was employed).

The final asymptotic expression for the width is:

$$\Gamma(R) = 2\pi D_2^2(R)\gamma, \tag{24}$$

$$\gamma = (A_0 + B_0)^2 + \sum_{l=1}^{\infty} \frac{A_l^2}{2l + 1}, \tag{25}$$

where the coefficients A_l describe the contribution of the electron-electron interaction and B_0 is related with the shake-off type mechanism (Ivakin *et al* 1990b). We give here only expression for A_l:

$$A_l = (R_{1s} \mid R_{1s^*}) \left(R_{el}(r_2) R_{1s}(r_1) \mid \frac{r_<^l}{r_>^{l+1}} \mid R_{2s}(r_2) \frac{f_l(r_1)}{r_1} \right). \tag{26}$$

The round brackets again symbolise radial matrix elements, $r_> = \max[r_1, r_2]$, $r_< = \min[r_1, r_2]$.

Thus the asymptotic theory of the Penning process reduces calculations to *one-center two-electron* matrix elements. It should be stressed that the radial matrix elements entering the expression (26) do not depend on the internuclear distance R. They depend only on the electron binding energy $(\frac{1}{2}\beta^2)$ in the atom B (and the factor $A^{(b)}$ in Eq.(18)). This allows the construction of A_l, B_0 as the universal functions of β and the use of interpolation procedures for applications of the theory to various species B.

Our calculations (Ivakin *et al* 1990b) were performed for the partners He(2^3S) - K, Na, Zn, Cd, Hg. The functions R_{el}, f_l were obtained by numerical integration. The summation over the ejected electron orbital momentum l in Eq.(25) exhibits rapid convergence. The crude estimate of the Penning process cross sections was made using the asymptotic expressions for $\Gamma(R)$ in the whole range of R and the straight line approximation for the trajectories. Satisfactory agreement with the experiment was obtained. For close collisions, the ionisation probability is large and the details of $\Gamma(R)$ behaviour are less important. However assumption of straight line trajectories should be dropped for thermal collisions. Direct comparison of $\Gamma(R)$ can be made with the recent numerical calculations by Pedial *et al* (1989). The shape of $\Gamma(R)$ curve for He(2^3S)-Na pair agrees for large R, but the asymptotic values are higher by the factor about 2, the difference increasing for small R.

The present asymptotic theory is applicable only for the Penning process with triplet helium metastables. In the case of singlet metastables the mechanism of two-electron transition differs since the constraints induced by the Wigner rule are absent. We only mention here that the asymptotic theory in this case can be constructed reducing the problem to the calculation of the superelastic electron scattering He(2^1S) $+ e \longrightarrow$ He(1^1S) $+ e$ for *negative* incident electron energies (Toivonen 1989). The long range contribution to the quasimolecule width also exists leading to the power dependence ($\Gamma(R) \sim R^{-12}$, see Manakov *et al* (1984), Ostrovsky and Toivonen (1987)). It is interesting that numerically the Penning cross sections for singlet and triplet metastables are close although the transition mechanisms differ.

Experimental data testify that in many cases the cross sections for the Penning process with excitation of the ion (13) are comparable in magnitude to those for the Penning process (12) (Baltayan *et al* 1985a,b; Inaba *et al* 1982a,b, 1983; Vogel and Tolmachev 1982). The process (13) may be governed by two-electron transitions (if the inner shell electron is ejected from the atom B). Then the asymptotic theory described above can be employed. However the process (13) is efficent also when only three-electron transitions contribute (the latter circumstance can be seen from the comparison of the electron configurations in the atom B and ion B^{+*}).

An asymptotic theory of the three-electron process (13) can be constructed along the scheme similar to that presented above (Ivakin *et al* 1991). The additional (third)

Table 1. Cross sections and rate constants for the Penning process with excitation of the ion: $He(2^3S) + B \longrightarrow He_0 + B^{+*} + e$. σ: present theoretical cross section in the units 10^{-15} cm^2; K_t: present theoretical estimate for the rate constants in the units 10^{-10} cm^3/s; K_e: experimental data by Baltayan *et al* (1985a,b) (in the same units).

B $\Rightarrow$ Cd				B $\Rightarrow$ Zn			
Final state of Cd$^+$	σ	K_t	K_e	Final state of Zn$^+$	σ	K_t	K_e
$5p^2P_{1/2}$	0.090	0.11	0.78	$4p^2P_{1/2}$	0.088	0.11	0.08
$5p^2P_{3/2}$	0.091	0.11	0.93	$4p^2P_{3/2}$	0.089	0.11	0.27
$4d^95s^2\ ^2D_{5/2}$	3.3	4.1	1.75	$3d^94s^2\ ^2D_{5/2}$	3.1	3.9	0.23
$4d^95s^2\ ^2D_{3/2}$	3.3	4.1	1.04	$3d^94s^2\ ^2D_{3/2}$	3.1	3.9	0.15
$5s\ ^2S_{1/2}$	0.63	0.79	3.16	$4s\ ^2S_{1/2}$	0.61	0.76	6.4
Total	7.4	9.3	7.7	Total	7.0	8.8	7.1

active electron performs a transition (excitation) on the center B. This distinction is reflected in the initial and final states of the system Ψ_i, Ψ_f. A non-zero result in the expression for the width $\Gamma(R)$ is obtained when the third (excited) electron is coupled by the electron-electron interaction with the first (transferred) or second (ejected) electron. The residual electron performs a transition due to shake off. The first and the second electrons perform transitions close to the helium core whereas the third electron is located near the core B. Thus *two-center* expansion of the electron-electron interaction $1/r_{12}$ should be employed in contradistinction to the process (12).

The table shows the results of our calculations (Ivakin *et al* 1991) for He(2^3S) - Cd and He(2^3S) - Zn collisions compared with the experimental data by Baltayan *et al* (1985a,b). The approximations used in evaluation of the cross sections are formulated above. The ionic states ^2P_J are populated by three-electron transitions. Since ^2D_J states can be populated by two-electron transitions we assume three-electron mechanism negligible for them. The ground ionic states ^2S_J are populated by the process (12). The total rate constants summed over final states are reproduced well. For the partial rates the theory overestimates two-electron mechanism of the process (13) and underestimates three-electron mechanism and the process without excitation of the ion. No others theoretical works on this process exist to our knowledge.

6. Conclusion

The advent of powerful computers made many atom-atom collision problems accessible for direct calculations. However many systems remain too complex for *ab initio* study. For instance, the quasimolecules (HeHg)$^+$ and (HeCd)$^+$ considered in the Sections 3 and 4 correspond to highly excited states with appreciable relativistic effects and the inner shell vacancies involved. For such situations even the model potential methods are not developed to any extent. Additional difficulties appear in the analysis of the quasimolecular states embedded into continuum (this is the case for the processes (11)-(13)). The two-electron (correlation) effects generally are more difficult for treatment than one-electron processes.

The asymptotic approach implies analytical or simple numerical calculations. Its advantage is that it allows simple and appealing mechanisms of the processes to be revealed. The applicability of the asymptotic approach is limited to large internuclear separations. It can be said that it directly treats the threshold behaviour in the impact parameter for inelastic processes. However even in the situations when the asymptotic approach is not applicable directly it gives the useful guidelines in the analysis.

References

Alexandrov M S, Bochkova O P, Ivanov V S, Kuligin A V, Piotrovsky Yu A and Tolmachev Yu A 1990 *Optika i Spektrosk.* **69** 518-22

Baltayan P, Pebay-Peyroula J C and Sadeghi N 1985a *J.Phys.B: At.Mol.Phys.* **18** 3615-28

Baltayan P, Pebay-Peyroula J C and Sadeghi N 1985b *J.Phys.B: At.Mol.Phys.* **19** 2695-702

Belyaev A K, Zagrebin L A and Tzerkovnyi S I 1987 *Optika i Spektrosk.* **63** 968-73

Belyaev A K, Zagrebin L A and Tzerkovnyi S I 1989 *Khimicheskaya Fizika* **8** 435-41

Belyaev A K and Tzerkovnyi S I 1989 *Optika i Spektrosk.* **66** 778-83

Bochkova O P, Kuligin A V, Oginets O V, Piotrovsky Yu A, Sergeev Yu N and Tolmachev Yu A 1987 *Optika i Spektrosk.* **63** 1184-6

Bochkova O P, Ivakin I A, Oginets O V, Ostrovsky V N, Piotrovsky Yu A, Sergeev Yu N, Tolmachev Yu A and Kuligin A V 1988 *Optika i Spektrosk.* **65** 786-92 [*Opt. & Spectrosc.* **65** 464-7]

Bochkova O P, Ivakin I A, Ostrovsky V N, Tolmachev Yu A and Kuligin A V 1989 *Optika i Spektrosk.* **67** 510-6 [*Opt. & Spectrosc.* **67** 298-301]

Bochkova O P, Kuligin A V and Tolmachev Yu A 1990 *Optika i Spektrosk.* **68** 1222-4

Bochkova O P, Ivakin I A, Kuligin A V, Ostrovsky V N, Piotrovsky Yu A and Tolmachev Yu A 1991 *Optika i Spektrosk.* **70** 19-25

Chibisov M I and Janev R K 1988 *Phys. Rep.* **166** 1-87

Collins G J 1973 *J.Appl.Phys.* **44** 4633-52

Devdariani A Z and Ostrovsky V N 1986 *Optika i Spektrosk.* **60** 904-10 [*Opt. & Spectrosc.* **60** 558-61]

Demkov Yu N and Ostrovsky V N 1975 *Zh.Eksp.Teor.Fiz.* **69** 1582-93 [*Sov.Phys.-JETP* **42** 806-11]

Dowek D, Houver J C, Pommier J, Richter C and Royer T 1990 *Phys. Rev. Lett.* **64** 1713-6

Duman E L, Tyshchenko N P and Shmatov I P 1983 *Doklady AN SSSR* **271** 853-6

Inaba S, Goto T and Hattori S 1982a *J.Phys.Soc.Japan* **51** No 2

Inaba S, Goto T and Hattori S 1982b *J.Phys.D: Appl.Phys.* **15** No 1

Inaba S, Goto T and Hattori S 1983 *J.Phys.Soc.Japan* **52** No 4

Ivakin I A 1989 *Thesis*, Leningrad State University

Ivakin I A, Karbovanets M I and Ostrovsky V N 1987 *Optika i Spektrosk.* **63** 494-500

Ivakin I A, Kereselidze T M, Ostrovsky V N and Tolmachev Yu A 1990a *Optika i Spektrosk.* **69** 1224-7

Ivakin I A, Ostrovsky V N and Toivonen N R 1990b *Optika i Spektrosk.* **69** 728-33

Ivakin I A, Ostrovsky V N and Toivonen N R 1991 *Optika i Spektrosk.* **70** 1177-81

Jonsen R, Leu M T and Biondi 1973 *Phys.Rev.A* **8** 1808-13

Karbovanets M I, Lazur V Yu and Chibisov M I 1984 *Zh.Eksp.Teor.Fiz.* **86** 84-93 [*Sov.Phys.-JETP* **59** 47-52]

Kuligin A V 1991 *Thesis*, Leningrad State University

Manakov N L, Ovsyannikov V D, Ostrovsky V N and Yastrebkov V N 1984 *Optika i Spektrosk.* **56** 222-7

Olson R E, Smith F T and Bauer E 1971 *Appl.Opt.* **10** 1842-55

Ostrovsky V N 1983 *Zh.Eksp.Teor.Fiz.* **84** 1323-28 [*Sov.Phys.-JETP* **57** 766-9]

Ostrovsky V N 1991 *J.Phys.B: At.Mol.Opt.Phys.* (submitted for publication)

Ostrovsky V N and Toivonen N R 1987 *Vestnik Leningradskogo Universiteta, Ser. fiz.-him.* No 4 7-12

Ostrovsky V N and Tolmachev Yu A 1983 *Optika i Spektrosk.* **55** 1064-5 [*Opt. & Spectrosc.* **55** 646-7]

Ostrovsky V N and Sheinerman S A 1989 *Optika i Spektrosk.* **67** 16-9 [*Opt. & Spectrosc.* **67** 8-10]

Ostrovsky V N, Adjouri C, Andersen N, Gaboriaud M N, Roncin P and Barat M 1991 *Proc. I Int.Conf. on Atomic and Molecular Dynamics*, Taipei

Ovchinnikova M Ya and Shalashilin D V 1988 *Khimicheskaya Fizika* **7** 175-86

Padial N T, Sohen J S, Martin R L and Lane N F 1989 *Phys.Rev.A* **40** 117-24

Piotrovsky Yu A, Tolmachev Yu A and Kasyanenko S V 1982 *Optika i Spektrosk.* **52** 754-6

Roncin P, Adjouri C, Gaboriaud M N, Guillemot and Barat M 1990 *Phys.Rev.Lett.* **65** 3264-7

Soskida M-T I and Shevera V S 1975 *Pis'ma Zh.Eksp. Teor.Fiz.* **22** 545-49

Toivonen N R 1989 *Thesis*, Leningrad State University

Turner-Smith A R, Green J M and Webb C E 1973 *J.Phys.B: At.Mol.Phys.* **6** 114-30

Vogel D and Tolmachev Yu A 1982 *Beitr.Plasmaphys* **22** 387-400

Woerlee P H, El Sherbini T M, de Heer F J and Saris F M 1979 *J.Phys.B: At.Mol.Phys.* **12** L235-41

The dissociative ionization of simple molecules by ion impact

C J Latimer

School of Mathematics and Physics, The Queen's University of Belfast,
Belfast BT7 1NN, UK

ABSTRACT: The energy and angular distribution of fragment ions and
ion pairs produced in simple diatomic gases by 3–30 keV ion beams. The
results provide detailed information on the collision mechanisms and
precursor states involved in the dissociative charge transfer
processes studied.

1 INTRODUCTION

The process of dissociative ionization in heavy particle collisions has been
the subject of very little study despite its fundamental nature and its
significance in interstellar clouds and planetary atmospheres. Therefore
many important features of even the most basic processes, involving
collision between protons and hydrogen molecules, are still not understood.

In contrast dissociative ionization by electrons (Dunn and Kieffer 1963,
Crowe and McConkey 1973, Burrows et al 1980) and photons (Dehmer and Dill
1978, Strathdee and Browning 1979) has been extensively studied. In these
experiments mass analysis of the product ions coupled with studies of their
energy and angular distribution have shown for example that the protons
produced in hydrogen exhibit many features which cannot be accounted for by
simple direct excitation to repulsive states of H_2^+. Rather they arise from
autoionizing states of H_2, although it is not easy to accurately identify all
the states involved or the mechanism by which the H^+ ions are produced (Kirby
et al 1981, Guberman 1983).

In recent experiments in our laboratory we have investigated the energy and
angular distribution of fragment ions and fragment ion pairs produced in the
single and double ionization of simple diatomic molecules (including H_2) by
2–30 keV ions. Previous experimental work on these processes, which has
been limited simply to a charge and mass analysis of the product ions,
indicates that within this energy range fragment ions and ion pairs arise
predominantly through the charge transfer processes (Afrosimov et al 1969).

$$A^+ + BC \rightarrow A + B^+ + C$$
$$\rightarrow A + B^+ + C^+ + e$$

A series of similar but complimentary experiments has been performed in high
energy ($\sim$ 2 MeV) regime where pure ionization processes dominate by Edwards,
Wood and coworkers (Edwards et al 1990 and references therein).

2 EXPERIMENTAL APPROACH

A schematic diagram of the apparatus is shown in figure 1. A 2–30 keV ion

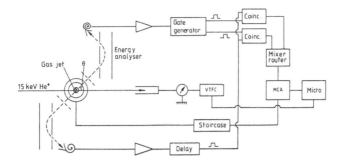

Figure 1 A schematic diagram of the apparatus

beam was crossed at 90⁰ by a low—pressure gas jet of target molecules in the ground vibrational state at the centre of a ramp voltage labelled region defined by two parallel circular plates. A weak radial draw out field of a few vols per cm allowed the observation of all secondary ions produced without significantly perturbing their angular distribution. Fragment ions, appearing at a prescribed energy and angle, θ, at right angles to the gas jet and in a plane parallel to, and midway between, the circular plates, were selected using two identical parallel plate analysers which viewed the interaction region from opposite directions to allow the detection of fragment ion pairs produced with equal and opposite momenta in a Coulomb explosion. Fragment ions could be identified by time of flight. Coincident spectra (corresponding to proton—pair production plus random coincidences) and non—coincident spectra (corresponding to random coincidences and therefore total proton production) were accumulated simultaneously using a coincidence/mixer router/MCA system. The technique used ensured that neither sin θ or analyser transmission factors were necessary and that the thermal motion of the H_2 gas had a negligible effect.

3 RESULTS AND DISCUSSION

3.1 Total Fragmentation Processes

Total and differential charge transfer cross sections have been obtained for collisions between H^+ and He^+ ions and hydrogen and also the well known He^+—O_2 process.

3.1.1 H^+, He^+—H_2 Collisions

An example of the diffuse fragment proton spectra taken at 75⁰ with 15 keV H^+ ions is shown in figure 2. This can be successfully interpreted as arising

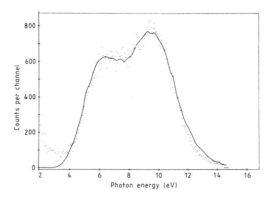

Figure 2 An energy spectrum of fragment protons produced at 75⁰ in 15 keV H^+—H_2 collisions. The full curve is a theoretical fit based on the reflection approximation and the curves shown in figure 3.

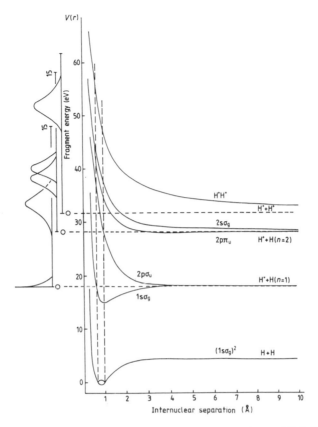

Figure 3 Potential energy curves for selected states of H_2 and H_2^+ (Sharp, 1971).
Also shown are the reflection approximation predictions of fragment proton
energy spectra produced by Franck—Condon excitation (Wood et al, 1977)

from Franck—Condon transitions, shown in figure 3, to four repulsive states
$2p\pi_u$, $2s\sigma_g$, $2p\pi_u$ and H^+H^+. The detailed shapes of the energy distribution are
defined by the ground—state wavefunction of H_2 in accord with the reflection
approximation (Coolidge et al 1936, Yousif et al 1988, Tellinghuisen 1985)
and the shape of the repulsive states of the molecular ion. It is apparent
that although the fit is not perfect in the region 2.5–5.0 eV where protons
from autoionizing states of H_2 have been observed with electrons (Crowe and
McConkey 1973, Burrows et al 1980) and photons (Strathdee and Browning 1979)
it would appear that these states are relatively unimportant in keV H^+ charge
transfer collisions.

However as can be seen from figure 4 this conclusion is not valid in the case
of 15 keV He^+/H_2 collisions. The higher energy group around 9 eV is similar
to that obtained using 15 keV H^+ projectiles and can readily be described in a
similar manner as arising from excitation of $2p\sigma_u$ and the H^+H^+ states.
However the lower energy group differs considerably from that observed with
H^+ projectiles, having a broad maximum around 4 eV rather than 5.5 eV. It
cannot therefore be simply described in terms of excitation to the $2p\pi_u$ and
$2s\sigma_g$ states. The only satisfactory explanation is that there is now a
significant contribution from dissociative autoionization of doubly
excited states giving rise to 2.5–5.0 eV fragments.

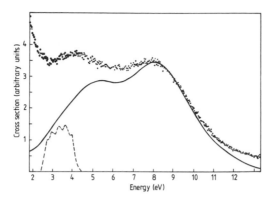

Figure 4 An energy spectrum of fragment protons produced at 90^0 in 15 keV He^+-H_2 collisions. Full curve: equivalent results for 15 keV H^+-H_2 collisions (Lindsay et al 1987) which correspond to transitions to the $2p\pi_u$, $2p\sigma_u$ and H^+H^+ states. Broken curve: proton kinetic—energy distribution arising from the $^1\Sigma_u{}^+$, two—electron excited state of H_2 (Kanfer and Shapiro 1983)

Since dissociation occurs in a time that is short compared with the period of molecular rotation, the fragment ion trajectory will indicate the orientation of the target molecule in space. In general it can be shown that the fragment angular distribution in the axial recoil approximation, which applies when the fragments have kinetic energies much larger than the rotational spacing, is given by the expression (Zare 1972)

$$I(\theta) = (\sigma/4\pi)(1 + \beta P_2(\cos \theta)) \qquad (1)$$

where σ is the total cross section, β is the anisotropy parameter which depends on the nature of the molecular orbital and P_2 $(\cos \theta)$ is given by $\frac{1}{2}$ $(3 \cos^2 \theta - 1)$ in the dipole approximation. The β values provide important information on the symmetry of the final state which comprises both a molecular ion and an ejected electron which, in the present experiment is captured by the projectile.

A typical angular distribution is shown in figure 5 for the $2p\pi_u$ state where three degenerate channels are available. The β value (~ 0.97) indicates that 90^0 or more transitions are $\Sigma \to \Pi$. This result is remarkably similar to the conclusions of molecular photoionization work (Dehmer and Dill 1978). The symmetry about 90^0 indicates that there is negligible momentum transfer in the charge transfer collision process.

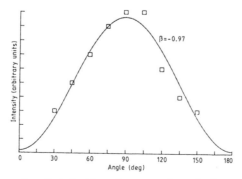

Figure 5 The angular distribution of protons from the $2p\pi_u$ state of $H_2{}^+$ produced in 15 keV H^+-H_2 collisions. The full curve is a fit of equation (1) to the data with $\beta = -0.97$.

3.1.2 He⁺–O₂ Collisions

The fragment O⁺ ions produced in 3--25 He⁺/O₂ charge transfer collisions (see figure 6) display mainly discrete, rather than diffuse, energy spectra which are a result of transitions to bound states which rapidly predissociate. The peaks at 2.97 and 1.94 eV are easily identified as arising from near resonant charge transfer excitation to the $c^4 \Sigma_u^-$ state of O_2^+ followed by predissociation to form $O(^3P) + O^+(^4S)$ and $O(^1D) + O^+(^4S)$ fragments (Yousif et al 1987). The angular distributions are non isotopic with $-0.2 < \beta < -0.6$ indicating that the predissociation lifetime must be less than the rotational period ($\sim 10^{-11}$ secs) and in accord with multiconfigurational SCF calculations (Tanaka and Yoshimine, 1979).

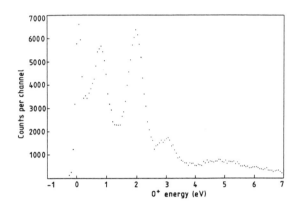

Figure 6 An energy spectrum of secondary ions produced at 150^0 in 5 keV He⁺–O₂ collisions.

Clearly however other non resonant channels are important. The peak at 0.73 eV arises from an exoergic process producing O_2^+ ($B^2 \Sigma_g^-$, v) ions which predissociate to form $O(^3P) + O^+(^4S)$. The broad feature at ~ 4.6 eV can be ascribed to direct dissociation via a Franck--Condon transition to the repulsive $^2\pi_u$ state of O_2^+.

3.2 Ion Pair Fragmentation Processes

3.2.1 The Double Ionization of Hydrogen

This is a special situation, since only in this case is there a single exactly known (pure Coulomb) doubly charged repulsive curve. Thus the fragment proton energy distribution (see curve (4) in fig 3) is directly related to the ground state wavefunction of H_2. An example of an energy spectrum taken at 90^0 with 15 keV H⁺ ions is shown as figure 7 and can be seen to consist of a single broad peak centered at 9.8 eV, which is clearly consistent with Franck--Condon transitions from the ground state of H_2. The transformation of this energy distribution using the simple reflection approximation to give the square of the ground state wavefunction is shown in figure 8. Theory and experiment are clearly in good accord although there is a small as yet unexplained displacement.

The energy spectrum of fragment proton pairs also taken at 90^0 but with 15 keV He⁺ ions (figure 9) is not a single broad peak at 9.8 eV in accordance with the above picture. An additional group of ion pairs is observed centered around an energy of 4.5 eV. This lower energy group is discrete and found to be insensitive to projectile velocity and so a simple breakdown of the

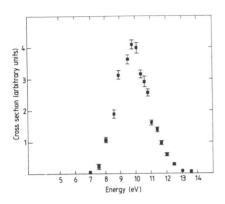

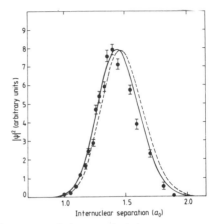

Figure 7 An energy spectrum of fragment proton pairs produced at 90^0 in 15 keV H^+–H_2 collisions.

Figure 8 The square of the ground state wavefunction of H_2. Points are a transformation of the data in figure 7. The full and broken curves are for the $J = 0$ and $J = 6$ rotational levels respectively.

Franck–Condon principle cannot provide an explanation. A two–step process provides a possible explanation (Savage et al 1990). The $H_2^+(^2\Sigma, v)$ state has the correct internuclear separation for an intermediate step, however given the collision timescale, a satisfactory description of events can only be obtained if this state is populated via autoionization of one of the doubly excited autoionizing state of H_2 eg $^1\Sigma_g$ ($2p\sigma_u^2$) (Hazi 1987, Kanfer and Shapiro 1983) which, as we have already seen, are significantly excited in such collisions.

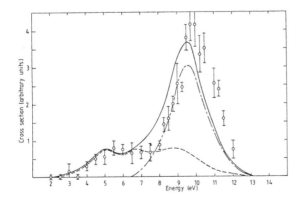

Figure 9 An energy spectrum of proton pairs produced in 15 keV He^+–H_2 collisions. The full curve is the sum of direct transitions to the H^+H^+ potential and two step transitions via the two electron excited state $^1\Sigma_g$ $(2p\sigma_u)^2$ and H_2^+ ($1s\sigma_g$) state (dashed curve).

3.2.2   The Double Ionization of Deuterium

In order to explore further the phenomena already described for a hydrogen target, the double ionization of deuterium has also been investigated. The heavier deuterium molecule will change the picture in two ways. Firstly the ground state potential well is some 16% narrower than in H_2 and, in the simple reflection approximation, this should be reproduced as a similar narrowing of the fragment ion pair energy distribution. Secondly the increased mass of the nuclei means that nuclei on a repulsive curure will separate more slowly.

This in turn implies that in a two step process of the type described above in the case of He$^+$–H$_2$ collisions there should be an enhancement of the intermediate step which requires the repulsive doubly excited state to autoionize into D$_2^+$ rather than dissociate.

Figure 10 shows an energy spectrum of fragment D$^+$ ion pairs produced at 90^0 in 15 keV H$^+$–D$_2$ collisions. The above expectations are clearly apparent. The direct double ionization process, centered as usual at 9.8 eV is indeed proportionally narrower than in H$_2$. Furthermore the two step process giving ~ 5 eV fragments which was unobservable in H$^+$–H$_2$ collisions (figure 7) is now clearly apparent. A further enhancement of such processes has also been observed in 15 keV He$^+$–D$_2$ collisions.

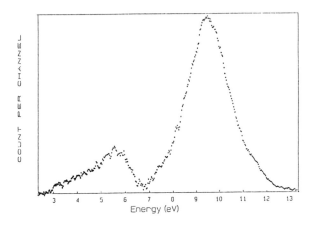

Figure 10 An energy spectrum of fragment D$^+$D$^+$ ion pairs produced at 90^0 in 15 keV He$^+$–D$_2$ collisions.

3.2.3 The Formation of Quasibound States of N$_2^{++}$ in H$^+$–N$_2$ collisions

In the case of collisions with non–hydrogenic targets we have the opposite situation in that the ground state wavefunction of the target molecule is much better known than the potential energy curves of the doubly ionized states. The reflection approximation procedure can now be reversed to provide information about the potential energy curves of the doubly charged molecular ion.

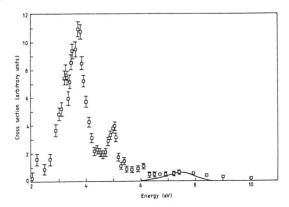

Figure 11 An energy spectrum of N$^+$N$^+$ ion pairs produced at 90^0
in 15 keV H$^+$–N$_2$ collisions. The full curve is the expected distribution
arising from the repulsive N$_2^{++}$ ($^2\Delta_u$) state (Whetmore and Boyd 1986).

An example of an energy spectrum taken at 90^0 in 15 keV H^+–N_2 collisions is shown in figure 11 and can be seen to show considerable structure due to the production of quasi–bound predissociating states of N_2^{++}. Six dissociation channels are observed and the precursor electronic states of N_2^{++} tentatively identified (Yousif et al, 1990). The angular distributions of the fragment ion pairs exhibit strong energy dependent anisotropies characteristic of short predissociation lifetimes.

REFERENCES

Afrosimov V V, Leiko G A, Mamaev Yu A and Panev M N, 1969, Sov Phys – JETP <u>29</u>, 648.

Burrow M D, McIntyre L C, Ryan S R and Lamb W E, Phys Rev A <u>21</u>, 1841.

Coolidge A S, James H M and Present R D, 1936, J Chem Phys <u>4</u>, 193.

Crowe A and McConkey J W, 1973, Phys Rev Lett <u>31</u>, 192.

Dehmer D and Dill D, 1978, Phys Rev A <u>18</u>, 164.

Dunn G H and Kieffer L J, 1963, Phys Rev <u>132</u>, 2109.

Edwards A K, Wood R M, Davis J L and Ezell R L, 1990, Phys Rev A <u>42</u>, 1367.

Guberman S L, 1983, J Chem Phys <u>78</u>, 1404.

Hazi A V, 1974, J Chem Phys <u>60</u>, 4358.

Kanfer S and Shapiro M, 1983, J Phys B: At Mol Phys <u>16</u>, L655.

Kirby K, Uzer T, Allison A C and Dalgarno A, 1981, J Chem Phys <u>76</u>, 2820.

Savage O G, Lindsay B G and Latimer C J, 1990, J Phys B: At Mol Opt Phys <u>23</u>, 4313.

Sharp T E, 1971, At Data <u>2</u>, 119.

Strathdee S and Browning R, 1979, J Phys B: At Mol Phys <u>12</u>, 1789.

Tanaka K and Yoshimine M, 1979, J Chem Phys <u>70</u>, 1626.

Tellinghuisen J, 1985, Advances in Chemical Physics <u>LX</u>, ed K P Lawley (New York: Wiley–Interscience) pp 299–369.

Whetmore R W and Boyd R K, 1986, J Phys Chem <u>60</u>, 5540.

Wood R M, Edwards A K and Steuer M F, 1977, Phys Rev A <u>15</u>, 1433.

Yousif F B, Lindsay B G and Latimer C J, 1987, J Phys B: At Mol Phys <u>20</u>, 5079.

— , 1988, J Phys B: At Mol Opt Phys <u>21</u>, 4157.

— , 1990, J Phys B: At Mol Opt Phys <u>23</u>, 495.

Zare R N, 1967, J Chem Phys <u>47</u>, 204.

Spectroscopic and dynamical studies of chemiluminescent ion–molecule reactions

Th. Glenewinkel–Meyer, A. Kowalski, B. Müller, Ch. Ottinger,
D. Rabenda, P. Rosmus, and H.-J. Werner

Max–Planck–Institut für Strömungsforschung
Bunsenstraße 10 D–3400 Göttingen / Fed. Rep. Germany

This brief report summarizes recent experimental and theoretical investigations of low–energy (1–1000 eV) ion–neutral collisions which result in optical emission of the products. Three studies have been selected. In the first of these the emphasis was on spectroscopic problems, while the other two concentrated on dynamical aspects.

The apparatus is described in [1]. Basically it consists of a mass spectrometer with a target gas cell or target molecular beam positioned behind its exit slit. A powerful optical detection system collects and analyzes light which is emitted from the collision zone. In particular a sophisticated position–sensitive photomultiplier ("Mepsicron") enabled optical multichannel detection to be employed, a technique which proved indispensable especially for the first of the experiments to be described.

A) Group IIIa ion/halogen molecule reactions

A prototype of this class is the chemiluminescent exchange reaction

$$Al^+ + Cl_2 \rightarrow AlCl^{+*} + Cl \ , \tag{1}$$

where the asterisk indicates electronic excitation of a radiating state of the product. Interest in this reaction arose from the fact that ions such as $AlCl^+$ are isoelectronic to $MgCl$ and the corresponding alkaline earth halides, of which the spectroscopy is very well documented. By contrast, no information on optical emission from the group III halide ions, e.g. $AlCl^+$, existed prior to our work. Since our apparatus had previously demonstrated its ability to produce otherwise inaccessible spectra we undertook a comprehensive study of reactions of the type (1), with the ions B^+, Al^+, Ga^+, In^+ colliding with F_2, Cl_2 and Br_2 molecules [2].

In the case of reaction (1), two new band systems were discovered at collision energies of a few eV_{cm}. One is very broad and relatively unstructured, the other is much more compact and with vibrational structure. Interestingly, in the corresponding reaction with F_2 only the first of these appeared.

The analysis of these emissions may be outlined as follows: The narrow band system can successfully be interpreted on the basis of analogies with the isoelectronic species MgCl . Here it is known that the ground state is ionically bound, Mg^+Cl^- , and dissociates diabatically into $Mg^+(^2S)+Cl^-(^1S)$. Likewise an excited state $MgCl(^2\Pi)$ is known which dissociates diabatically into the excited ion pair, $Mg^+(^2P)+Cl^-(^1S)$. A survey of all known spectroscopic information on the alkaline earth halides reveals a striking correspondence between the wavelengths of the atomic ion $^2P \rightarrow {}^2S$ resonance line and of the corresponding molecular $^2\Pi \rightarrow \Sigma$ band system for the same metal. Using this rule as a guide, the narrow $AlCl^+$ emission system (and others) could be identified as the ionic counterparts: Here a similar parallelism exists between the (well-known) atomic-ion resonance lines such as $Al^{++}(^2P \rightarrow {}^2S)$ and the (newly found) molecular band systems, e.g. $AlCl^+(^2\Pi \rightarrow {}^2\Sigma)$. Further confirmation of this assignment came from the $^2\Pi_{1/2,3/2}$ spin–orbit splitting in those cases where it could be resolved ($GaCl^+$, $GaBr^+$, $InCl^+$, $InBr^+$). Again the corresponding neutral systems provide a valuable clue; here the $^2P_{1/2,3/2}$ splitting of the atomic (diabatic) dissociation product runs parallel to the known molecular $^2\Pi_{1/2,3/2}$ splitting. A very similar relationship was found to be obeyed between the known Ga^{++} and In^{++} spin–orbit splittings on the one hand and the corresponding $^2\Pi_{1/2,3/2}$ splittings of the newly observed, above–mentioned spectra on the other.

The broad, featureless emission spectra of AlF^+, $AlCl^+$ etc. had to be assigned on a different basis. They do not have a neutral counterpart. However, we had earlier observed similar broad spectra from BH^+ and AlH^+ and had interpreted them in analogy with the isoelectronic species BeH, MgH . In all these cases an avoided crossing exists between the repulsive diabatic $^2\Sigma$ ground state, arising from the ^1S metal atom and $H(^2S)$, and the attractive $^2\Sigma$ state formed from ^3P metal plus $H(^2S)$. Substituting a halogen ^2P atom for $H(^2S)$, a quite similar situation is expected for

$AlCl^+$ etc.: A weakly bound excited $^2\Sigma$ state is expected to arise adiabatically from the asymptote $Al^+(^3P) + Cl(^2P)$, which would have its minimum displaced towards larger internuclear distances compared to the $^2\Sigma$ ground state. Thus the $^2\Sigma$–$^2\Sigma$ emission spectrum will extend far to the red and will be rather unstructured, such as was observed.

These and other conjectures were put on a firm quantitative basis by high–level ab initio calculations [3] for AlF^+ and $AlCl^+$. They confirmed all the expected relationships outlined above and gave excellent agreement between the observed and calculated emission wavelengths. In addition, they revealed the reason for the absence of a (narrow) $^2\Pi$–$^2\Sigma$ emission band system in AlF^+: The upper state is here predissociated by a repulsive $^2\Pi$ state from $Al^+(^1S) + F(^2P)$. Such a state exists also for $AlCl^+$, but here it perturbs the excited $^2\Pi$ state only far from its minimum, and therefore does not interfere with the emission.

In summary, the chemiluminescence experiments have led to the discovery of a whole class of about 20 new band systems. Although the resolution was very poor, the way is now open to high–resolution (e.g. LIF) experiments aimed specifically at the spectral regions concerned here.

B) Isotopic branching ratio in the reaction $P^+ + HD \rightarrow PH^{+*}(PD^{+*}) + D(H)$

The chemiluminescent exchange reaction between P^+ and H_2, D_2 had been studied previously by us [4]. The emission bands were found to be quite narrow, as a result of predissociation setting in above certain J levels. The PH^+ and PD^+ bands, being slightly displaced in wavelength as a result of the spectroscopic isotope shift, are therefore well separated. This suggests a study of the branching ratio between the two isotopic product channels in the case of the $P^+ + HD$ reaction, as a function of collision energy [5].

As a reference, the energy dependence of the luminescence cross section was first measured with H_2 and D_2 targets. In both cases the light yield was found to increase from the thermodynamic threshold energy (~ 4.4 eV$_{cm}$) to a peak at 6.2 eV$_{cm}$, followed by a rather rapid decrease. With HD , on the other hand, a strong isotope effect was observed, favoring PD^+ emission at low and PH^+ emission at high energies. It could be explained on the basis of a very simple kinematic model.

The assumption was made that the quantity controlling the light yield is not the CM energy, but the energy contained in the relative motion of the P^+ ion with respect to only that constituent target atom, H or D, with which it ultimately reacts. This "pairwise" relative energy, E_p, is then very nearly $\frac{1}{2}$ of the CM energy for the H_2 and D_2 reactions, such that the peak cross section occurs here at $E_p = 6.2/2 = 3.1$ eV . Assuming that also for the two branches of the HD reaction the peak will lie at $E_p=3.1$ eV , this would correspond to $E_{cm}\approx 3 \times 3.1 = 9.3$ eV for the PH^+ branch (because the reduced mass of $P^+ + HD$ is about 3 times that of $P^+ + H$), and to $1.5 \times 3.1 = 4.6$ eV_{cm} for the PD^+ branch (from the ratio of the reduced masses $(P^+ + HD)/(P^+ + D) \approx 1.5$). Comparing with the experimental result, the expectation is borne out very well in the first case: The PH^+ branch does indeed peak at 9.3 eV_{cm} . In fact, transforming the entire cross section curve measured with $P^+ + H_2$ by the factor 1.5 on the energy scale, yields excellent agreement with the entire cross section curve for the $P^+ + HD \rightarrow PH^+ + D$ branch.

For the other branch, however, the corresponding transformation is only partly successful; it can explain the decreasing (high–energy) flank of the cross section curve very well, but is fails completely as regards the increasing, low–energy flank. In particular, the "predicted" peak position at 4.6 eV_{cm} disagrees totally with the observation (~ 6.1 eV_{cm}). However, in the low–energy regime the pairwise energy model cannot possibly obtain. In its pure form it would transform the low–energy part of the cross section curve into values of E_{cm} below the thermodynamic threshold, which is, of course, energetically forbidden. The model has to be modified, therefore, for the $P^+ + HD \rightarrow PD^+ + H$ branch in the region just above the (fixed) threshold. Here we assumed that the cross section increases with the CM energy at the same rate as measured with both H_2 and D_2 targets, except, of course, scaled down by a factor of two to allow for the fact that in HD only half the molecule is reactive as regards the PD^+ branch. With this additional prescription the rising flank of the observed PD^+ branch cross section curve was then also reproduced almost perfectly.

Thus it appears that the pairwise energy model can explain the observed, very pronounced isotope effect very well on purely kinematical grounds. Further support for this model comes from the rotational excitation of the PH^+ and PD^+ reaction products which can be deduced from the observed spectra by means of a computer simulation. It is found that the PD^+ product is much more rotationally excited than PH^+. This, too, can be quantitatively explained on the basis of the greater reduced mass (assuming similar impact parameters) for the interacting pair $P^+ + D$ compared to $P^+ + H$. Thus this rather rudimentary model appears to be capable of explaining quantitatively not only the relative cross sections, but also fine details such as internal product excitation.

C) Luminescent charge transfer $H^+ + HCl \rightarrow HCl^+(A\ ^2\Sigma^+) + H$

This reaction, yielding an electronically excited product by transfer of an electron rather than by exchange of an atom as in (A) and (B), is attractive both from an experimental and a theoretical point of view. The emission spectra of $HCl^+(A)$, due to the $A\ ^2\Sigma \rightarrow X\ ^2\Pi$ transition, are very well structured and display clearly a $v',v''=0$ progression ($v'=0-6$). This facilitates an accurate determination of the $HCl^+(A)$ vibrational population distribution by means of a computer simulation. It was found, for example, that at low collision energies $v'=0$ is the most strongly populated level, while at higher energies (~ 100 eV) $v'=1$ dominates. At the same time this system is also simple enough to be amenable to an accurate ab initio calculation. Using state–of–the–art MR–SCF CI codes, the lowest three electronic surfaces were calculated for a wide range of geometries (300 points) [6]. The entrance surface of the reaction is asymptotically the middle one of the three. A very interesting finding was its behaviour as a function of the angle of approach of the proton to the HCl molecule. For near–collinear arrangements, the entrance surface is attractive, thus approaching the ground state surface, which leads asymptotically to $H+HCl^+(X\ ^2\Pi)$ products. Thus in such geometries the predominant charge transfer product will be the non–radiating ground state HCl^+. At near–perpendicular directions of the collision, however, the entrance channel surface "switches" upwards to become repulsive, thereby approaching the third surface, which yields the optically observable product $HCl^+(A)$. It is therefore in these geometries that our reaction will preferentially occur. It proceeds actually through an avoided intersection between the two topmost surfaces, as a detailed configuration analysis of the wavefunction has shown.

The ab initio points were fitted with continuous functions, providing a complete, accurate determination of the electronic potential energy surfaces, in particular in the vicinity of the important avoided crossing. This enabled a dynamical treatment of the non–adiabatic transition between the entrance and exit surface to be performed [7]. It was based on the well-known Landau–Zener formula for the transition probability, as applied to a straight–line trajectory model with the orientation of the HCl target assumed fixed during the collision. The latter assumption is fully justified for the collision energies considered. The important characteristic of the present Landau–Zener treatment is that the parameters ΔE and $F_1 - F_2$ (energy gap between the two surfaces and difference of their slopes) were not simply considered as adjustable fit parameters, but were taken directly from the ab initio surfaces. With proper averaging over the contributing HCl orientation angles and internuclear distances, partial absolute cross sections were then obtained for the excitation of the individual vibrational levels v'=0–5 of HCl$^+$(A) , as a function of the collision energy. These could be compared with the experimental results, which were also put on an absolute basis by means of calibration experiments, using another, well documented luminescence reaction as a reference standard. The agreement between the calculated and the experimental cross section curves was very satisfactory as regards their shape and relative magnitudes up to a collision velocity of about 200 km/sec. Above this velocity the Landau–Zener theory is not expected to be valid because the crossings between the potential energy surfaces can then no longer be considered as localized. The absolute magnitude of the calculated and measured cross sections differed by a factor of two , which is, however, within the estimated uncertainty of the calibration procedure.

In summary, the three types of reactions considered demonstrate the power of the luminescence method in the investigation of rather complex collision processes. The detailed data obtained allow fruitful comparisons with the results from modern quantum chemical and dynamical calculations.

References

[1] Ch. Ottinger, in "Gas Phase Ion Chemistry", M.T. Bowers, ed., Academic Press, New York 1984, Vol. 3, p. 249.

[2] Th. Glenewinkel–Meyer, A. Kowalski, B. Müller, Ch. Ottinger, and W.H. Breckenridge, J. Chem. Phys. **89** (1988) 7112.

[3] Th. Glenewinkel–Meyer, B. Müller, Ch. Ottinger, P. Rosmus, and H.-J. Werner, J. Chem. Phys. 1991, in press.

[4] B. Müller and Ch. Ottinger, J. Chem. Phys. **85** (1986) 243.

[5] D. Rabenda, Diplomarbeit 1991.

[6] Th. Glenewinkel–Meyer, Ch. Ottinger, P. Rosmus, and H.-J. Werner, Chem. Phys. **152** (1991) 409.

[7] Th. Glenewinkel–Meyer and Ch. Ottinger, submitted to J. Chem. Phys. .

Site-specific electronic excitation in ion–molecule collisions

Shigetomo Kita, Hajime Tanuma[*], and Noriyuki Shimakura[**]

Department of Physics, Nagoya Institute of Technology, Gokiso, Showa-ku, Nagoya 466, Japan
[*]Central Research Laboratory, Hitachi, Ltd., Kokubunji, Tokyo 185, Japan
[**]General Education Department, Niigata University, Ikarashi, Niigata 950-21, Japan

ABSTRACT: Site-specific electronic excitation observed in the differential energy-transfer measurements for Li^+-CO collisions is discussed on the basis of knowledge about energy-transfer mechanisms in molecular collisions. The site dependence of the electronic excitation can be qualitatively explained by taking the electron-density distribution of the CO molecule into account.

1. INTRODUCTION

Investigation of the energy transfer in ion-molecule collisions is of fundamental interest in conjunction with chemical reaction dynamics, plasma physics, and surface science. The energy-transfer mechanisms in molecular collisions are complicated because of the rotational and vibrational degrees of freedom in the molecules, in contrast to atomic collisions. Impulsive molecular collisions generally cause rotational, vibrational, and electronic excitations, and even molecular dissociation.

Differential energy-transfer spectroscopy has been proved to be a powerful technique for the study of molecular collision dynamics as well as of atomic collisions. The rotational and vibrational excitation mechanisms have been elucidated by means of energy-transfer measurements (Faubel and Toennies 1978, Schepper et al 1979, Buck et al 1980, Vedder et al 1981, Faubel et al 1982, Hasegawa et al 1985, Buck et al 1985). Differential scattering experiments have also recently been carried out to study the electronic excitation mechanisms in ion-molecule collisions (Nieder et al 1987, Dhuicq and Sidis 1987, Quintana et al 1989, Kita et al 1990). In this paper we discuss a specific electronic excitation mechanism in the Li^+-ion impact on the heteronuclear CO molecule (Tanuma et al 1989, Kita et al 1990). Our discussion is limited to the impulsive collisions under repulsive force. The electronic excitation mechanism is closely related to the character of the short range repulsion.

The repulsive potentials between atoms (or ions) have a simple character and are represented by the atomic parameters (radius and softness), which characterize the electron-density distributions $\rho(r)_{i,j}$ in the outer region of the interacting partners i and j (Gilbert 1968, Carlson 1973, Kita et al 1976). As is already well-known, electronic excitation in moderate-energy atomic collisions takes place through potential-curve crossing, and is characterized by the crossing radius R_x and the interaction energy V_{12} at the crossing point between two potentials V_1 and

V_2 of the electronically ground and excited states. The crossing radius R_x is related to the atomic radii of the colliding partners (Bisgaard et al 1980, Heydenreich et al 1983). This indicates that the electron-density distributions $\rho_{i,j}$ of the colliding partners also govern the potential curve crossing. For a strongly asymmetric heteronuclear system, as it will be discussed here, the repulsive potential as well as the crossing radius R_x is mostly determined by the electron-density distribution ρ_j for one of the interacting particles having a larger radius and softness (Kita et al 1976, Esbjerg and Nørskov 1980, Laughlin 1982). One can, therefore, explain electronic excitation in such an asymmetric system by taking account of only the electron density ρ_j.

Electronic excitation in impulsive molecular collisions is accompanied by a huge momentum transfer. Therefore, the electronic excitation mechanisms can only be understood on the basis of knowledge about rotational and vibrational excitations. We will first briefly discuss the vibro-rotational excitation mechanisms.

2. ROTATIONAL AND VIBRATIONAL EXCITATION

In collisions at moderate energies between alkali ions and rare-gas atoms, electronic excitation is observed only in some specific systems (Kita et al 1983, Kita et al 1984, Inouye et al 1984). This is the case for ion-molecule collisions (Hasegawa et al 1985, Kohlhase et al 1985, Tanuma et al 1989, Kita et al 1989). The Na^+-N_2 and Na^+-CO systems are suitable for the explanation of rotational and vibrational excitation mechanisms in collisions, because electronic excitation is not observed in these colliding systems (Hasegawa et al 1985, Kita et al 1987, Nakamura et al 1987, Kita et al 1988, Tanuma et al 1988).

As an example, angular dependence of the energy-transfer spectra at the laboratory (lab) energy $E_{lab} = 100$ eV is shown in Fig. 1 for the Na^+-N_2 and Na^+-CO collisions (Hasegawa et al 1985, Nakamura et al 1987). Energy dependence of the spectra at the lab angle $\theta = 20°$ is also exhibited in Fig. 2. The abscissa in the figures indicates the relative energy-transfer $\Delta E/E$ in the center-of-mass (CM) frame. A single peak is observed at small angles, while at angles larger than $15°$ the spectra for the Na^+-N_2 and Na^+-CO collisions are composed of double and triple peaks, respectively. The spectra, thus, depend strongly on the scattering angle, but are almost independent of the collision energy at a certain angle. Very similar structures in the spectra have been found for the first time in the velocity distributions of K atoms scattered from the N_2 and CO molecules at the low CM energy $E \simeq 1$ eV (Schepper et al 1979), and have also been observed in Li^+-N_2 and Li^+-CO collisions at $E_{lab} \simeq 10$ and 20 eV (Gierz et al 1984).

Classical trajectory (CT) calculations provide valuable information on the rotational and vibrational excitation mechanisms (Karplus et al 1965, LaBudde and Bernstein 1971). The theoretical analyses of the energy-transfer spectra of Na^+-N_2 and Na^+-CO with the CT method as well as with the quantal infinite-order-sudden approximation (Schinke and Bowman 1983) confirm that the peaks located at larger energy transfers, peak (2) for Na^+-N_2 and peaks (2) and (3) for Na^+-CO, are attributed to rotational excitation of the molecules, and are due to the rotational rainbow effect (Bosanac 1980, Beck et al 1981, Jones et al 1982, Schinke and Bowman 1983). This suggests that rotational excitation plays an important role in the energy-transfer mechanisms even when the energy is high enough to lead to other inelastic processes. The rotational rainbows originate from the collisions around the molecular orientations $\gamma_r = 50$ and $130°$, where γ is the angle between the molecular axis and the intermolecular distance vector **R** at the turning point in the collisions. The energy-transfer spectra of Na$^+$-N$_2$ have one rotational rainbow peak because N$_2$ is a homonuclear diatomic molecule. The spectra of Na$^+$-CO have, on the other hand, two rotational rainbows, because CO is heteronuclear. The two peaks (2) and (3) for Na$^+$-CO are due to the rainbows arising from the collisions of Na$^+$ ions with the CO molecules on the O-atom and C-atom sides, respectively. Vibrational excitation takes place exclusively around the orientations $\gamma = 0$ and 90°. Classical excitation cross section $\sigma(\Delta E, \theta)$ at a certain angle θ is approximately proportional to $\sin\gamma$, and so contribution from the orientation around $\gamma = 0°$ is very small. Peak (1) in the spectra of Figs. 1 and 2 is attributed to vibrational excitation around the orientation $\gamma = 90°$.

The arrows N, C and O in Fig. 1 illustrate the energy transfers evaluated with the spectator model, in which the ion is assumed to interact with one of the constituent atoms of the molecules. This model fairly reproduces the energy-transfer locations of the rotational rainbow peak (2) in Na$^+$-N$_2$, and of the rainbow peaks (2) and (3) in Na$^+$-CO. Around the rotational rainbow angles $\gamma_r = 50$ and 130°, the short range repulsion of the ion-molecule system is given exclusively by the ion-atom interaction of the closest pair. Then, the ion-molecule collision around γ_r can be regarded approximately as an ion-atom interaction. This is why the rotational rainbow position can be well explained by the spectator model. Peak (1) in the spectra, however, originates from the three-body interaction, and can not be reproduced at all with this simple model.

As discussed above, the energy transfer due to rotational and vibrational excitation is very big at larger angles. All the peaks in the energy-transfer spectra observed at larger angles can be ascribed to different types of rainbow effects by taking the orientation dependence of the

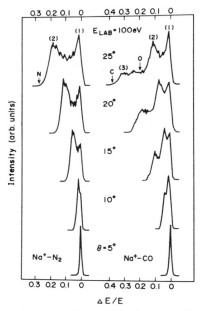

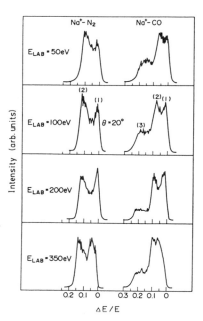

Fig. 1. Angular dependence of the energy-transfer spectra for Na$^+$-N$_2$ and Na$^+$-CO collisions at $E_{lab} = 100$ eV.

Fig. 2. Energy dependence of the energy-transfer spectra for Na$^+$-N$_2$ and Na$^+$-CO collisions at $\theta = 20°$.

energy transfer $\Delta E(\gamma)$ into account (Tanuma et al 1988). Since the momentum transfer depends strongly on the molecular orientation, differential energy-transfer measurements allow us to obtain information on the orientation dependence of electronic excitation.

3. ELECTRONIC EXCITATION

By means of differential energy-transfer spectroscopy, we have studied the inelastic scattering of alkali ions (Li^+, Na^+ and K^+) from N_2 and CO molecules at moderate collision energies of $50 \lesssim E_{lab} \lesssim 350$ eV. According to the measurements, electronic excitation was observed in the Li^+- and K^+-ion impact, while noticeable electronic excitation could not be found for the Na^+ ions. This system dependence of electronic excitation is very similar to that in the alkali ion impact on Ar atoms. The electronic excitation in Li^+-N_2 and Li^+-CO collisions has also been investigated previously at higher energies $E_{lab} \gtrsim 500$ eV and smaller angles $\theta \lesssim 7°$ (Sato et al 1976, Kobayashi and Kaneko 1976). Here we will discuss electronic excitation mechanisms in moderate-energy Li^+-N_2 and Li^+-CO collisions (Tanuma et al 1989, Kita et al 1990).

3.1. Li^+-N_2 collisions

Electronic excitation in the Li^+-ion impact on the heteronuclear molecule CO is distinctly site-specific, while no site-dependence of electronic excitation is observed in the scattering of Li^+ ions from the homonuclear molecule N_2. Electronic excitation accompanied by a big momentum transfer has never been investigated and is not familiar to us. For the sake of our discussion we will first present the experimental findings in impulsive Li^+-N_2 collisions.

In Fig. 3 angular dependence of the energy-transfer spectra measured at $E_{lab} = 120$ eV is shown. Signal A in addition to signal X is seen in the spectra measured at larger angles. The prominent signal X has a single peak at small angles, but splits into double peaks X_D and X_N at larger angles. Measurements of the angular and energy dependence of the spectra indicate that peak X is attributed to electronically elastic collisions, i.e., rotational and vibrational excitation. The double peaks X_D and X_N correspond to peaks (1) and (2) in the Na^+-N_2 collisions, respectively. Figure 4 shows the differential cross sections (DCS's) measured at $E_{lab} = 120$ eV. The excitation DCS's $\sigma(\theta)_A$ (open triangles in the figure) begin to appear at $\theta \simeq 27.5°$ and have their maximum value around $\theta = 40°$. The total DCS's (open circles) have a hump around $\theta = 32.5°$, which is close to the threshold angle $\theta \simeq 27.5°$ for signal A. The signal A is observed only at reduced angles $\tau = E_{lab}\theta$ larger than 3 keV deg ($= \tau_N$), independent of the collision energy. For higher energies of $E_{lab} \gtrsim 250$ eV, neutral Li atoms produced in the charge exchange reactions, furthermore, are observed at angles $\tau > \tau_N$. All these findings are the same as for the angular and energy dependence of electronic excitation in atomic collisions (Lorents and Conklin 1972, Brenot et al 1975, Kita et al 1987). Peak A observed in the spectra is ascribed to electronic excitation.

The energy difference T_N of peak A from the rotational rainbow peak X_N, $T_N = \Delta E(A) - \Delta E(X_N)$, extracted from the experimental spectra for $E_{lab} = 160$ eV is shown with the open circles in Fig. 5. The difference T_N is almost independent of the scattering angle, while the difference $T_D = \Delta E(A) - \Delta E(X_D)$ depends strongly on the angle, which can be seen from Fig. 3. The difference T_N is close to the vertical excitation energy $\Delta E = 9.16$ eV for the $a\ ^1\Pi_g \leftarrow X\ ^1\Sigma_g^+$ transition of the N_2 molecule (Lassettre et al 1968), where $a\ ^1\Pi_g$ is the lowest-lying excited electronic (singlet) state. These features are almost the same at other collision energies. This result

suggests that rotational excitation in electronically inelestic
scattering is almost the same as that in electronically elastic
scattering, and the difference T_N is nearly equal to the vertical energy
for the electronic transition. These experimental findings in Li^+-N_2
collisions are considered to be the general feature of electronic
excitation in impulsive molecular collisions.

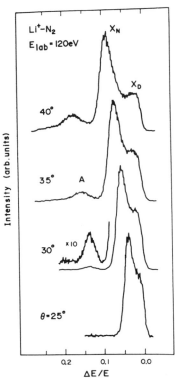

Fig. 3. Angular dependence of the
energy-transfer spectra for Li^+-N_2
collisions measured at E_{lab}=120 eV.

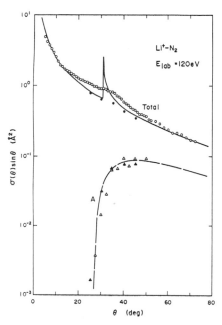

Fig. 4. Angular dependence of the
DCS $\sigma(\theta)$ weighted with $\sin\theta$ in the
lab system for Li^+-N_2 at E_{lab}= 120
eV. O and △ , experimental total and
excitation DCS's, respectively. ——
and — — , total and excitation
DCS's, respectively, calculated by
assuming spherical potentials. ● and
▲ , total and excitation DCS's,
respectively, calculated with the
CTSH method.

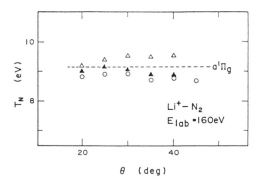

Fig. 5. Angular dependence of the
energy difference $T_N = \Delta E(A) - \Delta E(X_N)$
evaluated from the experimental
and theoretical energy-transfer
spectra of Li^+-N_2 collisions at
E_{lab} = 160 eV. O , experimental
results; ▲ , calculations with the
rigid-rotor model; and △ , vibrotor model calculations. ---- ,
excitation energy for the vertical
$a\ ^1\Pi_g \leftarrow X\ ^1\Sigma_g^+$ transition of the
N_2 molecule.

3.2. Li⁺-CO collisions

The knowledge about electronic excitation accompanied by a huge momentum transfer for a homonuclear molecule system makes it possible to obtain information on the site dependence of electronic excitation in an ion impact on a heteronuclear molecule. The experimental results for Li⁺-CO collisions will be presented here.

Angular dependence of the energy-transfer spectra for Li⁺-CO is exhibited in Fig. 6. The spectra have a peak A in addition to the prominent peaks X_0 and X_C. The two peaks X_0 and X_C correspond to the electronically elastic peaks (2) and (3) in the spectra for the Na⁺-CO collisions. The quasi-elastic peak, which corresponds to peak X_D for Li⁺-N₂ and signal (1) for Na⁺-N₂ and Na⁺-CO, is observed separately from peak X_0 only at larger angles of $\theta \gtrsim 35°$. Peak A can be seen at angles $\theta > 20°$, and is attributed to electronic excitation of the CO molecules. The evaluation of the energy differences of peak A from peaks X_C and X_0 shows that the difference T_C, $\Delta E(A)-\Delta E(X_C)$, is almost independent of the scattering angle, which is similar to T_N for Li⁺-N₂, and has the averaged value of 8.2 eV. On the other hand, the quantity T_0, $\Delta E(A)-\Delta E(X_0)$, depends strongly on the angle. This result suggests that peak A is attributed to electronic excitation in the Li⁺-impact on the C-atom side of the CO molecule.

At higher collision energies $E_{lab} \gtrsim 160$ eV and larger scattering angles, a double peak structure due to electronic excitation is observed in the energy-transfer spectra illustrated in Fig. 7. As can be seen in the

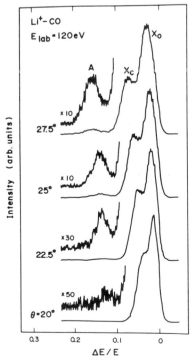

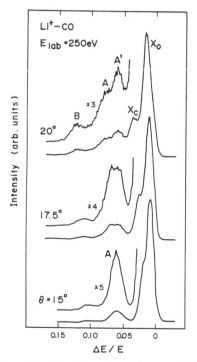

Fig. 6. Angular dependence of the energy-transfer spectra for Li⁺-CO collisions measured at E_{lab}=120 eV.

Fig. 7. Angular dependence of the energy-transfer spectra for Li⁺-CO collisions measured at E_{lab}=250 eV.

figure, the electronically inelastic peak A measured at E_{lab} = 250 eV changes its structure drastically around θ = 17.5°. At θ = 20°, the spectrum has electronically elastic peaks X_0, X_c, and electronically inelastic peaks A, A', and B. Peak B is assigned to two-electron excitation of the CO molecules, but we do not discuss about it here. According to the evaluation of four energy differences of peaks A and A' from peaks X_c and X_0, only the differences $T_c = \Delta E(A) - \Delta E(X_c)$ and $T'_0 = \Delta E(A') - \Delta E(X_0)$ are independent of the scattering angle. The two differences T_c and T'_0 have almost the same value of 8.6 eV, which is nearly equal to T_c = 8.2 eV obtained at E_{lab} = 120 eV. These energy differences, furthermore, correspond to the vertical excitation energy $\Delta E \simeq 8.4$ eV for the A $^1\Pi$ ← X $^1\Sigma^+$ transition of the CO molecule (Skerbele et al 1967, Tilford and Simmons 1972). These results suggest that the two peaks A and A' belong to an identical electronic transition, i.e., CO(A $^1\Pi$) ← CO(X $^1\Sigma^+$), and are attributed to excitation in the Li$^+$ impact on the C-atom and O-atom sides, respectively. The measurements of the angular and energy dependence of the electronic excitation show that the threshold angle for peak A, which is attributed to the Li$^+$-C interaction, is approximately τ_c = 2.1 keV deg. The onset of signal A', which is due to the Li$^+$-O interaction, is observed around the reduced angle τ_0 = 4.2 keV deg. Electronic excitation in Li$^+$-CO collisions, is thus distinctly site-specific.

The DCS's of Li$^+$-N$_2$ and Li$^+$-CO collisions at E_{lab} = 200 eV are exhibited in Fig. 8, where the excitation DCS $\sigma(\theta)_A$ for Li$^+$-CO means $\sigma(\theta)_A + \sigma(\theta)_{A'}$. The excitation DCS for Li$^+$-CO could be evaluated only at small angles, because the separation between peaks A' and X_c becomes worse at larger angles. The total DCS's for the two systems show small differences only around θ = 20°.

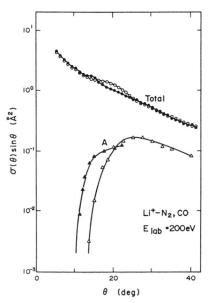

Fig. 8. Angular dependence of the exper-imental DCS $\sigma(\theta)$ weighted with sinθ of Li$^+$-N$_2$ and Li$^+$-CO collisions at E_{lab} = 200 eV. ○ and △ are the total and exci-tation DCS's for Li$^+$-N$_2$, respectively. ● and ▲ are the total and excitation DCS's for Li$^+$-CO, respectively.

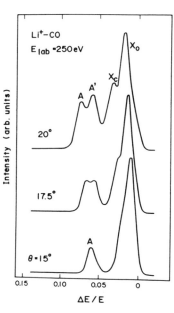

Fig. 9. Angular dependence of the energy-transfer spectra for Li$^+$-CO collisions at E_{lab} = 250 eV calculated with the CTSH method based on the vibrotor model.

The excitation DCS's of the two systems show distinctly different angular dependence, while these have almost the same values around $\theta = 20°$.

3.3. Potential-crossing characteristics

The potential parameters at the crossing have to be evaluated in order to discuss the electronic excitation mechanisms. Information on the electronic excitation in molecular collisions can be obtained by calculations using the classical trajectory surface hopping (CTSH) approach (Tully and Preston 1971). In such a calculation one needs the interaction $V(R, \gamma) = V(r_{12}, r_{23}, r_{31})$ of the colliding system, where the subscript 1 means ion, 2 and 3 are constituent atoms of the molecules, and r_{12}, r_{23} and r_{31} are the interatomic distances, respectively. Ground-state potential surfaces for the Li$^+$-N$_2$ and Li$^+$-CO systems computed with a self-consistent-field (SCF) approximation are available (Staemmler 1975, 1976). Theoretical investigations of rotational and vibrational excitation in low-energy Li$^+$-N$_2$ collisions, however, show that the SCF surface overestimates the vibrational component of the intermolecular forces (Vilallonga and Micha 1987). In our calculations, the interaction potentials for the colliding systems were approximated by the additive binary interaction. The results calculated with the statistical electron gas model were used as the ion-atom interactions $V(r_{12})$ and $V(r_{31})$ for the ground state (Ishikawa et al 1976). The ground-state potentials $V(R, \gamma)$ at the equilibrium intramolecular distance $r_{23} = r_e$ used in our study agree well with the ab initio potentials.

The intermolecular potentials for the electronically excited state of Li$^+$-N$_2$ (a $^1\Pi_g$)   and Li$^+$-CO(A $^1\Pi$) were estimated from a curve fitting procedure of the DCS's. The intramolecular potentials $V(r_{23})$ of the ground and excited states of the molecules were extracted from spectroscopic data (Huber and Herzberg 1979). The transition probability at the surface crossing was given by the Landau-Zener expression. Examples of the best fit result of the DCS's for Li$^+$-N$_2$ at $E_{lab} = 120$ eV and the computed energy-transfer spectra for Li$^+$-CO collisions at $E_{lab} = 250$ eV   and $15 \le \theta \le 20°$ are demonstrated in Figs. 4 and 9, respectively. The energy difference T_N evaluated from the theoretical spectra for Li$^+$-N$_2$ collisions at $E_{lab} = 160$ eV is also shown in Fig. 5. The interaction V_{12} at the crossing was estimated to be 1.6 eV for both systems, Li$^+$-N$_2$ and Li$^+$-CO. Other potential parameters, i.e., crossing radius R_x and potential height $V(R_x)$,   at the crossing point deduced from our experiments are tabulated in Table I for the equilibrium intramolecular distance $r_{23} = r_e$. The crossing parameters R_x and $V(R_x)$ were also evaluated with the SCF-CI (CI denotes configuration interaction) approximation. The theoretical values of R_{x1} and $V(R_{x1})$ are also given in Table I. The theoretical distances R_{x1} adequately explain the experimental values of R_x. All the potential heights $V(R_{x1})$ calculated with the SCF-CI method are approximately a

Table I. Crossing parameters for the intramolecular distance $r_{23} = r_e$.

System	γ(deg)	R_x[a] (Å)	$V(R_x)$[a] (eV)	R_{x1}[b] (Å)	$V(R_{x1})$[b] (eV)	R_{x2}[c] (Å)
Li$^+$-N$_2$	0	1.32	27.5	1.23	55	1.26
	90	0.75	27.5			
Li$^+$-CO	0(C)	1.47	21.4	~1.35	~45	1.48
	90	0.75	26.7			
	180(O)	1.22	32.6	~1.15	~60	1.10

[a] Experiments. [b] SCF-CI calculations. [c] Evaluations with the binary model.

factor 2 higher than the experimental values. Nevertheless, the calculations reproduce qualitatively the system and site dependence of the potential height $V(R_x)$.

3.4. Site-specific electronic excitation

The intermolecular potentials for Li^+-N_2 and Li^+-CO can be favorably approximated by additive binary interactions. As mentioned above, the ion-molecule potentials at the orientations apart from $\gamma = 90°$ can be approximated with ion-atom interactions. The crossing distances of the ion-molecule systems at the specific linear configuration ($\gamma = 0$ and $180°$) can, therefore, be approximately evaluated by assuming binary interactions. The crossing distances R_{x2} given in Table I are the values evaluated from ionic and atomic radii defined by the outermost atomic orbital. The distances R_{x2} adequately reproduce the experimental values R_x and the calculated values R_{x1}. The system and site dependence of the crossing distance R_x is considered to originate from the electron density distributions of the N_2 and CO molecules.

The site-specific electronic excitation in Li^+-CO collisions can also be qualitatively explained in another manner by taking account of the electron-density distribution of the molecular orbital which is to be promoted. Excitation from the ground-state $X\ ^1\Sigma^+$ to the $A\ ^1\Pi$ of the CO molecule corresponds approximately to the transition of an electron from the 5σ molecular orbital to the 2π orbital. The 5σ and 2π orbitals are the highest occupied and lowest unoccupied molecular orbitals of the CO molecule, respectively (Huo 1965). The 5σ molecular orbital has a component of the atomic orbital of C-atom larger than that of O-atom, so the electron density of the 5σ orbital is much higher on the C-atom side than on the O-atom side (Huo 1965). In collisions of small reduced angles, the intermolecular distance R_t at the turning point is large and the ion interacts exclusively with the 5σ electron of the CO molecule on the C-atom side. Electronic excitation of the CO molecule is, then, considered to take place only in the Li^+-ion impact on the C-atom side for such small angle scattering. The Li^+ ion is expected to interact with the 5σ electron on the O-atom side only in the violent collisions of larger reduced angles, which correspond to smaller R_t. This would be the origin of the site-specific electronic excitation in Li^+-CO collisions.

ACKNOWLEDGMENTS

We are very grateful for longstanding dialogue with Professor H Inouye (Bunri University Tokushima, Tokushima, Japan) in Sendai. We wish to thank Professor N Kobayashi (Tokyo Metropolitan University, Tokyo, Japan), Professor I Kusunoki and Professor Y Sato (Tohoku University, Sendai, Japan) for valuable discussions. One of the authors (S K) also gratefully acknowledges to Professor M Izawa (National Lab. High Energy Phys., Tsukuba, Japan), Dr A Kohlhase (Siemens AG, Munich, Germany), Mr T Hasegawa (Asahi Glass, Ltd., Tokyo, Japan) and Dr M Nakamura (RIKEN, Wako, Japan) for their contributions to our subsequent studies.

REFERENCES

Beck D, Ross U and Schepper W 1981 Z. Phys. A **299** 97-104
Bisgaard P, Andersen T, Sørensen B V, Nielsen S E and Dahler J S 1980 J. Phys. B **13** 4441-52
Bosanac S 1980 Phys. Rev. A **22** 2617-22
Brenot J C, Dhuicq D, Gauyacq J P, Pommier J, Sidis V, Barat M and Pollack E 1975 Phys. Rev. A **11** 1245-66
Buck U, Huisken F and Schleusener J 1980 J. Chem. Phys. **72** 1512-23

Buck U, Otten D and Schinke R 1985 J. Chem. Phys. **82** 202-16
Carlson E H 1973 J. Chem. Phys. **58** 1905-7
Dhuicq D and Sidis V 1987 J. Phys. B **20** 5089-105
Esbjerg N and Nørskov J K 1980 Phys. Rev. Lett. **45** 807-10
Faubel M, Kohl K H and Toennies J P 1982 Faraday Discuss. Chem. Soc. **70** 205-20
Faubel M and Toennies J P 1978 Adv. At. Mol. Phys. **13** 229-314
Gierz U, Toennies J P and Wilde M 1984 Chem. Phys. Lett. **110** 115-22
Gilbert T L 1968 J. Chem. Phys. **49** 2640-2
Hasegawa T, Kita S, Izawa M and Inouye H 1985 J. Phys. B **18** 3775-82
Heydenreich W, Menner B, Zehnle L and Kempter V 1983 Z. Phys. A **312** 285-92
Huber K P and Herzberg G 1979 *Molecular Spectra and Molecular Structure IV. Constants of Diatomic Molecules* (New York: Van Nostrand Reinhold)
Huo W M 1965 J. Chem. Phys. **43** 624-47
Inouye H, Izawa M, Kita S, Takahashi K and Yamato Y 1984 Bull. Res. Inst. Sci. Meas. Tohoku Univ. **32** 41-66
Ishikawa T, Kita S and Inouye H 1976 Bull. Res. Inst. Sci. Meas. Tohoku Univ. **24** 101-8
Jones P L, Hefter U, Mattheus A, Witt J and Bergmann K 1982 Phys. Rev. A **26** 1283-301
Karplus M, Porter R N and Sharma R D 1965 J. Chem. Phys. **43** 3259-87
Kita S, Hasegawa T, Kohlhase A and Inouye H 1987 J. Phys. B **20** 305-15
Kita S, Izawa M, Hasegawa T and Inouye H 1984 J. Phys. B **17** L885-90
Kita S, Izawa M and Inouye H 1983 J. Phys. B **16** L499-504
Kita S, Noda K and Inouye H 1976 J. Chem. Phys. **64** 3446-9
Kita S, Tanuma H and Izawa M 1987 J. Phys. B **20** 3089-103
Kita S, Tanuma H and Izawa M 1988 Chem. Phys. **125** 415-24
Kita S, Tanuma H and Izawa M 1989 unpublished
Kita S, Tanuma H, Kusunoki I, Sato Y and Shimakura N 1990 Phys. Rev. A **42** 367-82
Kobayashi N and Kaneko Y 1976 unpublished
Kohlhase A, Hasegawa T, Kita S and Inouye H 1985 Chem. Phys. Lett. **117** 555-60
LaBudde R A and Bernstein R B 1971 J. Chem. Phys. **55** 5499-516
Lassettre E N, Skerbele A, Dillon M A and Ross K J 1968 J. Chem. Phys. **48** 5066-96
Laughlin R B 1982 Phys. Rev. B **25** 2222-47
Lorents D C and Conklin G M 1972 J. Phys. B **5** 950-62
Nakamura M, Kita S and Hasegawa T 1987 J. Phys. Soc. Jpn. **56** 3161-75
Niedner G, Noll M and Toennies J P 1987 J. Chem. Phys. **87** 2685-94
Quintana E J, Andriamasy A, Schneider D J and Pollack E 1989 Phys. Rev. A **39** 5045-7
Sato Y, Niurao K, Takagi H and Inouye H 1976 J. Chem. Phys. **65** 3952-7
Schepper W, Ross U and Beck D 1979 Z. Phys. A **290** 131-41
Schinke R and Bowman J M 1983 *Molecular Collision Dynamics* ed J M Bowman (Berlin: Springer-Verlag) pp 61-115
Skerbele A, Dillon M A and Lassettre E N 1967 J. Chem. Phys. **46** 4162-4
Staemmler V 1975 Chem. Phys. **7** 17-29
Staemmler V 1976 Chem. Phys. **17** 187-96
Tanuma H, Kita S, Kusunoki I and Sato Y 1989 Chem. Phys. Lett. **159** 442-6
Tanuma H, Kita S, Kusunoki I and Shimakura N 1988 Phys. Rev. A **38** 5053-65
Tilford S G and Simmons J D 1972 J. Phys. Chem. Ref. Data **1** 147-87
Tully J C and Preston R K 1971 J. Chem. Phys. **55** 562-72
Vedder M, Hayden H and Pollack E 1981 Phys. Rev. A **23** 2933-40
Vilallonga E and Micha D A 1987 J. Chem. Phys. **86** 760-75

Slow ion–surface interactions studied by means of ion-induced electron emission statistics

HP Winter

Institut für Allgemeine Physik, Technische Universität Wien
Wiedner Hauptstraße 8-10, A-1040 Wien, Austria

ABSTRACT: Electron emission due to impact of slow (impact velocity below 1 a.u.) singly, doubly and multiply charged ions on clean metal surfaces has been studied by means of the related electron emission statistics. The applied method is described and their capability demonstrated for delivering fairly detailed informations on the various electronic transitions involved in slow ion-surface interaction.

1. INTRODUCTION

Bombardment of the surface of a solid, in particular a clean metal, by slow neutral or ionized atoms or molecules can result in emission of electrons, due to transfer of potential energy (potential emission - PE) and/or kinetic energy (kinetic emission - KE) from the projectiles onto the target electrons. Such processes are both of fundamental interest and considerable practical importance and have therefore already been investigated for long time. The principal paper on PE is still by Hagstrum (1954), and more recent reviews on PE have been given by Varga (1987, 1988, 1989) and Andrä (1989). KE has recently been reviewed by Hasselkamp (1988) and Hofer (1990). Both classes of processes depend very critically on target surface conditions, because of which experimental data on electron emission from slow ion - surface interaction should principally be taken with a grain of salt.

As a general distinction, PE results primarily from electronic transitions between projectile and surface before the impact has already taken place, whereas KE can only be initiated after the particle has made its close contact with the surface. In a more detailed view, PE arises from Auger-type processes involving time intervals of the order of 10^{-15} s. The latter are comparable to the flight times of relatively slow ($v \leq 10^5$ m/s) particles within the distances where electronic transitions between approaching projectiles and a metal surface are probable (typically several 10^{-10} m for singly charged ions, cf. Hagstrum (1954), and increasing in proportion to the ion charge state q for multicharged ions, Varga 1989). Consequently, for such projectiles a complete neutralisation/deexcitation, which can involve a large number of electronic transitions, may not be completed until surface impact. Therefore, PE becomes the more efficient the lower the impact energy, and no impact energy threshold is obeyed.

KE is related to the stopping power of projectiles within the uppermost atomic layers of the solid. Electrons which are freed with kinetic energies in the ten eV range in this region may escape across the surface barrier into vacuum. Consequently, KE processes which usually last not longer than about 10^{-12} s, are subject to an impact energy threshold.

From these simple considerations we conclude that even for slow ion-induced electron emission ("sIEE") the PE and KE processes will generally be interrelated, which clearly questions the physical relevance of their still quite common distinction.

The present work is based on a new experimental approach (cf. chapter 2), which in contrast to the so far used methods can better distinguish among different contributions to sIEE and thus clarify the relative importance of potential and kinetic projectile energy in such processes. We determine the probabilities for emission of a given number n = 0, 1, 2, etc. electrons after ion impact on the surface or, in other words, the statistics of ion-induced electron emission ("ES"). Results from such ES studies are demonstrated and discussed for bombardment of a clean gold surface by slow singly charged ions in chapter 3, slow doubly charged ions in chapter 4 and finally slow multicharged ions in chapter 5.

2. EXPERIMENTAL METHODS AND DATA EVALUATION

sPIE from solid surfaces has so far almost exclusively been studied by regarding the total electron yield (i.e the mean number of electrons emitted per projectile particle, usually denoted by γ) and/or the ejected electron energy distribution dN_e/dE_e. Figs. 1a,b show standard experimental arrangements for such measurements with sketched results.

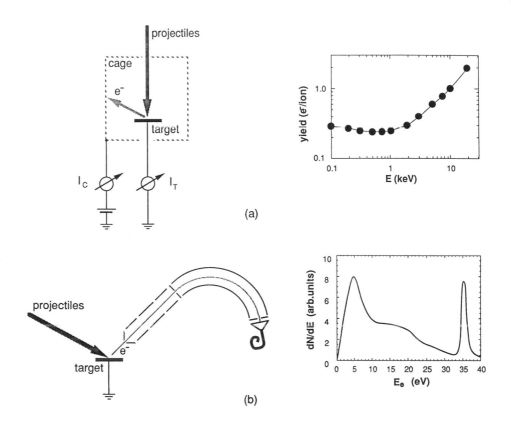

Fig. 1: (a) Measurement of total sIEE yields, (b) spectroscopy of sPIE-ejected electrons

However, both kinds of data give merely qualitative informations and do not provide a straightforward access to the mechanisms responsible for electron emission. In this situation it is of interest to measure the actual number of electrons resulting from particular emission events. The socalled emission statistics (ES) corresponds to the set of probabilities W_n for ejection of any given number of n electrons and is related to the total emission yield γ.

$$\gamma = \sum_{n=0}^{\infty} n \cdot W_n \; ; \quad \sum_{n=0}^{\infty} W_n = 1 \qquad (1)$$

We have recently described in detail how such ES can be precisely determined for both ionized and neutral projectiles (Lakits et al. (1989a), Aumayr et al. 1991). A short explanation of the essential features will now be given with reference to fig. 2, which shows a setup especially well suited for measurements at low ion impact energy ($E \geq 30$ q eV, Lakits et al. 1990).

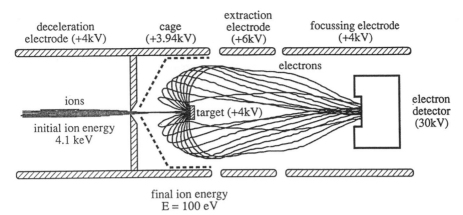

Fig. 2: Experimental setup for measuring slow ion-induced ES and γ data for impact of slow (≥ 30 eV) ions on a clean metal surface (Lakits et al. 1990).
The given electrode potentials refer to production of 100 eV singly charged ions by deceleration of a primary 4,1 keV ion beam.

The selected primary ions are directed into a detection unit within UHV environment, where they can be decelerated just before hitting the sputter-cleaned metal target situated inside a highly transparent conical electrode. The latter causes backward deflection of emitted electrons into an acceleration lens setup and finally a solid state detector in the geometrical shadow of the target. Emitted electrons thus hit the detector surface with an energy of typically 25 keV. Within the time resolution of the detector electronics (typically ≥ 1 ns), an emission of n electrons cannot be resolved into n single events but will just be registered as one single detector pulse corresponding to the n-fold energy of a single 25 keV electron, thus giving rise to a typical pulse height distribution as shown below in fig. 3. From such raw ES distributions relative values for W_n can be derived after proper corrections for detector resolution and background contributions (Aumayr et al. 1991). Absolute values for W_n including n = 0 can then be obtained by means of equ. 1, if the corresponding value for γ is already known. Otherwise, this data can be measured with the present setup in the following way.
The deceleration lens in front of the ES detector target has been designed such that the energy of primary ions can be varied without altering their flux, as could be proven in numerous trial experiments. Independent from ion deceleration inside the detector unit the ion beam intensity can be reproducibly attenuated by a selectable factor of up to 10^6, via a magnetic quadrupole lens in the ion beam line far upstream of the ES detector. Applying at first ion currents of the order of several nA with such impact energies that $\gamma \gg 0.1$ can be assured, the total electron yields can be determined from ion- and electron current measurements as indicated in fig. 1a. Subsequently the ion flux will be strongly attenuated to typically less than 10^3 particles/s, but can still be precisely measured by counting all electrons emitted from the target and using the already determined value of γ (see above). Since a once set ion flux can be kept fixed besides even strong ion deceleration, the measurement of emitted electron fluxes for various ion impact energies with the ion flux known, permits the corresponding γ data to be determined rather precisely. This can be achieved down to quite low impact energies where measurements by the standard current measurement techniques are unfeasible because of far too small ion and electron currents. In the here described way total electron yields of less than 10^{-3} could still precisely be measured (Lakits et al. 1990).

3. ELECTRON EMISSION INDUCED BY SINGLY CHARGED IONS - STUDYING PE BELOW AND ABOVE THE KE THRESHOLD

Raw ES data for impact of Ne+ ions on clean gold are shown in fig. 3, with the impact energy varied from 100 eV up to 16 keV. Note that Ne+ has no long-lived highly excited state (Winter 1982), because of which only Ne+ ground state ions can have been involved.

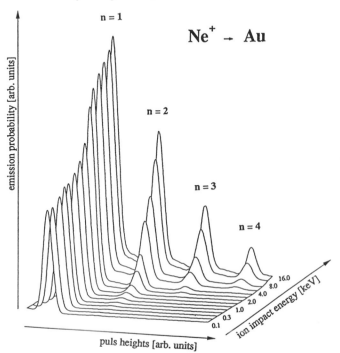

Fig. 3: Raw experimental data of ES for impact of Ne+ on clean polycrystalline gold

At the lowest impact energy (E ≤ 15 eV/amu) obviously only one electron is ejected, whereas at higher E also emission of 2, 3, etc. electrons is gradually appearing (Lakits et al. 1990). Consequently, as long as KE is not possible, only PE can contribute to the total electron yield by ejecting one electron, although the transferrable potential energy (W ≤ 21,6 eV) would suffice to emit up to three electrons. The here demonstrated results are supported by similar measurements for impact of He+, Ar+ (see below) and other singly charged ions on clean gold (Lakits et al. 1990). There is a fundamental difference between PE- and KE-related ES.

PE involves reasonably well defined transitions between electronic states of the target surface and the approaching projectiles, and therefore gives rise to emission of a correspondingly well defined number of electrons. KE, on the other hand, involves dissipation of projectile kinetic energy among a relatively large number of target electrons which are facing similar chances of being ejected. The number of these electrons increases with transferred kinetic energy, and thus with the projectile velocity, whereas the potential energy carried by the approaching ion is transferred via some Auger type electronic transitions to one electron only.

We stress that such measurements can be extended toward rather small γ values. For example, consider impact of Ar+ at low impact energy (cf. fig. 4a; a pure ground state Ar+ beam has been assured by adequate operation of the applied plasma ion source, see Hofer et al. 1982).

To identify the exact location of the KE threshold, the course of γ vs. impact velocity E as apparent from fig. 4a is not very revealing. However, in fig. 4b the ratio of relative ES probabilities for emission of n = 2 and 1 electrons respectively has been plotted. From the latter diagram the KE threshold impact velocity can be found in a straightforward manner, since the PE leads to emission of one electron only (see above discussion).

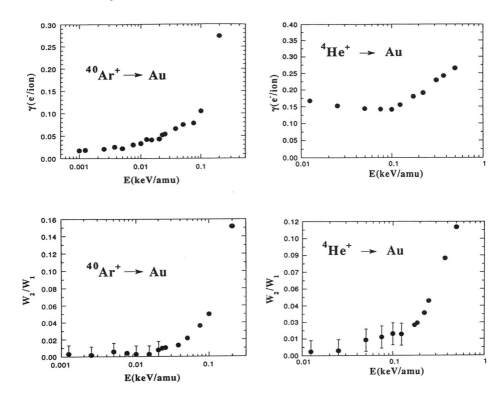

Fig. 4: (a) Total electron emission yields for impact of Ar+ ground state ions on clean gold vs. impact velocity. (b) Ratio of emission probabilities W_2/W_1 for impact of Ar+ on clean gold vs. impact velocity. Data from Töglhofer et al. (1991).

Fig. 5: As for fig. 4, for impact of He+ on clean gold. Data from Töglhofer et al. (1991)

An upper limit for the KE threshold should principally be found for that impact velocity where the kinetic energy transfer in head-on collisions from projectile particles onto quasi-free metal electrons in the solid surpasses the metal surface work function Φ. This threshold impact velocity would be about 300 eV/amu for a clean gold surface (Lakits et al. 1990) and thus considerably higher than the here determined KE threshold of about 15 eV/amu for Ar+, cf. fig. 4b. Similar results for He+ impact on clean gold are shown in figs. 5a,b and reveal a KE treshold impact velocity of less than 50 eV/amu.

The precise origin of kinetic electron emission at such rather low impact velocities is not well understood. We believe that it might be caused by autoionisation of highly excited quasimolecules which have been transiently formed in close encounters of projectile ions with single surface atoms.

Similar measurements have also been carried out with singly charged molecular ions.

As an example, figs. 6a,b show total yields and W_2/W_1 ratios for impact of atomic and molecular hydrogen ions on clean gold (Winter et al. 1991).

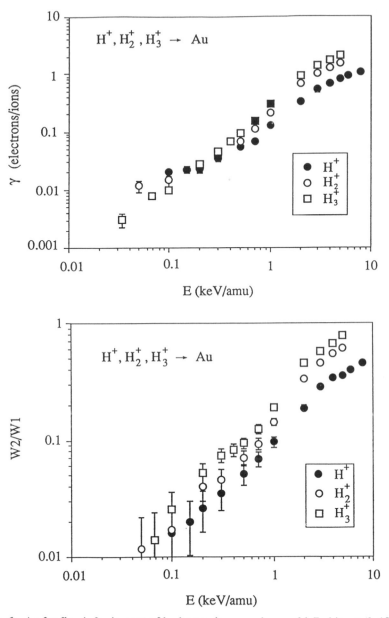

Fig. 6: As for fig. 4, for impact of hydrogen ions on clean gold (Lakits et al. 1989b).

For all three projectile species the PE is negligibly small. The KE gradually disappears around 50 eV/amu. We like to mention that the measured total electron yields as well as the ES for molecular ions could be well reconstructed by respectively summation or convolution of corresponding data measured for equally fast neutral and singly charged hydrogen atoms, if the same number of projectile electrons was taken care of (Lakits et al. (1989b), Winter et al. 1991). This is clear evidence for sizeable electronic shielding effects in KE induced by impact of molecular projectile ions, which can be seen as just another signature of the socalled "molecular effects" in KE processes (Hasselkamp 1988).

4. PE STATISTICS FOR IMPACT OF DOUBLY CHARGED IONS

Potential emission may become rather strong for impact of slow multicharged ions ("MCI") on a metal surface, due to the enhanced potential energy content of such projectiles (Delaunay et al. 1987a). ES measurements are very useful to elucidate the interplay of various electronic transitions governing such PE processes. According to Hagstrum (1954), electronic transitions involved in PE processes can engage one or more electrons. The only one-electron transition of interest is resonance neutralisation ("RN"), which does not directly result in electron emission but often acts as precursor for the then following electron-emitting two- or more-electron transitions. Auger deexcitation ("AD") can cause electron emission if the involved deexcitation energy is larger than the metal surface work function Φ. Auger neutralisation ("AN") may give rise to electron emission if the potential energy change in the neutralisation step surpasses 2Φ. Autoionisation ("AI") will take place after formation of a doubly (multiply) excited projectile particle via double (multiple) RN transitions ("DRN", "MRN"). The last process has first been identified by Hagstrum and Becker (1973) for impact of He^{2+} on Ni single crystal surfaces.

Similar processes could later be demonstrated for impact of multicharged ions (Zehner et al. (1986), Delaunay et al. (1987b), de Zwart et al. (1989), Folkerts and Morgenstern 1990).

Varga et al. 1982 have precisely measured ejected energy distributions for impact of doubly charged rare gas ions Ne^{2+} - Xe^{2+} on clean W. They identified for both Ne^{2+} and Ar^{2+} a group of very slow electrons which they related to two successive deexcitation steps on the way from X^{2+} toward the X^{+o} ground state along the two conceivable paths

$$X^{2+} \to RN \to X^{+*} \to AD(e^-) \to X^{+*'} \to AD(e^-) \to X^{+o} \qquad (2a)$$

or

$$X^{2+} \to AN(e^-) \to X^{+*} \to AD(e^-) \to X^{+o} \qquad (2b)$$

For actual occurence of electron-emission in the steps labelled by "(e^-)", sufficiently large deexcitation energies have to be made available (see above), which can be assured for the collision systems Ne^{2+}, Ar^{2+} - W (see also Varga 1988,1989).

However, recently measured electron energy distributions for keV grazing collisions of He^{2+}, Ne^{2+} and Ar^{2+} on clean Cu (Wouters et al. 1989) provided no evidence for transitions as described by equs. (2). These authors assumed that only the following transitions according to equs. (3), which each can give rise to at most one electron, are taking place in the X^{2+} - X^{+o} neutralisation steps.

$$X^{2+} \to DRN \to X^{o**} \to AI(e^-) \to X^{+o} \qquad (3a)$$

or

$$X^{2+} \to AN(e^-) \to X^{+o} \qquad (3b)$$

To clarify this situation, we have measured ES for impact of 100 eV He^{2+}, Ne^{2+} and Ar^{2+} on clean gold, cf. fig. 7. At 100 eV impact energy for all three collision systems an exclusive PE situation is given (see figs. 3 - 5). Emission of up to three electrons can unambiguously be seen for both He^{2+} and Ne^{2+}, but not for Ar^{2+} (note that a third electron can be emitted via AN of X^{+o} to the X^o ground state after conclusion of the transitions given by equs. 2, cf. previous chapter). Furthermore, for impact of $^3He^{2+}$ the dependences of W_2/W_1 and W_3/W_1 vs. ion velocity are shown in figs. 8a,b. With falling E both data go through a minimum and then rise again toward the lowest impact velocities covered by these measurements. This behaviour has to be caused by improving probabilities for the transitions described by equs. 2 in comparison to those covered by equs. (3).

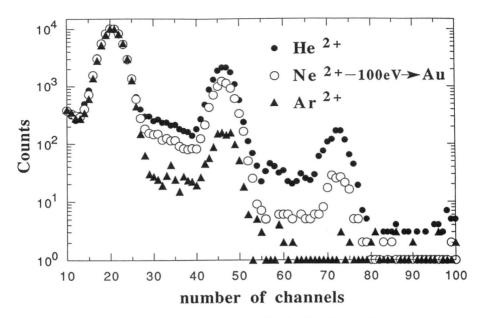

Fig. 7: Raw ES measured for impact of 100 eV $^3He^{2+}$, $^{20}Ne^{2+}$ and $^{40}Ar^{2+}$ on clean gold (from Töglhofer et al. (1991), all W_1 peaks have been normalized).

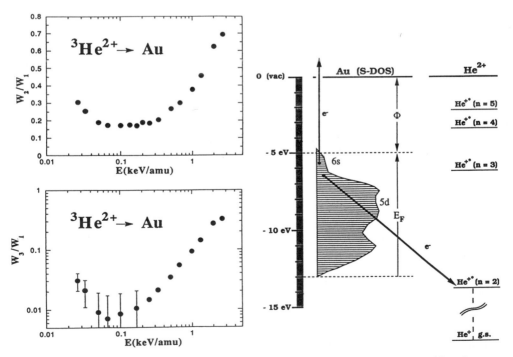

Fig. 8: Course of W_2/W_1 (a) and W_3/W_1 (b) vs. projectile velocity for impact of $^3He^{2+}$ on Au (Töglhofer et al. 1991).

Fig. 9: Density of states for a gold surface (Eastman 1970, Evans et al. 1974) vs. HeII binding energies

Further evidence for this behaviour is provided from the trend for the three projectile species apparent in fig. 7. For He^{2+} neutralisation (see also fig. 9), the first deexcitation step according to equs. (2) can involve at most 13,6 eV and for both Ne^{2+} (deexcitation into Ne^+ $2s2p^6\,^2S$) and Ar^{2+} (deexcitation into Ar^+ $3s3p^6\,^2S$) at most 14,1 eV (cf. Varga et al. 1982). While these first deexcitation steps are almost equally large for all three ion species, binding energies of the related X^{+o} ground states differ strongly (54,4 eV for He, 41,0 eV for Ne and 27,6 eV for Ar, respectively). In first approximation, the transition matrix elements for electron emitting processes described by equs. (3) should be the larger, the smaller the involved X^{2+} - X^{+o} energy differences. Keeping this in mind, we can explain the decreasing probability for W_3/W_1 by the competition from processes following equs. (3), which is increasingly successful when changing the projectile species from He^{2+} via Ne^{2+} to Ar^{2+}.

Whereas Varga et al. (1982) identified a slow electron group also for Ar^{2+} - W collisions, we could not find an equivalent in the emission of up to three electrons for impact of Ar^{2+} on Au. This may be caused by the difference in surface-density of states ("S-DOS") for both target species. Tungsten features a surface work function of $\Phi = 4,5$ eV and a high S-DOS just below the Fermi level (Feuerbacher and Christensen 1974), whereas for gold (cf. fig. 9) Φ is about 5 eV and the S-DOS much less pronounced in the vicinity of the Fermi level. These differences could explain a relative decrease of the probabilities for electron emission due to the first deexcitation processes described by equs. (2) in comparison with the electron emission probabilities covered by equs. (3).

As soon as the total electron yields have been precisely determined (respective measurements are in progress), we can isolate transitions for the various electron-emitting processes discussed above. For this purpose, balance equations for all identified transitions will be used in a similar way as for the analysis of ejected electron energy distributions (Wouters et al. (1989), Zeijlmans van Emmichoven and Niehaus (1990), Folkerts and Morgenstern (1989), Schippers et al. 1991). Informations from impact velocity-dependent ES measurements will be especially attractive, since the number of involved electron-emitting processes can be directly determined.

5. PE STATISTICS FOR MULTICHARGED ION IMPACT

For impact of slow multicharged ions on metal surfaces both a large number of PE-related electronic transitions and a partial interrelation of PE- and KE processes may take place. In such a situation ES studies can again be very useful.

Fig. 10a shows ES measured for impact of Ne^{q+} (q= 0,1,2,3) on clean gold (Lakits 1989). If an MCI can become completely neutralized before hitting the surface, the then initiated KE will be produced by the already neutral projectile and therefore the ES of the previously completed PE sequence will not be entangled with the KE-related ES. In such a situation the latter can be precisely determined with originally neutral projectiles. With this information at hand, the ES for PE can then be easily obtained by folding an ES measured for the neutral species out of the ES measured for equally fast ions in the charge state of interest. Fig. 10b shows, however, that such a deconvolution procedure works only properly for Ne^{q+} impact energies up to 6 keV (corresponding to a projectile velociy limit $v_l \leq 2,4 \times 10^5$ m/s). At higher v, the appearance of "negative probabilities" in the deconvoluted ES signals a beginning interrelation of PE- and KE contributions or in other words, at $v > v_l$ the MCI can no more become completely neutralized before hitting the surface. Therefore, determination of v_l delivers the necessary neutralisation time $t_n \geq d_q/v_l$, where d_q is the distance from the surface within which electronic transitions between projectile and surface are possible with nonvanishing probability. Apell 1987 has derived for d_q about 3q (Angstroms). Recently, Bardsley and Penetrante 1991 performed classical Monte Carlo calculations on electron emission due to impact of highly charged ions on metal surfaces, from which they concluded that the calculations of Apell 1987 may predict somewhat too large values for d_q.

Similar ES measurements as for Ne^{q+} have also been performed for Ar^{q+}, leading to comparable values for v_l. It is of special interest that ES measured for a q-fold charged projectile could not be meaningfully unfolded with ES measured for equally fast ions in any lower charge state $q' \geq 1$. This is a clear hint that the involved neutralisation sequences circumvent the lower charged ground states, but rather undergo MRN transitions with subsequent electron production via multiple AI processes.

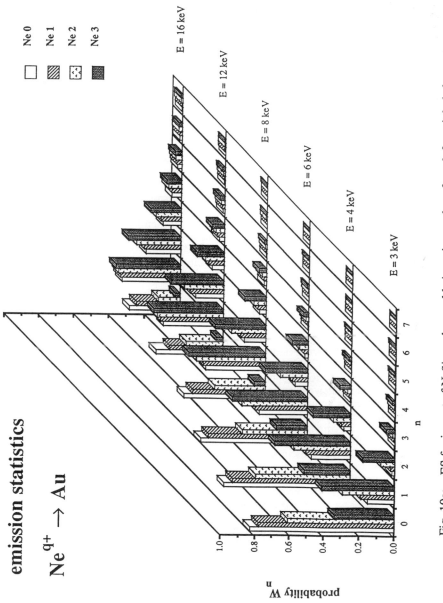

Fig. 10a: ES for impact of Ne^{q+} on clean gold. At a given n, bars from left to right belong to q = 0, 1, 2, 3, respectively. Data from Lakits (1989).

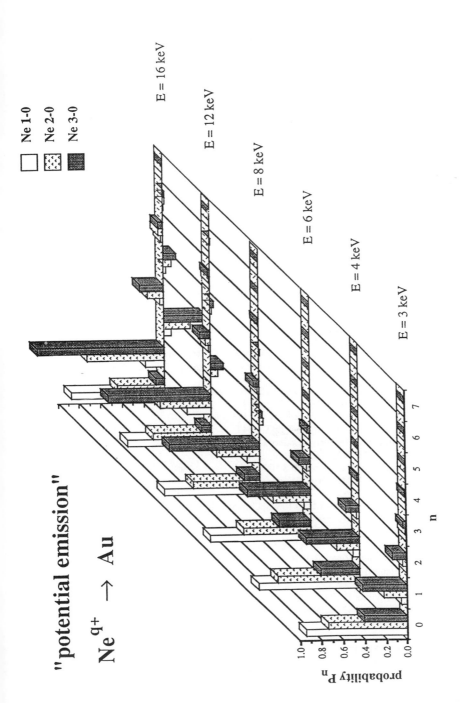

Fig. 10b: ES for exclusive potential emission due to differently charged ions, as obtained by an unfolding procedure. At a given n, bars from left to right belong to q = 1, 2, 3, respectively. Note the onset of "negative probabilities" above E = 6 keV.

6. SUMMARY AND OUTLOOK

In the present paper we have demonstrated recent progress in studying slow ion-induced electron emission from a clean metal surface, which could be achieved by measuring and evaluating the statistics of emitted electrons (ES). Such ES measurements permit the correct separation of potential and kinetic effects and show their fundamental difference quite clearly. The threshold impact velocity for onset of kinetic emission can be precisely determined and it has also become possible to study PE at impact energies above the KE threshold. Armed with this new ES technique we could also clarify some open questions regarding potential emission induced by doubly charged rare gas ions. The most interesting applications are still to come when electron emission by slow highly charged ions will be investigated, for which purpose the ES method is especially well suited because of its sensitivity and supplementary informations to the ones already obtained by techniques commonly applied in this field.

Acknowledgments

This work was supported by Austrian Fonds zur Förderung der wissenschaftlichen Forschung, project no. P 6381, and by Kommission zur Koordination der Kernfusionsforschung in Österreich at the Austrian Academy of Sciences. Useful discussions with Drs. F. Aumayr and G. Lakits and with Mrs. K. Töglhofer are gratefully acknowledged.

References

Andrä H J 1989 Nucl.Instrum.Meth.Phys.Res. B **43** 306
Apell P 1987 Nucl.Instrum.Meth.Phys.Res. B **23** 242
Aumayr F, Lakits G and Winter H 1991 Appl. Surface Sci. **47** 139
Bardsley J N and Penetrante B M 1991 Proc. Symposium on Surface Science, Obertraun/Austria, February 10 - 16, 1991, p. 3 and private communication
Delaunay M, Fehringer M, Geller R, Hitz D, Varga P and Winter H 1987a Phys.Rev. B **35** 4232
Delaunay M, Fehringer M, Geller R, Varga P and Winter H 1987b Europhys. Letters **4** 377
Eastman D E 1970 Phys.Rev. B **2** 1
Evans S, Evans E L, Parry D E, Tricker M J, Walter M J and Thomas J M 1974 Proc. Faraday Disc.Chem.Soc. **58** 97
Feuerbacher B and Christensen N E 1974 Phys.Rev. B **10** 2373
Folkerts L and Morgenstern R 1989 J. de Physique **50** C1-541
Folkerts L and Morgenstern R 1990 Europhys. Letters **13** 377
Hagstrum H D 1954 Phys.Rev. **96** 325; 336
Hagstrum H D and Becker G E 1973 Phys.Rev. B **8** 107
Hasselkamp D 1988 Comments At.Mol.Phys. **21** 241
Hofer W O 1990 Scanning Micr. Suppl. **4**
Hofer W, Vanek W, Varga P and Winter H 1982 Rev.Sci.Instrum. **54** 150
Lakits G 1989 Thesis, Technische Universität Wien (unpublished)
Lakits G, Aumayr F and Winter H 1989a Rev.Sci.Instrum. **60** 3151
Lakits G, Aumayr F and Winter H 1989b Europhys. Letters **10** 679
Lakits G, Aumayr F, Heim M and Winter H 1990 Phys.Rev. A **42** 5780
Schippers S, Oelschig S, Heiland W, Folkerts L, Morgenstern R, Eeken P, Urazgil´din I F and Niehaus A 1991 Proc. Symposium on Surface Science, Obertraun/Austria,February 10 - 16, 1991, p. 285 and to be published
Töglhofer K, Aumayr F and Winter H 1991, to be published
Varga P, Hofer W and Winter H 1982 Surface Sci. **117** 142
Varga P 1987 Appl.Phys. A **44** 31
Varga P 1988 in Electronic and Atomic Collisions, ed H.B. Gilbody et al, Elsevier, p. 793
Varga P 1989 Comments At.Mol.Phys. **23** 111
Winter H 1982 Rev.Sci.Instrum. **53** 1163
Winter H, Aumayr F and Lakits G 1991 Nucl.Instrum.Meth.Phys.Res. B (in print)
Wouters P A A F, Zeijlmans van Emmichoven P A and Niehaus A 1989 Surface Sci. **211/212** 249
Zehner D M, Overbury S H, Havener C C, Meyer F W and Heiland W 1986 Surface Sci. **178** 359
Zeijlmans van Emmichoven P A and Niehaus A 1990 Comments At.Mol.Phys. **24** 65
de Zwart S T, Drentje A G, Boers A L and Morgenstern R 1989 Surface Sci. **217** 298

Neutralization of ions at metal surfaces

I.Urazgil'din[#] , P.Eeken, A.Niehaus

University of Utrecht, Princetonplein 5, 3508 TA Utrecht, NL
[#]Moscow State University, Moscow 119899, USSR

ABSTRACT: It is shown that direct information on electronic processes leading to neutralization of ions when they slowly approach metal surfaces can be obtained from the energy spectra of electrons emitted in the various possible spontaneous ionization processes. This is demonstrated for the example systems He^{++} / Pb(111) and Ar^{++} / Pb(111). The information contained in the spectra obtained at vertical energies of the order of electron volts is retrieved by comparison with model calculations. The analysis in terms of the model leads to a complete reconstruction of the neutralization during the ion-surface approach.

INTRODUCTION

When atomic ions approach a metal surface at velocities small compared to the Fermi-velocity of the metal electrons, the ion- surface interaction leads to various electron exchange- and Auger-type electron emission processes. This is known to be true for singly charged ions since the early work of Hagstrum (1954), and has been found recently to also hold for multiply charged ions (Varga et al 1982, de Zwart 1987, Zeijlmans van Emmichoven et al 1988). The electrons are "spontaneously" ejected, at a certain projectile- surface distance, and their energies, therefore, are determined by the "vertical" energy separation of the relevant initial- and final electronic states of the projectile-metal system. This description suggests the formulation of a quantitative model for the calculation of electron spectra, based on rate equations, using model functions of the distance for the quantities mentioned. Such a model has recently been formulated by us (Zeijlmans van Emmichoven et al 1988, Zijlmans van Emmichoven and Niehaus 1990), and has been applied to the analysis of He^{++}/ Cu(110) spectra. Later measurements for this system at lower "vertical" collision energy, and measurements for the systems discussed in this contribution, however, suggested the necessity of a further refinement of the model. In what follows, we will give an outline of the analysis and of experimental electron spectra using such a refined model.

ANALYSIS OF ELECTRON SPECTRA

Crucial for the application of the model is a scheme of the various transitions considered to be important. For the case of collisions of singly charged ions with metal surfaces, the general scheme is

$$A^+ + M \xleftarrow{\text{RT}} A^* + M^+ \tag{1}$$
$$\xrightarrow{\text{AC}} A + M^{++\#} + e^-$$

with (RT) indicating a resonance one electron transition between ion (A^+) and metal (M), and (AC) and (PI) indicating the "Auger-type" two electron transitions which, in principal, can lead to electron emission. The Auger capture (AC) involves two metal electrons, while the "Penning-ionization" involves one metal electron and one atomic electron. The energy of an ejected electron depends on the recombination energy of the atomic "hole" filled in the transition, and on the energy by which the metal electrons were bound before the transition. In case of (PI), therefore, the spectrum of emitted electrons directly reflects the density of electronic states of a certain binding energy, the so called "SDOS", while the spectrum caused by (AC) approximately reflects a "self convolution" of the (SDOS), because two metal electrons are involved.

The number of possible reaction pathways increases considerably for multiply charged ions approaching a metal surface. Apart from the fact that more "holes" and therefore more successive steps of the Auger-type processes (PI) and (AC) are possible, also the process of atomic auto ionization (AU) following the capture of two or more electrons becomes possible. The energy of the electrons ejected in this process is, in contrast to the processes (AC) and (PI), well defined, leading to lines in the spectra.

For the analysis of the Ar^{++}/ Pb- spectra we use the following scheme:

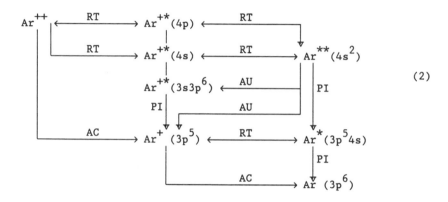

$$(2)$$

The selection of states considered is mainly dictated by energy considerations: population of an atomic state by (RT) is only possible, if the one electron atomic binding energy is at resonance, at least for part of the collision, with one electron binding energies in the metal for which the (SDOS) is non zero. Instead of the many states deriving from the configurations Ar^{+*} $(..3s^2 3p^4 (^3P)3d,4s,4p)$, only the two symbols Ar^{+*} (4p) and Ar^{+*} (4s) are indicated as representative for the two energy groups which may roughly be distinguished. (RT)'s that correspond to electron loss, indicated by arrows pointing to the left, are assumed to occur when the atomic binding energy is resonant with empty one electron metal states above the Fermi-level.

For the He^{++}/ Pb- system, we use a scheme which differs from the one used in our earlier analysis of spectra of the He^{++}/ Cu(110) system. In analogy to scheme (2), we now also consider contributions from states He^{+*}, which, in the the He^{++}/ Cu(110) spectra obtained at higher vertical energies, were not identified and therefore omitted in the analysis. Only He^{+*} (n=2) states need be considered, because higher states will decay at large distances by (RT). The nearly degenerate He^{+*} (n=2) levels are split by the first order Stark-effect in the field of the image charge into three states, which we label as $(2s,p)\Sigma,\Pi,\Sigma^*$. The scheme used for the present analysis is the following:

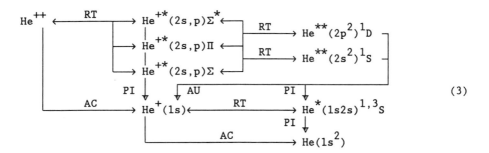

$$\text{(3)}$$

Measured electron spectra for four different vertical energies are shown in figures 1 and 2., together with theoretical spectra calculated on the basis of the schemes (2) and (3). The spectra are obtained at a detection angle perpendicular to the projectile beam and in the plane containing the surface normal. The energy resolution of the electron detector used was chosen low enough to not influence the widths of the observed structures. The transmission function of the detector is believed to be approximately constant above ca.5 eV, and to decrease strongly towards zero energy due to spurious fields.

Fig.1: Measured and calculated electron spectra for the system Ar^{++}/ Pb(111). Measurements and calculations are performed for the Ar^{++}- beam energies given, and for an angle of incidence between beam and surface of two degrees. Below ca.5eV the measured spectra are influenced by the strongly decreasing transmission of the detector. The main structure, and its dependence on collision energy, is seen to be well reproduced in the calculations in which the processes of scheme (2) are considered. The peaks are due to (AU), the edge at ca. 8eV is due to (PI) by $Ar^{+*}(3s3p^6)$, and the hump around 4 eV is caused by (AC) of Ar^+. The towards low electron energies rising background is due to (AC) of Ar^{++}.

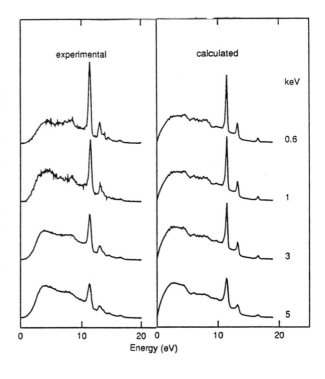

In principle, the model we used to calculate electron spectra is the same as the one developed earlier in our group (Zeijlmans van Emmichoven et al 1988, Zeijlmans van Emmichoven and Niehaus 1990). We only indicate its general structure. The basic ingredients are the following.

Fig.2: Measured and calculated electron spectra for
the system He^{++}/ Pb(111).
Measurements and calcula-
tions are performed for
the angles of incidence
between the He^{++}- beam and
the surface given, and for
a beam energy of 1 keV.
The vertical energies that
correspond to the angles
of 2, 10, 20, and 45
degrees are, respectively,
1.2, 30, 117,and 500 electron volt. The calcula-
tions are based on the
scheme (3). The peaks are
due to (AU), the low
energy hump is caused by
(AC) of He$^+$, and the
quasi-continuous back-
ground is due to (AC) of
He^{++} and to (PI) of He^{+*}.
For increasing vertical
collision energy the (AC)-
processes become dominant.
(AU) at low vertical
energy is largely due to
decay of He** after
reflection.

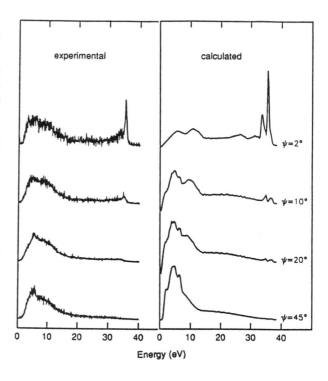

(i) The projectile-surface motion is described using a planar potential
V(z), yielding, for a given vertical collision energy $E_\perp$ - E_{beam} * $sin^2(\psi)$,
a defined vertical velocity v(z).

(ii) The various energy levels $E_k(z)$ relevant for the description of the
time evolution of the electronic system are taken to be well defined
functions of the distance (z).

(iii) The electronic transitions among the various electronic states are
described by transition probabilities $G_{ik}(z)$, which also are assumed to be
well defined functions of (z).

(iv) The kinetic energy of the emitted electron is taken to be given by the
"vertical" difference between electronic states: $\epsilon_{ik}(z)$ - $E_i(z)$- $E_k(z)$,
with $E_k'(z)$ the final state electronic energy for the system with one
electron less than in the initial state.

(v) As in the original version of the model, the possibility of a "non
adiabatic response" of the metal to the sudden charge changes arising in
the Auger-type processes is incorporated by interpolating between an
"adiabatic" final state energy, leading to an electron energy ϵ^a, and a
"diabatic" final state energy,leading to an electron energy ϵ^d,as

$$\epsilon = \epsilon^a - (\epsilon^a - \epsilon^d) * \exp(-t/\tau).\tag{4}$$

In contrast to the original version, however, the "response time" (τ) is not introduced as a free parameter, but is estimated from the surface plasmon frequency (ω) as $\tau = \omega^{-1}$. We approximate the "ionization time" as $t = r_s[2\epsilon^a]^{-1/2}$, with r_s the radius of the spherical volume available to one electron of the conduction band of the maetal. This leads to the estimate

$$t/\tau \cong \left(\frac{3}{4\, r_s\, \epsilon^a} \right)^{1/2}.\tag{5}$$

Based on the simplified description implied by (i)-(v), the electron spectrum can be calculated if the functions $v(z)$, $E_k(z)$, and $G_{ik}(z)$ are known. The procedure we followed to calculate spectra is briefly outlined below.

First, the distance dependent probabilities for the population of the various electronic states, $N_k(z)$, are calculated numerically by solving the relevant set of coupled differential equations,

$$\frac{dN_i(z)}{dz} = \frac{\sum_{k \neq i} (\, G_{ki}(z)\, N_k(z) - G_{ik}(z)\, N_i(z))}{v(z)}.\tag{6}$$

Then, the electron spectrum due to the subset of Auger-type transitions, l $\longrightarrow$ m, say, is obtained by adding contributions from all distances and from all processes (lm), taking into account the individual transition energies $\epsilon_{lm}(z)$. Due to the presence of a *distribution* of binding energies (E_b) in the metal, different consequences concerning the transition probabilities arise: (i) In case of (RT), transitions are possible in a *range* of distances, in contrast to (RT) in atomic collisions. (ii) In case of (PI), the probability becomes an *integral* over binding energies:

$$G_{lm}(z) = \int_0^{E_F} g_{lm}(z, E_b)\, P(E_b)\, dE_b\ ,\tag{PI} \qquad (7)$$

with $P(E_b)$ being the (SDOS). (iii) In case of (AC), where two electrons with binding energies E_b^1 and E_b^2 are involved, the probability is a *double integral*:

$$G_{lm}(z) = \int_0^{E_F} g_{lm}(z, E_b^1, E_b^2)\, P(E_b^1)\, P(E_b^2)\, dE_b^1\, dE_b^2\tag{AC} \qquad (8)$$

The contribution of a process l $\rightarrow$ m to the electron spectrum at the well defined energy $\epsilon(z, E_b^1, E_b^2)$ is given by

$$P(\epsilon)_{lm} = \left| \frac{g_{1m}(z, E_b^1, E_b^2) \; P(E_b^1) \; P(E_b^2)}{v(z) \; |d\epsilon_{1m}(z)/ \; dz|} \right|_{z \; = \; z_{1m}(\epsilon, E_b^1, E_b^2)}, \tag{8}$$

and the spectrum is obtained by appropriately summing over contributions belonging to different binding energies. In the original version of the model, the g_{1m} were taken to be independent of (E_b). As a refinement, we now allow g_{1m} for the processes (PI) and (AC) to depend on the binding energy of the "down" electron, i.e., the electron that recombines with the core hole of the atom.

For the actual numerical calculations shown in figures 1 and 2, the following model functions were used:

(i) $V(z) = 2\pi n Z_1 Z_2 a* \; 1.17 \; \exp(-0.3(z+2.5)/a) - f*q^2/4$ \hfill (9)

$a = 0.8853/(\; \sqrt{Z_1} + \sqrt{Z_2} \;)^{2/3}$, $f = (1-\exp(-z))/z$

with Z_1 and Z_2 the atomic numbers of projectile and target atoms, (n) the surface density of target atoms, q the projectile charge state, and (f) a factor that replaces the reciprocal distance from the image plane(z^{-1}), and accounts for the "saturation" at small (z) of the image charge induced by Z_p. The distance between first atomic layer and image plane is taken to be 2.5 atomic units.

(ii) $E_i(z) = E_i(\infty) - Z_{eff}^2*f/4- \dfrac{(\alpha/2)}{(1+\exp(-(z-1)/1\;)} * Z_{eff}^2*(f/2)^4$ \hfill (10)

$Z_{eff} = q + r*\exp(-z/ \; l_r)$

with (α) the polarizability of the (r) excited electrons, which are polarized in the field caused by the image charge of the effective charge Z_{eff}. The screening distance l_r for the excited electrons is estimated from "Slaters rules".

(iii) $G_{ik}(z)_{RT} = const(RT)*P(E_b)*\exp(-\gamma(z)_{RT}* \; z)$ \hfill (11)

$\gamma(z)_{RT} = \sqrt{2E_b(z)}$

with $E_b(z)$ the binding energy at which a resonant transition is possible, $P(E_b)$ the probability for E_b in the (SDOS), and const(RT) a free parameter of the model.

(iv) $g_{1m}(z)_{PI} = const(PI)*\exp(-\gamma(z)_{PI}* \; z)$ \hfill (12)

$\gamma(z)_{PI} = \left(\dfrac{1}{2\sqrt{2E_b}} + \dfrac{1}{2\sqrt{2RE(z)}} \right)^{-1}$

with RE(z) the recombination energy of the atomic core hole filled in the transition, and const(PI) a free parameter.

(v) $g_{1m}(z)_{AC}$= const(AC)*exp($-\gamma(z)_{AC}$* z)   (13)

   $\gamma(z)_{AC}$= $\gamma(z)_{PI}$

withconst(AC) a free parameter.

(vi) $G_{1m}(z)_{AU}$ = const(AU)   (14)

const(AU) is a free parameter for the Ar^{**}-stateconsidered, and is taken to be given by the known atomic lifetimes for He^{**}.

(vii) ϵ^a - ϵ^d= f*$\left(Z_{eff,m}q_m + \dfrac{(Z_{eff,1}-Z_{eff,m})^2}{4}\right)$ (15)

with q_m the number of positively charged "holes" created in the metal conduction band in the final state of the transition.

(viii) The Stark splitting of the He^{+*}(n=2) states in the field of the image charge we take to be given by

$\Delta(\Sigma-\Pi)$ - $\Delta(\Pi-\Sigma^*)$ = $(3/8)*f^2$ (16)

(ix) The (SDOS) was taken from literature.

In the case of Ar^{++} projectiles, in the calculation it was taken into account that the beam contained, in addition to $Ar^{++}(3s^23p^4)^3P$, also small amounts of Ar^{++} in the metastable states ^1D, and ^1S. The same reaction scheme (2) was assumed to be valid for all three states.

In the numerical calculations the free parameters defined above were varied and determined by trial and error to yield best agreement with the measured spectra, and with the change of the spectra with vertical collision energy. Two changes of the "physics" incorporated into the model had to be made to avoid obvious discrepancies.(i) the diabatic correction of the emitted electron energy implied in relations (4,5,15) led, in the case of the (PI) Ar^{+*}(3s3p^6), to a contribution to the spectrum not having the observed sharp high energy edge. In the calculations, therefore, the diabatic correction was omitted in this case. A possible explanation for the different behaviour is that, the ionization is less localized due to a dominance of the "radiative" part of the matrix element. (ii) For the He^{++}/ Pb system, the assumed "distance between the first surface layer and the image plane", introduced in connection with the potential (9),had to be reduced.In the calculations a value of 1.5 atomic units was used. The values of the free parameters determined are given in table 1 (atomic units are used).

System	constant factors of probability functions detd.			
	RT	PI	AC	AU
He^{++}/ Pb	0.01	0.001	0.10	-
Ar^{++}/ Pb	0.01	0.015	0.20	$5*10^{-4}$

Table 1.
Together with the corresponding functions, the determined constants allow to describe the rather complex neutralization events during the ion-surface approach. As an example we show in figure 3. for the Ar^{++}/ Pb system the

Fig.3: The population of states, considered in the analysis (see scheme (2)), as it results from adapting the free parameters in the model functions by comparison with the measured spectra. The sum of the population numbers shown as curves, is normalized to unity. For clarity two panels are used. The curves represent the population "on the way in" for a collision at 1 keV, and at an angle of incidence between beam and surface of 2 degrees. At these conditions the neutralization is seen to be complete at the turning point, which is seen to lie slightly above the image plane. The rather complex multistep neutralization takes place in a rather narrow region around 5(a.u) above the image plane. This region shifts inward with vertical energy.

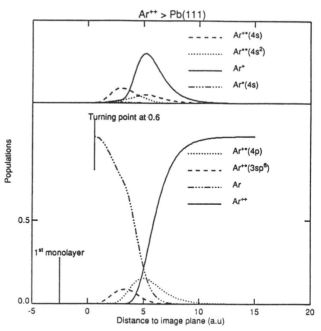

Fig.4: Transition probabilities for electronic processes occurring in Ar^{++} / Pb collisions. The probabilities follow from adapting calculated- to measured electron spectra, using the model described in the text. (RC) indicates one electron resonance capture. (RC $4s^2$) indicates capture of the "second" electron, namely, $Ar^{+*}(4p) \longrightarrow Ar^{**}(4s^2)$. As expected, the (AC) process involving two metal electrons is considerably less important than the other processes at large ion-metal distances, while close to the image plane (AC) and (AU) have similar probabilities, atomic- and metal electrons becoming indistinguishable.

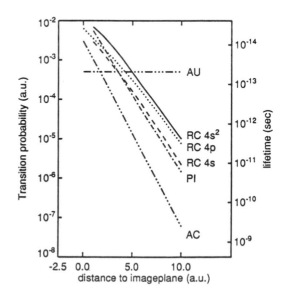

distance dependent population of the various states for the case of a vertical energy of 1.2 eV (corresponding to 1 keV and 2°). In figure 4. the distance dependent transition probabilities for the Ar^{++}/ Pb system are plotted. Finally, we show in figure 5. how the spectrum due to Ar^{++} in the ground state (^{3}P) is composed of contributions from the various spontaneous ionization events occurring in such collisions.

Fig.5: Decomposition of the calculated spectrum for Ar^{++}(3s^{2}3p^4) ^{3}P/ Pb at 1 keV and 2° incidence angle into the contributions from the various spontaneous ionization processes. It is clearly seen that the (PI) processes reflect the (SDOS) of lead, containing the fine structure split band below the Fermi-level due to the (6p^2) electrons, and the band due to the (6s^2) electrons at higher binding energies. The distortion of the reflected (SDOS) by distance dependent level shifts is smallest for the contribution from (PI) by Ar^{+*}(3s3p^6). All structure in the spectra becomes less pronounced at higher vertical energies due to the fact that the processes occur at smaller distances where the distortion by level shifts becomes more important.

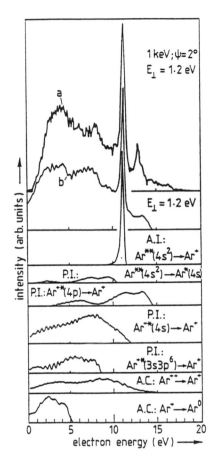

CONCLUSION

The agreement between calculated and measured spectra, as well as between their calculated and measured energy dependence, we consider to be very good in view of the amount of observed "structure" that has to be reproduced, and the number and type of parameters that are free to adapt. We therefore believe that the "physics" which underlies the model is essentially correct. On the other hand it should be clear that, in view of the complexity of the systems studied, the model as formulated here still has to be regarded as a first attempt of a description. More, and more accurate and specific experimental studies are necessary and may well lead to further refinements and modifications. The main value of the model is certainly that, its successful application to the analysis of experimental spectra yields all the information necessary for a detailed description of the time evolution of real ion-metal systems. For instance, once the free parameters for the He^{++} / Pb(111) system have been determined, electron spectra arising for collisions with Pb(111) can be predicted for any , not too high, vertical collision energy, and for any initial condition, namely, for collisions of He^+, He^{+*} (n=2), or $He^*(2^3,^1S)$.

In connection with practical applications it is especially interesting that it appears to be possible to retrieve from experimental spectra in well selected cases the (SDOS) of metals and its modification by adsorbates. In contrast to the information on the (SDOS) obtainable from photo- or X-ray induced electron spectra, which reflect the electronic states averaged over the first layer and several deeper layers ,this information would be related to electronic states of the first layer only, and therefore more relevant for the description of gas phase- surface interactions.

REFERENCES

- H.D.Hagstrum, Phys. Rev. **96** (1954) 336
- P.Varga, W.Hofer, and H.P.Winter, Surf. Sci. **117** (1982) 142
- P.A.Zeijlmans van Emmichoven, P.A.A.F.Wouters, and A.Niehaus, Surf. Sci. **191** (1988) 115
- P.A.Zeijlmans van Emmichoven and A.Niehaus, CAMP **24** (1990) 65
- S.T.de Zwart, Nucl. Instr. Meth. **B 23** (1987) 239

Multiphoton processes in atomic hydrogen

R M Potvliege† and Robin Shakeshaft‡

† Physics Department, University of Durham, Durham, DH1 3LE, England

‡ Physics Department, University of Southern California, Los Angeles, CA 90089-0484

Abstract. We present results of recent nonperturbative Floquet calculations of a.c. Stark shifts and rates for multiphoton ionization by intense laser fields.

The development of powerful pulsed lasers has spurred considerable interest in the study of the interaction of atomic systems with strong oscillating fields, [1] and in particular the study of phenomena that involve the absorption and emission of many photons and cannot be analyzed properly in perturbation theory, such as above-threshold ionization (ATI), [2] multi-ionization, [3] Stark shift induced resonances, [4] high-order harmonic generation, [5] or beyond the multiphoton regime, barrier suppression ionization. [6] Observations of these processes in experiments has stimulated the rapid progress in theory and the computational techniques that has been made these last few years on the way towards accurate *ab initio* nonperturbative calculations of measurable quantities.

[1] For a representative cross-section of current research in the field, see the April 1990 issue of Journal of the Optical Society of America B (a special issue on the Theory of high-order processes in atoms in intense laser fields, edited by K. Kulander and A. L'Huillier), and the Proceedings of the 5th International Conference on Multiphoton Processes (1990) edited by G. Mainfray and P. Agostini (write to Dr. G. Mainfray, DSM, Service de Physique Atomique, CEN Saclay, 91191 Gif-sur-Yvette Cedex, France, to order copies of these Proceedings). The forthcoming special issue of Advances in Atomic, Molecular, and Optical Physics devoted to multiphoton ionization (edited by M. Gavrila, 1991) will also contain reviews of the earlier literature.

[2] Stimulated free-free photon absorption by the photoelectron, leading typically to a photoelectron energy spectrum consisting of a sequence of peaks separated by $\hbar\omega$, where ω is the angular frequency of the laser field. This process (Agostini *et al* 1979) is similar to free-free absorption by scattered electrons (Andrick and Langhans 1976, Weingartshofer *et al* 1977).

[3] Multicharged ions are easily produced in rare gases by surprisingly weak fields; for example, Xe^{5+} ions are formed at 532 nm at intensities below 10^{13} W/cm^2 (L'Huillier *et al* 1983).

[4] Rates for multiphoton ionization, at a *fixed* frequency, can be dramatically enhanced by resonances which appear at intensities where the detuning between the Stark-shifted energy levels is an integral multiple of the photon energy (Gontier and Trahin 1973, Freeman *et al* 1987).

[5] For example, up to the 33rd harmonic of the Nd:YAG laser (1064 nm wavelength) can be produced in argon in the 10^{13} W/cm^2 range, and up to the 53rd harmonic of a Nd-Glass laser (1053 nm wavelength) has been detected for an incident intensity of about 5×10^{14} W/cm^2 (L'Huillier *et al* 1991).

[6] This process occurs in intense fields when the saddle point of the potential barrier is so lowered that the valence electron(s) can tunnel through it or flow over its top (Augst *et al* 1989, Meyerhofer 1990, Shakeshaft *et al* 1990).

We are mainly concerned, here, with the Floquet approach to multiphoton processes, in which the time-dependent Schrödinger Equation is reduced to an eigensystem of time-independent equations, the temporal variations of the intensity of the incident radiation being neglected in the first instance. [7] Theoretical results obtained within this framework can be compared to experimental data through a two-stage process. First of all, rates for ionization by a monochromatic field are calculated for (many) different intensities. These rates are then integrated over the temporal and spatial intensity distribution of a typical laser pulse, to produce a macroscopic ionization yield or a photoelectron spectrum (figure 1). [8] Clearly, it is not possible, or is very difficult, in the Floquet approach to take into account effects that depend critically on the temporal variation of the intensity, or to describe the atom if it ionizes within only a few cycles of the laser field. [9] Fully time-dependent calculations, which can accommodate realistic temporal profiles of the femtosecond pulses used in certain experiments, are in order in these cases. [10]

Let us consider a one-electron atom or ion with reduced mass μ undergoing multiphoton ionization under the effect of a classical, single-mode monochromatic [11] field of polarization vector $\hat{\epsilon}$, that is homogeneous over atomic dimensions, and has the electric field vector $\mathbf{F}(t)$ and vector potential $\mathbf{A}(t)$:

$$\mathbf{F}(t) = F_0 \, \mathrm{Re}[\hat{\epsilon} \exp(-i\omega t)] \qquad \mathbf{A}(t) = (c/\omega) \, F_0 \, \mathrm{Im}[\hat{\epsilon} \exp(-i\omega t)]. \tag{1}$$

In the Floquet approach, one seeks solutions to the Schrödinger Equation of the form

$$\Psi(\mathbf{r}, t) = e^{-iEt/\hbar} \sum_{N=-\infty}^{\infty} e^{-iN\omega t} \psi_N(\mathbf{r}). \tag{2}$$

[7] Reviews of the theory and of applications of this method in atomic physics can be found e.g. in Chu (1989) and Manakov *et al* (1986). The ansatz (2) is implicitly used in time-independent perturbation theory, and in any nonperturbative approach that is cast within a time-independent framework. Please see Potvliege and Shakeshaft (1991) for a more complete review of our work.

[8] When calculating realistic angular and energy distributions, one should allow, unless the laser pulse is very short, for the ponderomotive distortion in the trajectories of the ejected electrons as they pass through the focal region on their way to the detector. Indeed, a free, "slow", classical electron moves through an inhomogeneous optical field as if accelerated away from the regions of large intensity I by a repulsive force $-\nabla P$, where $P = 2\pi e^2 I/(\mu c \omega^2)$, the ponderomotive energy, is the kinetic energy, averaged over one cycle, of a free electron that has zero drift velocity in the field (e.g. Bucksbaum *et al* 1987 and Freeman *et al* 1987).

[9] Rapid variations in the intensity can give rise to population trapping in a metastable superposition of Rydberg states (Burnett *et al* 1991, and references therein), population transfer to excited states via non-diabatic transitions at quasienergy-curve crossings, rainbow features in the energy spectrum due to interference between the photoelectron amplitude generated on the rising and falling edge of the pulse (Reed and Burnett 1991), and a breakdown of the usual inversion symmetry if significant ionization occurs during one cycle (Kulander *et al* 1991).

[10] Various methods have been proposed to solve numerically the 3D time-dependent Schrödinger Equation, for example working on a lattice, or using a convenient discrete basis set (in the position or in the momentum space), or treating the system in a density-matrix theory combined with multi-channel quantum defect theory (e.g. Kulander 1987, X. Tang *et al* 1989 and 1990, Collins and Merts 1990, DeVries 1990, Pindzola and Dörr 1991, LaGattuta 1991, Pont *et al* 1991a). ATI spectra and energy-resolved angular distributions can now be obtained (Schafer and Kulander 1990) and the time-integration be carried out over several tens of optical cycles.

[11] The method is readily generalized to polychromatic fields (Chu 1989, Dörr et al 1991b). Two cases should be distinguished for two-color experiments. In the most usual "incoherent" case, the two frequencies are either incommensurable, or, if they are not, there is no well defined relative phase relationship between the fields. In the "coherent" case, one of the two laser beams is a (low) harmonic of the other one with a well defined relative phase relationship (Muller *et al* 1990).

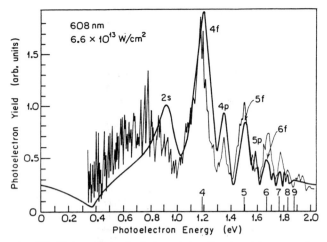

Figure 1. Yield in photoelectrons, into the lowest open channel, versus photoelectron energy, for ionization of H(1s) by a 608 nm pulse whose peak intensity is 6.6×10^{13} W/cm^2 and whose duration (FWHM) is 0.5 psec. The bold curve is the result of Floquet calculations (Dörr *et al* 1990a); the thin curve represents experimental data (Rottke *et al* 1990). Some of the subpeaks are labelled by the dominant configuration of the resonant Floquet state.

Working in the velocity gauge, where the interaction of the atom with the field is

$$V(t) = (-e/\mu c)\mathbf{A}(t) \cdot \mathbf{p} \equiv V_+ e^{-i\omega t} + V_- e^{i\omega t}, \qquad (3)$$

the harmonic components are solutions of the time-independent coupled equations

$$[E + N\hbar\omega - H_a]\psi_N = V_+ \psi_{N-1} + V_- \psi_{N+1} \qquad (4)$$

that are regular at $r = 0$ and behave as a superposition of outgoing waves (and damped ingoing waves) at $r \sim \infty$:

$$\psi_N(\mathbf{r}) \sim \sum_M f_{MN}(\hat{\mathbf{r}}) \, r^{iZ/k_M} \, e^{ik_M r}/r. \qquad (5)$$

In equation (5), H_a denotes the field-free atomic Hamiltonian. The wavenumber k_M of the M-th channel is defined as

$$k_M = [(2\mu/\hbar^2)(E + M\hbar\omega)]^{1/2} \qquad (6)$$

and E is a complex eigenvalue, the ac-quasienergy, which depends on the parameters of the radiation field but not on time. The quasienergy may be expressed, modulo $\hbar\omega$, as $E = E_0 + \Delta - i\Gamma/2$, where E_0 (the eigenenergy of the initial state), Δ (the ac-Stark shift), and Γ (the induced width of the state, which is positive) are real. The total ionization rate (integrated over all directions of the emergent photoelectron and summed over all channels) is $\Gamma/\hbar$. Hence calculating the total yield in ions or photoelectrons, and to a limited extent a photoelectron spectrum, requires finding the relevant eigenvalues of equation (4), while ATI spectra or angular distributions are obtained by extracting the amplitudes f_{MN} from the harmonic components ψ_N. [12]

There are two choices for the sign of the channel wavenumber k_M, owing to the the square-root branch in equation (6). Hence there are infinitely many solutions (since there are infinitely many channels and two sign choices per channel). The *dominant* eigenvalues, the ones that are physically meaningful, are the solutions that correspond to taking $\text{Im}\, k_M > 0$ for all closed channels and $\text{Re}\, k_M > 0$ for all open channels. In a closed channel, we have $\text{Re}(E + M\hbar\omega) < 0$, and we therefore require $\exp(ik_M r)$ to decay as r increases (the atom absorbs an insufficient number of photons to ionize), while in an open channel we have $\text{Re}(E + M\hbar\omega) > 0$, and we therefore require $\exp(ik_M r)$ to behave as an outgoing wave as r increases (the photoelectron moves outwards as it escapes). However, there are infinitely many *shadow* eigenvalues for which at least one k_M has the "wrong" sign, and these shadow eigenvalues follow different trajectories in the E plane as the intensity varies. [13] Most of them remain shadow eigenvalues, without much physical significance, at any relevant intensity, but certain ones may emerge as a dominant eigenvalue upon crossing a multiphoton threshold (where the real part of one channel wavenumber vanishes) at a certain appearance intensity I_{app}. The state corresponding to this shadow eigenvalue becomes a new dominant state whose properties cannot be distinguished from those of a decaying bound state. [14] Conversely, upon crossing a multiphoton threshold a dominant eigenvalue would become a shadow eigenvalue: as the minimum number of photons required to ionize the atom varies with the intensity, the system is described by a succession of solutions to the Floquet equations that differ by the choice of the square-root branch for the different channel wavenumbers. [15]

The approach we adopted to solve the coupled equations (4) is similar in many respects to the numerical method described by Maquet, Chu, and Reinhardt (1983), though with some important technical differences. We expand the harmonic components $\psi_N(\mathbf{r})$ on a discrete basis set consisting of spherical harmonics and radial "Sturmian" functions $S_{nl}^{\kappa}(r)$:

$$S_{nl}^{\kappa}(r) = N_{nl}^{\kappa}\, r^{l+1} e^{i\kappa r}\, {}_1F_1(l+1-n, 2l+2, -2i\kappa r) \qquad (7)$$

[12] Note, from equations (2) and (5), that the harmonic component ψ_N, with photon index N, represents the absorption of N real or virtual photons; f_{MN} is the amplitude for absorbing M real photons and $N - M$ virtual photons. The M-photon ionization amplitude is a coherent sum of the amplitudes f_{MN} (Potvliege and Shakeshaft 1988b, 1990).

[13] Shadow poles of the S-matrix were first discussed in the context of elementary particle physics; it has long been recognized that they play also a role in processes such as electron impact excitation of the 2^3S state of He (in the threshold behaviour of the cross sections), dissociative attachment, or collisional electronic excitation of molecules (Burke 1968, Chapman and Herzenberg 1972, Herzenberg and Ton-That 1975, Ohno and Domcke 1983). It is in the domain of multiphoton processes that shadow poles may perhaps be of greatest interest (Ostrovskii 1977, Potvliege and Shakeshaft 1988b).

[14] The appearance of "new bound states" has been seen in model atom calculations (Bhatt *et al* 1988) and interpreted as the emergence of stray shadow eigenvalues as dominant ones (Dörr and Potvliege 1990).

[15] All the dominant and shadow eigenvalues can be considered to be different branches of a single multivalued function of the field strength F_0. The analytical properties of this function have been investigated by Manakov and Fainshtein (1981) and by Pont and Shakeshaft (1991a). Its only physically significant singularities at finite F_0 are branch points where two branches (originating from the same bound state at $F_0 = 0$ or from different bound states) meet. The radius of convergence of the Rayleigh-Schrödinger perturbation series (which is in general non-zero in the a.c. case, in contrast with the d.c. case, Graffi *et al* 1985) is the distance from $F_0 = 0$ to the nearest branch point (Alvarez and Sundaram 1990, Pont and Shakeshaft 1991a).

where κ may be complex and where N_{nl}^κ is a normalization factor. Projecting equations (4) onto this basis yields a large, sparse, complex non-Hermitian matrix eigenvalue equation for the quasienergy. The boundary conditions (5) are implemented implicitly through the choice of $\arg(\kappa)$. Indeed, any well-behaved function of r which vanishes as r^{l+1} for $r \sim 0$ and behaves as $r^\nu \exp(ikr)$ for $r \sim \infty$ can be expanded in terms of the S_{nl}^κ with expansion coefficients that vanish for $n \sim \infty$ provided that [16]

$$|\arg(\kappa) - \arg(k)| < \pi/2. \tag{8}$$

An eigenvalue which is stable with respect to changes in the basis set, and in particular to changes in $\arg(\kappa)$, then corresponds to a physical quasienergy.

Quasienergy spectra, total and partial ionization rates, and rates for harmonic generation by a single atom have been calculated, for atomic hydrogen, over wide ranges of wavelengths, intensities, and initial states, and for linear as well as for circular polarizations. As an illustration of typical results, we show in figure 2 the quasienergy spectrum for a rather long wavelength (1064 nm) and intensity not exceeding 7×10^{12} W/cm^2, and in figure 3 the total rate for ionization of H(1s) at the same wavelength for intensity up to 3×10^{13} W/cm^2. In our choice of gauge [equation (3)] the Rydberg states close to the continuum (such as states 5, 20, and 10, which are predominantly 8j, 7i, and 6h atomic states, respectively) and the continuum threshold do not shift in the field, while the ground state (state 15 in the figure) and the other deep lying states shift downward, approximately by the ponderomotive energy P; the energy difference between the quasienergies of two states is of course gauge-invariant. Clearly, the dressed initial state shifts in and out resonance with the excited states as the intensity increases. For example, the first group of resonances peaks in figure 3 corresponds to intermediate 11- or 12-photon transitions to Rydberg sublevels associated with the crossings, in the intensity range 6×10^{12} W/cm^2 to 10×10^{12} W/cm^2, of the 1s level with the group of levels labelled 20-25, 27, and 1 in figure 2. [17] The ionization potential of the atom being different at each crossing, the Stark shift induced resonances manifest, prominently, in the energy spectrum of the ejected electrons (Freeman *et al* 1987, figure 1). We see also from figure 3 that perturbation theory is utterly inadequate to describe the yield for ionization of H(1s) by infrared radiation at any intensity where ionization might be significant. It is interesting to note not only that in general the total rate, for multiphoton

[16] Shakeshaft (1986). Examples of the application of this method to the evaluation of multiphoton matrix elements and of phaseshifts in field-free potential scattering can be found in Potvliege and Shakeshaft (1988a,c). The complex Sturmian basis is also very convenient for calculating rates of high-order multiphoton ionization or harmonic generation for one-electron systems in perturbation theory (Potvliege and Shakeshaft 1989a,b). Note that rather than use a complex basis set we could use, like e.g. Maquet *et al* (1983), a real basis set but with the Hamiltonian transformed by the familiar complex scaling $r \rightarrow r \exp i\theta$.

[17] Whether or not a crossing in the real part of the quasienergy of two states gives rise to resonance enhancement, and whether or not the crossing is true or avoided, depends critically on the width of the states. Analysis of a curve crossing within a two-level model is illuminating (Gontier and Trahin 1979, Holt *et al* 1983, Crance 1988, Potvliege and Shakeshaft 1989c). The atom should be described by a superposition of two dressed states, with time-dependent coefficients, in the vicinity of an avoided crossing, and one should evaluate the probability of jumping from one adiabatic curve to the other as the intensity sweeps through the resonance. However, as long as the passage through the resonance is fast on the time scale set by the quasienergy gap, and this is often the case at long wavelength for the resonances involving a many-photon transition between the ground state and an excited state, a reasonable approximation is to represent the system by just one "diabatic" Floquet state. This approximation may be inappropriate at shorter wavelengths (e.g. Pindzola and Dörr 1991).

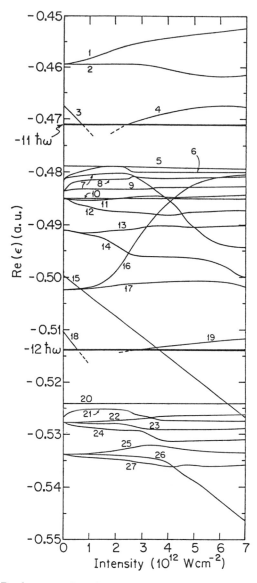

Figure 2. Real parts of various quasienergies for hydrogen irradiated by linearly polarized light of wavelength 1064 nm. (An integral multiple of the photon energy, different for each curve, has been substracted from the quasienergy, in such a way that a multiphoton dipolar transition between two dressed states is allowed at any true or avoided crossing.) The curves are labelled from 1 to 27 for convenience. All those quasienergies corresponding to dominant states that reduce in the zero-field limit to atomic states with principal quantum number ≤ 6 are shown, and a few more besides. (From Potvliege and Shakeshaft 1989c.)

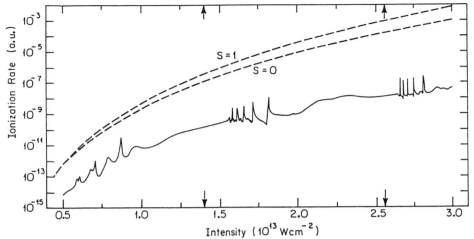

Figure 3. Total ionization rate vs intensity for H(1s) irradiated by linearly polarized light of wavelength 1064nm. Dashed curves are partial rates for (12+S)-photon ionization obtained within lowest-order perturbation theory. The arrows indicate the intensities at which the the minimum number of photons required to ionize the atom increases from 13 to 14 and from 14 to 15. (From Potvliege and Shakeshaft 1989c.)

ionization from the ground state level, obtained in perturbation theory increases much too rapidly with increasing intensity at long wavelengths, [18] but also that in general the nonperturbative rate increases, far from resonances, less and less rapidly. This overestimation of the perturbative rate is due to the neglect of both the increase in the ionization potential and the (oscillating) polarization of the atom – the effect of the latter is that the electron spends less time near the atomic nucleus where it can absorb photons.

The atom actually tends to stabilize against ionization at sufficiently large intensity when the frequency of the field is substantially larger than the threshold frequency for one-photon ionization (Gavrila 1985). [19] This theoretical prediction is supported by the Floquet total ionization rate shown in figure 4 : it increases with the intensity until an intensity I_{max} (about 1.1×10^{16} W/cm^2) beyond which it decreases toward zero. Similar

---

[18] This is not true for certain Floquet states issued from Rydberg states with large l but small m. They are actually superpositions of several atomic states, dominated, however, by a particular one, in proportions that vary with the field strength; therefore, their widths may increase more rapidly with the intensity, in weak fields, than the linear power law expected for single Rydberg states (Potvliege and Shakeshaft 1989c).

[19] This is due to the difficulty of satisfying energy and momentum conservation simultaneously (Pont and Shakeshaft 1991). There is no stabilization if the frequency of the field is much below the characteristic atomic-orbital frequency, because field-ionization can occur when the intensity is large but the frequency is low. The a.c. shift and a.c. rate approach the cycle-averaged d.c. shift and d.c. rate, respectively, as either the wavelength or intensity increases; H(1s) ionizes within roughly one cycle at and above about 5×10^{14} W/cm^2, for wavelengths in the range 616-1064 nm (Shakeshaft et al 1990). In the low frequency regime, it is possible to represent the eigenenergy in a simple and convenient way by a power-series of ω^2 truncated after the first few terms, which are relatively easy to calculate for linear as well as for circular polarization (Pont et al 1990b, 1991b).

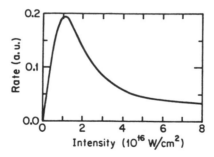

Figure 4. Total rate for ionization of H(1s) by linearly polarized 70-nm ($\omega = 0.65$ a.u.) light. (From Dörr *et al* 1990b; see Shakeshaft *et al* 1990 and Dörr *et al* 1991a for more detailed results.)

calculations for ionization from other levels of hydrogen confirm that this behaviour is indeed general; I_{max} is found to be approximately proportional to ω^3, and is therefore smaller at smaller frequency (provided $\hbar\omega > |E_0|$). [20] The ionization rate is quite large in the conditions of figure 4 (the atom would ionize in less than 1 fs), but it is much smaller for ionization from *excited* states, in particular from states with $n = l + 1$. For example, the rate for ionization from the 4f(m=3) state is 1.9×10^{-4} a.u. at $I_{max} = 1.1 \times 10^{14}$ W/cm^2 and 2.4×10^{-5} a.u. at 10^{15} W/cm^2 when $\omega = 0.1$ a.u. As shown in table 1, an atom prepared in this state can easily sustain a short but realistic very intense laser pulse, with a survival probability which *increases* with the peak intensity. This suggests that stabilization may indeed be achieved in experimental conditions.

Table 1. Survival probability at the end of a pulse of frequency 0.1 a.u. (456 nm wavelength) whose temporal intensity profile is gaussian with a duration (FWHM) of 20 cycles and a peak intensity I_0. The initial state is H(4f,m=3). TD : results of a fully time-dependent calculation (Pont and Shakeshaft 1991b); Fl : results of Floquet calculations. I_0 is expressed in units of 10^{14} W/cm^2.

	$I_0 = 0.5$	$I_0 = 1.0$	$I_0 = 2.0$	$I_0 = 4.0$	$I_0 = 6.0$	$I_0 = 8.0$	$I_0 = 10.0$
TD	0.79	0.71	0.67	0.68	0.70	0.72	0.74
Fl	0.76	0.66	0.62	0.63	0.65	0.67	0.69

[20] As can be seen from table 1, the high-frequency Floquet results are in agreement with the results of computer simulations (see also Kulander *et al* 1991). The shift also approaches the result of the high frequency theory (Pont *et al* 1988, 1990a) as the intensity increases beyond I_{max} (Bardsley and Comella 1989, Dörr et al 1991a). The rise and fall of the rate, in the vicinity of I_{max}, seems to be associated with an interaction between a shadow state and the dominant state (Dörr et al 1991a). It is worth noting that a "stable" atom still oscillates in the field, and would therefore provide a continuous source of high-frequency radiation through harmonic generation.

Acknowledgments

Many of the results mentioned here were obtained in collaboration with M. Dörr, M. Pont, D. Proulx, P.H.G. Smith, and Z. Teng. This work is being supported by the NSF, Grant No. Phy-9017079, by the UK Science and Engineering Research Council, and by a RIC grant from the University of Durham.

References

Agostini P, Fabre F, Mainfray G, Petite G and Rahman N K 1979 *Phys. Rev. Lett.* **42** 1127

Alvarez G and Sundaram B 1990 *Phys. Rev. A* **42** 452

Andrick D and Langhans L 1976 *J. Phys. B: At. Mol. Phys.* **9** L459

Augst S, Strickland D, Meyerhofer D D, Chin S L and Eberly J H 1989 *Phys. Rev. Lett.* **63** 2212

Bardsley J N, Szöke A, and Comella M J 1988 *J. Phys. B: At. Mol. Opt. Phys.* **21** 3899

Bhatt R, Piraux B and Burnett K 1988 *Phys. Rev. A* **37** 98

Bucksbaum P H, Freeman R R, Bashkansky M and McIlrath T J 1987 *J. Opt. Soc. Am. B* **4** 760

Burke P G 1968 *J. Phys. B: At. Mol. Phys.* **1** 586

Burnett K, Knight P L, Piraux B R M and Reed V C 1991 *Phys. Rev. Lett.* **66** 301

Chapman C J and Herzenberg A 1972 *J. Phys. B: At. Mol. Phys.* **2** 790

Chu S-I 1989 *Adv. Chem. Phys.* **73** 739

Collins L A and Merts A L *J. Opt. Soc. Am. B* **7** 647

Crance M 1988 *J. Phys. B: At. Mol. Opt. Phys.* **21** 2697

DeVries P 1990 *J. Opt. Soc. Am. B* **7** 517

Dörr M and Potvliege R M 1990 *Phys. Rev. A* **41** 1472

Dörr M, Potvliege R M and Shakeshaft R 1990a *Phys. Rev. A* **41** 558

Dörr M, Potvliege R M and Shakeshaft R 1990b *Phys. Rev. Lett.* **64** 2003

Dörr M, Potvliege R M, Proulx D and Shakeshaft R 1991a *Phys. Rev. A* **43** 3729

——1991b *Phys. Rev. A* **44** 574

Freeman R R, Bucksbaum P H, Milchberg H, Darack S, Schumacher D and Geusic M E 1987 *Phys. Rev. Lett.* **59** 1092

Gavrila M 1985 *Fundamentals of Laser Interactions I*, ed F Ehlotzky (New York : Springer Verlag) p 3

Gontier Y and Trahin M 1973 *Phys. Rev. A* **7** 1899

——1979 *Phys. Rev. A* **19** 264

Graffi S, Grecchi V and Silverstone H J 1985 *Ann. Inst. Henri Poincaré - Phys. Théor.* **42** 215

Herzenberg A and Ton-That D 1975 *J. Phys. B: At. Mol. Phys.* **8** 426

Holt C R, Raymer M G and Reinhardt W P 1983 *Phys. Rev. A* **27** 2971

Kulander K C 1987 *Phys. Rev. A* **35** 445

Kulander K C, Schafer K J and Krause J L 1991 *Phys. Rev. Lett.* **66** 2601

LaGattuta K J 1991 *Phys. Rev. A* **43** 5157

L'Huillier A, Lompré L A, Mainfray G and Manus C 1983 *J. Phys. B: At. Mol. Phys.* **16** 1363

——1991 *Adv. At. Mol. Opt. Phys.* in press

Maquet A, Chu S-I and Reinhardt W P 1983 *Phys. Rev. A* **27** 2946

Manakov N L and Fainshtein A G 1981 *Teor. Mat. Fiz.* **48** 375 [*Theor. Math. Phys.* **48** 815]

Manakov N L, Ovsiannikov V D and Rapoport L P 1986 *Phys. Rep.* **141** 319

Meyerhofer D D, Augst S, Peatross J and Chin S L 1990 in *Multiphoton Processes – Proc. 5th Int. Conf. on Multiphoton Processes*, ed G Mainfray and P Agostini (Saclay : CEA) p 317

Muller H G, Bucksbaum P H, Schumacher D W and Zavriyev A 1990 *J. Phys. B: At. Mol. Opt. Phys.* **23** 2761

Ohno M and Domcke W 1983 *Phys. Rev. A* **28** 3315

Ostrovskii V N 1977 *Teor. Mat. Fiz.* **33** 126 [*Theor. Math. Phys.* **33** 923]

Pindzola M S and Dörr M 1991 *Phys. Rev. A* **43** 439

Pont M and Shakeshaft R 1991a *Phys. Rev. A* **43** 3764

Pont M and Shakeshaft R 1991b to be published

Pont M, Offerhaus M J and Gavrila M 1988 *Z. Phys. D* **9** 297

Pont M, Walet N R and Gavrila M 1990a *Phys. Rev. A* **41** 477

Pont M, Shakeshaft R and Potvliege R M 1990b *Phys. Rev. A* **42** 6969

Pont M, Proulx D and Shakeshaft R 1991a *Phys. Rev. A* in press

Pont M, Potvliege R M, Shakeshaft R and Teng Z-j 1991b to be published

Potvliege R M and Shakeshaft R 1988a *Phys. Rev. A* **38** 1098

——1988b *Phys. Rev. A* **38** 6190

——1988c *J. Phys. B: At. Mol. Opt. Phys.* **21** L645

——1989a *Phys. Rev. A* **39** 1545

——1989b *Z. Phys. D* **11** 93

——1989c *Phys. Rev. A* **40** 3061

——1990 *Phys. Rev. A* **41** 1609

——1991 *Adv. At. Mol. Opt. Phys.* in press

Reed V C and Burnett K 1991 *Phys. Rev. A* **43** 6217

Rottke H, Wolff B, Brickwedde M, Feldmann D and Welge K H 1990 *Phys. Rev. Lett.* **64** 404

Schafer K J and Kulander K C 1990 *Phys. Rev. A* **42** 5794

Shakeshaft R 1986 *Phys. Rev. A* **34** 244 and 5119

Shakeshaft R, Potvliege R M, Dörr M and Cooke W E 1990 *Phys. Rev. A* **42** 1656

Tang X, Lambropoulos P, L'Huillier A and Dixit S N 1989 *Phys. Rev. A* **40** 7026

Tang X, Rudolph H and Lambropoulos P 1990 *Phys. Rev. Lett.* **65** 3269

Weingartshofer A, Holmes J K, Caudle G, Clarke E M and Kruger H 1977 *Phys. Rev. Lett.* **39** 269

Direct and indirect processes in single photon double ionization

Pascal LABLANQUIE

LURE, Bâtiment 209D, Université Paris Sud, 91405 ORSAY Cédex, France
and Institute for Molecular Science, Myodaiji, Okazaki, 444, Japan

ABSTRACT: Double ionisation in the valence region is shown to proceed from both direct and indirect processes. Investigations of the electronic correlations at hand in the first case are presented for the model system He. Existence of indirect processes through neutral or ionic excited intermediates is shown in Ar and Ne; their mechanisms are discussed by a comparison with the innershell indirect processes.

1. INTRODUCTION

Double ionization following the absorption of a single photon by an atom or a molecule is a specially interesting phenomenon because it is forbidden in a simple model: if we describe the target as composed of independent electrons in frozen orbitals, the photon absorption, represented by a monoelectronic operator, can only induce the ejection of one electron. Experiments have long shown that double photoionization is far from being a negligible process. The question then arises: how can we understand this mechanism? In fact two different cases have to be considered:
- the innershell region, where the photon energy is high enough to access the innershell electrons. In this case relaxation of the highly excited species thus created often takes place through the ejection of a secondary electron by the well documented Auger effect (see for example Carlson 1975). We deal in this case with a two step mechanism which is satisfactorily reproduced if we no longer consider the molecular orbital to be frozen (autoionisation process).
- the valence region, where the photon energy is low enough to enable interaction with the valence electrons only. In this case the clear 2 step process which prevails in the innershell zone is no longer possible, but yet doubly charged ions can be created as soon as it is energetically allowed. The intensity of this process, depending on the photon energy, can even be quite significant for a forbidden one (for instance A^{++} / A^{+} can reach up to 40% in Xe (El Sherbini and Van der Wiel 1972)). Theoretical calculations (Chang *et al* 1975, Carter *et al* 1977) have first suggested that we deal here with a direct ejection of the 2 electrons, which gains intensity from the electronic correlations; the He case was especially extensively studied, and experimental evidences confirmed this direct process. But recent observations (Becker *et al* 1989, Price and Eland 1989) strongly suggest that it is not the hole story, and that indirect processes also exist; in other words the 2 electron ejection can here also proceed in sequence, in a way similar to the one occurring in the innershell region.

In this paper, we shall detail this idea of direct and indirect double ionization occurring in the valence region. Firstly, the direct process will be considered; the origin of its strength lies in the electronic correlations, as will be shown by the total cross section for He^{++} formation; we will then present detailed investigations of these electronic correlations. They have been carried out essentially in the threshold region, where the Wannier theory and its extensions (ref) predict characteristic behavior originating from the electronic correlations in the final (A^{++} + 2e) state. We will take the example of the excess energy sharing between the two electrons in He (Lablanquie *et al* 1990), and review the other experimental work that has recently appeared on this subject . Secondly, we will deal with the indirect double ionization process: an analogy with the indirect single ionization one will help to define what is meant here by indirect process; we will see that it can be of two different types, involving either an excited neutral A or ionic A^+ intermediate. Evidence for the existence of both will be presented in the atoms Ne, Ar (Lablanquie 1989, Becker *et al* 1989), and Hg (Price and Eland 1990). Thirdly, the indirect processes involving innershell electrons will be considered. As they are much more intense than the ones occurring in the valence region, more insight into the mechanisms for these indirect double ionisation processes can be gained in this way. The decay of the resonant Kr $3d_{5/2} \rightarrow 5p$ will be taken as an example, and compared to the equivalent situation occurring in Xe (Okuyama *et al* 1990, Heimann *et al* 1987).

2. DIRECT VALENCE DOUBLE PHOTOIONIZATION

2.1 Origin of the intensity of valence double photoionization: electronic correlations.

The first experimental observation of the surprisingly high intensity for double electron ejection in the valence region was performed on the rare gases by Carlson (1967). Further experiments (Holland *et al* 1979 and references included) confirmed this point and showed that photon absorption can lead to double electron ejection with a probability that can be as high as 40% (in Xe at 60eV, El Sherbini and Van der Wiel 1972). The case of molecules was solved much more recently; as the loss of two electrons gives a doubly charged species which is generally not stable and usually rapidly dissociates into a pair of cations, we had to wait until coincidence methods (Mc Culloh *et al* 1965) became sufficiently sensitive to be applied in conjunction with photon excitation sources (Dujardin *et al* 1984, Curtis and Eland 1985, Lablanquie *et al* 1985). A number of molecules has been investigated so far (see Eland 1991 for a review), and a similar probability for the double electron ejection phenomenon was also found. Finally the case of the open shell atoms was recently investigated (see Samson and Angel 1990 for the case of N and O, Adam et al (1989) for the Ag one and a compilation of other related experiments).

However, calculations aimed to reproduce these observations were performed in a limited number of systems, namely He Ne Ar and H_2. The best understood case is naturally He. Figure 1 shows our measurements of the double photoionization cross section of helium (open circles, Lablanquie *et al* 1990); they agree very well with a compilation of experimental values (solid thick line, Bizau 1981). The first attempt at interpretation was proposed by Carlson (1967), using a "shake off", independent particle model: he considered that photon absorption leads to the ejection of a first 1s electron, then the remaining electron has to adapt its wave function $|\Psi(1s)\rangle_{He}$ to the eigenfunctions $|\Psi''(..)\rangle_{He^+}$ of the new He^+ system (monopole transitions). Satellite states $He^+(n)$ as well as the double continuum He^{++} can gain intensity in this way, but the calculations significantly differ from the experiment: only a forth of the double ionization intensity can be reproduced (note that this model is sufficient to explain the gross features of the double ionization process following innershell ionization: this shows how different is the valence region as far as the double ionization mechanism is concerned). Carlson suggested that electronic correlations, neglected in his simple picture were mainly responsible for the intensity of the double ionization process. This idea was confirmed by calculations taking into account these correlations; the first one included them in the initial state only (Byron and Joachain 1967) and could reproduce half of the measured

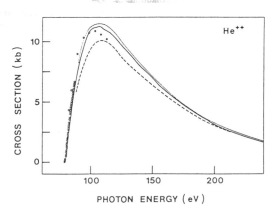

Figure 1: Double photoionization cross section in helium (see text).

intensity. Excellent agreement with experiment was obtained only when all correlations were included, either through the many body perturbation theory -dashed line in figure 1 (Carter and Kelly 1981)- or with a wave function description -thin solid line in figure 1 (Tiwary 1982).

As far as the double ionization probability is concerned, an interesting experimental observation was presented by Samson (1990): he found that in a number of cases the branching ratio for double photoionization σ^{++}/σ (where σ and σ^{++} stand for the total and double ionization cross section respectively) was proportional to the cross section for the electron impact ionization of the corresponding ion. This can be qualitatively explained if we consider that the photon absorption generates an energetic electron inside the target A, that will interact with the remaining electrons in a similar way as the electron impinging on the A^+ ion. Though slight differences exist in both cases, electron correlation effects must be similar and explain the observed proportionality. We should notice at this point that this "internal electron impact" model for the double photoionization doesn't necessarily imply an indirect process as defined below: this intermediate state after photon absorption should rather be considered as virtual: its lifetime is too short to be observed experimentally, as in the indirect processes described in §3.

The conclusion is that the origin of the intensity for double photoionization in the valence region is the electronic correlations involved, whether explicitly considered in the calculations, or experimentally taken into account by comparison with a similar system such as proposed by Samson (1990). But detailed understanding is still not available. Let us give examples of open questions:

- calculations in the more complicated Ne and Ar case were also successful in reproducing the correct order of magnitude for the double electron ejection (Chang and Poe 1975, Carter and Kelly 1977) but they only took into account the direct process (the only one possible in He or H_2), and neglected any indirect process (see below §3) which were not discovered at that time. What are the relative importance of the two contributions?

- the available calculations on H_2 (Le Rouzo 1988) include correlation in the initial state only, but in this case, the velocity formulation of the cross section already exceeds the experimental estimates (Dujardin *et al* 1987, Kossmann *et al* 1989); let us recall that for He the equivalent calculation (Byron and Joachain 1967) gave only half of the observation. Is it some kind of molecular effect?

- Is there some final state selectivity? in other words what are the branching ratios between the different existing A^{++} electronic states? Only a few experiments have yet tried to answer this question, as selective techniques have to be used. The first class used Fluorescence techniques to measure A^{++} partial cross sections (namely Ar^{++} $3s3p^5$ ($^{1,3}P$) one, Mobus *et al* (1991). The second class of experiments uses an electron-electron coincidence technique; Lablanquie *et al* (1987) first reported that at 56eV photon energy, the Ar^{++} $3p^{-2}$ (^3P) is the dominant species, it was interpreted as a consequence of the "kinematically disallowed" character of the 2 electron wavefunction associated with the other states (Greene and Rau 1983); but further investigations questioned this point: Price and Eland (1989, 1990) in a more sensitive experiment on Xe and Ar found that all final states can also be populated; although the ^3P state is still found to be dominant it can in some case (Ar at 48eV) be of equal intensity as the ^1D one; they also found intense indirect double ionisation processes which could be a reason for this discrepancy. Finally more careful theoretical examination (Huetz *et al* 1991) also showed that other effects can compensate the disallowed character of the final state.

- Physical parameters other than the total cross section can reflect these electronic correlations; however calculations mainly concentrated on this point, possibly because of the difficulty to handle double continua, and the lack of experimental data. The dependence on angles and energies of the photoelectrons have been investigated for some time in the threshold region (see below §2.2) but the whole photon energy scale was just examined recently on a theoretical point of view, in He by Le Rouzo and Dal Cappello (1991). More progress can be anticipated along these lines.

2.2 Tests of the electronic correlations in the threshold (Wannier) zone.

Since the direct valence double ionization has its origin in the electronic correlations, it is ideally suited to study them. In fact much effort has recently been undertaken to study a specific case: the threshold zone; it can be shown that the system is then governed by the correlations in the final state, between the two departing electrons; this comes from the classical Wannier theory (Wannier 1953) and its quantum generalizations (Peterkop 1971, Rau 1971), and means that the system behavior is that of 3 point particles $(A^{++} + 2e^-)$ in Coulomb interaction. This 3 body problem is especially interesting since it is directly connected to a number of analog physical situations, the most studied one being the electron impact ionization: $e^- + A \rightarrow A^+ + 2e^-$ (see Selles *et al* 1990 and references included). Though it is of low intensity (especially in this threshold zone) and hence experimentally hard to investigate, the threshold double ionization has a double interest: test the scale effects predicted by theory (compared to the other related systems) and provide a well defined system, thanks to the selection rules imposed by the dipolar transitions.

We wish to give here 2 examples of correlation effects we observed in the model He system (Lablanquie *et al* 1990).

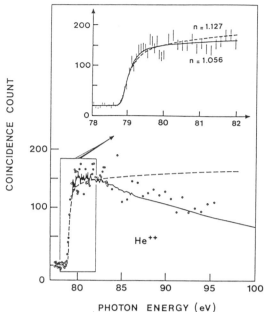

Figure 2: Yield of coincidences between low energy (less than 200meV) electrons and He^{++} ions, compared to the experimental quantity $\sigma^{++}(E)/E$ (solid line) and the theoretical E^{n-1} law, with n=1.056 (dashed line). Insert shows extension of the threshold zone, and comparison with two E^{n-1} laws.

First, the energy distribution of the departing electrons is predicted to be flat. In order to test this point, we measured the yield of low energy electrons (less than 200meV) associated to double ionization events; the experimental set up consists of a 127° electrostatic analyser facing a time of flight spectrometer, and uses monochromatized synchrotron radiation from the ACO storage ring in Orsay. This yield, obtained through e^- / He^{++} coincidences is displayed in figure 2; it is compared with the quantity $\sigma^{++}(E)/E$, where E and $\sigma^{++}(E)$ stand for the excess energy above threshold and the double ionization cross section respectively. If the energy distribution is flat, the partial cross section for ejecting an electron within a fixed bandwith (0-200meV) should be proportional to $\sigma^{++}(E)/E$; this is what is observed here within 20% error bars in a 15eV range above threshold. The predicted flatness of the energy sharing between the electrons is thus confirmed; but the surprising point here is the energy range over which this flatness holds true: Wannier theory and its extensions cannot predict the energy range over which they apply, this has to be determined experimentally, but the observed validity zone is here larger than expected, if we compare to the related electron impact ionization of He; in that case, the flatness of the electron distribution was found to be limited to the 3.6eV above threshold (Pichou *et al* 1978). Reason for this different behavior is not clear; we may observe here influence of the creation process, which is neglected in the Wannier theory; the recent calculations by Le Rouzo and Dal Capello (1991) of this energy distribution also find a flat behavior near threshold; as they do not include correlation in the final state, this suggests that another mechanism than the electron correlation analyzed in Wannier theory also contributes to this flatness.

Second, the same experiment can be used to test another theoretical prediction, namely that the cross section $\sigma^{++}(E)$ should be proportional to E^n with n=1.056. Since the energy distribution is found to be flat, we should observe the low energy electron yield to be proportional to $\sigma^{++}(E)/E=E^{n-1}$ Figure 2 shows that such a relation is limited to a 3eV excess energy range, and the best fit to our data gives n=1.05±0.05. Two other experiments came to a similar conclusion, and a more precise measurement of the Wannier exponent n, by use of different techniques: He^{++}/He^{+} branching ratio measurement with n=1.05±0.02 (Kossmann *et al* 1988) or threshold electron yield with n=1.060±0.007 (Hall *et al* 1991). The scale effect is here checked, as n was known to be 1.127 for 2 electrons leaving a singly charged center.

Recently, thanks to the always improved experimental performances, much work appeared on this subject: measurement of the anisotropy of the electron ejection with respect to the polarisation vector of the light (Hall *et al* 1991); observation of a sideways dissociation of the proton pair with respect to the polarization vector in H_2 (Kossmann *et al* 1989); use of electron-electron coincidences resolved in energy (Price and Eland 1990) or in energy and angle (Mazeau *et al* 1991). This last type of experiment is especially promising, as angular information is expected to give the best insight view of the processes involved; an extensive theoretical analysis for the noble gas case was recently performed by Huetz *et al* (1991). They predicted for instance a very sensitive angular behavior for the kinematically disallowed states, whose wavefunction presents a node for a 180° angle between the ejected electrons, whereas electron correlations precisely tend to eject them in opposite directions.

3. INDIRECT VALENCE DOUBLE PHOTOIONISATION

As seen above, after a photon absorption, 2 electrons can be directly released: both electrons are then equivalent and no intermediate step can be experimentally detected. Helium is the simplest example; but other mechanisms can occur in more complex systems. The first remark to show this, is that the double ionisation process is not an isolated phenomenon, but should rather be considered as the extension of the ionisation + excitation one (satellite formation: one electron ejected and another excited); the smooth behavior of threshold electron spectra across the double ionisation energy (see for example King *et al* 1988 for He) suggests that the double continuum and its associated Rydberg series of satellite single continua should be considered as a same entity. Analogy with the single ionisation case is then obvious: in that case, an ionisation channel is usually defined as composed of the ionisation continuum and the neutral Rydberg series that converge to it; autoionisation or indirect single ionization then arises from the interaction between two such ionization channels. In the same way, we can define indirect double ionization as the result of the interaction between two such "double ionization channels"; the only difference will be that we deal here with a two dimension problem: the double ionization continuum can be considered not only as the limit of Rydberg series of ionic satellite states A^{+*} but also of doubly excited neutral states A^{**} (another way to present this is to consider the double ionization continuum as the limit of a Rydberg series of single ionization channels). The consequence is that 2 kinds of indirect processes can now be expected:

$$h\nu + A \rightarrow A^{++} + 2e \qquad \text{direct process}$$
$$A^{+*} + e \qquad \text{indirect process 1)}$$
$$A^{**} \qquad \text{indirect process 2)}$$

The first clear observation of type 1 processes in atoms was performed in Xe by Price and Eland (1989), and simultaneously in Ne by Becker *et al* (1989), whereas in molecules such a process was suggested to occur at threshold in H_2O (Winkoun *et al* 1988) and CO (Lablanquie *et al* 1989); different names have been used to design them: "two step double ionisation" (Winkoun *et al* 1988), "valence Auger", "participator Auger" (Becker *et al* 1989), "non resonant Autoionisation" (Price and Eland 1990). Type 2 processes were observed in Ar by Lablanquie *et al* (1987); an appropriate denomination for them could be resonant double ionisation, whether the decay of these highly excited neutral state A^{**} proceeds directly in one step to $A^{++} + 2e$ (resonant one step double autoionisation) or in two steps through an intermediate ($A^{+*} + e$) step (resonant double autoionisation). Experimental evidence for these 2 indirect processes will be presented in the following.

Figure 3 shows the Ne^{++} cross section, together with the yield of low energy (less than 200meV) electrons. The structure in the Ne^{++} yield around 85eV indicates the presence of a

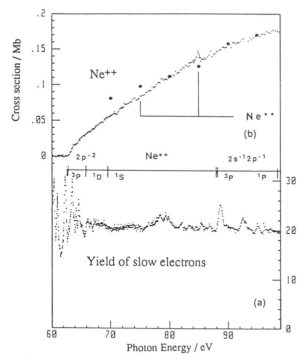

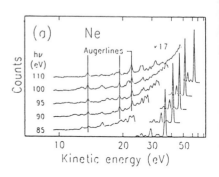

Figure 4 Neon : series of photoelectron spectra showing the valence Auger lines (figure 4 from Becker *et al* (1989)

Figure 3 Neon: (a) Yields for low energy electrons (b) cross section for Ne++ production , normalized at 80eV to the values of Holland *et al* (1979) - thick points-

weak type 2 process; this highly excited Ne* state was already observed in absorption measurements by Codling *et al* (1967) and we can see here that it can evolve to the double ionisation continuum. But type1 indirect paths appear to be more intense in Ne, as shown in the slow electron yield fig 3b: the observed peaks correspond to the release of a slow photoelectron associated with the creation of highly excited Ne+* states; a number of them is seen to be embedded in the double continuum, to which they can evolve by autoionisation; the observation of the electron emitted at that time has been performed by Becker *et al* (1989), as shown in Fig 4: we clearly see, when changing the photon energy, peaks corresponding to electrons of fixed kinetic energy, as it is expected for such second step process (valence Auger). Recently, Armen and Larkins (1991) calculated in detail the deexcitation paths of these highly excited Ne+* states; they found that radiative decay is very weak, and calculated which Ne++ state is preferentially produced; they showed that two different decay mechanisms are expected: inner-valence and valence multiplet participator Auger transitions, they correspond to decays where the Ne+* core changes either its electronic configuration [for instance: Ne+ $2s2p^5$ (^3P) $3p$ (^2S) $\rightarrow$ Ne++ $2s^22p^4$ (^1S or ^1D) + e] or its multiplet coupling [such as Ne+ $2s2p^5$ (^1P) $3p$ (^2S) $\rightarrow$ Ne++ $2s2p^5$ (^3P) + e], relative probability of these two processes were estimated, and await experimental confirmation.

In order to test this prediction, and gain a better insight into the mechanisms for these non resonant autoionisations, more sophisticated experiments have to be developed; one of the key point is undoubtedly the detection in coincidence of the two emitted electrons, as performed by Price and Eland (1989, 1990). Figure 5 shows one of their results on the Hg atom: it represents the kinetic energy distribution of the electrons associated with a given Hg++ state, namely Hg++ $5d^{10}$ (^1S) - fig 5a - and Hg++ $5d^9$ $6s$ (^3D) - fig 5b -; it was obtained by the coincidence detection of the two electrons as a function of the kinetic energy of the first one, E_1;

the kinetic energy of the second one , E_2, was adjusted so that their sum was kept constant and identified a given doubly charged state of energy IP according to: $h\nu = IP + E_1 + E_2$ (the photon energy $h\nu$ was 40.8eV -HeII radiation-). The first striking observation is the structured behaviour. Direct double ionisation process is expected to give a monotonous energy distribution, flat near the double ionisation threshold (see § 2.2); this is clearly not the case for the Hg^{++} (^1S) state, which is seen to be essentially created through non resonant autoionisation. On the opposite, the flat spectrum for the Hg^{++} (^3D) state suggests a dominant direct double ionisation path.

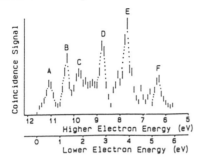
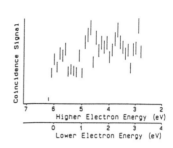

Figure 5 Mercury: Constant dication spectrum for the formation of the ^1S -left- and ^3D -right-states of Hg (figures 3 and 6 from Price and Eland 1990a)

Evidence for indirect double ionisation in molecules is less clear; the nuclear motion and the predominant dissociative character of the doubly charged states significantly complicate the problem. However, in H_2O (Winkoun *et al* 1988) and CO (Lablanquie *et al* 1989) molecules, the unexpected observation of the creation of a $(A^{+}+B^{+})$ pair below the threshold for double ionisation (as determined in the Franck Condon zone) strongly suggests a nonresonant autoionisation route. Recently, Becker and al (1991) observed the valence Auger electrons associated with this process.

4. INDIRECT DOUBLE IONISATION IN THE INNERSHELL REGION

We can see that the discovery of these indirect paths for double ionisation is very recent. Up to now, experiments have been essentially limited to their qualitative observation, much remains to be done in order to understand fully their mechanism. Limitation arises from the low intensity of this mechanism and the dominant underlying single ionisation path, which makes the use of coincidence techniques unavoidable. However indirect double ionisation happens to be the dominant process in the innershell region, and investigation of this zone, apart from its interest per se, can give some clues on what is occurring in the valence region. The equivalent of process 1) or non resonant Autoionisation, is here the well known Auger effect, whereas process 2) corresponds to the resonant Auger one.

We will give here an example of this last type, where a detailed picture of the different mechanisms could be obtained: namely the deexcitation of the highly excited $Kr^*3d^{-1}(^2D_{5/2})5p$ state. Much work has been devoted to this resonance since it was first observed by Codling and Madden (1964). Hayaishi *et al* (1984) determined by ion spectroscopy that its decay leads essentially to a Kr^{++} channel; our measurement (Lablanquie and Morin 1991) confirmed this and gave the following branching ratios between the possible ionisation states $1^{+} / 2^{+} / 3^{+}$: $17\pm1 / 73\pm1 / 10\pm1$.

Figure 6 compares a Photoelectron Spectrum taken on this resonance (Fig 6c) and out off resonance (Fig6b); many new peaks show up; their interpretation can be done simply in the framework of the spectator electron model: it states that the outer 5p electron remains uncoupled with the core, as it relaxes through an Auger effect similar to the one observed above the 3d

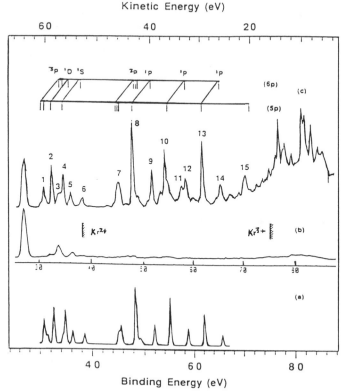

Figure 6 Krypton: Electron spectra taken on the $3d_{5/2} \rightarrow 5p$ resonance at 91.2eV (6c) and off resonance at 89.3eV (6b) compared to the reconstructed spectrum obtained with the model described in the text (figure 2 of Lablanquie and Morin 1991)

threshold; the presence of this 5p electron will just alter (increase) the kinetic energy of the Auger electron; if we assume a constant value ΔE for this increase, irrespective of the Auger transition involved, we should observe an image of the M_5NN Auger peaks [decay of the Kr^+ $3d^{-1}$ ($^2D_{5/2}$) hole]; this simple model effectively explains the most intense peaks (peaks n° 1, 2, 4, 7, 8, 10, 13); a second group of peaks appears as a weak and shifted image of the former and corresponds to final states with a similar Kr^{2+} core but with promotion of the 5p electron into the 6p orbital; ratio between this participator and spectator lines was estimated to be 25±7%. Figure 6a summarizes this discussion, and presents a reconstructed resonant Auger spectrum: inputs are position and intensities of the M_5NN Auger lines (Aksela *et al* 1984) and our measurement for ΔE=4.9eV (increase of the kinetic energy of the Auger electrons due to the presence of the 5p electron) and for the participator / spectator branching ratio. A very good agreement is found with the experiment. The second step of the analysis is to correlate this to the Kr^+ or Kr^{++} final states; this can be done if we notice that peaks n° 1 to 5 lie below the double ionisation threshold, and consequently correspond to Kr^+ formation, whereas on the opposite peaks 6 to 15 are embedded in the double continuum to which they will evolve by emission of a second electron; these electrons will contribute to the photoelectron spectrum as structure on the low energy side; estimation of the kinetic energy we expect agrees with the structure observed, but a definite proof needs a coincidence detection of the two electrons, as performed recently by Von Raven *et al* (1990). This shows that one decay mode of the neutral $Kr^*3d^{-1}(^2D_{5/2})5p$ state to the double ionisation channel is through a sequential mechanism (mediation through a type 1 path); we proposed here the "two step Auger" denomination as the first step is exactly an Auger decay, in this case. But another decay mode is possible: the direct simultaneous emission of the two electrons; this was named as "resonant shake off"; we prefer the denomination "resonant one step double autoionisation" as the denomination "shake off" is confusing and would suggest a very specific origin for the direct double ionisation process, which is well known to be insufficient (Carlson 1967). Estimation of the intensity for this one step double autoionisation path could be performed by combining the ratio between peaks 1 to 5 and 6 to 15, and the 17% value for Kr$^+$ formation; it gives a 51% estimate for the two step Auger path, the remaining of Kr^{++} formation (22%) can be ascribed to the one step 2 electron ejection. The following table summarizes these decay paths:

	17% Kr$^+$		
decay of the	73% Kr^{++}	51% two step Auger	33% 5p spectator
Kr*3d^{-1}(^{2}D$_{5/2}$)5p			18 % 5p participates $\rightarrow$ 6p
state		22% one step double ionisation	
	10% Kr^{+++}		

In the similar Xe case (decay of the Xe*4d^{-1}(^{2}D$_{5/2}$)6p state), experiment suggests a dominant one step double ionisation path (Heiman *et al* 1987). This is surprising in view of the present results, more work is needed to clarify the origin of this discrepancy. Okuyama *et al* (1990) recently developed a new experimental electron electron coincidence technique that gives a complete two dimension spectrum of the electrons associated with the double ionisation route; though of yet low resolution, this kind of experiment is expected to give more insight into this phenomenon.

5. CONCLUSION

The double photoionisation process in the valence region is seen to be very rich and complex. Electronic correlations being at the origin of its significant intensity, it is specially suited to investigate them. Much work is emerging on a very well defined case: the threshold zone, where behaviour of the system can be understood if we consider only one type of such electronic correlations: that between the two departing electrons; but experimental and theoretical analysis of all the various electronic correlations is still at its beginning. Double ionisation is also seen to proceed through direct and indirect paths; the existence of these last ones in the valence region was recently discovered, their relative intensity as well as the detailed mechanisms they sustain are yet unknown. Comparison with indirect double ionisation in the innershell zone, as presented here for Krypton, is very informative at this point, because it is conceptually the same physical problem; however, a major difference exists: in the innershell region, the intermediate species is created through a one electron transition involving the innershell electron, and is consequently much more intense than the "forbidden" direct path. On the opposite, in the valence region, the intermediate species imply excitation of two electrons in the target: they originate from electronic correlations, in the same way as the direct double ionisation does. Consequently, one can expect comparable intensities and stronger interference effects in this case, that could require a unified description, as the one introduced by Tulkki *et al* (1987) to treat the PCI effect. An indication for such an interference process is found in the Ne photelectron spectrum as measured at 1487eV by Svensson *et al* (1988): the Ne$^+$ 2s2p^5 (^{3}P and ^{1}P) 3p (^{2}S) lines are found to be slightly asymmetric, as a possible manifestation of their interference with the direct double ionisation path. Much work is still needed; a very powerful tool will undoubtedly be the electron electron coincidence experiments which appeared recently.

ACKNOWLEDGEMENTS

My warmest thanks go to my Orsay colleagues Irene Nenner and Paul Morin for their enthusiastic support and essential contributions to this work. It is a pleasure to thank also John Eland, Jacques Delwiche, Marie-Jeanne Hubin and Kenji Ito, for their collaboration. Financial support by CNRS and CEE (scientific training programme in Japan) is gratefully acknowledged.

REFERENCES

Adam M Y, Hellner L, Dujardin G, Svensson A, Martin P and Combet Farnoux F 1989 *J. Phys. B* 22, 2141
Aksela H, Aksela S and Pulkkinen H, 1984 , *Phys. Rev. A* 30, 2456
Armen G B and Larkins F P 1991 *J.Phys.B* 24, 741
Becker U, Wehlitz R, Hemmers O, Langer B and Menzel A 1989 *Phys. Rev. Lett* 63, 1054
Becker U, Hemmers O, Menzel A and Wehlitz R 1991 this conference poster n° Th107
Bizau J M, 1981 Thèse de troisième cycle, Orsay, unpublished
Byron F W and Joachain C J, 1967 *Phys.Rev.* 164, 1
Carlson T A 1967 *Phys.Rev.* 156, 142
Carlson T A 1975 "Photoelectron and Auger Spectroscopy" ed. Plenum Press, New York
Carter S L and Kelly H P   1977 *Phys.Rev.A* 16, 1525   1981 *Phys.Rev.A* 24, 170
Chang T N and Poe R T 1975 *Phys.Rev.A* 12,1432
Codling K and Madden R P 1964 *Phys.Rev.Lett.* 12, 106
Codling K, Madden R P and Ederer D L, 1967 *Phys.Rev.* 155, 26
Mc Culloh K E, Sharp T E and Rosenstock H M, 1965 *J.Chem.Phys.* 42, 3501
Curtis D M and Eland J H D, 1985 *Int. J. Mass Spect. and Ion Proc.* 63, 241
Dujardin G, Leach S, Dutuit O, Guyon P M and Richard-Viard M, 1984 *Chem.Phys.* 88, 339
Dujardin G, Besnard M J, Hellner L and Malinovitch Y, 1987 *Phys.Rev.A* 35, 5012
Eland J H D, 1991 in "VUV photoionisation of molecules and clusters" edited by Ng C Y
Greene C H and Rau, A R P, 1983 *J.Phys.B* 16, 99
Hall R I, Avaldi L, Dawber G, Zubek M, Ellis K and King G C 1991 *J.Phys.B* 24, 115
Hayaishi T, Morioka Y, Kageyama Y, Watanabe M, Suzuki I H Mikuni A, Isoyama G, Asaoka
    S and Nakamura M, 1984 *J.Phys.B*, 17, 3511
Heimann P A, Lindle D W, Ferret T A, Liu S H, Medhurst L J *et al* 1987 *J.Phys.B* 20, 5005
Holland D M P, Codling K, West J B and Marr G V, 1979 *J.Phys.B* 12, 2465
Huetz A, Selles P, Waymel D and Mazeau J 1991 *J. Phys. B* 24, 1917
King G C, Zubek M, Rutter P M, Read F H, Mac Dowell A A, *et al* 1988 *J.Phys.B* 21, L403
Kossmann H, Schwarzkopf O, Kammerling B and Schmidt V 1989 *Phys. Rev. Lett* 63, 2040
Lablanquie P, Nenner I, Millie P, Morin P, Eland J H D, Hubin-Franskin M J and Delwiche J,
    1985 *J.Chem.Phys.* 82, 2951
Lablanquie P, Eland J H D, Nenner I, Morin P, Delwiche J and Hubin-Franskin M J 1987
    Phys. Rev. Lett. 58,992
Lablanquie P 1989 Thèse d'Etat, Orsay, unpublished
Lablanquie P, Delwiche J, Hubin Franskin M J, Nenner I, Morin P, Ito K, Eland J H D,
    Robbe J M, Gandara G, Fournier J and Fournier P G, 1989 *Phys.Rev.A* 40, 5673
Lablanquie P, Ito K, Morin P, Nenner I and Eland J H D 1990 *Z. Phys. D* 16, 77
Lablanquie P and Morin P 1991 to be published in *J.Phys.B*
Mazeau J, Selles P, Waymel D and Huetz A, 1991 to be published
Mobus B, Schartner K H, Magel B and Wildberger M 1991 this conference poster n° We113
Okuyama K, Eland J H D and Kimura K 1990 *Phys. Rev.A* 41, 4930
Pichou F, Huetz A, Joyez G and Landau M, 1978 *J. Phys. B* 11, 3683
Price S D and Eland J H D 1989 *J. Phys. B* 22, L153
Price S D and Eland J H D 1990a *J. Phys. B* 23, 2269 1990b *J. El. Spec. Rel. Phen.* 52, 649
Le Rouzo H 1988 *Phys.Rev.A* 37,1512 and Dal Cappello C 1991 *Phys. Rev. A* 43, 318
Peterkop R 1971 *J. Phys. B* 4, 513
Rau A R P, 1971 *Phys. Rev. A* 4, 207
Samson J A R and Angel G C 1990 *Phys. Rev. A* 42, 5328
Samson J A R 1990 *Phys. Rev. Lett* 65, 2861
Selles P, Mazeau J and Huetz A 1990 *J. Phys. B* 23, 2613
Svensson S, Eriksson B and Gelius U, 1988 *J. Elect. Spect. Rel. Phen.* 47, 327
El Sherbini Th M and Van der Wiel M J 1972 *Physica* 62, 119
Tiwary S N 1982 *J. Phys. B* 15, L323
Tulkki J, Armen G B, Aberg T, Crasemann B and Chen M H, 1987 *Z.Phys.D* 5, 241
Von Raven E, Meyer M, Pahler M and Sonntag B 1990 *J.Elect. Spect. Rel. Phen.* 53, 677
Wannier G H, 1953 *Phys.Rev.* 90, 817
Winkoun D, Dujardin G, Hellner L and Besnard M J, 1988 *J.Phys.B* 21, 1385

Photodissociation of hydrogen: a subtle probe of collision physics

Wim J. van der Zande, Laurens D.A. Siebbeles, Juleon M. Schins and Joop Los
FOM-Institute for Atomic and Molecular Physics, Kruislaan 407, 1098 SJ Amsterdam
The Netherlands

Using fast-beam photofragment spectroscopy the photodissociation of molecular hydrogen is studied. The theoretical description of the measured differential photodissociation cross section is strongly related to that of an atomic collision process with well-defined energy and angular momentum. The analogue of an orbiting resonance, photodissociation of hydrogen by tunneling through a potential barrier is studied in detail. The accuracy of the molecular hydrogen potential curves makes it possible to recognize and quantify non-adiabatic effects.

1. INTRODUCTION

The scientific field of atomic and molecular collisions has received a lot of attention over many decades (Mott and Massey 1933). Understanding of atomic and molecular collisions is of considerable importance. In collision theory the interaction potential takes a central position. The Born-Oppenheimer approximation provides a well-proven prescription for the calculation of interaction potentials. The large difference between the masses of an electron and a nucleus forms the rationalization of the BO-approximation which assumes that the 'fast' electrons can easily follow the motion of the nuclei.

In many types of experiments the BO-approximation has been observed to break down. From the mentioned assumption it follows that collision processes at high center-of-mass velocities are a rich source of non-BO, also known as non-adiabatic, effects. Also, excited electronic states with loosely bound electrons can make non-adiabatic effects important even at very low collision energies. In the low energy regime two conditions have to be fulfilled to interpret consequences of the breakdown of the BO-approximation. BO-potential curves have to be known very accurately to exclude errors in these potentials as source of *apparent* non-adiabatic effects and experiments need to be very well-defined.

Photodissociation studies are such experiments. Photodissociation can be viewed as a half-collision. In photodissociation of a single quantum state the total energy is accurately known. The total angular momentum is exactly known or known up to one unit of angular momentum. Time reversal of an angular-resolved photodissociation event represents a collision with approximately known impact parameter. Molecular hydrogen is a system for which BO calculations of many of its electronic states exist with accuracies of better than 1 cm⁻¹. The theoretical description of the photodissociation of hydrogen is expected to be very accurate and discrepancies with experiment should be reconciled by invoking non-adiabatic effects. Part of the results presented will appear in different form (Siebbeles *et al.* 1991a,1991b). In this contribution the differential cross section of partly overlapping resonances is measured and described. The interference due to the coherent excitation of different rotational levels and the effects of non-adiabatic couplings are discussed.

1.1 Molecular hydrogen

The photodissociation of triplet molecular hydrogen is studied. Experimentally it is relatively simple to obtain a beam of metastable hydrogen in the $c^3\Pi_u^-$ state. Starting from this c-state a number of molecular Rydberg (n=3) states can be excited. Figure 1 shows the potential energy diagram of molecular hydrogen. The n=3 states, which are of importance for the present study, are the $g^3\Sigma_g^+$, the $h^3\Sigma_g^+$ (Rychlewski 1989) the $i^3\Pi_g$ (Kolos and Rychlewski 1977) and the $j^3\Delta_g$ (Rychlewski 1984) states, all characterized by a H_2^+ core with a (3sσ), or (3dλ, λ=δ,π,σ) excited electron attached. Radial coupling exists between the two $^3\Sigma_g^+$ states (Schins *et al* 1991a,1991b), while molecular rotation is known to affect the spectroscopy of the three (3dλ) states (Keiding and Bjerre 1987). The occurrence of barriers in the g-state and i-state potential curves is the result of strongly avoided crossings between these states and higher lying (partly double excited) ones (Lembo *et al* 1990). A collision between H(2p) and H(1s) atoms at energies of 330 meV which is just below the top of the i-state barrier can give rise to shape resonances. The reverse process, barrier tunneling is studied.

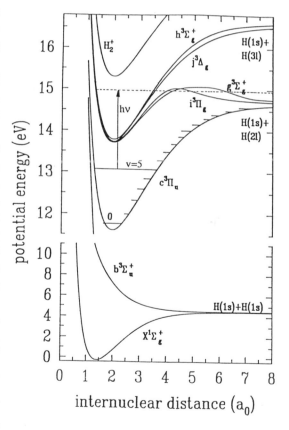

Fig. 1 Potential energy diagram of triplet hydrogen. The excitation step is indicated. Tunneling through the i-state barrier gives H(1s)+H(2p) (dashed line).

2. EXPERIMENTAL

In this section a brief discussion will be given of our fast beam photofragment set-up (de Bruijn and Helm 1986, Helm *et al* 1984). A fast (4-6 keV) beam of metastable H_2 is generated by charge transfer between (keV and mass selected) H_2^+ ions and cesium vapor. Fig. 2 shows the principle of the method (de Bruijn and Los 1981). The fast H_2^* beam passes through the cavity of a CW-dye laser, the polarization vector is chosen parallel to the fast beam. At this point metastable hydrogen is excited to dissociative n=3 levels. The photofragments scatter out of the beam due to the kinetic energy imparted to them in the dissociation process. These photofragments are intercepted on a time- and position-sensitive detector. The impact position of the fragments and the difference in arrival time are measured. The obtained spatial distance (R, resolution 140 μm) and distance in time (τ, resolution 1.5 ns) can be converted into a kinetic energy release value (KER or ε) and dissociation angle with respect to the laser polarization (θ).

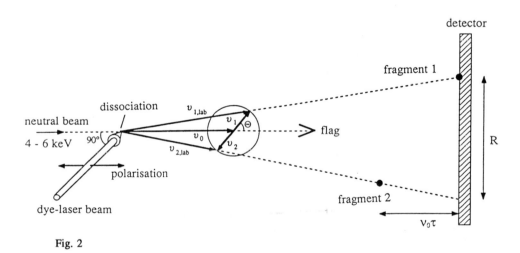

Fig. 2

Fig. 2 shows that a simple relation exists between the observables and the calculated quantities. The advantage of measuring ε and θ for each event is obvious; photon energy and the observed KER value (usually) identify the lower level unambiguously by invoking conservation of energy. Thus, if more than one lower level is photodissociated at a certain laser wavelength still quantum state specific differential photodissociation cross sections can be obtained.

3. THEORY OF PHOTODISSOCIATION

The original account of molecular photodissociation processes given by Zare (1972) is strongly rooted in collision physics. In a photodissociation experiment the yield of photofragments in a well-defined laboratory direction Θ (and Φ) is wanted. Thus, plane waves are used to describe continuum states. However, a photon carries one unit on angular momentum and the initial state angular momentum is well-defined. Therefore, few angular momentum states are needed. From this point of view a more direct way is provided by considering the asymptotic molecular states directly (Siebbeles *et al* 1991b). Be $|\Psi_0\rangle$ the initial state and H the total Hamiltonian consisting of H_0 (the Hamiltonian of the isolated molecule and the non-interacting photon) and H_{int} (the interaction operator). At time t the system is described by:

$$|\Psi(t)\rangle = \exp(\frac{-i\,H\,t}{\hbar})\ |\Psi_0\rangle \tag{1}$$

The probability of the molecule to be photodissociated into the solid angle $d\Omega = \sin\theta d\theta d\phi$ is equal to the square of the amplitude of the wavefunction at large R and at θ, ϕ:

$$\frac{dw_M(\theta,\phi)}{d\Omega} = \lim_{R\to\infty} \int |\,|\Psi(t)\rangle|^2 R^2 dr \tag{2}$$

The integration is performed over the coordinates of all electrons, denoted by **r**. If H_{int} is brought into account up to first order, then

$$|\Psi(t)\rangle = -\frac{1}{2i\,\pi} \int \frac{dE\,e^{-i\,Et/\hbar}}{E - E_0 + i\,\eta} \left[|\Psi_0\rangle + (E - H_0 + i\,\eta)^{-1} H_{int}\,|\Psi_0\rangle \right], \tag{3}$$

with $\eta \to 0^+$. This equation can be evaluated by using the closure relation:

$$I = |\Psi_0\rangle\langle\Psi_0| + \int d\varepsilon \sum_{N',M',p'} |\Phi_{N'M'\Lambda'p'\varepsilon}\rangle\langle\Phi_{N'M'\Lambda'p'\varepsilon}| \tag{4}$$

In Eq. 4 the kets $|\Phi\rangle$ contain all (accessible) continuum eigenstates of the molecular Hamiltonian H_0. N' is the total angular momentum and M', Λ' the projections on the laser polarization axis and the molecular axis. The symmetry index p' is related to parity and ε is the ker value. As we will see later these eigenstates can be pure or coupled BO-states. We note that quasibound states which are separated from the dissociation continuum by a barrier in the electronic potential are treated as true continuum states. Insertion of the closure relation and taking the limit of R$\to\infty$ gives:

$$\Phi(t)_{R\to\infty} \sim \sum_{N'\,M'\,p'} \Phi_{N'M'\Lambda'p'} (-i)^{N'} e^{i\,\delta_{N'}} \langle\Phi_{N'M'\Lambda'p'\varepsilon}|\,H_{int}\,|\Psi_0\rangle \frac{e^{i\,\kappa R}}{R} e^{-i\,E_0 t} \tag{5}$$

κ is the momentum associated with the kinetic energy of the photofragments, ε, via $\kappa = \sqrt{2\mu\varepsilon}/\hbar$. E_0 is the total energy. Bound excited states can be added to Eq. 4 too; however in taking the limit of R$\to\infty$ these bound states as well as Ψ_0 disappear. The magnitude of the matrix element $\langle\Phi_{N'M'\Lambda'p'\varepsilon}|\,H_{int}\,|\Psi_0\rangle$ is found by evaluating the transition dipole moment, and the Franck-Condon overlap between the lower initial and the excited continuum state wave functions. The angular momentum selection rules follow from the integration over the angular parts of the wavefunctions. A numerical determination of the stationary excited state(s) continuum wave functions also gives the phase shifts $\delta_{N'}$ of Eq. 5. These phase shifts strongly sense the presence of resonances. Finally, the differential cross section can be expressed in the form:

$$\frac{d\sigma(\theta)}{d\Omega} = \sum_M \frac{dw_M}{d\Omega} = \sigma_0[1 + \beta\,P_2(\cos\theta)] \tag{6}$$

Where σ_0 is the total cross section and $P_2(\cos\theta)$ is the second Legendre polynomial. Isotropy of the initial state is assumed; all its M_N magnetic sublevels equally populated. The anisotropy parameter, β, depends on the phase shifts and the magnitudes of the matrix elements. For direct photodissociation via a repulsive potential curve the β parameter is given by $\beta = 2$ if Λ equals Λ' (parallel transition) and $\beta = -1$ is $|\Lambda - \Lambda'| = 1$ (axial recoil, Zare 1972).

4. RESULTS

The photodissociation laser is tuned such that $c^3\Pi_u$ v=5, N levels are excited to energies about 350 cm^{-1} below the top of the barrier in the $i^3\Pi_g$ state potential curve, as has been indicated in Fig. 1. Fig. 3a shows a KER spectrum. The angular resolved spectrum in Fig 3b pertains only to dissociations of c-state, v=5, N=1 molecules as deduced from the ker value. These spectra are cuts through the recorded N(ϵ,θ) spectrum with the laser at a fixed energy of 15600 cm^{-1}. Fig 3a shows that simultaneously the v=5 N=1, 2 and 3 levels contribute to the total photodissociation signal. Dissociation of the N=1 level is resonance enhanced since the intensity of the N=1 signal is large with respect to its contribution in the fast beam. The angular spectrum shown in Fig. 3b shows the differential photodissociation cross section. The dotted line gives the least square fit to Eq. 6. The resulting β is β=0.48.

Fig. 4 shows the total (state specific, $c^3\Pi_u$, v=5, N=1)) photodissociation cross section results as function of the excitation wavelength. The crosses are the experimental results. Two broad features can be recognized as a Q- and R- transition to the $i^3\Pi_g$ "v'''=5, N'=1 and 2 states. The quotes around the vibrational assignment stress that a vibrational level is formally not defined for a quasi-bound continuum state. The width of the resonances of approximately 30 cm^{-1} reflect the efficient tunneling process.

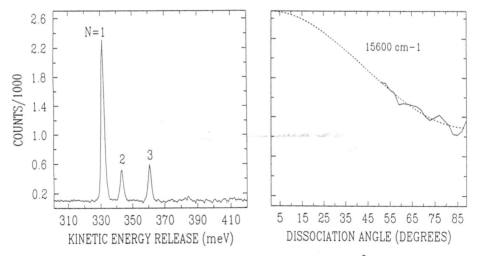

Fig. 3. a) kinetic energy release spectrum. Photodissociation signal is due to $c^3\Pi_u^-$, v=5, N=1, 2 and 3 levels; b) angular-resolved spectrum for photodissociation of the $c^3\Pi_u^-$, v=5, N=1 level only. The dashed curve is a fit to a 1+βP_2 function giving β=0.48

The v=5,N=1 differential photodissociation cross section is given in Fig. 5 in the form of the β-parameters extracted at each photon wavelength. Note how dramatically the differential cross section changes. On the resonances the β-parameters are $\beta \approx \frac{1}{2}$. Note that the β-parameters almost span the full range allowed from β=-1 (sin$^2\theta$ distribution) to $\beta \approx 2$ (cos$^2\theta$ distribution). In the following sections we will give a quantitative explanation for the results and illustrate the need to incorporate non-adiabatic effects to arrive at a correct description.

5. DISCUSSION

5.1 Quasi-classical picture

In this short section classical arguments will be given to rationalize the observed changes in the anisotropy parameter. The studied electronic transition is a Π-Π transition. In a classically treatment Jonah (1971) has calculated the effect of an 'orbiting' resonance on the anisotropy of photofragments. He showed that for a parallel transition a direct dissociation yields $\beta=2$. In the limit of dissociation after many molecular rotations the anisotropy is reduced to the limiting value of $\beta=\frac{1}{2}$. All β-values with $\beta<\frac{1}{2}$ for a parallel transition are non-classical.

It is tempting to reverse the argument of Jonah (1971) and extract an orbiting lifetime from the observed β-values. This implies that the lifetime *decreases* with decreasing excitation energy below 15526 cm^{-1}. The physical barrier that is constituted by the potential barrier only increases slightly. Thereby, the tunneling probability is slightly reduced. Thus the lifetime decrease is not caused by a changed reflection coefficient. Quantum mechanically, one is not allowed to extract a tunneling probability from a measurement at one excitation energy. At a single excitation energy no flux related to tunneling events can be measured since the eigenstate is necessarily stationary. The fact that the β-parameter below the resonance approaches the axial recoil value of $\beta=2$ is therefore hard to interpret in semi-classical terms.

5.2 Quantum-mechanical calculations

We start out with a one-state calculation. The c-state and i-state potential curves are taken from Kolos and Rychlewski (1977). Koch and Rychlewski (in Schins *et al* 1991b) provide various dipole moments. The Franck-Condon factors were evaluated by solving numerically the Schrödinger equation for a fixed total angular momentum (N'=1 or 2). In this particular case only a Q- and R-transition are possible since N'=0 does not exist in a Π−state. The two corresponding matrix elements and phase shifts are M_1 (=$M_{N'=1}$), M_2 and δ_1, δ_2 resp.. All these quantities depend strongly on the excitation energy (E). The expressions for the total cross section and the anisotropy parameter now read:

$$\sigma_0(E) \approx \left[M_1{}^2 + M_2{}^2 \right] \tag{7}$$

$$\beta(E) = \frac{1}{2} + \frac{3M_1M_2}{M_1{}^2 + M_2{}^2} \cos(\delta_1 - \delta_2) \tag{8}$$

From these equations one appreciates that coherence in the simultaneous excitation of the N'=1 and N'=2 character does not manifest itself in the total photodissociation cross section. Only the anisotropy parameter $\beta(E)$ is affected by coherence through the product M_1M_2. Note that the $\cos(\delta_1 - \delta_2)$ only starts to play a role if M_1 and M_2 are of the same order of magnitude. The dotted lines in Figs. 6 and 7 are the results of this calculation. The calculated curve is scaled to the first resonance. The excitation energy out of the calculation had to be shifted over 15 cm^{-1} to match the positions of the resonances. Unfortunately, the observed resonances are more narrow than the calculated ones. Also, the amplitude of the second resonance is not-well reproduced.

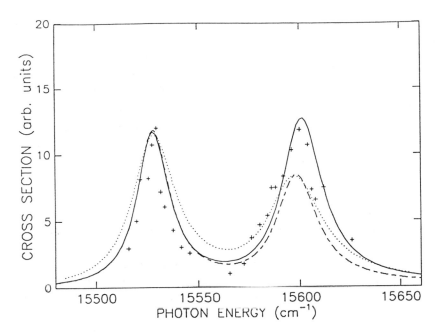

Fig. 4 Relative photodissociation cross section as function of excitation energy. The error in the experimental points (+) is 10%. The different lines are the results of various calculations (see text).

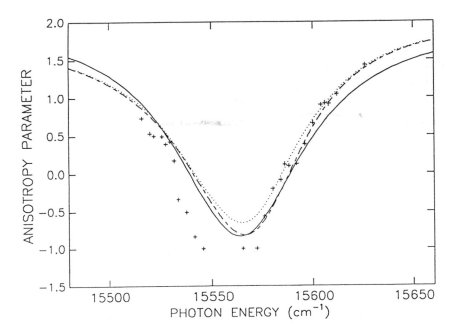

Fig. 5 Anisotropy parameters as function of the excitation energy. The error in the experimental points (+) is ±0.25. The curves represent different calculations (see text).

The calculated anisotropy parameter as function of excitation energy qualitatively agrees with the experiment. By looking at Eq. 8 we can appreciate the quantum-mechanical effects between the resonances. The phase shift δ_1 senses the first resonance and changes with almost π, δ_2 is still slowly changing. Thus, $\cos(\delta_1-\delta_2)$ changes sign resulting in a negative β-parameter. The discrepancy in the total cross section implies that the calculation overestimates the tunneling efficiency. The discrepancy observed in the β-parameters leads to the same conclusion. In the case of a broad resonance the phase shift $(\delta_1(E))$ changes less rapidly with E, thus $|\delta_1-\delta_2|<<\pi$ for all E. In the limit of axial recoil $M_1(E)$ equals $M_2(E)$ and $\delta_1(E)$ equals $\delta_2(E)$ giving $\beta=2$ in Eq. 8.

The quality of the BO-potential curves is not questioned. Nevertheless Rychlewski (1991) has recently calculated adiabatic corrections being the effect of the velocity of the nuclei on the i-state potential curve. The correction peaks at the top of the barrier (about 59 cm^{-1}). The magnitude of the correction reflects the change of electronic character over the barrier (see also section 6). As an increased barrier height reduces the tunneling probability, the dashed curves agree better with the respective experimental ones. More importantly the necessary shift in the first calculation of 15 cm^{-1} now turns out to be superfluous. Apart from scaling the magnitude of the first resonance no single parameter enters in the calculation.

Unfortunately the total cross sections still disagrees. In the following we include intramolecular couplings. In the general formalism given in section 3 intramolecular couplings can be incorporated; H_o changes by inclusion of the coupling operator, the new eigenstates $\phi_{N'M'p'\epsilon'}$ are now expressed as sum over the coupled $\phi_{N'M'\Lambda'p'\epsilon'}$ BO-states. As mentioned in section 1.1 the three $3d\lambda$ states are coupled through rotational coupling. In molecular coupling the parity of the state as well as the total angular momentum is conserved. The initial state is of positive total parity ($c^3\Pi_u^-$, N=1). In photo-excitation the total parity changes; via a Q-transition one therefore reaches the $i^3\Pi_g^+$, N'=1 and via an R-transition the $i^3\Pi_g$, N'=2 level. Thus, the N'=2 level only can interact with the $j^3\Delta_g^-$, N'=2 state. The N'=1 level can interact with both Σ_g^+ states. Taking into account the coupling with the Σ_g^+ -states is at the moment beyond our computational capacities. Fig. 1 shows that the $g^3\Sigma_g^+$ state also has a barrier, thus a second open channel enters. The coupling of the i-state with the j-state is taken into account. The rotational coupling can be described in the pure precession model (Van Vleck 1929) which yields a very good estimate for the coupling matrix elements. The $\Phi_{N'M'\Lambda'p'}$ of Eq. 4 are now combinations of (i- and j-) Born-Oppenheimer eigenstates. As a consequence, the matrix element M_2 is composed of two contributions:

$$M_2 = \int \chi_i(R) <i\,|u_z|c> \chi_{vN}^c(R)\,dR + \int \chi_j(R) <j\,|u_x+u_y|c> \chi_{vN}^c(R)\,dR \quad (9)$$

χ_i and χ_j are solutions from the two coupled equations that take rotational coupling into account. $<i\,|u_z|c>$ and $<j\,|u_x+u_y|c>$ are the transition dipole moments. Taking the limit of $R \to \infty$ removes the bound j-state part from the wavefunctions $\Phi_{N'M'\Lambda'p'}$ in Eq. 5. The anisotropy is thus affected only through the magnitude of the matrix elements but not by the Wigner rotation functions of the j-state; at large R the continuum is a pure i-state. The total cross section is influenced via M_2 by the transition dipole moment to the j-state contribution.

Figs. 4 and 5 show these final results (solid curves). We note that coupling with the *j*-state shifts the calculated position of the second resonance by less than 2 cm-1 but increases its intensity by a factor of two. The agreement between the zero-parameter calculation and the experimental total cross section is very good. The change in the calculated anisotropy parameter is, as anticipated, much smaller. In fact, the agreement with the experimental curve seems to worsen. However, the remaining discrepancies are very close to the experimental uncertainties.

6. 'VELOCITY' DURING BARRIERTUNNELING

The barrier in the *i*-state potential curve is due to a strongly avoided crossing. A doubly excited repulsive state of $2p\pi_u, 2p\sigma_u$ -character is responsible (Lembo *et al* 1990). The adiabatic corrections to the *i*-state potential curve are indications of this avoided crossing. We recently have performed experiments on the photo-dissociation of hydrogen through excitation of the $i^3\Pi_g(v'=4,N')$ levels. These quasi-bound levels lie 1350 cm-1 below the v'=5 resonances and are therefore much narrower ($\approx 10^{-2}$ cm-1). Large as yet unexplained discrepancies are found between the calculations and observations of their dissociation dynamics.

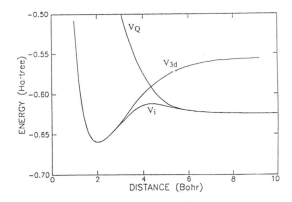

Fig. 6 Diabatic potentials V_Q (doubly excited) and V_{3d} (Rydberg) illustrating the change in character of the adiabatic *i*-state, V_i. The second $^3\Pi_g$ adiabatic potential curve is not shown.

These discrepancies formed the motivation to estimate the influence of the second adiabatic $^3\Pi_g$ state. In view of the accordance reached with the v'=5 resonances just described this influence should manifest itself more in the v'=4 than in the v'=5. Fig. 6 shows a set of *diabatic* $^3\Pi_g$ potential curves; V_{3d} has pure Rydberg($1s\sigma,3d\pi$) character, the repulsive V_Q (Guberman 1983) is doubly excited ($2p\pi,2p\sigma$) throughout. These states are coupled through the electron energy operator H_{el}. The off-diagonal coupling matrix element is $H_{el} = \r((V_{3d} - V_i)(V_Q - V_i))$. The formulation of this coupling ensures that the adiabatic *i*-state curve is reproduced upon transformation of the two diabatic states into their adiabatic counterparts. The energy solutions of the two-state coupled equations are calculated to be very close to the values found in the *i*-state adiabatic calculation. The important observation is that the width of the v'=5,N'=1 resonance is not affected but that the width of the v'=4,N'=1 resonance is reduced with 30% from 0.0028 to 0.0019 cm-1. The two-state diabatic calculation is equivalent to a two-state adiabatic calculation in which the two states are coupled by radial coupling. The results suggest that the second high lying adiabatic state has more influence on the v'=4 than on the v'=5. The radial coupling (by the $\frac{d}{dR}$ -operator) is related to the momentum of the nuclei.

We conclude that the increasing effect for the v'=4 level implies that the 'penetration velocity' through a potential barrier increases for energies deeper in the classically forbidden region of the barrier.

7. CONCLUSIONS

We have given a progress report on the photodissociation of molecular hydrogen. The photodissociation by tunneling through the potential barrier of the *i*-state potential has been stressed. The time reversal of this process can be viewed as an atomic collision process in which the energy is known exactly and angular momentum is known very well. To reproduce the observed total and differential photodissociation cross sections through v'=5 levels non-adiabatic couplings had to be taken into account. The effect of the coherent excitation of two partial waves is observed as a large change in anisotropy over the resonances. In a two-state diabatic calculation it is suggested that the diabatic character does not only increase with energies above the crossing but also below the crossing in the classically forbidden region.

This work is part of the research program of the Stichting voor Fundamenteel Onderzoek der Materie (Foundation for the Fundamental Research on Matter) and was made possible by the financial support of the Netherlands Organisation for the Advancement of Research.

8. REFERENCES
de Bruijn D P and Los J 1982 Rev. Sci. Instrum. **53** 1020
de Bruijn D P and Helm H 1986 Phys. Rev. **A34** 3855
Guberman S L 1983 J. Chem. Phys. **78** 1404
Helm H, de Bruijn D P and Los J 1984 Phys. Rev. Lett. **53** 1642
Jonah C 1971 J. Chem. Phys. **55** 1915
Keiding S R and Bjerre N 1987 J. Chem. Phys. **87** 3321
Kolos W and Rychlewski J 1977 J. Mol. Spectrosc. **66** 428
Lembo L J, Huestis D L, Keiding S R, Bjerre N and Helm H 1988 Phys. Rev. **A38** 3447
Mott N F and Massey H S W 1933 *The theory of atomic collisions* (Oxford; Clarendon Press)
Rychlewski J 1984 J. Mol. Spectrosc. **104** 253
Rychlewski J 1989 J. Mol. Spectrosc. **136** 333
Rychlewski J 1991 personal communication
Schins J M, Siebbeles L D A, Los J and van der Zande W J 1991a, Phys. Rev. A in press
Schins J M, Siebbeles L D A, Los J, van der Zande W J , Rychlewski J. and Koch H. 1991b Phys Rev in press
Siebbeles L D A, Schins J M, Los J and Glass-Maujean M 1991a Phys. Rev A in press
Siebbeles L D A, Schins J M, van der Zande W J, Beswick J A 1991b in preparation
Van Vleck J H 1929 Phys. Rev. **33** 467
Zare R N 1972 J. Mol. Photochem. **4** 1

Experimental determination of accurate van der Waals interaction potentials

Laura Beneventi, Piergiorgio Casavecchia, and Gian Gualberto Volpi

Dipartimento di Chimica, Universitá di Perugia, 06100 Perugia, Italy

ABSTRACT: The contribution of molecular beam scattering to the knowledge of intermolecular forces for van der Waals systems is examined with reference to recent work from this laboratory. Examples are shown of high-resolution measurements, in the thermal energy range, of elastic and total differential cross sections with well resolved diffraction and/or rainbow oscillations for Ne-Ne, Ar, Kr, and Xe, and for He, Ne, Ar-N_2, O_2, NO, CO_2, and Cl_2, respectively. The derivation of accurate interaction potentials, within a multiproperty analysis of microscopic and macroscopic properties, is discussed.

1. INTRODUCTION

A knowledge of the forces between atoms and molecules is fundamental to an understanding of the macroscopic properties of matter. Their determination is a very active field of research since accurate intermolecular potentials are essential for the ultimate success of the current endeavor of developing molecular descriptions of condensed matter behavior, and for a detailed understanding of the mechanism of chemical reactions and of the energy transfer in laser media and interstellar clouds. Despite their importance, up until two decades ago, our knowledge of interaction potentials for weakly bound van der Waals molecules (i.e., molecules whose binding energy ranges from about 0.005 to 0.05 eV) was surprisingly meager. The only source of information were traditional measurements of gaseous bulk properties, as transport and second virial coefficients, which are mainly sensitive to isotropically averaged quantities (Hirschfelder et al 1964). Simple spherical models of intermolecular potential, such as the Lennard-Jones model, were adequate for the description of the observed properties.

In the last 20 years, a growing explosion has occurred with the advent of new methods for characterizing molecular interactions. The methods which have led to a rapid recent progress in our knowledge of intermolecular forces involve molecular beam scattering and spectroscopic techniques (Maitland et al 1981). Initially, accurate potentials were limited to a few atom-atom interactions (Scoles 1980, Aziz 1984), while recently the combination of beam scattering, spectroscopy, as well as theoretical work has made it possible to determine accurate anisotropic potentials for atom-molecule interactions (Buck 1987, Hutson 1990), and also for open-shell atom interactions (Casavecchia et al 1982, Aquilanti et al 1976, 1990). For the latter, essential has been the implementation of the magnetic selection technique and of the proper theoretical framework by Aquilanti et al (1980, 1984, 1990). Detailed experimental data on the interaction involving polar systems and polyatomic molecules start also to

appear and the routes to understanding more complex systems have become
clearer (Rigby et al 1986). At the same time, it has become increasingly
obvious that the simple Lennard-Jones potential model is inadequate for the
description of many physical properties. As a result, more elaborate
potential models have been introduced, initially for atom-atom interactions
(Ahlrichs et al 1977, Tang and Toennies 1978), and, more recently, for
anisotropic atom-molecule systems (Rodwell and Scoles 1982, Tang and
Toennies 1980, Fuchs et al 1984, Bowers et al 1988).

From a theoretical point of view, the calculation of reliable potentials
for van der Waals systems is still a big problem, with the exception of a
few light cases. Generalization to atom-molecule interactions of atom-atom
modeling procedures has recently been applied beyond the H_2-rare gas
systems with various degrees of success (Fuchs et al 1984, Bowers et al
1988). Our most detailed knowledge of intermolecular forces comes from
experiment. The molecular beam scattering method opened a new era in the
determination of interaction potentials: it represents a direct tool, in
principle the most general one since it can probe both the attractive and
repulsive regions of the interaction, and both the isotropic and
anisotropic components (Buck 1987). For atom-molecule systems, in which the
molecular species possesses a permanent dipole or quadrupole moment,
spectroscopic techniques, such as molecular beam electric resonance (Novick
et al 1973, Muenter 1987), mid-infrared (IR) (Nesbitt 1988) and
far-infrared (IR) (Saykally 1989) laser rovibrational spectroscopy, can
successfully be applied for obtaining detailed information on the
intermolecular potential near the equilibrium configuration of the van der
Waals molecule and, in some cases, also far from equilibrium. Information
away from the minimum and on the repulsive part of the interaction still
relies on scattering experiments, which are in general complementary to
spectroscopic studies. For systems where the molecular species is a
homonuclear diatomic, such as N_2 and O_2, interacting with a rare gas,
scattering methods are practically the only source of detailed information
on the intermolecular potentials, since these systems are still out of
reach of the state-of-the-art techniques of mid-IR and far-IR laser
rovibrational spectroscopy.

The objective of this Progress Report is to review recent work from our
laboratory in which high-resolution differential scattering experiments
have been exploited, within a multipropery scheme of analysis, to derive
accurate interaction potentials for a variety of simple van der Waals
molecules. Being aware that in order to derive a potential energy surface
(PES) which is reliable over a wide range of internuclear distances, a
multiproperty analysis of several independent and complementary
experimental properties which are sensitive to different features of the
potential is necessary, we have always followed a rather general procedure
of analysis that can be outlined as follows. Information on the anisotropy
of the repulsive wall is extracted from state-to-state rotationally
inelastic differential cross sections, if available (Buck et al 1980,
Faubel 1983). When single rotational transitions cannot be resolved,
differential energy loss spectra still provide a sensitive probe of the
repulsive anisotropy, especially if rotational rainbows are observed.
However, the above type of data give information only on the "relative"
anisotropy of a system, i.e., the difference between potential curves of
different orientations (Buck 1986). The determination of the absolute scale
of the potential relies on the precise measurement of diffraction
oscillations in the total (elastic + inelastic) differential cross sections
(DCS). The origin of the diffraction oscillations in DCS is well understood
(Buck 1987). Their angular positions allow a direct measure of the diameter

σ (the distance at which V(R)=0) of the repulsive wall to better than 1%, and this represents one of the most detailed information among those provided by scattering experiments. Total DCS data also carry a significant amount of information on the anisotropy, since the anisotropy of the potential well depth and minimum position produces a quenching of the rainbow and diffraction oscillations, respectively (Pack 1978). In favorable cases this quenching can be accurately evaluated and reliable information on the anisotropy obtained (Beneventi et al 1986b, Danielson et al 1987, Buck 1987). Measurements of the rainbow structure in the total DCS, and/or of the glory structure in the total integral cross section (ICS), or of the second virial coefficient down to low temperatures, allow the determination of the other relevant features of the attractive well of the full PES. The long-range part of the interaction can be obtained experimentally from the absolute value of the ICS measured as a function of collision velocity, if available, or from calculated ab initio or semiempirical van der Waals coefficients. Transport data can then be used to check the validity of the potential so derived and, if available up to very high temperature, to extend its validity in the region of the repulsion. Transport and virial data are not very sensitive to fine details of the potential nor, especially, to its anisotropy. Bulk relaxation data and transport property field-effects provide additional data for testing the anisotropic components of the PES, especially in the region of the repulsive wall (Wong et al 1990, Beneventi et al 1991a).

Benefitting considerably from the very detailed information content of precise differential cross section data, measured under high-resolution conditions and presenting well resolved diffraction oscillations, and following the prescription outlined above, we have been able to derive or assess accurate potentials for a large variety of atom-atom (Ne-Ne, Ne-Ar, Ne-Kr, Ne-Xe) and atom-molecule (He-N_2, He-O_2, He-NO, He-CO_2, He-Cl_2, Ne-N_2, Ne-O_2, Ne-NO, Ne-Cl_2, Ar-N_2, Ar-O_2, Ar-NO, Ar-Cl_2, and Kr-NO) systems. Final PES for some of these van der Waals molecules have been recently published (Casavecchia et al 1984, Beneventi et al 1986ab, 1988ab, 1991ab), while PES for others are currently under fine-tuning. Before some typical results are reviewed and the method of analysis to derive the potential discussed, the experimental technique used to measure differential cross sections will be described.

2. EXPERIMENTAL

Measurements of differential cross sections were carried out by using a high-resolution crossed molecular beam apparatus built in our laboratory, and which has been described in detail elsewhere (Beneventi et al 1986). Briefly, two well collimated, differentially pumped, supersonic nozzle beams are crossed at 90° in a large scattering chamber kept at 10^{-7} mbar in operating conditions. The in-plane scattered lighter particle is detected by a rotatable triply differentially pumped ultra-high-vacuum quadrupole mass spectrometer detector maintained in the 10^{-11} mbar pressure range in the ionization region by extensive ion-, turbo-, and cryo-pumping.

Since the absolute accuracy of the measurements depends on the calibration of the velocity distributions of the two beams, as well as of the angular positions, great care was devoted to all these aspects. The beam velocities were measured by absolute time-of-flight analysis to better than 0.6%, the angular locations were determined to better than 0.03°, and the geometrical alignments were carried out to within 0.05 mm. The primary beam (He, Ne, or Ar) has an angular divergence of 0.4°, while the secondary target beam of 1.8°. While He and Ne beams were produced by high pressure expansion to get high speed ratios (s=36-59), the Ar, O_2, N_2, NO and Cl_2 beam pressures were chosen as to give the maximum signal intensity while avoiding possible problems from condensation effects (typical velocity spreads range from 7% for Ar to 20% for Cl_2). The detector angular resolution, which is variable in our machine, was set to 0.5° for a point collision zone. The same geometrical arrangement has been maintained for all atom-atom and atom-molecule experiments which have been performed over the past few years. This permits us to put on a very precise relative basis the different results. The narrow divergences in angle and velocity of the colliding beams, as well as the high angular resolution of the detector proved to be mandatory for the observation of a well resolved diffraction structure in the differential cross sections. The molecular species in the supersonic beams are expected to be in their lowest rotational levels. The angular dependence of the elastic and total (elastic + inelastic) DCS was obtaind by taking at least four scans of 30-90 s at each angle. The secondary target beam was modulated at 160 Hz by a tuning fork chopper for background subtraction. For most of the systems, we have also measured the DCS at negative angles with respect to the primary beam, which provides higher resolution (Beneventi et al 1988a, 1991b).

3. SURVEY OF EXPERIMENTAL RESULTS AND DISCUSSION

3.1 Rare gas-rare gas potentials

A high accuracy in the knowledge of the two body forces is fundamental for making advances in the knowledge of the three body forces. Benchmark systems for this endeavor are, as in other areas of physics, the rare gas systems. The interaction potentials for the like and unlike rare gas pairs are presently considered to be within one, two percent of the true potentials in the attractive and low repulsive regions, with the exception of some unlike pairs, as for instance Ne-heavier rare gases, which have still larger uncertainties (Aziz 1984). This is due to the fact that the Ne-heavier rare gas systems are characterized by a reduced mass which is too light for permitting, in the thermal energy range, the observation of a well developed rainbow structure, but too heavy for allowing the observation of diffraction oscillations which have very small angular spacing. As a result, the two type of oscillations were not resolved in the past in Ne-rare gas scattering because of the limited experimental resolution. We recall that the rainbow scattering angle is probably the most direct and unique measure of the potential well depth, ε, while the diffraction oscillations give a direct measure of the diameter, σ, of the repulsive wall (and then of the minimum position, R_m) (Buck 1987).

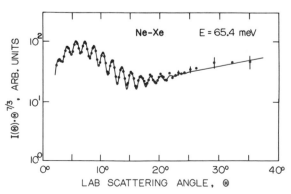

Fig. 1. Elastic differential cross section data for Ne-Xe at a collision energy of 65.4 meV. Continuous line: calculation with a multiproperty best-fit Ne-Xe potential (see text).

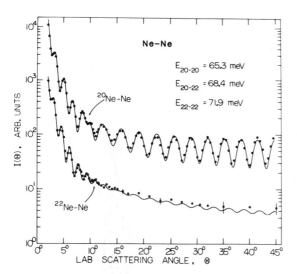

Fig. 2. Differential cross section data for Ne-Ne obtained by detecting ^{20}Ne (upper plot) and ^{22}Ne (lower plot). The continuous line is the prediction of the HFD-C2 potential of Aziz et al (1983). The natural isotopic abundance of the Ne target beam is taken into account in the calculation of the cross sections.

We have measured the elastic DCS under very high resolution conditions for Ne-Ne, Ne-Ar, Ne-Kr and Ne-Xe, which has permitted us to resolve clearly all the quantum diffraction oscillations superimposed on the main rainbow oscillation at collision energies corresponding to room temperature beams. As example of the data quality, we report in Figure 1 the angular dependence of the DCS for Ne-Xe at E=65.4 meV. In a previous measurement at a comparable collision energy the diffraction structure was not resolved at all (Ng et al 1974). In Ne-Ne the symmetry and diffraction oscillatory patterns have been clearly distinguished by performing isotope resolved angular distribution measurements. The results are shown in Figure 2. These new high quality data have allowed us to determine σ with an accuracy of better than 0.6% and ε to about 1%, and so they represent a stringent test of the available potentials for the systems in question. The HFD-C2 potential of Aziz et al (1983) is confirmed to represent the most accurate description of Ne$_2$ to present (see Figure 2). Instead, for Ne-Kr and Ne-Xe none of the available potentials (Aziz 1984) affords a very good fit to the experimental data. The present DCS data permit a sophisticated refinement of the Ne-Kr and Xe potentials within a multiproperty analysis, as shown for Ne-Ar (Candori et al 1986). This work is currently under way. The continuous line in Fig. 1 represents a calculation using a new improved Ne-Xe potential (Beneventi et al 1991c).

3.2 Rare gas-molecule potentials

After that the benchmark rare gas-rare gas potentials were understood, attention was devoted to the H$_2$-rare gas potentials. These represent the atom-molecule systems better characterized from both theoretical and experimental point of view (Buck 1982, Hutson 1990). Recently, much attention has been devoted to rare gas (Rg)-molecule potentials, as Rg-N$_2$, O$_2$, NO, CO$_2$, Cl$_2$: N$_2$ is a good prototype of other diatomics, and O$_2$ and N$_2$ are molecules of importance in aerodynamics; He-CO$_2$ is relevant in laser systems; Rg-Cl$_2$ have attracted large attention as model systems in vibrational predissociation studies (Cline et al 1989). Total DCS have been measured for a variety of the above systems (see Introduction). In the Figures 3 and 4 we summarize the total DCS data for the He-diatom and Ne-diatom (diatom = O$_2$, N$_2$, NO) systems, respectively, compared to the elastic DCS of the corresponding isotropic systems, namely He-Ar and Ne-Ar. On the right-hand-side (RHS), the spherical average potentials, derived

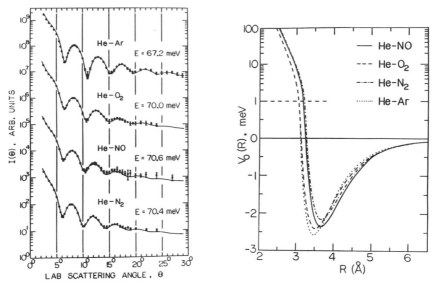

Fig. 3.   Left: Total differential cross section data for scattering of
He from O_2, NO, and N_2, compared with the elastic DCS data for the
corresponding   isotropic   He-Ar   system.   The continuous lines are IOS
calculations with the best-fit multiproperty potentials. For He-Ar the
accurate   He-Ar   potential   (Aziz   1984)   is   used.   Right: Comparison
between the spherical average potential, $V_0(R)$, for the same He-diatom
systems and the He-Ar potential.

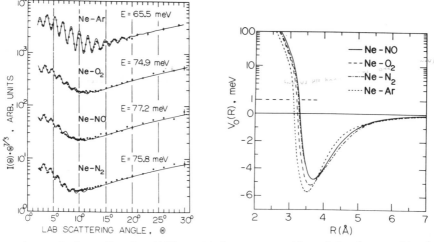

Fig. 4.   Left: Total differential cross section data for scattering of
Ne from O_2, NO, and N_2, compared with the elastic DCS data for the
corresponding   isotropic   Ne-Ar   system.   The continuous lines are IOS
calculations with the best-fit multiproperty potentials. For Ne-Ar the
curve is calculated from the potential of Candori et al (1986). Right:
Comparison between the spherical average potential, $V_0(R)$, for the same
Ne-diatom systems and the Ne-Ar potential.

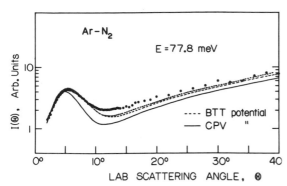

Fig. 5. Total differential cross section data, multiplied by $\Theta^{7/3}$, for Ar-N$_2$ compared with the predictions of the BTT (Bowers et al 1987), CPV (Candori et al 1983) and present multiproperty potential (dotted line) in the IOS approximation.

from a multiproperty analysis (see below), are also depicted. As an example of the total DCS for an Ar-containing system, in Figure 5 we report the data for Ar-N$_2$. Since for the He-containing systems the well depth is small compared to the collision energy, only diffraction oscillations are observable. For Ar-N$_2$, which has a much larger well depth, the rainbow structure is well observable, but because of a much larger reduced mass the diffractions superimposed are not resolved. The Ne-containing systems represent an intermediate case, in which the closely spaced diffractions superimposed on the fall-off of the rainbow are resolved. Both rainbow and diffraction oscillations are partially quenched due to the anisotropy of the potential well depth and minimum position, respectively. As far as Cl$_2$-rare gases are concerned, total DCS at two energies with well resolved diffraction oscillations have been measured for He-Cl$_2$. The diffraction structure superimposed on the main rainbow has also been resolved for Ne-Cl$_2$. The rainbow and two supernumerary rainbows have been observed for Ar-Cl$_2$. These data are being analyzed simultaneously with the vibrational predissociation data of Janda and coworkers (Beneventi et al 1991d).

Under our experimental conditions, it has been shown that the total DCS can be analyzed by using the powerful and simple infinite-order-sudden (IOS) approximation (Gianturco and Palma 1985, Bowers et al 1987, Beneventi et al 1988a). Analysis of the total DCS simultaneously with other available experimental properties has permitted us to derive accurate PES for He-N$_2$, O$_2$, NO (Beneventi et al 1986b, 1991a), He-CO$_2$ (Beneventi et al 1988b), and Ne-N$_2$ (Beneventi et al 1988a). More recently, an accurate PES has been obtained for Ne-O$_2$ by including in the analysis also the Zeeman spectrum (Beneventi et al 1991b). On the same line, work is now concentrated on Ne-NO (Casavecchia 1990, Beneventi et al 1991e), Ar-N$_2$ and Ar-O$_2$ (Beneventi et al 1991f), and He, Ne, Ar-Cl$_2$ (Beneventi et al 1991d). For Ar-N$_2$ an extensive multiproperty analysis is currently under way, in which up to ten different properties enter our determination (Beneventi et al 1991f).

In the data analysis for all systems, a potential model is chosen in which the anisotropy is described by making the size parameters ε and R_m angle-dependent, while the reduced form is taken to be the same for all orientations. The analytical form of the potential model usually used is:

$$V(R,\gamma) = \varepsilon(\gamma) \, f(x) \qquad\qquad x = R/R_m(\gamma)$$

where

$$\varepsilon(\gamma) = \overline{\varepsilon} \; [1 \; + \; A_1 \; P_1(\cos\gamma) \; + \; A_2 \; P_2(\cos\gamma) \; + \; \dots]$$

$$R_m(\gamma) = \overline{R}_m \; [1 \; + \; B_1 \; P_1(\cos\gamma) \; + \; B_2 \; P_2(\cos\gamma) \; + \; \dots] \tag{1}$$

Here R designates the distance between the center-of-mass of the molecule and the atomic partner, and γ the angle between R and the internuclear molecular axis. For homonuclear diatomics only even terms appear in Eq.(1), and a P_2 term is sufficient to describe the anisotropy effects experimentally observed. The reduced curve $f(x)$ is usually chosen to be the piecewise analytic exponential-spline-Morse-spline-van der Waals (ESMSV) or Morse-Morse-spline-van der Waals (MMSV) form. The explicit expressions can be found in Beneventi et al (1986b, 1991b). Only a few potential parameters are actually varied during the simultaneous best-fit procedure: essentially $\overline{\varepsilon}$, $\overline{R}_m$, the β parameter of the Morse describing the attractive and repulsive regions, and the anisotropy parameters A_2 and B_2 (Beneventi et al 1991b).

Table. Characteristic potential parameters

System Parameters	He-N_2	He-NO	He-O_2	Ne-N_2	Ne-NO	Ne-O_2
ε_0 (meV)	2.15	2.35	2.40	4.79	4.80	5.23
R_{m0} (Å)	3.69	3.67	3.52	3.72	3.70	3.62
$\varepsilon_\perp$ (meV)	2.97	3.39	3.26	6.70	7.10	7.20
$\varepsilon_\parallel$ (meV)	1.32	1.30	1.55	2.60	2.72	3.42
$R_{m\perp}$ (Å)	3.42	3.38	3.27	3.46	3.39	3.36
$R_{m\parallel}$ (Å)	4.11	4.14	3.93	4.15	4.16	4.03

In the Table we report the main potential characteristics for the He and Ne-diatom series. The spherical component $V_0(R)$, i.e., the first term of the Legendre expansion of the potential, for the same series which are depicted in the RHS of Figures 3 and 4, were obtained by Gauss-Legendre quadrature from the full surface (Beneventi et al 1986b). It is interesting to compare the various systems along a series. The location and spacing of the diffraction oscillations permitted us to establish the position of the low repulsive wall of $V_0(R)$, that is the σ_0 and R_{m0} parameters, with an accuracy of about $\pm 1\%$ for all the He and Ne-diatom systems. In particular, the location of the diffraction extrema mirrors the σ_0 value, as one can appreciate from the Figures 3 and 4. The relative precision between the various σ_0 values is much higher than 1%. The well depths, ε_0, could also be determined to high accuracy ($\approx \pm 2\%$-5%, depending on the system and the properties available for the multiproperty analysis). It should be noted that these precisions start to be comparable to those achieved for the rare gas-rare gas systems. As can be seen from the Figures 3 and 4 (see also the Table), the σ_0 and R_{m0} values increase in the order Rg-O_2, Rg-NO, and Rg-N_2 (Rg= He, Ne), that is they follow the same trend as the molecular polarizabilities ($O_2 \leq$ NO < N_2), while they correlate inversely with the equilibrium bond lenghts of the diatomic molecules ($O_2 >$ NO > N_2). The well depths are very similar to that of the corresponding isotropic rare gas-rare gas system for both He and Ne series, because it is known that well depth values depend mainly on the polarizability of the less polarizable partner, which is the rare gas for all cases. The weakest interaction along a series is that of Rg-N_2, as one would expect from the fact that N_2 has the larger polarizability and consequently Rg-N_2 the larger R_m. The R_{m0} value of Ar-N_2 is much larger ($R_{m0} \simeq 4$ Å) than for the lighter members of the series, which have almost identical R_{m0}. Obviously, also the well depth ε_0 increases significantly going towards Ar-N_2 ($\varepsilon_0 \simeq 10$ meV), because of the much larger polarizability of Ar with respect to Ne and He. The subtle interplay between attractive and repulsive interactions determines the similar R_{m0} for He-N_2 and Ne-N_2. An analogous situation occurs in the H_2-rare gas series (Buck 1982).

Comparisons with previous empirical and theoretical potentials for He-diatom and Ne-N_2,O_2 systems have been extensively discussed (Beneventi et al 1986b, 1988ab, 1991ab). For Ne-NO, the only potential previously proposed compare poorly with our results (Casavecchia 1990). For Ar-N_2 and Ar-O_2, comparison with previous empirical potentials (Candori et al 1983) shows significant disagreement, as can be seen in Figure 5 for Ar-N_2. Interestingly, comparison of our multiproperty potentials with the model potentials of Bowers et al (1988) for the Rg-N_2 series (Rg=He, Ne, Ar) shows good agreement, especially as far as $V_0(R)$ is concerned (Wong et al 1990, Beneventi et al 1991a). This is rather important, since supports the predictive power of the BTT potential model, which is indeed on a good theoretical basis. The model worked well also for the weakly anisotropic potentials of H_2-rare gases (Tang and Toennies 1980). The spherical component $V_0(R)$ of atom-molecule interactions has been recently shown (Liuti and Pirani 1985) to follow the same correlation found to be valid for the rare gas-rare gas systems. The possibility of predicting anisotropic potentials with some confidence appears to be feasible.

ACKNOWLEDGMENTS

We thank all our colleagues of the Molecular Beam Group of Perugia for useful discussion and collaboration on various parts of this work. Financial support from ECC Science Program, CNR Bilateral Agreements and "Progetto Finalizzato Chimica Fine", and ENEA is gratefully acknowledged.

REFERENCES

Ahlrichs R, Penco R and Scoles G 1977 **Chem.Phys.** 19 119
Aquilanti V, Liuti G, Pirani F, Vecchiocattivi F and Volpi G G 1976 **J.Chem.Phys.** 65 4751
Aquilanti V, Luzzatti E, Pirani F and Volpi G G 1980 **J.Chem.Phys.** 73 1181
Aquilanti V, Grossi G and Pirani F 1984 **Electronic and Atomic Collisions** eds J Eichler, I V Hertel and N Stolterfoht (Amsterdam: Elsevier) pp 441-450
Aquilanti V, Candori R, Cappelletti D, Luzzatti E and Pirani F 1990 **Chem.Phys.** 145 293
Aziz R A, Meath W J and Allnatt A R 1983 **Chem.Phys.** 78 295
Aziz R A 1984 in **Inert Gases** ed M L Klein (Berlin: Springer) ch. 2
Beneventi L, Casavecchia P and Volpi G G 1986a **J.Chem.Phys.** 84 4828
Beneventi L, Casavecchia P and Volpi G G 1986b **J.Chem.Phys.** 85 7011
Beneventi L, Casavecchia P and Volpi G G 1987 in **Structure and Dynamics of Weakly Bound Molecular Complexes** ed A Weber (Dordrecht: Reidel) pp 441-454
Beneventi L, Casavecchia P, Vecchiocattivi F, Volpi GG, Lemoine D and Alexander M H 1988a **J.Chem.Phys.** 89 3505

Beneventi L, Casavecchia P, Vecchiocattivi F, Volpi G G, Buck U, Lauenstein C and Schinke R 1988b **J.Chem.Phys.** 89 4671

Beneventi L, Casavecchia P and Volpi G G 1990 in **Dynamics of Polyatomic van der Waals Complexes** ed N Halberstadt and K C Janda (New York: Plenum) pp 399–407

Beneventi L, Casavecchia P, Volpi GG, Wong C C K, McCourt F R W, Corey G C and Lemoine D 1991a **J.Chem.Phys.** in press

Beneventi L, Casavecchia P, Pirani F, Vecchiocattivi F, Volpi GG, Brocks G, van der Avoird A, Heijmen and Reuss J 1991b **J.Chem.Phys.** in press (July)

Beneventi L, Casavecchia P and Volpi G G 1991c unpublished

Beneventi L, Casavecchia P, Volpi G G, Bieler C R and Janda J C 1991d work in progress

Beneventi L, Casavecchia P and Volpi G G 1991e to be published

Beneventi L, Casavecchia P, Volpi G G, Wong C C K and McCourt F R W 1991f **13th Int.Symp. on Molecular Beams** (El Escorial, 2–7 June 1991) Book of Abstracts E10

Bowers M S, Faubel M and Tang K T 1987 **J.Chem.Phys.** 87 5687

Bowers M S, Tang K T and Toennies J P 1988 **J.Chem.Phys.** 88 5465

Buck U, Huisken F, Schleusener J and Schaefer J 1980 **J.Chem.Phys.** 72 1512

Buck U 1982 **Faraday Discuss.Chem.Soc.** 73 187

Buck U 1986 **Comments At.Mol.Phys.** 17 197

Buck U 1987 in **Atomic and Molecular Beams Methods** ed G Scoles (New York: Oxford) vol. 1

Candori R, Pirani F and Vecchiocattivi F 1983 **Chem.Phys.Lett.** 102 412

Candori R, Pirani F and Vecchiocattivi F 1986 **J.Chem.Phys.** 84 4833

Casavecchia P, He G, Sparks R K and Lee Y T 1982 **J.Chem.Phys.** 77 1878

Casavecchia P, Laganá A and Volpi G G 1984 **Chem.Phys.Lett.** 112 445

Casavecchia P 1990 in **Dynamics of Polyatomic van der Waals Complexes** ed N Halberstadt and K C Janda (New York: Plenum) pp 123–141

Cline J I, Sivakumar N, Evard D D, Bieler C R, Reid B P, Halberstadt N, Hair S R and Janda K C 1989 **J.Chem.Phys.** 90 2605

Danielson L J, McLeod K M and Keil M 1987 **J.Chem.Phys.** 87 239

Faubel M 1983 **Adv.At.Mol.Phys.** 19 345

Fuchs R R, McCourt F R W, Thakkar A J and Grein F 1984 **J.Phys.Chem.** 88 2036

Gianturco F A and Palma A 1985 **J.Phys.B** 18 L519

Hirschfelder J O, Curtiss C F and Bird R B 1964 **Molecular Theory of Gases and Liquids** (New York: Wiley)

Hutson J M 1990 **Ann.Rev.Phys.Chem.** 41 123

Liuti G and Pirani F 1985 **Chem.Phys.Lett.** 122 245

Maitland G C, Rigby M, Smith E B and Wakeham W A 1981 **Intermolecular Forces: their Origin and Determination** (Oxford: Oxford)

Muenter J S 1987 in **Structure and Dynamics of Weakly Bound Molecular Complexes** ed A Weber (Dordrecht: Reidel) pp 3–21

Nesbitt D J 1988 **Chem.Rev.** 88 843

Ng C Y, Lee Y T and Barker J A 1974 **J.Chem.Phys.** 61 1996

Novick S E, Davies P, Harris S J and Klemperer 1973 **J.Chem.Phys.** 59 2273

Pack R T 1978 **Chem.Phys.Lett.** 55 197

Rigby M, Smith E B, Wakeham W A and Maitland G C 1986 **The Forces between Molecules** (Oxford: Oxford)

Rodwell W R and Scoles G 1982 **J.Phys.Chem.** 86 1053

Saykally R J 1989 **Acc.Chem.Res.** 22 295

Scoles G 1980 **Ann.Rev.Phys.Chem.** 31 81

Tang K T and Toennies J P 1978 **J.Chem.Phys.** 68 5501

Tang K T and Toennies J P 1980 **J.Chem.Phys.** 74 1148

Wong C C K, McCourt F R W and Casavecchia P 1990 **J.Chem.Phys.** 93 4699

The role of initial orbital alignment and orientation in charge transfer processes induced in atomic collisions

D. Dowek, J.C.Houver and C. Richter

Laboratoire des Collisions Atomiques et Moléculaires, Bat. 351, Université Paris-Sud, 91405 Orsay, FRANCE

ABSTRACT: The role of initial orbital alignment and orientation in charge transfer processes is studied in the keV energy range, where the collision velocity and the orbital velocity of the active electron are comparable. For electron capture by protons impinging on aligned and oriented states of Na(3p), strong anisotropy effects are observed in differential measurements at extremely small scattering angles corresponding to grazing incidence collisions. The data are compared with theoretical predictions based on elaborate calculations and on simple models.

1. INTRODUCTION

The investigation of initial orbital alignment and orientation effects on the outcome of binary collision events has in recent years developed into a powerful tool for a deeper understanding of fundamental processes. The experimental technique common to all these studies has been the use of laser polarized light to excite an atomic beam, thereby controlling the shape and the inherent angular momentum of the prepared target. It has been applied first to study electron impact excitation of atoms by Hertel and Stoll (1977) and a review of this topic may be found in Andersen et al (1988). One controls the orbital alignment of the excited atom by varying the linear polarization of the laser, whereas the orientation i.e. the dynamics of the electron in its circular orbit is defined by the use of right or left hand circular polarized light.

In the field of heavy particle collisions, significant progress in the description of collision mechanisms taking place at low energies has been achieved. The study of alignment effects in reactions like associative ionization, reactive scattering or resonant charge transfer processes, summarized for example in the reviews by Hertel et al (1985) and Leone (1988), has put in evidence the role of molecular states with defined symmetry. These results have motivated the development of powerful concepts such as the orbital following or the locking radius. The latter defines the R distance between the two encounters inside which the electronic cloud approximately follows the rotation of the internuclear axis, while staying space fixed outside. If the alignment effects can be studied experimentally by measuring total and differential cross sections for the reaction of interest, the investigation of the role of initial orbital orientation requires to analyze the right-left asymmetry in the angular scattering of the projectile by the polarized target. Here also the first crossed beam experiments with heavy particles were performed at thermal energies for the investigation of rovibrational energy transfer and fine structure transitions, then extended to higher velocities for electronic excitation and resonant energy transfer in the Na^+-Na(3p) collisions. These data probed in particular the role of coherence in the scattering process and the sensitivity on the long range part of the potentials (see e.g. Campbell et al (1988) for a recent review).

However, for all the reactions cited above the collision velocity v_c is quite small compared to the velocity of the active electron v_e. Our motivation is to study the role of initial orbital alignment and orientation in the region where the probability for outer shell electronic transitions are maximum, that means when the two characteristic velocities v_c and v_e are of the same order of magnitude. This task can be achieved in two types of collision experiments : in the first one the incoming particles are in the ground state and the electronic state of a collisionally excited atom is investigated by polarization analysis of the photons emitted after the collision ; in the second type, which corresponds to the time-reverse scheme, the incident channel is prepared in a defined quantum state by means of a polarized laser as discussed above. The first technique has been used in the last 15 years for the investigation of electronic transition mechanisms in experiments with planar symmetry and the data have been analyzed in the framework of the original ideas of Macek and Jaecks (1971) and Fano and Macek (1973). Reviews of the progress made can be found e.g. in the preceding ICPEAC publications (Andersen et al 1985, Hippler 1987). One important outcome of these studies has been the derivation of propensity rules for orientation in direct excitation processes taking place at large internuclear distances by Andersen and Nielsen (1986) and their confirmation in several experiments (Panev et al 1987). However, in this velocity range several inelastic channels are usually open and populated after the collision. Therefore, except in cases where the excitation mechanism is strongly selective, it is often difficult to analyze without ambiguity the decay photons due to the cascade effects. From this point of view the second type of experiment is preferable since the aligned or oriented state is uniquely defined. For this purpose we have developed an experiment where different mechanisms taking place in collisions between an ion or a neutral beam and a polarized excited target in the velocity range where $v_c / v_e \approx 1$ can be studied by performing differential measurements and we report the first data of this type. Our main interest until now has been centered on the production of neutral particles in charge transfer or electron detachment processes, and the first investigations have concerned the H^+, He^+, H_2^+-Na(3p) and H^--Na(3p) collisions respectively. Since the mean velocity of the 3p electron in the Na(3p) atom is $v_e \approx 0.47$ a.u., one may satisfy the above velocity matching condition in the keV energy range for such light projectiles. In this paper I will focus the discussion on the quasi-one-electron H^+-Na(3p) system chosen as a prototype for the investigation of non-resonant charge transfer and for which strong polarization effects have been measured.

Different theoretical approaches have been undertaken for the study of alignment and orientation in H^+-Na(3p) neutralization reactions. Detailed calculations for this system have been performed by Allan et al (1986,1990) and Courbin et al (1990,1991) with a 19-state molecular basis and by Fritsch (1988) with a 49-state atomic basis ; on the other side Nielsen et al (1990) have extended the propensity rule approach for orientation in charge transfer processes and performed also detailed calculations for the H^+-Na(3p) collision using a 19-state atomic basis (Dubois et al, 1991). All these calculations include electron translation factors for the description of electron momentum transfer subsequent to the capture process and secure Galilean invariance of the reference frame. A molecular picture modelizing the B^{3+}-He collision, for which orientation effects were observed recently (Roncin et al 1990) and will be discussed below, has been proposed by Ostrovsky (1991). It enlights the role of crucial parameters for the prediction of orientation. The comparison between the theoretical predictions and the experimental findings is discussed, some conclusions drawn, and remaining questions pointed out.

2. COLLISION SPECTROSCOPY

The spectrometer used for the experimental investigation (Royer et al (1988)) combines (i) high resolution time-of-flight technique for the kinetic energy analysis of the scattered projectiles, (ii) optical pumping of alkali atoms, (iii) and position sensitive detection to perform the angular scattering analysis of the products in a defined sector $\Delta\theta$. The well-collimated ion beam of energy E is chopped and crosses at right angle, in the horizontal plane, the Na beam produced in a supersonic expansion. The single mode dye laser is locked on the [3 $^2S_{1/2}$ F=2 $\rightarrow$ 3 $^2P_{3/2}$ F=3] hyperfine transition and the beam travels along the vertical direction as schematized in Figure 1. The polarization is controlled by means of a Pockels cell. The effective fraction α of excited Na(3p) atoms is measured directly as described by Houver et al (1989) and is typically 10%. In the future the use of a two mode laser as described by Campbell et al (1990) is planned in order to achieve a fraction α of about 30 to 40%. The channel-plate detector is located at a distance L from the collision volume, which can be varied between 2.5 and 7.50 m depending on the required energy resolution : typically for L = 4.50 m the energy resolution is $\Delta E < 1$ eV for a proton of $E_{H^+} = 1$ keV, and the total angular acceptance of the detector is $\Delta\theta = 0.5$ deg.

The rotation of the ion gun ensemble in the horizontal plane enables angular scattering analysis over a range of few degrees.

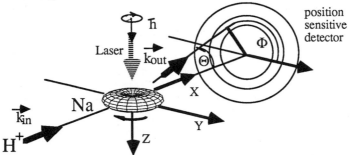

Fig. 1. Schematic diagram of the experiment, here with circularly polarized light excitation.

Two experimental difficulties specific to the energy range presently studied must be outlined. (i) the requirements of maintaining a low density of Na atoms, $n < 10^{11}$ at./cm^3 at the collision volume, in order to avoid radiation trapping and keep the target polarized, is severe with respect to the number of counted events. The processes studied first should then be chosen also to have large cross sections. (ii) Large cross sections are usually associated with large impact parameters, i.e. small scattering angles, especially in the velocity range where the collision velocity becomes comparable to the velocity of the active electron : therefore, the investigation of the orientation effects demands differential angular analysis in a very small scattering angular range, typically $\pm$ 0.1 or 0.2 deg. For this purpose, the angular aperture of the incident beam must be drastically limited, resulting again in a decrease of the number of measured events. These findings have rendered the development of position sensitive detection crucial. In the first series of experiments where total cross sections for the H$^+$-Na(3p) collision have been measured, the angular aperture of the incident beam was about $\Delta\theta = 0.2$ deg. whereas it has been reduced to 0.04 deg for the determination of the differential cross sections.

The first step of such an experiment consists in the identification of the channels populated. Since the target is composed of α% of Na(3p) and (1-α)% of Na(3s) atoms when the laser is shining on the collision region, the energy defect spectra for the H$^+$-Na(3p) collision are obtained from the weighted differences $I_{on}^P - (1-\alpha)I_{off}$, where I_{off} and I_{on}^P are the intensities for spectra accumulated without and with laser for light polarization defined by the index P. For both systems the following electron capture reactions have been identified :

$$H^+ + Na(3p) \rightarrow H(n=2) + Na^+ \quad \Delta E = 0.36 \, eV \quad (1)$$
$$H^+ + Na(3p) \rightarrow H(n\geq3) + Na^+ \quad \Delta E \leq -1.52 \, eV \quad (2)$$
$$H^+ + Na(3s) \rightarrow H(n=2) + Na^+ \quad \Delta E = -1.74 \, eV \quad (3)$$

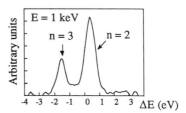

Fig. 2. Energy defect spectrum

ΔE is the energy defect of the reaction defined here positive for an exothermic process. Figure 2 displays typical energy defect spectrum at E = 1 keV for the H$^+$-Na(3p) collisions in arbitrary units. A number of measurements and much theoretical work has been devoted to the study of charge exchange processes in H$^+$-Na(3s) collisions showing large total cross sections (50 Å^2 at 1 keV proton energy) for reaction (3) which is completely dominant in the neutralization process (see e.g. Shingal and Bransden 1987 and references therein). Throughout the results reported here for the H$^+$-Na(3p) collision, this reaction has consequently been used as a reference for absolute calibration of the cross sections measured for reactions (1) and (2). The strong enhancement of the electron capture probability from the Na(3p) target compared to Na(3s) reveals the larger electron capture total cross sections, which are found to reach their maximum value of 200 Å^2

at $E = 0.5$ keV for reaction (1) and 80 Å^2 at $E = 3$ keV for reaction (2) (Royer et al 1988). The detection of Ly_α photons emitted after the collision has been measured by Finck et al (1988) and more recently by Gieler et al (1991) giving total cross sections for the production of H(2p) in reaction (1) consistent with the present data.

3. INTEGRAL ALIGNMENT EFFECTS

For the H^+-Na(3p) collision we have first measured the integral alignment for reaction (1) and (2) leading to the production of H(n=2) and H(n≥3) respectively, as a function of collision energy in the 0.5 to 2 keV range (Dowek et al, 1990). For this purpose, energy defect spectra are accumulated with a linear polarization of the laser light parallel (∥) and perpendicular (⊥) to the collision velocity in alternation, ensuring that the angular acceptance of the detector is larger than the angular spread of the scattered atomic beam. In order to analyze the observed anisotropy in terms of atomic orbitals, one must take into account that the two polarization directions (∥ and ⊥) do not correspond to the preparation of pure $m_L = 0$ p_x and p_y orbitals, which we label σ and π respectively, consistent with the notation of molecular physics. The hyperfine structure causes a scrambling of the orbitals which has been discussed by Fisher and Hertel (1982) in the case of the optical pumping of sodium. When the F=2 - F=3 transition is used, the two situations yield (∥) $\frac{5}{9}\sigma + \frac{4}{9}\pi$ and (⊥) $\frac{2}{9}\sigma + \frac{7}{9}\pi$. Using these coefficients, the dependence of the electron capture probability on the initial alignment of the Na(3p) orbital can be discussed in terms of $\sigma_\Sigma(n)$ and $\sigma_\Pi(n)$ cross sections for each H(n) + Na^+ channel populated, where Σ and Π label the molecular symmetry of the entrance channel in the $R \to \infty$ asymptotic limit (Richter et al, 1990). Figure 3 displays the variation of the anisotropy parameter A(n), $-1 \leq A \leq 1$, defined by :

$A(n) = [\, \sigma_\Sigma(n) - \sigma_\Pi(n)\,] / [\, \sigma_\Sigma(n) + \sigma_\Pi(n)\,]$ as a function of collision energy. For electron capture to the H(n=2) level the $\sigma_\Sigma / \sigma_\Pi$ cross section ratio remains close to 1.5 (A(2) $\approx$ 0.15) at all energies investigated. This result is comparable to what was obtained by Bähring and coworkers (1984) for Na^+-Na(3p) resonant charge transfer at much lower velocities ($v_c \approx 0.01$ a.u.). The measured polarization dependence of the cross sections was found consistent with semi-classical calculations predicting a rather constant

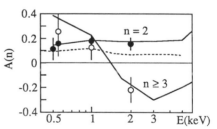

Fig. 3. Anisotropy parameter dots : our results, lines : calculations

(- - - Courbin et al, — Fritsch)

$\sigma_\Sigma / \sigma_\Pi$ cross section ratio of 1.7 (A(Na(3p)) $\approx$ 0.25). The different behavior of capture into the H(n≥3) levels, where the situation is similar to the H(n=2) case at 1 keV, but reversed at 2 keV with a preferential capture for the Π symmetry compared to the Σ symmetry, prevents an interpretation based on simple models to be proposed at this stage. The results compared with the values extracted from elaborate calculations of Fritsch (1989) and of Courbin et al (1990) show a rather good agreement as illustrated in Figure 3. However, despite this overall agreement no clear interpretation of these findings emerges in terms of simple physical picture. Further understanding of the charge transfer processes is expected from the angular differential analysis, since the impact parameter dependence of the various processes can then be studied.

In addition, the degeneracy of the Π^+ and Π^- molecular states is suppressed since the planar symmetry of the scattering process is restored.

4. DIFFERENTIAL ALIGNMENT EFFECTS

In sections 4 and 5 we report the first differential cross sections for electron capture from a polarized target in the intermediate energy range, where the collision velocity and the electron velocity become comparable. Since the neutral atoms produced after electron capture are forwardly scattered in a very narrow cone, $\Delta\theta < 0.2$ deg., the angular aperture of the incident beam has been reduced to $\Delta\theta_i \approx 0.04$ deg. and a position sensitive device developed by Brenot et al (1991) is used. The number of scattered particles is subsequently smaller and the differential analysis will be focused on reactions (1) and (3) for charge transfer into the H(n=2) state. The detector consists of three channel plates and a resistive square anode. Its angular discrimination is better than 0.001 deg. The characteristics of this device, the detailed experimental procedure and treatment of data required for such measurements will be described in a forthcoming paper. For each detected event three coordinates are recorded : the time of flight (T), the position (x,y) of the scattered particle, determined from the charges collected at the four corners of the anode, and an index i characterizing the collision (laser on/off, polarization, sodium beam on/off). A first selection of the events by the index i provides histograms for the number of counts I as a function of position for the H+ - Na(3s) and for the H+ - Na(3p) collisions after the weighted subtractions $I_{on}^P - (1-\alpha)I_{off}$ as explained in section 2. A second selection of the events is provided by using the time of flight information.

For the investigation of alignment effects, the double differential cross sections $\sigma(\theta,\phi)$, measured for a linear polarization direction of the laser light successively parallel, perpendicular, at 45° and 135° with respect to the collision velocity, can be parametrized as follows :

$$\sigma^{\parallel}(\theta,\phi) = A_0^{\parallel}(\theta)$$

$$\sigma^{\perp}(\theta,\phi) = A_0^{\perp}(\theta) + A_2^{\perp}(\theta)\cos(2\phi) \qquad (4)$$

$$\sigma^{45°/135°}(\theta,\phi) = A_0(\theta) \pm A_1(\theta)\cos(\phi) + A_2(\theta)\cos(2\phi)$$

The parameters $A_i(\theta)$ are extracted by fitting the data with the appropriate function of the azimuth angle ϕ. They are known functions of the scattering amplitudes $f_{\lambda\Sigma}(\theta)$, $f_{\lambda\Pi^+}(\theta)$ and $f_{\lambda\Pi^-}(\theta)$ for the collision process where, in a molecular representation, the initial state is Σ, Π^+ or Π^- and the final state λ is one of the $2s\Sigma$, $2p\Sigma$, $2p\Pi^+$ and $2p\Pi^-$ final states of the H(n=2) + Na+ channel :

$$A_0^{\parallel}(\theta) = \frac{5}{9}\sigma_\Sigma(\theta) + \frac{4}{9}\sigma_\Pi(\theta) \quad \text{and} \quad A_0^{\perp}(\theta) = \frac{2}{9}\sigma_\Sigma(\theta) + \frac{7}{9}\sigma_\Pi(\theta)$$

where the averaged Π cross section is : $\sigma_\Pi(\theta) = \dfrac{(\sigma_\Pi^+(\theta) + \sigma_\Pi^-(\theta))}{2}$

$$A_2^{\perp}(\theta) = \frac{1}{6}(\sigma_\Pi^+(\theta) - \sigma_\Pi^-(\theta))$$

$$A_0(\theta) = \frac{1}{2}[\frac{7}{9}\sigma_\Sigma(\theta) + \frac{11}{9}\sigma_\Pi(\theta)] \qquad A_2(\theta) = \frac{1}{12}(\sigma_\Pi^+(\theta) - \sigma_\Pi^-(\theta)) \qquad (5)$$

$$A_1(\theta) = \frac{1}{3}\text{Re}\left(\sum_\lambda f_{\lambda\Sigma}(\theta) f_{\lambda\Pi^+}^*(\theta)\right)$$

$\sigma_\Sigma(\theta)$, $\sigma_\Pi{}^+(\theta)$ and $\sigma_\Pi{}^-(\theta)$ are the differential cross sections for the production of H(n=2),

where $\sigma_\Sigma(\theta) = \sum_\lambda |f_{\lambda\Sigma}(\theta)|^2$, etc. These parameters can be alternatively expressed as functions of the density matrix elements for the scattering process (Andersen et al 1988)). Figure 4 (a) summarizes the experimental data obtained at E = 1 keV for reactions (1) and (3) in H+-Na(3p) and H+-Na(3s) respectively. We extract the $\sigma_\Sigma(\theta)$ and $\sigma_\Pi(\theta)$ cross sections on one side, and the difference $\sigma_\Pi{}^+(\theta) - \sigma_\Pi{}^-(\theta)$ on the other side, from the measurements with the polarization direction parallel and perpendicular to the collision velocity. The latter difference is found to be weak ($A_2(\theta) < 0.1\ A_0(\theta)$), then only the mean average $\sigma_\Pi(\theta)$ cross section is displayed in Figure 4 (a).

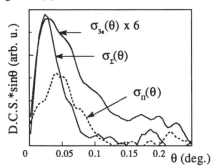

Fig. 4.(a) Differential cross sections :
present results

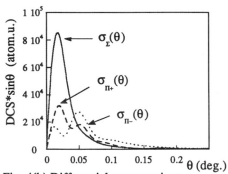

Fig. 4(b) Differential cross sections :
atomic calculations

The measured $\sigma_\Sigma(\theta)$ and $\sigma_\Pi(\theta)$ cross sections show similar shape. The integral alignment discussed above is clearly attributed to the smallest scattering angles. The $\sigma_{3s}(\theta)$ cross section reveals that scattering in the H+-Na(3s) collision occurs at relatively larger angles than in the H+-Na(3p) case. A rather good agreement is found with the results of the atomic calculation by Dubois et al (1991) at same energy displayed in panel (b), if one compares the mean average of the $\sigma_\Pi{}^+(\theta)$ and $\sigma_\Pi{}^-(\theta)$ curves shown in Figure 4 (b) with the $\sigma_\Pi(\theta)$ measured DCS. When the laser light polarization is directed at 45° or 135° with respect to the x axis the molecular state associated with the entrance channel includes a coherent superposition of Σ and Π^+ states and the real part of the product of scattering amplitudes $\Sigma(f_{\lambda\Sigma}(\theta)f^*_{\lambda\Pi+}(\theta))$ is extracted from the differential measurements according to eq. (4) and (5). A rather constant negative value of about 0.25 is measured for the ratio $\dfrac{A_1(\theta)}{A_0(\theta)+A_2(\theta)}$. Another set of parameters is used in the literature to characterize the collision, which describes the shape of the 3p electronic cloud that would be excited in the time inverse process after electron capture from an isotropically populated H(n=2) state. Namely the length, width of the electronic cloud and its alignment angle γ with respect to the collision velocity are directly related to the $\sigma_\Sigma(\theta)$, $\sigma_\Pi{}^+(\theta)$,

$\mathrm{Re}\Sigma(f_{\lambda\Sigma}(\theta)\,f^*_{\lambda\Pi+}(\theta))$ terms. Its height in the direction perpendicular to the collision plane is proportional to $\sigma_\Pi{}^-(\theta)$. Such a representation of our results is shown in Figure 5.

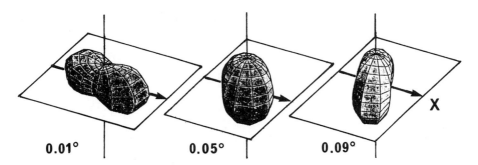

Fig. 5. Charge cloud after electron capture from an isotropically populated H(n=2) state in the time inverse process, as derived from the alignment parameters for selected scattering angles θ.

Here again, the differential alignment data present similarities with the Na+-Na(3p) collision at low velocities, since the pσ atomic orbital prepared at large internuclear separation was shown to be most efficient for the charge transfer process in the measurements reported by Witte et al (1987). However, even in the molecular regime characterizing the Na+-Na(3p) investigation where the concept of locking radius had appeared to be a reasonable approximation to interpret alignment effects in inelastic and superelastic transitions, no simple qualitative interpretation based on similar concepts has been proposed for the charge transfer process.

5. ORIENTATION EFFECTS

5.1 Orientation propensity for electron capture in H+-Na(3p) collisions.

The determination of the $\sigma^{\pm}(\theta,\phi)$ differential cross sections for a target excited with circularly polarized light left (+) or right (-) handed requires the same experimental procedure as described in the preceding section. The $\sigma^{\pm}(\theta,\phi)$ differential cross sections are here parametrized as :

$$\sigma^{\pm}(\theta,\phi) = C_0(\theta) \pm C_1(\theta) \cos\phi + C_2(\theta) \cos2\phi \qquad (6)$$

where the parameters $C_i(\theta)$ are accordingly defined by the following equations :

$$C_0(\theta) = 1/2 \, [\sigma_\Sigma(\theta) + 1/2 \, (\sigma_\Pi{}^+(\theta) + \sigma_\Pi{}^-(\theta))] \; ; \quad C_2(\theta) = 1/4 \, (\sigma_\Pi{}^+(\theta) - \sigma_\Pi{}^-(\theta))$$

$$C_1(\theta) = \mathrm{Im} \, (\sum_\lambda f_{\lambda\Sigma}(\theta) \, f_{\lambda\Pi+}^*(\theta)) \qquad (7)$$

Similarly to what was done for the analysis of alignment effects, the parameters $C_i(\theta)$ are extracted by fitting the data with the appropriate function of the azimuth angle ϕ (Houver et al, 1991). Figure 6 (a) displays the differential cross sections measured with a left handed circular polarization of the laser light, at the collision energy E = 1 keV ($v_c / v_e = 0.4$). The scattering in the half planes $\phi = 0$ and $\phi = \pi$, which display the strongest right-left asymmetry are represented.

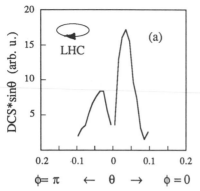 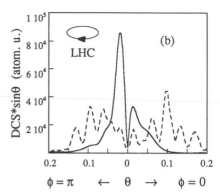

Fig. 6. Differential cross sections multiplied by sinθ for scattering in the plane $\phi = 0$ and $\phi = \pi$ (a) experiment, (b) molecular (--) (Courbin et al) and atomic (—) (c) (Dubois et al) calculations.

The data can be discussed through two different aspects. (i) The first question raised at the starting point of the present investigation was formulated in terms of a "velocity matching argument" suggested by Kohring et al (1983) in the classical description of charge transfer for protons colliding with Rydberg hydrogenic atoms. This simple picture predicts that electron transfer is favored when the projectile and the active electron, having comparable velocities, move in the same direction on the same side of the nucleus. The extension of the propensity rule approach to charge transfer process developed by Nielsen and coworkers (1990) predicts a maximum transition probability when the two phase terms in the following equation compensate :

$$\Delta \varepsilon \, \frac{a}{v_M} + \Delta m \, \pi = 0,$$ for an energy change $\Delta \varepsilon$ and effective interaction range a. When the treatment includes electron translation factors, this condition is modified by addition of an energy term $1/2 \, v^2$. Applied to electron capture to the H(n=2) state in H^+-Na(3p) collision at slightly higher velocities ($v_c = 0.35$ a.u.), this leads to a preferred orientation which is in agreement with the velocity matching image. Finally the differential cross sections calculated by Courbin et al (1990) and Dubois et al (1991) are shown in Figure 6 (b) : although they differ rather significantly from one another, they comfort the above predictions in terms of number of particles scattered in the $\phi=0$ or $\phi=\pi$ half-planes. The experimental results shown in Figure 6 for E = 1 keV exhibit indeed a strong right-left asymmetry, but in a sense which is opposite to these various theoretical approaches.

(ii) The second aspect in the analysis of scattering from an oriented target is the determination of the asymmetry parameter A(θ), termed orientation parameter in the following. A(θ) is equal to the angular momentum transfer $L_\perp$ in the time reverse process, often referred to in the literature. Using eq. (6) and (7) one obtains :

$$A(\theta) = L_\perp (\theta) = \frac{I^+(\theta,0) - I^+(\theta,\pi)}{I^+(\theta,0) + I^+(\theta,\pi)}$$

$$A(\theta) = \frac{I^+(\theta,0) - I^-(\theta,0)}{I^+(\theta,0) + I^-(\theta,0)} = \frac{C_1(\theta)}{C_0(\theta) + C_2(\theta)}$$

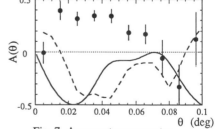

Fig. 7. Asymmetry parameter
dots : our results, lines : calculations
(- - - Courbin et al, — Dubois et al)

Consistently with the measured DCS, at E=1 keV, the orientation parameter A(θ) is positive at small scattering angles, reaches a maximum value of 0.4 and decreases to become negative at θ = 0.085 deg (see Figure 7). It is significantly larger than the one measured at low velocities in the Na^+-$Na(3p)$ resonant process (Witte et al 1987). The same behavior is found at E = 0.75 and 1.5 keV. The orientation parameters calculated by Courbin et al (1990) and Dubois et al (1991) are also presented in Figure 7, reflecting the difference found by comparison of the DCS. This drastic discrepancy between experiment and calculations, which both exhibit a large orientation effect, but of opposite sign at small scattering angles, has not been elucidated yet.

5.2. Orientation propensity for electron capture in B^{3+}-He collisions

The investigation of orientation effects in charge transfer reactions involving a multiply charged ion has been undertaken recently by Roncin et al (1990). It presents some advantage from the experimental side since the long range Coulombic repulsion in the outgoing channel leads to reasonably large scattering angles, even if capture occurs at large internuclear distances. The B^{3+}-He system, where only the two capture channels leading to $B^{2+}(n=2)$ are effective, has been selected to study the orientation of the $B^{2+}(2p)$ electronic cloud after electron capture.

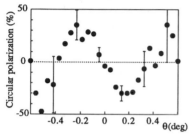

Fig. 8. Circular polarization

There, by detecting in coincidence the scattered B^{2+} ions collected on a position sensitive detector with the photons emitted in the $2p \rightarrow 2s$ decay (λ = 207 nm), the circular polarization of the 2p state is measured as a function of the scattering angle θ. This experiment is performed in the 1-12 keV collision energy range for which the v_c / v_e velocity ratio varies from 0.045 to 0.15 a.u. A strong orientation propensity, predicted by the theoretical analysis carried out by Hansen et al (1990), has been observed. Figure 8 illustrates the clear right-left asymmetry of the circular polarization defined here as (LHC - RHC) / (LHC + RHC) (equal to the angular momentum transfer $L_\perp$) as a function of scattering angle θ measured at 1.5 keV, the same tendency being found at higher energies. At small positive scattering angles the circular polarization is negative, revealing the dominant excitation of the m=-1 substate, in the standard geometry. When one considers the time reverse process this result fits the picture of velocity matching i.e. a capture favored when electron and projectile travel in the same sense of direction. However, for larger scattering angles the circular polarization changes sign and presents an oscillatory structure, even more visible with a Ne target (Adjouri et al, 1991) than with He. A simple interpretation of this behavior has been proposed recently by Ostrovsky (1991) in the frame of the molecular representation of the collision process. It can be outlined with the help of the diagram represented in Figure 9.

Electron capture occuring at the crossing radius R_C preferentially populates the $2p\sigma$ orbital either on the way in (1) or on the way out (2) of the collision. In this velocity range the $p\sigma$ orbital populated in (1) does not lock to the rotation of the internuclear axis, but rather remains space-fixed transforming partially into a $p\pi$ orbital. At the second passage of the system through the crossing region R_C the coherent superposition of the $p\pi$ component and of the $p\sigma$ orbital populated in (2) will give rise to a p_{+1} or p_{-1} state. The resulting circular polarization κ primarily depends on the rotation angle 2β and on the

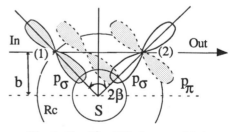

Fig. 9 : For $2\beta = 90°$, the $p\sigma$ orbital populated in (1) slips to a $p\pi$ orbital in (2)

Stückelberg phase ψ_0 developed between the two paths defined inside the crossing radius, both terms being impact parameter dependent.

Approximately, one finds : $\kappa(\theta) = \dfrac{\cos^2(1/2\psi\text{-}\beta) - \cos^2(1/2\psi\text{+}\beta)}{\cos^2(1/2\psi\text{-}\beta) + \cos^2(1/2\psi\text{+}\beta)}$

At the low velocities discussed here the inclusion of electron translation factors does not influence the orientation effect (Ostrovsky 1991).The combined effect of the rotational coupling responsible for the "slippage" of the electronic cloud, and of the radial coupling which induces electron capture is also clearly seen in the theoretical analysis of Hansen et al (1990).

6. CONCLUSION

In this experimental investigation of the role of initial orbital alignment and orientation for charge transfer in keV H^+-Na(3p) collisions we have reported strong anisotropy effects observed in differential measurements in a very narrow scattering angle range. The elaborate calculations performed with large molecular and atomic basis sets, including electron translation factors, account well for the alignment effects observed in total cross sections ; a better agreement is found with the atomic data concerning the shape of the $\sigma_\Sigma(\theta)$ and $\sigma_\Pi^\pm(\theta)$ differential cross sections. However the right-left asymmetry put in evidence in the angular scattering of the H atoms impinging on circular states of Na(3p) is opposite to the theoretical predictions resulting from the above calculations as well as from propensity rules and from the classical concept of velocity matching. Recent data have been reported for orientation propensity in s-p electron capture involving a multiply charged ion. There the reaction occurs at a well defined curve crossing and favors a simple description of the collision in the molecular representation. It is worth noticing that in this case, the model predicts that the circular polarization resulting from electron capture will be governed by the sum and difference of two phase terms, the angle which characterizes the rotation of the internuclear axis during the collision and the semi-classical phase developed in the potential scattering, similarly to what appears in the formalization of propensity rules in the atomic representation. In particular the circular polarization will be maximum (± 1) if $(\psi\pm2\beta) = 0$, whereas it will cancel if ψ or β is negligible with respect to the other. The search for explicit correspondences in the two approaches will certainly be fruitful. For the rather low velocities explored there, the effect of electron momentum transfer is expected to be negligible. This is not the case for the H^+-Na(3p) collision at the velocities where $v_c / v_e \approx 0.5$ and the specific role of the inclusion of translation factors in the orientation effect should be investigated (see e.g. Campbell et al 1991). At the present step for this system where p-p and p-s capture processes take place it would be interesting on the theoretical side to study more closely distinct regions of internuclear distances, in order to localize where the angular momentum transfer takes place. Whether such a capture process, where several states are coupled and where electron transfer is not very localized, can be modelized in order to favor the emergence of simple general physical pictures, remains an open question.

On the experimental side, the H^+-Na(3p) collision will be studied in an extended velocity range. Detection of Ly_α photons in coincidence with the scattered H atoms is in progress (Roller-Lutz et al 1991). Investigation of other collision systems will be performed in order to explore the general character of the strong effects observed presently. Complete information for the p-p electron capture process can be gained by analyzing the polarization of the final state : this will be attempted by detecting in coincidence the scattered neutral and the decay photon emitted after collision. The Li^+-Na(3p) system is a good candidate for such an experiment since the photons emitted by the Li(2p) excited atom lie in the visible range and the polarization analysis can be performed easily. Other systems where a s-p capture process involving a multiply charged ion occurs are currently studied by Roncin et al. The idea of colliding multiply-charged ions with a polarized target combines the advantage of the experimental method described in this report with reasonably large deflection angles and should be developed in the future. This is in progress since first measurements of total single electron capture cross sections in He^{2+}-Na(3p) collisions have been performed recently by Aumayr et al (1991).

ACKNOWLEDGEMENTS

The authors gratefully acknowledge N. Andersen for suggesting this study as well as for cordial and stimulating collaboration. Financial support by the Committee for the Development of European Science and Technology and Institut Français (Copenhagen) is gratefully acknowledged. We would like to thank the participants to this European programme from the Universities of Copenhagen, Freiburg, Madrid, Bergen and Orsay and their colleagues for fruitful discussions.

REFERENCES

Adjouri C, Ostrovsky V N, Andersen N, Gaboriaud M N, Roncin P and Barat M 1991
  Colloque : Dynamique des Ions, des Atomes et des Molécules Bourges, 3-5 juil.1991
Allan R J, Shingal R and Flower D R 1986 *J. Phys. B* **19** L251
Allan R J, Courbin C, Salas P and Wahnon P 1990 *J. Phys. B* **23** L461
Andersen N 1985 *XIV ICPEAC Book of Invited Papers* (North-Holland) pp 57-76
Andersen N and Nielsen S E 1986 *Europhys. Lett.* **1** 15
Andersen N, Gallagher J W and Hertel I V 1988 *Phys. Rep.* **165** 1
Aumayr F, Gieler M, Unterreiter E and Winter H 1991 *submitted to Europhysics Letters*
Bähring A, Hertel I V, Meyer E and Schmidt H 1984 *Phys. Rev. Lett.* **53** 1433
Brenot J C and Durup-Ferguson M 1991 *Adv. Chem. Phys.* **1** 168 in press.
Campbell E E B, Hülser H, Witte R and Hertel I V 1990 *Z. Phys. D* **16** 21
Campbell E E B, Schmidt H and Hertel I V 1988 *Adv. Chem. Phys. LXXII* 37
Campbell E E B, Hertel I V and Nielsen S E 1991 *J. Phys. B* in press.
Courbin C, Allan R J, Salas P and Wahnon P 1990 *J. Phys. B* **23** 3909 and Courbin C, private
  communication.
Courbin C, Allan R J, Salas P and Wahnon P 1991
Dowek D, Houver J C, Pommier J, Richter C, Royer T, Andersen N and Palsdottir B 1990
  Phys. Rev. Lett. **64** 1713
Dubois A, Hansen J P and Nielsen S E 1991, to be published
Fano U and Macek J 1973 *Rev. Mod. Phys.* **45** 553
Finck K, Wang Y, Roller-Lutz Z and Lutz H O 1988 *Phys. Rev. A* **38** 6115
Fisher A and Hertel I V 1982 *Z. Physik A* **304** 103
Fritsch W 1987 *Phys. Rev. A* **35** 2342 and private communication 1988.
Gieler M, Aumayr F, Hütteneder M and Winter H 1991, *submitted to J. Phys. B*.
Hansen J P, Kocbach L, Dubois A and Nielsen S E 1990 *Phys. Rev. Lett.* **64** 2491
Hertel I V and Stoll W 1977 *At. Mol. Phys.* **13** 113
Hertel I V, Schmidt H, Bähring A and Meyer E 1985 *Rep. Prog. Phys.* **48** 375
Hippler R 1987 *XV ICPEAC Book of Invited Papers* (North-Holland) pp 241-253
Houver J C, Dowek D, Pommier J and Richter C 1989 *J. Phys. B* **22** L 585
Houver J C, Dowek D, Richter C and Andersen N 1991 *submitted to Phys. Rev. Lett.*
Kohring G A, Wetmore A E and Olson R E 1983 *Phys. Rev. A* **28** 2526
Leone S R 1988 *Selectivity in Chemical Reactions* ed J C Whitehead
  (Kluwer Academic Publ.) 245
Macek J and Jaecks D H 1971 *Phys. Rev. A* **4** 2288
Nielsen S E, Hansen J P and Dubois A 1990 *J. Phys. B* **23** 2595
Ostrovsky V N 1991, to be published
Panev G S, Andersen N, Andersen T and Dalby P 1987 *Z. Phys. D* **5** 331
Richter C, Dowek D, Houver J C and Andersen N 1990 *J. Phys. B* **23** 3925
Roller-Lutz Z, Finck K, Wang Y and Lutz H O 1991 *XVII ICPEAC Abstracts of Contributed
  Papers*
Roncin P, Adjouri C, Gaboriaud M N, Guillemot L, Barat M and Andersen N 1990
  Phys. Rev. Lett. **65** 3261
Royer T, Dowek D, Houver J C, Pommier J and Andersen N 1988 *Z. Phys. D* **10** 45
Shingal R and Bransden B H 1987 *J. Phys. B* **20** 4815
Witte R, Campbell E E B, Richter C, Schmidt H and Hertel I V 1987 *Z. Phys. D* **5** 101

Application of the hyperspherical method to the calculation of reactive cross sections

J M Launay and M Le Dourneuf

UPR 261 du CNRS, Observatoire de Paris, 92195 Meudon, France

ABSTRACT : We describe the essential features of the hyperspherical method and display recent results obtained on the $F + H_2 \rightarrow FH + H$ and $He + H_2^+ \rightarrow HeH^+ + H$ reactions. For the first reaction, we find that the effect of H_2 rotational excitation on total integral and differential cross sections is non negligible. However, the discrepancy found between experimental and theoretical vibrational branching ratios (Launay and Le Dourneuf 1990) does not disappear with the inclusion of $j \neq 0$ initial states. For the second reaction, we find several structures in the energetic dependence of integral cross sections.

1. INTRODUCTION

A rigorous and accurate theoretical description of chemical reactions at the state–to–state level is obviously founded on the quantum scattering theory. However, when solving the nuclear Schrödinger equation for a chemical rearrangement process, one faces two types of difficulties. A formal difficulty concerns the choice of coordinates to represent nuclear motions. For example, in an atom–diatom reaction $A + BC \rightarrow AB + C$, the Jacobi coordinates $\vec{R}_\alpha$ ($\equiv \vec{r}_{\text{A–BC}}$) and $\vec{r}_\alpha$ ($\equiv \vec{r}_{\text{BC}}$) are adequate to describe the arrangement α of the reactants, but not the arrangement γ of the products, for which another set of Jacobi coordinates $\vec{R}_\gamma$ ($\equiv \vec{r}_{\text{C–AB}}$) and $\vec{r}_\gamma$ ($\equiv \vec{r}_{\text{AB}}$) is better suited. Moreover, neither set of Jacobi coordinates of the three arrangements $\lambda = \alpha, \beta, \gamma$ allows a convenient parametrisation of the intermediate ABC molecule during the reaction. A computational difficulty arises from the large number of vibrational, rotational and magnetic vjm levels which are usually populated in one or all the arrangements λ.

The first accurate calculations of cross sections for a chemical reaction were performed some fifteen years ago, in the threshold energy range of the $H + H_2 \rightarrow H_2 + H$ reaction (Schatz et al 1976, Elkowitz et al 1975, Walker et al 1978). However, lacks of generality in the methodology as well as limitations in computer power prevented the study of other systems or even of higher energies. A variety of new accurate quantum mechanical methods were developed some ten years later, when large memory vector computers became available. In linear algebraic variational methods, the wavefunction (Zhang and Miller 1987, Manolopoulos and Wyatt 1988) or the amplitude density

(Haug *et al* 1986) is expanded over elementary basis functions written in Jacobi coordinates of each arrangement. The expansion coefficients are determined once asymptotic boundary conditions are imposed. In hyperspherical coordinate methods (Hipes and Kuppermann 1987, Pack and Parker 1987, Launay and Lepetit 1988, Schatz 1988, Linderberg *et al* 1989), the strong interaction region is represented by a single set of coordinates. The wavefunction is propagated along the hyperradius and matched at large distances to asymptotic solutions expressed in the Jacobi coordinates of each arrangement. Accurate reaction cross sections were obtained using both variational and hyperspherical approaches, first for the $H + H_2 \rightarrow H_2 + H$ reaction (Zhang and Miller 1988, Mladenović *et al* 1988, Manolopoulos and Wyatt 1989, Launay and Le Dourneuf 1989), and then for the $D + H_2 \rightarrow DH + H$ reaction (Zhang and Miller 1989, Zhao *et al* 1990) and the $H + D_2 \rightarrow HD + D$ reaction (D'Mello *et al* 1991). Quantum mechanical calculations on these prototype reactions, which involve three light atoms and a thermoneutral potential energy surface, are relatively easy because a modest number of channels is accessible.

The application of quantum methods to more complex systems is of prime importance for the understanding of real chemical reactions and is currently an active field of research. The present report describes recent results obtained in our group. Section 2 briefly summarizes the salient features of the hyperspherical method. In sections 3 and 4, we present results for the $F + H_2 \rightarrow FH + H$ and $He + H_2^+ \rightarrow HeH^+ + H$ reactions. In section 5, we give some concluding remarks.

2. THEORY

The reaction region for a triatomic system can be described with hyperspherical body–frame coordinates (Pack and Parker 1987, Launay and Le Dourneuf 1989) which are closely related to the usual democratic coordinates (Whitten and Smith 1968). Three Euler angles (denoted collectively by ϖ) specify the orientation in space of the principal axis frame, with the z axis pointing along the least inertia axis and the y axis perpendicular to the molecular plane. The hyperradius $\rho = \sqrt{R_\lambda^2 + r_\lambda^2}$, where R_λ and r_λ are now mass–weighted Jacobi coordinates, characterises the size of the system, while two parameters θ and ϕ characterise its shape. θ is related to the ratio of the smallest to the largest inertial moments and to the area S of the ABC triangle through the relations $\theta = 2 \arctan \sqrt{I_z/I_x} = \arcsin (4S/\rho^2)$. It is equal to 0 for linear configurations and to $\pi/2$ for the non linear symmetric top configurations where $I_z = I_x = I_y/2$. The cyclic parameter ϕ given by $\phi = \arctan [2\vec{R}_\alpha . \vec{r}_\alpha/(R_\alpha^2 - r_\alpha^2)]$ characterises the arrangement in which the system fragments.

The exact nuclear hamiltonian can be split as :

$$H \equiv -\frac{1}{2\mu\rho^5} \frac{\partial}{\partial\rho} \rho^5 \frac{\partial}{\partial\rho} + \mathcal{H} + \mathcal{C} \tag{1}$$

where μ is the three–body reduced mass and $\mathcal{H}$ a reference surface hamiltonian which conserves the body–frame projection $J_z \equiv \Omega$:

$$\mathcal{H} = \frac{2}{\mu\rho^2} \left(-\frac{1}{\sin 2\theta} \frac{\partial}{\partial\theta} \sin 2\theta \frac{\partial}{\partial\theta} - \frac{1}{\cos^2 \theta} \frac{\partial^2}{\partial\phi^2} + \frac{J_z^2}{\sin^2 \theta} \right) + V(\rho; \theta\phi) \tag{2}$$

The sum of the first two terms in the right–hand side of equation (1) represents the $J = 0$ nuclear hamiltonian, augmented by a J_z–dependent centrifugal potential which includes the major part of the energy arising from rotation around the axis of least inertia. The term C in (1) contains the residual part of the rotational energy and the Coriolis coupling :

$$C = \frac{1}{2\mu\rho^2} \left(\frac{J_x^2 - J_z^2}{\cos^2 \theta / 2} + \frac{J_y^2}{\cos^2 \theta} - \frac{4i \sin \theta J_y}{\cos^2 \theta} \frac{\partial}{\partial \phi} \right) \tag{3}$$

C vanishes when the total angular momentum J is 0. It is minimal for linear configurations and diverges for the symmetric top configurations ($\theta = \pi/2$) because of the $\cos^2 \theta$ factors in the denominator of the last two terms in equation (3).

Potential–adapted surface states $\varphi_{k\Omega}$ are defined as the eigenfunctions of the reference hamiltonian $\mathcal{H}$ for fixed hyperradius ρ_p and body–frame component Ω :

$$\mathcal{H}^\Omega(\rho_p) \; \varphi_{k\Omega}(\rho_p; \theta\phi) = \epsilon_{k\Omega}(\rho_p) \; \varphi_{k\Omega}(\rho_p; \theta\phi) \tag{4}$$

Equation (4) is solved by variational expansion over *pseudo*–hyperspherical harmonics which diagonalise $\mathcal{H}$ with $V = 0$, by contrast with the *true* hyperspherical harmonics which diagonalise $\mathcal{H} + C$ with $V = 0$. At large ρ, we can transform the harmonics to a set of functions localised in θ–space and keep in the expansion only the ones which are localised at small θ. The surface functions φ are independent of the total angular momentum J. This property reduces substantially the computer time needed for their evaluation.

For given total angular momentum J and space–fixed component M, the solutions of the exact nuclear hamiltonian (1) can be expanded inside a small sector centered on ρ_p as :

$$\Psi^{JM}(\rho\theta\phi\varpi) = \frac{1}{\rho^{5/2}} \sum_{k\Omega} \varphi_{k\Omega}(\rho_p; \theta\phi) \, D_{M\Omega}^{J*}(\varpi) \, f_{k\Omega}^J(\rho_p; \rho) \tag{5}$$

where the $D_{M\Omega}^J$ are rotation matrix elements and the hyperradial functions $f_{k\Omega}^J(\rho_p; \rho)$ satisfy a set of coupled second–order differential equations :

$$\left(-\frac{1}{2\mu} \frac{d^2}{d\rho^2} + \frac{15}{8\mu\rho^2} - E \right) f_{k\Omega}^J(\rho_p; \rho) + \sum_{k'} \mathcal{H}_{kk'}^\Omega(\rho_p; \rho) f_{k'\Omega}^J(\rho_p; \rho)$$
$$+ \sum_{k'\Omega'} C_{k\Omega, k'\Omega'}^J(\rho_p; \rho) f_{k'\Omega'}^J(\rho_p; \rho) = 0 \tag{6}$$

The couplings originate, firstly, from the variation of $\mathcal{H}^\Omega(\rho)$ inside a sector and, secondly, from the J–dependent operator C. For the systems which we study here, the matrix elements of C are well behaved, since the large θ region of configuration space where they become large is not accessible.

Inside each sector, these equations are solved by standard numerical methods (De Vogelaere 1953, Johnson 1973, Manolopoulos 1986). At the boundary between sectors, the functions and derivatives are transformed to allow for the change of basis. The sector width is an important parameter of the calculation. If too small, the time needed to compute the surface states is too large. If too large, the flexibility of the surface basis is reduced and too many states are needed to get convergence. In practice, we found that a uniform width of $0.2a_0$ was adequate, in terms of both accuracy and efficiency.

After propagating the hyperradial components f to some large ρ, an asymptotic analysis is performed. $\mathcal{H}$ has been chosen such that its kinetic energy part is separable when expressed in terms of the Fock internal coordinates $\omega_\lambda = \arctan(r_\lambda/R_\lambda)$ and $\eta_\lambda = \arccos(\hat{R}_\lambda.\hat{r}_\lambda)$ (Launay and Le Dourneuf 1989). At large ρ in each arrangement λ, the potential V becomes independent on the bending angle η_λ. As a consequence, each surface state $\varphi_{k\Omega}$ converges to a particular $vj\Omega$ separable vibration–rotation wavefunction in a given arrangement λ :

$$\varphi_{k\Omega}(\rho;\theta\phi) \rightarrow \chi_{\lambda vj}(\rho;\omega_\lambda)P_{j\Omega}(\eta_\lambda) \tag{7}$$

The wavefunction (5) is then transformed from the body–fixed hyperspherical representation to the standard space–fixed Arthurs–Dalgarno (1960) representation in Jacobi coordinates. Finally, the reactance K and scattering S matrices are extracted to produce integral and differential cross sections (DCS).

The computer code is divided into three stages. The first stage, which concerns mainly the calculation of the φ basis of surface states is independent of the total angular momentum J and of the total energy E. The second stage, which concerns the evaluation of the C matrix elements and of the body–fixed to space–fixed transformation, depends on J but remains independent of E. The third stage, which computes the hyperradial components for each J and E values, forms the bulk of the computational effort. Fortunately, it involves mainly matrix multiplications and matrix inversions and uses highly vectorised routines.

Convergence is monitored by varying separately the number of k and Ω components in the wavefunction expansion (5). We first perform a calculation for $J = 0$ to determine the number of k components. We then perform calculations for $J \neq 0$ to determine the number of Ω components. In general, it is necessary to include more Ω components to converge detailed observables (such as differential cross sections) than averaged ones (such as integral cross sections). On one processor of a CRAY–2 computer, typical times for all partial waves are about 2 minutes per energy for the $H + H_2 \rightarrow H_2 + H$ reaction with about one hundred channels, and ten hours for the $F + H_2 \rightarrow FH + H$ with about one thousand channels.

3. THE $F + H_2 \rightarrow FH + H$ REACTION

This reaction has been extensively studied experimentally since the beginning of the 1970's because of its importance in the design of infrared chemical lasers. On the theoretical side, it has long been considered as a challenge because the $v'=0, \ldots, 3$ manifolds of the product arrangement are energetically accessible at threshold, due to its large exothermicity (31.8 kcal/mole). Integral and differential cross sections have already been computed for the reaction with an initial H_2 in the $(v=0, j=0)$ state (Launay and Le Dourneuf 1990, Launay 1991). Very recently, total integral cross sections which are in satisfactory agreement with the present ones have been obtained both with a wavepacket (Zhang 1991) and a variational method (Neuhauser et al 1991).

We report here recent results for rotationally excited initial $H_2(v=0, j)$ states, in relation with the crossed–beam experiments of Neumark et al 1985. In all this work, we used the T5A potential energy surface (Steckler et al 1985), which combines ab initio and empirical informations. This surface is adequate to describe the reaction from the $F(^2P_{3/2})$ ground state. Reaction from the fine–structure excited $F(^2P_{1/2})$ state involves a non–adiabatic transition to the lower state which becomes important at energies higher than the ones studied here. We used a basis of surface states with $\Omega \leq 7$, which dissociates asymptotically to (29, 26, 22, 18, 12) in the FH + H arrangement (in this notation, each number designates the maximum j quantum number for each vibrational manifold). In the $F + H_2$ arrangement, the states with even permutation symmetry for the two H nuclei (para–H_2) dissociate as (10, 4) while the states with odd permutation symmetry (ortho–H_2) dissociate as (9, 5). Four symmetry blocks at each J (even and odd permutation symmetry, even and odd parity) are needed to obtain cross sections out of initial $j \neq 0$ H_2 states. Partial waves with J up to 35 were considered to get convergence at the highest energy. The number of channels grows from about 120 for $J = 0$ to 700 or 800 for $J \geq 7$, depending on the symmetry block considered.

Table 1 shows that the dependence of integral cross sections on the initial rotational state of H_2 is fairly large, in contrast with the $D + H_2 \rightarrow DH + H$ reaction (Zhao et al 1990). In particular, an increase of initial rotational excitation produces lower vibrational states and decreases the total reaction cross section. Figure 1 shows that an increase of translational energy enhances the forward peak in the $v' = 3$ DCS, while an increase of initial rotational excitation reduces it. Table 2 shows that the forward to backward DCS ratio for production of FH($v'=3$) is in semi–quantitative agreement with experiment. However, no forward product $v'=2$ is detected (Neumark et al 1985), whereas our calculations predict a small but non negligible $v' = 2$ forward DCS.

Table 1. Integral cross sections (in a_0^2) for $F + H_2(v = 0,j) \rightarrow FH(v'=2, 3) + H$ for several collision energies and initial rotational states j. The total reaction cross section is also given at each energy.

		$j = 0$	$j = 1$	$j = 2$
	$v' = 2$	2.65	2.41	3.10
1.84 kcal/mole	$v' = 3$	8.20	5.40	3.36
	total	10.89	7.88	6.62
	$v' = 2$	3.96	5.41	4.85
2.74 kcal/mole	$v' = 3$	11.36	9.58	5.74
	total	15.40	14.52	10.87
	$v' = 2$	4.87	6.48	6.25
3.42 kcal/mole	$v' = 3$	11.81	10.40	7.06
	total	16.84	17.19	13.73

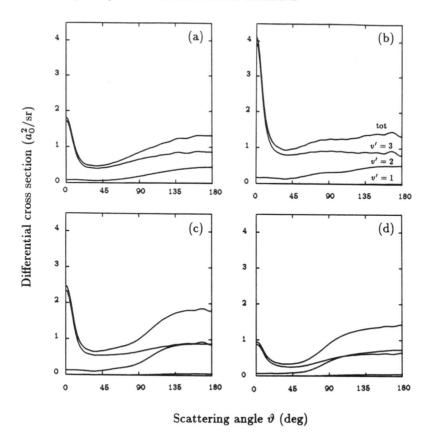

Differential cross section (a_0^2/sr)

Scattering angle ϑ (deg)

Figure 1. Differential cross sections for $F + H_2(v=0, j) \rightarrow FH(v') + H$ for several collision energies ϵ and initial rotational state j. **a** : $j = 0$, $\epsilon = 1.84$ kcal/mole. **b** : $j = 0$, $\epsilon = 2.74$ kcal/mole. **c** : $j = 1$, $\epsilon = 2.74$ kcal/mole. **d** : $j = 2$, $\epsilon = 2.74$ kcal/mole.

Table 2. Comparison between present and experimental results (Neumark *et al* 1985). ϵ is the collision energy in kcal/mole. r is the forward/backward DCS ratio for production of FH in the $v'=3$ state. $R_{v'}$ is the ratio of the integral cross sections for formation of FH in the v' vibrational state over the one for formation of FH in the $v'=2$ state. The results at 1.84 kcal/mole are for para-H_2 with 80 % H_2 in the $j=0$ rotational state and 20 % in $j=2$, while the results at 2.74 and 3.42 kcal/mole are for normal-H_2.

ϵ	r	r^{exp}	R_1	R_1^{exp}	R_3	R_3^{exp}
1.84	1.9	2.5	0.023	0.20	2.63	0.68
2.74	3.3	3.7	0.038	0.23	1.88	0.53
3.42	4.0	6.1	0.045	0.33	1.77	0.48

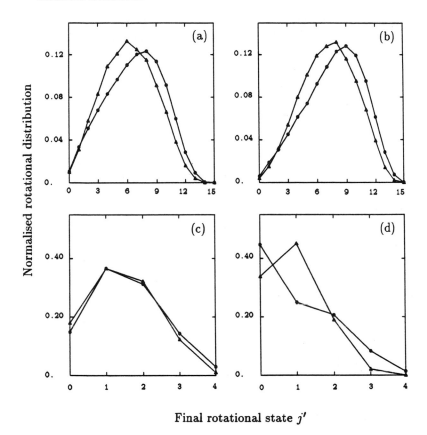

Figure 2. Rotational populations in the product vibrational state v' of HF(v') at fixed scattering angle. **a** : $v'=2$, $\vartheta = 140°$, $\epsilon = 1.84$ kcal/mole. **b** : $v'=2$, $\vartheta = 140°$, $\epsilon = 2.74$ kcal/mole. **c** : $v'=3$, $\vartheta = 100°$, $\epsilon = 1.84$ kcal/mole. **d** : $v'=3$, $\vartheta = 0°$, $\epsilon = 1.84$ kcal/mole. Open circles indicate the present results and upper triangles the experimental ones (Neumark *et al* 1985). Panels (a) and (b) are for normal-H_2, while panels (c) and (d) are for para-H_2 with 80 % H_2 in the $j = 0$ state. Rotational populations for each vibrational state have been normalised so that their sum is 1.

Figure 2 compares experimental and theoretical rotational distributions for the production of FH ($v'=2$, 3). The $v'=2$ distributions are in rather good agreement, the theoretical distribution beeing slightly too hot. The $v' = 3$ distributions are in quantitative agreement at $\vartheta = 100°$, but not in the forward direction.

Table 2 points out a discrepancy between theory and experiment in the magnitude of the R_1 and R_3 vibrational branching ratios. R_1 is smaller and R_3 bigger than the corresponding experimental values. This discrepancy is present at all scattering angles (figure 1). This already noted difference (Launay and Le Dourneuf 1990) between theory and experiment on the branching ratios for the $j = 0$ initial state is reduced by introducing the contributions of $j \neq 0$ initial H_2 states, but it remains sizeable.

Recently, the T5A potential surface has been corrected in the reactant region of configuration space using the scaled external correlation method, but it did not reduce the R_3 branching ratio in the $J = 0$ contribution for the initial $j = 0$ state (Lynch *et al* 1991b). A full calculation including all J and j contributions is necessary to ascertain this discrepancy, but our previous experience on the T5A surface, leads us to believe that these contributions will not reconcile theory and experiment. The discrepancy may rather originate from an inaccuracy of the potential energy surface in the product region of configuration space.

4. THE He + H$_2^+$ → HeH$^+$ + H REACTION

An accurate *ab initio* potential energy surface for the three–electrons HeH$_2^+$ system is available (Mc Laughlin and Thompson 1979, Joseph and Sathyamurthy 1987). The reaction is endothermic by 17.5 kcal/mole and there is a 5 kcal/mole well in the reactant valley. Recent quantum scattering calculations (Kress *et al* 1990, Lepetit and Launay 1991) have shown the existence of a rich spectrum of long lived resonances in $J = 0$ reaction probabilities. The lifetime of these resonances is of the order of 1 ps, an order of magnitude longer than the lifetimes found in the H + H$_2$ (Hipes and Kuppermann 1987) and F + H$_2$ (Bačić *et al* 1990) reactions. While integral cross sections in the last two systems show no structure as a function of energy because of the partial wave summation, it is interesting to know whether resonances can be observed in the He + H$_2^+$ reaction. This led us to extend our previous calculations (Lepetit and Launay 1991) and to calculate integral cross sections for 161 closely spaced energies in the range from 0.95 eV to 1.15 eV. We used a basis of surface states with $\Omega \leq 6$ which

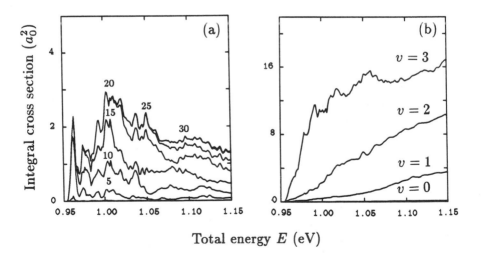

Figure 3. a : Integral cross sections for He + H$_2^+$(v, j=0) → HeH$^+$(v'=0, $j' = 0$) + H as a function of the total energy. b : Integral cross sections for He + H$_2^+$(v, j=0) → HeH$^+$($v' = 0$) + H. In panel (a), each curve represents the summation of all partial waves $J \leq J_{\rm m}$ with $J_{\rm m} = 0, 5, 10, 15, \ldots$. Total energy is measured with respect to the bottom of the He + H$_2^+$ valley at infinite separation. The opening of the v'=0 manifold of HeH$^+$ lies at 0.9549 eV.

dissociates asymptotically to (11, 7) in the $HeH^+ + H$ arrangement and to (20, 18, 16, 14, 10, 8) in the $He + H_2^+$ arrangement for the even permutation symmetry block which is adequate to get reaction cross sections from the $j = 0$ initial states. We included partial waves with J up to 50 at the highest energy considered, with 369 channels for $J \geq 6$.

Figure 3a shows that, contrarily to the $H + H_2$ and $F + H_2$ reactions, the partial wave summation does not wash out the resonance structure present on $J = 0$ reaction probabilities. Even total reaction cross sections for an initial $H_2^+(v=3, j=0)$ show prominent structures, as do cross sections obtained with the Bending Corrected Rigid Linear Model (Kress *et al* 1990), although these last approximate calculations are not in quantitative agreement with the present ones in the threshold region. These structures, which are the fingerprint of resonances on integral cross–sections, would be difficult to detect experimentally because of the finite energy resolution. An experimental study of differential cross sections (Pollard *et al* 1991) has been recently performed : it might be interesting to ascertain whether resonances can be seen on these observables, given the current energy resolution.

5. CONCLUSION

We have shown that the hyperspherical method provides a way to compute reaction cross sections for several triatomic chemical reactions. Although reduced space prevents us to report it here, we just mention that the $Cl + H_2 \rightarrow HCl + H$ has also been treated (Launay and Padkjær 1991). Using present day supercomputers like the CRAY-2, reactions involving about one thousand channels can be studied. Moreover, our computer code has been recently adapted to a RISC workstation which produced the $He + H_2^+$ results presented here.

One can predict that reactions between heavy species involving about 10^4 channels will be within reach in a not too distant future when teraflops computers will be available. It will be necessary to adapt our code to take full advantage of the new computer architectures which will probably be massively parallel. However, since most of the computer time is spent in general purpose linear algebra routines, one will benefit of the optimised libraries which will be available. With extensions in the methodology, triatomic reactions with several electronic surfaces and some four–atom reactions on a single adiabatic surface will be treated (Wu *et al* 1991).

Improvements in the methodology may also hasten progress in the field. In the present method, direct algorithms are used to diagonalise and invert matrices. As a consequence, the whole S–matrix which contains information on all possible initial states is obtained. This results in a growth of computational effort which scales as the cube of the number of channels. The implementation of reliable iterative algorithms or of wavepacket methods in which one selects a few initial states will reduce this growth and prove interesting when the number of channels exceeds about one thousand. In any case, one can conclude that, after the pioneering studies of the $H + H_2$ system, we have now reached an era of detailed quantum studies on real chemical reactions.

ACKNOWLEDGMENTS

The calculations reported here have been partly performed on a CRAY–2 computer through grants of the "Conseil Scientifique du Centre de Calcul Vectoriel pour la Recherche" (Ecole Polytechnique, Palaiseau). Financial support from the Groupement de Recherches 87 "Dynamique des Réactions Moléculaires" is also acknowledged.

REFERENCES

Arthurs A M and Dalgarno A 1960 *Proc. R. Soc.* A **256** 540
Bačić Z, Kress J D, Parker G A and Pack R T 1990 *J. Chem. Phys.* **92** 2344
De Vogelaere R 1955 *J. Res. Natl. Bur. Std* **54** 119
D'Mello M, Manolopoulos D E and Wyatt R E 1991 *J. Chem. Phys.* **94** 5985
Elkowitz A B and Wyatt R E 1975 *J. Chem. Phys.* **62** 2504
Haug K, Schwenke D W, Shima Y, Truhlar D G, Zhang J Z H and Kouri D J 1986 *J. Phys. Chem.* **90** 6757
Hipes P G and Kuppermann A 1987 *Chem. Phys. Lett.* **133** 1
Johnson B R 1973 *J. Comput. Phys.* **13** 445
Joseph T and Sathyamurthy N 1987 *J. Chem. Phys.* **86** 704
Kress J D, Walker R B and Hayes E F 1990 *J. Chem. Phys.* **93** 8085
Launay J M and Lepetit B 1988 *Chem. Phys. Lett.* **144** 346
Launay J M and Le Dourneuf M 1989 *Chem. Phys. Lett.* **163** 178
——1990 *Chem. Phys. Lett.* **169** 473
Launay J M 1991 *Theor. Chim. Acta* **79** 183
Launay J M and Padkjær S B 1991 *Chem. Phys. Lett.* **181** 95
Lepetit B and Launay J M 1991 *J. Chem. Phys.* in press
Linderberg J, Padkjær S B, Öhrn Y and Vessal B 1989 *J. Chem. Phys.* **90** 6254
Lynch G C, Steckler R, Schwenke D W, Varandas A J C and Truhlar D G 1991a *J. Chem. Phys.* **94** 7136
Lynch G C, Halvick P, Zhao M, Truhlar D G, Yu C H, Kouri D J and Schwenke D W 1991b *J. Chem. Phys.* **94** 7150
Manolopoulos D E 1986 *J. Chem. Phys.* **85** 6425
Manolopoulos D E and Wyatt R E 1988 *Chem. Phys. Lett.* **152** 23
——1989 *Chem. Phys. Lett.* **159** 123
McLaughlin D R and Thompson D L 1979 *J. Chem. Phys.* **70** 2748
Mladenović M, Zhao M, Truhlar D G, Schwenke D W, Sun Y and Kouri D J 1988 *J. Phys. Chem.* **92** 7035
Neuhauser D, Judson R S, Jaffe R L, Baer M and Kouri D J 1991 *Chem. Phys. Lett.* **176** 546
Neumark D M, Wodtke A M, Robinson G N, Hayden C C and Lee Y T 1985 *J. Chem. Phys.* **82** 3045
Pack R T and Parker G A 1987 *J. Chem. Phys.* **87** 3888
Pollard J E, Syage J A, Johnson L K and Cohen R B 1991 *J. Chem. Phys.* **94** 8615
Schatz G C and Kuppermann A 1976 *J. Chem. Phys.* **65** 4642
Schatz G C 1988 *J. Chem. Phys.* **90** 3582
Steckler R, Truhlar D G and Garrett B C 1985 *J. Chem. Phys.* **82** 5499
Walker R B, Stechel E B and Light J C 1978 *J. Chem. Phys.* **69** 2922
Whitten R C and Smith F T 1968 *J. Math. Phys.* **9** 1103
Wu Y M, Cuccaro S A, Hipes P G and Kuppermann A 1991 *Theor. Chim. Acta* **79** 225
Zhang J Z H and Miller W H 1987 *Chem. Phys. Lett.* **140** 329
——1988 *Chem. Phys. Lett.* **153** 465
——1989 *J. Chem. Phys.* **91** 1528
Zhang J Z H 1991 *Chem. Phys. Lett.* **181** 63
Zhao M, Truhlar D G, Schwenke D W and Kouri D J 1990 *J. Phys. Chem.* **94** 7074

Photoinitiated reactions in weakly bound complexes: Ca* + HX → CaX* + H

B. Soep and J.C. Whitham
Laboratoire de Photophysique Moléculaire
Université Paris-Sud, Orsay 91405, France

J.P. Visticot and A. Keller
Service des Photons atomes et Molécules
CEN Saclay, Gif sur Yvette 91191, France

Abstract

The Potential energy surfaces of the excited state reactions of Calcium ($(4s4p)^1D$, $(4s4p)^1P$) with halogen halide (HCl, HBr) have been explored by the optical excitation of a Ca-HX Van der Waals complex prepared in a supersonic expansion. Resonances and intense vibrational progressions have been analysed on a local mode basis.

The Ca-HX complex can be prepared in various electronic states, corelating to the ^1D or ^1P calcium states. The excitation of these different electronic states, in the case of Ca-HBr system, leads to a striking effect on the branching ratio for the two allowed excited electronic states ($A(^2\Pi)$ and $B(^2\Sigma)$) of the CaBr product.

INTRODUCTION

Numerous collision experiments have been performed to study chemical reactions involving excited electronic states of the reactants.

In this type of experiments, the initial state of well separated reactants is prepared, as selectively as possible, and the product states population is measured, after the chemical reaction has been completed. Only the asymptotic states are known, it is thus difficult to extract accurate information on the dynamics and the potential energy surface in the intermediate domain where the atoms exchange during the chemical reaction.

In a full collision experiment, two important parameters are difficult or can not be accessed: the relative reactants orientation and the impact parameter. Indeed, several experiments have been developed to orient the reactants[1-3] or align the orbital of an excited atom involved in the chemical reaction[4]. The impact parameter average and the passage from the space fixed initial condition to the molecular intermediate state, during the collision may destroy, in part, the orientation or the alignment defined at the beginning, and make intricate the interpretation of experimental results.

To avoid these depolarisation or reorientation effects, in the entrance channel of the collision, our goal has been to initiate the chemical reaction with the reactants near each other, so as to prepare the reactants in the excited electronic state directly in the molecular frame.

To achieve this goal we prepare the A + BC reactants within a stable A...BC van der Waals complex, in its ground state. Then, the chemical reaction is initiated by optical excitation of this complex to an excited electronic state (A...BC)*. The important fact is that the excited state of this complex corresponds to the local excitation of A within the A...BC complex: (A*...BC). We therefore prepare A* within the reach of BC, in the molecular field of the complex. This type of experiment looks like a photodissociation experiment, the main difference lies in the fact that within the weakly bond complex the two monomers, A* and BC, keep their individuality. For that reason, initiating a chemical reaction within a van der waals complex can be related to a collision experiment.

In that way, the parameters which are averaged in a collision experiment or partially destroyed by the collision dynamics in the entrance channel of the reaction are here determined:

- The reactants relative orientation is determined by the equilibrium geometry of the van der Waals complex in its ground state.

- The impact parameter of the full collision experiment

is related to the overall rotation of the van der Waals complex. The angular momentum correponding to the A-BC orbital motion is low, because the temperature achieved in the supersonic beam, used to form the complex, is low (about 5K).

- The preparation of selective intramolecular vibration mode of the complex, can give insight into the influence of the relative kinetic energy and geometrical arrangement of reactants, on the chemical reaction.

-These three last points are related to the mechanical relative arrangement preparation of the reactants. The interest of studying chemical reaction involving excited electronic state is that another type of arrangement can be controlled: the electronic cloud alignment. Indeed, specific electronic states of the complex will be excited, as A* is an excited atom, the molecular field lifts the orbital degeneracy of the atomic state. for instance, for a P atomic state, specific Σ, Π states will be selected by the optical excitation, allowing the determination of the atomic orbital alignment in the frame of the reactive complex. The specificity wich can be attained in the electronic excitation of the complex is related to orbital alignment collision as observed by Leone and coworkers[5], in the electronic relaxation of calcium (5p) and the reaction of calcium (4p) with HCl by Rettner and Zare[4].

The chemical reactions studied here are :

$$Ca* + HX \rightarrow CaX* + H \text{ with } X = Cl, Br$$

in which, a calcium atom in an excited electronic state $((4s3d)^1D$ or $(4s4p)^1P)$ reacts with the HX molecule to form the CaX molecule in an excited electronic state. Two excited electronic states are energetically allowed for the CaX product : the $A(^2\Pi)$ state and the $B(^2\Sigma)$ state.

Initiating the chemical reaction within the Ca...HX excited van der Waals complex, we have characterized the electronic and vibrational states of this complex. This allows us to probe the potential energy surface corresponding to the entrance channel of the chemical reaction. Beeing able to prepare the reactive system in a defined vibrational and electronic state we have studied the influence of these different initial conditions on the internal state distribution of the CaX products formed by the chemical reaction.

EXPERIMENTAL

Set-up

The reaction of the complex $Ca^*-HX \rightarrow CaX^*(A^2\Pi, B^2\Sigma^+)$ with X=Cl, Br, is observed in the silent zone of a pulsed supersonic expansion after excitation with a pulsed, tunable dye laser of the Ca-HX complex. The complex is formed by laser evaporation of a rotating calcium rod in a classical arrangement[6-10]. The resulting emission is dispersed through a small monochromator (50 Å resolution) and collected on a phototube. The tunable laser is generated in the range 4000-4300 Å by frequency mixing the fundamental frequency of a YAG laser with the output of a red dye laser (DCM ,LDS 698) pumped by the same Yag. In the range 4300-5000Å, direct pumping of coumarin dyes by the third harmonic of the YAG laser is used. Great care is taken to eliminate superradiance light in the case of direct pumping, while frequency mixing acts almost completely as a filter of this superradiance.

A delay of a few microseconds between the laser vaporization pulse (created by the 2nd harmonic of another synchronized Yag laser) and the laser excitation pulse is needed to allow the calcium atoms or complexes to reach the excitation zone. This delay depends upon the carrier gas (10 μs for He and 35 μs for Ar) and its accurate adjustment is crucial for the observation. A temperature of 5 K is achieved in argon expansion allowing the formation of complexes. We found, as in our study of the calcium free radicals, that helium was less efficient a coolant in our beam configuration[9]. The expansion has been characterized by the weak appearance of the Ca_2 bands at 4840Å; the rotational contour of the B-X(13-0) band in the expansion of calcium in pure argon corresponds to a temperature of 5 K.

The calcium evaporation power was maintained low, typically at 0.7 mJ in order to optimize the calcium metal formation with respect to metal cluster and to obtain the lowest beam temperatures. A 1% mixture of HX in argon is further diluted in argon in a continuous fashion allowing the concentration range of HX to be varied between 10^{-3} and 10^{-2}.

RESULTS

Several types of experiment have been conducted :
- action spectra, where the reaction product fluorescence is monitored as a function of the excitation laser.
- dispersed fluorescence of the product CaX[*]
- pump and probe experiments, where the excitation laser is tuned to a particular resonance of the action spectrum and the state distribution of CaX is probed with a second laser.

Only the two first experiments will be reported here, we thus shall present the action spectra which have been recorded with the two system: Ca...HCl and Ca...HBr, in a first part. In a second time, we shall present the emission spectra obtained with the CaBr system for different preparations of the reactive complex Ca...HBr.

ACTION SPECTRA: Characterization of the Ca-HX complex.

We have made three observations. First, the chemiluminescence signal of CaX resulting from the complex excitation is linear with the HX concentration. Second, it is only observed intensely in the expansion of argon (with HX). Third, when HCl, HBr, or HF are added in minute quantities to the pure argon expansion, the Ca_2 dimer signal disappears. We conclude that the chemiluminescence that we observe origins from a Ca-HX complex (linear dependence on HX and cold expansion), whose the binding energy is greater than that of Ca_2 (1075 cm^{-1})[11,12]. Two other observations support the absence of contributions from higher clusters in our spectra. First, all vibrational distributions of the products go up to the same energetics limit characteristic of a single complex. Secondly, a strong H/D isotopic effect has been observed for most bands of the Ca-HX complexes and can only arise from a single HX complexation.

Action spectra have been recorded in long scans between 4000 and 5000 Å monitoring the CaX A-X (or B-X) emission. This region covers the $Ca(^1S_0-^1D_2)$ and $(^1S_0-^1P_1)$ transitions and does not overlap with CaX transitions. No chemiluminescence signal is observed at longer wavelengths than 5000 Å which results from the energetical threshold for CaX^* formation

Ca-HCl Action spectrum

The action spectrum presented in figure 1a reveals two sets of bands localized near each of the two calcium transitions. Both sets extend over about 1000 cm^{-1} with broad structures in their red part and narrow structures in their blue part . A close-up of the band corresponding to the $Ca(^1S_0-^1P_1)$ transition is presented in figure 3a and the pattern of a vibrational progression merges from the figure. This pattern is slightly complicated by satellites within the bands which either appear as non symmetrical in shape or display shoulders. The ultimate width of the transitions may be given by one band, located at 24360 cm^{-1}, which appears sharp and therefore has been recorded under a higher laser resolution (0.1 cm^{-1}). It exhibits a quasi lorenztian line shape with no fine structure indicative of rotational transitions, therefore the bandwidth depends only on the diffuseness of the transition. This allow a determination of a lower limit for the lifetime of the excited Ca...HCl complex: about 1 ps.

Ca-HBr action spectra

The spectra have been recorded in the same conditions as for HCl and present a very similar outlook. In figure 1b, an overview of the action spectrum is presented. As for Ca-HCl, the spectrum consists in two regions close to the calcium atomic transitions and also these regions can be further divided in two sets of bands.

The main differences with Ca-HCl spectra appear in the $^1D_2$ region where the red part is now structured with a regular separation of about 60 cm^{-1}, decreasing slightly towards the blue. In the $^1P_1$ region, although the bands exhibit the same spacing as in the Ca-HCl spectrum, their width is somewhat broader. A spectrum has also been recorded with DBr, the structures of the spectrum have vanished throughout the spectrum of the deuterated compound, except for the band at 23400 cm^{-1} which splits up in three sub-bands.

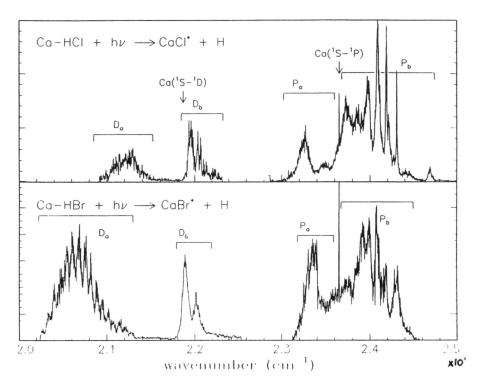

Figures 1a, b: CaHCl and CaHbr Action spectra.

Localization of the excitation energy

The Ca-HX complex is a weakly bound species in the ground state as we have seen in the preceding. How can we then describe the optically accessed states when we know that these excited states are reactive and, most likely, through the harpoon mechanism? This mechanism has been postulated[4,10] by analogy of the excited alkaline earth with the alkali (ground state) + HX reactions. The open shell Ca(4s4p) of the ionisable calcium can easily yield an electron jump to HX,[1] the crossing radius being 3.5 Å for Ca^{1}P+HCl between the covalent and Ca$^+$,HCl$^-$ surfaces. Therefore the vertical excitation of the complex should lead directly to the reaction domain.

The ionic curve is neither directly attained for Ca-HCl nor for Ca-HBr, by the optical excitation, as can be seen by

----

[1]The HCl$^-$ isolated ion is unstable but easily stabilised in the Ca$^+$ field in the range of distances of the complex (3 to 4 Å)

the inspection of the action spectra in figures 1a, 1b. If it were, the spectrum should be continuous and unstructured as are charge transfer spectra; this results from the rapidly varying nature of the ionic potential ($\approx$ 1 eV/Å). Rather, the spectra display sets of bands close to the calcium singlet transitions at strikingly similar positions for both the Ca-HCl and Ca-HBr complex. We therefore think that the observed transitions derive their oscillator strength from the calcium transitions. In such conditions, the reaction surfaces correlating at infinite distance with Ca(^{1}P,^{1}D) + HX should split up in several Λ components under the molecular field[14,15]. We thus expect two classes of transitions "Σ" and "Π" as allowed by electric dipole transitions for each excited atomic configuration.

In that way, the main groups of transitions on the action spectra has been interpreted as due to electronic states of the complex correlating at infinite separation with the Ca(^{1}S$_0$-^{1}D$_2$) and Ca(^{1}S$_0$-^{1}P$_1$) transitions; these states have been labeled as D$_a$, D$_b$ and P$_a$, P$_b$, respectively.

Vibrational motion of the reactive complex

We shall describe in the following the nature of the observed transitions in the action spectra in terms of local excitation and local modes. Before this discussion, it seems necessary to summarize what can be stated about the structure of the Ca-HCl van der Waals complex and what can be inferred on the energetics of the reaction.

Structure of the Ca-HCl complex

The geometries of the known complexes of an atom with HCl are linear with the hydrogen atom lying between the two heavy partners.[16] This applies also to the Hg-HCl van der Waals complex[17] and should be also true for Ca-HCl in the ground state as the calcium atom is even more polarisable (25 Å^3)[18] than Hg (5.1 Å^3)[19] and interacts strongly with the HCl quadrupole. Following Buckingham's developments, Hutson calculated the anisotropy of rare gas-HX complexes and found that the major contribution arose from the induction forces.[20] In this model, the HCl dipole imposes a linear geometry and the HCl quadrupole the H central position. Therefore, in calcium-HCl with a 3.5 Å center of mass separation, the barrier to linearity is high ($\approx$700 cm^{-1}) and the second well shallow. This value is not inconsistent with the existence of a deep well as stated in the experimental section. We can not make such definitive statements in the case of HBr because of the possible failure of a point charge electrostatic model for a bulkier atom like Br, nevertheless, we can safely assume a linear geometry for Ca-HBr, this complex being also a strongly bounded complex.

In conclusion, for HCl, the structure of the complex is a linear Ca-HCl geometry which is further confirmed by the initial results of ab-initio calculations[21] in accordance with the intuitive partial charge transfer between the calcium and the HX. This charge transfer will involve the S orbital of Ca and the antibonding orbital of HX located mainly on the H atom. Experiments would be welcome to confirm this structure.

Vibrational excitation

A closer examination of the previous spectra reveals vibrational structures that we shall analyze in the following in term of local modes of the excited complex Ca-HX. We therefore assume that the optical excitation process allows us to make one dimensional cuts of the excited potential energy surface along the van der Waals stretch, bend and the HCl coordinates. We shall combine the information contained in both Ca-HCl, Ca-HBr spectra provided their close resemblance.

The $^1D_2$ Ca-HBr stretching progression

At the long wavelength end of the CaHBr (^1D region) we notice a long progression of closely spaced peaks (≈ 70 cm^{-1}). These peaks which have a width of ≈ 40 cm^{-1} can be described by a one dimensional Morse oscillator with $\omega_e = 71$ cm^{-1} $\omega_e x_e = 0.3$ cm^{-1} (assuming that the first band observed corresponds to the origin). This long low frequency progression suggests a van der Waals mode. The first observed band is located at 1580 cm^{-1} to the red of the ^1D line thus the potential energy surface in the complex excited state is deeper by at least 1580 cm^{-1}. In such case the Ca-HBr distance should be appreciably reduced in the excited state and the progression must pertain to the van der Waals stretching mode. The resulting dissociation energy is $D_o \approx \omega_e^2/4\omega_e x_e \approx 4200$ cm^{-1} in the excited state. To reproduce the intensity of these transitions, we assume also that the ground state potential surface can be described by a morse potential with the parameters: $w_e = 35$ cm^{-1}, $w_e x_e = 0.2$ cm^{-1}, and with a difference between the excited and ground state equilibrium bond length of 0.5 Å. The experimental and the calculated spectrum are presented in the figure 2a. The order of magnitude of the stretch frequency in excited states of Ca-HX will be in the following kept as <70 cm^{-1} (depending on the well depth of the relevant potential).

The same portion of spectrum have been recorded with the CaDBr complex. To our disappointment, it does not display the same stretching progression that should be expected in a local mode analysis. Rather, a quasicontinuum is obseved. The onset of a continuum can be due to a greater number of bands possibly of increased breath due to an increased reactivity. The increase of bending level density from Ca-HBr to Ca-DBr may cause the appearance of a great number of bend-stretch combination bands. Such an increase of the complexity of the spectra due to deuteration has already been observed at a certain level of excess energy in the Hg-NH$_3$ and Hg-H$_2$O van der Waals complexes.[22,23]. This complexity could be combined to a larger width of the individual peaks. The reaction can have a very strong angular dependence with a maximum at equilibrium distance. As the vibrational amplitude will be smaller for Ca-DBr than for Ca-HBr, one could expect a greater reactivity for the deuterated compound. In any case, the long progression (1100cm^{-1}) with a small positive anharmonicity, observed in Ca-HBr has to be assigned to the Ca-HBr stretch.

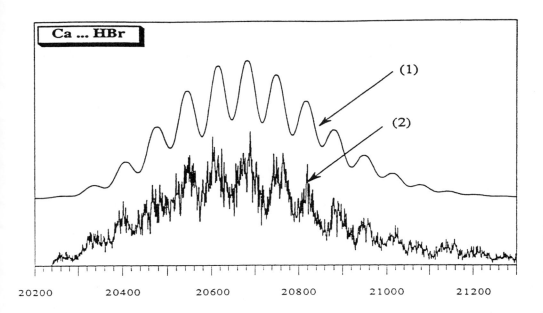

Figure 2: (1) Calculated spectrum assuming Morse potentials to describe the van der Waals streching motion in both excited and ground electronic states of the CaHBr complex. (2) CaHBr Experimental Action Spectrum, in the red side of the $^1S \rightarrow {}^1D$ calcium transition.

The ^1P Ca-HX bending progressions

We shall now analyze the spectra displayed in figure 2a progression in the blue part of Ca-HCl spectrum is observed with an irregular spacing of c.a. 200 cm^{-1}, a value implying a van der Waals mode progression. In order to assign these bands, we have substituted DCl for HCl. The corresponding spectrum is displayed in figure 2b. A strong isotopic effect is observed resulting in an important decrease of the band spacing. Thus, it is obvious that the motion of the H/D atom is involved in these progressions Furthermore, the relative positions of the 4 bands in the bluemost part of the spectrum closely match the energy separations of free HCl rotor lines for j=4,5,6,7 (E_j=b j(j+1), b is the HCl ground state rotational constant). This is confirmed by the observation in the corresponding Ca-DCl spectrum of 4 transitions matching the j=5,6,7,8 line positions of a free DCl rotor. This correspondence indicates that these ensembles of transitions lead to upper levels beyond the barrier to free rotation of HCl/DCl within the excited state of the complex. On the contrary, the other members of the progression relate to the bending motion associated with the anisotropy of the excited Ca–HCl potential. The observation of quasi free rotation has also been reported recently in the transition region of IHI[24].

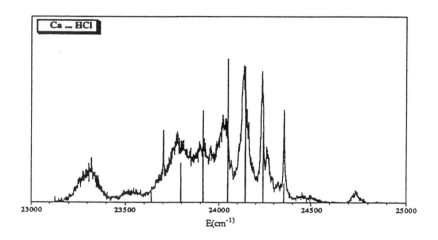

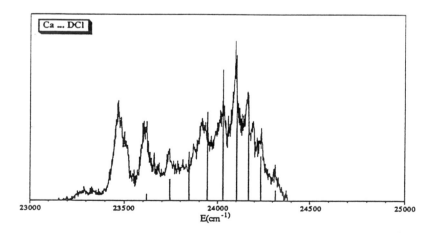

Figures 3a, b: CaHCl and CaDCl Action spectra around the $^1S \to ^1P$ calcium transition and the corresponding calculated spectrum with the hindered rigid rotor model.

To be more quantitative, we have used a bending model described in an previous article[10]. In this model the relevant angular momentum operators are:

- $J = L + l + j$: the total angular momentum
- j, m_j: the angular momentum of HCl molecule and its projection on the Ca-Cl axis (z axis of the molecular frame), respectivly.
- L, Λ: the electronic angular momentum and its projection on the z axis.
- l: the angular momentum corresponding to the overall rotation of the complex.
- Ω: The projection of the total angular momentum, J, on the z axis in the molecular frame.

The transitions in the Ca-HCl complex arise from the vibrationless state of the linear geometry, without angular momentum Ω along the complex axis. The selection rules allow only the access of excited $m_j = 0$ states, that for $\Sigma \rightarrow \Pi$ transitions the $\Delta\Lambda = 1$ change is cancelled by the change in the total angular momentum projection, $\Delta\Omega=1$. Therefore these $\Delta m_j =1$ transitions should be minor as only due to the Coriolis coupling that we have neglected. Such transitions have been observed by Nesbitt et al. in the case of strong Coriolis coupling of degenerate levels.[25]

We have fitted the bending potential $V(\theta)$ of the excited state to reproduce the positions and intensities of most of the transitions of the Ca-HCl ^1P region. The result of the fit is presented in figure 3a and the best potential in figure 4. The calculated levels and intensities reproduce correctly the pattern of the blue bands in figure 2a, with the origin taken on the calcium line. The details of the fit have been given in a previous article[10]. Changing only the H atom mass, we could directly obtain the Ca-DCl transitions wich are presented in figure 3b. This assesses that the main progressions in the blue part of the spectra in figure 3 are bending progressions of the Ca-HCl complex. This agreement between experimental and calculated peaks positions for Ca-HCl and Ca-DCl implies also that the origin of this progression changes only by $30cm^{-1}$ upon deuteration. This is compatible with a local mode description where the HCl bond frequency reduces in the excited state only by $\approx 200cm^{-1}$.

This model model was designed to give the framework of the transitions showing that one dimensional local mode excitation was promoted by the optical pulse. But it is seen that the model does not account for all the transitions observed in figure 3. If we look more closely in the blue part of figure 3, we see some non assigned transitions appearing as satellites or shoulders displaced by $20-50cm^{-1}$ from the main peaks of the bending progression. This separation is compatible with a stretch mode spacing in a shallower potential ($1500 \ cm^{-1}$) than the potential of Ca-HBr in $^1D_2$ region. Therefore a possible explanation for the existence of such bands relies on a combination of bend-stretch modes which are not easily predicted in the model.

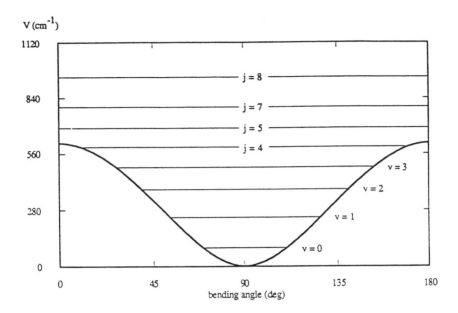

Figure 4: Excited electronic state bending potential for the CaHCl complex.

The inspection of the Ca-HBr spectrum in the same ^1P region shows a very similar pattern for the transitions, similar spacings but a more diffuse appearance. We have not attempted a bending analysis because the Ca-DBr spectrum is almost continuous as in the $^1D_2$ domain and probably for the same reason, i.e. a greater bend stretch coupling.

The previous analysis clearly confirms the existence of a second group of transitions which appear most distinctly in the red of the Ca-DCl spectrum (see figure 2) and bear no relation with the group of the bending transitions. We can exclude the contribution of Ca-$(HX)_n$, n>1, as these bands are sensitive to the H/D substitution, in both the HCl and HBr spectra. An other interpretation for this group of bands would be the existence of a Ca-XH isomer in the ground state which, on the basis of electrostatics, seems unlikely to be formed. The correlation of atomic and complex states implies in the case of calcium ^1P the presence of a second electronic state. We have already noticed the separation in two groups of the transitions observed in Ca-HBr and Ca-DBr spectra, fitting in this interpretation.

Reactivity of the Ca-HX states.

The purpose of the previous analysis has been to show that one could represent the excited complex by one dimensional modes on surfaces correlated to excited atomic calcium. This description may appear as surprising, for we have been probing a reactive surface , and we could conclude from the observation of structured spectra and resonances, that we had only access to the entrance valley of this reaction. On the other hand, the chemiluminescent reaction of excited calcium with HCl is known to occur with high cross sections ranging from 25 $Å^2(^1D_2)^{13}$ to 68 $Å^2(^1P)^4$. These cross sections agree with a crossing of the ionic and covalent surfaces without barrier at 3.5 $Å^4$. Hence, we have to expect a direct and fast reaction. In addition , in figure 2a the bending bands become increasingly sharp with excitation, so that the bending Ca-HCl vibration displaces the system away from a favourable conformation. We thus interpret the observation of distinct progressions and not featureless continua, in the action spectra as the excitation of local modes perpendicular to the reaction coordinate.

In this previous discussion we have shown that we have identified the electronic and vibrational states of the Ca...HX complex. We now shall present how different preparations of the reactive complex influence the internal states distribution of the CaX product. We report here only the results obtained on the Ca...HBr system.

Ca...HBr CHEMILUMINESCENCE SPECTRA

These spectra are characteristic of the CaBr $A^2\Pi-X^2\Sigma^+$ and $B^2\Sigma^+-X^2\Sigma^+$ emission. Each spectrum consists in two series of short vibrational progressions well separated. The spectral resolution is however insufficient to resolve either the band sequences (a few cm^{-1}) or the spin-orbit separation in the $A^2\Pi$ state (60 cm^{-1}).

The emission spectra resulting from the excitation of Ca-HBr are very dependent upon the excited state. This behavior is exemplified in figure 5. We found that the shape of the emission spectra was independent of the vibrational excitation within one electronic state of the Ca-HBr complex except in the ^1D region where the two peaks at 4568 and 4537Å do not have exactly the same emission profile. The shape of the emission spectra results from the intensity distribution over the A to X and B to X transitions and this distribution was found very different for the excitation of each each electronic state of the complex as exemplified in figure 5. We shall relate in the following these variations in pattern to the population of the CaBr A and B states formed by the reaction.

These emission spectra have been recorded throughout the excitation spectrum of the deuterated complex Ca-DBr. The same tendency has been observed: within what we assign to a single state of the complex, the emission pattern is constant, and the evolution in pattern from one electronic state to another is comparable but with different ratios (B/A) as we shall see later.

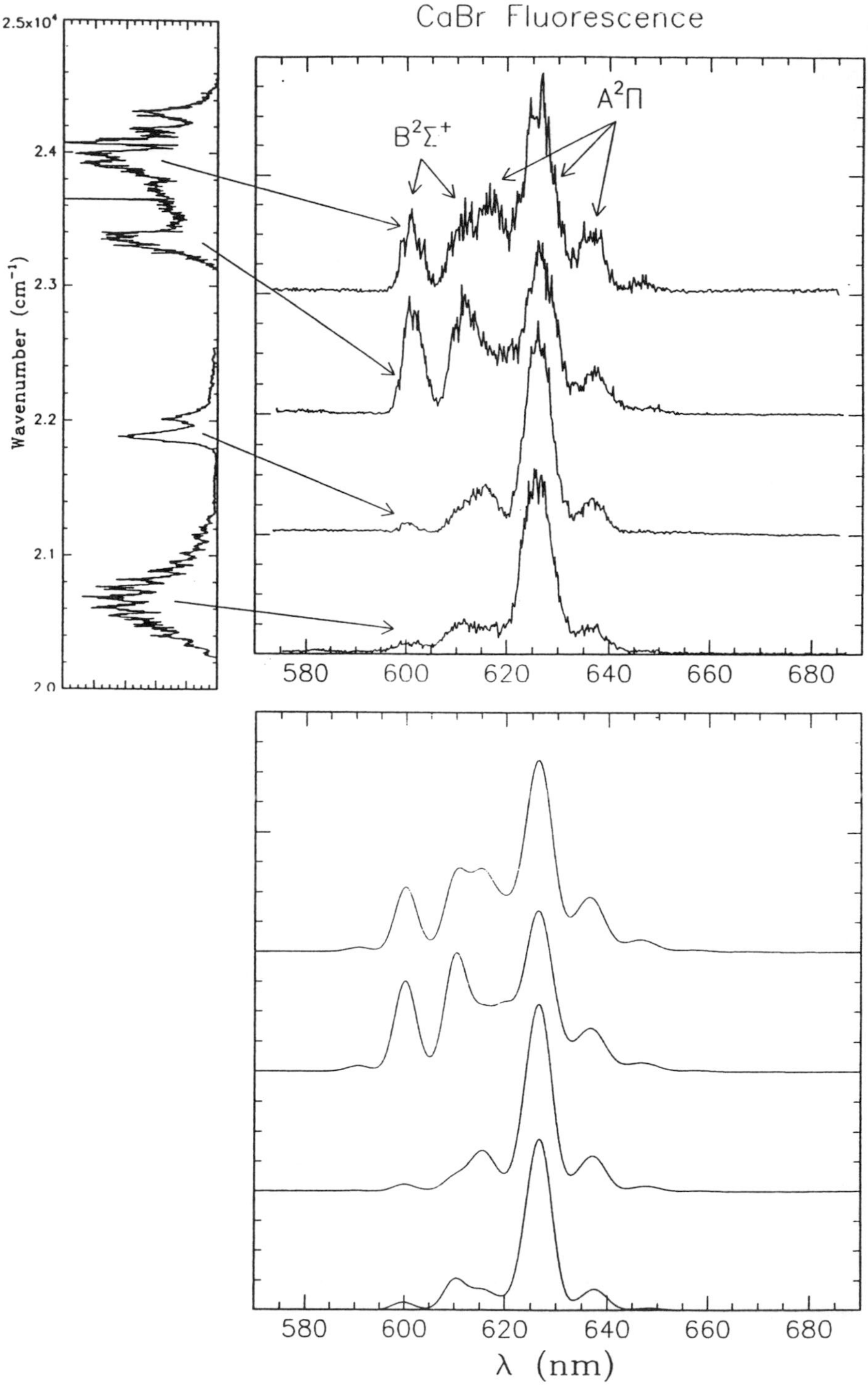

Figures 5, 6: *Fluorescence spectra of the CaBr molecule formed by the reaction within the CaHBr complex, for different excitation frequencies, and the corresponding simulated spectra.*

SIMULATION OF THE EMISSION SPECTRA

The CaX A,B to D transitions have been probed by a tunable red laser after excitation of selected peaks in the action spectra of Ca...HBr, in the region correlating to ^1P and ^1D calcium states. These pump and probe experiments allow a determination of the vibrational population of the CaX produced by the reaction, for the A or B states depending on the excitation frequency. these results will be reported in a forthcoming paper.

Combining the vibrational population that we have derived and Franck-Condon simulations, we have reproduced the experimental emission spectra and obtain the relative A to B branching ratios. The CaX molecule is very ionic in its lower states (X,A,B). Consequently, the sequences are strongly congested. However, the Franck-Condon factors corresponding to sequences $\Delta v=\pm 1$ or ± 2 vary very rapidly with the vibrational quantum number. The relative intensities of these sequences are a sensitive indication of the vibrational energy content in the CaX product, as it is seen in the emission spectra presented in figures 5, where these ratios increase with the excitation energy.

To be more quantitative, we have simulated the emission spectra using a standard program to calculate the Franck-Condon factors and we approximate the potentials by Morse functions. We have used this approximation because of the lack of accurate data for the whole range of vibrational levels populated in the A,B states; a very accurate rotational analysis of the CaBr X to A,B transitions has been performed but only for the low v's (v<2) and the parameters cannot be extrapolated to higher v[26]. The ω_e, $\omega_e x_e$, and R_e values used are presented in table I. The consistency of these values with the experimental data has been checked for the observed transitions A,B to D (1 photon) and also for a two photon excitation spectrum of X to D, all the band heads positions being reproduced within 1 Å.

CaBr	$X\ ^2\Sigma$	$A\ ^2\Pi$	$B\ ^2\Sigma$
T_e (cm^{-1})	0	15958.41$\pm$ 29.6 [c]	16383.13
ωe (cm^{-1})	284.94	287.95	285.
$\omega \chi_e$(cm^{-1})	0.8218	0.8066	0.902
r_e(Å)	2.5935	2.5764	2.5676

Table I : Spectroscopic constants for the CaBr molecule[26].

With these parameters and the vibrational distributions, we could reproduce the experimental spectra, the fit are given in figure 6. The A/B branching ratios deduced from these fits are summarized in table II.

λ (Å)	4855	4568	4537	4282	4185
Label	D_a	D_b	D_b	P_a	P_b
B/A	0.19	0.09	0.25	0.92	0.5

Table II : Branching ratios B/A for different excited electronic states of the Ca...HBr van der Waals complex.

A to B branching ratios

We have described the nature of the observed transitions in the action spectra in terms of local excitation and local modes. In our previous study, the analysis of the vibration progressions and of their spectral locations has led us to conclude to the existence of different electronic states correlating to the ^1P and ^1D calcium atomic states and corresponding to the different alignement of the excited electronic cloud with respect to the metal molecule axis. In this study, we have recorded chemiluminescence spectra resulting from the excitation of these electronic states. In the case of Ca-HBr, the emission spectra have been found to be specific of a given excited electronic state of the complex. We have observed that each group of transitions assigned to a single electronic state of Ca-HBr yields a single CaBr emission pattern of constant A to B relative branching ratio.

This is in support with the existence of different ("Σ" and "Π" in the quasi-diatomic limit) electronic states, each one having a specific branching to the excited states of CaBr. Different authors have inferred the conservation of the electronic momentum Λ during the chemical reaction. In the collisions of excited calcium atoms with HCl, the optical preparation of the pπ or pσ calcium states, with respect to the reagents approach, resulted in the preferential production of CaCl $A^2\Pi$ of $B^2\Sigma$, respectively. This was interpreted in the framework of the harpoon mechanism, as a single electron jump from $Ca(4s4p^1P)$ to HCl. Various crossings are accessible involving the jump of either the 4s or the 4p electron of Ca and the selectivity was assigned to the jump of the 4s electron leaving an oriented Ca[**] ion[4]. The role of the inner crossings with excited ionic configurations are essential in explaining the high chemiluminescence cross sections of excited alkaline-earth atoms with halogenated molecules and this is the basis of a model developed by Menzinger[27].

The model is represented in the extended correlation

diagram in figure 7. Along the approach coordinate, the crossings between the covalent and ionic curves are shown, the diagram is extended along the reaction coordinate to the various accessible states of the product. We have displayed the various electronic states of Ca-HBr which correlate at infinite separation with Ca(^{1}P) and Ca(^{1}S) and the ground and excited ionic surfaces Ca$^+$(4s,3d,4p)-HCl$^-$. This is done of course in a very schematic manner in the quasi-diatomic limit. After the excitation of the complex into a state of well defined symmetry, the reaction can proceed either by a jump to the ground state ionic surface (outer crossing) or to one of the various excited ionic surfaces (inner crossings). The selectivity results from the competition between the various covalent-ionic couplings which are governed by the symmetry of the involved states. In this respect, the observed variation of the A/B branching ratio can be either due to an increase of the X ground state yield at the expense of one of the excited states populations or to a competition between the two excited A and B states. Both cases are possible but we favor the latter explanation where a state of the complex of Σ symmetry will lead preferentially to the B^{2}Σ state of CaBr and, equivalently, a Π state of the complex will form the A^{2}Π CaBr state preferentially.

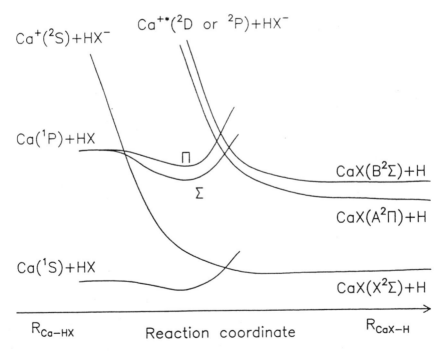

Figure 7: simplified corelation diagram for the Ca + HX chemical reaction.

Thus, the electronic states labeled D$_a$ and P$_a$ are assigned to Σ electronic states of the complex while D$_b$ and P$_b$

to Π electronic states. This is consistent with a possible linear structure of the excited Σ electronic state which should be stabilized by an overlap of the dσ or pσ Ca orbital with the σ* HCl antibonding orbital. On equal grounds, the P_b progression having been assigned to a bending progression, this implies a bent equilibrium geometry of the upper "Π" state and the same σ-σ* overlap favors a T-shape geometry. The other coupling scheme where the ground state competes with the state of Σ symmetry as the exit channel, would result in an opposite Σ, Π assignment for the complex states, therefore difficult to intepret with the above simple arguments.

CONCLUSION

We have shown that initiating a chemical reaction Ca* + HX -> CaX* + H within the Ca...HX Van der Waals complex, allows a direct acces onto the potential energy surface and to the dynamics of the reactive system.

This half collision type experiment allows a precise preparation of the reactive system in the molecular frame, corresponding to the interacting reactants.

We have observed a very dramatic effect on the branching ratio of the A, B channels of the chemiluminescent reaction Ca*+HBr→CaBr*+H by initiating the reaction from a van der Waals Ca-HBr complex selectively excited in various electronic states. This effect can be related to alignement selectivity, in full collision experiment[4]. But here the precise preparation of the system leads to incomparable higher selectivity effects than in full collision experiments, where the variation of the branching ratio A/B does not exceed a few percent, due in part to the depolarisation of the orbital alignment by the collision dynamics in the entrance channel of the reaction.

The selectivity seems to be very sensitive to the details of the surfaces which we intend to investigate both experimentally by looking at other compounds and theoretically through ab initio calculations which are under way.

REFERENCES

1.   K.H. Kramer, R.B. Bernstein, J. Chem. Phys. 42, 767 (1965)
      R.J. Beuler, R.B. Bernstein, K.H. Kramer, J. Am. Soc. 88, 553 (1966)
      S.R. Gandhi, J.J. Curtis, Q.X. XU, S.E. Choi and R.B. Bernstein, Chem. Phys. Lett, 132, 6 (1966)
      R.B. Bernstein, Chem. Phys. Lett, 132, 6 (1986)

2.   D.H; Parker, H. Jalink, and S. Stolte, J. Phys. Chem, 91, 5427 (1987)
      M.H.M. Jansen, D.H. Parker, and S. Stolte, J. Chem. Soc. Farad. Trans., 2, 85 (1989)

3.   M. Hoffmeister, R. Steying, and H. Loesh, J. Phys. Chem, 91, 5441 (1987)

4.   C.T. Rettner and R.N. Zare, J. Chem. Phys. 77, 2416 (1982)

5.   W. Bussert, D. Neushäfer, and S.R. Leone, J. Chem. Phys., 87, 3833, (1987)

6.   R.E. Smalley, Laser Chem. 2, 167 (1983)

7.   V.E. Bondybey, Science 227, 125 (1985)

8.   J.P. Visticot, B. Soep, and C.J. Whitham, J. Phys. Chem. 92, 4574 (1988)

9.   C.J. Whitham, B. Soep, J.P. Visticot, and A. Keller, J. Chem. Phys. 93, 991 (1990)

10.   A. Keller, J.P. Visticot, S. Tsuchiya, T.S. Zwier, M.C. Duval, C. Jouvet, B. Soep, and C.J. Whitham, *Dynamics of polyatomic van der Waals complexes*, NATO ASI Series, ed. N. Halberstadt and K.C. Janda, Plenum, New York, 1990. pp103

11.   W.J. Balfour and R.F. Whitlock, Can. J. Phys. 53, 472 (1975)
12.   V.E. Bondybey and J.H. English, Chem. Phys. Lett. 111, 195 (1984)

13.   U. Brinckmann and H. Telle, J. Phys. B 10, 133 (1977)

14   C. Jouvet, M. Boivineau, M.C. Duval, and B. Soep, J. Phys. Chem. 91, 5416 (1987)

15.   C. Jouvet, M.C. Duval, B. Soep, W.H. Breckenridge, C.J. Whitham, and J.P. Visticot, J. Chem. Soc. Faraday Trans. 2 85, 1133 (1989)

16.   S.E. Novick, K.R. Leopold, and W. Klemperer, in *Atomic and Molecular Clusters*, ed. R.E. Bernstein, Elsevier, Amsterdam, 1990, pp 359

17.   J.A. Shea and E.J. Campbell, J. Chem. Phys. **81**, 5326 (1984)

18. T.M. Miller and B.J. Bederson, Adv. At. Mol. Phys. **13**, 1 (1977)

19. R.R. Teachout and R.T. Pack, At. Data 3, 195 (1971)

20. J.M. Hutson, J. Chem. Phys. **91**, 4448 (1989)

21. J.P. Daudey and A. Keller, to be published

22. M.C. Duval, B. Soep, R.D. van Zee, W.B. Bosma, and T.S. Zwier, J. Chem. Phys. **88**, 2148 (1988)

23. M.C. Duval and B. Soep, to be published

24. S.E. Bradford, A. Weaver, D.W. Arnold, R.B. Metz, and D.M. Neumark,J. Chem. Phys. **92**, 7205 (1990)

25. D.J. Nesbitt, C.N. Lovejoy, T.G. Lindeman, S.V. ONeil, and D.C. Clary, J. Chem. Phys. **91**, 722 (1989)

26. P.F. Bernath and R.W. Field and B. Pinchemel, Y. Lefebvre, and J.S. Champs, J. Mol. Spect., 88, 175-193 (1981)

27. M. Menzinger, Acta Phys. Pol., Vol A73 (1988)
     M. Menzinger in "Selectivity in Chemical Reactions", Edited by J.C. Whitehead, Nato ASI Serie C, Vol 245, Cluwer Accademic, p 457 (1988)

Reactive scattering from aligned and oriented molecules

H.J.Loesch, A.Remscheid, E.Stenzel, F.Stienkemeier, and B.Wüstenbecker

Fakultät für Physik, Universität Bielefeld, D-4800 Bielefeld 1, Germany

ABSTRACT: Rotationally cold beams of CH_3I and ICl molecules have been oriented in a homogeneous static electric field and used to determine the influence of the collision geometry on reactions with K. Significant geometry related (steric) effects have been found for both systems. A beam of HF molecules has been state prepared and aligned using the method of infrared radiation pumping in a Stark field. Large steric effects in reactions with K and Li atoms have been observed.

1. INTRODUCTION

The rearrangement of atoms in a bimolecular collision plays, as the elementary microscopic step, a key role in all macroscopic gas phase reactions such as combustions or chemical processes in the atmosphere. A prerequisite for understanding and controlling macroscopic reactions is to understand them on the molecular level as comprehensive as possible. A great number of experimental and theoretical investigations has been devoted to the study of energetic properties of bimolecular reactions, like the influence of initial translational and internal energy on the reaction or the distribution of energy into the various degrees of freedom of the products. But besides energy, there is also a geometrical aspect in a reactive collision (Bernstein et al. 1987, Stolte 1982, Brooks 1976). As the approaching molecules have internal structure, the relative orientation of their axes (collision geometry) is important e.g. for the initiation of the reaction. The direct determination of such geometry related (steric) effects in reactions requires the preparation and variation of the collision geometry. In this paper we discuss two novel methods for the latter and add new results to the rather small supply of steric effects known to date.

2. ORIENTATION AND ALIGNMENT OF THE MOLECULAR AXIS

To facilitate the presentation, we restrict ourselves to polar linear or symmetric top molecules and denote their symmetry axis as molecular axis. The permanent electric dipole moment of these molecules, $\vec{d}$, lies parallel to the axis. In a gas at thermal equilibrium and without external fields the axes are distributed isotropically. But this may change, if one or both of these prerequisites are not met. In a given coordinate system the spatial distribution of the molecular axes of an ensemble of molecules is described by the probability density $\tilde{A}(cos\theta, \phi)$ of the axis ($\vec{d}$) to point into the direction given by the polar and azimuthal angles θ, ϕ. $\tilde{A}$ is normalized according to

$$\int_{4\pi} \tilde{A}(cos\theta, \phi) \cdot dcos\theta \cdot d\phi = 1 . \tag{1}$$

If $\tilde{A}$ is symmetric with respect to an inversion of the axis the ensemble of molecules is called "aligned" if not, that is, when one direction is preferred, the molecules are called "oriented".

Alignment and orientation are always prepared with respect to a vector quantity whose direction (quantization axis) is typical for the preparation technique. On the other hand, the typical vector quantity of the collision process is the relative velocity $\vec{V}$ of the approaching reagents and, in general, the direction of $\vec{V}$ deviates from the one of the quantization axis. This situation is illustrated by Figure 1. The quantization axis is denoted by $\vec{E}$ and the prepared axis distribution (an orientation) is characterized by a polar diagram; the most likely direction of the axis is $\|\vec{E}$. The collision geometry is described by the angle of attack, γ_a, between $\vec{V}$ and the molecular axis. If $\vec{E}$ is set $\perp \vec{V}$, the most likely γ_a is $90°$. The atom A attacks preferentially the side of BC. Now, and this is essential to experiments on steric effects, without changing the prepared distribution just by rotating the quantization axis, the collision geometry changes. If $\vec{E}$ is rotated clockwise (counterclockwise) by $90°$, A attacks most likely the C-end (B-end) of BC. In this way, we can select any arbitrary collision geometry. It is quite intuitive that the flux of products measured as a function of the field angle $\beta = \angle(\vec{E}, \vec{V})$ or of the most probable angle of attack will inform us about geometry related properties of the chemical forces.

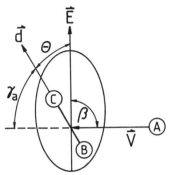

Fig. 1. Definition of angles

3. ORIENTATION DEPENDENT CROSS SECTIONS

To become more quantitative, we express the scattered flux of products in the center of mass (CM) frame by the orientation dependent, double differential reaction cross section $I(cos\gamma_a, \phi_a, \vartheta, \varphi; E_{tr})$ which depends on the direction of the axis (γ_a, ϕ_a); ϑ and φ are the polar and azimuthal scattering angles of the detected product, that is, of the recoil velocity $\vec{u}$ of the product. E_{tr} denotes the relative translational collision energy. I gives the probability that in a collision of an atom with a molecule whose axis is fixed products are scattered by ϑ, φ into the solid angle $d\omega = dcos\vartheta \cdot d\varphi$. The Cartesian frame to which these angles refer, is defined by $\vec{V}$ as z-axis and an x-axis which stands perpendicular on the plane defined by the velocities of the approaching reagents (scattering plane).

In a realistic experiment, the direction of the axes is not fixed but distributed and the observable double differential cross section is related to I by

$$J_A(\vartheta, \varphi, u; E_{tr}) = \int_{4\pi} I(cos\gamma_a, \phi_a, \vartheta, \varphi; E_{tr}) \cdot \tilde{A}(cos\gamma_a, \phi_a) \cdot dcos\gamma_a \cdot d\phi_a. \tag{2}$$

In cases, where $\tilde{A}$ is cylindrically symmetric around $\vec{V}$, it is useful to define a mean cross section which is independent on φ

$$\bar{I}(cos\gamma_a, \vartheta, u; E_{tr}) = \frac{1}{2\pi} \int_0^{2\pi} I(cos\gamma_a, \phi_a, \vartheta, \varphi, u; E_{tr}) \cdot d\phi_a \tag{3}$$

and the observable quantity is then

$$J_A(\vartheta, u; E_{tr}) = \int_{-1}^{1} \bar{I}(cos\gamma_a, \vartheta, u; E_{tr}) \cdot A(cos\gamma_a) \cdot dcos\gamma_a. \tag{4}$$

$A(cos\gamma_a) \cdot dcos\gamma_a$ is the probability for finding γ_a in the interval $dcos\gamma_a$. Usually A can be represented by a short series of Legendre polynomials, $P_i(cos\gamma_a)$. As a consequence, the measured cross sections contain only limited information on the orientation dependence of $\bar{I}$. This becomes immediately clear, if one expands also $\bar{I}$ in a series of the P_i

$$\bar{I}(cos\gamma_a, \vartheta, u; E_{tr}) = \sum_{i=0}^{\infty} J_i(\vartheta, u; E_{tr}) \cdot P_i(cos\gamma_a). \tag{5}$$

Due to the orthogonality of the P_i, the highest moment J_i which contributes to the observable quantity J_A is J_n, where n is the highest order of P_i in the expansion of $A(cos\gamma_a)$. The steric information resulting from experiments with aligned/oriented molecules is thus a usually small number of moments.

In case the experiments provide an integral reaction cross section, the observed quantities $\sigma(E_{tr})_A$ can be expressed by the orientation dependent integral reaction cross section, $\sigma(\cos\gamma_a; E_{tr})$, in analogy to Eq. 4. Following the same procedure as before one finds that again only a small number of moments, $\sigma(E_{tr})_i$, can be determined.

4. EFFECTS OF AXIS ORIENTATION

4.1 Orientation by Brute Force

The Brute Force (BF) method (Loesch and Remscheid 1990) bases on the simple fact that a homogeneous static electric field $\vec{E}$ exerts a torque onto a polar molecule which tries to rotate the dipole moment $\vec{d}$ towards $\vec{E}$, the direction of lowest potential energy. An initially non rotating molecule will start angular oscillations around $\vec{E}$, but when the field-strength increases temporally, the amplitudes become smaller and smaller and thus the direction of $\vec{d}$ is increasingly better defined. If the molecule rotates initially, higher fields are required to disrupt the rotation first and then force $\vec{d}$ (brutally) towards $\vec{E}$. The BF method is not restricted to symmetric top molecules, like the frequently used hexapole field technique, but is generally applicable to all kinds of polar molecules. The significance of the BF technique has recently been emphasized also by Friedrich and Herschbach (1991).

To become a little more quantitative, let us consider the Schrödinger equation for the angular motion of a rigid polar molecule in an external field $\vec{E}$

$$[\mathbf{H_{rot}} - w \cdot cos\theta - W_i]\Psi_i = 0. \tag{6}$$

$\mathbf{H_{rot}}$ denotes the Hamiltonian of the free rotational motion devided by a typical rotational constant B of the molecule, $w = E \cdot d/B$ is the reduced field coupling parameter, $\theta = \angle(\vec{E}, \vec{d})$, W_i is the reduced energy eigenvalue and Ψ_i denotes the wave function of the angular motion which is a function of all three Eulerian angles in the general case. The axis distribution is related to the square modulus of Ψ_i and given for symmetric top molecules by

$$\tilde{A}(cos\theta) = \{2\pi\}|\Psi_i|^2; \tag{7}$$

for linear molecules $\{2\pi\}$ has to be replaced by 1. To orient a supersonic beam of molecules, we generate at the site we need the preparation a strong homogeneous field. When the molecules reach the fringing field, the rotational states (e.g. $Y_{J,M}$) in the field free region adiabatically transform into the strong field states following simple correlation rules such as ground state to ground state. In a beam the rotational states are distributed according to a rotational temperature T_r and each of these states is transformed individually. The resulting total axis distribution, $A(cos\theta)$, is then a sum of properly weighted $|\Psi_i|^2$ ($i \rightarrow M$, $k = J - |M|$). We have calculated A for various E and T_r and found the simple expression

$$A(cos\theta) = \frac{1}{2} \cdot [1 + A_0(E, T_r) \cdot cos\theta]. \tag{8}$$

The extent of orientation, characterized by A_0, depends on E and T_r according to

$$A_0(E, T_r) = a_0 \cdot E(kV/cm)/T_r(K) \tag{9}$$

provided $A_0 << 1$ holds. For CH_3I and ICl we found $a_0 = 0.0366$ and 0.0307, respectively. Eqs. 8, 9 demonstrate that high fields and low rotational temperatures are prerequisites for a sensible application of the BF technique.

4.2 Apparatus

We have investigated the exoergic reactions

$$K + CH_3I \rightarrow KI + CH_3, \quad \Delta D_{00} = -0.9 \ eV \tag{R1}$$

$$K + ICl \rightarrow KI + Cl, \ KCl + I, \quad \Delta D_{00} = -1.2, \ -2.2 \ eV \tag{R2}$$

A schematic diagram of the experimental configuration is displayed in Figure 2. The two reagent beams intersect perpendicularly. The reaction products are detected by surface ionization on a Re-ribbon which is housed in a UHV-chamber. For measuring angular distributions of scattered products the detector can be rotated around the intersection volume in the scattering plane. The laboratory (LAB) scattering angle is denoted by Θ. The velocity distributions of products are measured via the time of flight (TOF) method using a rotating chopper wheel with four narrow slits. The electrodes generating the orientation field are placed symmetrically around the intersection volume. Basically, these are two parallel plates standing perpendicular on the scattering plane. They have a slit in this plane to let the beam particles and products freely pass. By applying suitable potentials, $\pm U$, a field which lies in the plane can be generated; reversing the sign of the potential reverses also the direction of $\vec{E}$. By rotating the plates, any direction of $\vec{E}$ in the plane can be chosen.

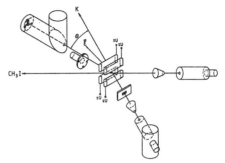

Fig. 2. Molecular beam apparatus

To determine the influence of the orientation field on the reactions, we have first rotated the plates such that $\vec{E}$ is parallel to the mean relative velocity $\vec{V}$ of the reagents. Then, we measure the flux of products at the deflection angle Θ, $I_{LAB}(\Theta, E)_+$, for a given period of time. Subsequently, the field is reversed and the flux of products, $I_{LAB}(\Theta, E)_-$, is measured again. From these fluxes we define the *steric effect*, S, by

$$S(\Theta, E) = \frac{I_{LAB}(\Theta, E)_+ - I_{LAB}(\Theta, E)_-}{I_{LAB}(\Theta, E)_+ + I_{LAB}(\Theta, E)_-} \tag{10}$$

Qualitatively, $S(\Theta, E)$ represents the difference of reactivity between the I-end and the CH_3-end of CH_3I or the Cl-end and the I-end of ICl in reaction $R1$ or $R2$, respectively, with respect to the formation of products which are scattered by the angle Θ. It has been shown (Loesch and Remscheid 1991) that the angular distribution of products measured for $E = 0$, $I_{LAB}(\Theta)$, and $S(\Theta, E)$ are related to the zeroth moment J_0 and to the ratio of first and zeroth moment J_1/J_0 of the expansion Eq. 5, respectively.

4.3 Experimental results for $K + CH_3I$ and discussion

The results are well documented in two previous papers (Loesch and Remscheid 1990 and 1991); therefore, it appears sufficient to give here only a brief summary and refer to these papers for more details. Experiments have been performed at two relative translational energies, $E_{tr} = 0.79$ and 1.24 eV. First, angular and velocity distributions of products (KI) have been measured without orientation field. From these data we have deduced the zeroth moment of the cross section which can be represented by a product of an only ϑ-dependent function, $g_0(\vartheta, para)$ and an only u^*-dependent function, $f_0(u^*, para)$,

$$J_0(\vartheta, u; E_{tr}) = g_0(\vartheta, para) \cdot f_0(u^*, para)/u_{max}. \tag{11}$$

u_{max} denotes the maximal recoil velocity of KI and $u^* = u/u_{max}$. The argument *para* symbolizes a freely floating set of parameters which has been fixed by a least squares fit procedure. The angular distribution $g_0(\vartheta)$, obtained for $E_{tr} = 1.24$ eV is displayed in Figure 3. The most intense flux of products occurs at $\vartheta = 180°$, that is, KI is predominantly scattered contrary to the approaching K-atoms. This behavior has been found also at lower (Rulis and Bernstein 1972) and higher E_{tr} (Rotzoll et al. 1975).

The result of our experiments with the orientation field turned on is presented in Figure 4. Plotted is the steric effect versus the scattering angle Θ for the two collision energies. S features an oscillatory behaviour. It is positive at large angles (backward scattering) and negative at small angles (forward scattering). Qualitatively, this means more products are scattered into the backward direction in the $\vec{E}$ parallel to $\vec{V}$ situation, where the K atoms encounter preferentially the I-end of the CH_3I molecule, while the CH_3-end forms more products into the forward direction. As most products are scattered backward, the *I*-end is called the *head* (most reactive end) and the CH_3-end the *tail* of the molecule (Brooks 1976).

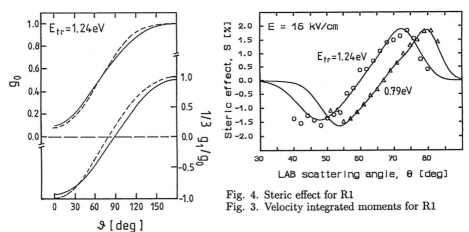

Fig. 4. Steric effect for R1
Fig. 3. Velocity integrated moments for R1

To determine the first moment $J_1(\vartheta, u; E_{tr})$ we choose for J_1 the same ansatz as before (Eq. 11) with an u^* dependent function $f_1(u^*, para)$ and a ϑ dependent function

$$g_1(\vartheta, para) = g_0(\vartheta, para) \cdot [3 \cdot a \cdot cos(\vartheta - \alpha)] \tag{12}$$

where $g_0(\vartheta, para)$ is the result of the measurement without orientation field. Eq. 12 reflects the observed oscillatory form of S; the amplitude a, the phase α and the width of f_1 are free parameters. These parameters have been varied until the numerically simulated steric effect fits the data (solid line in Figure 4). The resulting ratio of velocity integrated moments, $1/3 \cdot g_1(\vartheta)/g_0(\vartheta) = a \cdot cos(\vartheta - \alpha)$ is displayed in Figure 3. It should be emphasized that the amplitude a is very close to unity ($0.88 \leq |a| \leq 1$) and α is roughly zero. For more details see Loesch and Remscheid (1991).

Amplitude and phase of the first moment can be nearly quantitatively rationalized by a simple dynamical model called the DIPR-model (Kuntz 1972). The main assumption is that in the moment the atom reaches a critical distance to the center of mass of the molecule (reaction shell) CH_3I dissociates immediately along the symmetry axis ($\vec{d}$), where CH_3 gets an impulse into the direction of $\vec{d}$, $\vec{p}_0 = p_0 \cdot \vec{d}/d$ and I the same impulse into the opposite direction. The initial momentum of CH_3, $\vec{p}_C$ and $\vec{p}_0$ form the final momentum of the product CH_3, $\vec{p'} = \vec{p}_C + \vec{p}_0$ and conservation of momentum in the CM frame requires that KI moves with $-\vec{p'}$. Consequently, there is a strong correlation between $\vartheta = \angle(-\vec{p'}, \vec{V})$ and the orientation of the molecule. Is the initial momentum of CH_3 small compared to p_0 the relation $\vartheta = \pi - \gamma_a$ holds. The ratio of moments $1/3g_1/g_0$ predicted by the DIPR-model is given as dashed line in Figure 3. The good agreement suggests that the reaction is indeed characterized by the explosion of the reagent molecule and the resulting strong correlation of ϑ and γ_a.

In the framework of this model the angular distribution of the zeroth moment $g_0(\vartheta)$ is governed by the probability, W_γ, for a trajectory with impact parameter b to be reactive. This probability depends in the molecular frame on the angle, γ_c (see Figure 5). With the ansatz $W_\gamma = [(1 + cos\gamma_c)/2]^n$ and $n = 3$ it is possible to simulate $g_0(\vartheta)$ almost perfectly as shown by the dashed curves in Figure 3. W_γ is displayed in Figure 5. The atoms approaching the CM of CH_3I collinearly to the $I - C$-line have a reaction probability of 100 % and at least 50 %, if they approach it within a *cone of acceptance* centered around this line with a full apex angle of 108°.

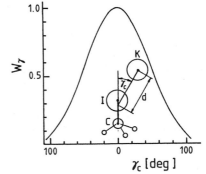

Fig. 5. Reaction probability for R1

4.4 Experimental results for $K + ICl$ and discussion

The BF technique is the only one to date which can be used to orient polar $^1\Sigma$-molecules. The reaction $R2$ represent thus the first step into a completely new field. We have recently completed a series of experiments involving angular and velocity distributions of scattered products measured with and without orientation field (Loesch and Möller 1991). The reaction is particularly interesting, since two product channels are open which can be investigated seperately. Unlike reaction R1 the analysis of the raw data is in a very early state.

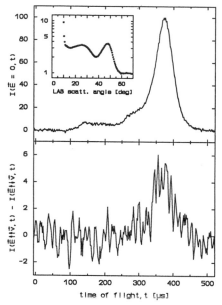

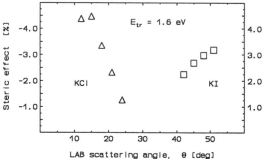

Fig. 7. Steric effect for R2: KCl (left scale); KI (right scale)

Fig. 6. TOF profiles for R2: E=0 (upper panel); difference of TOF profiles for the two field directions (lower panel). Insert: total angular distribution

An angular distribution of scattered particles measured at $E_{tr} = 1.6\ eV$ without orientation field is inserted in the upper panel of Figure 6. The various structures are caused by the fact that surface ionization detects elastically scattered K and the products KI and KCl with the same efficiency. The sharp decrease of intensity at small angles is due to elastic scattering off the seeding gas He, the subsequent peaks reflect the forward scattered KCl molecules and the KI products, respectively. Superimposed to the peak regions is the flux resulting from an unknown channel, most likely KCl recoiling from I in the electronically excited $^2P_{1/2}$ state. Position and shape of the structures as well as the mean velocities suggest that the products are predominantly forward scattered in the CM-frame.

The different mean velocities of the various products can be utilized for their identification. For this purpose, we have measured TOF distributions of scattered particles at several LAB scattering angles. The result obtained at 48° is presented in the upper panel of Figure 6. The fast flat peak (small times) is associated with the products K, KCl, the adjacent slower shoulder is presumably due to KCl recoiling from $I(^2P_{1/2})$ and the marked peak results from KI. To determine the steric effects for each component individually we have measured TOF distributions for two field directions, $\vec{E}$ parallel and antiparallel to $\vec{V}$. The lower panel of Figure 6 shows the difference of the two TOF distributions at 48°. There is a clear steric effect for the KI flux, but no significant difference is observable for the other components. The steric effect is positive and amounts to 5 % of the signal measured without field.

The result of a preliminary analysis of the presently available TOF distributions is presented in Figure 7. Shown is the steric effect (Eq. 10) as a function of the scattering angle for KI (right scale) and KCl (left scale). The angles cover the region of largest product flux and correspond to (near) forward scattering in the CM-frame. We find a positive steric effect for the formation of KI of $\approx$ 3% and a negative one for KCl which decreases from -4 % at small angles to -1 % at larger angles. With the dipole moment pointing from Cl to I these results mean qualitatively that forward scattering takes place preferentially, when the reacting atom of ICl points *away* from the approaching K. Using the idea of the harpooning model the electron is accommodated by the reacting atom although it is farther away from the K^+ than the non reacting one. The magnitude of the effect suggests that - like in the case of $K + CH_3I$- a strong correlation between the deflection angle and the direction of the molecular axis exists.

5. EFFECTS OF AXIS ALIGNMENT

5.1 Alignment by radiation pumping in a Stark field

The principles for aligning linear molecules in a linearly polarized radiation field have been described by Zare (1982). The method has a serious disadvantage in case the nuclei of the molecule have spin which couple to the rotation. After excitation, the randomly oriented nuclear spins, $\vec{I}$, and the angular momentum of rotation, $\vec{J}$, couple and form total angular momenta with various directions around which $\vec{J}$ precesses. This leads to a deterioration of the axis alignment which is particularly serious for $J \leq I$ (Altkorn et al. 1985).

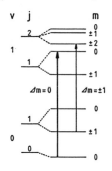

Fig. 8. Stark splitting of low v,j - levels of HF

A way to avoid nuclear spin depolarization is to excite the molecule in a homogeneous electric field $\vec{E}_S$. If the molecule features a Stark effect, which leads at feasible field-strengths to an interaction energy substantially larger than nuclear spin-rotation coupling, rotation is coupled to the space fixed Stark field rather than to the various directions of the total angular momentum. To preserve the decoupling until the reaction volume is reached, the Stark field must accompany the molecules with more or less the same magnitude, however, the *same direction* is not required. To present the technique a little more detailed, let us consider the schematic diagram of lowest lying ro-vibrational levels of HF splitted in a Stark field (Figure 8). By tuning the radiation field to one of the Stark components of the $v' = 0, j' \rightarrow v = 1, j = j' \pm 1$ transitions one excites the molecules vibrationally and -most important for creating an alignment- specifies completely (except for the sign of M) the rotational state. The axis distribution of the vibration marked subgroup of molecules is then given by the square modulus of the prepared *rotor wave function*. In the experiments presented below we use the $R_1(0)$ line which populates the $v = 1, J = 1, M = 0$ state. The axis distribution is then given by the expression

$$\tilde{A}(cos\theta) = |Y_{J=1,M=0}(cos\theta)|^2 = \frac{3}{4\pi} \cdot cos^2\theta \tag{13}$$

where θ is the angle between the quantization axis, $\vec{E}_S$, and the axis (dipole moment $\vec{d}$). Eq. 13 describes HF molecules, whose axes are preferentially aligned along $\vec{E}_S$ but never perpendicular to $\vec{E}_S$. We have used also the $\Delta M = \pm 1$ component of the $R_1(1)$ line which populates the $J = 2, M = \pm 2$ state. The axis distribution is given by $\tilde{A}(cos\theta) = 15/(32\pi) \cdot sin^4\theta$ with zero and maximal probability for the axis parallel and perpendicular to $\vec{E}_S$, respectively.

5.2 Apparatus

We have investigated the reactions

$$K + HF(v = 1, J = 1) \rightarrow KF + H, \quad \Delta D_{00} = +0.75 \ eV \tag{R3}$$

$$Li + HF(v = 1, J = 1) \rightarrow LiF + H \quad \Delta D_{00} = -0.04 \ eV. \tag{R4}$$

The molecules have been state prepared and aligned using the technique described above. The experimental set up is very similar to the one shown in Figure 2 except for the components used for the preparation. As radiation source, we use a color center laser which is operated in the single mode. Laser beam and molecular beam intersect perpendicularly a few cm upstream from the reaction volume. The Stark field $\vec{E}_S$ is generated by two parallel plates standing perpendicular to the scattering plane. The polarization vector is rotated into a direction parallel to $\vec{E}_S$ by a half wave retardation plate. The field between the excitation zone and the reaction volume is generated by four rods which are bent such that at the reaction volume the guiding field, $\vec{E}_g$, is either parallel or perpendicular to the relative velocity $\vec{V}$ of the approaching reagents. The direction of $\vec{E}_g$ is controlled by the potentials applied to the rods. The field-strength over the entire length is set to $\approx 10kV/cm$. This guarantees that during the flight of the molecule from the excitation zone to the reaction volume the prepared state follows adiabatically the local quantization axis which rotates from $\vec{E}_S$ to $\vec{E}_g$. Furthermore, the field dependence of the observed steric effects indicates that complete decoupling of nuclear spin and rotation is already obtained for $E_g > 4$ kV/cm, so 10 kV/cm is sufficient.

The two directions of the guiding field $\vec{E}_g$ give rise to two substantially different distributions of the axis with respect to $\vec{V}$, that is, to two different reaction geometries. In a Cartesian frame, where $\vec{V}$ is the z-axis and x-axis stands perpendicular on the scattering plane, the two distributions which follow from Eq. 13 are given by

$$\tilde{A}_{\parallel} = \frac{3}{4\pi} \cdot cos^2\gamma_a \qquad \tilde{A}_{\perp} = \frac{3}{4\pi} \cdot sin^2\gamma_a \cdot cos^2\phi_a \qquad (14)$$

for $\vec{E}_g \| \vec{V}$ and $\vec{E}_g \perp \vec{V}$, respectively. The polar diagrams of the distributions displayed in Figure 9 illustrate the preferred collision geometry. For the $\|$ situation the atoms attack the two ends of the molecule while for the $\perp$ situation side-on attacks prevail. It should be noted that a preparation according to $\tilde{A}_{\perp}$ destroys the usual cylindrical symmetry of the scattering around $\vec{V}$.

The infrared laser beam is on-off modulated with a frequency of 17 Hz. The detector signal at an angle Θ is rooted to a set of two phase sensitive scalers which eventually provide the difference of the laser *on* and *off* signals. This laser correlated difference signal $I_{LCD}(\Theta)$ is in all cases studied so far proportional to the flux of products resulting from reactions with HF in the $v=1$, $J=1$, $M=0$ state (Hoffmeister et al. 1989). In the following we suppress these quantum numbers and characterize I_{LCD} by the direction of $\vec{E}_g$ with respect to $\vec{V}$ namely by $\|$ or $\perp$.

5.3 Experimental results for $K + HF$ and discussion

Figure 9 shows a typical example for angular distributions of scattered products measured for the two collision geometries at the same E_{tr} (0.62eV). First of all, there is a drastic steric effect: end-on collision lead to a substantially larger flux of products than side-on collisions. The curves have nearly the same peak position (a slight shift of $\approx 0.5°$ appears in the slopes) and exactly the same shape which is to a great extent given by the centroid distribution (solid line). The latter results, if the recoil velocity of the product in the CM frame is assumed zero. The near congruence of the two distributions is in the first place a consequence of the special kinematics of the reaction $R3$ which allows only very small recoil velocities of the heavy product. Thus the angular distribution reflects only the velocity profiles of the parent beams without information on the dynamics.

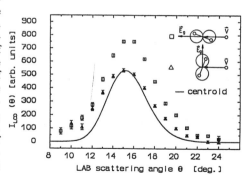

Fig. 9. Angular distributions of products for R3 measured for the indicated collision geometries.

It is therefore useful to define the steric effect independently of the scattering angle by

$$S(E_{tr}) = \frac{I_{LCD,\|}(\Theta_p) - I_{LCD,\perp}(\Theta_p)}{I_{LCD,\|}(\Theta_p) + I_{LCD,\perp}(\Theta_p)} \qquad (15)$$

where Θ_p denotes the angle of peak intensity. The steric effect (squares) is plotted versus the collision energy E_{tr} in Figure 10. It is largest at the lowest E_{tr} (0.38 eV) which is close to threshold (0.29 eV) and decreases with E_{tr}. For comparison, we have also included the data of our previous experiments (circles)

(Hoffmeister et al.1987). The substantially smaller values reflect the much less marked alignment of the $J = 2$ state prepared without Stark field.

As the recoil velocity of KF in the CM frame is nearly zero the detector collects most of the flux scattered out of plane and thus the integral over the measured in plane angular distribution is proportional to the total flux of products, that is, to the integral reaction cross section, $\sigma(E_{tr})$. As the shape of the distributions are independent of the field direction the peak intensities are a measure for σ and thus S can be interpreted as

$$S(E_{tr}) = \frac{\sigma_\parallel(E_{tr}) - \sigma_\perp(E_{tr})}{\sigma_\parallel(E_{tr}) + \sigma_\perp(E_{tr})}. \qquad (16)$$

To extract the steric information, we describe here the observed quantities by the orientation dependent integral reaction cross section $\sigma(cos\gamma_a, E_{tr})$ in analogy to Eq. 4. Expansion of $\sigma(cos\gamma_a, E_{tr})$ and the prepared axis distributions (Eq. 14) in series of Legendre polynomials shows that $S(E_{tr})$ is a simple function of the ratio of the second and zeroth moment, σ_2/σ_0, of the $\sigma(cos\gamma_a, E_{tr})$ series.

Both sets of data in Figure 10 can be quantitatively rationalized using the *angle dependent line of centers* model (Smith 1982; Levine and Bernstein 1984). In this model, reaction occurs with unit probability, whenever the fraction of the translational energy along the line connecting the atom and the center of mass of the molecule exeeds the barrier height, $V_b(cos\gamma_c)$, at the critical distance d (reaction shell); γ_c is illustrated in the insert of Figure 10. Assuming only straight trajectories and no reorientation during the approach, the model provides a simple expression for $\sigma(cos\gamma_a, E_{tr})$ from which the ratio of moments σ_2/σ_0 and hence $S(E_{tr})$ can be deduced. We have carried out calculations using the following ansatz for $V_b(cos\gamma_c)$

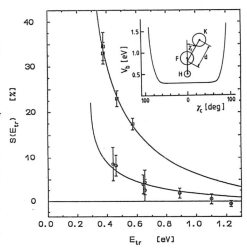

Fig. 10. Energy dependence of steric effect for R3. Insert: angle dependent barrier height

$$V_b(cos\gamma_c) = v_0 \cdot |1 - cos\gamma_c|^n + E_{th} \qquad (17)$$

where v_0 and n are free parameters. The solid lines for both data sets in Figure 10 have been obtained for *one* set of parameters: $v_0 = 1.33 \ eV$ and $n = 6.69$. The threshold was set to $E_{th} = 0.29 \ eV$. The resulting angle dependent barrier is displayed in the insert of Figure 10. It has a flat minimum for a collinear conformation, most likely $K - F - H$ and rises steeply beyond 80°. The full apex angle of the *cone of accceptance* is thus roughly 160°.

5.4 Experimental results for $Li + HF$ and discussion

Figure 11 shows an angular distribution of LiF measured at $E_{tr} = 0.42 \ eV$ for $\vec{E}_g \parallel$ and $\perp \vec{V}$. First of all, there is a huge influence of the prepared reaction geometry on the product flux and second, the distributions differ signficantly from the centroid distribution. For this reason, the detected flux of products measured in plane is by no means representative for the total flux and reflects the differential rather than the integral reaction cross section and an angular dependent steric effect in analogy to Eq. 15 can be determined. The result is displayed in Figure 12 as a function of the scattering angle in the CM frame. The LAB-CM transformation is based on a primitive fixed velocity procedure. For large angles (backward scattering) $S(\Theta)$ is positive and reaches the enormous value of 0.6. Towards the forward direction the effect decreases and turns negative.

To find a rationale for the exceptional magnitude of the effect we have recently performed a trajectory calculation (Loesch et al. 1991) based on the ab-initio potential energy surface of Chen and Schaefer (1980). To simulate the prepared axis distributions we set the initial angular momentum of HF to zero and generate $cos\gamma_a$ and ϕ_a such that these quantities are distributed according to Eq. 14. The considerable difference in the peak heights in Figure 11 as well as the difference in the peak positions are recovered and so are sign and magnitude of the steric effect. The latter can be judged from a comparison with the histogram shown in Figure 12. As a more detailed analysis of the trajectories shows, the geometry related differences in the cross sections -which eventually cause sign and magnitude of S- are mainly due to the propensity of the products to separate into opposite directions *along the axis of HF*. As for $\vec{E}_g \perp \vec{V}$ the axis stands preferentially perpendicular to the scattering plane, a substantial fraction of products is scattered above or below the plane and is thus lost for the detection. In case $\vec{E}_g \| \vec{V}$, the preference for out of plane scattering vanishes as the axis are symmetrically distributed around $\vec{V}$. Consequently, more products appear in the plane which is reflected by the significantly larger cross section. The preferred backward scattering for $\vec{E}_g \| \vec{V}$ is a consequence of the asymmetric reactivity of the two ends of HF. Only attacks on the F-end form products. After initiation of the reaction the H-atom is ejected along the HF- axis into the forward direction of the CM frame and the detected LiF into the backward direction.

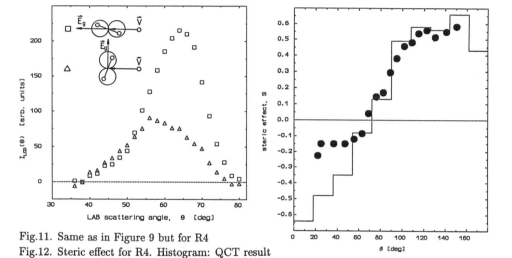

Fig.11. Same as in Figure 9 but for R4

Fig.12. Steric effect for R4. Histogram: QCT result

ACKNOWLEDGMENT

Support of this work by the Deutsche Forschungsgemeinschaft (SFB 216, P5) is gratefully acknowledged.

REFERENCES

Altkorn R, Zare R N and Greene C H 1985 Mol.Phys. **55** 1
Bernstein R B, Herschbach D R and Levine R D 1987 J.Phys.Chem. **91** 5365
Brooks P R 1976 Science **193** 11
Chen M M L and Schaefer III H F 1980 J.Chem.Phys. **72** 4376
Friedrich B and Herschbach D R 1991 Z.Physik D **18** 153
Hoffmeister M, Schleysing R and Loesch H J 1987 J.Phys.Chem. **91** 5441
Kuntz P J 1972 Mol.Phys. **23** 1035
Levine R D and Bernstein R B 1984 Chem.Phys.Letters **105** 467
Loesch H J and Möller J 1991 Contributed paper to this ICPEAC
Loesch H J and Remscheid A 1990 J.Chem.Phys. **93** 4779
Loesch H J and Remscheid A 1991 J.Phys.Chem. accepted
Loesch H J, Stenzel E. and Wüstenbecker B 1991 J.Chem.Phys. submitted
Rotzoll G, Viard R and Schügerl K 1975 Chem.Phys.Letters **35** 353
Rulis A M and Bernstein R B 1972 J.Chem.Phys. **57** 5497
Smith I W M 1982 J.Chem.Educ. **59** 9
Stolte S 1982 Ber.Bunsen-Ges.Phys.Chem. **86** 413
Zare R N 1982 Ber.Bunsen-Ges.Phys.Chem. **86** 422

Transition from molecular structure to collective excitations in alkali metal clusters

J. CHEVALEYRE

Laboratoire de Spectrométrie Ionique et Moléculaire (associé au C.N.R.S. n°171)
Bat 205, Université Claude Bernard LYON I
43, boulevard du 11 Novembre 1918 69622 Villeurbanne Cedex, France

ABSTRACT: Visible spectra from trimer to octomer of alkali metals are shown to be a source of very valuable information. They allow to understand as well the geometric structure as the electronic properties of the involved clusters. A Jahn–Teller treatment leads to the determination of potential surface, rotational constants and isotopic shift for trimer. For larger cluster sizes, comparison with SCF–CI molecular calculations allows geometric structures with very close calculated energy to be discriminated as well as the progressive appearance of giant resonances due to collective excitations is evidenced.

1. INTRODUCTION

Small metal clusters, generally produced in supersonic beams, offer the opportunity to study in the vapor phase, e.g. with negligible external interactions, electronic properties and geometric structure of regularly increasing size homonuclear molecules. Especially, those clusters are very convenient to describe how their properties evolve with molecular size, the final purpose being to examine how these properties extrapolate to the bulk ones. As an example, it is clear that in atoms or diatomic molecules, valence electrons are localized on well determined orbitals while they are free in the bulk. A progressive delocalization of such electrons has been demonstrated by Bréchignac *et al* (1988) on mercury clusters where the s–p hybridization which leads to the conduction band in the bulk gives rise to a broadening of the 5d–6p core–valence transition observed in the U.V. range for cluster size larger than n=12 .

On the theoretical side, the electronic shell model, firstly developed for predicting the stability of nucleus in nuclear physics (Mayer *et al* 1955) was successfully adapted to clusters stability (Knight *et al* 1984). It assumes that the valence electrons are confined in a spherical jellium of the homogeneously distributed charge of the remaining ions. Those electrons are then calculated to occupy the successive shells:

$1s^2$, $1p^6$, $1d^{10}$, $2s^2$, $1f^{14}$, $2p^6$, $1g^{18}$, $2d^{10}$, $3s^2$, ...

characterized by radial and angular quantum numbers; the superscript represents the $2(2l+1)$ degeneracy of a given shell. The specific stability assigned to closed shell clusters was confirmed by the "magic numbers" that appeared in the early experiments on alkali metal clusters: relative abundances in the sodium clusters mass spectra by Knight *et al* (1984) or ionization potentials of potassium clusters determined by photoionization by Knight *et al* (1986).

Though extensions to ellipsoidal distorsions (Clemenger 1985) or attempts to allow for the geometric structure through effective core potential (Manninen 1986) have been developped with some success, it is clear that electronic shell model which basically ignores the cluster structure can not account for properties closely related to this structure. Examples of this kind such as the structure in itself or relative intensities in photoabsorption spectra will be developed thereafter.

Among recent attempts to include geometric structure in quantum chemistry calculations, theoretical ab initio calculations using the SCF—MRD—CI procedure (Boustani *et al* 1987) were very promising as they allowed to predict ground state structure in the size range $2 \leqslant n \leqslant 13$ as well as relative intensities in optical absorption spectra of the smallest alkali metal clusters: $2 \leqslant n \leqslant 8$ (Bonacic—Koutecky *et al* 1988a, b and 1990a, b). It is well known in spectroscopy of small molecules that relative intensities are very sensitive to the electronic wave functions; as will be shown below, any change in the geometric structure induces variation of electronic wave functions which result in drastically different patterns for the optical absorption spectra. This possibility challenged new experiments of photoabsorption spectroscopy of small alkali metal clusters by Broyer *et al* (1990) and Blanc *et al* (1991) with the obvious purpose to reach both the geometric stucture and a better understanding of electronically excited states of these molecules.

We present in this paper some recent results on sodium and lithium clusters photoabsorption spectroscopy which illustrate how the strong demand for experimental results that raised up to check ab initio molecular calculations has been partially answered in the last few years by means of optical spectroscopy. The discussion of these results will be limited to three cluster sizes that represent very typical illustrations of the usefullness and efficiency of photoabsorption spectroscopy:

i) the trimer: Na_3 can be considered as a text book example of Jahn—Teller effect applied to the interpretation of spectra of non linear triatomic molecules; due to the lightness of lithium, new results related to rotational structure and isotopic effect have been obtained,

ii) the hexamer: its structure is at the frontier between planar and three-dimensional geometry, it demonstrates the efficiency of molecular calculations combined with experimental photoabsorption spectroscopy to determine unambiguously a cluster structure,

iii) the octomer: it illustrates the concept of giant resonances resulting from collective excitation of free electrons.

2. EXPERIMENT

Alkali clusters (Na_n or Li_n) are produced in supersonic molecular beams by coexpansion of alkali metal vapor with argon as inert carrier gas. To reach reasonable metal vapor pressures (10-100 Torr), high temperatures were required: up to 1200 °C for lithium. For this reason, a special oven heated by tungsten filaments and made of molybdenium alloy (TZM) was built. With argon pressures varying in the range 1-5 bars, small and very cold metal clusters were produced due to both low metal-metal and high argon-metal collision rates.

The laser excitation is performed by two pulsed lasers crossing at right angle the alkali clusters supersonic beam. It leads to the detection of ions which is both a very sensitive and easy to operate procedure. To allow for the various stabilities of excited states, two different excitation techniques were used.

Small clusters ($n \leqslant 3$) were excited through the two photon ionization (TPI) technique: a first tunable dye laser is scanned over the cluster excited states while a second fixed wavelength laser brings the excited molecule up to the ionization region. This technique, very convenient to study dimer or trimer is no longer valid for larger size clusters; their excited states are generally dissociative with very short dissociative times in the ps time scale. We made then use of the depletion spectroscopy (DS) technique: the first laser, frequency doubled by second harmonic generation, directly ionizes clusters from their ground state and gives a reference ion signal. The second laser is then scanned through excited dissociative states inducing a depletion in the ground state population and consequently in the reference ion signal. Ions originating from those two excitation techniques are then mass selected through time of flight or quadrupole mass spectrometers and the ion signal of a given mass is recorded as a function of the exciting laser frequency.

Li_n spectra recorded by TPI (Li_3) or DS (Li_{4-8}) are reported in Fig 1 after processing to give photoabsorption cross sections.

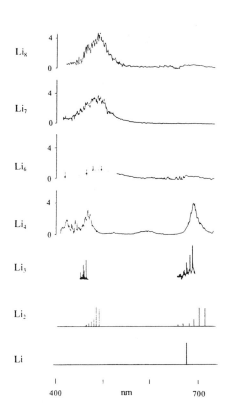

Fig 1. Absorption spectra of Li_n clusters $1 \leqslant n \leqslant 8$. Cross sections are in arbitrary units for Li, Li_2 and Li_3 and in sq. Angstroms for larger clusters

3. Na₃ AND Li₃

Before starting with Li₃ analysis, we must recall that interpretation of Na₃ optical spectra in the last few years made this molecule the most completely analyzed trimer through the dynamical Jahn–Teller effect. After the pioneering first observation by Herrmann *et al* (1979), the vibronic structures of five excited states labelled A, B, B', C and D have been recorded by Delacretaz *et al* (1985 and 1986) and Broyer *et al* (1986 and 1989). A, B and C states are now completely analyzed, but D and ground states present linear configuration that precludes analysis via the second order Jahn–Teller hamiltonian, they are not yet fully understood.

Distorsion of the equillibrium D₃ₕ triangle induces a vibronic coupling between the degenerate electronic states of a trimer.   At second order, the Jahn–Teller hamiltonian is written (Thomson *et al* 1985 and Longuet–Higgins *et al* 1988):

$$H_{JT} = \begin{pmatrix} q^2/2 & 0 \\ 0 & q^2/2 \end{pmatrix} + \begin{pmatrix} 0 & kq\,e^{-i\varphi} + g/2q^2\,e^{2i\varphi} \\ kq\,e^{i\varphi} + g/2q^2\,e^{-2i\varphi} & 0 \end{pmatrix}$$

where q and φ are related to the normal coordinates:
$q^2 = Q_X{}^2 + Q_Y{}^2$ and $\tan\varphi = Q_X/Q_Y$,
Let us remark that the symmetric vibrational mode Q_Z remains uncoupled to the two non symmetric modes Q_X and Q_Y; k and g are the 1st and 2nd order coefficients, they determine the potential surface of the distorted molecule.

i) if g=0 and k≠0, a pseudo rotation motion takes place in the bottom valley of a "mexican hat" potential surface with a stabilization energy:
$E_S = k^2\,\omega_0\,/\,2(1-g)$ where ω_0 is the zero–order vibrational frequency. The nicest example of quasi free pseudo-rotating molecule is the B $^2E'$ state of Na₃ whose spectrum exhibits the fractional quantization due to the quantum adiabatic phase (Delacretaz *et al* 1986) with the single energy pattern:
$E_{u,j} = (u+1/2)\,h\omega_0 + Aj^2$

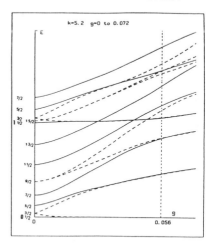

Fig 2. Dependence of energy levels on quadratic Jahn–Teller parameter
---- doubly degenerated levels
——non degenerated levels
k=5.2 and g=0.056 fit the Na₃ A state

ii) More generally, g is non zero, the bottom valley of the "mexican hat" is then distorted with three minima and saddle points whose energy difference is the localization energy:
$E_{loc} = gk^2\,\omega_0\,/\,(1-g^2)$

Indeed, in this energy range the vibronic states are triply degenerated and localized between two consecutive saddle points. As an example, in Na_3 A state this degeneracy appears in the calculation as far as g value increases (Fig. 2).

In Li_3 spectra, two excited states labelled A and C have been recorded. For C state, our data are not yet sufficient to get a complete interpretation. Ground (deduced from hot bands) and A states have been calculated through second order Jahn–Teller hamiltonian. Their molecular constants (in cm^{-1}) are collected below with Na_3 analyzed states; ω_s is the totally symmetric mode frequency

	E_{loc}	E_s	ω_0	ω_s
Na_3 A (Dugourd *et al* 1990a)	1847	196	103.7	150
Na_3 B (Delacretaz *et al* 1986)	1050	26	127	137
Na_3 C (Broyer *et al* 1989)	180	40	127	135
Li_3 X (Dugourd *et al* 1990b)	268	26	150	301
Li_3 A (Blanc *et al* 1991)	57	17	191	326

As lithium is the lightest metal, we benefited from this property to go further in spectroscopic investigations: both rotational structures and isotopic effect that depend on the reduced mass through $1/\mu$ or $1/\sqrt{\mu}$ laws have been also analyzed (Dugourd Ph *et al* to be published). Figs 3a and 3c show the rotational structure of the origin bands in A←X and C←X transitions; they are the first example of completely resolved rotational structure in a metal trimer transition.

The symmetries of these transitions are responsible for the different structures that appear in the spectra. The A←X transition is perpendicular (E'←E') and the selection rules $\Delta J=0,\pm1$, $\Delta K=0,\pm1$ lead to a more congested spectrum than in the C←X parallel transition (E"←E') with selection rules $\Delta J=0,\pm1$, $\Delta K=0$. In these spectra, the structures appeared nearly symmetric (especially for the A←X transition) which suggests that rotational constants of the three states involved could be rather close together. A numerical fit led to the A and B rotational constants of the isoscele triangle and the ξ Coriolis coupling for those three states (cm^{-1}).

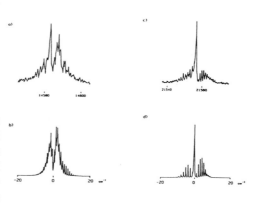

Fig 3. Rotational structure of Li_3 vibronic bands
a) recorded A←X origin band
b) CI–calculated A←X origin band
c) recorded C←X origin band
d) CI–calculated C←X origin band

	A	B	ξ
Li_3 X	0.285	0.584	−0.43
Li_3 A	0.285	0.57	−0.2
Li_3 C	0.28	0.56	−0.36

Calculated spectra deduced from these constants showed the best agreement with experimental ones for a rotational temperature T_{rot} = 10K (Figs 3b and 3d). Moreover rotational constants, directly related to inertia moments, when combined to Jahn–Teller analysis allow to completely determine the geometry and potential surfaces. For instance, these results for the Li_3 ground state are shown in Fig 4.

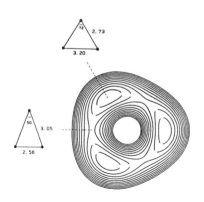

Fig 4. Li_3 ground state potential surface projection with a 10 cm^{-1} difference between subsequent contours. Geometries of the molecule at the minima and saddle points are also shown with distances in Angstroms and angles in degrees

On the other hand, A←X spectra of $^{21}Li_3$, $^{20}Li_3$ and $^{19}Li_3$ have been recorded and isotope shifts have been measured. Theoretical interpretation is based on the consideration that charge center and mass center fly apart when ^{7}Li atom(s) is exchanged with ^{6}Li atom(s). Charge center is always considered as the origin as, in first approximation, electric field and consequently potential surfaces are not altered in the exchange of isotopes. Then, an additional kinetic energy due the motion of the mass center around the charge center must be taken into account. As a result, the totally symmetric Q_Z vibrational mode is now coupled to Q_X and Q_Y non symmetric modes; we can then demonstrate that, depending on the energy position with respect to stabilization and localization energies, isotope shift obeys a ρ or ρ^2 law where $\rho = \sqrt{(\mu/\mu_i)}$. This further information is of great interest when line positions or intensities are not sufficient to completely determine a given state. For example, it could be used to remove remaining ambiguity in the C state spectrum (Dugourd 1991) if new spectra with enriched lithium could be recorded.

4. Li_6

Ab initio calculations have been performed for Li_6 ground state with the SCF–MRD–CI procedure by Koutecky *et al* (1990) starting with a (13s, 3p/6s, 3p) atomic orbitals basis set. The most striking feature of the results is the calculation of three energetically close lying structures: a planar D_{3h} structure, a C_{5v} pentagonal pyramid with only one atom slightly out of the plane of the pentagonal basis and a completely non planar C_{2v} tripyramid. Their respective atomization energies: 4.10, 4.18 and 4.23 ev exhibit very small differences and we may think that competition between those three geometries could be studied with thermally excited hexamers. Indeed, Li_6 clusters were produced at low temperatures in our seeded molecular beam (10 K for Li_3 rotational temperature (Dugourd 1991)). The experimental absorption cross sections are shown in Fig. 5; they exhibit two broad bands: the most intense at 2.5 ev and a weaker one at 1.8 ev.

To record those spectra, it appeared necessary to take great care of the Li_6 daughter fragments originating from Li_8 and Li_7 under the action of the depleting laser. By shifting the ionizing laser downstream by a few cms, the fragments will be removed from the beam because they fly apart due to their transverse velocities; indeed geometric constraint limited the shift between the two lasers to about 1 cm and Li_6 fragments were discriminated in views of their different velocities distribution. In Fig. 5, the experimental absorption cross sections are compared to oscillator strengths calculated for vertical transitions from the three different ground state geometries. It appears that the 3-D tripyramidal C_{2v} geometry is the only one to give a spectrum in good agreement with experimental calculations. For the two other structures, calculated oscillator strengths in the 500 nm region is completely negligible while an intense band, expected in the 600 nm region is in complete disagreement with our observed spectra. This result allows to conclude that the geometry of the Li_6 clusters formed in our seeded molecular beam is unambiguously the C_{2v} tripyramid; it is the first non planar structure in increasing size lithium clusters.

This result must be compared to similar calculations with spectra carried out with sodium clusters (Selby *et al* 1991). Na_6 cluster spectrum exhibits the intense 600 nm band calculated only for planar D_{3h} and 3-D C_{5v} pentagonal pyramid. This difference in ground state geometric structure between sodium and lithium appears only for hexamers Na_6 and Li_6, all other clusters with $n \leqslant 8$ having similar structure. This illustrates

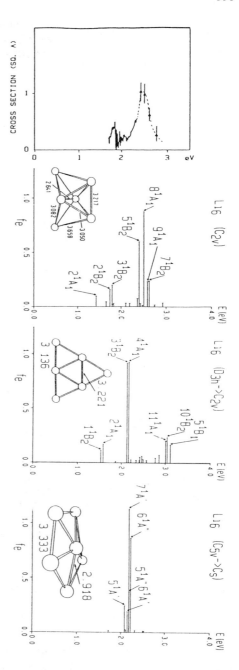

Fig 5. Comparison between experimental spectra and CI-calculated transitions of C_{2v}, D_{3h} and C_{5v} geometries of Li_6

competition between the different close energy geometries and how hexamers are the crucial size when proceeding from planar to 3–D geometries in alkali metal clusters.

5. Li_8

The Li_8 excitation spectrum shown in Fig. 6 exhibits an intense broad band centered at 490 nm and a very weaker one at 670 nm. As for Li_6, this result must be compared to ab initio calculations by the Koutecky group (to be published). The minimum energy for the ground state is calculated for a core tetrahedron surrounded by four electrons (one facing each face) with T_d symmetry. Three transitions with significant oscillator strengths are calculated at 460, 520 and 670 nm, they match very satisfactorily our experimental spectra.

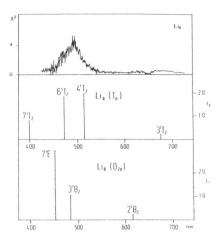

Fig 6. Comparison between experimental spectra and CI–calculated transitions of T_d, and D_{2d} geometries of Li_8

Nevertheless, a second, very close equilibrium geometry is calculated 0.1 ev higher in energy (section of the fcc lattice with D_{2d} symmetry); it leads to two intense transitions at 470 and 480 nm which also could be partially responsible for the broad band observed at 490 nm. Consequently, if we can assure that Li_8 T_d clusters which are the only one to account for the weak 670 nm band are present in our beam, we cannot exclude the presence of the D_{2d} isomer.

These results must be compared to the semi–classical Mie–jellium model for surface plasmon resonance (Selby *et al* 1989). Under the action of the positive ions restoring force fully correlated electrons can oscillate resonantly with the electromagnetic field of an exciting laser provided that laser frequency is equal to the plasmon frequency: $\omega_i^2 = N e^2 / m_e \alpha_i$ (1)
the polarizability α_i is determined along each of the three principal axis of the metal ellipsoid. As octomers are closed shell clusters with spherical geometry, only one plasmon frequency is calculated leading to one broad band in the photoabsorption spectrum. In sodium octomers, a giant resonance has been observed at 2.52 ev for Na_8 by de Heer *et al* (1987) and Wang *et al* (1990) and at 2.62 ev for Na_8^+ by Brechignac *et al* (1991) in agreement with the Mie theoretical value: 2.59 ev; the difference between the two experimental results can be attributed to the change in the restoring force when one electron is removed. This very satisfactory result is not encountered with Li_8 where two unexpected features must be outlined: i) while a broad band is well observed, the weaker one observed at 670 nm is not predicted by the Mie theory and ii) the 2.60 ev energy of the broad band differs very significantly from the calculated value: 3.45 ev

We must remark that, some years ago, similar discrepancies between measured and calculated surface plasmon frequencies were observed in solid lithium by Kloos (1971). A theoretical treatment by Ham (1962a, b) in terms of distorsions of the Fermi surface accounted for the fact that electrons in lithium are more tightly bounded than in other alkali metals. As a result, an effective mass m^* must be introduced in equation (1). Using Ham's formulation a ratio $m^*/m = 2.05$ is deduced; it leads for Li_8 calculated giant resonance to a modified value : 2.44 ev in better agreement with the observed value: 2.60 ev. This kind of behavior, where the motion of valence electrons is not completely free, could be more general as similar results in silver ions clusters by Meiwes–Broer (1991) have been reported recently and could be assigned to interaction with d electrons.

As for Li_8, the observation of weak transitions in addition to the plasmon frequency has been also reported for Na_8 by Wang *et al* (1990) in disagreement with the outcomes of the Mie–jellium theory. Indeed, molecular calculations including both geometric structure and electron correlation account for this result (CI–calculations by Bonacic–Koutecky to be published). Electron correlation can also be introduced through interaction of hole–particle pairs according to the random phase approximation (Eckardt 1985 and Yannouleas *et al* 1990), recent calculations have been successfully carried out for Na_8 by combination with an effective core potential (Gatti *et al* 1990). Anyway, this approach shows clearly how interferences between hole–particle excitations lead to very high oscillator strength. While a very large number of excited states are calculated, only a few of them, corresponding to a constructive interference of single excitations, gather all the oscillator strength and lead finally, as cluster size increases, to the observation of the dominant plasmon frequency. From this point of view, comparison of spectra from Li_3 to Li_8 perfectly illustrates this evolution: the number of calculated states increases very rapidly with the cluster size while, schematically, two bands in the blue and red regions remain observed in all spectra, the blue band that converges to the plasmon frequency becoming more and more dominant. We can then think, by extrapolation of this behavior, that molecular calculations, well convenient to describe these small clusters, could lead to results very close to the Mie–jellium theory for larger size ones.

6. CONCLUSIONS

We have measured the absorption cross sections of small sodium and lithium clusters up to n=8. These results are shown to be without any equivalent to determine geometric structure, elctronic properties and, in views of the lightness of lithium, rotational structure and isotopic shift. Nevertheless, some aspects of the spectra remain questionable. For example, what is the origin of the broadening of molecular bands observed for the smallest clusters, typically n=4 to 6: spectral congestion or dissociative lifetimes? New experiments with sub-picosecond lasers, similar to the ones developped recently by Gerber (1991) with diatomic molecules, should be applied with interest to alkali metal small clusters.

REFERENCES

Blanc J, Broyer M, Chevaleyre J, Dugourd Ph, Kuhling H, Labastie P, Ulbricht M, Wolf J P and Woste L 1991 *Z. Physik* **D19** 7

Boustani I, Pewerstof W, Fantucci P, Bonacic–Koutecky V and Koutecky J 1987 *Phys. Rev.* **B35** 9437
Bonacic–Koutecky V, Fantucci P and Koutecky J 1988a *Chem. Phys. Letters* **146** 518
Bonacic–Koutecky V, Kappes M M, Fantucci P and Koutecky J 1988b *Chem. Phys. Letters* **170** 26
Bonacic–Koutecky V, Fantucci P and Koutecky J 1990a *Chem. Phys. Letters* **166** 32
Bonacic–Koutecky V, Fantucci P and Koutecky J 1990b *J. Chem. Phys.* **93** 3802
Bréchignac C, Broyer M, Cahuzac Ph, Delacretaz G, Labastie P, Wolf J P and Woste L 1988 *Phys. Rev. Letters* **60** 275
Bréchignac C, Cahuzac Ph, Carlier F, de Fructos M and Leygnier J 1991 *Z. Physik.* in press
Broyer M, Delacretaz G, Labastie P, Whetten R L, Wolf J P and Woste L 1986 *Z Physik* **3** 131
Broyer M, Delacretaz G, Ni G Q, Whetten R L, Wolf J P and Woste L 1989 *J. Chem. Phys.* **90** 843
Broyer M, Chevaleyre J, Dugourd Ph, Wolf J P and Woste L 1990 *Phys. Rev.* **A42** 6954
Clemenger K 1985 *Phys. Rev.* **B32** 1359
de Heer W A, Selby K, Kresin V, Masui J, Vollmer M, Chatelain A and Knight W D 1987 *Phys. Rev. Letters* **59** 1805
Delacretaz G and Woste L 1985 *Surface Science* **156** 770
Delacretaz G, Grant E R, Whetten R L, Woste L and Zwanziger J W 1986 *Phys. Rev. Letters* **56** 2598
Dugourd Ph, Chevaleyre J, Perrot J P and Broyer M 1990a *J. Chem. Phys.* **93** 2332
Dugourd Ph, Chevaleyre J, Broyer M, Wolf J P and Woste L 1990b *Chem. Phys. Letters* **175** 555
Dugourd Ph 1991 *Thesis* Lyon
Eckardt W 1985 *Phys. Rev.* **B31** 6360
Gerber G 1991 *Proceedings ICPEAC XVII Brisbane*
Gatti C, Polezzo S and Fantucci P 1990 *Chem. Phys. Letters* **175** 645
Ham F S, 1962a *Phys. Rev.* **128** 82
Ham F S, 1962b *Phys. Rev.* **128** 2524
Herrmann A, Hoffmann M, Leutwyler S, Schumacher E and Woste L 1979 *Chem. Phys. Letters* **62** 216
Kloos T, 1973 *Z. Physik* **265** 225
Knight W D, Clemenger K, de Heer W A, Saunders W A, Chou M Y and Cohen M L 1984 *Phys. Rev. Letters* **52** 2141
Knight W D, de Heer W A and Saunders W A 1986 *Z. Physik* **D3** 109
Koutecky J, Boustani I and Bonacic–Koutecky V 1990 *Int. J. Quant. Chem.* **38** 149
Longuet–Higgins H C, Opik V, Pryce M H L and Rack S A 1988 *Proc. Roy. Soc.* **A244** 1
Manninen M 1986 *Solid State Comm.* **59** 281
Mayer M G and Jensen H D 1955 *Elementary theory of nuclear structure* (Wiley New York)
Meiwes–Broer K H 1991 *Proceedings ICPEAC XVII Brisbane*
Selby K, Vollmer M, Masui J, Kresin V, de Heer W A and Knight W D 1989 *Phys. Rev.* **A40** 5417
Selby K, Kresin V, Masui J, Vollmer M, de Heer W A, Scheidemann A and Knight W D 1991 *Phys. Rev.* **B** in press
Thomson T C, Truhlar D G and Mead C A 1985 *J. Chem. Phys.* **82** 2393
Wang C R C, Pollack S and Kappes M 1990 *Chem. Phys. Letters* **166** 26
Yannouleas C, Pacheco J M and Broglia R A 1990 *Phys. Rev.* **B41** 6088

Attachment of slow electrons to molecular clusters

T. Kraft, M.-W. Ruf, H. Hotop

Fachbereich Physik, Universität D-6750 Kaiserslautern, FRG

Abstract: In this report, we discuss the attachment of slow electrons to molecular clusters with emphasis on recent work involving state-selected Rydberg atoms. The electron attachment behaviour of the clusters is discussed on the basis of that for the monomers for three different prototype molecules, namely SF_6 (strong capture at zero energy), O_2 (capture into short-lived states at energies above zero), and N_2O (dissociative attachment through repulsive surfaces). For N_2O clusters, a very strong variation of the negative ion mass spectra with principal quantum number n of the Rydberg electron (10 < n < 50) is found; at high n, negative ion formation proceeds with a remarkable size-selectivity.

1. INTRODUCTION

The investigation of clusters, which provide the link between gas phase and solid state physics, is an active and important area of current research. Increasing interest has been paid over the last years to the formation (Haberland et al. 1987) and spectroscopy of negatively charged clusters, especially molecular clusters (Arnold et al. 1989). The energy dependence of free electron attachment was studied by Klots and Compton (1987), Märk and colleagues (1985, 1986), Knapp and colleagues (1985, 1987), Vostrikov et al. (see Märk 1991) and Illenberger et al. (Lotter and Illenberger 1990, Hashemi et al. 1990); recent survey articles were presented by Hashemi et al. (1990) and by Märk (1991). Even though these experiments were carried out with a relatively poor electron energy resolution of 0.5 - 1 eV, they revealed as an interesting new phenomenon the "zero energy resonance" (Märk et al. 1985, Stamatovic 1988) in the electron attachment to clusters made of molecules, which as isolated species do not attach very slow electrons (e.g. O_2, CO_2, H_2O). This phenomenon is important in connection with electron transport in dense gases (Hatano 1986), and it is desirable to study electron attachment to clusters at very low energies with high resolution. As an interesting and efficient alternative to slow free electrons, Rydberg electrons can be used to investigate the attachment process at energies $\leq$ 0.1 eV (Dunning 1987). This idea was first applied to clusters by Kondow and colleagues (1987), who studied the mass spectra of negative cluster ions formed in collisions of van-der-Waals clusters with electron-beam excited rare gas Rydberg atoms Rg** (principal quantum numbers n $\cong$ 25 - 35). Kraft et al. (1989, 1990) and, independently, Desfrancois et al. (1989) used laser-excited, state-selected rare gas Rydberg atoms Rg**(nl) (n $\cong$ 8 - 50, l = 0,2,3) with the aim to investigate the dynamics of electron attachment to neutral clusters at very low energies and to look for effects due to postattachment interactions of the Rg^+ ion with the negative cluster ions. It is of special interest to look for changes in the attachment behaviour, when going from the monomer XY to the clusters $(XY)_m$, to study variations with cluster size ($m \geq 2$) (for instance with regard to the branching ratio for dissociative and non-dissociative processes) and to reveal size-selective behaviour (Kraft et al. 1990). We note that size-selected neutral clusters so far have not been used in electron attachment work. In this (selective) report we discuss the present status of electron attachment at low energies and concentrate on molecular clusters $(XY)_m$ built from XY = SF_6, O_2 and N_2O. The neutral clusters are characterized by weak van-der-Waals attractions; the resulting negative cluster ions are mostly electrostatically bound aggregates. These systems are prototypes for which the monomer XY exhibits a) strong capture at zero energy with formation of a long-lived XY^- ion (XY = SF_6), b) capture at energies above zero through $XY^-(v>0)$ resonances with short autodetachment lifetime (XY = O_2), c) dissociative electron attachment producing $X^- + Y$ through repulsive potential surfaces (XY = N_2O).

The paper is organized as follows: in section 2, we describe some experimental aspects. In section 3, we summarize the essential physics of Rydberg electron attachment to molecules XY at high and low n. In section 4, we discuss the attachment of slow free electrons and of state-selected Rydberg electrons to the chosen cluster systems $(XY)_m$ and add some concluding remarks.

2. EXPERIMENTAL

The basic apparatus for electron attachment studies to molecular clusters consists of a) a source of free electrons or Rydberg atoms; b) a nozzle beam, containing the molecular clusters, as formed in a supersonic expansion of a high pressure gas through an orifice with typically 0.05 - 0.1 mm diameter; c) a mass spectrometer with sufficient mass range and a suitable ion detector. We briefly describe our crossed-beams apparatus and the method for production of state-selected rare gas Rydberg atoms $Rg^{**}(nl)$ (Kraft et al. 1989). A well-collimated supersonic metastable rare gas atom beam from a differentially-pumped dc discharge source is crossed in the reaction chamber by a skimmed, differentially-pumped nozzle beam (nozzle diameter 50 μm) and two single mode dye laser beams; the latter provide efficient excitation of $Rg^*(ms\ ^3P_2)$ atoms via the closed transition involving the intermediate $Rg^*(mp\ ^3D_3)$ level to the respective (ns,J=2) or (nd,J=4) Rydberg level of interest. The resulting positive or negative ions are analyzed by a quadrupole mass spectrometer (m/q up to 2000 amu/e), mounted coaxially with the supersonic beam and detected by an off-axis dual microchannel plate electron multiplier. The metastable flux is some 10^9 /s with an average velocity of about 560 / 800 m/s for Ar / Ne. For the neutral cluster production, stagnation pressures p_0 up to 6 bar and nozzle temperatures in the range T_0 = 200 - 300 K have been used. The presence of clusters is probed by measuring positive cluster ion spectra due to Penning ionization of $(XY)_m$ by metastable Ne* (or He*) atoms. The overall performance of the apparatus can be summarized by quoting typical product ion intensities. Using $Ar^{**}(20d,J=4)$ Rydberg atoms, an SF_6 count rate of about 10^5/s from the calibration reaction

$$Rg^{**}(nl) + SF_6 \rightarrow Rg^+ + SF_6^-$$

(rate constant around $2 \cdot 10^{-7}$ cm³/s for n = 20) was observed at a diffuse SF_6 density of about 10^9 /cm³. In the negative cluster ion spectra of CO_2 and N_2O clusters, the intensities of the most abundant ions $(CO_2)_{16}^-$ and $(N_2O)_6 \cdot O^-$ were about 2000/s for neat expansions at $T_0 = 300$ K and $p_0 = 5$ bar.

3. ATTACHMENT OF RYDBERG ELECTRONS TO MOLECULES

Many molecules are known to exhibit very large electron attachment cross sections $\sigma_e(\varepsilon)$ at meV energies ε, among them SF_6. The threshold law for s-wave electron capture predicts $\sigma_e(\varepsilon) \sim \varepsilon^{-0.5} \sim v^{-1}$ (Chutjian 1991), and the corresponding rate constant $k_e = \sigma_e \cdot v$ is independent of velocity v. Within the quasi-free electron model (Matsuzawa 1972) the rate constant k_{nl} for electron attachment Rg**(nl) + XY → Rg$^+$ + XY$^-$ is connected with free electron attachment through

$$k_{nl} = \int k_e(v) \, |G_{nl}(v)|^2 \, d^3v,$$

where $G_{nl}(v)$ is the velocity distribution of the Rydberg electron, i.e. k_{nl} is obtained as the velocity average of the free electron rate constant with respect to the Rydberg electron wave function. For s-wave attachment, $k_e(v) = const = k_0$, and one obtains $k_{nl} = k_0$. The quasi-free electron model assumes that the Rydberg core Rg$^+$ can be neglected and that the Coulombic ion pair Rg$^+$ + XY$^-$ will always separate. One may intuitively take this as granted at high n in view of the large extension of the Rydberg orbital (for high n and small l, the average radius amounts to $<r>_{nl} \cong (3/2) n^2 a_0$ with $a_0 =$ Bohr radius = 52.9 pm). Qualitatively, one may view s-wave Rydberg electron attachment to occur as a sudden electron transfer from Rg** to XY at an average Rg** - XY distance R_t close to $<r>_{nl}$ such that the positions and velocities of the nuclei remain unchanged. The relative kinetic energy of the Rg** - XY system at these large R_t should be very close to the asymptotic collision energy E_{rel}. Therefore after electron transfer, the Coulomb pair Rg$^+$ - XY$^-$ suddenly finds itself bound by

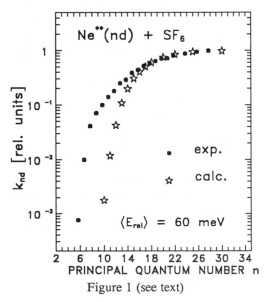

Figure 1 (see text)

$V_c(R_t) = - E_0 a_0 / R_t \cong - 2 E_0 / 3 n^2$
$\cong - 18$ eV / n^2 with a local kinetic energy close to E_{rel}. As long as $E_{rel} > |V_c(R_t)| \cong 18$ eV / n^2 dissociation of the Coulomb pair is expected to occur. On the contrary, observation of the ionic products Rg$^+$ and XY$^-$ is not granted, if

$n < n_c = (18 \text{ eV} / E_{rel})^{1/2}$. At $E_{rel} = 60$ meV (typical nozzle beam), $n_c = 17$; at these collision energies, one therefore expects postattachment effects to be negligible at least for $n > 30$, as verified by Dunning and colleagues. They found the rate constant k_{nl} for several molecules including SF_6 to be constant for $n > 35$ (as expected for quasi-free s-wave attachment) and showed the velocity distribution of Xe^+ ions resulting from $Xe^{**}(nf) + SF_6$ attachment collisions at $T = 300$ K to be essentially identical with that of the Xe^{**} atoms for high n, but perturbed for low n (Dunning 1987). Figure 1 illustrates the n-dependence of $k_{nl}(SF_6)$, as measured by Harth et al. (1989) for $Ne^{**}(nd) + SF_6$ in the range $6 \leq n \leq 30$. For comparison, we show the results of a calculation, which assumes the $Ne^+ - SF_6^-$ formation to occur with a distribution of electron transfer distances R_t according to the radial density distribution of the respective nd-orbital and which allows for separation of the Coulomb pair on the basis of the escape criterion $E_{rel} > |V_c(R_t)|$ applied to the relevant collision energy distribution $g(E_{rel})$. One observes that this simple Coulomb escape model, first applied by Zollars et al. (1986) to rationalize the n-dependence of $k_{nl}(SF_6)$ for $K^{**}(nd) + SF_6$ collisions, is not able to properly describe the experimental data at low n. Harth et al. (1989) therefore suggested that at low n dissociation of bound $Rg^+ - SF_6^-$ pairs through conversion of vibrational energy of SF_6^- becomes important. Results of Pesnelle et al. (1990) and Desfrancois et al. (1991) point into the same direction. We conclude that the quasi-free electron model provides a good description at sufficiently high n (dependent on collision energy E_{rel}), whereas towards lower n, postattachment effects become progressively more important.

The influence of the Rydberg core becomes very clear in attachment reactions with small molecules, which have short autodetachment lifetimes τ_A. Viewed within the two-step model, the Rydberg core ion Rg^+ has a chance to stabilize the intermediate XY^{-*} ion, if stabilizing momentum exchange occurs between Rg^+ and XY^{-*} within τ_A. For $XY = O_2$, the lifetime of the lowest-lying resonance state $O_2^-(^2\Pi_g, v=4)$ is around 10^{-11} s; for Rg^+ ions with velocities of about 10^5 cm/s, only those born with transfer distances $R_t < 100 \, a_0$ and with a sufficiently small impact parameters $b \leq b_c$ (to allow for a close, momentum-exchanging encounter !) can contribute to stabilization of O_2^-. One therefore expects cross sections of order πb_c^2 for Rydberg states with radial extensions $< 100 \, a_0$ ($n < 8$). O_2^- production was in fact observed by Walter et al. (1986) in $K^{**}(nd) + O_2$ and by Harth et al. (1989) in $Ne^{**}(ns,nd) + O_2$ collisions at low n with maximum rate constants around 10^{-12} cm³/s (cross section $\cong 10^{-17}$ cm²) for $n^* = 8$. Correspondingly and as expected, the effective impact parameter range ($b_c \cong 2 \cdot 10^{-9}$ cm) is only a small fraction of the orbital radius.

It should be noted, however, that at very low n the above two-step model of the ion pair formation process becomes questionable, and a quasi-molecular description involving curve crossings may be more adequate.

4. ATTACHMENT OF SLOW ELECTRONS TO MOLECULAR CLUSTERS

4.1 SF_6 clusters

In Figure 2, we compare the positive and negative cluster ion mass spectra due to He* Penning ionization and Ar**(22d) Rydberg electron transfer collisions with a seeded SF_6 beam ($T_0 = 300$ K, $p_0 = 5$ bar, 20 % SF_6 in Neon carrier gas). In accord with the respective ion formation for collisions with monomers (SF_5^+ / SF_6^-), the positive / negative cluster ion spectra consist of $(SF_6)_{q-1}*SF_5^+$ / $(SF_6)_q^-$ ions. The respective mass spectra have a similar general appearance; the ion intensities monotonically decrease with rising q with a sharp drop from q = 1 to 2.

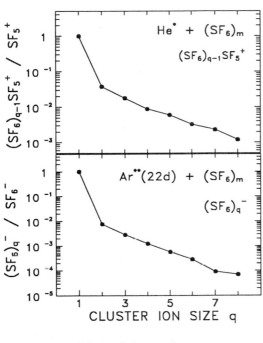

Figure 2 (see text)

Since Penning ionization is expected to probe the effective surface area exposed by the cluster $(XY)_m$ (Ohno et al. 1984, Siddiqui et al. 1984), which scales as $m^{2/3}$, the ionization cross section should slowly rise with m (by about a factor of four from m = 1 to 8). Let us assume for simplicity that the $(SF_6)_{q-1}*SF_5^+$ and $(SF_6)_q^-$ ions stem from the neutral cluster with q = m (i.e. SF_6-evaporation is a weak process). The fact that the relative cluster intensities for q ≥ 2 decrease less rapidly with q for positive than for negative ions, indicates that the ratio $\sigma_{PI}(m)/\sigma_{nl}(m)$ of the m-dependent Penning ionization to electron attachment cross section (nl = 22d) increases with m. Since the electron attachment cross section for the SF_6 monomer is very large (not far from the s-wave de Broglie cross section $\pi \lambda^2$), it is plausible that the attachment cross section for $(SF_6)_m$ clusters (m ≥ 2) does not change significantly with m. Then, the m-dependence of $\sigma_{PI}(m)/\sigma_{nl}(m)$ would mainly mirror the m-dependence of σ_{PI}. So far it is not well known, however, which role SF_6-evaporation may play in the formation of the positive and negative ion mass spectrum. We also mention that apparatus-dependent discrimination effects in the detection of different cluster sizes (especially for positive ions) may influence the mass spectra and could account for differences with regard to previous work (Mitsuke et al. 1986). Nevertheless, it is justified to state that the positive Penning ionization mass spectra provide a very useful reference for judging the neutral cluster size distribution and thereby to draw conclusions on the m-dependence of the electron attachment process.

In measurements of negative cluster ion spectra for different n = 16 - 28 we found the normalized ion intensity ratios $(SF_6)_q^- / SF_6^-$ (q $\geq$ 2) to be very nearly independent of n (for q = 2 the intensity ratio is constant within $\pm$ 15 %); this behaviour is compatible with the idea that the primary electron attachment process is dominated by that of the SF_6 monomer. More recently, Desfrancois et al. (1991) have provided evidence for an increase of $(SF_6)_2^- / SF_6^-$ towards very low n (about a factor of 2 in going from n = 25 to n = 9) and for an evaporative process due to the influence of the Rg^+ ion core upon the negative SF_6 cluster.

Apart from SF_6 clusters, we also studied clusters of the zero-energy attaching molecules CCl_4, C_6F_6 and $1,1,1-C_2Cl_3F_3$ and found no or only weak variations of the normalized negative cluster ion mass spectra with n. For $(C_6F_6)_m$, only $(C_6F_6)_q^-$ ions are detected, for $(CCl_4)_m$, the mass spectrum is dominated by the dissociated $(CCl_4)_q*Cl^-$ ions. For $(C_2Cl_3F_3)_m$, for which the monomer yields exclusively Cl^- (as is the case for CCl_4), intact $(C_2Cl_3F_3)_q^-$ ions were observed besides the dominant dissociated $(C_2Cl_3F_3)_{q-1}*Cl^-$ ions (q $\geq$ 1) with a relative intensity around 30 % (for CCl_4 and other molecules, see also Kondow 1987).

4.2 O_2 clusters

The attachment cross section of free electrons to isolated O_2 molecules yields exclusively O^- ions and is characterized by a broad peak centered at $\varepsilon \cong 7$ eV due to dissociative attachment via the repulsive $O_2^-*(^2\Pi_u)$ state. The electronic ground state of the O_2^- ion $(^2\Pi_g)$ is subject to autodetachment for v > 3 and the $O_2^-(^2\Pi_g, v \geq 4)$ resonance levels strongly influence the electron scattering processes from O_2 (see, e.g., Field et al. 1988). In the first study of free electron attachment to O_2 clusters, Märk et al. (1985, 1986) found both $(O_2)_{q-1}*O^-$ and $(O_2)_q^-$ ions (q $\geq$ 1) with substantially different energy dependences. A peak is observed at energies close to 0 eV ("zero energy resonance") which exclusively yields $(O_2)_q^-$ ions; the width of the peak is found to decrease with rising q and appears limited by the experimental energy resolution (about 0.5 eV) at high q. The contribution to $(O_2)_q^-$ formation from the high-lying repulsive states (for clusters, at least three repulsive states contribute to attachment, see Märk 1991) strongly decreases towards higher q. The "zero energy resonance" has later been observed in other systems, e.g. CO_2 and H_2O (Märk 1991). Tsukada et al. (1987) theoretically treated electron attachment to CO_2 clusters and correlated the zero energy resonance with attachment into an "extended affinity state" of the clusters above a minimal size. For O_2 clusters such calculations have not been carried out, and the experimental data do not allow definite conclusions on the m-dependence of the primary attachment process (the minimal size has to be at least m = 2). In the first study of Rydberg electron attachment to O_2 clusters (T_0 = 200 K, p_0 = 5.5 bar), Kraft et al. (1989) observed $(O_2)_q^-$ cluster ions up to q = 13, see figure 3. The appearance of the positive ion mass spectrum, resulting from Ne* Penning ionization,

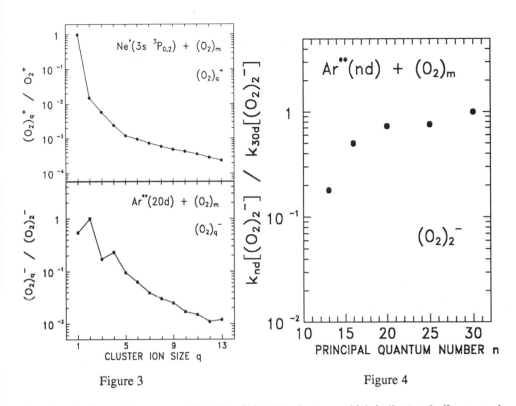

Figure 3   Figure 4

is quite similar to that presented in Fig. 2 for SF$_6$ clusters, which indicates similar neutral cluster size distributions in the two cases. In contrast, the negative ion spectrum, which exclusively consists of $(O_2)_q^-$ ions, exhibits structure with the dimer ion ($q = 2$) as the most abundant species. The dimer to monomer ratio for the negative ions is about 150 times larger than for the positive ions. This finding can be mainly attributed to the small cross section for O_2^- production from monomers, as discussed in section 3. The signal rate of the $(O_2)_q^-$ cluster ions (at $q = 2$, we observe about 1000 ions /s for Ar**(20d)) suggests that the Rydberg electron attachment process to O_2 clusters is efficient with an average rate constant estimated to be $\geq 10^{-8}$ cm^3/s by comparison with $(CO_2)_q^-$ production, for which Kraft et al. (1991a) determined an average value of $(4.5 \pm 3) \cdot 10^{-8}$ cm^3/s. The observed structure with peaks at $q = 2$ and 4 reflects negative cluster ion stability, which is particularly high for $(O_2)_2^-$ (Hiraoka 1988; see also Kraft et al. (1989) for more details). Clearly, O_2 evaporation takes place and plays an important role in the synthesis of the negative ion mass spectrum. It is therefore difficult to judge on the m-dependence of electron attachment from a comparison of the positive and negative ion spectrum. It is desirable to study in detail the evolution of the cluster ion mass spectra as a function of expansion conditions in order to learn more about the dynamics of Rydberg electron attachment to these important clusters.

The relative $(O_2)_q^-$ intensities (including $q = 1$) are essentially independent of n in the range 16 - 30. The n-dependence of the $(O_2)_2^-$ intensity yields an effective n-dependent rate constant similar to that for SF_6^- production, as illustrated in Fig. 4. The substantial decrease towards lower n is likely due to postattachment effects. At high n the rate constant tends to become constant; this suggests s-wave-type attachment, whereas for O_2 monomers, the $O_2^-(^2\Pi_g)$ resonances involve d-waves. In the clusters, the symmetry selection rules are relaxed (Hatano 1986, Märk 1991), and s-wave attachment to $(O_2)_m$ clusters (m $\geq$ 2) becomes possible. As a final remark, we note that the n-independence of the $(O_2)_2^- / O_2^-$ intensity ratio is a clear indication that in the covered n-range the O_2^- production predominantly results from attachment to $(O_2)_m$ (m > 2) clusters.

4.3 N_2O clusters

Neutral N_2O molecules are linear in their ground state, whereas the anions N_2O^- are bent by about 135° (adiabatic electron affinity 0.22 eV (Hopper et al. 1976)). The barrier of N_2O^- against dissociation into O^- is rather low ($\cong$ 0.4 eV). Attachment of free electrons to N_2O monomers proceeds by dissociative attachment (O^- formation) with a broad peak at $\varepsilon = 2.25$ eV (Knapp et al. 1985). Attachment to N_2O clusters yields predominantly $(N_2O)_{q-1}*O^-$, but also $(N_2O)_q^-$ ions (q $\geq$ 1). The excitation function of $(N_2O)_{q-1}*O^-$ ions shows a single broad peak with the maximum shifting to lower energies with increasing q. For $(N_2O)_q^-$ ions, the energy dependence exhibits at least two peaks with q-dependent maximum positions. No clear "zero energy resonance" was observed as for O_2 or CO_2 clusters (Knapp et al. 1987).

Transfer of unselected Rydberg electrons to N_2O clusters was studied by Kondow (1987) and by Yamamoto et al. (1991). Attachment of state-selected Ar**(ns,nd) Rydberg atoms to N_2O clusters was investigated by Kraft et al. (1990, 1991b), who discovered a very strong n-dependence of the mass spectrum. In Figure 5 we compare the positive ion spectrum, due to Ne* Penning ionization and yielding almost exclusively $(N_2O)_q^+$ ions, with the spectrum of $(N_2O)_{q-1}*O^-$ ions (q $\geq$ 5), which represent the major negative product ions resulting from electron transfer from Ar**(20d) Rydberg atoms to N_2O clusters; in addition, $(N_2O)_q^-$ (q $\geq$ 5) ions are found. Figure 6 illustrates the amazing n-dependence of the negative cluster spectrum, documented and discussed in some detail by Kraft et al. (1990). These data were obtained with a neat N_2O expansion at $T_0 = 300$ K and $p_0 = 5$ bar.

$(N_2O)_q^-$ and $(N_2O)_{q-1}*O^-$ ions with q = 1-3 are missing in the Rydberg attachment mass spectrum at all n (11 $\leq$ n $\leq$ 45, i.e. electron energies < 0.1 eV), while these ions, especially $(N_2O)_{q-1}*O^-$, are clearly detected at higher energies using free electrons (Knapp et al. 1987, Yamamoto et al. 1991). Although the adiabatic electron affinity of N_2O molecule is positive, the strong geometry change from the linear N_2O molecule and the corresponding negative values for the vertical electron affinity of small $(N_2O)_m$ clusters prevent the attachment of very slow free and Rydberg electrons at low m. The state-selected Rydberg data indicate that attachment of very slow electrons becomes possible for

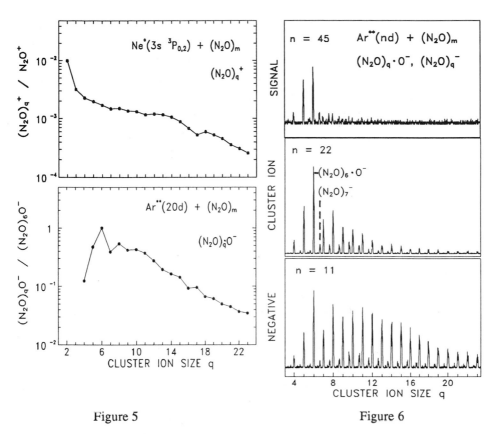

Figure 5   Figure 6

m $\geq$ 5, yielding predominantly $(N_2O)_{q-1}*O^-$, but also $(N_2O)_q^-$ ions with a q- and n-dependent branching ratio. At high n, attachment proceeds in a rather size-selective way, yielding mainly $(N_2O)_{q-1}*O^-$ and $(N_2O)_q^-$ ions with $5 \leq q \leq 9$. Test experiments showed that Ar**(nd) atoms are efficiently ionized also in collisions with larger clusters at high n, suggesting that electron transfer leads to autodetaching negative ion states for larger m. From our results obtained so far it is likely that the change of the negative cluster spectrum with n (Fig. 6) reflects a size-selective resonance effect in negative ion formation associated with the variation of the energy of the (bound) Rydberg electron.

It is of obvious interest and highly desirable to investigate the negative ion formation from N_2O and other molecular clusters by using monoenergetic free electrons with energies $\leq$ 0.1 eV. Based on the recent achievement of laser photoelectron attachment to SF_6 at sub-meV resolution (Klar et al. 1991), we plan to extend this method to cluster targets. Future progress in the understanding of electron attachment to clusters will depend on an improved characterization of the neutral cluster distribution. Calibrated positive cluster ion mass spectra, produced in electron, photon or metastable atom impact, could be used as a reference at low m, once the fragmentation pattern is known from studies such as those carried out, e.g. for Ar and CO_2, by Buck (1988).

This work has been supported by the Deutsche Forschungsgemeinschaft through Sonderforschungsbereich 91. We gratefully acknowledge K. Harth, A. Fras and D. Klar for their fruitful cooperation. We thank A. Chutjian, E. Illenberger, T.D. Märk and J.P. Schermann for providing manuscripts prior to publication.

Arnold, S.T., Eaton, J.G., Patel-Misra, D., Sarkas, H.W., Bowen, K.H., 1989, in "Ion and Cluster-ion Spectroscopy and Structure", Maier, J.P. (ed.), p 417, Elsevier, Amsterdam
Buck, U., 1988, J. Phys. Chem. **92**, 1023
Chutjian, A., 1991, Review paper in this volume
Desfrancois, C., Khelifa, N., Schermann, J.P., 1989, J. Chem.Phys. **91**, 5853
Desfrancois, C., Khelifa, N., Lisfi, A., Schermann, J.P., 1991, J. Chem. Phys. (xxx)
Dunning, F.B., 1987, J. Phys. Chem. **91**, 2244
Field, D., Mrotzek, G., Knight, D.W., Lunt, S., Ziesel, J.P.. 1988, J. Phys. B **21**, 171
Haberland, H., Ludewigt, C., Schindler, H.-G., Worsnop, D.R., 1987, in "Large finite systems", Jortner, J. et al. (eds), p 195, Dordrecht:Reidel
Harth, K., Ruf, M.-W., Hotop, H., 1989, Z. Phys. D **14**, 149
Hashemi, R., Kühn, A., Illenberger, E., 1990, Int. J. Mass Spectrom. Ion Proc. **100**, 753
Hatano, Y., 1986, Electronic and Atomic Collisions (Lorents, D.C. et al., eds.), p 153, Elsevier, Amsterdam
Hiraoka, K., 1988, J. Chem. Phys. **89**,3190
Hopper, D.G., Wahl, A.C., Wu, R.L.C., Tiernan, T.O., 1976, J. Chem. Phys. **65**, 5474
Knapp, M., Kreile, D., Echt, O., Sattler, K., Recknagel, E., 1985, Surf. Sci. **156**,313
Knapp, M., Echt, O., Kreisle, D., Märk, T.D., Recknagel, E., 1987, in "Physics and Chemistry of Small Clusters", Jena, P. et al. (eds), p 693, Plenum Press, New York and London
Klar, D., Ruf, M.-W., Hotop, H., 1991, Joint Symposium on Electron and Ion Swarms and low Energy Electron Scattering, Abstracts of Papers, p 2, Bond University, Gold Coast, Australia
Klots, C.E., Compton, R.N., 1978, J.Chem. Phys. **69**, 1636
Kondow, T., 1987, J. Phys. Chem. **91**, 1307
Kraft, T., Ruf, M.-W., Hotop, H., 1989, Z. Phys. D **14**, 179
Kraft, T., Ruf, M.-W., Hotop, H., 1990, Z. Phys. D **17**, 37
Kraft, T., Ruf, M.-W., Hotop, H., 1991a, Z. Phys. D **18**, 403
Kraft, T., Ruf, M.-W., Hotop, H., 1991b, Z. Phys. D **20**, 13
Lotter, J., Illenberger, E., 1990, J. Phys. Chem. **94**, 8951
Märk, T.D., Leiter, K., Ritter, W., Stamatovic A., 1985, Phys. Rev. Lett. **55**, 2559
Märk, T.D., Leiter, K., Ritter, W., Stamatovic A., 1986, Int. J. Mass Spectrom. Ion Proc. **74**, 265
Märk, T.D., 1991, Int. J. Mass Spectrom. Ion Proc. **107**, 143
Matsuzawa, M., 1972, J. Phys. Soc. Japan **32**, 1088, **33**, 1108
Mitsuke, K., Kondow, T., Kuchitsu, K., 1986, J. Phys. Chem. **90**, 1552
Ohno, K., Matsumoto, S., Harada, Y., 1984, J. Chem. Phys., **81**, 4447
Pesnelle, A., Ronge, C., Perdrix, M., Watel, G., 1990, Phys. Rev. A **42**, 273
Siddiqui, H.R., Bernfeld, D., Siska, P.E., 1984, J. Chem. Phys. **80**, 567
Stamatovic, A., 1988, Electronic and Atomic Collisions (Gilbody, H.B. et al., eds.), p 729, Elsevier, Amsterda
Tsukada, M., Shima, N., Tsuneyuki, S., Kageshima, H., Kondow, T., 1987, J. Chem. Phys. **87**, 3927
Walter, C.W., Smith, K.A., Dunning, F.B., 1989, J. Chem. Phys. **90**, 1652
Yamamoto, S., Mitsuke, K., Misaizu, F., Kondow, T., Kuchitsu, K., 1991, J. Phys. Chem. **94**, 8250
Zollars, B.G., Walter, C.W., Lu, F., Johnson, C.B., Smith, K.A., Dunning, F.B., 1986, J. Chem. Phys. **84**, 5589

Optical and electronic properties of small metal clusters

K.H. Meiwes-Broer

Fakultät für Physik, Universität Bielefeld, D-4800 Bielefeld 1, FRG

Abstract: Physical properties of small metal clusters significantly deviate from those of the respective bulk. Experimental tools like photoelectron and photoabsorption spectroscopy are suitable to monitor details of the electronic level structure of these species.

I. Introduction

The physical and chemical properties of materials may change upon reduction of their sizes. This phenomenon is well-known in the study of, e.g., thin films where the electrical conductivity is influenced by the confinement of the charge carriers to two dimensions. A three-dimensional confinement to a few Å in diameter will induce descrete energy levels. Experimental and theoretical evidence suggests a picture of the valence electrons in an alkali cluster as a system of Fermi particles, being quantized in a mean field made up by the positive atomic cores. The order and spacing of the levels depends on the geometry of the confining volume as well as the details of the effective potential seen by the electrons. Such 'quantum size effects' are well known from, e.g., nuclear physics. They are expected to play an important role in the transition regime between atom and solid-state physics [1]. So far, the main experimental tool for the investigation of cluster properties was mass spectrometry. Discontinuouties in mass spectra revealed special features of the electronic structure. Only recently, photoelectron-spectroscopy as well as photoabsorption spectroscopy have been applied to gas-phase clusters in order to map the metal cluster level structure directly.

In this contribution we first discuss the basic theoretical concepts of small metal clusters. Next, the experimental technique for the investigation of photoelectrons and photofragments from clusters in a beam is described. Some photoelectron spectra will be presented and discussed. Finally, complementary photodissociation results serve as a cross-check of the model discussed.

II. Theoretical Concepts

Cluster systems are supposed to have properties intermediate between atoms and solids. Therefore, theoretical concepts employ methods from both atomic and solid state physics or chemistry. In addition, the shell nature gives rise to apply nuclear physics methods.

At present, a number of *ab initio* approaches exist to solve the Schrödinger equation in the Hartree-Fock framework. Correlation effects may also be included by, e.g., the configuration interaction (CI) technique [2]. *Ab initio* methods require an exact calculation of all integrals contributing to the elements of the Fock matrix. The investigation of systems with more and more electrons and nuclei, however, leads to an enormous number of integrals, driving the cost of the calculation out of present reach. This has led to efforts to find sensible and systematic simplifications where, e.g., the differential overlap between atomic orbitals is neglected. Another simplification is the application of pseudopotentials which can greatly reduce the computational demands of a given problem. An explicit treatment of core orbitals is replaced by a potential function [3].

Quantum chemistry theories have been applied to clusters for a long time. Already an alkali trimer is found to appear in different arrangements, depending on the charge state [4]. There seems to be no simple rule to build up larger clusters by smaller ones. Correspondingly, changing the charge state by, e.g., ionization or electron detachment may lead to a significant change in the geometry. Therefore, for a thorough interpretation of spectroscopic experiments the geometric cluster structure has to be known.

In contrast with quantum chemistry methods, the theoretical study of solids uses translational symmetry. Nevertheless, approaches derived from solid state physics have successfully been applied to alkali clusters [5]. The computational method is based on local-spin-density (LDA) and pseudopotential approximations. Again, with increasing cluster size ($N \geq 10$), the complexity of the calculation and the number of degrees of freedom in arranging the atomic positions dramatically increase. Hence, to examine clusters over a large range of sizes, it is necessary to simplify the problem.

Experimental results on sodium [6] or on noble metal clusters (see below) suggested the validity of an electronic shell theory. The underlying idea resembles the shell theories for atoms and nuclei: Fermions are confined to lengths of about 10^{-14} m in nuclei, and to about 10^{-10} m in atoms. In spite of different interactions involved, both fermion systems can be explained by a quantization in a global mean field with spherical symmetry. The nucleus in an atom acts almost like a point charge resulting in a spherically symmetric field. The eigenstates of the atomic Hamiltonian can be characterized by the orbital angular momentum, i.e., the electronic states can be ordered in shells which is reflected in the periodic table.

In nuclei, the fermions are strongly interacting nucleons. The resulting shell structure is qualitatively similar to that of the atoms [7]. A Fermi gas model of the nucleus assumes that the nucleons move in an effective potential (e.g., Woods-Saxon-Potential), thus leading to the well-known magic numbers. Note that in the nucleus the spin-orbit-interaction significantly influences the level ordering: e.g., the shell closing for 28 nucleons cannot be explained without this interaction.

Is it possible to apply the shell model also to metal clusters? To establish a smooth mean field, the (valence) electrons have to be delocalized over the whole cluster. The electron mean free path should be large compared to the particle diameter. In the case of, e.g., bulk sodium, a mean free path of around 200 Å can be estimated, which makes the probability of a collision with an ion in a small sodium cluster very low.

Consequently one can use the independent particle, mean field approximation as in nuclei. The positions of the ionic cores are not known. However, it turns out that these are not very sensitive to the level ordering. Therefore, the structure of the lattice is left out and replaced by a uniform positively charged background. This is the jellium approximation. The effective field acting on each conduction electron is the average effect of the attraction from the ions and the repulsion from the other electrons. Such fields have been calculated self-consistently by Ekardt et al. [8] and Brack et al. [9] in the density functional formalism.

One result for spherical Na_{34} is given in fig. 1 [8]. The single particle potential turns out to have a flat bottom and a smoothly rising spherical rim. The calculations assume a radius that reproduces the bulk density; in principle, however, the experimentally confirmed lattice shrinkage could be incorporated by changing the bulk Wigner-Seitz radius, r_s, into a corrected bond length. These and similar calculations have been performed on various cluster sizes and materials [8-11]. Generally, the widths of the potentials increase with increasing N, whereas the potential depths increase with increasing number of valence electrons per atom.

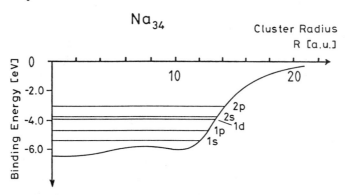

Fig. 1: *Effective single particle potential of Na_{34}, calculated in the local density approximation in the spherical jellium model. After Ekardt [8].*

In an atom, the spherically symmetric field from the nucleus assures spherical symmetry. There is no similar external field in the cluster. Like nuclei, metal clusters with partially filled shells will tend to deform as the electronic energy can be lowered that way [12]. The distortion from a spherical shape changes the symmetry and thus the energy level splittings of the system. The simplest appropriate deformation is to change the geometry from a sphere to a spheroid with axial symmetry. One approach of this kind is the Clemenger-Nilsson model [12] where a simple harmonic-oscillator single particle Hamiltonian is used. As a result, many subshells arise in the energy level spectrum which in part qualitatively coincide with structures in mass and electron spectra.

Different refinements of the jellium model have been discussed taking into account the structure of the ionic background [13,14]. We do not expect the shell theory to be applicable to transition metal cluster. Although there exist valence s and p electrons which are needed to establish the electronic shell structure, the unfilled, localized d-orbitals cannot be ignored in the determination of structural and electronic properties [15]. Here, tight-binding types of electronic theories are used to calculate, e.g., the binding energies of iron clusters [16].

Finally, analytical cluster models are available [17]. They rely on very simplified assumptions, but are suitable for rough estimates of cluster properties like binding energies or melting temperatures as a function of size up to the bulk limits.

For the description of the optical activity of small particles, the mathematical framework of Mie [18] is suitable. It is based on Maxwell's equations and is not expected to work for nano- and subnanometer sized species. Only in the last years, the pioneering theoretical work of Ekardt [19] and experimental results of de Heer et al. [20] revealed that the optical activity of very small particles with only about ten atoms can – to a first order – be understood as a simple collective excitation of the valence electrons, called a "giant resonance". Such an excitation is equivalent to the first excitation mode in the Mie theory. Within the jellium model, the position of the plasmon energy is strongly influenced by the number of valence electrons N_e. The spill-out of the electrons beyond the positive jellium background leads to static polarizabilities above the classical value. The relative influence of the spill-out increases with decreasing cluster size, thus leading to a shift of the giant resonance to longer wavelengths (red-shift). Besides this crude picture, first quantum chemistry calculations have been done which explain the simple optical spectra of small alkali clusters in terms of the interference of several single particle excitations [21].

III. Experimental Section

a. Cluster production

Vaporization of metals by thermal heating is limited to low melting point materials. Sputtered clusters, on the other hand, are (at least vibrationally) excited. Here, we use a sputtering source to produce cluster ions for the study of the optical activity [22]: Clusters are generated by bombarding a silver target with 25 keV Xe^+. After acceleration to 1.8 keV and mass selection by a Wien filter, monodispersed cluster ion beams will be subject to experimental investigation.

One way to produce <u>cold</u> cluster ions and anions of high melting point materials is the vaporization of the metal by pulsed laser light, or by a pulsed arc discharge in a seeded beam source. The laser vaporization source LVS used in experiments discussed here is similar to the standard technique of laser vaporization of a fixed or rotating target rod of the desired metal mounted in a pulsed, high pressure supersonic nozzle. Different types of pulsed gas valves (piezo or magnetically driven) with backing pressures of $2 \ldots 15$ *bar* produce He or Ne pulses. Target vaporization is effected by (pulsed) excimer laser light (ArF, $XeCl$, 100 $mJ/pulse$), producing an intense plasma which is flushed by the carrier gas pulse through a channel of variable length into high vacuum. A number of channel designs (diameter $0.5 \ldots 2$ mm, length $5 \ldots 50$ mm) and nozzle geometries (cylindrical, conical) have been tested to optimize the intensity of (positively and negatively charged) clusters which emerge directly from the source without additional ionization.

In a refinement of the LVS, the vaporizing laser is replaced by a pulsed high-current arc between two electrodes in a pulse of carrier gas, which flushes the nascent plasma through a channel and conical nozzle into high vacuum [23,24]. The heart of this pulsed arc cluster ion source "PACIS" consists of a ceramic block containing two cylindrical electrodes with a spacing of about 2 mm and a diameter of 5 mm. A modified discharge circuit of an excimer laser or a home-built high power pulser serve to initiate the discharge. The operation of the PACIS strongly depends on the target material and desired cluster charge state as well as on the nozzle geometries. In some cases it is necessary to add a small mixing chamber between ceramic block and extender.

b. Photofragmentation spectroscopy

The desirable experimental method to study the optical activity is photo*absorption* in a beam of mass selected particles. However, the low intensities of such beams prevent a direct photoabsorption experiment. Instead, we use photo*fragmentation* to measure the optical activity as done in other publications [25,26]: Two collimators (1 mm diameter) confine the ion beam and define the interaction region. Here the clusters are irradiated collinearly with light from a pulsed excimer-pumped dye laser system (Lambda Physik LPX 110, FL2002). In order to extend the photon energy region, second harmonic generation (SHG) is achieved via frequency doubling in BB01 und BB02. Additionally, excimer laser lines (248 $nm, 308$ nm) are used. Behind the interaction region the ion beam is steered out of the laser beam axis with a static quadrupol deflector which acts as an energy selector field. Photodissociated clusters cannot reach the detector. Additionally, a grid can be used to repell cluster fragments. As an example, Fig. 2 gives the depletion spectrum of Ag_{21}^+ at a photon energy of 3.7 eV.

Care is taken to get optimal overlap of cluster and laser beams. A pyroelectrical detector serves to measure the laser intensity (typically $0.05 - 1.5$ mJ/cm^2 per pulse) before and after each run. Using this technique, depletion spectra are recorded for different photon

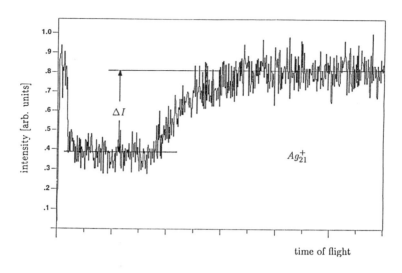

Fig. 2: *Depletion spectrum of Ag^+ at a photon energy of 3.7 eV. The length of the well and its overall shape reflect the degree of light and ion beam overlap, as well as details of the fragmentation dynamics. From [27].*

energies from $E = \hbar\omega = 2.3$ eV to 5.7 eV. Beer's law is then used to extract absolute photofragmentation cross sections.

c. Photoelectron spectroscopy

Principally, the standard methods as used in the photoelectron spectrocopy of molecules can also be applied to cluster [28]. One basic problem, however, is the need to work on charged species as the charge is necessary for a unique mass separation. Therefore it is desirable to have an electron spectrometer with a high angular acceptance. In 1986 Cheshnovsky et al. [29] and our group [30,31] independently developed electron TOF spectrometers with an acceptance of up to 4π sr basing on the principle of the magnetic bottle [32].

A schematic diagram of the Bielefeld apparatus is shown in Fig. 3. To obtain a high pressure gradient (source operation at 10^{-4} $mbar$, collision free acceleration of the cluster ions at 10^{-6} $mbar$, photodetachment at 10^{-8} $mbar$) the vacuum system consists of three differentially pumped subunits. The metal cluster beam source employs pulsed plasma formation either by a laser vapaorization source, or by the PACIS, see above. The cluster bunches are spatially and temporally focussed into the starting area of the TOF electron spectrometer where they are hit by the light pulse from, e.g., a nitrogen laser ($h\nu = 3.68$ eV, pulse lenght 10 ns). Detached electrons pass from a high into a low magnetic field and are detected after a drift distance of up to 3.5 m [31].

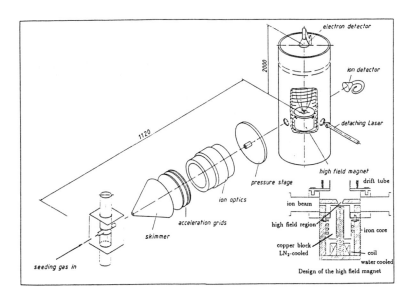

Fig. 3: *Schematic diagram of the photoelectron experiment. Metal cluster anions are produced in the pulsed arc cluster ion source (PACIS) and cooled by supersonic expansion (left). After pulse formation, electrons are photodetached from mass identified cluster bunches and undergo energy analysis in a magnetic-bottle TOF spectrometer (right). The shape of the high-field magnet is shown in the insert. Distances in mm. From ref. [33].*

The main action of the diverging magnetic field is an efficient collection of the photoelectrons. For a photoionized cluster at rest the degree of electron trajectory parallelization is given by the ratio of the high (B_h) to low (B_l) magnetic field. Angular momentum conservation determines the maximum single θ between the spectrometer axis and the helical motion in the weak field; from this the intrinsic energy resolution can by approximated by $\Delta E/E \approx B_l/B_h$. In contrast to Kruit et al. [32] who ionized (neutral) atoms at thermal velocities, our clusters are fairly fast (about 10^4 m/s) and thus give rise to noticeable Doppler broadening. In order to reduce this effect the cluster ion energy has to be kept as low as possible. Resulting electron TOF spectra are numerically transformed into energy spectra. We will plot the data as a function of electron binding energy, given by the photon energy minus the kinetic energy of the detached electron.

IV. Results and Discussion

Photoelectron spectra for a sequence of mass selected silver and thallium clusters are shown in Fig 4. Clearly, the evolution of the electronic structure of such small aggregates is not continuous, rather it happens in distinct steps. Addition of one atom may create considerably different electron spectra.

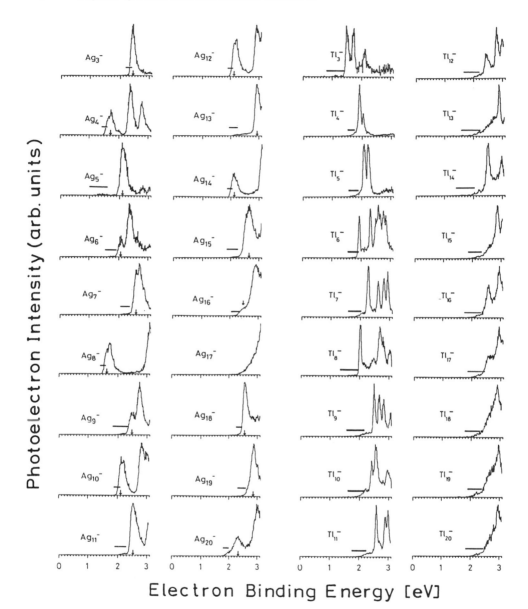

Fig. 4: *Photoelectron spectra of Ag_N^- and $T\ell_N^-$, $N = 3-20$, at a detachment energy of 3.68 eV (337 nm). After ref. [34,35].*

These spectra contain much information. A detailed discussion would be beyond the scope of this paper. A few aspects will be highlighted. First of all we may point out that the electronic band structure of the solid material is in general quite different from that of clusters. The monovalent metal Cu (or Ag) may serve as an example. Its surface UPS spectrum shows a flat contribution of the s/p-band close to threshold (i.e. around the Fermi edge), in contrast to the strong structures (caused by the quantum mechanical discrete level ordering of the delocalized electron states) in the corresponding cluster UPS spectra. This difference between the bulk and cluster UPS spectra is less evident in transition metals due to the high multiplicity and the more localized nature of the d states [31,34].

Furthermore, the detachment thresholds are expected to show a dependence on the particle radius (R). In a classical approach we have to take into account a correction to the image charge contribution due to the finite size of a cluster. For the electron affinity of a spherical cluster, a $\beta \cdot 1/R$-term has to be subtracted from the bulk work function [36]. The absolute value of β, however, can only be calculated taking into account quantum mechanical effects (exchange and correlation) [37]. Such density functional calculations reveal a higher order correction term which goes beyond the classical considerations. As an example, Fig. 5 compares results of the density functional calculations to measured threshold energies (see Fig. 4) for Ag_N. Due to the spherical jellium potential assumed in the calculations, only the mean threshold dependence is given. The measured data, however, show steps due to shell closings, especially at $N = 7$ (eight valence electrons), and at $N = 19$ (twenty valence electrons). At these sizes, the measured threshold values are close to the calculated ones.

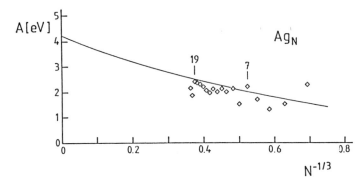

Fig. 5: *Results of density functional calculations (solid line, ref. [37]) compared to measured photodetachment energies (from Fig. 4) for Ag_N. Closed shell clusters are indicated by their numbers of atoms N.*

Obviously an extension of the calculations to deformed clusters is necessary. Nevertheless, the general trend is found as well as the high N limit. The curvature of the calculated behaviour vs. $N^{-1/3}$ depends on the assumed internuclear separation (Wigner Seitz radius r_s) and therefore on the material.

One further feature may be noted: Whenever a "shell" of delocalized electrons (e.g. in the jellium model) is closed and a spherical configuration of high stability is reached, the addition of one electron yields a particularly low ionization energy (since this electron has to be filled into a new "shell" of lower binding energy, c.f. Fig. 1). This behaviour can be seen directly between, e.g., Ag_7^- and Ag_8^-, or Ag_{19}^- and Ag_{20}^- (cf. Fig. 4). Ag_7^-, for example, contains eight valence electrons altogether, and thus produces a closed "shell" of electrons with p character. The addition of one more electron leads to occupation of the next higher state with lower binding energy E_B, and a new peak appears in the UPS spectrum (Ag_8^-, or Ag_{20}^-). The energy difference between this peak and the next one at higher binding energy could be called the HOMO-LUMO gap. Such gaps are also present at $n = 10, 12, 14$. Within the jellium model, these originate from deviations from spherical symmetry which break the degeneracy of angular momentum states and cause a splitting of energy levels.

The PE spectra at Tl_n^- (see Fig. 4, right side) display very rich and narrow structures. So far it is not clear whether thallium clusters can be calculated within the jellium model. The large s/p separation in Tl atoms might lead to an alkali-like behaviour. Indeed, the positions of the first lines in the PE spectra display an even/odd alternation, and a pronounced low E_B for $n = 8$, similar to the situation in Ag_n.

Now we aim for a "crosscheck" of the jellium model which – at least qualitatively – is suited to explain some features in the level structure of alkali and coinage metal clusters. Starting from the idea of a "delocalized electron gas" in small ideal metal particles, it should be possible to collectively excite the valence electrons vs. the positively charged core. With the photofragmentation spectroscopy described above, the optical activity of silver clusters is studied. Fig. 6 shows the photofragmentation cross section of Ag_9^+, Ag_{11}^+ and Ag_{21}^+. In the case of Ag_9^+ and Ag_{21}^+ (eight and 20 electron systems) single broad peaks dominate the spectra. As is indicated by the error bars, there might possibly be further structure which is not resolved. Additionally, a systematic trend is found for an increased "background" signal at higher photon energies. Nevertheless, there is no doubt about the existence and overall shapes of these collective resonances. In the case of Ag_9^+ the peak maximum is shifted towards higher energies (blue shift) with respect to Ag_{21}^+. The linewidth of Ag_{21}^+ appears to be smaller than that of Ag_9^+.

In contrast to the 8- and 20-electron systems, the optical spectrum of Ag_{11}^+ (10 valence electrons) shows two well-separated peaks at $3.44\ eV$ and $4.22\ eV$. Such splittings arise from dipole excitations in deformed clusters where – in the simplest case of an ellipsoid – two peaks are expected according to different resonance energies for the two main axes. Within the ellipsodial shell model [25,38] a 10-electron system undergoes prolate deformation which leads to an intensity distribution of 1/3 and 2/3 for the low and high energy peaks, respectively. For Ag_{11}^+ we find from the integrals of the two Lorentzians an intensity distribution of 0.24 and 0.76. The slight discrepancy might originate from shape isomerism due to the high internal energy.

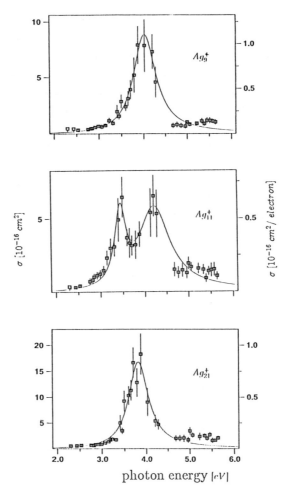

Fig. 6: *Absolute photofragmentation cross sections of silver cluster ions. Lorentz curves are used to fit the experimental values. Fit parameters for Ag_9^+ : $\hbar\omega_0 = 4.02eV$, $\Gamma = 0.62eV$, $\sigma = 8.84Å^2$; Ag_{21}^+ : $\hbar\omega_0 = 3.82eV$, $\Gamma = 0.56eV$, $\sigma = 16.79Å^2$. From ref. [27].*

From Mie theory [18], the resonance energy of a collective electron excitation is given by

$$\varepsilon(\omega_{Mie}) - 2 = 0, \tag{1}$$

$\varepsilon(\omega)$ being the dielectric function of the metal.

With the static polarizability $\alpha = Nr_s^3$, eq. 1 yields an expression for the resonance energy as function of N:

$$\omega_{Mie}^2 = \frac{Ne^2}{m\alpha}. \tag{2}$$

For a free electron metal, density functional theory reveals an electron spill-out, which leads to a red shift of the resonance energy with decreasing cluster size. Such a redshift is indeed observed for alkali clusters [25,26]. When using this approach for silver, however, resonances in the collective excitation are predicted at 2.61 eV and 2.72 eV, for Ag_9^+ and Ag_{21}^+, respectively. These values are far away from the experimental ones, cf. Fig. 6, which show a red shift. The deviation originates from the influence of the d-electrons in silver. Within the semiempirical approach of Apell et al. [39], the influence of the d-electrons can be accounted for by the use of the optical data of a silver surface. We find values of 3.93 eV for Ag_9^+ and 3.83 eV for Ag_{21}^+ in excellent agreement with the observed data. A complete discussion of this approach and of all investigated clusters up to Ag_{70}^+ will be given in a further publication [40].

In conclusion, experimental tools are developed for the detailed investigation of optical as well as electronic properties of mass-selected metal cluster. The optical activity of alkali and coinage metal clusters is basically concentrated in giant resonance-like absorption profiles which each shows one single peak in cases where the cluster is expected to be spherically shaped. For a cluster which is supposed to be deformed (e.g. oblate or prolate), the absorption profiles clearly split into two (or eventually three) components, being in qualitative accordance with collective oscillations of the valence electrons in different directions within the cluster.

The simple picture of delocalized valence electrons in a smooth potential turns out to – at least roughly – explain some features of the photoelectron spectra of Ag_N^-. Unlike photoemission of bulk silver, the clusters display a variety of discrete energy levels which usually do not smoothly develop with N. Nevertheless, for silver, the simple optical spectra as well as the complicated photoelectron spectra strongly support the picture of an "electron gas", confined to the volume of the cluster. Also for other materials, cf. thallium, richly structured photoelectron spectra show the distinct (and so far unexplained) character of the clusters. In a sense, small metal clusters may be regarded as "quantum dots" with completely new properties.

Acknowledgement

It is a pleasure to thank the coworkers devoted to the experimental studies, G. Ganteför, M. Gausa, R. Hector, L. Köller, and J. Tiggesbäumker. We are indebted to Professor Dr. H.O. Lutz for his continuous interest and support. The work has been supported by the Deutsche Forschungsgemeinschaft and the BMFT, which is also gratefully acknowledged.

References

[1] "Elemental and Molecular Clusters", Eds. G. Benedek, T.P. Martin, G. Pacchioni, Springer-Series in Materials Science 6, Springer, Berlin, Heidelberg, New York (1988);
"Small Particles and Inorganic Clusters", Eds. O. Echt, E. Recknagel, Springer, Berlin, Heidelberg, New York (1991)

[2] see e.g.: A. Szabo, N.S. Ostlund, "Modern Quantum Chemistry: Introduction to Advanced Electronic Structure Theory", MacMillian, New York (1982)

[3] "Semiempirical Methods of Electronic Structure Calculation", Ed. G.A. Siegal, Modern Theoretical Chemistry, Vol. 7 and 8, Plenum Press, New York (1977)

[4] I. Boustani, W. Pewestorf, P. Fantucci, V. Bonacić-Koutecký, and J. Koutecký, Phys. Rev. B 35, 9437 (1987);
B.K. Rao, S.N. Khanna, P. Jena, in "Physics and Chemistry of Small Clusters", NATO ASI Ser. B 158, p. 369, Plenum Press, New York (1987)

[5] J.L. Martins, J. Buttet, R. Car, Phys.Rev.Lett. 53, 655 (1984)

[6] W.D. Knight, K. Clemenger, W.A. de Heer, W.A. Saunders, M.Y. Chou, and M.L. Cohen, Phys.Rev.Lett. 52, 2141 (1984)

[7] T. Mayer-Kuckuk, "Kernphysik", Teubner, Stuttgart (1984)

[8] W. Ekardt, Phys.Rev. B 29, 1558 (1984);
Z. Penzar, W. Ekardt, Z.Phys. D 17, 69 (1990)

[9] M.E. Spina, M. Seidl, and M. Brack, Proc. of Symposium on Atomic and Surface Physics 90, Obertraun/Austria (1990)

[10] M.Y. Chou, M.L. Cohen, Phys.Lett. A 113, 420 (1986)

[11] M.Y. Chou, A. Cleland, M.L. Cohen, Solid State Commun. 52, 645 (1984)

[12] K. Clemenger, Phys.Rev. B 32, 1359 (1985)

[13] M. Manninen, Phys.Rev. B 34, 6886 (1986)

[14] L.C. Balbas, A. Rubio, J.A. Alonso, G. Borstel, Chem.Phys. 120, 239 (1988)

[15] W.A. de Heer, W.D. Knight, M.Y. Chou, M.L. Cohen, Solid State Phys. 40, 93 (1987)

[16] G.M. Pastor, J. Dorantes-Dávila, K.H. Bennemann, Chem.Phys.Lett. 148, 459 (1988)

[17] L. Skala, H. Müller, in "PDMS and Clusters". Eds. E.R. Hilf, F. Kammer, K. Wien, Lecture Notes in Physics 269, Springer, Berlin, Heidelberg, New York (1987)

[18] G. Mie, Ann.d.Physik 25, 377 (1908)

[19] W. Ekardt, Phys.Rev. B 31, 6360 (1985)

[20] W.A. de Heer, K. Selby, V. Kresin, J. Masui, M. Vollmer, A. Châtelain, W.D. Knight, Phys.Rev.Lett. 59, 1805 (1987)

[21] V. Bonačić-Koutecký, P. Fantucci, J. Koutecký, J.Chem.Phys. 93, 3802 (1990)

[22] J. Tiggesbäumker, L. Köller, H.O. Lutz, K.H. Meiwes-Broer, Chem.Phys.Lett., in press

[23] G. Ganteför, H.R. Siekmann, H.O. Lutz, K.H. Meiwes-Broer, Chem.Phys.Lett. 165, 293 (1990)

[24] H.R. Siekmann, Ch. Lüder, H.O. Lutz, K.H. Meiwes-Broer, Z.Phys. D 20, 417 (1991)

[25] K. Selby, M. Vollmer, J. Masui, V. Kresin, W.A. de Heer, W.D. Knight, Phys.Rev. B 40, 5417 (1989)

[26] C. Bréchignac, Ph. Cahuzac, F. Carlier, J. Leygnier, Chem.Phys.Lett. 164, 433 (1989);
H. Fallgren, T.P. Martin, Chem.Phys.Lett. 168, 233 (1990)

[27] J. Tiggesbäumker, L. Köller, H.O. Lutz, K.H. Meiwes-Broer, in "The Physics and Chemistry of Finite Systems: From Clusters to Crystals", Eds. P. Jena, S. Khanna, B. Rao, NATO ASI Ser. B (1992) (in press)

[28] D.G. Leopold, J.H. Ho, and W.C. Lineberger, J.Chem.Phys. 86, 1715 (1987)

[29] O. Cheshnovsky, P.J. Brucat, S. Yang, C.L. Pettiette, M.J. Craycraft, and R.E. Smalley, in "Physics and Chemistry of Small Clusters", NATO ASI Ser. B, Vol. 158, 1 (1987);
C.L. Pettiette, S.H. Yang, M.J. Craycraft, J. Conceicao, R.T. Laaksonen, O. Cheshnovsky, and R.E. Smalley, J.Chem.Phys. 88, 5377 (1988)

[30] G. Ganteför, K.H. Meiwes-Broer, and H.O. Lutz, Phys.Rev. A 37, 2716 (1988); and Z.Phys. D 9, 253 (1988)

[31] G. Ganteför, K.H. Meiwes-Broer, and H.O. Lutz, Faraday Discuss.Chem.Soc. 86, 197 (1988)

[32] P. Kruit, F.H. Read, J.Phys. E 16, 313 (1983)

[33] M. Gausa, J. Faehrmann, K.H. Meiwes-Broer, submitted

[34] G. Ganteför, M. Gausa, K.H. Meiwes-Broer, H.O. Lutz, J.Chem.Soc.Faraday Trans. 86, 2483 (1990)

[35] M. Gausa, G. Ganteför, H.O. Lutz, and K.H. Meiwes-Broer, Int.J.Mass Spectr.Ion Proc. 102, 227 (1990)

[36] J.M. Smith, Am.Inst.Aeronaut.Astronaut J. 3, 648 (1965);
D.M. Wood, Phys.Rev.Lett. 46, 749 (1981)

[37] M. Seidl, K.H. Meiwes-Broer, M. Brack, J.Chem.Phys. 95, 1295 (1991)

[38] K. Clemenger, Phys.Rev. B 32, 1359 (1985)

[39] P. Apell, Å. Ljungbert, Sol.St.Com. 44, 1367 (1982)

[40] J. Tiggesbäumker, L. Köller, K.H. Meiwes-Broer, to be published

Annihilation of positrons on large molecules

T J Murphy[*] and C M Surko

Physics Department, University of California, San Diego; La Jolla, California 92093 USA

ABSTRACT: We study the annihilation rate of positrons, confined in a Penning trap, in the presence of a variety of molecular species. The previously observed anomalously high annihilation rate of positrons on alkane molecules is shown to be a general property of large molecules. Measurements of the annihilation rate in alkanes, alkenes, and aromatic compounds suggest certain patterns in the annihilation rate per molecule.

1. INTRODUCTION

Measurements of the annihilation rate of positrons on alkanes (Paul and Saint-Pierre 1963, Heyland *et al.* 1982, Surko *et al.* 1988, Leventhal *et al.* 1990) show a rapid increase in this rate as a function of the size of the molecule. This increased rate has previously been attributed to resonance annihilation (Smith and Paul 1970) and to a bound state of the positron with the molecule (Surko *et al.* 1988, Leventhal *et al.* 1990).

Annihilation rates are generally expressed relative to the Dirac annihilation rate for annihilation of positrons on uncorrelated free electrons (Massey 1976), so that the annihilation rate is

$$\Gamma = \pi r_0^2 c \, n \, Z_{eff}, \tag{1}$$

where n is the number density of atoms or molecules, r_0 is the classical radius of the electron, and c is the speed of light. For free electrons, $Z_{eff}=1$. For atoms and molecules, Z_{eff} has been interpreted as the effective number of electrons per atom or molecule taking part in the annihilation process.

[*] Present address: Lawrence Livermore National Laboratory, Livermore, California 94550, USA.

However, measurements of Z_{eff} for alkane molecules (Surko *et al.* 1988, Leventhal *et al.* 1990) have yielded values of Z_{eff}/Z as great as 2 X 10^4. These large values of Z_{eff} were interpreted to indicate that the positron is attached to the molecule for times much longer than that of a simple collision, and they suggest that there can be binding between positrons and large hydrocarbons. Measurements have now been extended to include hexenes with one, two, or three double bonds, as well as other molecules, including benzene rings and some perfluorinated compounds for comparison. The results indicate that the anomalously high annihilation rate in large molecules is a general phenomenon, but they also illustrate distinct differences between the Z_{eff} in the hydrocarbons and the perfluorinated hydrocarbons.

2. DESCRIPTION OF THE EXPERIMENT

These measurements were performed in a positron trap (Surko *et al.* 1989) in a manner similar to previous measurements of the annihilation rate in hydrocarbons (Surko *et al.* 1988, Leventhal *et al.* 1990) and in xenon (Murphy and Surko 1990). Radial confinement of positrons in the trap is achieved using a magnetic field. Axial confinement is due to electrostatic potentials from an electrode structure. The trap allows the accumulation of positrons from a low-energy positron beam and the storage of room-temperature positrons for times up to several minutes.

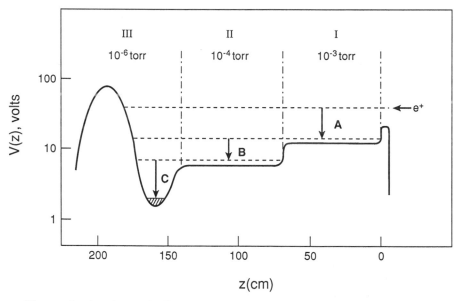

Figure 1: A schematic diagram of the trapping arrangement showing the background nitrogen pressure and the electrode potential, $V(z)$, as a function of distance along the trap. There is an applied magnetic field of 640 G in the z-direction. Positrons are trapped in region III.

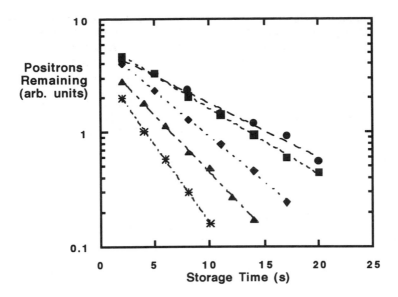

Figure 2: The dependence of positron lifetime on the presence of perfluorohexane. Shown is the number of positrons remaining after a 5 s fill and varying store times for C_6F_{14} pressures of ● 0.00, ■ 0.27, ◆ 0.61, ▲ 1.03, and ✳ 1.55 μtorr.

A principal advantage of using the positron trap is that substances which exist as liquids or solids at atmospheric pressure may be added in vapor form to the trap in quantities sufficient to measurably affect the annihilation rate in the trap. This has allowed measurement of Z_{eff} for alkanes up to $C_{16}H_{34}$, as well as for aromatic compounds which normally exist as solids and liquids. The use of low-pressure gases insures that one is studying a binary process involving one positron and one molecule (Charlton 1990). In addition, the positron temperature can be directly measured so that we have independent verification of the energy of the positrons involved in the interaction.

The positron trap is shown in Figure 1. Positrons from a 40-mCi ^{22}Na source are moderated using a 1-μm, single-crystal tungsten transmission moderator (Lynn *et al.* 1985, Zafar *et al.* 1988) to form a 40-eV positron beam. The positrons are then guided down a solenoid into the trap. The pressure of nitrogen gas in stage I is adjusted so that on average the positrons experience one inelastic (electronic excitation or ionization) collision upon traversing the length of stage I, thus becoming trapped within the electrode structure. Subsequent electronic excitation collisions result in the positrons becoming confined to stages II and III (in less than 1 ms) and finally in stage III (in a time of the order of 30 ms). The positrons then cool to room temperature (*i.e.* the temperature of the nitrogen gas) in a characteristic time of about 0.6 s.

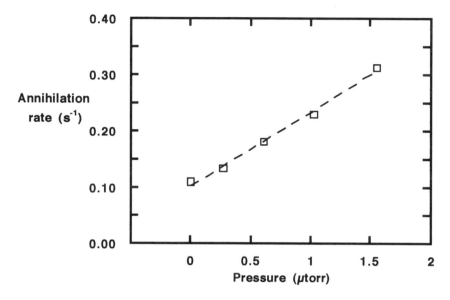

Figure 3: Annihilation rate vs. pressure of C_6F_{14} added to the trapping region (stage III) of the positron trap.

Various gases and vapors may be added directly to stage III in order to study the annihilation of positrons on these gases. Annihilation rates are determined by accumulating positrons for a fixed amount of time (filling) and then waiting a variable amount of time while the positrons annihilate (storing) before dumping the remaining positrons onto a metal plate and measuring the characteristic 511-keV annihilation radiation. This radiation is measured using a NaI(Tl) scintillator, the light pulse being proportional to the number of positrons remaining (Fig. 2). For a binary process, the annihilation rate is a linear function of molecule density (Fig. 3) and is given by

$$\Gamma = \Gamma_0 + AP_{N_2} + BP_X ,$$
(2)

where A and B are constants proportional to Z_{eff} of nitrogen and the molecule under study, respectively, P_{N_2} and P_X are the pressures of these gases, and Γ_0 is the annihilation rate on other large molecules present as contamination in the vacuum (Surko *et al.* 1988).

For the results presented here, the positrons are at room temperature. The molecules used in these studies were obtained from Aldrich Chemical Company, and were of the highest purity available for each. The liquids were degassed prior to use by repeated freezing and thawing under vacuum.

3. RESULTS AND DISCUSSION

The measured values of Z_{eff} are listed in Table 1 for alkenes, ring compounds, and fluorocarbons. These values are plotted in Fig. 4 *vs.* polarizability calculated using the method of Miller and Savchik (1979). The hydrocarbons, including those with double bonds and rings, fit a power law dependence between Z_{eff} and polarizability well (Surko *et al.* 1989). The fluorocarbons, however, fall short of this relation by a factor of 100.

Molecule	Formula	Z	Z_{eff}
Hexane	C_6H_{14}	50	98,000
1-Hexene	C_6H_{12}	48	185,000
trans 3-Hexene	C_6H_{12}	48	196,000
1,3-Hexadiene	C_6H_{10}	46	389,000
cis 2, *trans* 4-Hexadiene	C_6H_{10}	46	413,000
trans 2, *trans* 4-Hexadiene	C_6H_{10}	46	388,000
1,3,5-Hexatriene	C_6H_8	44	414,000
Benzene	C_6H_6	42	18,400
Toluene	C_7H_8	50	155,000
Naphthalene	$C_{10}H_8$	68	494,000
Anthracene	$C_{14}H_{10}$	94	4,330,000
Decahydronaphthalene*	$C_{10}H_{18}$	78	389,000
Perfluorohexane	C_6F_{14}	162	535
Hexafluorobenzene	C_6F_6	90	1200
Octafluorotoluene	C_7F_8	114	1240
Octafluoronaphthalene	$C_{10}F_8$	132	3080

Table 1: Measured values of Z_{eff} for hexenes, aromatics, and some perfluorocarbons.

The effect of adding one double bond to the six-carbon chain (hexane to hexene) is to approximately double Z_{eff}. Adding a second double bond (hexene to hexadiene) doubles the Z_{eff} again, but adding a third double bond (hexadiene to hexatriene) has no effect. Compared to the factor of 16,000 going from CH_4 to $C_{16}H_{34}$, these changes are not very significant. Apparently,

---

* From Leventhal *et al.* (1989).

the annihilation is not as sensitive to the presence of π-orbitals as to the size of the molecule.

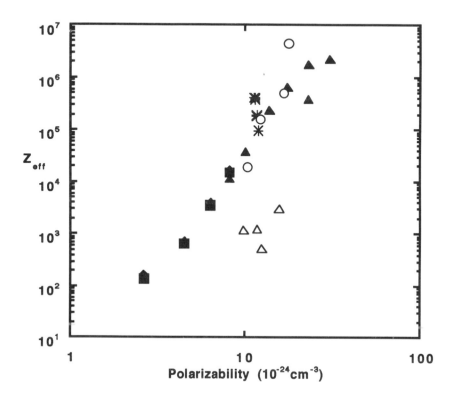

Figure 4: Measurements of Z_{eff} for a number hydrocarbons as a function of polarizability from this and previous work for alkanes, hexenes, aromatic compounds, and perfluorinated compounds: Present work: O rings, Δ perfluorocarbons, ✳ 6-carbon chains; Previous work:

● Wright *et al.* (1983), ■ Heyland *et al.* (1982), ◆ Paul and Saint-Pierre (1963), and ▲ Leventhal *et al.* (1990).

We note, however, that the Z_{eff} of benzene is a factor of ten lower than that of the hexenes. This may be due to the symmetry of benzene combined with its delocalized electrons. The addition of a methyl group to benzene to form toluene destroys this symmetry, and toluene has a Z_{eff} comparable to six- or seven-carbon chains.

Naphthalene and anthracene, which are double and triple ring molecules, have the delocalized orbitals that benzene has, but lack the symmetry of benzene. The Z_{eff} of naphthalene is 27 times that of benzene. That of anthracene is 235 time that of benzene. Since the Z_{eff} of naphthalene is very close to that of decahydronaphthalene (Leventhal *et al.* 1990), which lacks delocalized electrons, it appears that the symmetry is of greater importance in setting the lower value for benzene.

These trends do not hold if one considers the perfluorinated versions of these hydrocarbon compounds. Immediately noticeable is the much lower value of Z_{eff} for the perfluorinated compounds. For example, perfluorohexane has a Z_{eff} 200 times smaller than that of hexane, despite the fact that perfluorohexane contains over three times the number of electrons. And while hexane has a Z_{eff} five times larger than that of benzene, perfluorohexane has a Z_{eff} half that of hexafluorobenzene. The Z_{eff} of octafluorotoluene is nearly identical to that of hexafluorobenzene, and octafluoronaphthalene exceeds hexafluorobenzene by merely a factor of 2.6. Apparently the parameters which determine the annihilation rates in hydrocarbons and perfluorocarbons are different. This may be due to the very high electronegativity of fluorine.

4. CONCLUSIONS

The anomalously-high annihilation rate of positrons in the presence of alkane molecules has been found to be typical of the behaviour observed for a wide variety of large molecules. This rate is less sensitive to the presence of double bonds than to the size of the molecule, but it is greatly reduced for benzene and for perfluorinated molecules.

ACKNOWLEDGMENTS

We thank K. Brueckner, K. Fagerquist, J. Onuchic, S. L. McCall, and M. Charlton for stimulating discussions. This work is supported by the Office of Naval Research.

REFERENCES

Charlton M 1990, *Comments At. Mol. Phys.* **24**, 53

Heyland G R, Charlton M, Griffith T C, and Wright G L 1982, *Can. J. Phys.* **60**, 503

Leventhal M, Passner A, and Surko C M 1990, Annihilation in Gases and Galaxies, NASA Conference Pub. No. 3058 (National Aeronautics and Space Administration, Washington, D. C.)

Lynn K G, Nielsen B, and Quateman J H 1985, *Appl. Phys. Lett.* **47**, 239

Massey Sir Harrie 1976, *Phys. Today* **29**, No. 3, p. 42

Miller K J and Savchik J A 1979, *J. Am. Chem. Soc.* **101**, 7201

Murphy T J and Surko C M 1990, *J. Phys. B: At. Mol. Opt. Phys.* **23**, L727

Paul D A L and Saint-Pierre L 1963, *Phys. Rev. Lett.* **11**, 493

Smith P M and Paul D A L 1970, *Can. J. Phys.* **48**, 2984

Surko C M, Leventhal M, Crane W S, Passner A, Wysocki F J, Murphy T J, Strachan J D, and Rowan W S 1986, *Rev. Sci. Instrum.* **57**, 1862

Surko C M, Passner A, Leventhal M, and Wysocki F J 1988, *Phys. Rev. Lett.* **61**, 1831

Surko C M, Leventhal M, Passner A 1989, *Phys. Rev. Lett.* **62**, 901

Wright G L, Charlton M, Clark G, Griffith T C, and Heyland G R 1983, *J. Phys. B: At. Mol. Phys.* **16**, 4065

Zafar N, Chevallier J, Jacobsen F M, Charlton M, and Laricchia G 1988, *Appl. Phys. Lett.* **A47**, 409

Ionization of atoms by particle and antiparticle impact

H Knudsen

Institute of Physics, University of Aarhus, DK-8000 Aarhus C, Denmark

During the last decade, high-quality beams of low-energy antimatter particles have become available. This has given us the opportunity of measuring how reaction cross sections for charged-particle - atom collisions are affected by a change of the sign of the projectile charge or by a change of the projectile mass, while all other important parameters are fixed. The abundant experimental information obtained in this way has spurred a great deal of theoretical effort to explain the observed effects. In this paper, we present examples of the new experimental information on the ionization process.

1. INTRODUCTION

The process which leads to the liberation of target electrons, following the impact of a charged particle on an atom or a molecule, is one of the most important and fundamental phenomena studied in the field of atomic collisions. The interest in the ionization process goes back many decades, and it has been studied intensely since then. The recent availability of low-energy monoenergetic beams of antiparticles has made possible a comparison of data obtained with equivelocity p^+, p^-, e^+, and e^-, and hence a substantial amount of new information on the collision dynamics as well as on the influence of electron correlation on the process, has become available. This new experimental information, together with the theoretical advances spurred by it has recently been reviewed by Schultz et al (1991) and Knudsen and Reading (1991). In these papers can be found a thorough description of the new insight into the processes of multiple ionization and energy loss which have been acquired through measurements with antiparticles. Here we shall concentrate on the basic process of single ionization and further, restrict ourselves to its total cross section σ^+ which we define as the cross section for the release of one *free* electron from the target.

2. EXPERIMENTAL TECHNIQUE

It is outside the scope of this paper to discuss in detail the modern developments in the techniques by which slow, monoenergetic antiparticle beams are created and used for atomic collision experiments. A thorough description of this subject can be found in Knudsen and Reading (1991).

The most versatile antiparticle beams consist of positrons or antiprotons, although experiments have been performed with other particles such as muons and pions. In figure 1 is shown the magnetic positron beamline at Aarhus University, together with the experimental setup used to measure single- and double-ionization cross sections (Knudsen *et al* 1990). Here a $_{22}$Na source emits β^+ particles which are moderated in tungsten to give a slow, monoenergetic beam of $> 10^4$ e$^+$ sec^{-1} which is magnetically guided through the scattering gas cell. Ions created here due to the e$^+$ impact are extracted onto a ceratron detector, the signal of which, together with the signal from the beam detecting MCP, defines the time-of-flight of the created ions and hence their charge. The same scattering chamber has been used to measure ionization cross sections for impact of e$^-$ and p$^+$ coming from conventional sources, and for p$^-$ which were delivered from the LEAR at CERN (see, eg, Andersen *et al* (1990a,b)).

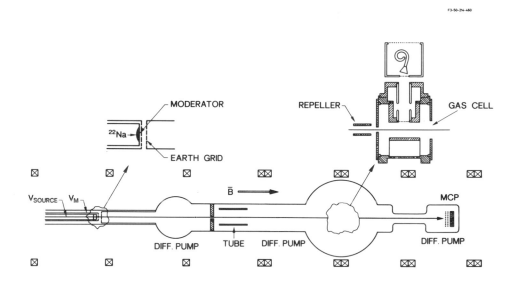

F3-50-214-460

Fig. 1. The experimental setup used by Knudsen *et al* (1990) to measure ionization by positron impact.

3. PROTON/ANTIPROTON MEASUREMENTS FOR LIGHT TARGETS

If we regard the impact of a point-like particle of charge q and velocity v_p on a light atom, in which the typical electron velocity is v_e, then at *very* high (but nonrelativistic) particle speed ($v_p > > v_e$), the interaction with the atomic electrons is short-lasting and, if q is low, also weak. In this case, the ionization can be described adequately by first-order perturbation theory. The first Born theory (Bethe 1930) gives

$$\sigma^+ \propto q^2 v_p^{-2} \ln v_p \tag{1}$$

and consequently, the ionization cross section does not depend on the sign of the projectile charge.

At somewhat lower particle speeds, $(v_p > v_e)$, we have to take into account that the interaction takes a longer time, and therefore the target-electron wave function may be modified during the first part of the collision: The density of electrons is relocated away from the target nucleus and towards the projectile if, for example, the latter is positive. It has been suggested that this modification of the target wave function should lead to an initial departure from the first Born result in such a way that the proton cross section becomes a few per cent smaller than the antiproton cross section as v_p decreases (Ermolaev 1990). This effect has not yet been seen experimentally. However, at somewhat lower v_p, it is clear that the *polarization* of the target-electron density towards a positive - and away from a negative projectile - leads to

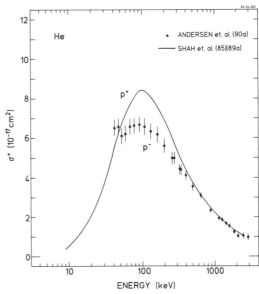

Figure 2.

$\sigma^+(p^+) > \sigma^+(p^-)$. This can be seen in figure 2, where the experimental single-ionization cross section for proton (Shah *et al* 1985, 1989) and antiproton (Andersen *et al* 1990a) impact on He is shown as an example. The difference between $\sigma^+(p^+)$ and $\sigma^+(p^-)$ has been reproduced by a number of calculations, see, eg, Andersen *et al* (1990a). It stems from the distant collisions which are dominating in the ionization at large v_p.

At medium values of the projectile speed $(v_p \sim v_e)$, the ionization cross section reaches its maximum value. Here, a number of mechanisms leads to a decrease of the proton cross section relative to that for antiprotons. They are all connected with close encounters which dominate at such impact velocities: Firstly, the advent of the projectile inside the orbit of the target electrons creates an *increased/decreased binding* of the electrons, which leads to an increase/decrease of σ^+, depending on whether the projectile is negative or positive. Secondly, in slow, close collisions, equivelocity projectiles of opposite charge do not follow the same trajectory: A proton is deflected away from the central, high-density part of the electron cloud, and this decreases σ^+ relative to that for negative particles. Furthermore, a positive particle is decelerated in the field of the target nucleus. At rather low v_p, where σ^+ decreases with decreasing v_p, this also leads to a smaller value of σ^+ relative to that for a negative particle. The combination of these mechanisms gives the so-called *Coulomb trajectory effects*. Finally, for $v_p \sim v_e$, *electron capture* becomes likely in the case of positive projectiles. This alternative reaction channel subtracts probability from the pure ionization channel and hence lowers the value of σ^+ for positive particles. All these mechanisms reduce σ^+ for positive particles relative to that for negative particles and result in the cross-over of $\sigma^+(p^+)$ and $\sigma^+(p^-)$ which can be seen in figure 2 at projectile velocities below the maximum in $\sigma^+(p^+)$.

At low projectile velocities ($v_p < < v_e$), the projectile-target-electron interaction becomes adiabatic and can be described in a quasimolecular picture. According to Kimura and Inokuti (1988), the evolvement of the quasimolecule leaves very little chance for ionization to happen in the case of proton impact (see figure 2). On the contrary, for antiproton impact there are no bound quasimolecular states if the projectile approaches the target nucleus closer than a certain critical distance which is in the order of magnitude of the electron-orbital radius. Therefore, at $v_p < < v_e$, it is expected that $\sigma^+(p^+) < < \sigma^+(p^-)$ and that $\sigma^+(p^-)$ stays fairly constant with decreasing v_p.

4. ELECTRON/POSITRON MEASUREMENTS FOR LIGHT TARGETS

During the last few years, a rather substantial amount of data on the total cross section for single ionization of atoms and molecules by positron impact has been obtained. We show as an example in figure 3 the results of Fromme *et al* (1986), Diana *et al* (1990), and Knudsen *et al* (1990) for the helium target. Also shown for comparison are the electron-impact data of Krishnakumar and Srivastava (1988) and Montague *et al* (1984). The arrow indicates where the projectile energy E is equal to the target-ionization potential I^+.

In general, for all targets investigated, the positron data agree well with the data for electron impact at the highest ($v_p > > v_e$) projectile velocities. This is as expected for this region, where perturbation theory should apply, and where consequently equation (1) should be valid.

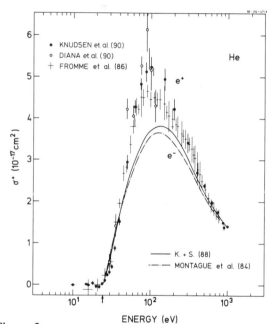

Figure 3.

At lower velocities, we see the effect of target polarization which makes $\sigma^+(e^+) > \sigma^+(e^-)$. A confirmation of the validity of the polarization picture is obtained if we plot $\sigma^+(e^+)/\sigma^+(e^-)$ for example at $E = 10\ I^+$, as a function of the target-polarizability parameter, as is done in figure 4 for all targets where such data are available. (For references, see Knudsen *et al* (1990) and Spicher *et al* (1990). As expected in this model, the larger the polarizability, the larger the ratio between the positron and the electron cross section.

At even lower impact velocities, where v_p approaches v_e, both $\sigma^+(e^+)$ and $\sigma^+(e^-)$ decrease steeply towards zero. This is due to the relatively small kinetic energy carried by these light particles, which reduces the accessible phase-space for final ionization states. For all targets investigated experimentally, there is a clear indication that at such velocities, the positron cross section decreases faster with decreasing energy than does the electron cross section. This is probably due to the same mechanisms (increased/decreased binding, Coulomb trajectory, and (for e^+ projectiles) electron capture)) that make the proton cross section become much smaller than that for antiprotons. However, the collision dynamics are considerably more complicated for light-particle impact in this velocity range than for heavy-particle impact. This is due to the fact that in the former case, the projectile may lose a substantial part

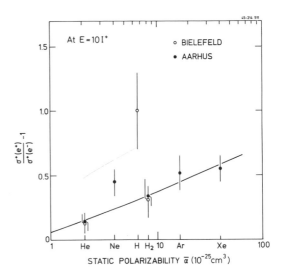

Figure 4.

of its energy in an ionization event, and therefore it typically leaves the collision with a velocity of the same magnitude as the ionized electron. This means that *mutual screening* of the outgoing particles is crucial for the ionization probability (Campeanu *et al* 1987), and that *electron exchange* is important in the case of electron impact.

For projectile energies just above the ionization threshold of the target, the excess energy $\epsilon = E - I^+$ is so small that only a very limited phase space is available for ionization. Wannier (1953) pointed out that for single ionization by impact of electrons on neutral atoms, the two active electrons will move away from the residual ion along a potential ridge at $r_1 = r_2$ and $\theta_{12} = \pi$, where r_i is the distance from the i'th electron to the ion, and θ_{12} the relative emission angle of the two electrons. Using a classical picture, which is justified because of the large value of $r_1 + r_2$ as the electrons escape, Wannier found that the distribution of electron energy is flat between zero and ϵ, that θ_{12} is near π, and that

$$\sigma^+ \propto \epsilon^\nu \qquad (2)$$

with $\nu = 1.127$. Later, these results were also obtained by a number of semiclassical calculations (see, e.g., Klar (1981a, 1984) and Grujic (1982)) as well as by Classical Trajectory Monte-Carlo (CTMC) calculations (Wetmore and Olson 1986).

The predictions of the Wannier theory for two-electron escape have been confirmed by experiments, both on electron-impact ionization of light atoms (see, eg, Cvejanovic and Read (1974)) and of inner shells of heavy atoms (see, eg, Hippler *et al* (1983)) as well as on photon-induced double ionization (Donahue *et al* (1982) and Kossmann *et al* (1988)). Generally, the predictions were found to be valid up to a few eV of excess energy, but the range of validity stretches up to several tens of eV when the residual ion possesses many passive electrons. According to Klar (1981a), this is due to the screening imposed by these electrons.

The Wannier theory for threshold ionization has been extended to positron impact on neutral targets by Klar (1981b, 1984) and by Grujic (1982). In their model, the electron and positron escape both in the same direction, with the electron trailing the positron so that $r_+ \sim 2.15\ r_-$. They find that the energy of an outgoing particle is still evenly distributed in the interval between zero and ϵ, and that

$$\sigma^+ \propto \epsilon^{2.65} \tag{3}$$

which should be valid in a larger ϵ range than that which applies for the electron-impact case (Grujic 1981).

The foundations of the Wannier theory for positron impact are not as solid as those for electron impact. There is a number of reasons for this: (i) For positron impact, in addition to the ionization- and the elastic/inelastic scattering channels, there is the very strong positronium-formation channel. This greatly complicates the theoretical approach. It may, for example, be necessary to take into account intermediate Ps states. (ii) For positron impact, the two light particles are much closer to one another in the final state than is the case for electron impact, this throwing doubt upon the non-over-lapping, classical orbits of the theory. (iii) The high degree of spatial symmetry inherent in the two-electron escape does not exist for positron impact; here neither the center of mass nor the center of charge coincides with the position of the residual ion.

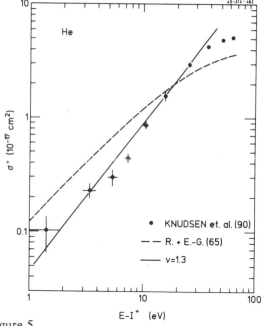

Figure 5

The collision system $e^+ + H$ has been investigated via CTMC calculations by Dimitrijevic and Grujic (1983) in the energy range $0.7 < \epsilon < 1.25$ eV. They point out that the assumption that the residual ion resides at the center-of-mass (which was employed by Klar (1981b, 1984) and Grujic (1982) in their extension of the Wannier theory to positron impact) may lead to large errors in the threshold law. They find that if the hydrogen nucleus is assumed infinitely heavy, then $v = 2.43$, in agreement with equation (3), but taking into account the proper recoil of the residual ion, they find $v = 1.64$. The same system was the subject of CTMC calculations by Wetmore and Olson (1986). In the energy range

5.5 < ε < 7.5 eV, they found the cross section σ^+ to be described by equation (2) with $\nu = 3$, in much better agreement with equation (3). It should be noted, however, that these two CTMC calculations do not cover the same energy range.

In figure 5, we show the experimental results of Knudsen *et al* (1990) for σ^+ for positron impact on He near threshold. Also shown are the experimental results for electron impact by Rapp and Englander-Golden (1965). At the lower energies shown in the figure, this latter cross section follows closely the threshold law, equation (2), with the Wannier exponent $\nu = 1.127$. As can be seen, the positron-impact data do not agree with the Wannier prediction $\nu = 2.65$, but rather follows a power law with $\nu \approx 1.3$ (solid line) - at least in the range 1 < ε < 20 eV. The same result was obtained by Knudsen *et al* (1990) for positron impact on H_2 and Ne.

Clearly, the experimental positron data are not accurate enough to allow with certainty an extraction of the threshold power law. However, there is an indication that either the Wannier theory for positron impact is not correct, or that it is valid only at smaller ε than was previously assumed. The positron data agree rather well with the energy dependence of σ^+ predicted by the CTMC calculation of Dimitrijevic and Grujic (1983) for 0.7 < ε < 1.25 eV, but not with the corresponding CTMC result obtained for 5.5 < ε < 7.5 eV by Wetmore and Olson(1986). In view of these findings, it seems worthwhile to readdress the question of the near-threshold behaviour of the ionization cross section for positron impact on atoms and molecules.

5. ELECTRON/POSITRON INNER-SHELL IONIZATION

Already in 1964, the first experimental measurements of inner-shell ionization by electron and positron impact was published (Hansen *et al* 1964). For the large impact energies of that work, no difference between the K-shell cross sections $\sigma_K(e^+)$ and $\sigma_K(e^-)$ could be discerned. Recently, similar measurements have been performed for Cu and Ag targets at impact energies approaching the K-shell-ionization threshold I_K. In figure 6, these data are shown, plotted as a function of E/I_K (Ito *et al* 1980, Ebel *et al* 1989, Shultz and Campbell 1985). Also shown are the data of Knudsen *et al* (1990) and of Krishnakumar and Srivastava (1988), for the He target. At large E, there is, as expected, no difference between the cross sections for positron and electron impact because of the perturbative nature of such encounters. At lower energies, however, K-shell ionization of heavy targets

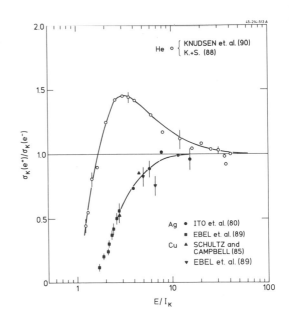

Figure 6.

happens mainly in close collisions, and the projectile-energy loss is relatively large. It was shown by Ito *et al* (1980) and Ebel *et al* (1989) that the decrease towards zero as E approaches I_K of $\sigma_K(e^+)/\sigma_K(e^-)$ can be reproduced by a calculation which takes into account the effects of exchange and of the Coulomb trajectory of the projectiles. Contrary to the behaviour of the heavy-target data, the He-target results show an initial increase above unity as the projectile energy decreases. We understand this as being due to target polarization which is important in the distant collisions which give a substantial contribution to the ionization cross section for the light target.

6. COMPARISON OF σ^+ FOR IMPACT OF HEAVY AND LIGHT PROJECTILES

If the single ionization cross sections measured for impact of equivelocity positrons and protons (or equivelocity electrons and antiprotons) are compared, then the effects of the large mass difference between the two kinds of particle can be studied. Figure 7 shows such data obtained from the measurements of Knudsen *et al* (1990), Shah *et al* (1985, 1989), Andersen *et al* (1990a), and Krishnakumar and Srivastava (1988). For the largest projectile velocities, there is no difference between the cross sections for light and heavy projectiles. This agrees with the expectation from the first Born approximation (equation (1)) which is independent of the projectile mass. Below projectile velocities corresponding to 1 MeV/amu, however, a general decrease of $\sigma^+(e^+)/\sigma^+(p^+)$ and of $\sigma^+(e^-)/\sigma^+(p^-)$ with decreasing projectile velocity sets in. This is due mainly to the low mass of the light particles: For projectile velocities, where the heavy particles possess kinetic energies in the order of 100 keV, the kinetic energy of an equivelocity light particle is near the target-ionization threshold, and the

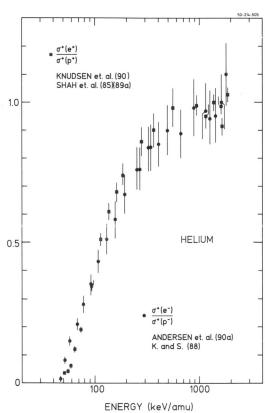

Figure 7.

available phase space for final ionization states for light-particle impact is considerably reduced. As this reduction is given by the mass of the light projectiles, but is independent of their charge, it should be the same for positrons and electrons. As can be seen in the figure, the overall trend of the data supports this picture.

On top of this effect, there are more subtle differences. As an example, let us note that the effects of the Coulomb trajectory are enhanced for positrons and electrons due to their small mass. The result of this fact can be seen in figure 8, where the data of figure

7 have been replotted as $\sigma^+(p^+)/\sigma^+(p^-)$ and $\sigma^+(e^+)/\sigma^+(e^-)$. For impact energies between 100 and 1000 keV/amu, we observe the effect of target polarization which makes $\sigma^+(p^+)>\sigma^+(p^-)$. Stemming from distant collisions, this mechanism should give the same relative difference between the cross sections for positive and negative projectile impact, irrespective of the projectile mass. The extra difference observed between $\sigma^+(e^+)$ and $\sigma^+(e^-)$ in this velocity range is probably due to the enhanced acceleration/ deceleration of the light particles during the encounters: The deceleration of the e^+ in the field for the positive target nucleus leads to a relatively larger σ^+ because, for the projectile velocities that we are concerned with

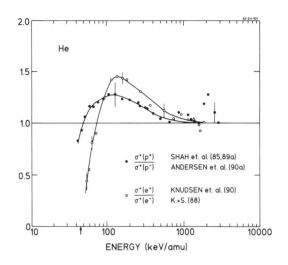

Figure 8.

here, σ^+ increases with decreasing velocity (see figure 3). This model actually agrees quantitatively with the data of figure 8 because, if we extract $\sigma^+(E-\Delta E)/\sigma^+(E+\Delta E)$ from figure 3, using the realistic value $\Delta E = 15$ eV, then this factor reproduces the difference between $\sigma^+(e^+)/\sigma^+(e^-)$ and $\sigma^+(p^+)/\sigma^+(p^-)$.

6. CONCLUSION

We have presented some of the new insight into the process of single ionization by charged-particle impact that has been achieved via measurements where the recently developed, low-energy antiparticle beams were employed.

REFERENCES

Andersen L H *et al* 1990a *Phys.Rev.A* 41 6536
Andersen L H *et al* 1990b *J.Phys.B* 23 L395
Bethe H A 1930 *Ann.Physik* 5 325
Campeanu R I, McEachran R P and Stauffer A D 1987 *J.Phys.B* 20 1635
Cvejanovic S and Read F H 1974 *J.Phys.B* 7 1841
Diana L M *et al*: *Annihilation in Gases and Galaxies* ed R J Drachmann (NASA Scient.Tech.Info.Div.) p 121
Dimitrijevec M S and Grujic P 1983 *J.Phys.B* 16 297
Donahue J B *et al* 1982 *Phys.Rev.Lett.* 48 1538
Ebel F *et al* 1989 *Positron Annihilation* eds L Dorikens-Vanpraaet, M Dorikens and D Segers (Singapore: World Scientific) p 15
Ermolaev A 1990 *Phys.Lett.A* 149 151
Fromme D, Kruse G, Raith W and Sinapius G 1986 *Phys.Rev.Lett.* 57 3031
Grujic P 1982 *J.Phys.B* 15 1913
Hansen H, Weigmann H and Flammersfelld A 1964 *Nucl.Phys.* 58 241

Hippler R *et al* 1983 *J.Phys.B* **16** L617

Ito S, Shimizu S, Kawaratani T and Kubota K 1980 *Phys.Rev.A* **22** 407

Kimura M and Inokuti M 1988 *Phys.Rev.A* **38** 3801

Klar H 1981a *J.Phys.B* **14** 3255Kimura M and Inokuti M 1988 *Phys.Rev.A* **38** 3801

Klar H 1981b *J.Phys.B* **14** 4165

Klar H 1984 *Electronic and Atomic Collisions* eds J Eichler, I V Hertel and N Stolterfoht
    (Amsterdam: Elsevier Science Publ.) p 767

Knudsen H, Brun-Nielsen L, Charlton M and Poulsen M R 1990 *J. Phys.B* **23** 3955

Knudsen H and Reading J F 1991 (submitted to *Phys.Repts.*)

Kossmann H, Schmidt V and Andersen T 1988 *Phys.Rev.Lett.* **60** 1266

Krishnakumar E and Srivastava S K 1988 *J.Phys.B* **21** 1055

Montague R G, Haarrison M F A and Schmidt A C H 1984 *J.Phys.B* **17** 3295

Rapp D and Englander-Golden P 1965 *J.Chem.Phys.* **43** 1464

Schultz D R, Olson R E and Reinhold C O 1991 *J.Phys.B* **24** 521

Schultz P J and Campbell J L 1985 *Phys.Lett.* **112A** 316

Shah M B and Gilbody H B 1985 *J.Phys.B* **18** 899

Shah M B, McCallion P and Gilbody H B 1989 *J.Phys.B* **22** 3037

Spicher G, Olson B, Raith W, Sinapius G and Sperber W 1990 *Phys.Rev.Lett.* **64** 1019

Wannier G H 1953 *Phys.Rev.* **90** 817

Wetmore A E and Olson R E 1986 *Phys.Rev.A* **34** 2822

High resolution studies of laser stimulated radiative recombination

F.B. Yousif[1], P. Van der Donk[1], Z. Kucherovsky[1], J. Reis[1],
E. Brannen[1], J.B.A. Mitchell[1], and T.J. Morgan[2].

[1] Dept. of Physics, University of Western Ontario, London,
Ontario, Canada N6A 3K7
[2] Physics Dept., Wesleyan University, Middletown, CT 06457

ABSTRACT.
Laser stimulated radiative recombination to the n=11 and
n=12 levels of atomic hydrogen has been demonstrated using
a merged electron-ion beam apparatus and a CO_2 laser.

INTRODUCTION.

When an electron collides with an atomic ion (charge q), it
can be captured by dropping down from the ionization continuum
into a bound state of the neutral atom (or ion of charge q-1).
To do this however it must lose energy and this can be achieved
by the emission of a photon. This process, known as radiative
recombination (RR) is actually very improbable as the
characteristic time for radiation emission (10^{-9}s) is much longer
than the collision time (10^{-15}s). Typical RR rate coefficients
are of the order of $10^{-11}cm^3s^{-1}$. The cross section for capture
to a specific n state of a bare nucleus can be obtained from the
formula (Bethe and Salpeter, 1977):

$$\sigma - 2.1x10^{-22}cm^2 \frac{E_0^2}{nE_e(E_0+n^2E_e)} \ldots (1)$$

where E_0 is the ionization potential of the resulting atom (or
ion), and E_e is the electron kinetic energy, (both in eV).
If the ion has attached electrons, then the free electron may
be captured through resonant dielectronic recombination (DER) if
it has the correct energy. If the electron and gas densities in
the vicinity of the ion are low then the likelihood of
recombination stabilization by collision with a third body is
small. If the ion is a bare nucleus the only means of
recombination is then RR. Neumann et al.(1983) have shown that
the rate of this process may be enhanced by stimulating the
stabilization transition by photon impact. This can be achieved
in principle by performing the electron-ion collision in the

photon field produced by a laser.

$$h\nu + e + A^+ \rightarrow A + 2h\nu$$

They showed that the recombination cross section can be enhanced by a factor G given by:

$$G = \frac{\sigma_{stim}}{\sigma_{spon}} = \frac{Pc^2}{F\Delta\nu 8\pi h\nu^3} \ldots (2)$$

where P is the laser power in watts, c, the velocity of light, F, the cross sectional area of the laser beam, ν the frequency of the laser photons (and of the photons emitted). $\Delta\nu$ is a measure of the match between the bandwidth of the laser, the electron energy spread and the width of the atomic state into which the recombination proceeds. The more closely these three quantities match, the more efficient the recombination process will be. In practice, $\Delta\nu$ will be dominated by the electron energy spread. The velocity spread of the electrons is related to $\Delta\nu$ by:

$$\Delta\nu = \frac{m\nu_e}{h}\Delta\nu_e \ldots (3)$$

It can be deduced from equation 2 that the gain is proportional to the principle quantum number (of the state into which the capture occurs) squared. Thus there are advantages in using a long wavelength laser to perform the stimulation to a high n state. A carbon dioxide laser can be used to populate states n≥11. (If doppler tuning was employed, n=10 could also be populated). This is particularly fortunate in our case since our electron-ion merged beam apparatus (Fig.1.) operates in a dc mode and high cw powers can be produced using a CO_2 laser. The low collision energy, high energy resolution capabilities of this apparatus have been demonstrated in previous measurements of dissociative recombination (Van der Donk et al., 1991). One of the major obstacles to the measurement of radiative recombination in this device is the background noise due to neutral hydrogen atoms being produced through charge exchange between the proton beam and the background gas in the apparatus. This can be minimized by

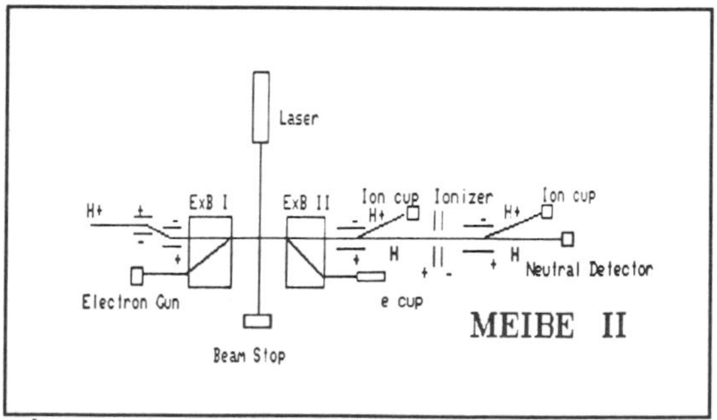

Figure 1. Schematic of the merged-beams apparatus showing the laser and field-ionizer arrangement.

maintaining the collision region at ultra-high vacuum (10^{-10} Torr) and by using a high ion energy. At 400keV, the charge exchange cross section for protons on hydrogen, (the major gas in the apparatus at these pressures) is $10^{-20} cm^2$. Neutrals produced from recombination are distinguished from background produced neutrals by modulating the electron beam and counting in and out of phase with the modulation. The true recombination signal is then the difference in the two count rates so produced.   Even with these provisions, this would still be a difficult measurement given the small size of the recombination cross section. An innovation that has greatly improved the signal to noise ratio in this experiment is the use of a field ionizer to detect the recombination products. This device is capable of generating an axial electric field of up to 150 KV/cm and this in turn can field ionize excited atoms with $n \geq 7$. The advantage of this is that, since the laser stimulated recombination process specifically populates a single energy level, only atoms in this state need be detected. The charge exchange reaction cross section leading to the production of atoms in state n is proportional to n^{-3} and so the number of neutral atoms so produced is less than 1/1000th the total charge exchange background. It was found in the experiment that the measured background rate was on average about 5 counts per minute. Most of this arose from collisions with surfaces during the passage of the neutrals through the ionizer. The signal count rate was also of this order of magnitude and typical measurement times of 24 hours per data point were necessary in order to achieve statistical accuracy.

EXPERIMENTAL APPARATUS.

The merged electron-ion beam apparatus used in this measurement (fig 1) has been described in detail elsewhere and so only a brief overview will be given here. The proton beam is produced using an rf source mounted in the terminal of a 400 KeV Van de Graaff accelerator. After mass analysis it is directed into the interaction region. The electron beam is produced using an indirectly heated cathode and magnetically confined by an axial field of 25 Gauss produced by Helmholz coils. This beam is merged with the ion beam using a trochoidal analyzer. After a distance of 86mm, it is demerged and collected in a Faraday cup. Within the interaction region, the overlap of the electron and ion beams is measured at two points using a pair of beam scanners. This allows the effective interaction area f to be determined. After leaving the interaction region, the ion beam passes through a pair of electrostatic deflection plates which deflect it into a second Faraday cup. The neutrals formed in the interaction region pass on towards the ionizer, operating with an axial electric field E_{ax}. Atoms in energy states with n greater than n_c where

$$n_c = \left(\frac{6.8 \times 10^{10}}{E_{ax}} \right)^{1/4} \ldots (4)$$

are ionized (Federenko et al., 1965). The ions thus formed are deflected onto a solid state detector. Those with $n < n_c$ pass on to be detected by a second solid state detector.   The recombination cross section is determined using the formula:

$$\sigma = \frac{Ce^2 f}{I_e I_i L} \left| \frac{v_e v_i}{v_e - v_i} \right| \ldots (5)$$

where C is the signal count rate, e, the electronic charge, I_e, I_i, v_e and v_i are the electron and ion beam currents and velocities respectively. In order to induce stimulated radiative recombination, the merged electron and ion beams are crossed by the beam from a 20 Watt CO_2 laser. This beam has a width of 1.5 mm and so only intersects 1.7% of the total interaction region. An ideal arrangement would be to allow the laser beam to pass coaxially through the interaction region.   This was deemed however to be too dangerous given the destructive power of this beam. The collision energy is varied by changing the electron beam energy, the ion energy remaining fixed. The energy in the centre of mass frame of reference, E_{cm} is given by

$$E_{cm} = (E_e^{1/2} - E_+^{1/2})^2 \ldots (6)$$

E_+ being the reduced ion energy given by $E_+ = (m_e/m_i) E_e$, where E_e, E_i m_e and m_i are the electron and ion energies and masses.

RESULTS AND DISCUSSION.

The electrostatic analyzer used to separate the primary ion beam from the neutrals has a field of 6KV/cm and so this will field ionize all atoms in states with n>19. These will not be subsequently detected.   During this series of measurements the field ionizer was set to ionize atoms with n>8. The experimental protocol is such that the electron and ion beams are merged and the laser beam is crossed with the merged beams.   The electron energy is varied across the energy region where stimulated recombination is

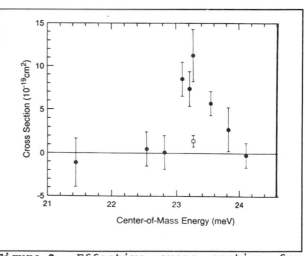

Figure 2. Effective cross section for stimulated radiative recombination to the n=11 state of H. ● Laser on. ■ Laser off.

expected to occur. Off resonance the resulting measured neutrals will be due to spontaneous recombination to states with 8<n<19. When the electron energy is such that it is in resonance with the laser beam energy to stimulate recombination, there will be a contribution to the cross section due to this process. The experimental results for stimulated capture into the n=11 state are shown in figs. 2. The circles represent cross sections measured with the laser on while the point represented by a square was taken with the laser off. It can be seen that laser stimulated recombination leads to an enhancement of the cross section by about a factor of 3. What is particularly striking is the very narrow width of the peak (0.5 meV) reflecting the very high energy resolution achievable with a merged beam experiment.

In figure 3, results taken at a higher electron energy are displayed. Here stimulated recombination occurs to the n=12 state. Again the data point represented by the square was taken with the laser off. As expected, the energy resolution here is not so good and the peak width is about 1 meV wide. These results have been reported elsewhere (Yousif et al., 1991) and an analysis of the measured gain has been given. To summarize this analysis, when the

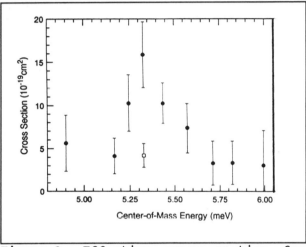

Figure 3. Effective cross section for stimulated radiative recombination to n=11 state of H. ● Laser on. ■ Laser off.

ratio of illuminated to total interaction region is taken account of the measured gain factors are 1720±860 for n=11 and 4790±2830 for n=12. This is of the order of the gains estimated using eqn.2.(2388 and 1139 respectively). Yousif et al. have also shown that in these measurements, the effect of neutral particle reionization due to the use of high laser power is negligible.

SUMMARY.

Laser stimulated radiative recombination has clearly been demonstrated for capture into high rydberg states of atomic hydrogen. This measurement and that reported by Schramm et al. at this conference (see also Schramm et al., 1991) represent the first experimental observations of this process. In future studies we intend to use a multi-pass cavity technique to more uniformly illuminate the interaction region and to examine other ions such as He^+, C^+, N^+ and O^+. Molecular systems will also be studied.

ACKNOWLEDGEMENTS.

The authors would like to acknowledge the generous financial support of the United States Air Force Office of Scientific Research and of the Canadian Natural Sciences and Engineering Research Council.

REFERENCES.

Bethe, H. and Salpeter, E. (1977) Quantum Mechanics of One- and Two-Electron Atoms. Plenum Publishing Co., New York.

Neumann, R., Poth, H., Winnacker, A. and Wolf, A. (1983) Z. Phys. A.313, 253.

Van der Donk, P., Yousif, F.B., Mitchell, J.B.A. and Hickman, A.P. (1991) Phys, Rev. Lett. 67, 42.

Federenko, N.V., Aukudinov, V.A. and Il'in, R.N. (1965) Sov. Phys. Tech. Phys. 10, 461.

Yousif, F.B., Van der Donk, P., Kucherovsky, Z., Reis, J.,Brannen, E., Mitchell, J.B.A. and Morgan, T.J. (1991) Phys. Rev. Lett.67, 26.

Schramm, U., Berger, J., Grieser, M., Habs, D., Jaeschke, E., Kilgus, G., Schwalm, D., Wolf, A., Neumann, R. and Schuch, R. (1991) Phys. Rev. Lett. 67, 22.

Laser-induced recombination in merged electron and proton beams

U.Schramm, J.Berger, M.Grieser, D.Habs, G.Kilgus, T.Schüßler, D.Schwalm, A.Wolf
Physikalisches Institut der Universität Heidelberg and
Max-Planck-Institut für Kernphysik, Heidelberg,
Federal Republic of Germany

R.Neumann
Gesellschaft für Schwerionenforschung, Darmstadt, FRG

R.Schuch
Manne-Siegbahn-Institute, Stockholm, Sweden

ABSTRACT: The enhancement of radiative recombination by induced transitions from the continuum into certain bound states has been observed in merged beams of electrons, protons and laser photons at the Heidelberg Test-Storage-Ring (TSR). Using this process of laser-induced recombination the photorecombination spectrum of transitions into the $n = 2$ state was measured with high resolution and compares well with theory if weak electric fields in the interaction region are taken into account.

1. INTRODUCTION

The recombination of a positive ion with an electron in a binary collision is made possible by the spontaneous emission of a photon. This radiative recombination (RR) process is usually treated [Stobbe 1930] as the inverse of photoionization. The basic theory of radiative transitions predicts that the recombination rate can be enhanced in an external resonant light field by stimulated photon emission. The concept of stimulated radiative recombination in a strong laser field (LIR) was proposed for positronium formation [Rivlin 1979], treated theoretically for the recombination of electrons and ions [Faisal 1981, Ritchie 1984], and discussed with regard to antihydrogen formation in a storage ring [Neumann 1983] and to short-wavelength lasers [Fill 86]. The direct observation of the enhancement of the recombination rate by external radiation has first been reported by our group [Schramm 1991] investigating the process $p^+ + e^- + h\nu \longrightarrow H^0_{(n=2)} + 2h\nu$ and later by Yousif et al. [1991].

In the experimental investigation of RR great progress was made by using merged electron and ion beams in a storage ring. Thus, the formation of hydrogen atoms was observed [Budker 1978, Bell 1981, Poth 1989] during electron cooling of a proton beam in a storage ring, where the neutral atoms leave the ring tangentially. Up to now, however, only the total recombination rate was measured, which represents the average over the center-of-mass (c.m.) energy distribution of the electrons and the sum over a large number of final states. Information about the recombination rate into specific final states and about the electron energy distribution is contained in the spectrum of the photons

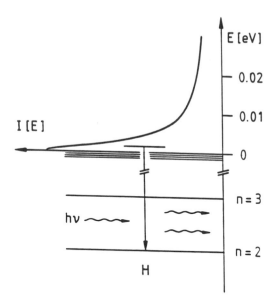

Figure 1: Schematic picture of the LIR-process of an electron and a proton to the bound state of a hydrogen atom with main quantum number $n = 2$.

emitted in the RR process; however, owing to its low intensity, this radiation could not yet be investigated using merged beams.

The process of laser-induced recombination, performed at a storage ring, offers new opportunities for the spectroscopic investigation of recombination processes [Neumann 1983]. As shown in the schematic picture (figure 1), LIR allows, by choosing the photon energy the determination of the c.m. energy of the captured electron, as long as it is small compared to the energy spacing of the final states. The induced recombination rate as a function of the laser wavelength represents the photorecombination spectrum. It can be measured by this method with high resolution, due to the narrow bandwidth of the laser and with high sensitivity, due to the complete detection of the recombination products. Moreover, the influence of external perturbations on the recombination process should manifest itself at energies close to the ionization limit of the recombined atom. In particular, external fields can lower the ionization threshold or the population of weakly bound states, caused by ternary collisions, might appear in the recombination signal if transitions to lower states are induced by the laser. Besides using LIR for studying the photorecombination spectrum, one can also consider its application for enhancing weak recombination processes and for producing the recombined system in a well-defined final state for further investigations.

2. EXPERIMENT

Recombination experiments between ions and electrons in a storage ring are generally based on the following scheme: The circulating ion beam is merged with the intense

electron beam of the electron cooling device. The charge of those ions which capture
an electron is reduced by one and therefore the recombined ions leave the closed orbit
of the stored beam after the next bending magnet. The detection of the recombination
products then takes place in a particle detector outside the closed orbit of the ring.

The first observation of laser-induced recombination was achieved at the Heidelberg
Test-Storage-Ring (TSR, [Krämer 1990]) in merged beams of cooled protons, electrons
and laser photons. Protons were chosen, because overlapping the three beams inside
the interaction region of the electron cooling device [Steck 1990] had been a serious
problem in previous beam times. The neutral H^0-atoms, moving straight through the
storage ring magnets, carried the information about the proton beam position inside the
electron cooler solenoid into areas of the TSR, accessible to scraper measurements. To
investigate the LIR process into the level with main quantum number $n = 2$ the light of
a pulsed dye laser tuned to the Balmer-series limit was shot into the interaction region
antiparallel to the proton beam. The increase of the hydrogen formation rate due to
stimulated free-bound transitions was observed in coincidence with the laser pulses.

Protons with an energy of 21.0 MeV were supplied by the Heidelberg MP-Tandem Van-
de-Graaf accelerator. Using a multiturn injection technic to accumulate the beam inside
the TSR, up to 5×10^9 protons were stored in the ring of 55.40 m circumference for several
hours. Inside the electron cooler the protons were merged with a velocity-matched electron beam (corresponding to an electron-energy of 11.4 keV) over a length of 1.5(1) m.
The electron beam diameter was 5.1 cm and the current 0.917 A, leading to an electron
density of 4.53×10^7 cm^{-3}. The electron beam was immersed in a longitudinal magnetic
field (30.2 mT) to counteract the electron space charge. Inside the electron beam the
space charge gave rise to a radial electric field vanishing on the beam axis and increasing linearly towards the electron-beam edge up to a strength of about 100 V/cm. By
Coulomb collisions in the overlapping beams, the protons acquired the same average
velocity as the electrons and, moreover, a beam diameter of only about 2 mm and a
longitudinal velocity spread below 10^{-4}. The precise proton velocity in the interaction
region was $\beta = 0.2081(2)$ in units of the velocity of light and varied during the experiment by less than $\pm 2 \times 10^{-4}$, as verified by observing the proton revolution frequency
at a Schottky noise pickup.

In the steady state, the relative velocity between protons and electrons is essentially
determined by the electron velocity in the rest frame of the beam. The electron velocity
distribution inside the interaction region is expected to be an anisotropic Gaussian distribution, characterized by the ratio of the longitudinal to the transversal velocity spread.
Mainly based on earlier dielectronic-recombination measurements [Kilgus 1990], a transverse velocity spread corresponding to the cathode temperature of 1220 K (0.105 eV) and
a longitudinal velocity spread corresponding to an energy of about 1 meV was expected.
The recombined atoms were separated from the circulating proton beam in a bending
magnet about 5 m downstream of the electron cooler. Behind this magnet they had
to pass through a 0.5-mm-thick quartz plate, serving as a mirror for inflecting laser
light antiparallel to the proton beam, and were then detected with an efficiency of more
than 96% on a 40-mm diam. micro-channel plate, covered by a suitable degrader foil.
The observed spontaneous-recombination rate was in good agreement with a theoretical
estimate [Bell 1981, Omidvar 1990] of the rate coefficient $\sum_n \langle v \sigma_n(\vec{v}) \rangle$. In this expression
σ_n denotes the recombination cross-section into level n and $\vec{v}$ the electron c.m. velocity;
the brackets indicate the average over the c.m. velocity distribution, and the sum extends
over those final states in which the hydrogen atoms are not field-ionized in the storage-
ring dipole ($n \leq 6$) and can thus reach the detector region. The formation rate of
hydrogen atoms by electron capture of protons in the residual gas was negligible.

Light pulses from a tunable dye laser (Coumarine 2), pumped by a XeCl excimer laser,
were focused to a cross section of ≈ 0.1 cm^2 in the interaction region. From the proton

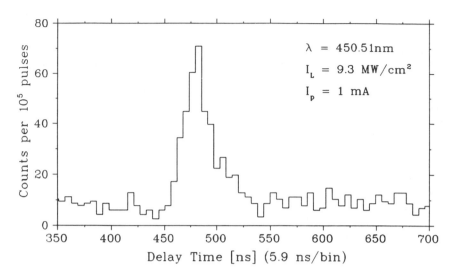

Figure 2: Time spectrum of the recombined hydrogen atoms observed in coincidence with the laser pulses.

beam velocity one determines the laboratory wavelength corresponding to the (field-free) binding energy of the hydrogenic $n = 2$ state ($E_{n=2} = -3.3996$ eV) to be 450.46(7) nm. The laser pulses were ≈ 20 ns long and had a peak power of 1 to 3 MW at the dye laser exit. The laser power was continuously monitored via the pulse height of a fast photodiode, hit by light reflected from one of the optical elements and calibrated versus the average laser output. A value of 0.7 is estimated for the transmission of the optical elements between the laser and the interaction region. The laser beam was aligned to the neutral hydrogen beam formed by RR, whose position was measured by scrapers behind the interaction region and in front of the micro-channel plate detector. This procedure ensured the overlap of the laser beam with the proton beam in the interaction region itself since the neutral beam, unlike an ion beam, is not deflected in the magnetic field of storage-ring elements.

The hydrogen atoms formed by the laser-induced process were detected in coincidence with the laser pulses to separate them from those formed by spontaneous radiative recombination. The repetition rate of the laser system was about 10 s^{-1}. The time spectrum which was recorded in a window of about 1 μs (figure 2) shows a flat background due to RR and a large enhancement in a time interval corresponding to the passage of the laser pulse through the overlap region. The width of this LIR peak approximately agrees with the sum of the travelling time of a hydrogen atom through the overlap region (25 ns) and the laser pulse length (20 ns). According to Neumann et al. [1983] a gain factor G_n is defined as the induced rate into the selected state with main quantum number n normalized to the total spontaneous rate: $G_n = R_n^{ind.} / R^{spon.}$. The gain factor was obtained by taking the ratio of the number of counts in the time-of-flight broadened LIR peak to the number of background counts in a time interval equal to the laser pulse duration.

3. RESULTS

To investigate the photorecombination spectrum, the gain factor was measured for different laser wavelengths. Making use of the expected and observed linear dependence of the gain factor on the laser intensity, G is normalized to an intensity of $I_{laser} = 17.5 \, \text{MW/cm}^2$. It is presented in figure (3) as a function of the relative c.m. electron energy E_e given by the difference of the photon energy E_γ and the field-free ionization threshold of the hydrogenic $n = 2$ level: $E_e = E_\gamma - |E_n|$. Via the Doppler-effect, the uncertainty in the absolute beam velocity was the principal contribution (4.6×10^{-4} eV) to the systematic error of the c.m. electron energy. The uncertainty of the laser calibration corresponded to less than 8×10^{-5} eV, whereas the spectral resolution was limited by the spread and the fluctuation of the proton velocity, corresponding to a bandwidth of below 5×10^{-5} eV. The gain factor $G(E_e)$ (figure 3) starts to rise already at -3.1(5) meV, clearly below the expected field-free threshold. It reaches a maximum value of 23(3) close to this threshold and falls off to half maximum within ≈ 6 meV and then decreases with E_e at a smaller rate.

The gain factor as a function of the laser intensity (see inset of figure 3) should saturate if the probability for recombining atoms to become re-ionized by the same laser pulse approaches unity. Atoms formed in the $2p$ state decay to the ground state with a mean lifetime of 1.6 ns; their ionization rate reaches the inverse of this lifetime at a laser intensity of $I_{laser} \approx 20 \, \text{MW/cm}^2$ [Neumann 1983]. For atoms formed in the $2s$ state the interaction time is given by the laser pulse length of 20 ns; accordingly, LIR into this subshell saturates at $\approx 2 \, \text{MW/cm}^2$. Since the laser intensity was varied above this value, the $2s$ contribution was always saturated and the observed linear increase of the gain factor is due to the $2p$ contribution only.

In order to understand the measured photorecombination spectrum we first consider LIR in a pure Coulomb field [Neumann 1983], denoting by E_e the asymptotic kinetic energy of an incident electron. The spectral shape of the gain factor is given by the energy distribution of the electrons, multiplied by the recombination cross-section $[\sigma_n(E_e) \propto 1/E_e]$ and by v $[\propto \sqrt{E_e}]$. Taking into account a flattened electron velocity distribution, characterized by $T_\parallel \ll T_\perp$, one obtains for $E_e \geq 0$:

$$G_{nl}^f(E_e) = \frac{c^2 I_{laser}}{8\pi\nu^3} \frac{e^{-E_e/kT_\perp}}{\sqrt{\pi}\sqrt{kT_\perp}\sqrt{E_e}} \, \text{erf}\left(\sqrt{\frac{E_e}{kT_\parallel}}\right) \frac{g_0(nl)}{n} \left[\sum_{n'} \frac{g_0(n')}{n'}\right]^{-1} \quad (1)$$

and $G_{nl}^f(E_e) = 0$ for $E_e < 0$. The last two factors represent the ratio of the spontaneous rate into a certain subshell nl to the total spontaneous rate. Due to field ionization in the bending magnet the sum over n' only runs from $n' = 1...6$. The $g_0(nl)$ and $g_0(n')$ are the Gaunt factors for a given (sub)shell at $E_e = 0$ [Eqs. (20) and (21) and Table 1 of Ref. [Omidvar 1990]], ν represents the laser frequency. The maximum gain at $E_e = 0$ is therefore given by

$$G_{nl}^f(0) = \frac{c^2 I_{laser}}{8\pi\nu^3} \frac{2}{\pi\sqrt{kT_\perp \, kT_\parallel}} \frac{g_0(nl)}{n} \left[\sum_{n'} \frac{g_0(n')}{n'}\right]^{-1} \quad (2)$$

Above the ionization limit the spectral shape of LIR is described by equation (1). The function $G_{nl}^f(E_e)$ was fitted to the data points at $E_e \geq 0$, yielding a longitudinal thermal energy of $kT_\parallel = 2.0(6)$ meV and a transverse thermal energy of $kT_\parallel = 0.08(5)$ eV in reasonable agreement with the expectations. Using these temperatures a gain factor of

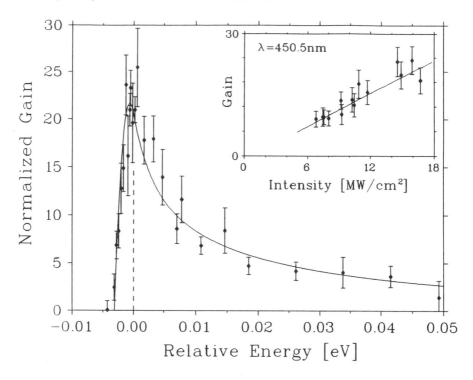

Figure 3: Normalized gain factor for laser-induced recombination as a function of the relative c.m. electron energy E_e at a fixed laser intensity of 17.5 MW/cm². Solid line: Fit according to equation (4). Inset: Maximum gain factor as a function of the laser intensity.

$G_{2p}^f(0) = 59$ was calculated for the intensity of $I_{laser} = 17.5\,\mathrm{MW/cm^2}$. The $2s$ contribution, obtained at the saturation intensity of $2\,\mathrm{MW/cm^2}$, only amounted to $G_{2s}^f(0) = 2.5$; this shows that its influence on the normalization of the data in figure (3) could be neglected. The observed maximum gain (see figure 3) is, however, much below the expected sum of $G_{2p}^f(0) + G_{2s}^f(0)$. Moreover, according to equation (1), the recombination spectrum should never extend below $E_e = 0$.

We explain these deviations by the influence of external static fields on the hydrogenic ionization limit [Schramm 91]. From measurements of the proton revolution frequency and of the hydrogen beam position, a small displacement of the proton beam from the electron beam axis by $(5 \pm 1)\,\mathrm{mm}$ can be inferred. Thus, an electric field of $F = (20\pm4)\,\mathrm{V/cm}$, due to the electron space-charge and directed radially towards the electron-beam axis, was present in the interaction region. The linear potential of this external field exceeds the Coulomb potential at a distance $\approx 1\,\mu\mathrm{m}$ from a proton. It leads to a saddle point of the potential at $E_{sp} = -3.4(3)\,\mathrm{meV}$ and reduces the ionization threshold of the hydrogenic $n = 2$ level by $|E_{sp}|$, in good agreement with figure (3). The perturbation due to the magnetic field can be neglected since for the present field strength, the diamagnetic potential, rising quadratically, reaches the size of the electric potential only at a much larger distance from the proton ($\approx 80\,\mu m$).

We are not aware of theoretical calculations concerning photorecombination in an electric field. However, photoionization of low-lying states of hydrogen in an electric field has been investigated both theoretically [Luc-Koenig 1980, Kondratovich 1984] and experimentally [Glab 1985, Rottke 1986]. In the theoretical models a complete quantum description is used only for the radiative transition occurring close to the nucleus, within the atomic radius of the low-lying state. Once the electron has been excited, a semiclassical description of its motion in the superimposed Coulomb and linear potentials is adequate. In particular, the calculation of classical trajectories [Kondratovich 1984] shows that the threshold excitation energy E_t at which an electron can just escape from the Coulomb potential is a function of the angle θ between the electron trajectory close to the nucleus and the direction of the electric field. For $E_{sp} \le E_e \le 0$ the threshold energy is given by:

$$E_t(\theta) = E_{sp} \sin(\theta/2) . \tag{3}$$

Therefore the relative c.m. electron energy relevant for the induced recombination process in the superimposed potentials has to be related to the equipotential surface of the threshold energy: $\hat{E}_e(\theta) = E_e + |E_t|$. Thus, we proceed by using the modified kinetic energy $\hat{E}_e(\theta)$ in equation (1). The average over the solid angle covered by the variation of θ then leads to the spectral shape:

$$\hat{G}_{nl}^f(E_e) = \frac{1}{2} \int_0^\pi d\theta \, \sin\theta \, G_{nl}^f(\hat{E}_e(\theta)) . \tag{4}$$

Using this approach, a surprisingly good fit to the data is obtained (solid line in figure 3). Moreover, the maximum gain factor for the field free case of 60(3) agrees with the prediction of $G_{2p}^f(0) + G_{2s}^f(0)$, as well as the saddle point energy of $E_{sp} = 0.0031(5)$ eV. The longitudinal temperature $kT_{\|} = 1.5(5)$ meV is slightly lower than the value given above and therefore closer to the expectation.

In conclusion, clear evidence of laser-induced electron–proton recombination was observed and the process was used to study the photorecombination spectrum with high resolution. The observed spectral shape reflects the energy distribution of the electrons in the c.m. frame of the protons and reveals the effect of a weak electric space-charge field. It is planned to extend the LIR studies of photorecombination spectra to highly-charged ions in order to investigate the influence of external perturbations, such as the deformation of the Coulomb potential by Debye screening of the ions in the electron beam, or the population of high Rydberg states by ternary recombination mentioned above.

This work has been funded by the German Federal Minister for Research and Technology (BMFT) under contract no. 06HD133I and by the Gesellschaft für Schwerionenforschung (GSI), Darmstadt.

4. REFERENCES

Bell M et al. 1981 Nucl. Instrum. Methods **190**, 237
Bell M and Bell J S 1982, Part. Accel. **12**, 49
Budker G I and Skrinsky A N 1978, Usp. Fiz. Nauk **124**, 561 [Sov. Phys. Usp. **21**, 277]
Faisal F H M, Lami A and Rahman N K 1981, J. Phys. B **14**, L569
Fill E E 1986, Phys. Rev. Lett. **56**, 1687

Glab W L, Nayfeh M H 1985, Phys. Rev. A **31**, 530
Kilgus G et al 1990, Phys. Rev. Lett. **64**, 737
Kondratovich V D et al. 1984, J. Phys. B **17**, 1981; **17**, 2011; **23**, 3785 (1990)
Krämer D et al. 1990 Nucl. Instrum. Methods **A287**, 268
Lami A, Rahman N K and Faisal F H M 1984, Phys. Rev. A **30**, 2433
Luc-Koenig E and Bachelier A 1980, J. Phys. B **13**, 1743 (1980); **13** 1769
Neumann R, Poth H, Winnacker A and Wolf A 1983, Z. Phys. A **313**, 253
Omidvar K and Guimaraes P T 1990, Astrophys. J. Suppl. Ser. **73**, 555
Poth H et al. 1989, Z. Phys. A **332**, 171
Ritchie B 1984, Phys. Rev. A **30**, 1849
Rivlin L A 1979, Sov. J. Quantum Electron. **9**, 353
Rottke H and Welge K H 1986, Phys. Rev. A **33**, 301
Schramm U et al. 1990, Phys. Rev. Lett. **67**, 22
Steck M et al. 1990, Nucl. Instrum. Methods **A287**, 324
Stobbe M 1930, Ann. Phys. (Leipzig) **7**, 661
Yousif F B et al. 1990, Phys. Rev. Lett. **67**, 26

Exchange in electron impact ionization: Triple differential measurements with polarized beams

L. Frost*, G. Baum, W. Blask, P. Freienstein, S. Hesse, W. Raith.

Fakultät für Physik, Universität Bielefeld, D–4800 Bielefeld 1

* Present address: Fachbereich Physik, Universität Kaiserlautern,
  D–6750 Kaiserslautern

ABSTRACT

We present measurements of the relative triple differential cross sections (TDCS) for 100 eV and 54 eV spin–polarised–electron impact ionisation of spin–polarised lithium atoms. Parallel or antiparallel alignment of the spins permits measurements of singlet and triplet scattering. These have been averaged over in all previous (e,2e) measurements. The goal of the experiment is to provide direct measurements of the importance of exchange scattering at low energies for different, well–defined, kinematics. Results are compared with a Distorted Wave Born Approximation. Experimental methods are discussed.

1. INTRODUCTION

The purpose of this report is to discuss the recent progress in the measurement and calculation of spin–dependent effects in low–energy electron impact ionisation of light atoms, in particular in triple differential cross sections. The experiment combines techniques from (e,2e) coincidence spectroscopy with polarised beam technology to fully determine the kinematics and initial spin state of each ionising collision.

Spin dependence of scattering occurs either through spin–orbit forces for the projectile electron or the bound target electron, or through scattering with exchange of co–ordinates of the projectile and target electrons, or both processes. For the experiments with light atoms discussed here, the spin–orbit interaction can be neglected (but see McClelland 1990).

Exchange scattering occurs independent of spin orientation, however, two electrons with antiparallel spins are distinguishable (the occurence of exchange can in principle be determined) whereas two electrons with parallel spins are not distinguishable. In the former case the scattering amplitudes sum in quadrature, whereas in the latter the two amplitudes interfere, leading to a different scattering intensity. Experiments with unpolarised particles average over these two possibilities, providing only an indirect test of theoretical calculations of the exchange amplitude.

The most detailed measurements of the ionisation process are provided by triple differential cross section measurements in which the two electrons from an ionising event are detected in coincidence at well–defined scattering angles and energies. Such (e,2e) measurements provide a very sensitive test of reaction theory, many examples of which are given in the reviews by McCarthy (1976) and Lahmam–Benanni (1991).

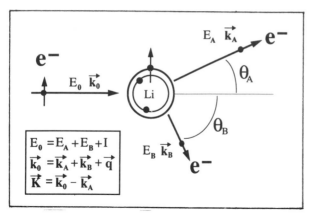

$$E_0 = E_A + E_B + I$$
$$\vec{k_0} = \vec{k_A} + \vec{k_B} + \vec{q}$$
$$\vec{K} = \vec{k_0} - \vec{k_A}$$

Figure 1 shows the coplanar asymmetric (e,2e) kinematics used in this experiment, together with the definitions of various collision parameters. The projectile and target electron spin polarisations are parallel to the lithium beam and hence perpendicular to the scattering plane. The incident electron is given the index 0, the faster of the two escaping electrons the index A.

Figure 1: Coplanar (e,2e) kinematics for spin polarised beams. E_0, E_A, E_B are the electron energies; k_0, k_A, k_B the momenta.

2. THEORETICAL BACKGROUND

Before beginning detailed discussions it is useful to gain insight into the size of the spin–dependent effects for ionising collisions. A useful parameter, which will be discussed in detail below, is the spin asymmetry 'A', which is just the relative difference in cross sections for scattering with projectile and target electron spins anti–parallel or with spins parallel. Figure 2 shows results as a function of energy for the spin asymmetry in the total ionisation cross section for a number of alkali atoms. The data were obtained by measuring ion count rates. It is seen that strong spin–dependent effects extend to 15 or 20 times the atom ionisation energy.

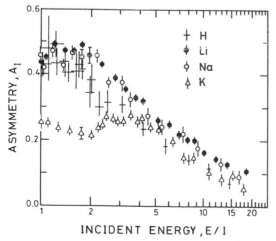

Figure 2: The spin asymmetry in the total ionisation cross section for alkali atoms (Baum 1989). The energy scale is normalized to each element's ionisation energy, I.

If 'f' is the amplitude for direct scattering of two electrons which leads to ionisation, and 'g' is the amplitude for scattering in which the coordinates of the projectile and the ejected electrons are exchanged, then the triple differential cross section for collisions with parallel or with antiparallel spins can be defined as:

$$(1) \quad \sigma_{\uparrow\uparrow} = |f-g|^2, \quad \sigma_{\uparrow\downarrow} = |f|^2 + |g|^2$$

If the collision partners are unpolarised then the cross section can be written in various forms:

$$(2) \quad \sigma = \frac{1}{2}\sigma_{\uparrow\uparrow} + \frac{1}{2}\sigma_{\uparrow\downarrow} = \frac{1}{4}|f+g|^2 + \frac{3}{4}|f-g|^2 = \frac{1}{4}\sigma_s + \frac{3}{4}\sigma_t$$

(3) $\sigma = |f|^2 + |g|^2 - \mathrm{Re}(f^*g) = \sigma_{\mathrm{DIR}} + \sigma_{\mathrm{EXCH}} - \sigma_{\mathrm{INT}}$

This last form for the cross section represents σ as the sum of direct, exchange and interference terms.

The spin asymmetry parameter 'A', which was mentioned in the introduction, is thus:

(4) $A = \dfrac{\sigma_{\uparrow\downarrow} - \sigma_{\uparrow\uparrow}}{\sigma_{\uparrow\downarrow} + \sigma_{\uparrow\uparrow}} = \dfrac{1}{P_e\, P_a} \cdot \dfrac{N_{\uparrow\downarrow} - N_{\uparrow\uparrow}}{N_{\uparrow\downarrow} + N_{\uparrow\uparrow}}$

where P_e and P_a are the polarisations of the projectile electron and atomic beams and $N_{\uparrow\downarrow}$, $N_{\uparrow\uparrow}$ are the measured count rates.

Substitution of the above expressions gives the following useful relations:

(5) $A = \dfrac{|f|^2 + |g|^2 - |f-g|^2}{|f|^2 + |g|^2 + |f-g|^2} = \dfrac{\mathrm{Re}(f^*g)}{\sigma} = \dfrac{|f| \cdot |g| \cdot \cos\, \delta}{\delta}$

where δ is the relative phase between the two amplitudes f and g.

Note that 'A' can be written using equation 2 in terms of the singlet and triplet scattering amplitudes σ_s and σ_t:

(6) $A = \dfrac{\sigma_s - \sigma_t}{\sigma_s + 3\sigma_t}$

This shows that if singlet scattering dominates, then the spin asymmetry parameter tends to unity. If triplet scattering dominates, then 'A' tends to $-1/3$.

Finally, equation 3 shows that 'A' is just the ratio of the 'interference' term from equation to the cross section for ionisation with unpolarised particles. This confirms the intuitive use in the introduction to this paper of 'A' as a measure of exchange scattering, but shows that it measures the interference of g with f rather than the magnitude of the g amplitude itself.

Bartschat (1990) has discussed the usefulness of the 'maximum interference approximation' for the calculation of the spin asymmetry for the total ionisation cross section. In this approximation it is assumed that f and g are exactly in phase ($\delta = 0^\circ$). This gives the best agreement in the absolute value of the cross section because it reduces the generally too–large results of Born–type perturbation calculations. However, when applied to a Distorted Wave Born Approximation (DWBA) using a local energy–dependent exchange potential for the target, it gave spin asymmetries significantly lower than the measurements shown in Figure 2 for below 30 eV and slightly too high at higher energies.

Joachain has estimated the role of exchange in calculating TDCS at intermediate energies (Joachain 1986 and references therein). He considered both coplanar asymmetric and also coplanar symmetric kinematics for a helium target, at energies of several hundred electron volts, and showed that g was negligible in the former case. In the latter case, and for studies which demonstrated the importance of second–order terms in the electron–electron interaction, he used the Ochkur approximation in which $g = (K^2/k_0^2)f$. This term is obviously negligible in the asymmetric kinematics where K is generally small and k_0 rather large, but is significant in coplanar symmetric kinematics.

Brauner, Briggs and Klar (Brauner 1989) have considered direct and exchange scattering for atomic hydrogen, using a T—matrix formulism in which the final state wavefunction has the correct asymptotic form for three charged particles in the continuum. Singlet and triplet final states were calculated, allowing the spin asymmetry A to be derived. They have recently extended the treatment to helium and work is in progress on an extension to lithium.

Walters, Whelan and Zhang have applied a DWBA to the calculation of f and g amplitudes for any light atom. Comparison with relative TDCS at 500, 200, and 100 eV for helium in the coplanar symmetric kinematics (Rösel 1991, Frost 1990) show good agreement. The model breaks down at lower energies (100 eV and below) in the coplanar symmetric kinematics. Zhang, Whelan and Walters (Zhang 1990) have recently published results for lithium which are compared in this report with measurements of the triple differential cross section and the spin asymmetry.

3. DESIGN OF THE EXPERIMENT

3.1 Introduction

Systematic errors in the determination of 'A' can occur through fluctuations or drifts in the coincidence count rate, through inclusion of extraneous noise pulses, or through errors in the determination of Pe and Pa. These will be discussed in the appropriate sections below. The statistical error ΔA in a determination of the spin asymmetry 'A' is quite large, due to the subtraction of quantities of similar magnitude and is given by:

$$(7) \ \Delta A = (P_e \cdot P_a)^{-1}(N)^{-1/2} \quad N = N_{\uparrow\downarrow} + N_{\uparrow\uparrow}$$

The minimum time required to achieve a given precision $\Delta A/A$ is:

$$(8) \ T \approx (P_e \cdot P_a)^{-2}(N)^{-1} (\Delta A)^{-2}L$$

where L depends on the signal to noise ratio (L $\approx$ 1 for negligible background count rates).

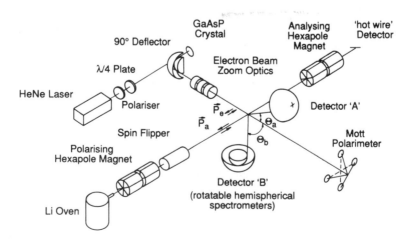

Figure 3: Overview of the experiment

Clearly it is advantageous to use the highest possible beam polarisations. For the experiments reported here Pe = 31 %, Pa = 28 %, and L = 1. Thus, to achieve a percentage error of e.g. 5 % requires approximately 600 times longer measurement times for the spin asymmetry than for the triple differential cross section itself. All possible steps have been taken to increase the coincidence count rate in the present experiments by designing large acceptance angle detectors, broadening the energy resolution and using high–transmission electron optics.

An overview of the apparatus is shown in Figure 3. It consists of the polarised electron beam line, the lithium beam line and the interaction region with associated detectors and electronics. Each will now be discussed in turn.

3.2 Polarised electron beam

The polarised electron source uses photoemission from the (100) surface of a GaAsP crystal treated with caesium and oxygen. The sense of spin polarisation depends on the sense of rotation of the circular polarisation of the HeNe laser beam used, and is changed from left– to right–hand polarisation automatically during measurements. The design is similar to that reported by Pierce (1980). The present system generates up to 10 microamperes of current, with a polarisation of 31 % measured to an accuracy of ±5 % by a Mott polarimeter operating at 100 keV.

A constant focus, constant beam–angle, constant magnification zoom lens (see Fig. 4) has been designed to permit experiments from threshold (5 eV) to several hundred electron volt. Within this range, the focal point and diameter of the beam at the interaction region can be kept nearly constant, and the beam–angle equal to zero.

3.3 Polarised lithium beam

The lithium beam is produced in a separate vacuum chamber by an effusive source heated to 770⁰ C. One filling of the oven of 150 g suffices for about 150 hours of operation. Lithium is explosive in air, so three or four days are needed to clean and load the oven. The lithium beam is spin–polarised using a hexapole magnet. The electron and nuclear spins (I = 1 for lithium–6) recouple in the weak guiding field (10^{-4} Tesla) following the hexapole magnet so that the theoretical maximum for the beam polarisation in the interaction region is 1/(2I+1) or 33 % for lithium–6. The sense of polarisation, parallel or antiparallel to the beam axis, can be flipped diabatically in a magnetic 'spin flipper' (Schroeder 1983).

The beam intensity is monitored in the following vacuum chamber using a hot–wire surface ionisation detector. A second hexapole magnet, prior to the detector, allows the beam polarisation to be measured automatically during spin asymmetry measurements. It was found to vary slightly between 28 % and 30 %.

3.4 (e,2e) Measurements

Figure 4 shows a drawing of the electron energy analysers, which are mounted coaxial with the lithium beam on two independently rotatable vertical wheels. The analysers can select scattering angles from approximately −40⁰ to 130⁰ in coplanar kinematics (with a minimum mutual angle of 60⁰). Their energy resolution is approximately 5 %. The entrance optics of the analysers to have large acceptance angles in order to improve the signal count rates. Calculations showed that a phase–space at the interaction region of approximately r = ±1.5 mm and $\phi = \pm8^0$ was achieved.

The magnetic guiding field of 10^{-5} Tesla in the interaction region causes a measured 0.6^0 shift in the trajectories of 50 eV electrons, which is negligible with respect to the experimental resolution.

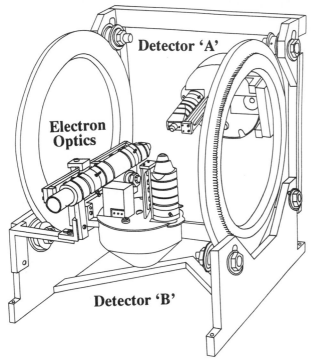

Detector 'A'

Electron Optics

Detector 'B'

Figure 4: Electron spectrometers and electron beam optics

The fast coincidence electronics and computer control uses standard methods. Pulses from one electron detector start a time—to—pulse—height converter which is stopped by pulses from the other detector. The two electrons from an ionising event have a fixed time interval, depending on fixed delays in the electronics as well as possible differences in energies and trajectories within the spectrometers, and always generate pulses which are recorded within a given small range on the multichannel analyser. Electrons from inelastic scattering, or two separate ionising events, have random time intervals. The latter generates a 'background' which is subtracted from the 'true coincidence' signal. Electrical noise picked up from the external environment is rejected by a 'noise gate'. Particular care was taken with the noise rejection, since counting times are long and signal rates low: measurements showed less than one noise count per 10 hours, which is less than 1/400th of the lowest signal rates.

A PC computer determines the incident and outgoing electron energies and angles, 'flips' the electron beam or atom beam polarisations and gates off the lithium beam, as required for each measurement. Twelve combinations of spin directions and beam—gating are used, instead of the nominal two, to detect drifts. For each combination, the signal and background counts in the separate timing windows are recorded. The PC also records the electron beam intensity, atomic beam intensity and the energy—loss signal represented by the count rates in the two detectors.

4. RESULTS AND DISCUSSION

The TDCS for an electron impact energy of 100 eV on lithium is shown in Figure 5 on a logarithmic scale. The detected electron energies were both 47.3 eV. One detector was fixed at $\Theta_A = 45^0$ while the other detector angle varied from 30^0 to 90^0. The relative cross section measurements have been normalized to the theoretical calculations of Zhang (1990) at $\Theta_A = \Theta_B = 45^0$. The dashed line represents Zhang's calculations cor··o-luted with the experimental resolution by averaging the calculated values over a range in Θ_A and Θ_B of $\pm 18^0$. The out—of—plane ($\pm 18^0$ azimuthal angle) cross sections should of course also be considered but this has not yet been possible. This point will be discussed later.

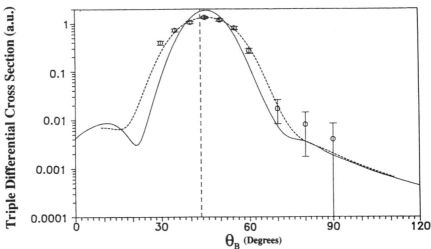

Figure 5: The TDCS for lithium at $E_0 = 100$ eV impact energy: $E_A = E_B = 47.3$ eV, $\theta_A = 45°$. Solid line, DWBA theory (Walters 1990); dotted line, angle-averaged DWBA (see text): data points are normalized to the dotted curve at 45°. The vertical dashed line is the direction for zero momentum transfer to the ion.

The shape of the peak is well reproduced by the DWBA calculations. The position of the peak is also well reproduced, within the precision of the measurements. Note that the cross section decreases by a factor of about 200 from the maximum near $45°$ to the point at $90°$.

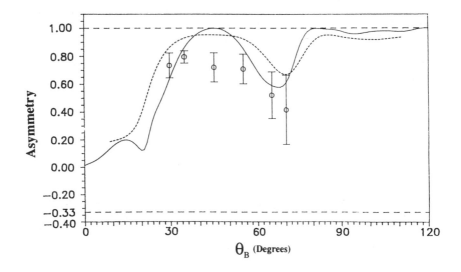

Figure 6: The spin asymmetry in the TDCS for lithium, with the same parameters as in figure 5.

The corresponding spin asymmetry measurements in the triple differential cross section are shown in Figure 6. The solid curve is the DWBA calculation and the dotted curve shows the average over the experimental resolution. The large error bars at 65^0 and 70^0 are due to the much smaller cross section at larger angles. The measurement time per point varied from 3 to 5 hours. The error bars show either the statistical error determined from the count rates, or the standard deviation derived from the 2 to 5 independent measurements at each angle, whichever was the greater.

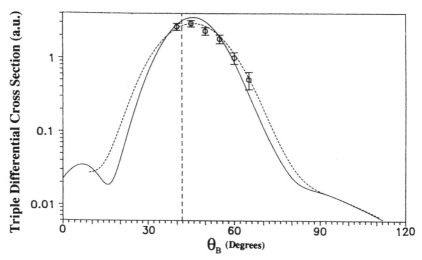

Figure 7: TDCS as in figure 5, for $E_0 = 54.4$ eV, $E_A = E_B = 24.5$ eV, $\theta_A = 45°$.

The calculations follow the trend of the data, but are about 20 % too high. This contrasts with the results of Bartschat (1991), where a DWBA calculation of the asymmetry in the total ionisation cross section was too low even when using the 'maximum interference' approximation which maximised this parameter.

The measured spin asymmetry at 45^0 is only 0.8, instead of 1.0 as required by symmetry considerations (both electrons have identical spatial wavefunctions so the total spin wavefunction must be antisymmetric, i.e. pure singlet, resulting in A = 1.0). A possible explanation is the contribution of triplet scattering from non–symmetric kinematics over the $\pm 8^0$ acceptance angles of the detectors.

The results for these kinematics show a large positive asymmetry over the whole of the forward ("binary") peak in the TDCS. This shows that the scattering is predominantly singlet in nature, with the exchange amplitude being approximately equal to the direct scattering amplitude in magnitude and phase (see equation). This must be true at $\Theta_A = \Theta_B = 45^0$, but it is surprising that it extends across the whole forward peak. The 'maximum interference' approximation is seen to be excellent within this range.

Further measurements at 75^0 and 80^0 are planned, to determine whether the asymmetry continues to decline as the geometry becomes more asymmetric, or whether the DWBA prediction is true, that singlet scattering (and hence 'A') should increase.

The TDCS for 54 eV impact is shown in Figure 7. Again agreement between (convoluted) theory and experiment is good.

Spin asymmetry measurements for 54 eV impact (Figure 8) have so far only been made at three angles, but are in full agreement (within the statistical errors) with the DWBA predictions. In contrast with the results at 100 eV, the asymmetry at $\Theta_A = \Theta_B = 45^0$ is consistent with a value of 1 after allowing for the angular resolution. This gives some support to the suggestion that the discrepancy at 100 eV is due to the angular resolution: the forward peak is broader at 54 eV, reducing such effects.

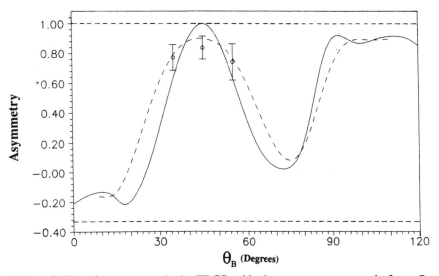

Figure 8: The spin asymmetry in the TDCS, with the same parameters as in figure 7.

5. IMPROVEMENTS

The main experimental difficulties are: (1) the required measurement times depend inversely on the square of the beam polarisations, which are only about 30 %; (2) the necessity of using large acceptance angles to achieve acceptable signal rates makes comparison with theoretical calculations much more difficult; (3) the choice of lithium as a target leads to low duty–cycles in the beam line and also limits the number of theoretical calculations available for comparison. Three improvements, now under development, will much reduce these problems.

Firstly, the polarisation of the atomic beam can be increased by at least a factor of 2 using the technique of radiofrequency transitions in a non–uniform magnetic field and a second hexapole magnet (Chamouard 1990). The required RF–resonator is being built. A reduction in measurement times by a factor of about 3 can be expected. Secondly, an atomic hydrogen (RF discharge) source is being constructed. This will improve the duty cycle of the experiment and also permit comparison with a much wider range of theoretical calculations. Thirdly, new electron spectrometers are being designed to provide improved angular resolution without sacrificing signal. This is particularly important for comparing measurements with calculations.

6. CONCLUSIONS

Relative TDCS for 100 eV and 54 eV electron impact ionisation of lithium have been measured using spin–polarised electron and lithium beams. The measured TDCS are in good agreement with DWBA calculations after accounting for the experimental angular resolution using a simple averaging technique. The measured spin asymmetries in the TDCS at 100 eV are reproduced in general form by the (averaged) DWBA theory, but theory is approximately 20 % higher. This could be due to contributions from asymmetric and out–of–plane scattering kinematics, which have not yet been properly accounted for in the approximate averaging method used. The spin asymmetry at 54 eV has so far only been measured at three angles, but shows excellent agreement with the DWBA.

As well as completing the measurements in Figure 8, other measurements at 54 eV would be of interest. Calculations recently received from Walters and Zhang for $\Theta_A = 40^0$ show that the forward peak is broader and the variation in the asymmetry stronger. The calculations predict that the asymmetry varies over the full range from $-1/3$ to $+1$ over a range of Θ_B from 20^0 to 40^0.

The feasibility of combining the (e,2e) coincidence spectroscopy technique with spin asymmetry measurements has been demonstrated. Spin asymmetry values approaching unity, i.e. pure singlet scattering, have been measured.

7. ACKNOWLEDGEMENTS

We would like to thank Professors Madison, Joachain and Bartschat for many valuable discussions. Professors Walters and Zhang kindly provided many extra tabulations of their calculations. Mr. M. Streun has collaborated in the 54 eV measurements. This work has been supported by the Deutsche Forschungsgemeinschaft SFB 216.

8. BIBLIOGRAPHY

Baum G, Freienstein P, Frost L, Granitza B, Raith W, Steidl H
  NIST Spec. Publ. 789, Proc. Int. Symp. Corr.&Polarization, Aug. 1989
Brauner M, Briggs J S and Klar H 1989 J Phys B 22 2265
Chamouard P A, Courtois A, Faure J et al 1990
  7th Int Conf Pol Phenom Nucl Phys, Paris, July 1990
Frost L, Freienstein P, Wagner M 1990 J Phys B 23 L715–L720
Joachain 1986 Comm Atomic & Mol Phys 17(5) 261
Lahmam–Bennani A 1991 J Phys B 24 2401–42
McCarthy I E, Weigold E 1976 Phys Rep 27(6) 275–371
McClelland J J, Buckman S J, Kelley M H, Celotta R J 1990 J Phys B 23 L21–L24
Pierce D T, Celotta R J, Wang G C 1980 Rev Sc Instr 51 478–9
Rösel T, Dupre C, Joeder J, Duguet A, Jung K, Lahmam–Bennani A, Ehrhardt H 1991
  J Phys B (in press)
Schroeder W, Baum G 1983 J Phys E 16 52–6
Zhang X, Whelan C T, Walters H R J 1990 "1990 (e,2e) Collisions and Related
  Problems" (H. Ehrhardt, ed.) pp 264–73

Coincidence measurements of Auger lineshapes

Birgit Lohmann

Division of Science and Technology, Griffith University
Nathan Qld AUSTRALIA 4111

ABSTRACT:

Coincident lineshapes of the L_3-$M_{23}M_{23}(^1S_0)$ and L_3-$M_{23}M_{23}(^1D_2)$ Auger lines in argon have been measured by detecting the Auger electron in coincidence with the scattered electron produced after electron impact ionization. The lineshapes were measured at excess energies of 2, 5 and 202eV. Post-collision interaction effects are evident at all three energies, and at 2eV excess energy the L_3-$M_{23}M_{23}(^1D_2)$ line exhibits strong interference effects.

1. INTRODUCTION

Electron impact measurements in which the scattered electron produced after an inner-shell ionization is detected in coincidence with a subsequently emitted Auger electron have been used in recent years to investigate a number of interesting phenomena (see the recent review by Lahmam-Bennani 1991, and references therein). In particular, they have been used to investigate the effect of post-collision interaction (PCI) on the coincident Auger lineshapes (Sewell and Crowe 1982, 1984a,b, Sandner and Völkel 1984, Stefani et al 1986).

Previous investigations of these effects have concentrated on the region of low excess energies (the excess energy refers to the kinetic energy of the ejected electron), as strong PCI between the Auger electron and the slow ejected electron is expected under these conditions. The post-collision interaction manifests itself as a shift in the energy of the Auger line and an asymmetrical broadening of the lineshape. In the case where the excess energy is low, the Auger electron can overtake the slow ejected electron and the resulting change in the ionic Coulomb field seen by the electrons results in the observed changes in the lineshape. The slow electron experiences a decrease in energy and the Auger experiences an increase.

A number of theoretical calculations exist for predicting the effect of PCI on Auger lineshapes after photoionization (Niehaus 1977, Russek and Mehlhorn 1986, Armen et al 1987), and recent calculations have been presented for particle impact ionization (Kuchiev and Sheinerman 1986, Russek and Mehlhorn 1986). Recent quantitative calculations by Sandner and Völkel (1989) of Auger lineshapes after electron impact ionization show that when *non-coincident* Auger lineshapes are measured, the PCI effects persist even at high impact energies, due to the presence of unobserved slow collision products. A number of the aforementioned calculations include the dependence on the angle between the ejected electron and the Auger electron, although in general when comparing with experimental data, an average over this angle is performed, since in non-coincidence experiments this angle cannot be explicitly determined.

In this paper some recent results of coincidence measurements of the L_3-$M_{23}M_{23}(^1S_0)$ and L_3-$M_{23}M_{23}(^1D_2)$ lines in argon will be presented, at excess energies of 2eV, 5eV (Lohmann 1991a) and 202eV. The observed lineshapes exhibit both PCI and interference effects.

2.   EXPERIMENTAL DETAILS

The apparatus which was used to perform these measurements is depicted schematically in figures 1 and 2. The incident electron beam is produced by a commercial electron gun and has an energy width of about 0.5eV. The electron beam crosses a gas jet at right angles, and the scattered electrons and Auger electrons are energy analyzed by identical hemispherical analyzers and detected by channeltrons. The two analyzers (mean radius 6.25cm) are coplanar with the incident beam and are preceded by 5-element retarding lenses. A more detailed description of the apparatus will be presented elsewhere (Lohmann 1991b).

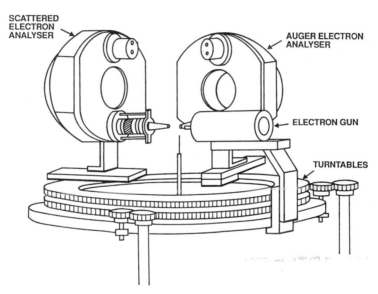

Fig 1.   The electron coincidence spectrometer.

The scattered electron analyzer is positioned at 21^0 and the Auger electron analyzer is positioned on the other side of the electron beam at 72^0. The lineshape measurements were performed with an energy resolution of either 250meV or 450meV. The scattered electron analyzer had an energy resolution of 1.8eV. The coincidence count rates in these experiments are very low, and the data acquisition is hence fully automated. The coincidence data is obtained by fixing the incident and scattered electron energies, and stepping the Auger analyzer detection energy over the range of interest. Each run consists of multiple scans over the entire energy range, in order to minimize the effect of any drifts. The counting time at each energy is normalized to the scattered electron count rate, to compensate for any variations in gas pressure or beam current. Many such runs were added to produce the final data sets.

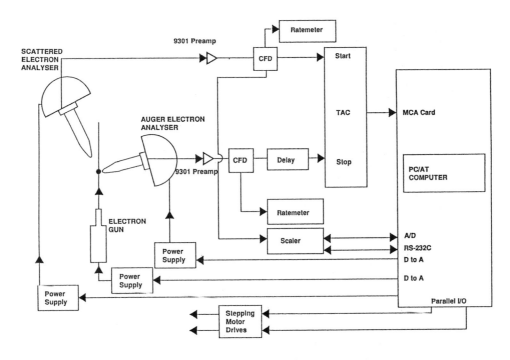

Fig 2. A schematic diagram of the apparatus
and associated electronics.

3. RESULTS AND DISCUSSION

The results are presented in figures 3-7. Figures 3 and 4 correspond to an excess
energy of 5eV and 2eV respectively, and were obtained with an Auger analyzer
energy resolution of 450 meV. The incident energy was 1000eV. The dashed line is
the non-coincident Auger signal and the solid line is a fit of two (symmetric) mixed
Gaussian-Lorentzian functions to the data. The 5eV results clearly show the shift to
higher energy of the Auger lines which is expected due to the post-collision
interaction between the fast Auger electron and the slow ejected electron, however
the lines do not appear significantly distorted. The spectrum also exhibits an
underlying coincidence background due to the presence of (e,2e) events. The energy
range covered in the measurements corresponds to a range of separation energies of
52.8-48.4eV, which is above the double ionization threshold. McCarthy and Weigold
(1976) and McCarthy et al (1989) have investigated this region of the (e,2e) spectrum
and have found configuration interaction satellites of the $3s^{-1}$ ionization which extend
into the Ar^{2+} continuum, leading to a final Ar^{2+} $3p^4(^1D)$ state of the ion. The
presence of this direct channel for the production of a doubly-ionized argon atom is
of interest since the Auger process also leaves the atom doubly-ionized. If the direct
and indirect processes leave the target in the same final ion state, then interference
between the two amplitudes is possible. Sandner and Völkel (1984), in their
measurements of the L_3-$M_{23}M_{23}(^1S_0)$ line found that at certain angles, in the presence
of this background, their measured $^1S_0$ lineshape could be fitted by a Beutler-Fano
profile, indicating that an interference process may be occurring.

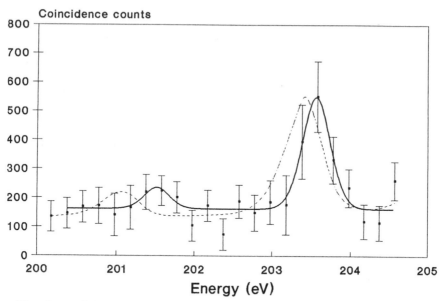

Fig 3.   Coincidence spectrum of the argon L_3-$M_{23}M_{23}(^1S_0)$ and L_3-$M_{23}M_{23}(^1D_2)$ Auger lines at an excess energy of 5eV and an energy resolution of 450meV. The broken curve is the non-coincident spectrum and the solid line is a fit to the data.

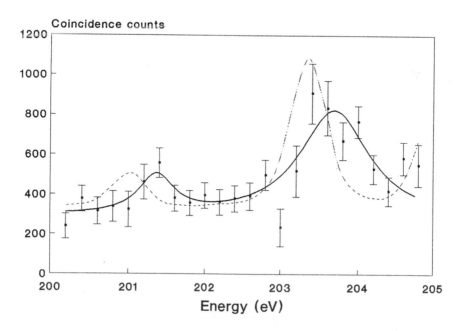

Fig 4.   As for figure 3, but measured at an excess energy of 2eV.

Figure 4 shows a coincidence scan of the same energy range, but at an excess energy of 2eV. This corresponds to scanning a range of separation energies of 49.8-45.2 eV, which results in an increase in the (e,2e) background. Both Auger lines are again shifted to higher energy due to the post-collision interaction, and the line is severely distorted. The solid line is an attempt to fit two mixed Gaussian-Lorentzians (G-L) to the data. As the G-L lineshapes are not asymmetric, and the $^1D_2$ lineshape exhibits a pronounced asymmetry, the fit is poor. In addition to a significant broadening on the high energy side, the lineshape exhibits a decrease in intensity below the background at 203 eV.

In order to investigate the details of the $^1D_2$ lineshape further, additional measurements were made over the energy range 202.7-204.7eV with improved resolution in the Auger analyzer (250meV). The results are presented in figure 5. Again the broken line is the non-coincident spectrum, and in this case the solid line is a fit of a Beutler-Fano type profile (Shore 1967, Comer and Read 1972) convoluted with a Gaussian instrument function of width 250meV. The data exhibit a clear decrease in intensity below the background on the low energy side, and a broadening on the high energy side. The Beutler-Fano profile provides a good fit to the data, and thus a persuasive argument in support of interpreting the distortion of the lineshape as evidence of interference between the Auger process and the direct process.

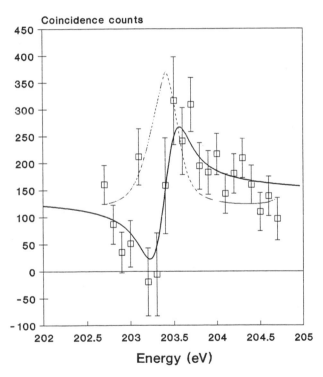

Fig 5. Coincidence spectrum of the argon L_3-$M_{23}M_{23}$($^1D_2$) Auger line at an excess energy of 2eV and an energy resolution of 250meV. The broken curve is the non-coincident spectrum and the solid curve is a fit of a Beutler-Fano type profile to the coincidence data (see text for details).

As mentioned earlier, a number of theoretical models exist which can be used to predict Auger lineshapes after photoionization. The post-collision interaction is then a result of the Coulomb interaction between the outgoing slow electron, the Auger electron and the residual ion. In the case of electron impact ionization, the situation is complicated by the presence of the scattered electron. A full calculation of the PCI should thus include the effect of the scattered electron. However, under conditions where this electron leaves the target region with a much higher energy than the ejected or Auger electrons, one can neglect the effect of the scattered electron, and the lineshapes should be fairly well modelled by calculations which only consider two electrons in the final state.

A useful analytical form for the Auger lineshape function due to Kuchiev and Sheinerman (1986) is given below. $P(\varepsilon)$ is the probability of detecting an Auger electron of energy ε.

$$P(\varepsilon) = \frac{\Gamma_i/2\pi}{(\varepsilon_A-\varepsilon)^2 + \Gamma_i^2/4} \; \kappa(\varepsilon)$$

$$\kappa(\varepsilon) = \frac{\pi\xi}{\sinh(\pi\xi)} \exp[2\xi \text{ arc tan } \frac{2(\varepsilon_A-\varepsilon)}{\Gamma_i}]$$

ε_A is the nominal Auger transition energy, and Γ_i is the initial-state width. The quantity ξ is given by

$$\xi = \frac{-1}{k_{ej}} + \frac{1}{|\underline{k}_{ej} - \underline{k}_A|}$$

where $\underline{k}_{ej}$ is the wave vector of the ejected electron and $\underline{k}_A$ is that of the Auger electron. Thus the lineshape formula includes a dependence on the angle between the ejected and Auger electrons. Armen et al (1987), using a fully quantum mechanical approach, have derived a lineshape formula which essentially coincides with the *angle-averaged* lineshape of Kuchiev and Sheinerman.

In figure 6, the measured coincidence spectrum of the L_3-$M_{23}M_{23}(^1S_0)$ line at 2eV excess energy and 250 meV energy resolution is compared with the above lineshape formula, calculated for the angle-averaged case (excess energy of 3eV, and convolved with the instrumental resolution). The lineshape calculated at an excess energy of 3eV has been used as it is in somewhat better agreement with the measured lineshape.

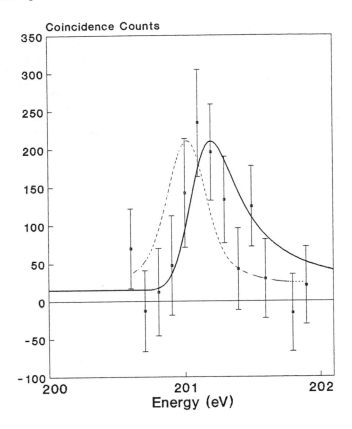

Fig 6. Coincidence spectrum of the argon L_3-$M_{23}M_{23}(^1S_0)$ Auger line at an excess energy of 2eV and an energy resolution of 250meV. The broken curve is the non-coincident spectrum and the solid curve is the angle-averaged Kuchiev and Sheinerman lineshape function (see text) calculated at an excess energy of 3eV.

Armen (1988) has investigated the behaviour of the *angle-dependent* Kuchiev and Sheinerman lineshape formula in detail, and has found that the effects of PCI on the lineshape can vary dramatically as the angle between the ejected electron and the Auger electron is varied. To summarize, Armen found that for $E_{ej} \ll E_{AUGER}$, the angle dependence of the lineshape is small. The lines are shifted to higher energy and broadened on the high energy side, with the effect being more pronounced at large angles (180^0) than at small angles (10^0). However, as the excess energy increases, this changes dramatically. For angles of 90^0 or greater, the shift is always to higher energy and vanishes as E_{ej} approaches infinity. When $E_{ej} \approx E_{AUGER}$, and the angle is less than 90^0, the shift and distortion can reverse themselves. Thus at 10^0, for example, the line is shifted to an energy below the nominal Auger energy and is strongly asymmetrical on the low energy side. This effect persists until $E_{ej} \simeq 1.5 E_{AUGER}$, after which the shift tends to zero again as E_{ej} increases. In contrast, the angle-averaged lineshape predicts a monotonic decrease of the (positive) energy shift as the excess energy decreases. The shift to lower energy of the Auger line when the ejected energy is in the range $E_{AUGER} \simeq E_{ej} \sim 1.5 E_{AUGER}$ can be classically explained as follows (Armen 1988). If the ejected and Auger electrons are moving in the same general direction, with the ejected electron moving just fast enough that it cannot be overtaken by the Auger electron, then the Auger electron will always experience a retarding force which will slow it down, resulting in a line shifted to lower energies.

In order to investigate these effects experimentally in the electron excited case, ideally one would need to perform a triple coincidence experiment in which the kinematics of all three outgoing electrons are fully determined. In the present measurements only the scattered electron and the Auger electron are detected, and the direction of the ejected electron is undetected. Hence it would appear to be a reasonable assumption to average over the ejected electron angle. However, measurements and calculations of the triple differential cross section for ionization of the inner and valence shells of argon (Stefani et al 1986, Avaldi et al 1989, McCarthy 1991) show that the distribution of the ejected electron can be far from isotropic. Generally, one observes a binary peak close to the momentum transfer direction, and a recoil peak approximately 180^0 from the binary peak. Thus by fixing the scattering angle and the scattered and ejected electron energies, we also fix the distribution of the ejected electron in space. In the kinematical arrangement used in these experiments, the position of the Auger electron analyzer coincides roughly with where one might expect the binary peak to occur, and hence the measured Auger lineshapes may exhibit evidence of a non-isotropic PCI.

In order to investigate such possible effects, the L_3-$M_{23}M_{23}(^1D_2)$ line was measured in coincidence with the scattered electron at a primary energy $E = 1200eV$ and an ejected electron energy of 202eV. The coincidence cross section drops significantly due to the decrease in the single differential cross sections as the energy sharing between the ejected electron and the scattered electron becomes more equal. Due to the low count rates the measurements were performed with an energy resolution of 450meV. The results are presented in figure 7. The broken line is the non-coincident spectrum. The coincidence spectrum is clearly broadened on the low energy side, and the peak appears displaced slightly to lower energies. The fine solid line is a calculation at 202eV excess energy of the angle-averaged Kuchiev and Sheinerman lineshape formula.

As expected, it exhibits essentially no deviation from the non-coincident lineshape, and does not explain the observed coincidence lineshape. The thick solid line is calculated using the angle-dependent Kuchiev and Sheinerman lineshape, and is the sum of a lineshape calculated at an angle of 5^0 between the ejected and Auger electrons, and a lineshape corresponding to an angle of 180^0. Essentially, this simulates a situation where there is a contribution to the lineshape from cases where the angle between the ejected electron and the Auger electron is close to zero (this produces a line shifted to lower energies and distorted on the low energy side) and a contribution from cases where there is a large angle between the ejected electron and the Auger electron (which produces an unshifted lineshape).

Such a simulation might correspond to the case where there is a binary lobe in the ejected electron distribution which is roughly in the direction of the Auger analyzer, and a recoil lobe in the opposite direction.

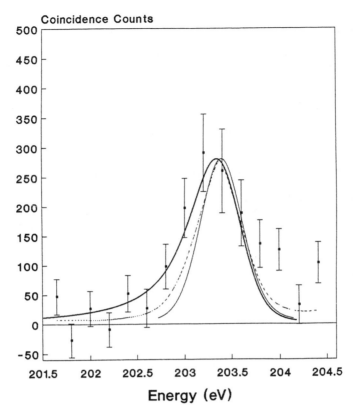

Fig. 7. Coincidence spectrum of the argon L_3-$M_{23}M_{23}(^1D_2)$ Auger line at an excess energy of 202eV and 450meV energy resolution. The broken line is the non-coincident spectrum and the thin solid line is the angle-averaged Kuchiev and Sheinerman lineshape function at an excess energy of 202eV. The thick solid line is the angle-dependent Kuchiev and Sheinerman lineshape function (see text for details).

It is apparent that the calculated line gives a very satisfactory fit to the coincidence data. The negative shift and the distortion on the low energy side of the Auger lineshape at an excess energy where the angle-averaged lineshape calculation predicts essentially an unshifted Lorentzian, indicates that an angle dependence of the PCI has been observed. An appropriate future experiment would be to measure the coincidence Auger lineshape under conditions where the *ejected* electrons are measured in coincidence with the Auger electrons at small angles between the two.

This work was supported by a grant from the Australian Research Council.

REFERENCES
Armen G B 1988 Phys. Rev. A **37** 995
Armen G B, Tulkki J, Åberg T and Crasemann B 1987 Phys. Rev. A. **36** 5606
Avaldi L, McCarthy I E and Stefani G 1989 J. Phys. B.:At. Mol. Opt. Phys. **22** 3305
Comer J and Read F M 1972 J. Phys. E.:Sci. Instrum. **5** 211.
Kuchiev M Yu and Sheinerman S A 1986 Zh. Eksp. Teor. Fiz. **90** 1680
  [Sov. Phys.-JETP 1986 **63** 986]

Lahmam-Bennani A 1991 J. Phys. B.:At. Mol. Opt. Phys. **24** 2401.
Lohmann B 1991a J. Phys. B.:At. Mol. Opt. Phys. **24** L249
- 1991b (to be published)
McCarthy I E 1991 (private communication)
McCarthy I E, Pascual R, Storer P and Weigold E 1989 Phys. Rev. A **40** 3041
McCarthy I E and Weigold E 1976 Phys. Rep. **27C** 277
Niehaus A 1977 J. Phys. B.:At. Mol. Phys. **10** 1845
Russek A and Mehlhorn W 1986 J. Phys. B.:At. Mol. Phys. **19** 911
Sandner W and Völkel M 1984 J. Phys. B.: At. Mol. Phys. **17** L597
Sandner W and Völkel M 1989 Phys. Rev. Lett. **62** 885
Sewell E C and Crowe A 1982 J. Phys. B.:At. Mol. Phys. **15** L357
Sewell E C and Crowe A 1984a J. Phys. B.:At. Mol. Phys. **17** L547
- 1984b J. Phys. B.:At. Mol. Phys. **17** 2913
Shore B W 1967 Rev. Mod. Phys. **39** 439
Stefani G, Avaldi L, Lahmam-Bennani A and Duguet A 1986
    J. Phys. B.:At. Mol. Phys. **19** 3787
van der Straten P and Morgenstern R 1986 J. Phys. B: At. Mol. Phys. **19** 1361.

Ionization in collisions between Na and Cs atoms

C.Gabbanini, S.Gozzini, A.Lucchesini and L.Moi*

Istituto di Fisica Atomica e Molecolare del C.N.R.
Via del Giardino 7, 56100   Pisa   Italy
*Dipartimento di Fisica - Universita' di Siena
Via Banchi di Sotto 55, 53100 Siena   Italy

ABSTRACT: A crossed-beams apparatus has been used to study collisions between Na and Cs atoms. Two kinds of collisional processes have been investigated: the associative ionization (AI) and the Hornbeck-Molnar ionization (HMI). Cross section values have been measured relative to others of known processes. HMI cross sections have been also calculated by using the Janev-Mihajlov model. The dissociation energy of the $NaCs^+$ molecular ion has been derived and compared with theoretical predictions.

1. INTRODUCTION

Selective laser excitation of atoms in well defined states may produce atomic   and/or molecular ions through collisional processes. Collisional processes with alkali atoms are so efficient that it is possible, under particular conditions, to create a plasma with resonant laser power densities that are orders of magnitude lower than those required in the process of multiphoton ionization (Lucatorto and McIlrath 1976). The study of elementary collisional processes leading to ionization is of great importance both to test theories and to understand complex and new physical situations. For instance the elementary process of associative ionization has opened the promising and new field of collisions between "cold" atoms (Gould et al 1988).

Here we present an investigation of two collisional processes between Na and Cs atoms:

a) the associative ionization in which $NaCs^+$ molecular ions are produced through the collision between Na(3p) and Cs(6p) atoms (Gabbanini et al 1991a);

b) the Hornbeck-Molnar ionization (Hornbeck and Molnar 1951) in which a   collision between a highly excited atom and a ground state atom produces a molecular ion. More precisely we studied the collision of Na(3s) atoms with cesium atoms in one of the 8d, 10s, 9d, 11s and 10d excited levels (Gabbanini et al 1991b).

We should remark that, while for the homonuclear collisions between alkali atoms there are many experiments both on AI and on HMI, for heteronuclear collisions only two quantitative experiments, one on Na-Li (Johnson et al 1988) and the other on K-Rb (Djerad et al 1987), are reported in literature. The reason for this lack of data is essentially due to the increasing experimental difficulties that AI and HMI heteronuclear collisions present and not to a minor interest of these processes. Infact the absolute determination of cross sections of such heteronuclear collisions requires the measurement of many independent parameters and systematic errors can be easily made. We followed a procedure that simplifies the acquisition of experimental data once that some reference values are available.

2. EXPERIMENTAL APPARATUS

The AI and HMI experiments have been made by using the same apparatus, shown in Fig.1. Two c.w. lasers have been used for atom excitation: an actively stabilized monomode diode laser (STC- model LT 50 A-03U) to excite cesium atoms to the $6p_{3/2}$ level; a frequency stabilized ring dye laser to excite sodium atoms to the $3p_{3/2}$ level in the AI experiment, or to excite cesium atoms to a high s or d level in the HMI experiment. The two laser beams are superposed on a dichroic mirror and coupled to an optical fiber whose output side is next to a window of the atomic beam chamber. Here, in a high vacuum region, two atomic beams, coming from two indipendently heated ovens, cross in a collision cell. The cell is used for three different reasons: it increases the atomic densities, it defines the potential at which ions are produced and, finally, it creates a vaporlike condition as a consequence of atom-wall and atom-atom collisions. This last point, that is demonstrated by the Doppler-broadened absorption linewidth, is essential for our quantitative data extraction, as will be shown in the next sections. The excitation region is internal to the collision cell and the volumes excited by the two lasers are equal within 5%. Ions are extracted by two electrostatic lenses, mass analyzed by a quadrupole filter and detected by an electron multiplier. A multichannel analyzer and a counter are alternatively used depending on the signal intensity. Ions can also be produced by impact with the electrons emitted by a hot tungsten filament located outside the collision cell at a well defined potential. The apparatus allows also to analyze the fluorescence emitted by the atoms in the collision cell.

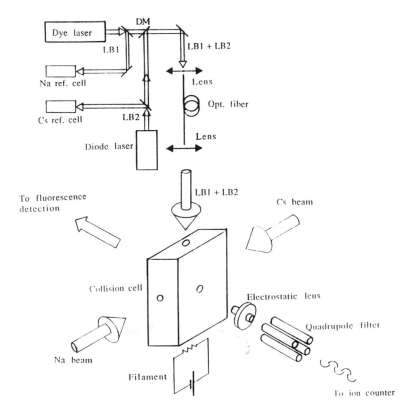

Figure 1. Sketch of the experimental apparatus. LB: laser beam; DM: dichroic mirror.

3. ASSOCIATIVE IONIZATION

The process studied in this case is:

$$Na(3p_{3/2}) + Cs(6p_{3/2}) \rightarrow NaCs^+ + e^- + \Delta E \qquad (1).$$

An example of the detected mass spectra is shown in Fig.2. $NaCs^+$ and Na_2^+ molecular ions are present together with atomic ions. The first ones are produced by process (1) and by the homonuclear AI:

$$Na(3p_{3/2}) + Na(3p_{3/2}) \rightarrow Na_2^+ + e^- + \Delta E \qquad (2).$$

The energy defects ΔE of reactions (1) and (2) are defined by the difference between the initial state energies, given by the sums of the atomic excited state energies, and the final state energies, related to the ground states of the molecular ions. Atomic ions cannot be produced by direct ionizing collisions because of the high energy defect involved in these reactions.   Therefore they are due to different processes like charge-transfer collisions with molecular ions, electron impact ionization and molecular ion dissociation. As a last remark, we observe that no evidence of Cs_2^+ is present in the mass spectrum.

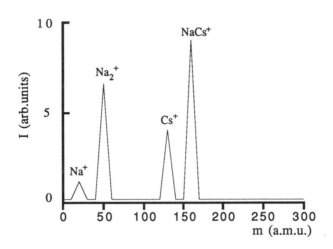

Figure 2. Typical mass spectrum in the AI experiment.

In order to both simplify the measurement and avoid systematic errors, we measured the rate coefficient of process (1) relative to reaction (2), whose rate coefficient has been measured by different groups. We adopted as a reference the value measured by Bonanno et al (1983) that is $K(Na_2^+) = (3.4 \pm 1.4) \times 10^{-11}$ $cm^3 s^{-1}$. In principle we could take as a reference also the homonuclear AI process in cesium, but this reaction is highly endothermic and its rate coefficient is very low and not well determined; this is demonstrated by the absence of Cs_2^+ in our mass spectra.

The detected ionic yields of processes (1) and (2) are:

$$I(NaCs^+)=eK(NaCs^+)n(3p_{3/2})n(6p_{3/2})\varepsilon(NaCs^+) V \qquad (3)$$

$$I(Na_2^+)=eK(Na_2^+)n^2(3p_{3/2})\varepsilon(Na_2^+) V/2 \qquad (4)$$

where e is the electronic charge, $n(3p_{3/2})$ and $n(6p_{3/2})$ are the laser-excited atomic populations, $\varepsilon(m)$ is the efficiency of the apparatus for ion mass m (in our case inversely proportional to the square root of m) and V is the interaction volume (the factor 1/2 in eq.(4) is due to statistical reasons). Dividing eq.(3) by eq.(4), we obtain:

$$K(NaCs^+) = 0.9 \ K(Na_2^+) \frac{I(NaCs^+)}{I(Na_2^+)} \frac{n(3p_{3/2})}{n(6p_{3/2})} \tag{5}.$$

The excited population ratio is determined through the collection of the fluorescence light. It depends on the excited state lifetimes that may be affected by radiation trapping. We evaluated this effect that, if not considered, would have introduced an important systematic error in the measurements. The evaluation has been made by applying the theory of Milne (1926) to our experimental situation (Gabbanini et al 1991a). This theory is valid in the range of low atomic densities as in our case; the absolute atomic densities have been estimated by the ratio between D_1 and D_2 fluorescence lines. The calculation shows that there is a condition where radiation trapping affects Na and Cs atom lifetimes in the same way, so that the lifetime ratio results unaffected; this condition corresponds to a Na/Cs density ratio of 3.9 that was used for the measurements.

By averaging over many ionic and fluorescence measurements, we obtained the rate constant value: $K(NaCs^+) = (5.4 \pm 3.2) \times 10^{-11} \ cm^3 s^{-1}$. The indetermination is mainly due to the reference rate constant indetermination. The cross section of $NaCs^+$ AI, by estimating the mean relative velocity between colliding atoms from Doppler width measurements, results equal to $\sigma(NaCs^+) = 8 \ Å^2$.

4. HORNBECK-MOLNAR IONIZATION

The studied HMI reaction involves the collision between a Rydberg state cesium atom and a ground state sodium atom. Rydberg atoms are prepared by two step laser excitation; by changing the second step laser wavelength, the 8d, 10s, 9d, 11s and 10d levels of cesium are alternately populated. A typical mass spectrum is shown in Fig.3. $NaCs^+$ and Cs_2^+ molecular ions, due to heteronuclear and homonuclear HMI collisions respectively, are present together with Cs^+ atomic ions. Cs^+ can be produced both by the processes already discussed in the AI section and by ion pair formation and photoionization of the Rydberg state (each laser has infact sufficient photon energy to keep excited cesium atoms above ionization threshold). Direct Penning ionization of Cs atoms has very low probability for these excited levels. Homonuclear HMI collisions in Cs have been studied by Korchevoi et al (1977) who measured cross sections of the process corresponding to several levels. For all the levels that we excited, homonuclear HMI reaction shows high cross section values. In this kind of reactions the cross sections increase with effective quantum number n*, starting from a threshold value related to the energy defect, up to n* ≈ 10, where they begin to decrease. At that point, infact, the Penning ionization process starts to be competitive with HMI. The cesium levels excited in the experiment are in the range $5.5 \leq n* \leq 7.5$ and are all above threshold for homonuclear HMI. The heteronuclear HMI threshold was not known in advance because the dissociation energy of $NaCs^+$ molecular ion, that defines its ground state energy, has never been measured. In the next section, we'll show how this experiment allows the extrapolation of the $NaCs^+$ dissociation energy.

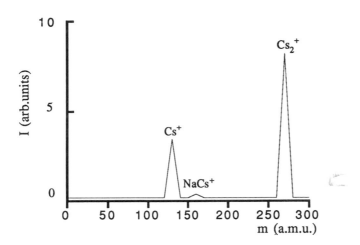

Figure 3. Typical mass spectrum in the HMI experiment after excitation of cesium to 9d level.

By exciting the nl level of cesium, we detected the ion yields:

$$I(NaCs^+) = e\ K_{nl}(NaCs^+)\ [Na(3s_{1/2})]\ [Cs(nl)]\ \varepsilon(NaCs^+)\ V \qquad (6)$$

$$I(Cs_2^+) = e\ K_{nl}(Cs_2^+)\ [Cs(6s_{1/2})]\ [Cs(nl)]\ \varepsilon(Cs_2^+)\ V \qquad (7).$$

Dividing eq.6 by eq.7, we obtain:

$$R = \frac{K_{nl}(NaCs^+)}{K_{nl}(Cs_2^+)} = 0.77 \frac{I(NaCs^+)\ [Cs(6s_{1/2})]}{I(Cs_2^+)\ [Na(3s_{1/2})]} \qquad (8).$$

It is then possible to get the ratio R between the two rate constants just with a measurement of the current and ground state density ratios. The knowledge of the cross sections $\sigma_{nl}(Cs_2^+)$ (Korchevoi et al 1977), together with the evaluation of the two mean atomic velocities, allows the derivation of absolute cross sections $\sigma_{nl}(NaCs^+)$, that are shown in

Figure 4. The measured values are: $\sigma_{8d}(NaCs^+)=0.04$ $Å^2$; $\sigma_{10s}(NaCs^+) = 1.0$ $Å^2$; $\sigma_{9d}(NaCs^+)=8.0$ $Å^2$; $\sigma_{11s}(NaCs^+)=3.6$ $Å^2$; $\sigma_{10d}(NaCs^+)=4.0$ $Å^2$.

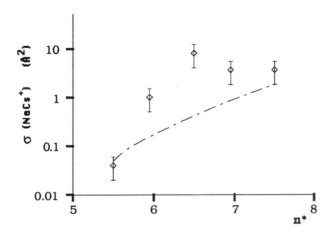

Figure 4. Cross sections for $NaCs^+$ HMI together with theoretical curve.

In the figure also a theoretical curve is plotted. It has been derived by applying the method of Janev and Mihajlov (1981) to $NaCs^+$ HMI and by using the molecular potential curves of Valence (1978). The Janev and Mihajlov method considers an initial channel $V_i(r)$, associated with a repulsive state, and a final state $V_f(r)$, associated with the ground state of the molecular ion. The difference $\varepsilon(r)=V_i(r)-V_f(r)$ is equal to the energy released to the electron after ionization. The ionization probability is defined depending on $\varepsilon(r)$ and on the photoionization cross section of the Rydberg level. The total cross section is calculated by an integration of the ionization probability over the impact parameters and over the kinetic energies (Gabbanini et al 1991b). The model, as previously demonstrated (Weiner and Boulmer 1986), gives the right order of magnitude but is unable to accurately reproduce the experimental data.

5. DISSOCIATION ENERGY

The HMI process can be used to experimentally determine the dissociation energy D_e of a molecular ion. Infact, passing from a Rydberg level to the successive one, it is possible to determine a threshold energy value where the HMI process becomes strongly effective. This energy value is then associated with a dissociation energy. The precision depends on the excited level spacings and on the kinetic energies of the colliding atoms. HMI has already been used by Solarz et al (1977) to determine the dissociation energies of Ca_2^+, Sr_2^+ and Ba_2^+. In our case we can identify the threshold energy value of $NaCs^+$ HMI between the 8d and 10s levels of cesium, where the associated cross section increases by more than one order of magnitude. This implies $3100 \ cm^{-1} < D_e(NaCs^+) < 3500 \ cm^{-1}$, that is in good agreement with the theoretical value of Von Szentpaly et al (1982) $D_e(NaCs^+) = 3209 \ cm^{-1}$.

6. CONCLUSIONS

Experiments on $NaCs^+$ formation in different collisional processes (AI and HMI) are presented. These processes involve atoms excited by lasers in different schemes. An experimental evaluation of $NaCs^+$ dissociation energy is reported. Work is in progress to excite higher Rydberg states of cesium.

ACKNOWLEDGEMENTS

The authors would like to thank M.Badalassi and M.Tagliaferri for technical assistance.

REFERENCES

Bonanno R, Boulmer J and Weiner J 1983 Phys.Rev.A **28** 604
Djerad M T, Cheret M and Gounand F 1987 J.Phys.B **20** 3789
Gabbanini C, Biagini M, Gozzini S, Lucchesini A and Moi L 1991a Phys.Rev.A **43** 2311
Gabbanini C, Biagini M, Gozzini S, Lucchesini A and Moi L 1991b J.Phys.B **24** XXX
Gould P L, Lett P D, Julienne P S, Philips W D, Thorsheim H R and Weiner J 1988 Phys.Rev.Lett. **60** 788
Hornbeck J A and Molnar J P 1951 Phys.Rev. **84** 621
Janev R K and Mihajlov A A 1981 J.Phys.B **14** 1639
Johnson C B, Wang M X and Weiner J 1988 J.Phys.B **21** 2599
Korchevoi Yu P, Lukashenko V U and Khil'ko I N 1977 Sov.Phys.Tech.Phys. **21** 1356
Lucatorto T B and McIlrath T J 1976 Phys.Rev.Lett. **37** 428
Milne E A 1926 J.London Math.Soc. **1** 40
Solarz R W, Worden E F and Paisner J A 1977 Opt.Eng. **21** 1356
Valence A 1978 J.Chem.Phys. **69** 355
Von Szentpaly L, Fuentealba P, Preuss H and Stoll H 1982 Chem.Phys.Lett. **93** 555
Weiner J and Boulmer J 1986 J.Phys.B **19** 599

Ejected electron spectra of doubly excited states from double capture in collisions of bare ions with helium atoms

CD Lin, Z Chen and R Shingal

Department of Physics, Kansas State University, Manhattan, Kansas 66506 USA

ABSTRACT: The ejected electron spectra of doubly excited states of ions resulting from double capture in collisions of bare ions with helium atoms in the energy range of a few keV/amu are calculated. The electron capture amplitudes are obtained using close-coupling expansion within the independent electron approximation. By combining these electron capture amplitudes with the Auger decay amplitudes and taking into account the post-collision-interaction effect the electron spectra as a function of the ejection angles are evaluated. Results from C^{6+}-He and O^{8+}-He collisions are compared to experiments.

1. INTRODUCTION

In recent years double electron capture between a multiply charged ion and a many-electron target atom or molecule has been studied in many laboratories. In the earlier studies, the energy gain spectroscopy method has been used to show that double capture processes are important in these collisions. With the advent of high-resolution Auger-electron spectroscopy, a large amount of absolute or relative cross sections to individual doubly excited states have been reported (Mack et al. 1989; Stolterfoht et al. 1990; Moretto-Capelle et al. 1989; Posthumus and Morgenstern, 1990). The electron spectra are often measured at one or more angles from which experimentalists have deduced the absolute or the relative cross sections for the formation of individual doubly excited states. To get such information, however, a number of assumptions about the formation and the decay of doubly excited states must be made. These assumptions are often not justified and the deduced "experimental" cross sections are not reliable. Only very recently have the experimentalists begun to measure the angular distribution of Auger electrons resulting from the collisions of multiply charged ions with atoms (Holt et al. 1990; Boudjema et al. 1991).

Despite the experimental activities in this area and the progress of ion-atom collision theory for single-electron processes made in the last decade, there have been very few theoretical calculations for such two-electron collision processes. This is partly due to the complexity of the processes involved. Consider, for example, O^{8+}-He collisions at a few keV/amu. It has been shown that double capture populates mostly $3\ell3\ell'$ and $3\ell4\ell'$ manifolds and smaller $3\ell n\ell'$ ($n\geq5$) manifolds. At such low collision energies, models based on the close-coupling method are obviously the most suitable. In principle, close coupling expansion based on the two-electron atomic or molecular orbitals can be formulated to solve the time-dependent Schrödinger equation, but the actual implementation for such a collision system is clearly very complicated. By including all the important channels, one needs to include close to 150 basis functions (counting the different magnetic states) in the calculation. Such a large calculation is not practical even for today's supercomputers.

In view of such practical difficulties, a simpler model based on the independent electron model is desirable. For ion-atom collisions where the projectile incident energy is of the order of a few keV/amu or greater, the collision time is short compared with the decay time of the doubly excited states. Thus the ejected electron spectra can be considered to be formed in a two-step process where doubly excited states are first collisionally populated, then followed by Auger emissions. In this article we discuss the angular distribution of the emitted electrons.

The angular distribution of the Auger electron of an isolated excited state has been studied over the years. For the doubly excited states populated by ion-atom collisions two other effects must be considered. First, the Auger emission occurs in the electric field of a receding ion such that the angular distribution is different from that of an isolated atom or ion. This has been called the post-collision-interaction (PCI) effect in general. Second, the Auger widths of some doubly excited states are comparable to the energy separations between neighboring states such that the Auger emission from diffferent states must be treated coherently. Since each doubly excited state can also decay radiatively, the branching ratio for the Auger emission of each state must be taken into account as well.

The theoretical model used for the calculation of the electron spectra is given in Section 2. In Section 3, the model is applied to analyze the electron spectra of $2\ell3\ell'$ doubly excited states of C^{4+} resulting from the collision of C^{6+} on He and of $3\ell3\ell'$ doubly excited states of O^{6+} resulting from the collision of O^{8+} on He. The angular distributions of the electron spectra are analyzed and the results compared to the experimental data. The conclusions are given in Section 4.

2. THEORY

2.1 The Independent Electron Approximation

Consider the collision of a bare ion of charge Z with a helium target atom.

The time-dependent Schrodinger equation is

$$\left[-\frac{1}{2}\nabla_1^2 - \frac{1}{2}\nabla_2^2 - \frac{2}{r_{1a}} - \frac{2}{r_{2a}} + \frac{1}{r_{12}} - \frac{Z}{r_{1b}} - \frac{Z}{r_{2b}} - i\frac{\partial}{\partial t} \right] \psi\left(r_1, r_2, t\right) = 0 \qquad (1)$$

where a and b refer to the helium and the bare ion nucleus, respectively. One version of the independent electron model is to replace the $1/r_{12}$ term with a screening potential. Under such a replacement, the resulting equation becomes separable, and the two-electron wave function can be written as

$$\psi\left(r_1, r_2, t\right) = \phi\left(r_1, t\right)\phi\left(r_2, t\right), \qquad (2)$$

where the one-electron wave function satisfies

$$\left[-\frac{1}{2} \nabla_1^2 + V(r_{1a}) - \frac{Z}{r_{1b}} - i\, \frac{\partial}{\partial t} \right] \phi(r_1, t) = 0 \tag{3}$$

and where the 'model potential' V includes the screening between the two electrons. Equation (3) can be solved by the close-coupling method using the two-center atomic orbital expansion method. For such a one-electron system, there is no difficulty to include up to 50-70 basis functions. According to this model, the two-electron wave function as t⇒∞ for each b, is

$$\psi(r_1, r_2, t \to \infty) = \sum_{n\ell m} \sum_{n'\ell'm'} a_{n\ell m}(b)\; a_{n'\ell'm'}(b)\; u_{n\ell m}(r_1)\; u_{n'\ell'm'}(r_2) \tag{4}$$

where the u's are the hydrogenic orbitals for nuclear charge Z. To extract the capture amplitude we project $\Psi(r_1, r_2, t \Rightarrow \infty)$ into each doubly excited state which is represented by a configuration interaction wave function. From the capture amplitudes the total double capture cross sections can be obtained.

In the model above one still needs to specify the potential V(r) to be used for the one-electron calculation. For the first electron, V(r) is chosen such that the binding energies of the ground state and the first few excited states of helium are in agreement with experimental values (Shingal 1988). For the second electron, we choose $V(r_a) = -2/r_a$ so that the total ionization energy to remove both electrons is correct. This is equivalent to assuming that the second electron is completely relaxed after the first electron has been removed, but it also shows that the second electron is captured from the smaller distances where the screening by the other electron can be neglected. By not treating the two electrons equivalently, the change of screening can be included. In this case the wave function (2) has to be properly antisymmetrized.

2.2   The Ejected Electron Spectra

For comparison with experimental electron spectra, the autoionization of doubly excited states after capture should be considered and several effects should be included: (1) Each doubly excited state populated is aligned in general, i.e., the cross section for each magnetic substate is not equal. This would result in unisotropic angular distributions; (2) Each state has a nonzero fluoresence yield which must be considered; (3) The ejected electron's spectrum for each state is modified by the PCI effect. This effect changes the electron's spectra from the symmetric Lorentzian shape to an asymmetric shape stretching toward the lower energy side; (4) When doubly excited states are not well separated, the ejected electron spectra from neighboring states must be added coherently.

The expression which describes the angular distribution of the ejected electron which includes the effects stated above is

$$I(k) = 2\pi \sum_{n\ell_1} \sum_M \left| \sum_\alpha f^\alpha(\epsilon, \epsilon_L)\, a_{LM}^\alpha(\theta_p)\, 4\pi \sum_{\ell_2} e^{i(\sigma_{\ell_2} + \ell_2 \pi/2)} \right.$$

$$\left. \langle \alpha LM | V_A | n\ell_1 \epsilon_k \ell_2 LM \rangle \sum_{m_1 m_2} \langle \ell_1 \ell_2 m_1 m_2 | LM \rangle\, Y_{\ell_2 m_2}^*(k) \right|^2. \tag{5}$$

In this expression, the electron is detected at an angle $k=(\theta,\phi)$ with respect to the incident direction. The electron capture amplitude to the doubly excited state $|\alpha LM\rangle$ is $a_{LM}^{\alpha}(\theta_p)$ where θ_p is the scattering angle of the projectile. The operator V_A is the electron-electron interaction. After autoionization the one-electron ion is left in the $|n\ell_1\rangle$ state and the ejected electron in the $|\epsilon_k\ell_2\rangle$ continuum ($\epsilon_k = k^2/2$) which has a Coulomb phase shift σ_{ℓ_2}. The amplitdue f^{α} describes the PCI effect. There are many theoretical models for the PCI effect (Baker and Berry, 1966; Devdariani et al. 1977; Niehaus, 1977; Arcuni, 1986; van der Straten and Morgenstern, 1986a,b; van der Straten et al. 1988; Barrachina and Macek, 1989). We follow the prescription of van der Straten and Morgenstern (1986a,b) for each isolated state. Note that the amplitudes are added coherently and thus the interference among the states are accounted for.

The equation above can be simplified for collisions where the projectile scattering angles are not measured. In such integral measurements, the electron spectra do not depend on the azimuthal angle ϕ. If the PCI effect and the interference can be neglected then each state has the angular distribution

$$I(\theta) = \sum_{\ell} A_{\ell} \, P_{\ell}(\cos\theta) \qquad (6)$$

where ℓ is limited to even integers, and A_{ℓ} is related to the alignment parameters. This simple situation occurs approximately, for example, in the decay of $2\ell3\ell'$ doubly excited states to the $1s\epsilon\ell''$ continuum.

3. RESULTS AND DISCUSSION

3.1 C^{6+} + He

In Table 1 we show the calculated single capture cross sections to n=2,3 and 4 states and double capture cross sections to $2\ell3\ell$, $3\ell3\ell'$ and $3\ell4\ell'$ states from the present independent electron model. The single capture cross sections can be compared to other theoretical calculations based on solving the two-electron Schrödinger equation (Fritsch and Lin, 1986; Shimakura et al. 1987) and with experiments (Dijkkamp et al. (1985); Liu et al. (1989)). For double capture cross sections there have been no other theoretical calculations based on the two-electron model. Electron spectra

Table 1.   Comparison of cross sections for single capture to n=2,3 and 4 states and double capture to doubly excited states in the $2\ell3\ell'$, $3\ell3\ell'$, $3\ell4\ell'$ manifolds for 60 keV C^{6+} + He collisions. The cross sections are in units of 10^{-17} cm^2.

	Experiment		Theory	
n=2			4.7[b]	
n=3	80[a]		110.0[b]	126[d]
n=4	16[a]		13.0[b]	13.8[d]
$2\ell3\ell'$	10.6 ± 0.3[c]		8.2[b]	
$3\ell3\ell'$	2.4 ± 0.2[c]		2.5[b]	
$3\ell4\ell'$	1.32 ± 0.1[c]		5.4[b]	

a   For O^{6+} + He from Dijkkamp et al. (1985);   b   present calculations.
c   Stolterfoht et al. (1990)   d   Fritsch and Lin (1986).

resulting from double capture to $2\ell n\ell'$ ($n\geq3$) and $3\ell n\ell'$ ($n\geq3$) doubly excited states have been observed experimentally at different angles (Stolterfoht et al. 1990; Mack et al. 1989; Sakaue et al. 1989, 1990). Stolterfoht et al. (1990) have attempted to deduce absolute double capture cross sections from the experimental electron spectra measured at a particular angle. In doing so, the Auger electron distributions were assumed to be isotropic and an average fluorescence yield was used for each manifold. We show in Table 1 their results and the comparison with the calculated values for double capture to the $2\ell3\ell'$, $3\ell3\ell'$ and $3\ell4\ell'$ manifolds. In general, the agreement between the calculated and experimental results is quite good, particularly if one recognizes that the cross section for double capture to the largest double capture manifold ($2\ell3\ell'$) is less than 10% of the cross section for single capture to the n=3 states. For the weaker $3\ell3\ell'$ and $3\ell4\ell'$ manifolds, the ratios are only a few percents.

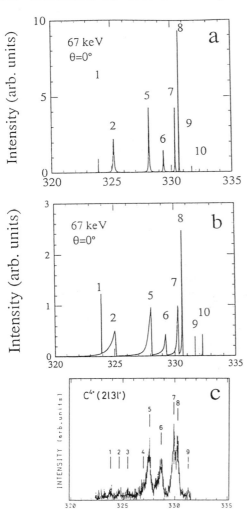

The ejected electron spectra in general depends largely on the angle of the emitted electron. Except under special circumstances it is not straightforward to deduce total double capture cross sections from measurements of electron spectra at a few angles. Thus it is important that theoretical electron spectra be compared directly with the experimental ones. In Fig. 1 the electron energy spectra resulting from the Auger decay of double capture to $C^{4+}(2\ell3\ell')$ states in C^{6+} collisions with He at 67 keV are shown for the electron ejection angle of $\theta=0°$. In Fig. 1a, the spectra are obtained by

Figure 1. Electron spectra at 0° for the $2\ell3\ell'$ manifold of C^{4+} resulting from the collision of bare carbon ions with helium atoms at 67 keV. (a) Theoretical spectra assuming that each state decays incoherently and no PCI effect is included. For narrow lines, the actual width may be smaller than what is shown. (b) Theoretical spectra including the PCI effect and that the Auger decay from all the states has been treated coherently. (c) The experimental electron spectra from Sakuae et al. (1989, 1990).

treating each state independently and neglecting the PCI effect.   In this
approximation, each resonance has a Lorentzian shape and the width of each
state reflects the Auger width of that state.   In Fig. 1b the effect of PCI
is included and the contribution to the electron spectra from each state is
treated coherently.   The scales in Fig.1a and Fig. 1b are identical.   It is
clear that the PCI effect broadens and shifts the strength of each line
toward lower energies.   To obtain the theoretical electron spectra in
Fig. 1, the atomic parameters for all the states in the (2,3) manifold must
be calculated.   One can compare Fig. 1b with the experimentally determined
electron spectra of Sakaue et al. (Fig. 1c) measured at 0°.   The overall
agreement between Fig. 1b and experimental data is satisfactory.   The
theoretical spectra have not been convoluted with the electron energy
resolution so that each line profile shows more pronounced skewness than the
experimental one.   The relative intensities among the lines also appear to
be in qualitative agreement with the experiment.   The Auger yield for each
state has been included in the calculated spectra.   Among the $2\ell 3\ell'$ states,
two of them, $^1P^e$ and $^1D^o$, (lines #3 and #4 in Fig. 1c) do not decay by
electron emission and are not seen in the theoretical electron spectra.

3.2 O^{8+} + He

For this system the cross section for double capture to doubly excited
states is about the same order as single capture cross sections for
collision energies at a few keV/amu.   This has been shown by Bliman et al.
(1983).   The calculation of Kimura and Olson (1985) disputed this statement,
but they did not include double capture channels in their basis set.   In
Table 2 we show the single and double capture cross sections calculated by
the present independent electron model and the comparison with the calculations of Bliman et al. (1983) and of the deduced experimental results.
Double capture cross sections to the (3,3) and (3,4) manifolds are also
compared to experimental data from different sources.

In this section we concentrate on the electron spectra resulting from double
capture to the $3\ell n\ell'$ (n≥3) manifolds. The spectra had been studied by Mack
et al. (1989) for 96 keV $^{18}O^{8+}$ ions on He where the electron spectra were
measured at 50°.   The same system has also been studied by Moretto-Capelle
et al. (1989) at 80 keV where the electrons were measured at 10°.   Both
experiments were able to separate most of the individual doubly excited

Table 2. Comparison of theoretical and experimental single (sc) and double
(dc) capture cross sections, and double capture to (3,3) and (3,4) manifolds
for O^{8+} + He collisions. The cross sections are given in units of 10^{-16} cm^2.
The theoretical results are from the present calculation at 96 keV.   The
experimental results are at different energies (see Table).

	Theory		Experiments
single capture	16[a]	13[b]; 22[c]	
double capture to:			
(3,3)	5.3[a]	1.8[b]	
(3,4)	9.3[a]	3.0[b]	
single + double capture	31.0[a]	18[b]	17±3.4[b]; 23±4.6[d]; 34.4±6.9[e]
ratio: $\sigma(3,3)/\sigma(3,4)$	0.57[a]	0.60[b]	0.85[f]; 0.64[g]

a. Present results. b. Bliman et al. (1983). c. Kimura and Olson (1985).
d. Afrosimov et al. (1982).   e.   Iwai et al. (1982) at 18 keV.   f.
Moretto-Capelle et al. (1990) at 80 keV. g. Barat et al. (1987) at 10 keV.

states of the $3\ell3\ell'$ manifold. The same system has also been investigated using the energy gain spectroscopy by Rocin et al. (1989) at the much lower energy of 10 keV where the angular distribution of the scattered projectiles has been measured. However, the latter method does not separate individual doubly excited states formed. This collision system has also been studied by Mann (1987) using electron spectroscopy.

The theoretical analysis of Auger electron energy spectra for the (3,3) doubly excited states of O^{6+} resulting from the double capture in slow O^{8+}-He collisions is more complicated. There are three important differences for this system as compared to the (2,3) states discussed above. First, each state can decay to 2s and 2p states of the O^{7+} ion. Second, the emitted Auger electrons have smaller energies; they are in the range of 34-50 eV. Third, the Auger widths of a number of states are comparable to the energy separations between neighboring states.

To calculate the electron spectra, the atomic parameters for all the eleven singlet states within the (3,3) manifold have to be calculated. For convenience, we number each state starting with the lowest one, although each state is also labelled by the K and T quantum numbers (Lin 1986). Since A=+1 for all the states in the manifold, this quantum number was not given explicitly. The Auger width and the double capture cross section to each state (for 96 keV O^{8+} ions on He) from our calculation are also needed. The branching ratios for Auger decay to 2s and 2p states were obtained from the calculations by Bachau as quoted in Chetioui et al. (1989). In calculating the electron spectra, we used the energies and widths from Ho (1981) in the PCI model but the Auger decay amplitudes needed in the calculation of the spectra are from our own CI calculations. Our calculated energies and widths do not differ significantly from Ho's.

The post-collision interaction effect is more important for slow Auger electrons. For each isolated state, the PCI effect shifts the intensity toward the low-energy side and broadens the apparent width of each state. It also has the effect of enhancing the overlap between neighboring states making it impossible to identify the area under each peak with the electron's intensity associated with that state. The spectra are also no longer symmetric with respect to $\theta=90°$ because of the interference among the states and the PCI effect.

We compare in Fig. 2 the theoretical electron spectra with the experimental data of Mack et al. (1989). The agreement between the theoretical spectra and experimental one is only qualitative. However, from a private communication (Morgenstern, 1991), the experimental spectra were found to be in error. In particular, the "new" experimental energies are now in agreement with those calculated by Ho. Since our spectra were calculated using Ho's energies, we expect that the locations of the peaks in Figs. 2a and 2b will be in agreement with the new data. Further comparison of experiment with this calculation has to await until the new experimental spectra becomes available. On the other hand, the "pileup" of electron intensity at lower energies is due to the PCI effect and is expected to remain the same.

Besides the independent electron approximation used, we should mention that the present treatment of the PCI effect is incomplete. The present model treats the PCI effect of each state individually by multiplying the amplitude of each state by a shape function $f^\alpha(\epsilon, \epsilon_L)$. In the (3,3) doubly excited states considered here, this treatment is inadequate. We note that

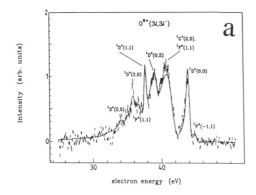

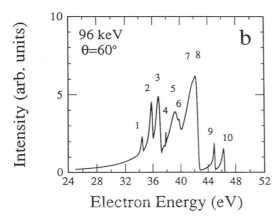

Figure 2. Comparison of the theoretical electron spectra at 60° in the emitter frame with the experimental electron spectra of Mack et al. (1989). The states are labelled in terms of the $(K,T)^A$ quantum numbers, and in the order of increasing energies are: 1. $(2,0)$ $^1S^e$; 2. $(2,0)$ $^1D^e$; 3. $(1,1)$ $^1P^o$; 4. $(0,2)$ $^1D^e$; 5 $(0,0)$ $^1S^e$; 6. $(2,0)$ $^1G^e$; 7. $(1,1)$ $^1F^o$; 8. $(0,0)$ $^1D^e$; 9. $(-1,1)$ $^1P^o$; 10. $(-2,0)$ $^1S^e$. One of the states in this manifold does not autoionize and A=+1 for all the states listed.

states #6, #7 and #8, for example, are easily mixed by the electric field from the receding He^{++} ions. In considering the PCI effect, these states should be treated together in a manner similar to the linear Stark effect as proposed in the literature (Stolterfoht et al. 1979; Maraglia and Macek 1990).

4. SUMMARY AND CONCLUSIONS

In this paper we analyze the electron energy spectra of doubly excited states populated by double electron capture processes in slow collisions between multiply charged ions with atoms. In the theoretical calculations, we assume that doubly excited states are first populated by double capture, the ejected electron spectra are obtained by following the electron emission from individual states. It is shown that one needs to include the post-collision interaction effect and that the emitted electrons from different states must be treated coherently. This is particularly true when the Auger electron energies are small and when the energy separations between neighboring states are comparable or smaller than the widths of the states.

Doubly excited states populated in slow collisions are usually strongly aligned and the resulting ejected electron spectra are not isotropic. We emphasize that it is extremely difficult to deduce double capture cross sections to individual doubly excited states by fitting the measured electron spectra at one angle only. Except for very special circumstances, the number of parameters to be fitted from the experimental data are too many and the fitted results may not be unique. When the PCI effect is large and/or when coherence is important, it is more desirable that the experimental electron spectra be compared to the theoretical electron spectra directly at different angles.

We have considered the electron energy distribution from the Auger decay of the $(2,3)$ doubly excited states of C^{4+}. These states decay to the 1s state of C^{5+} where the energies of the Auger electrons are in the order of 330 eV. In this case, the Auger decay of each doubly excited state can be treated independently and the PCI effect is negligible such that measurements of the electrons at a few angles would allow the determination of double capture cross sections to individual doubly excited states. We have also considered the Auger electrons from the $(3,3)$ states of O^{6+} following double capture in O^{8+}-He collisions. For this system the PCI effect is very large and the electron emissions from different states are all mixed such that the electron intensity under a peak cannot be attributed to individual states directly.

The major limitation of the present calculation is the use of the independent electron approximation in the calculation of double capture amplitudes. It is desirable that theoretical calculations based on the two-electron model be carried out. The latter may be quite complicated for $(3,3)$ and $(3,4)$ doubly excited states because of the large number of states in each manifold. On the other hand, such two-electron calculations for the $(2,2)$ and $(2,3)$ doubly excited states are possible, except that the PCI and the overlapping interference for these states are much smaller.

Acknowledgment

This work is supported in part by the US Department of Energy, Office of Basic Energy Sciences, Division of Chemical Sciences.

REFERENCES

Afrosimov VV et al. 1982 12th Int. Conf. on Physics of Electronic and Atomic Collisions
Barat M et al. 1987 J. Phys. **B20** 5771
Barker R B and Berry H W 1966 Phys. Rev. **151** 14
Barrachina R O and Macek J H 1989 J. Phys. **B20** 2151
Bliman S et al. 1983 J. Phys. **B16** 2849
Bordenave-Montesquieu A et al. 1987, XVIth International Conference on the Physics of Electronic and Atomic Collisions, Abstracts of Papers, edited by   Geddes et al. (Brighton, UK, 1987) p 551
Boudjema M et al. 1991 to be published in J. Phys. B (1991)
Chetioui A et al. 1989 J. Phys. **B22** 2865
Devdariani A Z, Ostrovski V N and Sebyakin Yu N 1977 Sov. Phys. JETP **46** 215
Dijkkamp D, Ciric D, Vlieg E, de Boer A and de Heer F J 1985 J. Phys. **B18** 4763
Fritsch W and Lin C D 1986 J. Phys. **B19** 2683
Ho Y K 1981 Phys. Rev. **A23** 2137
Holt R et al 1991 Phys. Rev. **A43** 607

Iwai T et al. 1982 Phys. Rev. **A26** 105
Kimura M and Olson R 1985 J. Phys. **B18** 2729
Lin C D 1986 Adv. At. Mole. Phys. **22** 76
Liu C J, Dunford R W, Berry H G, Pardo R C, Groeneceld K O, Hass M and
    Raphaelian M L A 1989 J. Phys. **B22** 1217
Mack M, Nijland J H, van der Straten, Niehaus A and Morgenstern R 1989
    Phys. Rev. **A39** 3846
Mann R 1987 Phys. Rev. **A35** 4988
Miraglia J and Macek J H 1990 Phys. Rev. **A43** 3971
Moretto-Capelle P et al. 1990 J. Phys. **B22** 271
Posthumus J H and Morgenstern R 1990 J. Phys. **B23** 2293
Roncin P, Barat M, Gaboriaud M N, Guillemot L and Laurent H 1989 J. Phys.
    B22 509
Sakaue H A et al. XVIth International Conference on the Physics of
    Electronic and Atomic Collisions, Abstracts of Papers, edited by Dalgarno
    A, Freund R S, Lubell M S and Lucatorto, T B (New York, 1989) p 570
Sakaue H A et al. 1990 J. Phys. **B23** L401
Shimakura N, Sato H, Kimura M, and Watanabe T 1987 J. Phys. **B20** 1801
Shingal R 1988 J. Phys. **B21** 2065
Stolterfoht N, Brandt D and Prost M 1979 Phys. Rev. Lett. 43 1654
Stolterfoht N et al. 1990 Phys. Rev. **A42** 5396
van der Straten P, Morgenstern R and Niehaus A 1988 Z. Phys. **D8** 35

Evidence for interference between resonant and nonresonant transfer and excitation

M. Benhenni[x+], S.M. Shafroth[x], J.K. Swenson[*♦], M. Schulz[*♣], J.P. Giese[*●],
H. Schone[*●], C.R. Vane[*], P.F. Dittner[*], S. Datz[*]

[x] Department of Physics and Astronomy, University of North Carolina
   at Chapel Hill, Chapel Hill, N.C. 27599-3255 USA.
[*] Physics Division, Oak Ridge National Laboratory, Oak Ridge,
   TN 37831 USA.

ABSTRACT: The first angular distribution measurements of Auger electrons
arising from transfer and excitation Auger (TE) processes were performed at O^{5+}
projectile energies of 8 MeV, where the resonant (RTE) and the nonresonant (NTE)
transfer and excitation have equal contributions in populating the $(1s2s2p^2)^3D$ state,
and at 13 MeV where RTE is predominant and at 6 MeV where NTE is more
significant. .The TE angular distributions are strongly forward peaked along the beam
direction. The data were fitted to $w(\theta_p)=C[1+a_2P_2(\cos\theta_p)+a_4P_4(\cos\theta_p)]$ and the
magnetic substate population probabilities were inferred.

1.INTRODUCTION

Resonant (RTE) and nonresonant (NTE) transfer and excitation are two well
known processes(1-18) which involve the transfer of a target electron to the
projectile and the excitation of a projectile electron. In RTE, the excitation and capture
are correlated since the captured target electron excites the projectile electron. On the
other hand, in NTE, the capture and excitation are uncorrelated since the
projectile electron is excited by the target nucleus. Another process that has
been identified recently (14,17) is the two electron transfer and excitation process
(2eTE). In 2eTE, the excitation and capture are also uncorrelated, but in this case,
the projectile electron is excited by the non captured target electron. Another
mechanism that can give rise to the same final state as RTE, NTE and 2eTE is the
direct ionisation of the target electrons known as the Binary Encounter
electrons (BE).Since all these four channels can form the same final state, they are
likely to interfere if the amplitudes for the individual channels are added
coherently (3,12,13,18). The interference betweeen RTE and BE that is evident
from the "Fano" lineshape of the doubly excited state have been discussed
previously (8,18,19). In earlier RTE and NTE experiments, isotropic
angular distribution of RTE and NTE was assumed and the RTE and NTE
amplitudes were added incoherently neglecting any interferences between
these processes. This paper reports a strong anisotropy of the transfer and
excitation processes angular distributions and gives evidence for possible
destructive interference between RTE and NTE.

2.EXPERIMENT AND RESULTS

The experiment was performed at the EN Tandem of Oak Ridge National Laboratory. Beam of O^{5+} was formed and directed toward the He target contained in a cylindrical gas cell of 25 mm and kept at 10 mTorr pressure. Linearity of electron yield versus pressure ensured single collision conditions at 10 mTorr. The emitted electrons were detected with the high resolution projectile electron spectrometer in the Lab angular range [5-55°] . The electrons were decelerated to one fourth of their lab energy to enhance their energy resolution. They were then energy analysed by a two stage refocusing 30° parallel plate analyzer and detected by an 8×50 mm^2 microchannel plate equipped with a resistive anode encoder at a position proportional to their energy. The Doppler broadenings caused by the variation of the observation angle $d\theta_L = \pm 0.4°$ permitted by the spectrometer entrance slit width, become more severe as the observation angle increases. They are eliminated to first order through refocusing of the projectile electrons on the detector which is remotely positioned along the shifted focal line. Details concerning the spectrometer and the refocusing technique have been reported previously (20).

Fig 1 shows the electron spectra at various lab emission angles at 8 MeV projectile energy. The state of interest is the $(1s2s2p^2)^3D$ state at 449 eV, known to be formed through the transfer and excitation processes (8). The other lines in the spectrum are Li-like lines formed through different excitation mechanisms. Among these lines, the $(1s2s^2)^2S$ excited state at 412 eV decays to the ground state $(1s^2)^1S$ and therefore it is an isotropic line. The ^3D Auger decay differential cross sections were normalized to the $(1s2s^2)^2S$ differential cross section. This normalization serves to eliminate errors on the beam current integration, target length, solid angle and gas pressure.

In fig. 2, the ratios $[d\sigma/d\Omega(^3D)]/ [d\sigma/d\Omega(^2S)]$ are plotted versus projectile emission angle θ_p at the projectile energies 6, 8 and 13 MeV, where $d\sigma/d\Omega(^3D)$ and $d\sigma/d\Omega(^2S)$ are the respective Auger ground state decay differential cross sections of the ^3D and the ^2S excited states respectively. The projectile angle θ_p was obtained from the lab emission angle θ_L by the kinematic frame transformations. The solid lines are linear least squares fits to the data represented by

$w(\theta_p) = C[1 + a_2 P_2(\cos\theta_p) + a_4 P_4(\cos\theta_p)]$ given by angular momentum theory,

where $P_2(\cos\theta_p)$ and $P_4(\cos\theta_p)$ are the second and fourth order Legendre polynomials respectively. Fig 2 shows that the ^3D angular distributions are forward peaked along the beam direction, more forward peaked at the resonance energy 13 MeV where RTE populates the ^3D state predominantly (8), than at the energies 6 and 8 MeV where NTE has a significant contribution (8). Furthermore, the ^3D angular distribution at 8 MeV projectile energy where RTE and NTE have equal magnitudes (8), is less forward peaked than the angular distribution at 13 MeV energy where RTE predominates or at 6 MeV where NTE predominates. This suggests that destructive interferences may be taking place between the RTE Auger (RTEA) and NTE Auger (NTEA) in the forward direction. This result is in agreement with previous calculations by T. Reeves (13), where a dip occurs in the ^3D energy dependence cross section after adding coherently the magnitudes for RTE and NTE. This destructive interference was also suggested by Itoh et al (7) to explain the positive shift of

8 MeV O^{5+} + He

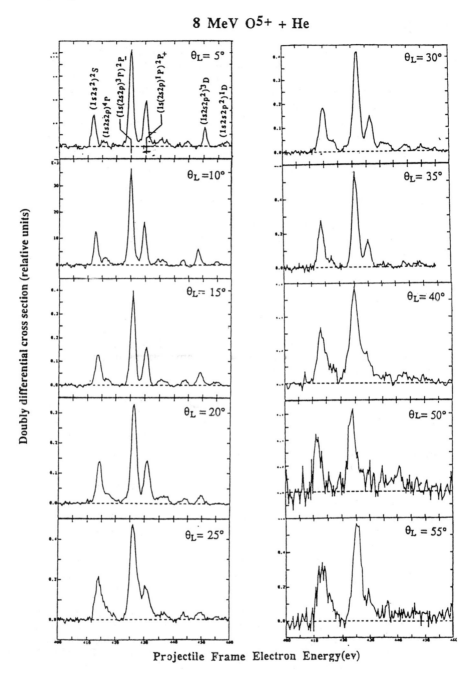

Fig. 1. Auger electron spectra in the projectile frame at various Lab angles θ_L following O^{5+} + He collisions at 8 MeV O^{5+} projectile energy.

the resonance energy and the smaller width of the resonance than the predicted value by the Impulse approximation. He claimed that this departure of the peak position and width from the theory can be understood if the RTE and NTE amplitudes interfere destructively and therefore the RTE and NTE amplitudes are added coherently.

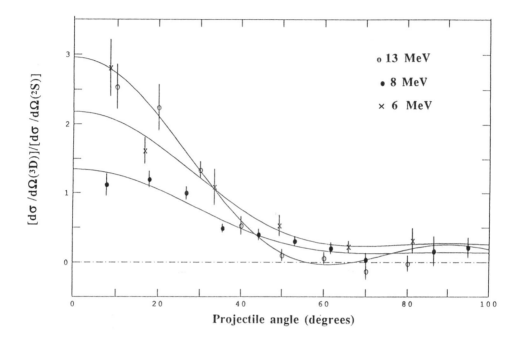

Fig. 2. Relative angular distributions of the $(1s2s2p^2)^3D$ Auger decay to ground state $(1s^22s)^2S$ at 6, 8 and 13 MeV O^{5+} energies. The solid lines are least squares fits to the data and given by $w(\theta_p)=C[1+a_2P_2(\cos\theta_p)+a_4P_4(\cos\theta_p)]$.

Furthermore, the magnetic substate population probabilities $Q(lm)$ were inferred from the anisotropy coefficients as follows:

$Q(20) = 0.2[1 + a_2 + a_4]$
$Q(2\pm1) = 0.2[1 + 0.5a_2 - 0.66a_4]$
$Q(2\pm2) = 0.2[1 - a_2 + 0.15a_4]$

Table. 1. Magnetic substate population probabilities of the $(1s2s2p^2)^3D$ formed following O^{5+} collisions with He.

E(MeV)	Target	$a_2 \pm \Delta a_2$	$a_4 \pm \Delta a_4$	Q(20) ± ΔQ(20)	Q(2±1) ± Δ(Q2±1)	Q(2±2) ± ΔQ(2±2)
13	He	3.43 ± 0.88	2.26 ± 1.18	1.33 ±0.50	0.24 ± 0.28	-0.40 ± 0.40
8	He	1.92 ± 0.34	0.97 ± 0.25	0.79 ± 0.08	0.26 ± 0.05	-0.15 ± 0.05
6	He	1.80 ± 0.33	1.03 ± 0.37	0.77 ± 0.11	0.24 ± 0.06	-0.13 ± 0.07

Table 1 shows that the ^3D state is exclusively populated with the magnetic substate $m_l=0$ at the resonance energy of 13 MeV where RTE predominates in agreement with Bhalla's calculations (18). However, at 6 and 8 MeV projectile energies where NTEA plays a significant role in populating the ^3D state, small contributions from the magnetic substates $m_l=\pm1$ are also observed. The presence of the magnetic substate $m_l= \pm1$ at these energies can be undestood since the electron excitation is caused by the target nucleus therefore changing the initial angular momentum $m_l=0$ whereas, in RTE the interaction is electron electron interaction of the type $1/r$ (18) and therefore conserves the initial angular momentum $m_l=0$. The negative values of Q(2±2) shown in table 1 may be attributed to contributions from close lying lines (21) which may have different angular distributions. These negative values may also be attributed to the destructive interferences between RTEA and NTEA which may introduce odd terms in the linear combination of Legendre polynomials. At the resonance energy, NTEA contribution can be neglected and the only interference contribution is the interference between RTEA and the binary encounter process which varies as $P_2(\cos\theta_L)$ in the angular range considered here (18). Therefore, the sum of the resonance term and the interference term is well fitted with a linear combination of only even order Legendre polynomials. Here, the interference of RTE with 2eTE can be neglected since the threshold energy for the 1s-2p excitation is 16.3 MeV.

3.CONCLUSIONS

In summary, we have measured the angular distributions of the ^{3}D Auger decay to ground state in collisions of O^{5+} with He. The data show that the angular distributions are forward peaked along the beam direction, more forward peaked at the resonance energy of 13 MeV where RTE is predominant than at 6 or 8 MeV projectile energies where NTE plays a significant role in the formation of the ^{3}D state. Furthermore, the ^{3}D angular distribution at 8 MeV projectile energy where RTE and NTE have equal contribution, is less forward peaked than at 13 MeV where RTE predominates or at 6 MeV where NTE predominates. This suggests that destructive interferences between RTEA and NTEA may be taking place in the forward direction in agreement with previous calculations. Moreover, the magnetic substate population probabilities have been inferred from the angular momentum coupling theory and show that the ^{3}D state is exclusively populated with the magnetic substate $m_l=0$ at 13 MeV. However, at 6 and 8 MeV projectile energies, where NTE is important, small contributions from the magnetic substates $m_l=\pm1$ are also observed. Moreover, the negative values of Q(2±2) at nonresonance, where NTE plays a significant role may be attributed to destructive interference between RTE and NTE.

ACKNOWLEDJMENTS: we acknowledge support by the U.S department of Energy, office of Basic Energy Sciences, Division of Chemical Sciences, under contracts No. DE-FGO5-87ER40361 with Martin Marietta Energy Systems, Inc. and the University of North Carolina Research Council.

+ Present address: LAGRIPPA, Centre d'études nucleaires de Grenoble 85X, 38041 Grenoble cedex, France

♦ Present address: L421, Lawrence Livermore National Laboratory, Livermore, CA 94450, USA

♣ Present address: University of Missouri, Rolla, Dept of Physics, Rolla, Mo 65401, USA

• Present address: Department of Physics, Kansas State University, Manhattan, KS 66506, USA

References

1)D. Brandt, Phys. Rev. A 27, 1314 (1983).
2)D. Brandt, Nucl. Inst. and Meth 214, 93 (1983).
3)J.M. Feagin, J.S. Briggs, and T.M. Reeves,
  At. Mol. Phys. 117, 1057 (1984) and J. Phys. B17, 1057 (1984).
4)M. Clark, D. Brandt, J.K. Swenson, and S.M. Shafroth,
  Phys. Rev. Lett 54, 544 (1985).
5)J.A. Tanis, E.M. Bernstein, M. Clark, W. G. Graham,
  R.H. McFarland, T.J. Morgan, B.M. Johnson, K.W. Jones, and
  M. Meron, Phys. Rev. A31, 4040 (1985).
6)P.L. Pepmiller, P. Richard, J. Newcomb, J. Hall, and T.R. Dillingham,
  Phys. Rev. A 31,734 (1985).
7)A. Itoh, T.J.M. Zouros, D. Schneider, U. Stettner, W. zeitz and
  N. Stolterfoht, J. Phys B: At. Mol. Phys. 18, 4581 (1985).
8)J.K. Swenson, Y. Yamazaki, P.D. Miller, P.F. Dittner, P.L. Pepmiller,
  S. Datz, and N. Stolterfoht, Phys. Rev. Lett. 57, 3042 (1986).
9)J.K. Swenson, J.M. Anthony, M. Reed, M. Benhenni, S.M. Shafroth,
  D.M. Peterson, and L.D. Hendrick, Nucl Inst and Meth B 24/25,
  184 (1987).
10)J.M. Anthony, S.M. Shafroth, M. Benhenni, E.N. Strait, T.J.M. Zouros,
  L.D. Hendrick, and D.M. Peterson, journal de Physique, colloque C9,
  supplement au N0 12, tome 48, 301 (1987).
11)M. Schulz, E. Justiniano, R. Schuch, P.H. Mokler, S. Reusch,
  Phys. Rev. Lett 58, 1734 (1987).
12)W. Fritch and C.D. Lin, Phys. Rev. Lett 61, 690 (1988).
13)T. Reeves, Electronic and Atomic collisions, Elsevier Science Publishers,
  Invited papers of the International Conference on the Physics of Electronic and
  Atomic Collisions, edited by H. B. Gilbody, W. R. Newell, F. H. Read,
  and A.C.H. Smith (North Holland, Amsterdam) 685 (1988).
14)M. Schulz, R. Schuch, S. Datz, E.L.B. Justiniano, P.D. Miller, and
  H. Schone, Phys. Rev A 38, 5454 (1988).
15)T.J.M. Zouros, D. H. Lee, J. M. Sanders, J.L. Shinpaugh, T.N. Tipping,
  P. Richard, and S.L. Varghese, Nucl. Inst and Meth,
  Phys. Res. B 40/41, 17 (1989).
16)T.J.M. Zouros, D.H. Lee, T.N. Tipping, J.M. Sanders, J.L. Shinpaugh,
  P. Richard, K.R. Karim, and C.P. Bhalla, Phys. Rev A 40, 6246 (1989).
17)M. Schulz, J.P. Giese, J.K. Swenson, S. Datz, P.F. Dittner,
  H.F. Krause, H. Schone, C.R. Vane, M. Benhenni, and S.M. Shafroth
  Phys. Rev. Lett 62, 1738 (1989).
18)C.P. Bhalla, Phys. Rev. Lett 64, 1103 (1990).
19)M. Benhenni, S.M. Shafroth, J.K. Swenson, M.Schulz, J.P. Giese, H. Schone,
  C.R. Vane, P.F. Dittner and S. Datz, Phys Rev Lett 65, 1849 (1990)
20)J.K. Swenson, Nucl. Inst. and Meth. B10/11, 899 (1985).
21)N. Stolterfoht, P.D. Miller, H.F. Krause, Y. Yamazaki, J.K. Swenson, R. Bruch,
  P.F. Dittner, P.L. Pepmiller and S. Datz, Nucl. Inst and Meth, Phys. Res. B,
  24/25, 168 (1987).

Author Index

Subject Index